AF395705

Development of Unconventional Reservoirs

Development of Unconventional Reservoirs

Special Issue Editor

Reza Rezaee

MDPI • Basel • Beijing • Wuhan • Barcelona • Belgrade • Manchester • Tokyo • Cluj • Tianjin

Special Issue Editor
Reza Rezaee
Curtin University
Australia

Editorial Office
MDPI
St. Alban-Anlage 66
4052 Basel, Switzerland

This is a reprint of articles from the Special Issue published online in the open access journal *Energies* (ISSN 1996-1073) (available at: https://www.mdpi.com/journal/energies/special_issues/development_unconventional_reservoirs).

For citation purposes, cite each article independently as indicated on the article page online and as indicated below:

LastName, A.A.; LastName, B.B.; LastName, C.C. Article Title. *Journal Name* **Year**, *Article Number*, Page Range.

ISBN 978-3-03928-580-8 (Pbk)
ISBN 978-3-03928-581-5 (PDF)

Contents

About the Special Issue Editor

Professor Reza Rezaee of Curtin's Department of Petroleum Engineering has a Ph.D. in Reservoir Characterization. He has over 27 years' experience in academia, being responsible for both teaching and research. During his career, he has been engaged in several research projects supported by major oil and gas companies; these commissions, together with his supervisory work at various universities, have involved a wide range of achievements. During his research career, he led several major research projects funded by various oil and gas companies. He has received a total of more than \$2.2 M funds through his collaborative research projects. He has supervised over 70 M.S. and Ph.D. students during his university career to date. He has published more than 170 peer-reviewed journal and conference papers and is the author of 4 books on petroleum geology, logging and log interpretation, and gas shale reservoirs. His research has mostly focused on integrated solutions for reservoir characterization, formation evaluation, and petrophysics. Currently, he is focused on unconventional gas including gas shale and tight gas sand studies. As a founder of the Unconventional Gas Research Group of Australia, he has established a unique and highly sophisticated research lab at the Department of Petroleum Engineering, Curtin University. This lab was established to conduct research on petrophysical evaluation of tight gas sands and shale gas formations.

Preface to "Development of Unconventional Reservoirs"

Unconventional energy resources will retain their importance for many years to come since conventional hydrocarbons will soon come to an end. From a technical viewpoint, the more expensive clean and sustainable energy sources cannot currently compete with the relatively cheap non-renewable fossil fuels. Unconventional resources, such as shale gas/oil, tight gas sand, coalbed methane, oil shale, and gas hydrate, have played important roles as leveraging resources that have led the oil and gas prices to soar in recent years. At this stage, many countries, including major oil and gas producing countries, are seriously trying to assess unconventional resources as their future commodities.

Assessment criteria for unconventional resources are different from those used to evaluate conventional hydrocarbon reservoirs. To be successful in the exploration and development of unconventional sources, many important factors have to be considered.

Although the subject of this Special Issue is broad, we tried to collect some studies that provide guidance on some of the major factors involved in evaluating and developing unconventional plays. The topics covered by the articles submitted to this Special Issue are diverse. However, they can be classified into several groups based on the type of unconventional plays covered by each paper.

The first group of the papers covers some aspects of shale gas plays and includes the following works:

- Comparative Porosity and Pore Structure Assessment in Shales: Measurement Techniques, Influencing Factors and Implications for Reservoir Characterization by Yujie and Rezaee;
- Total Organic Carbon Enrichment and Its Impact on Pore Characteristics: A Case Study from the Niutitang Formation Shales in Northern Guizhou by Liu et al.;
- Accumulation Conditions and an Analysis of the Origins of Natural Gas in the Lower Silurian Shiniulan Formation from Well Anye 1, Northern Guizhou Province by Guo et al.;
- Volumetric Measurements of Methane-Coal Adsorption and Desorption Isotherms — Effects of Equations of State and Implication for Initial Gas Reserves by Ekundayo and Rezaee;
- A Prediction Model for Methane Adsorption Capacity in Shale Gas Reservoirs by Zou and Rezaee;
- Investigation of Analysis Methods for Pulse Decay Tests Considering Gas Adsorption by Han et al.;
- Numerical Simulation of Gas Production from Gas Shale Reservoirs — Influence of Gas Sorption Hysteresis by Ekundayo and Rezaee;

- Numerical Analysis of Transient Pressure Behaviors with Shale Gas MFHWs Interference by Gao et al.; and
- Performance Evaluation of CO2 Huff-n-Puff Gas Injection in Shale Gas Condensate Reservoirs by Meng et al.

The second group of papers deals with some aspects of coalbeds, including the following works:

- Experimental and Simulation Studies on Adsorption and Diffusion Characteristics of Coalbed Methane by Kim et al.;
- Variation of Petrophysical Properties and Adsorption Capacity in Different Rank Coals: An Experimental Study of Coals from the Junggar, Ordos, and Qinshui Basins in China by Wang et al.; and
- Design and Evaluation of a Surfactant–Mixed Metal Hydroxide-Based Drilling Fluid for Maintaining Wellbore Stability in Coal Measure Strata by Chen et al.

The third group of papers focuses on tight gas/oil sands and includes the following works:

- Insight into the Pore Characteristics of a Saudi Arabian Tight Gas Sand Reservoir by Adebayo et al.;
- The Characteristics of Oil Migration due to Water Imbibition in Tight Oil Reservoirs by Yang et al.;
- Stress-Dependent Permeability of Fractures in Tight Reservoirs by Cao et al.;
- Applicability Analysis of Klinkenberg Slip Theory in the Measurement of Tight Core Permeability by Zou et al.;
- Pressure Transient Performance for a Horizontal Well Intercepted by Multiple Reorientation Fractures in a Tight Reservoir by Xing et al.; and
- Catalytic Effect of Cobalt Additive on the Low Temperature Oxidation Characteristics of Changqing Tight Oil and its SARA Fractions by Wang and Wang.

The last group are classified as miscellaneous, covering a variety of works dealing with some technical aspects of different unconventional plays:

- Experimental Investigation on Injection and Production Pattern in Fractured-Vuggy Carbonate Reservoirs by He et al.;
- Visual Experimental Study on Gradation Optimization of Two-Stage Gravel Packing Operation in Unconventional Reservoirs by Meng et al.;
- Study of Downhole Shock Loads for Ultra-Deep Well Perforation and Optimization Measures by Deng et al.;
- An Automatic Classification Method of Well Testing Plot Based on Convolutional Neural Network (CNN) by Chu et al.;
- Study on the Impacts of Capillary Number and Initial Water Saturation on the Residual Gas Distribution by NMR by Li et al.;

- Numerical Investigation of Downhole Perforation Pressure for a Deepwater Well by Deng et al.; and
- The Effect of Supercritical CO2 on Shaly Caprocks by Hadian and Rezaee.

Keywords: unconventional reservoirs; shale gas; shale oil; tight gas sand; coalbed methane

Reza Rezaee
Special Issue Editor

Article

Comparative Porosity and Pore Structure Assessment in Shales: Measurement Techniques, Influencing Factors and Implications for Reservoir Characterization

Yujie Yuan * and Reza Rezaee

Western Australian School of Mines, Minerals, Energy and Chemical Engineering, Curtin University, Perth, WA 6845, Australia; R.Rezaee@curtin.edu.au
* Correspondence: yujie.yuan@postgrad.curtin.edu.au

Received: 9 May 2019; Accepted: 29 May 2019; Published: 31 May 2019

Abstract: Porosity and pore size distribution (PSD) are essential petrophysical parameters controlling permeability and storage capacity in shale gas reservoirs. Various techniques to assess pore structure have been introduced; nevertheless, discrepancies and inconsistencies exist between each of them. This study compares the porosity and PSD in two different shale formations, i.e., the clay-rich Permian Carynginia Formation in the Perth Basin, Western Australia, and the clay-poor Monterey Formation in San Joaquin Basin, USA. Porosity and PSD have been interpreted based on nuclear magnetic resonance (NMR), low-pressure N_2 gas adsorption (LP-N_2-GA), mercury intrusion capillary pressure (MICP) and helium expansion porosimetry. The results highlight NMR with the advantage of detecting the full-scaled size of pores that are not accessible by MICP, and the ineffective/closed pores occupied by clay bound water (CBW) that are not approachable by other penetration techniques (e.g., helium expansion, low-pressure gas adsorption and MICP). The NMR porosity is largely discrepant with the helium porosity and the MICP porosity in clay-rich Carynginia shales, but a high consistency is displayed in clay-poor Monterey shales, implying the impact of clay contents on the distinction of shale pore structure interpretations between different measurements. Further, the CBW, which is calculated by subtracting the measured effective porosity from total porosity, presents a good linear correlation with the clay content (R^2 = 0.76), implying that our correlated equation is adaptable to estimate the CBW in shale formations with the dominant clay type of illite.

Keywords: gas shale; NMR; helium porosimetry; clay bound water; porosity; pore size distribution; low-pressure gas adsorption; MICP

---

1. Introduction

The increasing demand of unconventional energy resources raises the significance of shale reservoir investigation [1,2]. Shales are defined as the laminated fine-grained argillaceous sedimentary rock, which are essentially constituted by minerals involving silt-sized particles (4–62.5 μm) and clays (<4 μm) in couple with organic matter (OM) [3–6]. The porosity and pore size distribution (PSD), performing as the most fundamental pore structure parameters to estimate gas storage capacity and fluid transporting behaviour in shale complex pore structures [4,7–9], are significantly associated with clay minerals and the promising OM that significantly varies between different shale formations [10–14]. The clay mineral or OM develops the micropore (i.e., pores smaller than 2 nm per International Union of Pure and Applied Chemistry (IUPAC) classification [15]) and mesopore (i.e., pores ranging from 2 nm to 50 nm per IUPAC classification [15]) system, complicating shale pore structures and resulting in the extremely low permeability, low porosity and the large distinction of PSD in shales.

To date, three types of laboratory techniques are applied for pore characterization or quantification. Microscopy techniques, e.g., transmission electron microscopy (TEM) and scanning electron microscopy (SEM), perform as the helpful petrographic-imaging approaches for porosity estimation [16], however, provide objective results and are not adaptable to cover the full range of PSD in shales [12]. Radiation Scatterings, such as small angle neutron scattering (SANS) and ultra-small angle scattering (USANS) techniques, are capable to quantify the continuous PSD in tight sandstones [17] and coals [18,19]. However, the applications in shale systems are still under debate due to the limitation of neutron sources [2,20–22]. Fluid penetration methods, i.e., low-pressure (<18.4 psi) CO_2 gas adsorption (LP-CO_2-GA), low-pressure N_2 gas adsorption (LP-N_2-GA), mercury intrusion capillary pressure (MICP) and nuclear magnetic resonance (NMR), enable a wide range of pore structure detection and have been universally utilized in shale research studies [23–31]. However, MICP displays destructive disadvantages and is not approachable to the pore throat sizes smaller than 3.6 nm [3,11,32,33], merely inter-communicated pores are available for detection [34]. Helium expansion is attainable to the connected pore space corresponding to effective porosity, while the acquisition of PSD is not available [7]. LP-CO_2-GA coupled with LP-N_2-GA is approachable to the pore sizes ranging from 0.35 nm to 200 nm [35]. However, only interconnected pores are accessible [34], and the results are sensitive to measurement procedures and highly dependent on the sample pre-treatment such as the dewatering/ outgassing temperatures and the size of the smashed shale fragments [36–38]. NMR, which is acknowledged as a non-destructive technique, is adaptable for measuring the total porosity and PSD in shales [4,27,39–41].

Unlike the conventional rocks displaying consistent results in porosity and PSD among different fluid-penetration measurements [42], shales, however, tend to reveal significant discrepancies. For example, MICP porosity in Barnett shales exhibit ~25–50% lower when compared to helium porosity values [43]. Similar porosity inconsistencies up to 50% have also been found in previous studies [32,44,45]. To fully understand the variations of shale pore structure interpretation between different measurements, the comprehensive techniques are highly required to be combined and compared in parallel.

This paper discusses the discrepant results of different measurements for the two shales in typical composition (i.e., the Carynginia shales of the Perth Basin in Western Australia and the Monterey shales of San Joaquin Basin in the U.S). Porosity is compared based on MICP, NMR and helium expansion. PSD is interpreted based on MICP, NMR, and low-pressure gas adsorption. The influencing factors are discussed for result discrepancies. Implications are provided for shale gas reservoir characterization.

2. Materials and Methods

2.1. Shale Samples

Shale samples from two formations were analyzed and compared between different measuring techniques. Carynginia samples, by the name of "AC1-AC8", were collected from Arrowsmith well in the Perth Basin, Western Australia. Monterey shales, by the name of "M1-B-M10-B" and "M1-M6", came from well-1B and well-1, respectively, in the San Joaquin Basin, USA. Geological settings of Carynginia and Monterey shale formation were displayed in other studies [27,46].

Table 1 shows the mineralogical composition in Carynginia and Monterey shales. Carynginia shales are characterized by abundant clay minerals, constituting 31.1–50.8 wt % of the total mineral contents (e.g., the average value of Carynginia clay content is 36.6 wt %). The quartz contents occupy 35.6–53.2 wt % (e.g., quartz averages in 45.17 wt %), while the minorities are shown in K-feldspar, plagioclase and other minerals. Monterey shales present a low-clay content (e.g., the mean value of the clay content is around 9.0 wt %) but a relatively high proportion in quartz content. The clay type in both of Carynginia and Monterey shales have been identified as illite [27,47].

Table 1. XRD mineralogical composition for shales from Carynginia and Monterey formation. Some data were collected from the other studies.

Name	Formation	Depth (m)	Total Clay (wt %)	Quartz (wt %)	K-Feldspar (wt %)	Plagioclase (wt %)	Other Minerals (wt %)
AC1	Carynginia	2780.2	50.8	35.6	2.6	5.0	6.0
AC2	Carynginia	2781.7	43.2	40.3	3.6	7.6	5.3
AC3	Carynginia	2789.9	32.3	47.6	5.4	9.4	5.3
AC4	Carynginia	2794.4	31.1	53.0	3.3	8.1	4.5
AC5	Carynginia	2806.4	40.7	41.3	3.6	7.6	6.8
AC8	Carynginia	2825.3	32.3	53.2	1.4	10.6	2.5
M1-B [47]	Monterey	1633.7	7.3	83.6	1.6	0.7	6.8
M2-B [47]	Monterey	1658.1	4.9	55.2	0.0	0.5	39.4
M3-B [47]	Monterey	2409.7	11.1	59.2	4.1	1.8	23.8
M4-B [47]	Monterey	2539.9	6.8	77.5	2.2	1.3	12.2
M5-B [47]	Monterey	2602.7	N/A	N/A	N/A	N/A	N/A
M6-B [47]	Monterey	2631.0	24.2	50.4	3.3	5.2	16.9
M7-B [47]	Monterey	2723.4	8.4	77.0	2.5	1.8	10.3
M8-B [47]	Monterey	2772.8	8.5	71.0	1.4	3.3	15.8
M9-B [47]	Monterey	2802.0	14.7	72.6	2.2	3.6	6.9
M10-B [47]	Monterey	2879.4	5.6	10.6	0.0	0.0	83.8
M1 [47]	Monterey	1669.5	N/A	N/A	N/A	N/A	N/A
M2 [47]	Monterey	2200.9	10.0	69.0	4.0	6.0	11.0
M3 [47]	Monterey	2203.2	N/A	N/A	N/A	N/A	N/A
M4 [47]	Monterey	2362.4	N/A	N/A	N/A	N/A	N/A
M5 [47]	Monterey	2362.7	N/A	N/A	N/A	N/A	N/A
M6 [47]	Monterey	2485.3	7.0	68.0	5.0	6.0	14.0

2.2. Nuclear Magnetic Resonance (NMR)

Prior to NMR experiments, Carynginia shale plugs of 1.5′ diameter were cleaned with toluene/methanol mixture, and fully saturated with 30,000 ppm brine that are matched with the average formation salinity. The low-field NMR measurements were performed on saturated samples by using 2 MHz Magritek Rock Core Analyzer, which was set under 30°C with P54 probe and conducted under the constant magnetic resonance frequency. NMR T_2 spectrum was acquired by using the experimental parameters, i.e., 100 μs inter-echo spacing (TE), 10,000 ms inter-experiment delay, 10,000 number of echoes and the minimum 200 signal to noise ratio (SNR), coupled with the Carr-Purcell-Meilboom-Gill sequence [48–50].

Applying NMR T_2 spectrum to study shales pore structure is fundamentally established on the transverse relaxation dominated by surface relaxation mechanism [39]:

$$\frac{1}{T_2} = \rho_2\left(\frac{S}{V}\right) \tag{1}$$

where T_2 is the transverse relaxation time; ρ_2 is surface relaxivity, which is considered as a constant value representing the strength of surface relaxation; $\frac{S}{V}$ is the surface volume ratio that is closely intimated with pore sizes. Pore size distribution could be interpreted via T_2 spectrum, with smaller pore sizes corresponding to shorter relaxation times.

2.3. Low-Pressure Gas Adsorption (LP-GA)

Low-pressure N_2 gas adsorption (LP-N_2-GA) was applied to measure the pore size distribution (PSD) and the pore volume. Prior to the measurements, shale samples were crushed into fragments of 60 mesh sizes and degassed over 8h for pore surface cleaning. LP-N_2-GA was performed on Micromeritics® TriStar 3020 instrument at the bathing temperature of 77.4 K. N_2 was penetrated into the degassed samples under the constant temperature for the acquisition of adsorption isotherm. PSD is obtained by using the Barrett, Joyner and Halenda (BJH) theory based on N_2 adsorption isotherm [35]. The interpretations were carried out on the embedded TriStar II 3020 standard software.

2.4. Mercury Injection Capillary Pressure (MICP)

MICP measurements were performed on Micromeritics Autopore IV 9500 V1.09 porosimeter on chip samples weighing around 10 g. Prior to the test, shale chips were evacuated under the pressure of 70 µmHg for 10 min. The non-wetting mercury, as the working probe for pore access, follows the parameters applied as: Hg density of 13.53 g/mL; adv. and rec. contact angle of 130°; Hg surface tension of 485 dynes/cm. The mercury filling pressure was performed of 0.51 psia under 10 s for equilibration, followed by injection under high pressure, ranging from 0.1 MPa (14.5 psi) to the maximum 413.7 MPa (60,000 psi), which corresponds to the pore throat size from 3.6 nm to 1100 µm.

The pore throat size distribution of tested samples is obtained using Washburn equation assuming cylindrical pores (Equation (2)) [51]:

$$r_i = \frac{-2\sigma\cos\theta}{P_c} \tag{2}$$

where r_i is the pore throat radius calculated under mercury pressure of P_c (psi), µm; σ is the mercury surface tension (485 dynes/cm applied in the test); θ is mercury contact angle (130° applied in the test); p_c is the injection pressure ranging from 14.5 psi to the maximum 60,000 psi.

3. Results

3.1. Porosity

The porosity values obtained from three measuring techniques (i.e., MICP, Helium, and NMR) are shown for two different shale formations (i.e., Carynginia and Monterey) (Figure 1). An obvious porosity distinction is displayed in NMR between the clay-rich samples (i.e., Carynginia) and the clay-poor samples (i.e., Monterey). An overall higher NMR porosities are exhibited in Carynginia compared to Monterey. In addition, the porosity discrepancies are apparently exhibited between NMR and the other two measurements in Carynginia samples. Carynginia presents the highest porosity value in NMR, which is more than two times as MICP porosity, and about three times as helium porosity (i.e., the porosity measured by NMR, helium, and MICP ranges in 8.02–12.87%, 3.03–3.78%, and 1.93–4.15%, respectively). However, the Monterey exhibits a high porosity consistency in NMR, helium and MICP (Figure 1). The porosity measured from MICP and helium demonstrates high consistencies in both Carynginia and Monterey. As shown in Figure 2, the cross-plot of helium porosity versus MICP porosity generates a very good positive linear relationship, with the correlation coefficient (R^2) of 0.93. The porosity values for each sample are shown in Appendix A (Table A1).

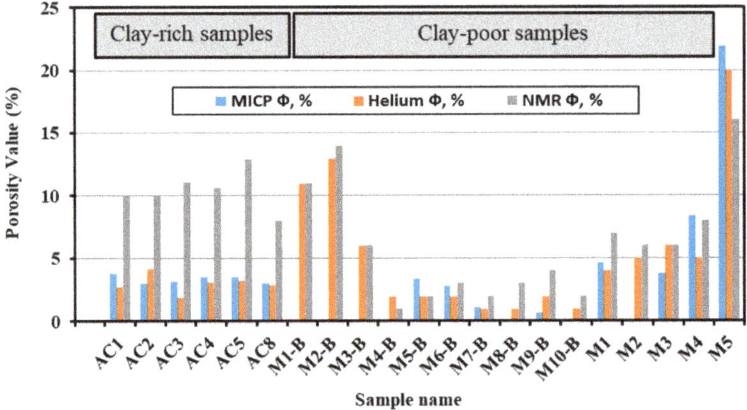

Figure 1. The porosity values obtained from mercury intrusion capillary pressure (MICP), Helium, and nuclear magnetic resonance (NMR) for two different shale formations.

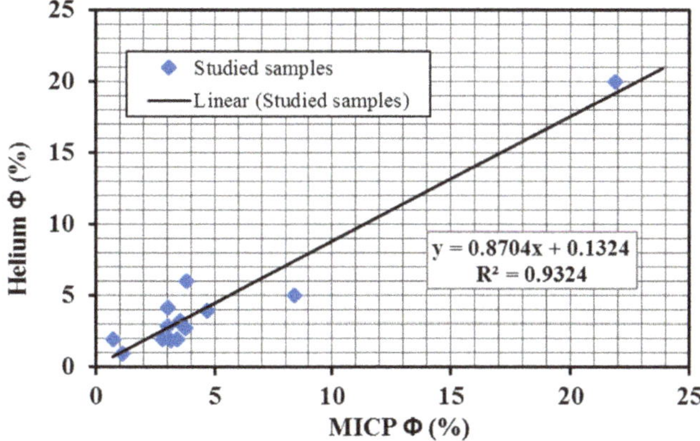

Figure 2. The cross-plot of Helium porosity (%) versus MICP porosity (%) for the studied samples.

3.2. The Pore Size Distribution from NMR

Figure 3 presents the NMR T_2 spectrum in Carynginia shales, with the majority of pores identified in small pore sizes. The peak values of T_2 curves correspond to the T_2 relaxation time around 0.3–1 ms. The samples of higher clay contents, e.g., AC1 and AC2 (i.e., 50.8% and 43.2%), exhibit larger amplitude and narrower spectrum with the peak value locating in smaller pore sizes. The samples of relatively lower clay contents, e.g., AC8 and AC4 (i.e., 32.3% and 31.1%), display smaller amplitude and wider distributions, presenting a general larger pore sizes. A uniform pore size distribution is commonly indicated in Carynginia shales.

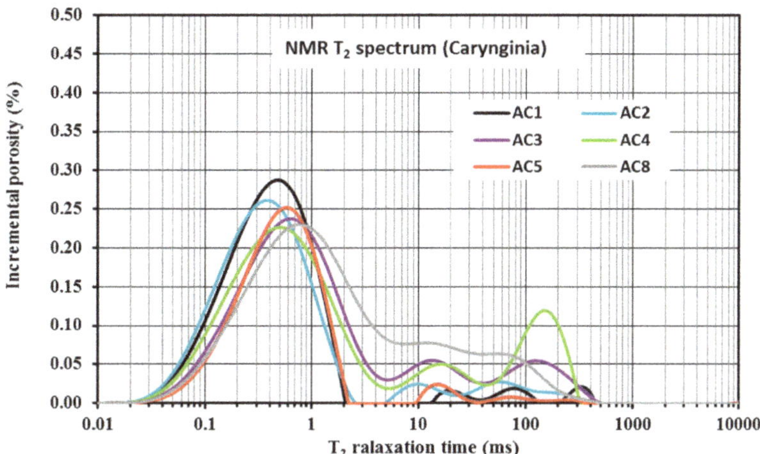

Figure 3. NMR T_2 spectrum for Carynginia shales (i.e., AC1-AC8).

NMR T_2 spectra of Monterey shales, i.e., M1-M6 and M1-B-M10-B, are displayed in Figures 4 and 5, respectively. Significant variations are demonstrated in Monterey shales compared to Carynginia shales. As shown in Figure 4, Monterey shales from Well 1 (except for M5) exhibit the major pores in larger pore size ranges. The peak locations of the spectrum correspond to T_2 relaxation time ~1–100 ms, coupled with an overall wider spectrum range, indicating a general uneven pore size distributions. The peak values (except for M5) correspond to the incremental porosity between 0.1% and 0.13%, displaying overall lower values than that in Carynginia shales. M5 shows the trimodal spectrum associated with three typical pore types. The spectrum peak of M5 occurs at T_2 relaxation time in 300–500 ms, representing the majority of larger pores or fractures.

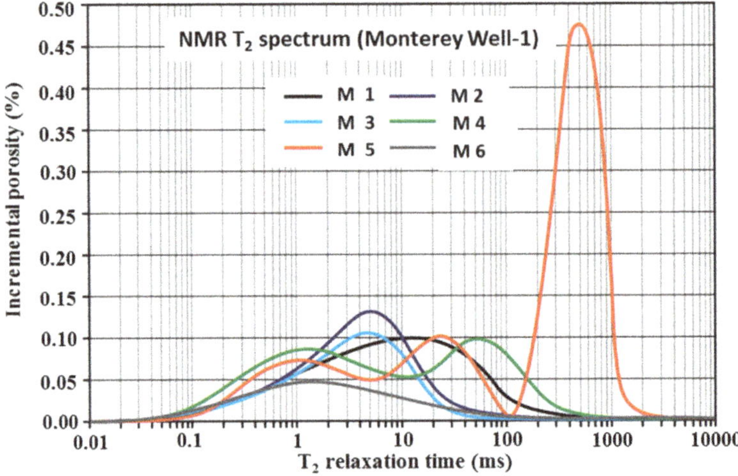

Figure 4. NMR T_2 spectrum for Monterey shales (i.e., M1–M6) collected from Well 1. Modified from Rivera [46].

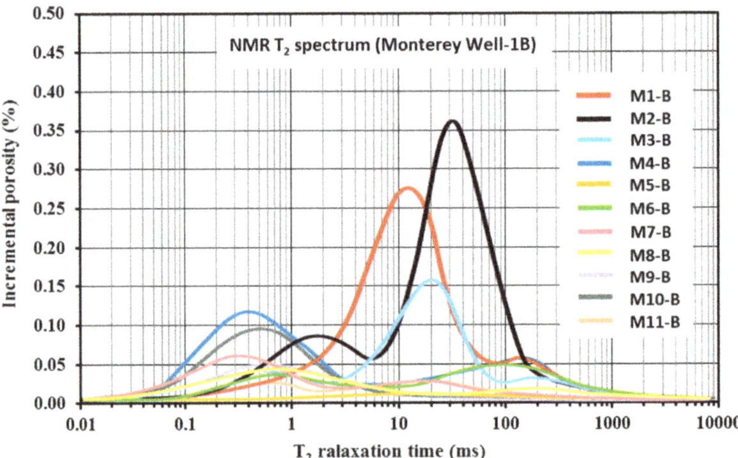

Figure 5. NMR T_2 spectrum for Monterey shales (i.e., M1-B–M6-B) collected from Well 1B. Modified from Rivera [46].

Figure 5 shows the T_2 spectrum of Monterey shales from Well 1B with multiple modal types. M4-B, M10-B, M7-B, M9-B exhibit a majority of small pore sizes, corresponding to the T_2 relaxation times of ~0.3–1 ms. M1-B and M2-B with semi-modal distributions present the PSD peaks at larger T_2 relaxation time (i.e., 14 ms and 40 ms respectively) and higher incremental porosity (i.e., 0.27% and 0.36% respectively). M3-B exhibits trimodal spectrum with the main pore size locates at ~20 ms. When compared to T_2 spectrum in clay-rich Carynginia, the pore sizes in Monterey are rather unevenly distributed and universally locating in larger pore sizes. Moreover, unlike Carynginia, no obvious correlations are observed between the clay contents and the NMR PSD amplitudes in Monterey shales.

3.3. The Pore Size Distribution from Gas Adsorption

Figure 6 displays Caryngina PSD obtained from LP-N_2-GA experiments. As can be seen, the PSD peak in Carynginia appears around 20 nm, implying the pore majority locating in fine mesopore sizes that dominantly controls the total pore volume. Monterey shales, by contrast, present a different scenario (Figure 7), showing PSD peak at ~50–100 nm with the pore majority in fine macropore ranges, which is intimately related to the high quartz content.

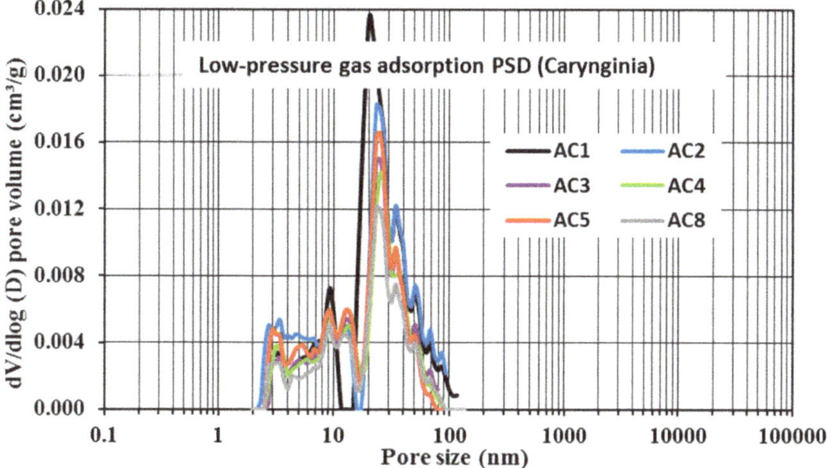

Figure 6. Carynginia pore size distribution (PSD) derived from low-pressure N_2 gas adsorption based on BJH theory.

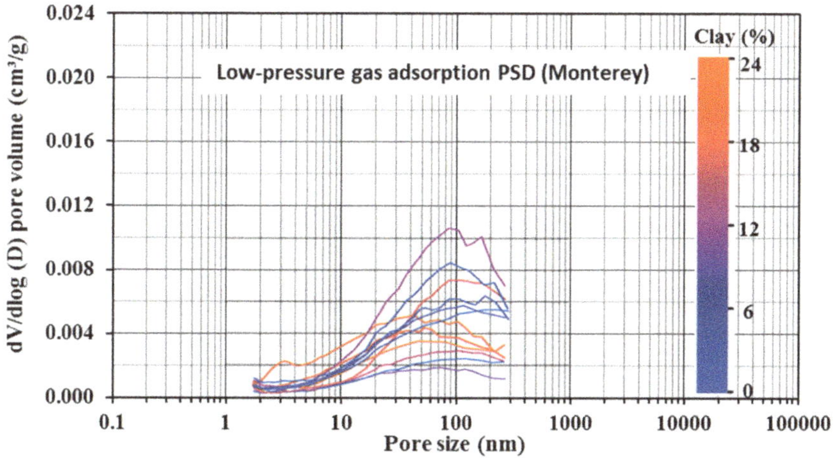

Figure 7. Monterey pore size distribution (PSD) derived from low-pressure N_2 gas adsorption. Modified from Saidian, Kuila [44].

3.4. The Pore Throat Size Distribution from MICP

Carynginia and Monterey pore throat size distributions (PTDs) measured by MICP are plotted in Figures 8 and 9. The peaks of MICP- derived PTD in Carynginia are commonly located in pore sizes ~4–5 nm, which are smaller compared to that interpreted by LP-N_2-GA. A wider range of the detectable large pores (i.e., pore sizes larger than 100 nm) is revealed by MICP technique compared to LP-N$_2$-GA. Consistent with the NMR and LP-N$_2$-GA interpretations for Carynginia samples, larger PTD amplitude is shown in the samples of higher clay (e.g., AC1, AC2), while the lowest PTD amplitude is found in samples of the lowest clay samples (i.e., AC8). The Monterey PTD, however, presents a weak interrelationship between the clay content and the amplitude of curve (Figure 9), which agrees with the behaviours of Monterey PSD (e.g., Figures 4, 5 and 7) that is most likely under the large influences of low-clay contents [44].

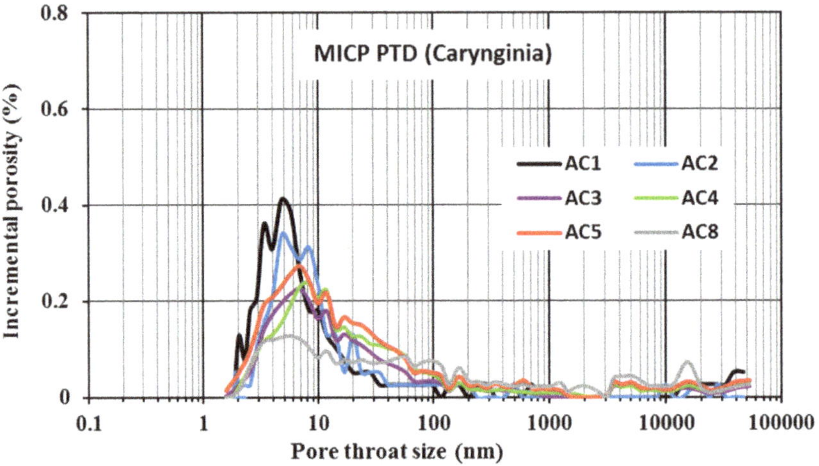

Figure 8. MICP pore throat size distribution (PTD) for Permian Carynginia shales.

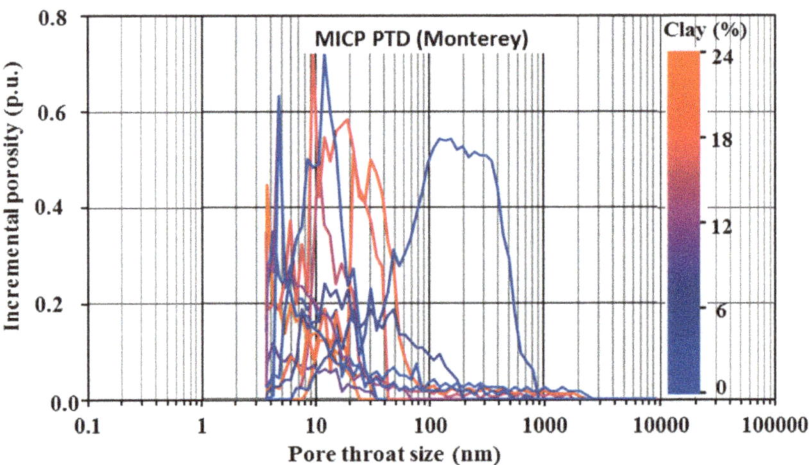

Figure 9. MICP pore throat size distribution (PTD) for Monterey shales. Modified from Saidian, Kuila [44].

4. Discussion

Carynginia samples are characterized by abundant clay contents, while Monterey shales are clay-poor (Table 1). NMR technique, which is highlighted by non-destructive measurement of total porosity, involves the detection of effective porosity and clay bound water (CBW), which is tightly bound on the surface area of clay minerals and universally quantified by cutting the effective porosity off total porosity [52,53]. Other penetration approaches, e.g., helium, MICP, and low-pressure gas adsorption, nevertheless, are merely approachable to the inter-connected pores, missing out the closed-pores or the ineffective porosity occupied by CBW. Under extreme circumstances, for shales containing very high clay contents and thus high CBW, the most of pore spaces could be nearly fully-occupied by the volume of CBW [8,54] that would influence the petrophysical properties in shales [27,55–57]. As helium porosimetry is able to obtain effective porosity by covering a wider pore size range (i.e., 0.1 nm–100 μm) than MICP (i.e., 3.6 nm–100 μm) (Figure 10) [34], the CBW is calculated by subtracting the helium porosity (i.e., effective effective) from NMR porosity (i.e., total porosity). Figure 11 cross-plots the calculated CBW versus the clay content in both Carynginia and Monterey shales. The CBW, which accounts for the porosity discrepancy between NMR and helium measurement, displays higher values in clay-rich Caryngnia shales, but lower values are found in Monterey shales. The correlation presents a good linear relationship ($R^2 = 0.76$), indicating that the correlation equation (Equation (3)) is adaptable for the estimation of CBW in the shale, whose clay type is dominantly contributed by illite:

$$CBW\ (\%) = 0.19 \times V_{sh}\ (\%) - 0.7 \tag{3}$$

where *CBW* is the volume of clay bound water (%), V_{sh} (%) is the clay contents (%). Moreover, the equation is most likely to fit into the formation with the brine salinity of 20,000–30,000 ppm that matches with our studied formations.

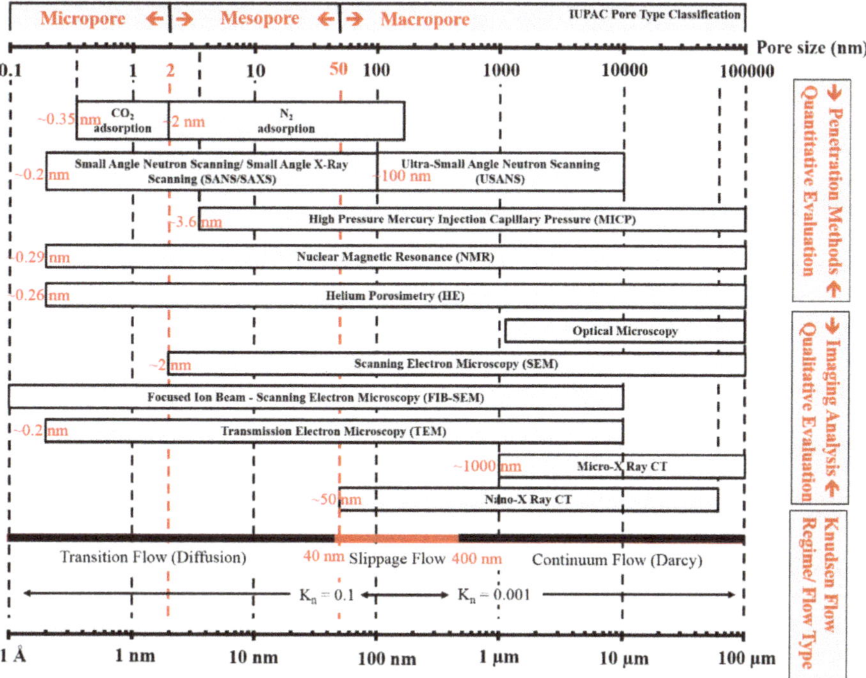

Figure 10. The multi-scaled methods for pore characterization in shales Modified from other studies [32,34,58,59].

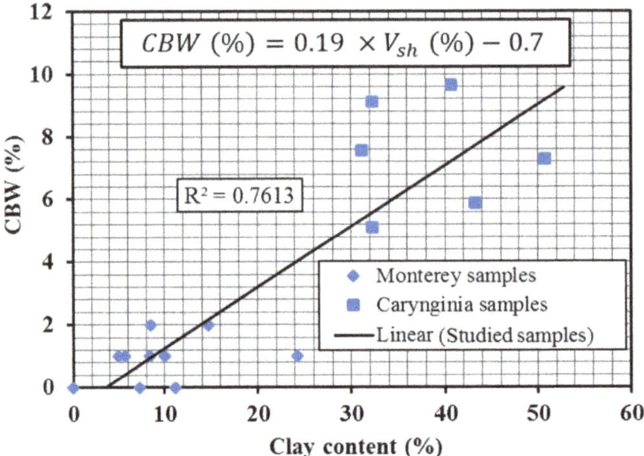

Figure 11. The cross-plot of clay bound water (CBW) (%) versus clay content (%) for studied shale samples.

Apart from the influencing factors associated with clays, the compatibility of the penetrated working fluid molecules with shale nanopore structure also causes the interpretation inconsistencies. Unlike NMR using H_2O as working fluid to access pore body, the working molecule involved in MICP is merely attainable to the limited pore throat size. The mineral-controlled geometrical pore shapes, which are highly intimated with the mineral compositions and assemblages, pose a large impact on the porosity discrepancies between NMR and MICP in Carynginia shales [60]. To summarize, the possible reasons for the higher NMR porosity over MICP are: (1) the volume of clay bound water; (2) the porosity contributed by pores smaller than 3.6 nm; (3) the different mechanisms involved in NMR pore body detection versus MICP pore throat detection (e.g., MICP assumes the pores are cylindrical in shape with a smooth surface, but the real pores are complicated with rough surfaces bound with water layers) [34]; (4) the pore shape combination that intimately related to shale compositions. When the comparisons are carried out between helium and MICP, theoretically, for shales containing high proportion of micropores, helium porosity is supposed to be higher than MICP due to its wider detection of pore size range [58]. However, the higher MICP porosity values are observed in some of the studied samples in both Carynginia and Monterey (e.g., AC1, AC3, M5-B, M4) (Figure 1). As the samples from both formations show a small proportion of micropores, the possible reasons could be explained by the increased mercury uptake induced by the high-pressure application (i.e., 60,000 psi) in MICP measurement [61]. Similar phenomenons have also been found in coals [62], which possess similar characteristics as shales [63,64].

5. Conclusions

The discrepancies in porosity or pore size distribution between MICP, NMR, and LP-GA porosimetry are largely controlled by shale compositions, particularly, the clay minerals. The clay-rich shales generate NMR porosity significantly higher than MICP and helium porosity, while the clay-poor shales exhibit a high porosity consistency between NMR, MICP and helium porosimetry.

The higher porosity values unveiled by NMR over MICP/helium technique are fundamentally attributed to CBW, meanwhile, the clay mineral compositions and assemblages, coupled with pore geometry also contribute to the discrepancies. The MICP and helium both detect intercommunicated pores and display consistent porosity for shales deficient in pores smaller than 3.6 nm. The shales of deficient micropores may possibly show higher helium porosity over MICP porosity, which essentially result from the high pressure application involved in MICP technique.

Author Contributions: Investigation and writing, Y.Y.; Supervision, reviewing and correcting, R.R.

Funding: This research was funded by China Scholarship Council. Grant number: 201606450018.

Acknowledgments: The authors acknowledge Unconventional Gas Research Group (UGRG), the discipline of Petroleum Engineering and the discipline of Chemical Engineering in Western Australian School of Mines (WASM), Curtin University for the facility assistance. The Department of Mines, Industry Regulation and Safety of the Government of Western Australia is acknowledged for their permission of core sample collections. The editor and anonymous reviewers are also appreciated for their comments to improve this work. Y. Yuan sincerely thanks China Scholarship Council-Curtin International Postgrad Research Scholarship (CSC-CIPRS) for their financial support.

Conflicts of Interest: The authors declare no conflict of interest.

Appendix A

Table A1. The porosity values obtained from MICP, Helium and NMR techniques for the studied samples in Carynginia and Monterey. Some data are collected from other studies [26,27,46].

Name	Formation	Depth (m)	MICP Φ, %	Helium Φ, %	NMR Φ, %
AC1	Carynginia	2780.2	3.78	2.78	10.06
AC2	Carynginia	2781.7	3.05	4.15	10.04
AC3	Carynginia	2789.9	3.17	1.93	11.05
AC4	Carynginia	2794.4	3.54	3.11	10.66
AC5	Carynginia	2806.4	3.56	3.22	12.87
AC8	Carynginia	2825.3	3.03	2.92	8.02
M1-B	Monterey	1633.7	N/A	11.0	11.0
M2-B	Monterey	1658.1	N/A	13.0	14.0
M3-B	Monterey	2409.7	N/A	6.0	6.0
M4-B	Monterey	2539.9	N/A	2.0	1.0
M5-B	Monterey	2602.7	3.4	2.0	2.0
M6-B	Monterey	2631.0	2.8	2.0	3.0
M7-B	Monterey	2723.4	1.1	1.0	2.0
M8-B	Monterey	2772.8	N/A	1.0	3.0
M9-B	Monterey	2802.0	0.7	2.0	4.0
M10-B	Monterey	2879.4	N/A	1.0	2.0
M1	Monterey	1669.5	4.7	4.0	7.0
M2	Monterey	2200.9	N/A	5.0	6.0
M3	Monterey	2203.2	3.8	6.0	6.0
M4	Monterey	2362.4	8.4	5.0	8.0
M5	Monterey	2362.7	21.9	20.0	16.0

References

1. Curtis, J.B. Fractured shale-gas systems. *AAPG Bull.* **2002**, *86*, 1921–1938.
2. Clarkson, C.R.; Solano, N.; Bustin, R.M.; Bustin, A.M.M.; Chalmers, G.R.L.; He, L.; Melnichenko, Y.B.; Radliński, A.P.; Blach, T.P. Pore structure characterization of North American shale gas reservoirs using USANS/SANS, gas adsorption, and mercury intrusion. *Fuel* **2013**, *103*, 606–616. [CrossRef]
3. Chalmers, G.R.; Bustin, R.M.; Power, I.M. Characterization of gas shale pore systems by porosimetry, pycnometry, surface area, and field emission scanning electron microscopy/transmission electron microscopy image analyses: Examples from the Barnett, Woodford, Haynesville, Marcellus, and Doig units. *AAPG Bull.* **2012**, *96*, 1099–1119.
4. Rezaee, R. *Fundamentals of Gas Shale Reservoirs*; John Wiley & Sons: Hoboken, NJ, USA, 2015.
5. Folk, R.L. *Petrology of Sedimentary Rocks*; Hemphill Publishing Company: Cedar Hill, TX, USA, 1980.
6. Javadpour, F. Nanopores and apparent permeability of gas flow in mudrocks (shales and siltstone). *J. Can. Pet. Technol.* **2009**, *48*, 16–21. [CrossRef]
7. Labani, M.M.; Rezaee, R.; Saeedi, A.; Hinai, A.A. Evaluation of pore size spectrum of gas shale reservoirs using low pressure nitrogen adsorption, gas expansion and mercury porosimetry: A case study from the Perth and Canning Basins, Western Australia. *J. Pet. Sci. Eng.* **2013**, *112*, 7–16. [CrossRef]

8. Topór, T.; Derkowski, A.; Kuila, U.; Fischer, T.B.; McCarty, D.K. Dual liquid porosimetry: A porosity measurement technique for oil-and gas-bearing shales. *Fuel* **2016**, *183*, 537–549. [CrossRef]

9. Jia, B.; Tsau, J.-S.; Barati, R. Different flow behaviors of low-pressure and high-pressure carbon dioxide in shales. *SPE J.* **2018**, *23*, 1452–1468. [CrossRef]

10. Sondergeld, C.H.; Ambrose, R.J.; Rai, C.S.; Moncrieff, J. Micro-structural studies of gas shales. In Proceedings of the SPE Unconventional Gas Conference, Pittsburgh, PA, USA, 23–25 February 2010.

11. Mastalerz, M.; Ambroseet, R.J.; Rai, C.S.; Moncrieff, J. Porosity of Devonian and Mississippian New Albany Shale across a maturation gradient: Insights from organic petrology, gas adsorption, and mercury intrusion. *AAPG Bull.* **2013**, *97*, 1621–1643. [CrossRef]

12. Furmann, A.; Mastalerz, M.; Bish, D.; Schimmelmann, A.; Pedersen, P.K. Porosity and pore size distribution in mudrocks from the Belle Fourche and Second White Specks Formations in Alberta, Canada. *AAPG Bull.* **2016**, *100*, 1265–1288.

13. Loucks, R.G.; Reed, R.M.; Ruppel, S.C.; Jarvie, D.M. Morphology, genesis, and distribution of nanometer-scale pores in siliceous mudstones of the Mississippian Barnett Shale. *J. Sediment. Res.* **2009**, *79*, 848–861. [CrossRef]

14. Curtis, M.E.; Cardott, B.J.; Sondergeld, C.H.; Rai, C.S. Development of organic porosity in the Woodford Shale with increasing thermal maturity. *Int. J. Coal Geol.* **2012**, *103*, 26–31. [CrossRef]

15. Rouquerol, J.; Avnir, D.; Fairbridge, C.W.; Everett, D.H.; Haynes, J.M.; Pernicone, N.; Ramsay, J.D.F.; Sing, K.S.W.; Unger, K.K. Recommendations for the characterization of porous solids (Technical Report). *Pure Appl. Chem.* **1994**, *66*, 1739–1758. [CrossRef]

16. Sigal, R.F. Pore-size distributions for organic-shale-reservoir rocks from nuclear-magnetic-resonance spectra combined with adsorption measurements. *SPE J.* **2015**, *20*, 824–830. [CrossRef]

17. Clarkson, C.; Freeman, M.; He, L.; Agamalian, M.; Melnichenko, Y.; Mastalerz, M.; Bustin, R.; Radlinski, A.; Blach, T. Characterization of tight gas reservoir pore structure using USANS/SANS and gas adsorption analysis. *Fuel* **2012**, *95*, 371–385. [CrossRef]

18. Radlinski, A.; Mastalerz, M.; Hinde, A.; Hainbuchner, M.; Rauch, H.; Baron, M.; Lin, J.; Fan, L.; Thiyagarajan, P. Application of SAXS and SANS in evaluation of porosity, pore size distribution and surface area of coal. *Int. J. Coal Geol.* **2004**, *59*, 245–271. [CrossRef]

19. Melnichenko, Y.; Radlinski, A.; Mastalerz, M.; Cheng, G.; Rupp, J. Characterization of the CO2 fluid adsorption in coal as a function of pressure using neutron scattering techniques (SANS and USANS). *Int. J. Coal Geol.* **2009**, *77*, 69–79. [CrossRef]

20. Radlinski, A.P. Small-angle neutron scattering and the microstructure of rocks. *Rev. Mineral. Geochem.* **2006**, *63*, 363–397. [CrossRef]

21. Bahadur, J.; Melnichenko, Y.B.; Radlinski, A.P.; Mastalerz, M.; Schimmelmann, A. Hierarchical pore morphology of cretaceous shale: A small-angle neutron scattering and ultrasmall-angle neutron scattering study. *Energy Fuels* **2014**, *28*, 6336–6344. [CrossRef]

22. Bahadur, J.; Melnichenko, Y.B.; Radlinski, A.P.; Mastalerz, M.; Schimmelmann, A. Small-angle and ultrasmall-angle neutron scattering (SANS/USANS) study of New Albany shale: A treatise on microporosity. *Energy Fuels* **2015**, *29*, 567–576. [CrossRef]

23. Knapp, L.J.; Nanjo, T.; Uchida, S.; Haeri-Ardakani, O.; Sanei, H. Investigating Influences on Organic Matter Porosity and Pore Morphology in Duvernay Formation Organic-Rich Mudstones. In Proceedings of the 24th Formation Evaluation Symposium of Japan, Chiba, Japan, 11–12 October 2018.

24. Kuila, U. *Measurement and Interpretation of Porosity and Pore-Size Distribution in Mudrocks: The Hole Story of Shales*; Colorado School of Mines, Arthur Lakes Library: Golden, CO, USA, 2013.

25. Rezaee, R.; Saeedi, A.; Clennell, B. Tight gas sands permeability estimation from mercury injection capillary pressure and nuclear magnetic resonance data. *J. Pet. Sci. Eng.* **2012**, *88*, 92–99. [CrossRef]

26. Al Hinai, A.; Rezaee, R.; Esteban, L.; Labani, M. Comparisons of pore size distribution: A case from the Western Australian gas shale formations. *J. Unconv. Oil Gas Resour.* **2014**, *8*, 1–13. [CrossRef]

27. Yuan, Y.; Rezaee, R.; Verrall, M.; Hu, S.-Y.; Zou, J.; Testmanti, N. Pore characterization and clay bound water assessment in shale with a combination of NMR and low-pressure nitrogen gas adsorption. *Int. J. Coal Geol.* **2018**, *194*, 11–21. [CrossRef]

28. Ross, D.J.; Bustin, R.M. The importance of shale composition and pore structure upon gas storage potential of shale gas reservoirs. *Mar. Pet. Geol.* **2009**, *26*, 916–927. [CrossRef]

29. Saidian, M.; Godinez, L.J.; Rivera, S.; Prasad, M. Porosity and pore size distribution in mudrocks: A comparative study for Haynesville, Niobrara, monterey, and Eastern European Silurian formations. In Proceedings of the Unconventional Resources Technology Conference, Denver, CO, USA, 25–27 August 2014.

30. Josh, M.; Esteban, L.; Piane, C.D.; Sarout, J.; Dewhurst, D.; Clennell, M. Laboratory characterisation of shale properties. *J. Pet Sci. Eng.* **2012**, *88–89*, 107–124. [CrossRef]

31. Wang, F.; Guo, S. Influential factors and model of shale pore evolution: A case study of a continental shale from the Ordos Basin. *Mar. Pet. Geol.* **2019**, *102*, 271–282. [CrossRef]

32. Sondergeld, C.H.; Newsham, K.E.; Comisky, J.T.; Rice, M.C.; Rai, C.S. Petrophysical Considerations in Evaluating and Producing Shale Gas Resources. In Proceedings of the SPE Unconventional Gas Conference, Pittsburgh, PA, USA, 23–25 February 2010.

33. Nelson, P.H. Pore-throat sizes in sandstones, tight sandstones, and shales. *AAPG Bull.* **2009**, *93*, 329–340. [CrossRef]

34. Caineng, Z.; Songqi, P.; Senhu, L.; Jinliang, H.; Songtao, W.; Dazhong, D.; Shasha, S.; Zhi, Y.; Yanyan, C.; Shufang, W.; et al. Shale Gas Formation and Occurrence in China: An Overview of the Current Status and Future Potential. *Acta Geol. Sin. Engl. Ed.* **2016**, *90*, 1249–1283. [CrossRef]

35. Barrett, E.P.; Joyner, L.G.; Halenda, P.P. The Determination of Pore Volume and Area Distributions in Porous Substances. I. Computations from Nitrogen Isotherms. *J. Am. Chem. Soc.* **1951**, *73*, 373–380. [CrossRef]

36. Holmes, R.; Rupp, E.C.; Vishal, V.; Wilcox, J. Selection of Shale Preparation Protocol and Outgas Procedures for Applications in Low-Pressure Analysis. *Energy Fuels* **2017**, *31*, 9043–9051. [CrossRef]

37. Figini-Albisetti, A.; Velasco, L.F.; Parra, J.B.; Ania, C.O. Effect of outgassing temperature on the performance of porous materials. *Appl. Surf. Sci.* **2010**, *256*, 5182–5186. [CrossRef]

38. Wang, G.; Ju, Y. Organic shale micropore and mesopore structure characterization by ultra-low pressure N 2 physisorption: Experimental procedure and interpretation model. *J. Nat. Gas Sci. Eng.* **2015**, *27*, 452–465. [CrossRef]

39. Coates, G.R.; Xiao, L.; Prammer, M.G. *NMR Logging: Principles and Applications*; Gulf Professional Publishing: Houston, TX, USA, 1999.

40. Morriss, C.; Rossini, D.; Straley, C.; Tutunjian, P.; Vinegar, H. Core analysis by low-field NMR. In *The Log Analyst*; Society of Petrophysicists and Well-Log Analysts: London, UK, 1997; Volume 38.

41. Schön, J.H. *Physical Properties of Rocks: Fundamentals and Principles of Petrophysics*; Elsevier: Amsterdam, The Netherlands, 2015; Volume 65.

42. Hossain, Z.; Grattoni, C.A.; Solymar, M.; Fabricius, I.L. Petrophysical properties of greensand as predicted from NMR measurements. *Pet. Geosci.* **2011**, *17*, 111–125. [CrossRef]

43. Sigal, R.F. Mercury Capillary Pressure Measurements on Barnett Core. *SPE Reserv. Eval. Eng.* **2013**, *16*, 432–442. [CrossRef]

44. Saidian, M.; Kuila, U.; Prasad, M.; Lopez, L.A. A Comparison of Measurement Techniques for Porosity and Pore Size Distribution in Shales (Mudrocks): A Case Study of Haynesville, Eastern European Silurian, Niobrara, and Monterey Formations. In *AAPG Memoir 110: Imaging Unconventional Reservoir Pore Systems*; AAPG: Tulsa, OK, USA, 2016.

45. Katsube, T.J.; Scromeda, N.; Williamson, M. *Effective Porosity of Tight Shales from the Venture Gas Field, Offshore Nova Scotia*; Geological Survey of Canada: Ottawa, ON, Canada, 1992; pp. 111–119.

46. Rivera, S. *Ultrasonic and Low Field Nuclear Magnetic Resonance Study of Lower Monterey Formation: San Joaquin Basin*; Colorado School of Mines, Arthur Lakes Library: Golden, CO, USA, 2014.

47. Rivera, S.; Prasad, M. Effect of Mineralogy on NMR, Sonic, and Resitivity: A Case Study of the Monterey Formation. In Proceedings of the Unconventional Resources Technology Conference, Denver, CO, USA, 25–27 August 2014.

48. Carr, H.Y.; Purcell, E.M. Effects of Diffusion on Free Precession in Nuclear Magnetic Resonance Experiments. *Phys. Rev.* **1954**, *94*, 630–638. [CrossRef]

49. Kenyon, B.; Kleinberg, R.; Straley, C.; Gubelin, G.; Morriss, C. Nuclear magnetic resonance imaging—Technology for the 21st century. *Oilfield Rev.* **1995**, *7*, 19–33.

50. Meiboom, S.; Gill, D. Modified Spin-Echo Method for Measuring Nuclear Relaxation Times. *Sci. Instrum.* **1958**, *29*, 688. [CrossRef]

51. Washburn, E.W. Note on a Method of Determining the Distribution of Pore Sizes in a Porous Material. *Proc. Natl. Acad. Sci. USA* **1921**, *7*, 115–116. [CrossRef]

52. Prammer, M.; Drack, E.; Bouton, J.; Gardner, J.; Coates, G.; Chandler, R.; Miller, M. Measurements of Clay-Bound Water and Total Porosity by Magnetic Resonance Logging. *SPE* **1996**, *37*. [CrossRef]

53. Coates, G.R.; Galford, J.; Mardon, D.; Marschall, D. A new characterization of bulk-volume irreducible using magnetic resonance. *Log Anal.* **1998**, *39*, 1.

54. Clavier, C.; Coates, G.; Dumanoir, J. Theoretical and Experimental Bases for the Dual-Water Model for Interpretation of Shaly Sands. *Soc. Pet. Eng. J.* **1984**, *24*, 153–168. [CrossRef]

55. Yuan, Y.; Rezaee, R. Fractal analysis of the pore structure for clay bound water and potential gas storage in shales based on NMR and N2 gas adsorption. *J. Pet. Sci. Eng.* **2019**, *177*, 756–765. [CrossRef]

56. Tan, M.; Mao, K.; Song, X.; Yang, X.; Xu, J. NMR petrophysical interpretation method of gas shale based on core NMR experiment. *J. Pet. Sci. Eng.* **2015**, *136*, 100–111. [CrossRef]

57. Yuan, Y.; Rezaee, R.; Tongcheng, H.; Verrall, M.; Si-Yu, H.; Jie, Z. Pore Characterization and Fluid Distribution Assessment of Gas Shale. In Proceedings of the 80th EAGE Conference and Exhibition 2018, Unconventional Resources I (EAGE-SPE), Copenhagen, Denmark, 11 June 2018.

58. Bustin, R.M.; Bustin, A.M.; Cui, A.; Ross, D.; Pathi, V.M. Impact of shale properties on pore structure and storage characteristics. In Proceedings of the SPE Shale Gas Production Conference, Fort Worth, TX, USA, 16–18 November 2008.

59. Busch, A.; Schweinar, K.; Kampman, N.; Coorn, A.; Pipich, V.; Feoktystov, A.; Leu, L.; Amann-Hildenbrand, A.; Bertier, P. Shale Porosity—What Can We Learn from Different Methods. In *Fifth EAGE Shale Workshop*; EAGE: Catania, Italy, 2016.

60. Yuan, Y.; Rezaee, R. Impact of Paramagnetic Minerals on NMR-Converted Pore Size Distributions in Permian Carynginia Shales. *Energy Fuels* **2019**, *33*, 2880–2887. [CrossRef]

61. Suuberg, E.M.; Deevi, S.C.; Yun, Y. Elastic behaviour of coals studied by mercury porosimetry. *Fuel* **1995**, *74*, 1522–1530. [CrossRef]

62. Yao, Y.; Liu, D. Comparison of low-field NMR and mercury intrusion porosimetry in characterizing pore size distributions of coals. *Fuel* **2012**, *95*, 152–158. [CrossRef]

63. Melnichenko, Y.B.; Rupp, J.A.; Mastalerz, M.; He, L. Porosity of Coal and Shale: Insights from Gas Adsorption and SANS/USANS Techniques. *Energy Fuels* **2012**, *26*, 5109–5120.

64. Mastalerz, M.; Wei, L.; Drobniak, A.; Schimmelmann, A.; Schieber, J. Responses of specific surface area and micro- and mesopore characteristics of shale and coal to heating at elevated hydrostatic and lithostatic pressures. *Int. J. Coal Geol.* **2018**, *197*, 20–30. [CrossRef]

Article

Total Organic Carbon Enrichment and Its Impact on Pore Characteristics: A Case Study from the Niutitang Formation Shales in Northern Guizhou

Li Liu [1,2,3], **Shuheng Tang** [1,2,3,*] **and Zhaodong Xi** [1,2,3]

[1] School of Energy Resource, China University of Geosciences (Beijing), Beijing 100083, China; liliu@cugb.edu.cn (L.L.); xizhaod@cugb.edu.cn (Z.X.)
[2] Key Laboratory of Marine Reservoir Evolution and Hydrocarbon Enrichment Mechanism, Ministry of Education, Beijing 100083, China
[3] Key Laboratory of Strategy Evaluation for Shale Gas, Ministry of Land and Resources, Beijing 100083, China
[*] Correspondence: tangsh@cugb.edu.cn

Received: 15 March 2019; Accepted: 15 April 2019; Published: 18 April 2019

Abstract: This study analyzes samples from the Lower Cambrian Niutitang Formation in northern Guizhou Province to enable a better understanding of total organic carbon (TOC) enrichment and its impact on the pore characteristics of over-mature marine shale. Organic geochemical analysis, X-ray diffraction, scanning electron microscopy, helium porosity, and low-temperature nitrogen adsorption experiments were conducted on shale samples. Their original TOC (TOC_o) content and organic porosity were estimated by theoretical calculation, and fractal dimension D was computed with the fractal Frenkel–Halsey–Hill model. The results were then used to consider which factors control TOC enrichment and pore characteristics. The samples are shown to be dominated by type-I kerogen with a TOC content of 0.29–9.36% and an equivalent vitrinite reflectance value of 1.72–2.72%. The TOC_o content varies between 0.64% and 18.17%, and the overall recovery coefficient for the Niutitang Formation was 2.16. Total porosity of the samples ranged between 0.36% and 6.93%. TOC content directly controls porosity when TOC content lies in the range 1.0% to 6.0%. For samples with TOC < 1.0% and TOC > 6.0%, inorganic pores are the main contributors to porosity. Additionally, pore structure parameters show no obvious trends with TOC, quartz, and clay mineral content. The fractal dimension D1 is between 2.619 and 2.716, and D2 is between 2.680 and 2.854, illustrating significant pore surface roughness and structural heterogeneity. No single constituent had a dominant effect on the fractal characteristics.

Keywords: Niutitang formation; TOC recovery; organic pores; porosity; pore structure

---

1. Introduction

Organic matter (OM) in shale refers to material abundant in organic carbon generated or retained in the shale through deep burial and increasing thermal maturity, including kerogen, bitumen, solid bitumen, residual OM, pyrobitumen, and char [1,2]. OM receives considerable attention in shale gas exploration and exploitation because of its essential role in hydrocarbon generation and in organic pore formation and development [3,4].

Total organic carbon (TOC) content is a key parameter for characterizing a shale reservoir and is an indicator of shale gas potential. However, previous studies adopted various values for the lower limit for TOC in a viable resource. For example, the principal gas-productive section of the Ohio Shale had at least 2.0% TOC, which enabled the shale gas system to become the first source of commercial gas production in North America [5]. The minimum TOC of an exploration target shale pay should be 2.5–3.0% [6]. The source rock in the core producing area of the Barnett Shale system had a present-day

TOC of over 1.0% [7]. In south China, shale gas "sweet spots" are characterized by high TOC content, usually higher than 3.0% [8]. However, TOC content decreases gradually during hydrocarbon generation and expulsion, and the TOC values mentioned above are present-day TOC or residual TOC and, thus, are lower than the original TOC (TOC_o) values [3,9,10]. Conclusions based on the present-day TOC cannot accurately reflect the original geochemical characteristics of thermally over-mature organic matter and may underestimate hydrocarbon generation potential [7,11]. Investigations were, therefore, undertaken into using TOC recovery to evaluate the effective shale gas layer [7,12,13]. Furthermore, previous studies did not investigate type-I kerogen in detail. Consequently, the original TOC value and distribution in marine shale with type-I kerogen and the factors controlling both still need to be clarified.

Organic matter starts to generate nanopores once it reaches a threshold thermal maturity. These nanopores are indispensable for shale gas adsorption [2,4]. Different methods are employed to calculate and analyze organic porosity and its contributions to total porosity. One category is the examination of images through processes such as two-dimensional (2D) image processing [14–16] and three-dimensional (3D) image reconstruction [17]. The other category is theoretical calculation, including low-field NMR [18,19], rock physics models [20,21], and calculations based on the mass balance principle [3,7,12,13,22]. However, the contributions of organic pores to total porosity vary significantly, and there is, thus, much discussion of the controlling factors of organic porosity and total porosity [14,23–25]. Although the results are not consistent, it is clear that there are complex relationships between the evolution of porosity and geological parameters. Moreover, pore structure parameters (specific surface area, pore volume, pore size distribution, and fractal dimension) and an understanding of their relationships with porosity are indispensable for fully characterizing pores [16,26,27].

The Lower Cambrian Niutitang Formation shale in north Guizhou Province, southwest China was identified as a promising shale gas target zone due to its remarkable TOC content and uniform distribution [28]. However, previous studies focused on the composition, pore structure, and gas adsorption of the shale [16,19,28–30] in the study area, and a limited number examined the characteristics of the original organic matter, its porosity evolution, its pore structure, and the relationships between them. Consequently, the objectives of this paper are (1) to conduct original TOC content recovery and organic porosity estimation, and (2) to characterize the pore structure and clarify how organic matter influences the pore characteristics. The understanding obtained with this study may be conducive to the evaluation of marine shale in the over-mature stage.

2. Geological Setting

The study area is in northern Guizhou, and well YX1 is located in northwestern Fenggang County (Figure 1). Tectonically, the northern Guizhou area evolved along with the Yangtze Platform, which experienced multiple periods of tectonic movement (from the Xuefeng movement to the Himalayan movement), resulting in complex geology [31]. These tectonic movements resulted in northeast (NE)- and north-northeast (NNE)-orientated folds and faults. The study area mainly features trough-like folds striking NE and NNE. The formation of S-shaped or Z-shaped single folds is attributed to tectonic deformation dominated by compression and facilitated by strike-slip. Their presence indicates that the geological conditions under which the area developed are complicated [31,32]. Differently trending faults intersect each other. The dip angles of faults are relatively steep, including some well-preserved upright fault planes [33].

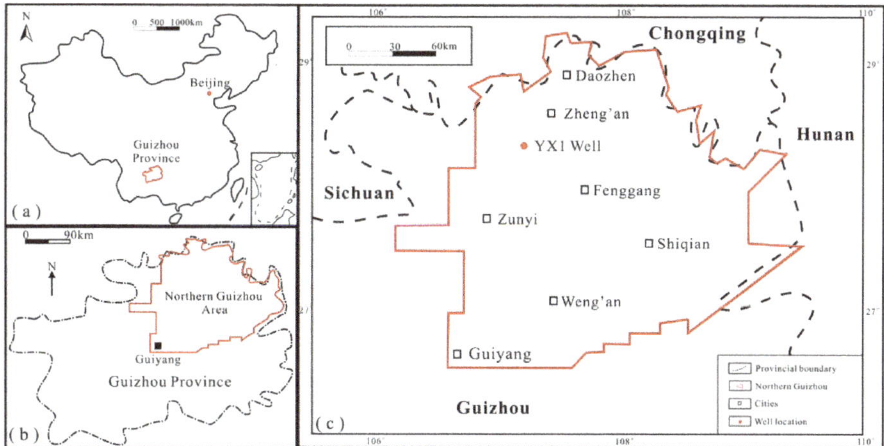

Figure 1. (**a**) Location of Guizhou Province; (**b**) location of Northern Guizhou area; (**c**) location of the study area and well location (Modified from Reference [34]).

The Lower Cambrian succession in the Upper Yangtze Platform was deposited during a major marine transgression from southeast to northwest. It is mainly composed of carbonaceous shale, and calcareous shale, carbonate, and siltstone. The black shale thickens southeastward; the present thick black shale in Sichuan, western Hunan and Hubei, Chongqing, and Guizhou province can be interpreted as being deposited in an open marine platform to marine shelf environment, while the deposits at the western margin of the Upper Yangtze Platform, which are dominated by terrestrial clastics and carbonates, are indicative of a littoral environment [28,35].

Occurrences of the Niutitang Formation in the study area can be subdivided into two members: (1) the lower member (LM), mainly consisting of carbonaceous shale, siliceous shale, and black shale with relatively high TOC content. It lies at depths of 1996–2031 m and is 35 m thick; (2) the upper member (UM), mainly consisting of organic-poor argillaceous mudstone and silty mudstone. It lies at depths of 1921–1996 m and is 75 m thick (Figure 2).

3. Samples and Methods

3.1. Samples and Experiments

Sixty-two samples were collected from the Lower Cambrian Niutitang Formation in YX1 well (Figure 1) with the depth from 1921 m to 2031 m (Figure 2), including 31 samples from the lower member (LM) and 31 samples from the upper member (UM). The core samples of the LM are dominated by black siliceous shale and carbonaceous shale, while the UM is mainly gray to dark-gray argillaceous shale and silty shale [34]. Measurements including organic geochemistry, optical microscopy, X-ray diffraction, scanning electron microscopy (SEM), nitrogen adsorption, and helium pycnometry were performed on samples. The methods used in this study were described thoroughly in Reference [36]; thus, they are not fully presented in this paper. Table 1 lists experiments and measurements conducted on each sample.

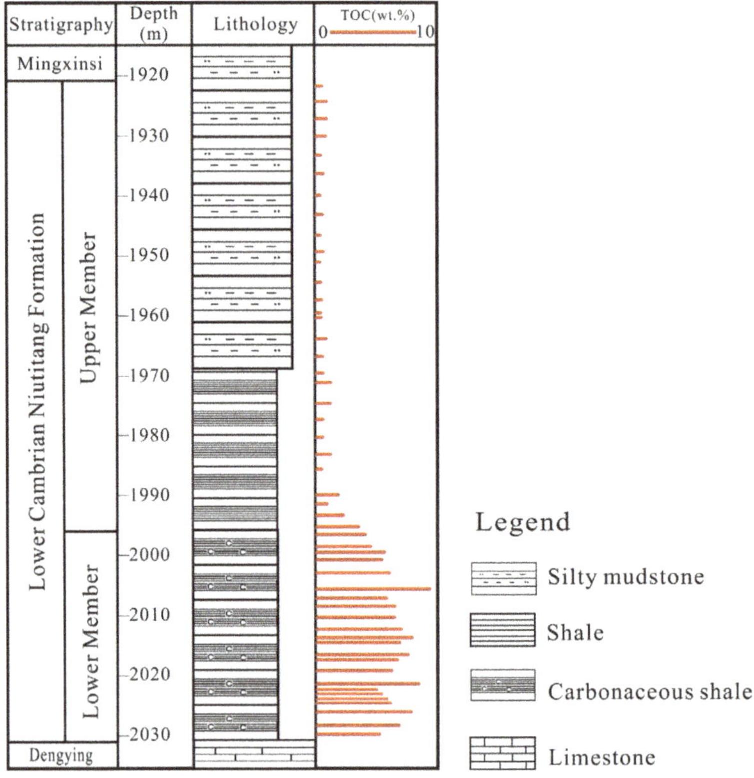

Figure 2. Lithological profile and total organic carbon (TOC) content of well YX1 section.

Table 1. Experiments and measurements conducted on samples.

Sample No.	Depth(m)	Organic Geochemistry				Mineralogy	Microscopy	Petrophysics	
		TOC	BR_o	MC	RE	XRD	SEM	NA	Porosity
1	1921.65	√	√	√	√	√	√		√
2	1924.32	√							
3	1927.22	√	√	√	√	√	√		√
4	1930.09	√							
5	1933.27	√	√		√	√	√		√
6	1936.30	√							
7	1939.32	√				√		√	
8	1939.94	√	√	√	√	√	√		√
9	1943.16	√							
10	1946.58	√		√	√	√	√		√
11	1949.36	√							
12	1951.14	√	√	√	√	√	√		√
13	1954.38	√							
14	1957.38	√							
15	1959.00	√				√		√	
16	1959.52	√	√	√	√	√	√		
17	1960.30	√							
18	1963.92	√	√	√	√	√	√		√
19	1966.73	√	√	√	√	√	√		√
20	1969.61	√							

Table 1. *Cont.*

Sample No.	Depth(m)	Organic Geochemistry				Mineralogy	Microscopy	Petrophysics	
		TOC	BR_o	MC	RE	XRD	SEM	NA	Porosity
21	1971.14	√	√		√	√	√		√
22	1974.76	√							
23	1977.42	√	√	√	√	√	√		√
24	1980.36	√							
25	1981.65	√							
26	1983.20	√							
27	1985.65	√	√	√	√	√	√		
28	1989.96	√							
29	1991.39	√	√	√	√	√	√		√
30	1993.35	√		√	√				
31	1995.31	√							
32	1996.57	√	√	√	√		√	√	√
33	1996.75	√				√			
34	1998.56	√							
35	1999.31	√	√		√	√	√		√
36	2000.81	√						√	
37	2002.05	√				√			
38	2003.02	√		√	√	√	√	√	√
39	2004.65	√				√			
40	2005.64	√						√	
41	2006.20	√							
42	2007.20	√	√	√	√	√	√		√
43	2008.57	√							
44	2010.33	√	√	√	√	√	√		√
45	2012.37	√							
46	2013.73	√							
47	2014.53	√	√	√	√	√	√	√	√
48	2014.80	√				√			
49	2016.80	√							
50	2017.21	√	√	√	√	√	√		√
51	2019.31	√							
52	2021.43	√	√	√	√		√	√	√
53	2021.50	√				√			
54	2022.38	√							
55	2023.15	√							
56	2023.98	√	√	√	√	√	√		√
57	2024.13	√						√	
58	2025.90	√				√			
59	2026.19	√	√	√	√	√	√		√
60	2028.36	√							
61	2029.86	√	√	√	√	√	√	√	√
62	2030.00	√				√			

Note: TOC: total organic carbon content; BR_o: bitumen reflectance values; MC: maceral composition; RE: Rock-Eval tests; XRD: X-ray diffraction; SEM: scanning electron microscopy; NA: nitrogen adsorption; Porosity: helium porosity.

3.2. Calculation for Original TOC and Organic Porosity

Claypool's Equation is considered to be useful [3,7,11], but the original production index (PI_o) used for calculation was assumed by the researchers; thus, the transfer ratio (TR) of kerogen was unreliable. Both hydrogen index (HI) and original hydrogen index (HI_o) were used to compute the TR, which was more reliable [13]. However, this calculation method ignores the mass change of source rock triggered by hydrocarbon generation and expulsion. Thus, rock mass ratio was taken into consideration [12], but the HI_o determination was not applicable for over-mature shale. As a result, the first step of their method was revised, and the method is presented in detail below.

Firstly, the original hydrogen index (HI_o) was calculated by Equation (1).

$$HI_o = (750 \times S + 450 \times L + 125 \times V + 50 \times I)/100, \tag{1}$$

where S, L, V, and I are volume percentages of sapropelinite, liptinite, vitrinite, and inertinite, respectively, on a mineral matter-free basis [11].

The TR refers to the degree of kerogen transformation due to hydrocarbon generation. Hydrogen index (HI) and original hydrogen index (HI_o) can be used to determine TR using Equation (2) [37].

$$TR = [1200 \times (HI_o - HI)]/[HI_o \times (1200 - HI)]. \tag{2}$$

The expulsion efficiency, f, refers to the ratio of the hydrocarbon expulsion amount to the total hydrocarbon generation amount, which can be expressed by Equation (3).

$$f = 1 - [S_1 \times (1 - TR)]/(S_2 \times TR). \tag{3}$$

The definition of mass conversion factor w is the ratio of present-day rock mass to original rock mass [12]; the factor can be obtained from Equation (4).

$$w = 1 - 0.833 \times TOC/100. \tag{4}$$

Then, TOC_o is determined using Equation (5).

$$TOC_o = TOC\,[1 - \alpha \times f \times TR \times (1 - 0.833 \times TOC/100)], \tag{5}$$

where the scaling parameter α refers to the proportion of convertible carbon to total carbon, and $\alpha = HI_o/1200$.

Finally, organic porosity Φ_{org} is obtained from Equation (6).

$$\Phi_{org} = t \times (TOC_o \times \alpha \times f \times w) \times \rho_b/\rho_k, \tag{6}$$

where t stands for the carbon equivalent mass of kerogen, $t = 1.2$; and ρ_b and ρ_k are the bulk density of source rock and kerogen density, respectively. The type-I kerogen density is 1.45 g/cm^3 [38].

3.3. Fractal Theory

A fractal dimension (D) can characterize pore geometry by evaluating complexity and irregularity of both pore surface and structure [39–41]. In this study, fractal dimensions can be determined using the Frenkel–Halsey–Hill (FHH) method, together with the N_2 adsorption isotherm data. The FHH model can be expressed by the following equation:

$$ln(V/V_0) = A \times [ln(ln(P_0/P)] + C, \tag{7}$$

where V is the volume of adsorbed gas molecules at the equilibrium pressure P, V_0 is the volume of monolayer coverage, A is the power-law exponent depending on D and the mechanism of adsorption, and P_0 is the saturation pressure of the gas. A is derived from the slope of linear regression equation in the plot of lnV versus $ln(ln(P_0/P))$. The fractal dimension D derives from the two following expressions: $D = A + 3$ and $D = 3A + 3$.

4. Results

4.1. Organic Geochemistry and Petrography

Figure 2 shows that the TOC contents of the Niutitang shale samples have a wide range of 0.29% to 9.36%; the upper and lower members average 0.69% and 5.93%, respectively. Only one sample from the upper member has a TOC value of over 2.0%. The bitumen reflectance (BR_o) value ranges from 1.47% to 2.56%, corresponding to vitrinite reflectance between 1.72% and 2.72% with a mean of 2.19% as calculated with the conversion equation [42]. The majority (60–84%) of the organic matter is sapropelinite, while liptinite accounts for 15–37%. The amounts of vitrinite and inertinite are quite small, 0–3% and 0–2%, respectively. Table 2 presents the calculated type index (TI) and shows that only two samples have a TI lower than 80, indicating that the Niutitang shale is dominated by type-I kerogen. This finding is consistent with previous studies [28,34].

Table 2. Sample thermal maturity and maceral.

Sample ID	Depth	BR_o	$EqVR_o$	S	L	V	I	TI	Type	HI_o
	m	%	%	%	%	%	%			mgHC/gTOC
1	1921.65	1.52	1.77	71	25	2	2	80	I	627.50
3	1927.22	1.47	1.72	76	21	1	2	83.75	I	653.75
5	1933.27	1.88	2.10	/	/	/	/	/	/	/
8	1939.94	1.53	1.78	70	27	2	1	81	I	649.50
10	1946.58	/	/	69	28	2	1	80.5	I	637.50
12	1951.14	2.34	2.52	80	18	1	1	87.25	I	682.75
16	1959.52	1.88	2.10	66	32	1	1	80.25	I	640.75
18	1963.92	1.96	2.17	72	27	1	0	84.75	I	662.75
19	1966.73	2.56	2.72	60	37	2	1	76	II_1	619.50
21	1971.14	2.06	2.27	/	/	/	/	/	/	/
23	1977.42	1.84	2.06	76	20	2	2	82.5	I	603.50
27	1985.65	2.47	2.64	68	29	2	1	80	I	640.50
29	1991.39	1.72	1.95	78	20	2	0	86.5	I	677.50
30	1993.35	/	/	76	22	1	1	85.25	I	670.75
32	1996.57	1.96	2.17	80	17	2	1	86	I	679.50
35	1999.31	2.19	2.39	/	/	/	/	/	/	/
38	2003.02	/	/	76	21	2	1	84	I	667.50
42	2007.2	1.68	1.92	66	32	1	1	80.25	I	634.25
44	2010.33	1.74	1.97	77	21	1	1	85.75	I	673.75
47	2014.53	2.00	2.21	76	23	1	0	86.75	I	674.75
50	2017.21	1.86	2.08	82	16	2	0	88.5	I	689.50
52	2021.43	2.43	2.61	71	26	2	1	81.5	I	652.50
56	2023.98	1.75	1.98	84	15	1	0	90.75	I	698.75
59	2026.19	2.33	2.51	70	26	3	1	79.75	II_1	646.25
61	2029.86	2.41	2.59	76	22	2	0	85.5	I	671.50

Note: ID = identifier; BR_o = bitumen reflectance; $EqVR_o$ = equivalent vitrinite reflectance, $EqVR_o$ = 1.09 × BR_o − 0.41; S = sapropelinite, L = liptinite, V = vitrinite, I = inertinite, TI = type index, TI = 100 × %S + 50 × %L − 75 × %V − 100 × %I, and TI > 80, 80 > TI > 0, and TI < 0 indicate Type I, Type II, and Type III, respectively. HI_o = original HI.

Table 3 illustrates some of the main characteristics of the relative mineral composition obtained with X-ray diffraction (XRD) analysis. The samples mainly contain quartz, feldspar, carbonate, pyrite, and clay (illite, mixed-layer illite/smectite, and chlorite). The quartz content ranges from 20.3% to 54.5% with an average of 37.8%. The clay mineral content ranges from 8.7% to 47.1% and comprises illite (50–100%), mixed-layer illite/smectite (0–29%), and chlorite (0–31%). In eight of the 11 samples from the lower member of the Niutitang Formation, the only clay mineral is illite, revealing that the lower member entered the post-mature stage, with vitrinite reflectance >2.0% [43]. The other samples contain illite, chlorite, and mixed-layer illite/smectite, in which 10% comprises smectite layers, indicating a highly mature stage [43,44]. The relationships between thermal maturity and clay mineral type and content ensure that the equivalent vitrinite reflectance ($EqVR_o$) values calculated by the fitting equation [42] are relatively reasonable. Furthermore, the pyrite content lies in the range of 1.3% to 13.8%; the mean pyrite contents of the upper and the lower member are 3.3% and 10.2%, respectively. Generally, pyrite forms in an anoxic environment [15]; thus, it can be taken as evidence that the lower member of Niutitang Formation had a stagnant, anoxic depositional environment [34].

Table 3. Mineralogical composition of the shale samples.

Sample	Depth	Q	F	Cal	Dol	Py	C	I/S	It	Ch	I/S Ratio
ID	m				%				Relative Percent (%)		
1	1921.65	37.9	9.6	2.1	/	1.7	47.1	/	73	27	/
3	1927.22	38.4	9.6	2.8	/	3.1	44.1	17	61	22	10
5	1933.27	40.2	10.4	/	/	2.7	46.7	/	69	31	/
7	1939.32	41.0	9.4	2.7	/	1.3	45.6	n	n	n	n
8	1939.94	39.4	9.7	2.6	/	1.3	47.0	/	73	27	/
10	1946.58	37.8	13.1	3.7	/	2.8	40.8	/	71	29	/
12	1951.14	39.4	9.6	1.5	/	2.9	46.6	/	69	31	/
15	1959.00	37.4	10	2.4	0	3.8	46.4	n	n	n	n
16	1959.52	39.7	9.4	2.3	/	2.6	44.5	12	60	28	10
18	1963.92	38.3	9.4	2.0	/	3.9	44.0	22	55	23	10
19	1966.73	36.1	9.6	9.6	2.1	3.0	39.7	17	57	26	10
21	1971.14	37.1	9.5	1.1	/	4.2	44.6	10	64	26	10
23	1977.42	36.9	10.4	0.9	/	2.5	46.1	24	50	26	10
27	1985.65	33.7	10.2	/	17.5	8.6	29.9	26	68	6	10
29	1991.39	33.2	13.4	/	11.6	5.0	36.8	23	77	/	10
33	1996.75	30.3	17.6	1.4	14.9	9.0	26.7	22	69	9	10
35	1999.31	32.6	16.1	2.8	12.0	12.3	24.2	/	100	/	/
37	2002.05	33.8	15.4	1.8	15.5	12.4	21.1	n	n	n	n
38	2003.02	27.1	17.4	9.9	13.3	12.3	20.0	/	100	/	/
39	2004.65	29.4	16.2	3.1	15.8	12.0	22.6	n	n	n	n
42	2007.20	36.4	12.3	5.0	6.1	13.8	21.9	/	100	/	/
44	2010.33	54.5	12.5	/	5.1	8.4	18.2	/	100	/	/
47	2014.53	36.3	16.1	1.0	26.9	8.4	10.3	/	100	/	/
48	2014.80	58.7	16.8	1.2	2.4	9.0	11.9	n	n	n	n
50	2017.21	43.2	23.2	/	9.3	10.3	14.0	/	100	/	/
53	2021.50	20.3	9.2	2.1	51.1	7.9	8.7	/	100	/	/
56	2023.98	42.0	16.6	8.1	3.8	8.5	21.0	29	71	/	10
58	2025.90	52.6	15.9	0.7	2.7	8.3	19.8	n	n	n	n
59	2026.19	52.3	16.3	/	5.7	7.8	18.0	/	100	/	/
61	2029.86	29.8	8.8	6.7	26.1	11.4	15.5	4	96	/	10
62	2030.00	33.1	8.5	5.7	23.5	12.2	15.4	n	n	n	n

Note: Q = quartz, F = feldspar, Cal = calcite, Dol = dolomite, Py = pyrite, C = clay minerals, I/S = illite/smectite mixed layer, It = illite, Ch = chlorite, n = no available data, "/" = not detected.

4.2. Estimation of Original TOC and Organic Porosity

Table 2 presents the results of original TOC reconstruction and organic porosity estimation. The HI_o ranges from 640.75 mg/gTOC to 689.5 mg/gTOC with a mean of 660.75 mg/gTOC. The TR values are over 95%, indicating that the organic matter present is approaching the end of hydrocarbon generation [3]. The low S2 values suggest that the source rock has almost no hydrocarbon generation potential (Table 4). Additionally, the high expulsion efficiency reveals that there is a minute quantity of remaining hydrocarbon. The TOC_o values for the upper member are between 0.64% and 1.91% with a mean of 1.15%, and those for the lower member range from 10.66% to 18.17% and average 14.02%, significantly higher than the upper member. S2 was between 4.14 mg/g (rock) and 125.28 mg/g (rock). The parameters in Table 5 demonstrate that a considerable amount of hydrocarbon was generated and expelled during thermal maturation. Furthermore, the recovery coefficient (the ratio of TOC_o to TOC_{pd}) is between 2.02 and 2.28 with an average of 2.16. TOC_o has a good positive correlation with TOC_{pd}, indicating that all of the samples experienced similar changes in TOC_o.

Table 4. Rock-Eval data and HI of samples.

Sample	Depth	T_{max}	S1	S2	S3	HI
ID	m	°C	mg/g	mg/g	mg/g	mgHC/gTOC
1	1921.65	442	0.01	0.06	0.46	13.33
3	1927.22	436	0.04	0.46	0.62	53.49
5	1933.27	593	0.01	0.01	0.57	2.78
8	1939.94	445	0.01	0.01	0.29	3.03
10	1946.58	597	0.01	0.01	0.43	3.33
12	1951.14	600	0.01	0.02	0.31	6.90
16	1959.52	573	0.01	0.01	0.42	2.70
18	1963.92	426	0.01	0.01	0.50	1.30
19	1966.73	364	<0.01	0.01	0.40	2.04
21	1971.14	600	<0.01	0.02	0.38	1.75
23	1977.42	524	0.01	0.02	0.47	4.00
27	1985.65	512	<0.01	0.02	0.46	5.00
29	1991.39	530	0.01	0.02	0.46	2.38
30	1993.35	/	/	/	/	/
32	1996.57	536	<0.01	0.01	0.76	0.25
35	1999.31	514	0.01	0.02	0.50	0.58
38	2003.02	523	<0.01	0.01	0.80	0.17
42	2007.2	309	<0.01	0.01	0.61	0.17
44	2010.33	572	0.01	0.01	0.65	0.16
47	2014.53	555	<0.01	0.01	0.58	0.15
50	2017.21	541	0.01	0.02	0.60	0.30
52	2021.43	586	0.01	0.01	1.21	0.12
56	2023.98	524	<0.01	0.01	0.93	0.17
59	2026.19	581	<0.01	0.01	0.81	0.13
61	2029.86	600	0.01	0.01	1.02	0.19

Note: T_{max}: maturity parameter based on the temperature at which the maximum amount of pyrolyzate (S2) is generated from the kerogen in a rock sample; S1: free hydrocarbons present in the rock (mg HC/g of rock); S2: remaining generation potential (mg HC/g of rock); S3: oxidizable carbon, (mg CO_2/g rock); HI: hydrogen index.

The organic porosity derived from TOC_o is shown in Table 5. Values for the upper member range from 0.85% to 2.67% and average 1.55%, whereas those for the lower member are between 12.84% and 22.59% with a mean of 16.94%. The two members had different capacities for generating organic pores due to the marked discrepancy in TOC_o between them. However, the calculated organic porosity is merely a theoretical value.

Table 5. TOC_o calculation, total porosity, and organic porosity estimation.

Sample	Depth	TR	α	f	ϱ_g	ϱ_b	TOC	TOC_o	Φ_{org}	Φ
ID	m				g/cm^3			%		
1	1921.65	0.990	0.540	0.998	2.82	2.79	0.45	0.93	1.28	1.07
3	1927.22	0.963	0.556	0.997	2.80	2.78	0.86	1.78	2.41	0.72
8	1939.94	0.998	0.541	0.998	2.82	2.75	0.33	0.71	0.94	2.49
10	1946.58	0.998	0.539	0.998	2.78	2.77	0.30	0.63	0.85	0.36
12	1951.14	0.996	0.569	0.998	2.79	2.78	0.29	0.66	0.93	0.36
16	1959.52	0.998	0.534	0.998	2.82	2.77	0.37	0.79	1.04	1.77
18	1963.92	0.999	0.552	0.999	2.78	2.77	0.77	1.70	2.31	0.36
23	1977.42	0.997	0.553	0.999	2.81	2.78	0.50	1.00	1.52	1.07
29	1991.39	0.999	0.565	0.999	2.82	2.79	0.84	1.91	2.67	1.07
44	2010.33	0.999	0.534	0.999	2.63	2.53	6.44	12.88	14.90	3.75
50	2017.21	0.999	0.562	0.999	2.64	2.58	6.67	14.22	17.44	2.26
52	2021.43	0.999	0.575	0.999	2.75	2.60	8.47	18.17	22.59	5.32
61	2029.86	0.999	0.539	0.999	2.76	2.61	5.17	10.66	12.84	5.43

Note: TR = transformation ratio, α = scaling parameter, f = expulsion efficiency, ϱ_g = grain density, ϱ_b = bulk density, Φ_{org} = organic porosity, Φ = total porosity.

4.3. Total Porosity

The helium porosity of the Niutitang shale in the study area ranges from 0.36% to 6.93% with a mean of 2.61%. That of the upper member lies in the range of 0.36% to 2.49% with a mean of 1.01%, while that of the lower member varies from 2.26% to 6.93% with a mean of 4.36%. Figure 3a shows how porosity changes with TOC. When the TOC is lower than 1.0%, the points are scattered; when the TOC is between 1.0% and 6.0%, porosity is positively correlated with TOC; when the TOC content exceeds 6.0%, the positive trend between these two parameters becomes negative, which is consistent with the findings of some previous studies [20,24,25,45]. However, it is noteworthy that sample 47 had a TOC content of 6.86% and a porosity of 6.93%, which differs from the negative trend illustrated by Figure 3a. A possible explanation for this phenomenon is that sample 47 has a relatively high brittle mineral content, as evidenced by the well-developed fractures seen in its SEM image (Figure 3b); fractures are less developed in the other samples.

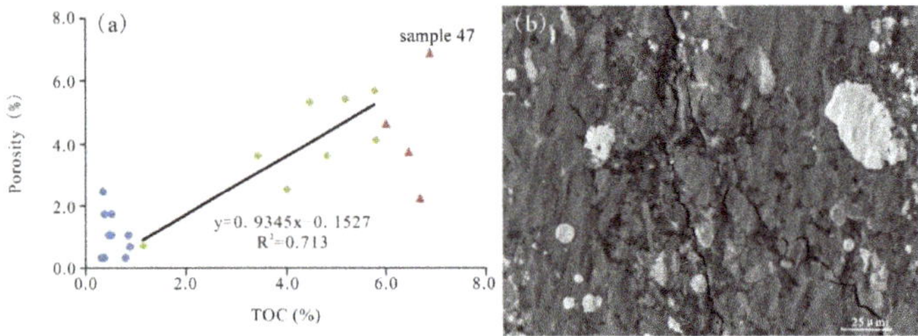

Figure 3. Relationship between total porosity and TOC content (**a**) and SEM image of sample 47 (**b**). In Figure 3a, red dots represent samples with TOC values over 6.0%, green dots represent samples with TOC values between 1.0% and 6.0%, and blue dots represent samples with TOC values below 1.0%.

4.4. N_2 Adsorption

4.4.1. N_2 Adsorption–Desorption Isotherm

The nitrogen adsorption–desorption isotherms from this study are presented in Figure 4. The isotherms of all the Niutitang Formation shale samples can be classed as type IV in the International Union of Pure and Applied Chemistry (IUPAC) classification system [46,47]. These isotherms have no plateau within the high relative pressure interval and are characterized by pronounced hysteresis loops due to capillary condensation within mesopores [46]. The adsorption volume is low and rises slowly in the low relative pressure range, but the isotherms became steep in the high relative pressure interval ($p/p_0 > 0.9$).

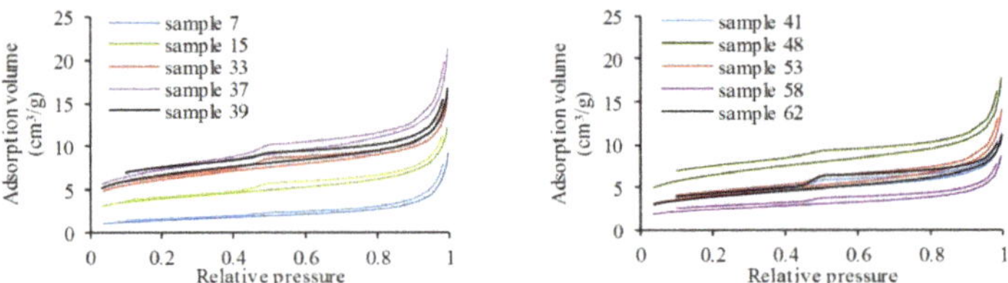

Figure 4. Nitrogen adsorption–desorption isotherms.

Although the degree of development of hysteresis loops varies, samples 7, 15, and 33 can be considered type H3 and the rest of samples type H4 (Figure 4) according to the IUPAC classification system, suggesting the existence of slit- or plate-like pores [46]. The hysteresis loops also indicate that ink-bottle pores are rare in the samples. Cylindrical pores, slit-like pores, and wedge-like pores are open pores that are conducive to gas flow, while ink-bottle pores can adsorb gas [47]. Consequently, the sorption capacity of these samples may be low.

4.4.2. Pore Structure

Values for pore structure parameters derived from the nitrogen adsorption isotherms are shown in Table 6. The Brunauer–Emmett–Teller (BET) specific surface area ranges from 5.08 m^2/g to 25.31 m^2/g with a mean of 16.21 m^2/g. The micropore surface area computed by the t-plot method ranges from 0.96 m^2/g to 7.78 m^2/g with an average of 4.85 m^2/g. The calculated Barrett–Joyner–Halenda (BJH) total pore volume varies between 0.013 cm^3/g and 0.029 cm^3/g and averages 0.019 cm^3/g, while the micropore volume lies in the range 0.0004–0.0037 cm^3/g with a mean of 0.0024 cm^3/g. The average pore diameter varies between 6.18 nm and 12.62 nm with an average of 7.45 nm, indicating that the Niutitang Formation shale is dominated by mesopores.

Table 6. Pore structure parameters for shale samples. BET—Brunauer–Emmett–Teller.

Sample ID	Depth	BET Surface Area	Micropore Area	Pore Volume	Micropore Volume	Average Pore Width
	m	m^2/g	m^2/g	cm^3/g	cm^3/g	nm
7	1939.32	5.0845	0.9653	0.014261	0.000436	12.6157
15	1959.00	13.8464	4.4582	0.0171	0.002103	7.3046
32	1996.75	21.4044	7.2142	0.021144	0.003405	6.1846
36	2002.05	25.3148	8.055	0.02947	0.00379	7.0011
38	2004.65	23.1018	7.7756	0.022171	0.003671	6.2249
40	2005.64	13.8811	4.0763	0.015734	0.001903	6.7046
47	2014.80	22.5701	6.507	0.024569	0.003036	6.4965
52	2021.43	14.5405	3.4514	0.020749	0.00159	7.5187
57	2025.90	8.7111	2.3449	0.012545	0.001091	8.036
61	2030.00	13.6746	3.6978	0.015642	0.001724	6.3833

Figure 5 shows all of the samples to have similar pore size distributions except for sample 7, which shows no significant peak and, thus, has a relatively uniform pore distribution. As a result, this sample exhibits the smallest BET surface area and pore volume and the largest average pore size. Other samples tend to be unimodal with a major peak in the pore size of 2–3 nm, revealing that they are dominated by mesopores. Additionally, a small but non-negligible amount of macropores are present.

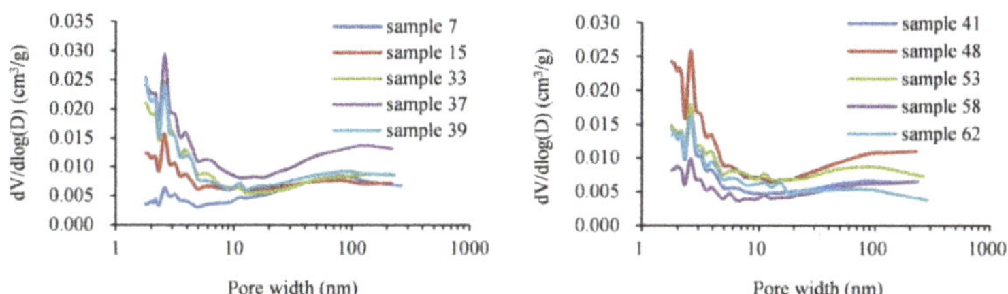

Figure 5. Pore width distribution of samples.

4.5. Fractal Characteristics

The FHH plots for the shale samples are presented in Figure 6. The presence of two distinct linear segments in these plots at P/P$_0$ ranges of 0–0.5 and 0.5–1 suggests that the fractal characteristics vary within those two relative pressure ranges; thus, the fractal dimensions D1 and D2 were acquired from the two linear intervals, respectively. Table 7 gives the slopes of the linear regression equations A_1 and A_2 and the fractal dimension values D1 and D2. By using the equation D = 3A + 3, the D1 values of samples 7 and 52 were calculated to be less than 2, which is unreasonable because the fractal dimension lies in the range 2 to 3 with increasing irregularity and complexity [39]. As a result, fractal dimensions calculated with the equation D = A + 3 are used for further analysis and discussion.

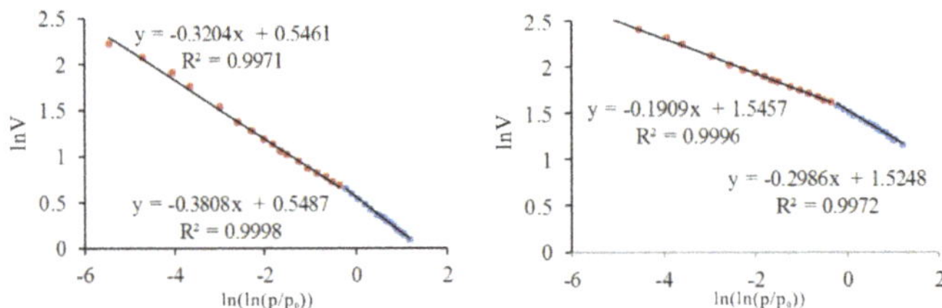

Figure 6. Plots of lnV vs. $\ln(\ln(P_0/P))$ obtained from the N_2 adsorption isotherms of two typical shale samples: (**a**) sample 7 and (**b**) sample 15. In Figure 6a,b, red dots indicate the P/P_0 range of 0.5–1, and blue dots indicate the P/P_0 range of 0–0.5.

Table 7. Fractal dimensions of pores obtained from N_2 adsorption of the shale samples.

Sample	Depth	P/P_0: 0–0.5			P/P_0: 0.5–1		
ID	(m)	R^2	A1	D1	R^2	A2	D2
				A + 3 3A + 3			A + 3 3A + 3
7	1939.32	0.9998	−0.381	2.619 1.858	0.9971	−0.320	2.680 2.039
15	1959.00	0.9972	−0.299	2.701 2.104	0.9996	−0.191	2.809 2.427
33	1996.75	0.9946	−0.289	2.711 2.133	0.9983	−0.146	2.854 2.563
37	2002.05	0.9951	−0.300	2.700 2.100	0.9979	−0.173	2.828 2.483
39	2004.65	0.9881	−0.284	2.716 2.147	0.9984	−0.149	2.851 2.553
41	2006.20	0.9912	−0.311	2.689 2.066	0.9994	−0.169	2.831 2.492
48	2014.80	0.9890	−0.311	2.689 2.068	0.9973	−0.162	2.838 2.515
53	2021.50	0.9957	−0.339	2.661 1.982	0.9983	−0.195	2.805 2.416
58	2025.90	0.9937	−0.323	2.677 2.030	0.9990	−0.215	2.785 2.354
62	2030.00	0.9952	−0.321	2.679 2.038	0.9906	−0.156	2.844 2.533

The D1 values range from 2.619 to 2.716 with an average value of 2.684, and the D2 values range from 2.680 to 2.854 with a mean of 2.813. The results illustrate that both the pore surfaces and structures of the shales are highly irregular and heterogeneous. Moreover, a good positive correlation between D1 and D2 can be observed in Figure 7, revealing that D1 and D2 were both suitable for characterizing pore structures in the Niutitang Formation shale. This finding is consistent with a previous study focusing on the Lower Silurian Longmaxi Formation shale [27]. The higher D2 values suggest that, with rising pore diameter, the pore surface of the shale becomes rougher and the pore structure more complicated.

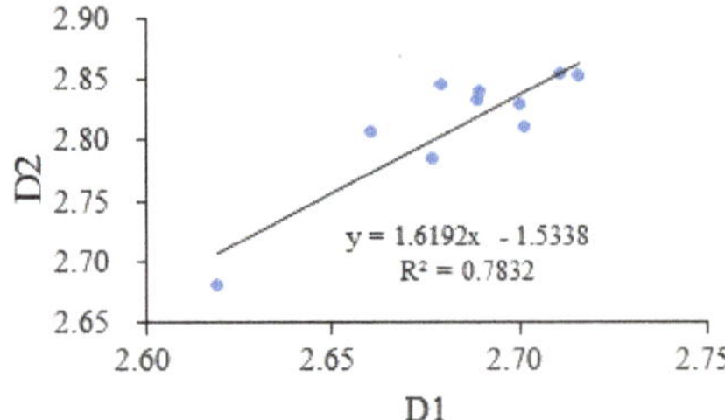

Figure 7. Relationship between D1 and D2.

5. Discussion

5.1. TOC Recovery

Hydrocarbon generation is a result of organic matter decomposition with increasing thermal maturity [6]; thus, original TOC decreases gradually due to the loss of the convertible part of the organic matter. As more than three-quarters of the samples entered the over-mature stage, the HI values of the LM samples dropped to below 1 mg/g (Table 4), illustrating that a large amount of hydrocarbon was generated and expelled.

In addition, 13 samples have a recovery coefficient between 2.02 and 2.28, showing little variation. However, the present-day TOC of the LM is markedly higher than that of the UM. Four LM samples with high present-day TOC had much higher TOC_o, while nine UM samples with low present-day TOC (<1.0%) had relatively low TOC_o. This indicates that TOC enrichment was dependent on the sedimentary environment. Previous studies characterized paleo-environments on the basis of productivity proxies and redox condition proxies [34,48,49]. For example, upwelling and hydrothermal events triggered a remarkable increase in paleo-productivity during LM deposition and showed that the redox conditions evolved from anoxic with euxinic intervals during LM deposition to oxic conditions with anoxic intervals upward [34]. In this study, mineral compositions are used to characterize the sedimentary environment. Firstly, quartz content is positively correlated with TOC in the LM samples but has a weak negative correlation with TOC in the UM samples (Figure 8). Quartz of biogenic origin has a positive correlation with TOC, whereas the correlation between terrigenous detrital origin quartz and TOC is negative [50,51]. Thus, the quartz source of the LM was dominantly authigenic, while that in the UM had both terrigenous inputs and an authigenic source. Secondly, the average pyrite contents of the UM and the LM shale are 3.4% and 10.3%, respectively. These results also indicate that the LM was deposited in a reducing environment with relatively stagnant water mass and that this transformed into an oxidizing environment upward.

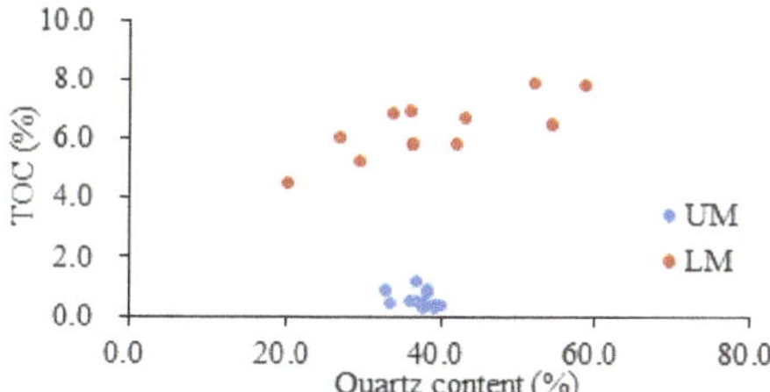

Figure 8. Relationship between TOC and quartz content.

5.2. Porosity and Controlling Factors

Organic matter decomposes during thermal maturation, leading to the formation of hydrocarbons and intraparticle organic pores [4], and migrated organic matter develops secondary pores as it cracks at high maturity [52]. Nanoporosity evolutionary sequence includes the formation stage (0.60% < VR_o < 2.0%), the development stage (2.0% < VR_o < 3.5%), and the conversion or destruction stage (VR_o > 3.5%) [53]. However, no general trend between organic porosity and maturity was observed when VR_o > 0.9% in the Woodford shale [14]. In this study, organoporosity was calculated with the method presented in Reference [12]. There is no obvious relationship between the calculated organoporosity and $EqVR_o$ (Figure 9). Because of the marked differences in the TOC of the two members, samples from the LM had higher organoporosity (over 10%) than those from the UM. Thus, both total porosity and organic porosity had no direct relationship with thermal maturity, and other factors like TOC content should be given more consideration when characterizing shale porosity.

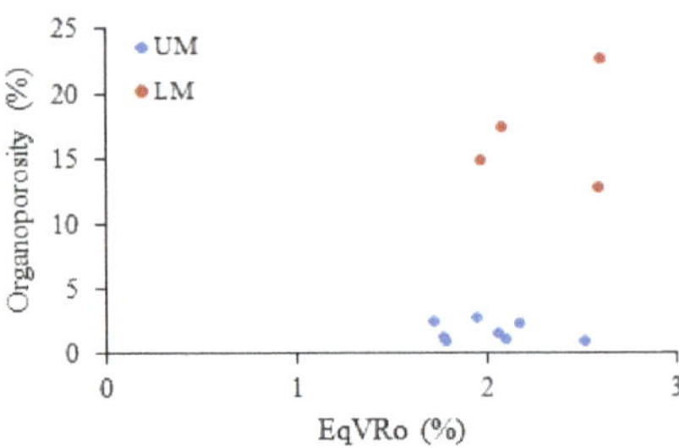

Figure 9. Relationship between total porosity and equivalent vitrinite reflectance (EqVR$_o$).

Figure 3a reveals the relationship between total porosity and TOC. Total porosity is positively correlated with TOC when 1.0% < TOC < 6.0%, after which the trend changes. A similar pattern was reported in other studies; for example, there was a change in a positive correlation at TOC = 5.5% in the Marcellus shale from North America [24], TOC = 5.0% in the Qiongzhusi Formation from southern Sichuan Basin [45], and TOC = 12% in the Permian shale from eastern China [25].

The contribution of per-gram organic carbon content to total porosity can be derived from the slope of the linear regression equation in a plot of total porosity versus TOC. For samples with 1.0% < TOC < 6.0%, the TOC contribution is 0.93% (Figure 3a), which is greater than in Lower Silurian shale in southern China (0.72%) [23] and higher than in Lower Cambrian shale in the southern Sichuan Basin (0.40%) [45]. However, these values may be invalid because they are based on limited samples; further study is needed to determine the contribution made by TOC content. For samples with TOC < 1.0% and TOC > 6.0%, organic pores make less of a contribution to porosity. Figure 10 shows that the recovery coefficient and calculated organoporosity have no obvious relationship with porosity, indicating that inorganic pores are the main contributors to porosity. When TOC < 1.0%, the volume of organic pores is too small to influence the porosity. The porosity decreasing when TOC > 6.0% may be attributable to mechanical compaction [24,25]. A previous study showed that Young's modulus values of kerogen, clay, and quartz were 10 GPa, 15–45 GPa, and 50–70 GPa, respectively [54], and an association between significant lowering of the kerogen modulus and organic pore development was illustrated. The immature kerogen modulus was 15–20 GPa, whereas kerogen in the gas window had a smaller modulus of 7–12 GPa [54]. This is one reason why fewer organic pores can be observed in the SEM images (Figure 11). Another possible explanation is that the organic matter seen in the SEM images is depositional organic matter, which mainly consists of kerogen and seldom yields pores [52]. Lastly, some micropores and fine mesopores cannot be clearly observed; thus, the insufficient magnification of the SEM images may be a factor [7].

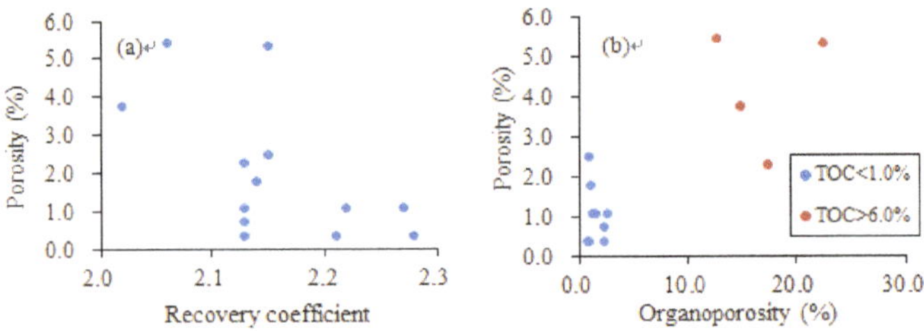

Figure 10. (**a**) Relationship between porosity and recovery coefficient; (**b**) relationship between porosity and organoporosity.

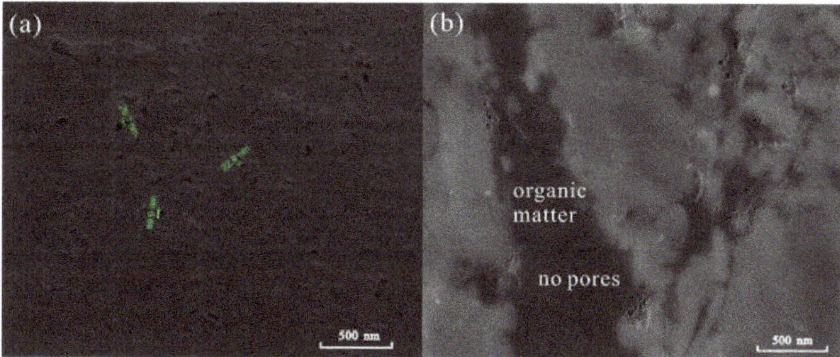

Figure 11. (**a**) Organic pores in sample 5; (**b**) Organic matter in sample 59.

The relationships between quartz, clay minerals, and total porosity are presented in Figure 12. Total porosity has a slight negative correlation with quartz content in the lower member but no obvious relationship with it in the Upper member. Similar findings were obtained from analysis of Devonian–Mississippian shale in western Canada, showing that total porosity had a negative correlation with biogenic quartz due to the loss of pores with a diameter of less than 10 nm [50]. Furthermore, quartz of detrital origin with angular grain boundaries is more likely to form interparticle pores, which are mainly macropores and mesopores, leading to a positive correlation between detrital quartz and total porosity [23,31].

The total porosity has a good negative correlation with clay content. Previous studies proposed that clay minerals increase total porosity [51,55]. Additionally, the lattice of randomly oriented clay mineral platelets forms a "house of cards" structure [14], contributing to the preservation of linear and triangular pores. However, for shale layers, higher clay content may weaken their capacity to resist compaction, which will cause a pronounced decrease in porosity. Clay-associated pores form with increasing clay content, but more pores, including organic pores and mineral pores, vanish during the mechanical compaction resulting from high clay content. As a result, the contribution of clay-associated pores to total porosity may be negligible [25]. Hypotheses regarding this phenomenon are preliminary and need further confirmation.

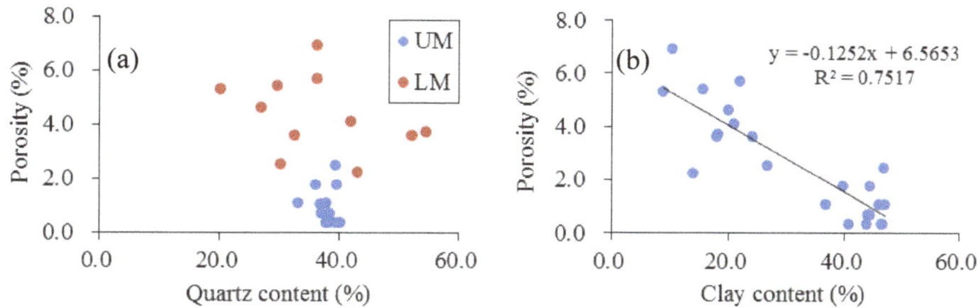

Figure 12. Effects of mineral composition on total porosity: (**a**) quartz; (**b**) clay minerals.

5.3. Pore Structure and Fractal Characterization

5.3.1. Factors Controlling Pore Structure

Figure 13 illustrates the relationships among the pore structure parameters of the shale samples. Specific surface area is significantly correlated with total pore volume ($R^2 = 0.81$) and micropore volume ($R^2 = 0.96$), which agrees with previous studies on highly mature marine shale [27,40,51,56]. Additionally, the average pore size exhibits a negative correlation with total pore volume and specific surface area (Figure 13a,b). Samples with smaller average pore diameters tend to possess more micropores and mesopores, leading to a higher pore volume and specific surface area. Micropores may be a more significant contributor to the specific surface area than are mesopores.

The relationships between shale constituents and pore structure parameters are shown in Figure 13. The slightly positive relationships between TOC content and total surface area, micropore surface area, total pore volume, and micropore volume seen in Figure 13c,d suggest that TOC is not the major contributor to these structural parameters in the samples. However, previous studies drew different conclusions, finding that specific surface area and pore volume have a significantly positive correlation with TOC content [27,40,57]. An examination of the SEM images of the samples used in those studies reveals that this correlation may be attributable to the presence of a large amount of well-preserved organic pores. Organic pores are limited in the samples considered in the current study, as seen in the SEM images, yielding a different outcome.

The moderate negative trends between quartz content and total surface area and total pore volume shown in Figure 13e,f reveal that quartz content does not exert a significant influence on pore structure in this study. Biogenetic quartz has smooth rims, which tend to form less interparticle porosity. In addition, quartz tends to provide less pore volume [58].

Specific surface area and total pore volume have a slightly negative relationship with clay content (Figure 13g,h), which is in agreement with some previous studies [25,26,58]. On the one hand, the specific surface area values of clay minerals vary considerably. For example, montmorillonite and mixed-layer illite/smectite have relatively high specific surface areas of 76.4 m^2/g and 30.8 m^2/g, respectively, but chlorite and illite have smaller surface areas of 11.7 m^2/g and 7.1 m^2/g [59]. As presented in Section 4.1, the clay content of these samples is dominated by illite (average 78%), which makes less of a contribution to the surface area. On the other hand, parts of clay-associated pores that are not protected from brittle minerals will be deformed by mechanical compaction. Thus, the specific surface area and total pore volume show little correlation with clay mineral content. To conclude, no simple relationship is seen between pore structure and shale composition in this study.

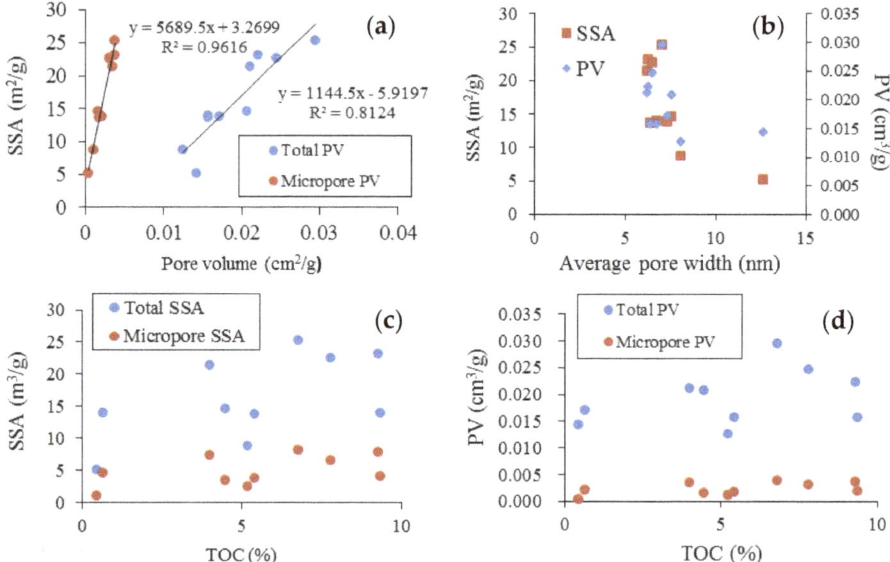

Figure 13. *Cont.*

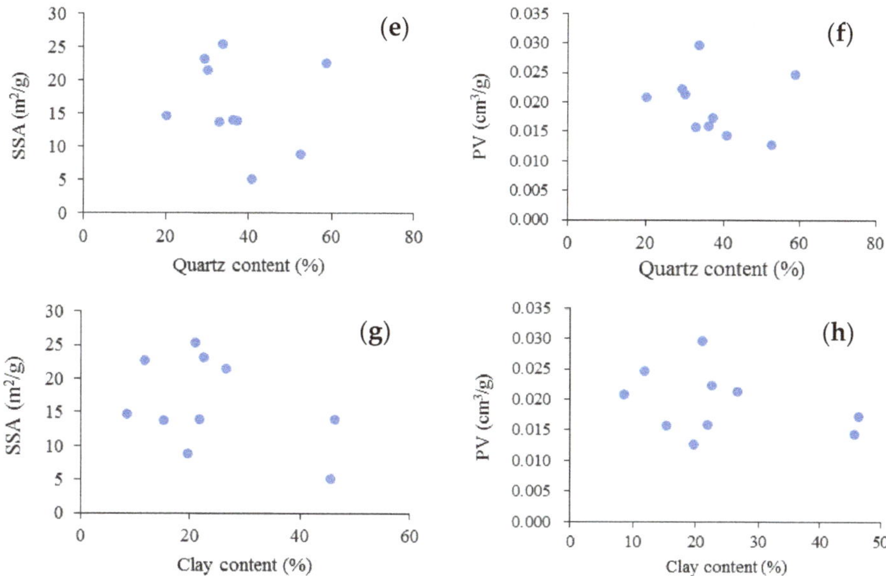

Figure 13. Relationship among structure parameters (**a,b**) and effects of shale composition on structure parameters (**c–h**); in (**b**) and (**e–h**), specific surface area (SSA) refers to total SSA, pore volume (PV) refers to total PV.

5.3.2. Characteristics and Controlling Factors of Fractal Dimension

Figure 14 shows the relationship between the fractal dimensions D1 and D2 and the constituents of the shale. Neither D1 nor D2 shows an obvious correlation with quartz content. This is because the quartz particles have smooth surfaces, resulting in a relatively smooth and regular overall surface and microstructure (Figure 14a). D1 and D2 exhibit no clear correlation with clay content (Figure 14b). The layered and flocculent structures of clay minerals increase the specific surface area of pores [58]; however, in this study, the samples have high illite content due to high thermal maturity, and the contribution is not significant. The lack of a relationship between mineral composition and fractal dimension indicates that mineral composition is not the major factor influencing fractal dimension.

Figure 14c shows a slight increase in D1 with rising TOC content and a moderately positive correlation between D2 and TOC content. Organic-rich shale tends to develop more nanopores due to kerogen decomposition during thermal maturation [3,52]. These nanopores form a complicated pore network, resulting in higher fractal dimensions [16,26]. However, limited organic pore development occurred in these samples due to their kerogen type; thus, the positive correlation between fractal dimension and TOC content is not significant in this study. In sum, the sedimentary environment and thermal maturation control the shale constituents, and no single constituent dominates the fractal characteristics.

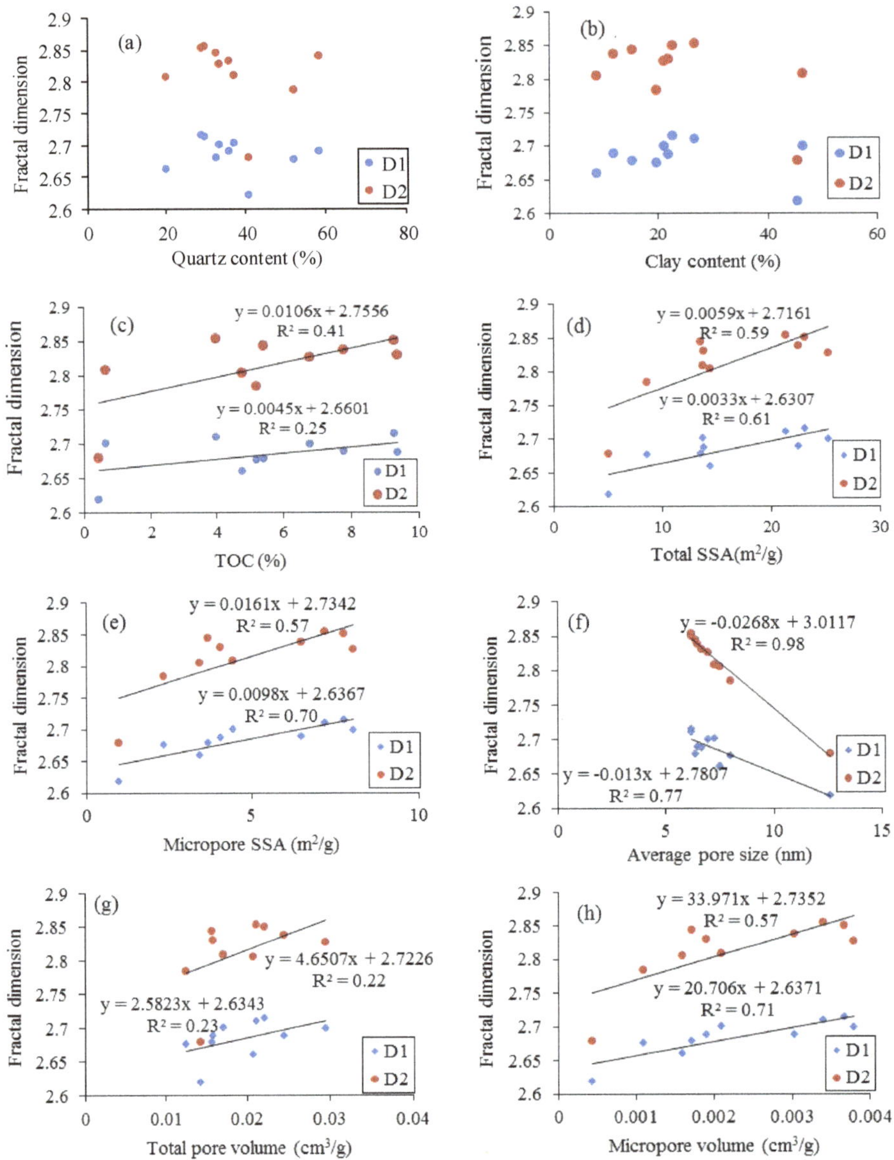

Figure 14. Relationship between fractal dimensions and shale composition (**a–c**) and relationship between fractal dimensions and structure parameters (**d–h**).

The relationship between the fractal dimensions and the pore structure parameters is also shown in Figure 14. D1 and D2 have positive linear correlations with surface area, and D1 has a better correlation with both total surface area (Figure 14d) and micropore surface area than does D2 (Figure 14e). This is because D1 represents the surface fractal dimension, which is dependent on the pore surface area [39]. Figure 14f exhibits a good linear correlation between fractal dimension and average pore size. As previous studies illustrated, a smaller pore size will not only increase the complexity and heterogeneity of the pore structure, but will also generate a rougher surface [16,27].

Moreover, the fractal dimension shows no obvious correlation with total pore volume (Figure 14g) but a better correlation with micropore volume (Figure 14h). On the one hand, the specific surface area is well correlated with pore volume; thus, samples with a higher fractal dimension have a larger pore volume. On the other hand, micropores are the major contributors to pore structure complexity and surface roughness. Thus, the correlation of the fractal dimension with micropores will be better than with other pores. Furthermore, the observation that both D1 and D2 have a pronounced correlation with the pore structure parameters reveals that fractal dimensions can be applied to characterize pore structure.

6. Conclusions

In this paper, the TOC distribution, pore structure, and evolutionary characteristics of Niutitang Formation shales from the northern Guizhou Province in southwest China were investigated using a series of experiments, mass balance calculations, and FHH theory. The following conclusions were drawn:

(1)   The organic matter was dominated by type-I kerogen with a total organic carbon (TOC) content of 0.29–9.36%, averaging 0.69% and 5.93% in the upper and the lower members, respectively. The equivalent vitrinite reflectance (EqVR$_o$) values ranged from 1.72% to 2.72% with a mean of 2.19%. The original TOC content values of the upper member were between 0.64% to 1.91% with a mean of 1.15%, while those of the lower member ranged from 10.66% to 18.17% and averaged 14.02%. The overall TOC recovery coefficient for the Niutitang formation was 2.16, indicating good hydrocarbon generation potential and that it is a promising exploration prospect. The sedimentary environment was shown to be the dominant factor in both present-day TOC and the original TOC distribution.

(2)   The total porosity of the Niutitang shale ranged from 0.36% to 6.93% with a mean of 2.61%. This ranged from 0.36% to 2.49% with a mean of 1.01% in samples from the upper member and 2.26% to 6.93% with a mean of 4.36% in samples from the lower member. When TOC was below 1.0%, porosity had no correlation with it; when TOC was between 1.0% and 6.0%, porosity was positively correlated with it; when TOC < 1.0% and TOC > 6.0%, porosity had no obvious correlation with it. This indicates that, when within a certain range, TOC was a significant controlling factor in porosity. The calculated organic porosity was higher than the measured total porosity, and the relationship between porosity and mineral composition was found to be indirect.

(3)   The nitrogen adsorption isotherms of all of the Niutitang Formation shale samples were type IV, with hysteresis loops identified as type H3 and type H4, suggesting the existence of slit- or plate-like pores. The BET specific surface area ranged from 5.08 m^2/g to 25.31 m^2/g, total pore volume from 0.013 cm^3/g to 0.029 cm^3/g, and average pore diameter from 6.18 nm to 12.62 nm. Pore structure parameters showed no obvious correlation with the composition of the shale.

(4)   The D1 values ranged from 2.619 to 2.716, with an average value of 2.684, and the D2 values ranged from 2.680 to 2.854, with a mean of 2.813. The irregularity of both the pore surface and structure indicated good gas sorption capacity. Sedimentary environment and thermal maturation were shown to control the constituents of the shale, and no single constituent dominated its fractal characteristics.

Author Contributions: L.L. and S.T. conceived and designed the experiments; L.L. and Z.X. performed the experiments; L.L. wrote the paper; Z.X. and S.T. revised the paper.

Funding: This work was financially supported by the National Science and Technology Major Project of China (grant No. 2017ZX05035001-003).

Acknowledgments: The authors gratefully acknowledge the China National Administration of Coal Geology for sample support and their permission to publish the results of this study.

Conflicts of Interest: The authors declare no conflicts of interest.

References

1. Jacob, H. Classification structure genesis and practical importance of natural solid oil bitumen ("migrabitumen"). *Int. J. Coal Geol.* **1989**, *11*, 65–79. [CrossRef]
2. Ko, L.T.; Ruppel, S.C.; Loucks, R.G.; Hackley, P.C.; Zhang, T.; Shao, D. Pore-types and pore-network evolution in Upper Devonian-Lower Mississippian Woodford and Mississippian Barnett mudstones: Insights from laboratory thermal maturation and organic petrology. *Int. J. Coal Geol.* **2018**, *190*, 3–28. [CrossRef]
3. Jarvie, D.M.; Hill, R.J.; Ruble, T.E.; Pollastro, R.M. Unconventional shale-gas systems: The Mississippian Barnett Shale of north-central Texas as one model for thermogenic shale-gas assessment. *AAPG Bull.* **2007**, *91*, 475–499. [CrossRef]
4. Loucks, R.G.; Reed, R.M.; Ruppel, S.C.; Jarvie, D.M. Morphology, Genesis, and Distribution of Nanometer-Scale Pores in Siliceous Mudstones of the Mississippian Barnett Shale. *J. Sediment. Res.* **2009**, *79*, 848–861. [CrossRef]
5. Curtis, J.B. Fractured shale-gas systems. *AAPG Bull.* **2002**, *86*, 1921–1938.
6. Bowker, K.A. Barnett Shale gas production, Fort Worth Basin: Issues and discussion. *AAPG Bull.* **2007**, *91*, 523–533. [CrossRef]
7. Jarvie, D.M. *Shale Resource Systems for Oil and Gas: Part 1—Shale-gas Resource Systems, In Shale Reservoirs—Giant Resources for the 21st Century, Breyer, J.A., Ed.*; AAPG: Tulsa, OK, USA, 2012; pp. 69–87.
8. Zou, C.; Dong, D.; Wang, Y.; Li, X.; Huang, J.; Wang, S.; Guan, Q.; Zhang, C.; Wang, H.; Liu, H.; et al. Shale gas in China: Characteristics, challenges and prospects (II). *Adv. Pet. Explor. Dev.* **2016**, *43*, 182–196. [CrossRef]
9. Tan, J.; Horsfield, B.; Mahlstedt, N.; Zhang, J.; Boreham, C.J.; Hippler, D.; van Graas, G.; Tocher, B.A. Natural gas potential of Neoproterozoic and lower Palaeozoic marine shales in the Upper Yangtze Platform, South China: Geological and organic geochemical characterization. *Int. Geol. Rev.* **2015**, *57*, 305–326. [CrossRef]
10. Li, Q.; Pang, X.; Li, B.; Zhao, Z.; Shao, X.; Zhang, X.; Wang, Y.; Li, W. Discrimination of effective source rocks and evaluation of the hydrocarbon resource potential in Marsel, Kazakhstan. *J. Pet. Sci. Eng.* **2018**, *160*, 194–206. [CrossRef]
11. Dang, W.; Zhang, J.; Tang, X.; Chen, Q.; Han, S.; Li, Z.; Du, X.; Wei, X.; Zhang, M.; Liu, J.; et al. Shale gas potential of Lower Permian marine-continental transitional black shales in the Southern North China Basin, central China: Characterization of organic geochemistry. *J. Nat. Gas Sci. Eng.* **2016**, *28*, 639–650. [CrossRef]
12. Chen, Z.; Jiang, C. A revised method for organic porosity estimation in shale reservoirs using Rock-Eval data: Example from Duvernay Formation in the Western Canada Sedimentary Basin. *AAPG Bull.* **2016**, *100*, 405–422. [CrossRef]
13. Modica, C.J.; Lapierre, S.G. Estimation of kerogen porosity in source rocks as a function of thermal transformation: Example from the Mowry Shale in the Powder River Basin of Wyoming. *AAPG Bull.* **2012**, *96*, 87–108. [CrossRef]
14. Curtis, M.E.; Cardott, B.J.; Sondergeld, C.H.; Rai, C.S. Development of organic porosity in the Woodford Shale with increasing thermal maturity. *Int. J. Coal Geol.* **2012**, *103*, 26–31. [CrossRef]
15. Tian, H.; Pan, L.; Zhang, T.; Xiao, X.; Meng, Z.; Huang, B. Pore characterization of organic-rich Lower Cambrian shales in Qiannan Depression of Guizhou Province, Southwestern China. *Mar. Pet. Geol.* **2015**, *62*, 28–43. [CrossRef]
16. Gu, Y.; Ding, W.; Yin, M.; Jiao, B.; Shi, S.; Li, A.; Xiao, Z.; Wang, Z. Nanoscale pore characteristics and fractal characteristics of organic-rich shale: An example from the lower Cambrian Niutitang Formation in the Fenggang block in northern Guizhou Province, South China. *Energy Explor. Exploit.* **2018**, *37*, 273–295. [CrossRef]
17. Bai, B.; Elgmati, M.; Zhang, H.; Wei, M. Rock characterization of Fayetteville shale gas plays. *Fuel* **2013**, *105*, 645–652. [CrossRef]
18. Jia, Z.; Xiao, L.; Chen, Z.; Liao, G.; Zhang, Y.; Wang, Z.; Liang, C.; Guo, L. Determining shale organic porosity and total organic carbon by combining spin echo, solid echo and magic echo. *Microporous Mesoporous Mater.* **2018**, *269*, 12–16. [CrossRef]
19. Sun, W.; Zuo, Y.; Wu, Z.; Liu, H.; Xi, S.; Shui, Y.; Wang, J.; Liu, R.; Lin, J. Fractal analysis of pores and the pore structure of the Lower Cambrian Niutitang shale in northern Guizhou province: Investigations using NMR, SEM and image analyses. *Mar. Pet. Geol.* **2019**, *99*, 416–428. [CrossRef]

20. Wang, Y.; Huang, J.; Li, X.; Dong, D.; Wang, S.; Guan, Q. Quantitative characterization of fractures and pores in shale beds of the Lower Silurian, Longmaxi Formation, Sichuan Basin. *Nat. Gas Ind. B* **2015**, *2*, 481–488. [CrossRef]
21. Xu, Z.; Shi, W.; Zhai, G.; Clay, C.; Zhang, X.; Peng, N. A rock physics model for characterizing the total porosity and velocity of shale: A case study in Fuling area, China. *Mar. Pet. Geol.* **2019**, *99*, 208–226. [CrossRef]
22. Yu, H.; Wang, Z.; Rezaee, R.; Zhang, Y.; Han, T.; Arif, M.; Johnson, L. Porosity estimation in kerogen-bearing shale gas reservoirs. *J. Nat. Gas Sci. Eng.* **2018**, *52*, 575–581. [CrossRef]
23. Tian, H.; Pan, L.; Xiao, X.; Wilkins, R.W.T.; Meng, Z.; Huang, B. A preliminary study on the pore characterization of Lower Silurian black shales in the Chuandong Thrust Fold Belt, southwestern China using low pressure N_2 adsorption and FE-SEM methods. *Mar. Pet. Geol.* **2013**, *48*, 8–19. [CrossRef]
24. Milliken, K.L.; Rudnicki, M.; Awwiller, D.N.; Zhang, T. Organic matter-hosted pore system, Marcellus Formation (Devonian), Pennsylvania. *AAPG Bull.* **2013**, *97*, 177–200. [CrossRef]
25. Pan, L.; Xiao, X.; Tian, H.; Zhou, Q.; Chen, J.; Li, T.; Wei, Q. A preliminary study on the characterization and controlling factors of porosity and pore structure of the Permian shales in Lower Yangtze region, Eastern China. *Int. J. Coal Geol.* **2015**, *146*, 68–78. [CrossRef]
26. Yang, R.; He, S.; Yi, J.; Hu, Q. Nano-scale pore structure and fractal dimension of organic-rich Wufeng-Longmaxi shale from Jiaoshiba area, Sichuan Basin: Investigations using FE-SEM, gas adsorption and helium pycnometry. *Mar. Pet. Geol.* **2016**, *70*, 27–45. [CrossRef]
27. Shao, X.; Pang, X.; Li, Q.; Wang, P.; Chen, D.; Shen, W.; Zhao, Z. Pore structure and fractal characteristics of organic-rich shales: A case study of the lower Silurian Longmaxi shales in the Sichuan Basin, SW China. *Mar. Pet. Geol.* **2017**, *80*, 192–202. [CrossRef]
28. Han, S.; Zhang, J.; Li, Y.; Horsfield, B.; Tang, X.; Jiang, W.; Chen, Q. Evaluation of Lower Cambrian Shale in Northern Guizhou Province, South China: Implications for Shale Gas Potential. *Energy Fuels* **2013**, *27*, 2933–2941. [CrossRef]
29. Zhang, J.; Fan, T.; Li, J.; Zhang, J.; Li, Y.; Wu, Y.; Xiong, W. Characterization of the Lower Cambrian Shale in the Northwestern Guizhou Province, South China: Implications for Shale-Gas Potential. *Energy Fuels* **2015**, *29*, 6383–6393. [CrossRef]
30. Li, T.; Tian, H.; Xiao, X.; Cheng, P.; Zhou, Q.; Wei, Q. Geochemical characterization and methane adsorption capacity of overmature organic-rich Lower Cambrian shales in northeast Guizhou region, southwest China. *Mar. Pet. Geol.* **2017**, *86*, 858–873. [CrossRef]
31. Wu, Z.; Zuo, Y.; Wang, S.; Chen, J.; Wang, A.; Liu, L.; Xu, Y.; Sunwen, J.; Cao, J.; Yu, M.; et al. Numerical study of multi-period palaeotectonic stress fields in Lower Cambrian shale reservoirs and the prediction of fractures distribution: A case study of the Niutitang Formation in Feng'gang No. 3 block, South China. *Mar. Pet. Geol.* **2017**, *80*, 369–381. [CrossRef]
32. Li, X.; Jiang, Z.; Wang, P.; Song, Y.; Li, Z.; Tang, X.; Li, T.; Zhai, G.; Bao, S.; Xu, C.; et al. Porosity-preserving mechanisms of marine shale in Lower Cambrian of Sichuan Basin, South China. *J. Nat. Gas Sci. Eng.* **2018**, *55*, 191–205. [CrossRef]
33. Wang, R.; Gu, Y.; Ding, W.; Gong, D.; Yin, S.; Wang, X.; Zhou, X.; Li, A.; Xiao, Z.; Cui, Z. Characteristics and dominant controlling factors of organic-rich marine shales with high thermal maturity: A case study of the Lower Cambrian Niutitang Formation in the Cen'gong block, southern China. *J. Nat. Gas Sci. Eng.* **2016**, *33*, 81–96. [CrossRef]
34. Li, J.; Tang, S.; Zhang, S.; Xi, Z.; Yang, N.; Yang, G.; Li, L.; Li, Y. Paleo-environmental conditions of the Early Cambrian Niutitang Formation in the Fenggang area, the southwestern margin of the Yangtze Platform, southern China: Evidence from major elements, trace elements and other proxies. *J. Asian Earth Sci.* **2018**, *159*, 81–97. [CrossRef]
35. Tan, J.; Horsfield, B.; Mahlstedt, N.; Zhang, J.; di Primio, R.; Vu, T.A.T.; Boreham, C.J.; van Graas, G.; Tocher, B.A. Physical properties of petroleum formed during maturation of Lower Cambrian shale in the upper Yangtze Platform, South China, as inferred from PhaseKinetics modelling. *Mar. Pet. Geol.* **2013**, *48*, 47–56. [CrossRef]
36. Xi, Z.; Tang, S.; Wang, J.; Yi, J.; Guo, Y.; Wang, K. Pore Structure and Fractal Characteristics of Niutitang Shale from China. *Minerals* **2018**, *8*, 163. [CrossRef]

37. Dahl, H.J.A.B. Quantitative hydrocarbon potential mapping and and organofacies study in the Greater Balder Area, Norwegian North Sea. In Proceedings of the 6th Petroleum Geology Conference; Dore, A.G., Vining, B.A., Eds.; Petroleum Geology: London, UK, 2005; pp. 1317–1329.

38. Rahman, R.R.F.K.A.M. Specific gravity as a kerogen type and maturation indicator with special reference to amorphous kerogens. *J. Pet. Geol.* **1983**, *6*, 179–194.

39. Yao, Y.; Liu, D.; Tang, D.; Tang, S.; Huang, W. Fractal characterization of adsorption-pores of coals from North China: An investigation on CH_4 adsorption capacity of coals. *Int. J. Coal Geol.* **2008**, *73*, 27–42. [CrossRef]

40. Yang, F.; Ning, Z.; Liu, H. Fractal characteristics of shales from a shale gas reservoir in the Sichuan Basin, China. *Fuel* **2014**, *115*, 378–384. [CrossRef]

41. Bu, H.; Ju, Y.; Tan, J.; Wang, G.; Li, X. Fractal characteristics of pores in non-marine shales from the Huainan coalfield, eastern China. *J. Nat. Gas Sci. Eng.* **2015**, *24*, 166–177. [CrossRef]

42. Charles, R.; Landis, J.R.C. Maturation and bulk chemical properties of a suite of solid hydrocarbons. *Org. Geochem.* **1995**, *22*, 137–149.

43. Dellisanti, F.; Pini, G.A.; Baudin, F. Use of T max as a thermal maturity indicator in orogenic successions and comparison with clay mineral evolution. *Clay Miner.* **2018**, *45*, 115–130. [CrossRef]

44. Ola, P.S.; Aidi, A.K.; Bankole, O.M. Clay mineral diagenesis and source rock assessment in the Bornu Basin, Nigeria: Implications for thermal maturity and source rock potential. *Mar. Pet. Geol.* **2018**, *89*, 653–664. [CrossRef]

45. Wang, F.; Guan, J.; Feng, W.; Bao, L. Evolution of overmature marine shale porosity and implication to the free gas volume. *Pet. Explor. Dev.* **2013**, *40*, 819–824. [CrossRef]

46. Sing, K.S.W. Reporting physisorption data for gassolid systems with special reference to the determination of surface area and porosity. *Pure Appl. Chem.* **1985**, *57*, 603–619. [CrossRef]

47. Labani, M.M.; Rezaee, R.; Saeedi, A.; Hinai, A.A. Evaluation of pore size spectrum of gas shale reservoirs using low pressure nitrogen adsorption, gas expansion and mercury porosimetry: A case study from the Perth and Canning Basins, Western Australia. *J. Pet. Sci. Eng.* **2013**, *112*, 7–16. [CrossRef]

48. Liu, J.; Yao, Y.; Elsworth, D.; Pan, Z.; Sun, X.; Ao, W. Sedimentary characteristics of the Lower Cambrian Niutitang shale in the southeast margin of Sichuan Basin, China. *J. Nat. Gas Sci. Eng.* **2016**, *36*, 1140–1150. [CrossRef]

49. Zhang, J.; Fan, T.; Algeo, T.J.; Li, Y.; Zhang, J. Paleo-marine environments of the Early Cambrian Yangtze Platform. *Palaeogeogr. Palaeoclimatol. Palaeoecol.* **2016**, *443*, 66–79. [CrossRef]

50. Bustin, R.M.; Cui, X.; Ross, D.J.K.; Murthy Pathi, V.S. Impact of shale properties on pore structure and storage characteristics. In Proceedings of the SPE Shale Gas Production Conference, Fort Worth, TX, USA, 16–18 November 2008; pp. 1–28.

51. Chalmers, G.R.L.; Ross, D.J.K.; Bustin, R.M. Geological controls on matrix permeability of Devonian Gas Shales in the Horn River and Liard basins, northeastern British Columbia, Canada. *Int. J. Coal Geol.* **2012**, *103*, 120–131. [CrossRef]

52. Robert, G.; Loucks, R.M.R. Scanning-Electron-Microscope petrographic evidence for distinguishing organic-matter pores associated with depositional organic matter versus migrated organic matter. *GCAGS J.* **2014**, *3*, 51–60.

53. Chen, J.; Xiao, X. Evolution of nanoporosity in organic-rich shales during thermal maturation. *Fuel* **2014**, *129*, 173–181. [CrossRef]

54. Zargari, S. Effect of Thermal Maturity on Nanomechanical Properties and Porosity in Organic Rich Shales (a Bakken Shale Case Study). Ph.D. Thesis, Colorado School of Mines, Ann Arbor, MI, USA, 2015.

55. Mastalerz, M.; Schimmelmann, A.; Drobniak, A.; Chen, Y. Porosity of Devonian and Mississippian New Albany Shale across a maturation gradient: Insights from organic petrology, gas adsorption, and mercury intrusion. *AAPG Bull.* **2013**, *97*, 1621–1643. [CrossRef]

56. Xi, Z.; Tang, S.; Li, J.; Zhang, Z.; Xiao, H. Pore characterization and the controls of organic matter and quartz on pore structure: Case study of the Niutitang Formation of northern Guizhou Province, South China. *J. Nat. Gas Sci. Eng.* **2019**, *61*, 18–31. [CrossRef]

57. Gu, Y.; Wan, Q.; Qin, Z.; Luo, T.; Li, S.; Fu, Y.; Yu, Z. Nanoscale Pore Characteristics and Influential Factors of Niutitang Formation Shale Reservoir in Guizhou Province. *J. Nanosci. Nanotechnol.* **2017**, *17*, 6178–6189. [CrossRef]

58. Li, A.; Ding, W.; He, J.; Dai, P.; Yin, S.; Xie, F. Investigation of pore structure and fractal characteristics of organic-rich shale reservoirs: A case study of Lower Cambrian Qiongzhusi formation in Malong block of eastern Yunnan Province, South China. *Mar. Pet. Geol.* **2016**, *70*, 46–57. [CrossRef]
59. Ji, L.; Zhang, T.; Milliken, K.L.; Qu, J.; Zhang, X. Experimental investigation of main controls to methane adsorption in clay-rich rocks. *Appl. Geochem.* **2012**, *27*, 2533–2545. [CrossRef]

Article

Accumulation Conditions and an Analysis of the Origins of Natural Gas in the Lower Silurian Shiniulan Formation from Well Anye 1, Northern Guizhou Province

Ruibo Guo [1], Jinchuan Zhang [1,2,3,]*, Panwang Zhao [1], Xuan Tang [1,2,3] and Ziyi Liu [1]

[1] School of Energy Resource, China University of Geosciences, Beijing 100083, China; 3006160043@cugb.edu.cn (R.G.); zhao_panwang@ctg.com.cn (P.Z.); tangxuan@cugb.edu.cn (X.T.); liuzy92@sina.cn (Z.L.)

[2] Key Laboratory of Shale Gas Exploration and Evaluation, Ministry of Land and Resources, China University of Geosciences, Beijing 100083, China

[3] Beijing Key Laboratory of Unconventional Natural Gas Geological Evaluation and Development Engineering, Beijing 100083, China

* Correspondence: zhangjc@cugb.edu.cn

Received: 18 September 2019; Accepted: 22 October 2019; Published: 26 October 2019

Abstract: The origin of natural gas and the mechanisms that lead to gas accumulation in the marine calcareous mudstone of the Lower Silurian Shiniulan Formation in northern Guizhou province are special and complicated. According to a combination of qualitative and quantitative methods, including the reconstruction of hydrocarbon generative potential and gas content's measurement, in the context of some geochemistry information—the origins of the natural gas of Shiniulan Formation is suggested to be mixed gas. Furthermore, the accumulation of the natural gas can be proposed combined with some geological information. Results indicated that the volume of the in-place gas content of Shiniulan samples, reinstated by the formulas' computation, reaches a yield of 3.67 $m^3 \cdot t^{-1}$ in rock. The theoretical gas content for Shiniulan Formation mudstone ranges from 1.6 to 5.8 $m^3 \cdot t^{-1}$ using the indirect calculation of gas content, and the total gas contents of those samples range from 0.065 to 0.841 m^3/t, according to the United States Bureau of Mines' (USBM) direct method. According to the combination of the reconstructed in-place gas content and the gas content, even mudstone in the Shiniulan Formation has potential to generating gas but could not satisfy the actual gas content in Shiniulan Formation. In addition, according to the composition, the carbon and hydrogen isotope charts of gaseous hydrocarbons further indicate that the gas origin of Shiniulan Formation is a mixed gas, which also means that the gas is not just generated in the layer, but has partly migrated from other formations, such as the Wufeng–Longmaxi Formation. The lower Shiniulan Formation in the study area is characterized by frequent interbed of limestone and calcareous mudstone. The geological information shows that well-developed fractures of mudstone and faults can be used as main pathways for the upward migration of gases from the underlying strata to the Shiniulan Formation. The limestone with fairly low porosity and permeability hinders the migration of natural gas as much as possible and keeps that efficiently reserved in the horizontal fractures of calcareous mudstone. This migration pattern implies that the interbedded rock association is also favorable for gas accumulation in the Shiniulan Formation.

Keywords: limestone and calcareous mudstone interbedding; gas content; source-mixed gas; fractures; northern Guizhou

---

1. Introduction

The great success of shale gas's exploration in North America has tremendously promoted the exploration and development of shale gas worldwide. The resources, such as net crude oil and petroleum-related production, are increasingly independent of imports, which has contributed mostly to the development of tight oil and shale gas reservoirs, as suggested by Jia et al. [1]. According to the Energy Information Administration (EIA) in 2017 [2], the new record for shale plays in the northeast extends to a longer-term trend of development, and the upward trends have continued in the days thereafter.

The exploration of shale gas is undergoing rapid development in China. In particular, the Lower Paleozoic marine facies rich in organic shale are an advantageous area for marine shale gas exploration. These facies are spread widely over the upper Yangtze region, with rich sources of rock layers, high organic matter abundance, rich organic matter types, high thermal maturity, a high content of brittle minerals, and moderate development of fractures having been demonstrated by Zhang and Zou and Wang and Tan [3–6]. Through the determination of the actual gas content of drilling cores in many areas of the southern Paleozoic, the shale gas content of the Lower Cambrian Qiongzhusi Formation was found to be distributed in the range of 0.13-5.02 $m^3 \cdot t^{-1}$, with an average of 2.14 $m^3 \cdot t^{-1}$. Accordingly, the shale gas content of the upper Ordovician Wufeng–lower Silurian Longmaxi Formation ranges from 0.29 to 6.50 $m^3 \cdot t^{-1}$ as pointed out by Zou et al. [7].

The northern Guizhou area is one of the research areas for the strategic investigation of shale gas resources in China; this area lies to the south of the Sichuan basin and has a generally low level of petroleum geological evaluation and exploration. The upper Ordovician Wufeng Formation–lower Silurian Longmaxi Formation and the Lower Cambrian Qiongzhusi Formation, which developed in the northern Guizhou area, have the most potential shale layers, with good hydrocarbon generation capabilities as effective formations in geological research and exploration for marine shale gas, as presented according to Han et al. and Zhao et al. [8,9].

The AY (AnYe)-1 well is located in Anchang Town, Zhengan County, in the north of Guizhou Province. The structural location belongs to the west wing of the Anchang syncline in the Wuling folded region. The purpose of drilling was to explore the upper Ordovician Wufeng Formation–lower Silurian Longmaxi Formation for shale gas. More remarkably, a number of high-pressure gas reservoirs were drilled in the Shiniulan Formation of the Silurian. This is the first time that high-yield gas flows have been obtained in the interbedded marine limestone and mudstone layers of the strata in southern China. The black shale was drilled for nearly 20 m from the upper Ordovician Wufeng Formation to the Lower Silurian Longmaxi Formation, which met the standard for the commercial exploration of shale gas. Furthermore, natural gas was prospected first in the Permian Qixia Formation, and then in the Baota Formation outside the Sichuan Basin. The drilling of the AY-1 well is a major breakthrough in the new area, which has a new formation system, and new types of natural gas and have opened up new areas for oil and gas exploration in southern China.

There is a fairly prominent gas logging abnormality in the gas source rock of the Shiniulan Formation, with 15 layers of gas logging abnormality drilled in the well depth of 2105-2204 m. The anomalous total hydrocarbon value rapidly soared from 0.35% to 85.40% while the abnormal methane value also rose from 0.25% to 80.5%. The initial production of the well AY-1 attained a maximum of 420×10^3 m^3 per day, as presented by Liu et al. [10]. It is expected that the output can reach 200×10^3 m^3 per day after acid fracturing. At present, the daily output reaches approximately 5000 cubic meters, and the cumulative gas production is nearly 600,000 cubic meters, signifying a principal breakthrough for gas exploration in Guizhou Province.

As a new stratum for natural gas exploration and trial in China, the investigation of gas potential and gas accumulation condition of reservoir is of vital importance to the next step of exploration and development in study area. This paper aims to evaluate the potential of the hydrocarbon generation of calcareous mudstone in the Shiniulan Formation by different methods which have not used in previous research, combined with the relationship between the stable isotopic compositions of gaseous hydrocarbons and the molecular composition of gases to confirm the origin of natural gas in lower part of the Shiniulan Formation due to its complexity and multiple solutions. Afterwards, characterizing the fault and fractures as controlling factors was used to estimate the accumulation model of natural gas in the reservoir with high yields.

2. Geological Setting

The Silurian in the upper Yangtze area has a good material foundation, which is not only widely distributed but also has a large thickness with rich organic matter, as presented by Wang. [11]. With the development of shale gas exploration, the resources of the Lower Silurian have achieved many breakthroughs in the Weiyuan, Changning, Fushun-Yongchuan, Fuling, and Pengshui areas in the southern Sichuan basin. Many wells have obtained high yields during the process of the gas testing suggested by Liang et al. [12].

The study area belongs to the South Sichuan–Northern Qianbei sub-region in the upper Yangtze (Figure 1). The Paleozoic strata have developed to a great extent, while Carboniferous and Devonian areas are lacking in the northern region of Guizhou Province. According to the lithology, sedimentary facies, and paleontological characteristics, the Sinian–Cambrian strata are divided into east and west regions. The study area is mainly exposed of the upper Cambrian–Silurian Formation, which is distributed in the anticlinoria. Among the strata, the Silurian strata are relatively developed with a large number of fossils that contact the underlying Ordovician but are in uncomfortable contact with the overlying Lower Permian. The Lower Silurian is divided into three groups from the bottom to top: the Silurian Longmaxi Formation, the Shiniulan Formation, and the Hanjiadian Formation.

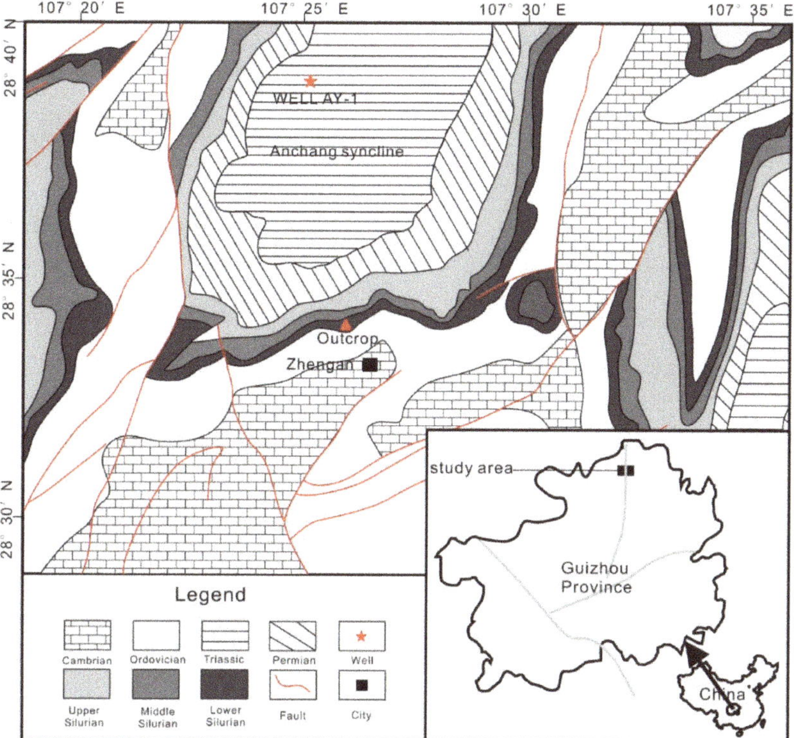

Figure 1. The geological map of the study area with the location of the sampling well and outcrop profile.

Silurian exists as an important source of rocks in the upper Yangtze region. The lithology of the Wufeng–Longmaxi Formation shows mainly black carbonaceous shale with a small amount of thin-layer of bioclastic limestone or bioclastic limestone strips, and the upper part is mainly a thin interbed of argillaceous limestone and mudstone refer to Wang et al. [13]. Furthermore, the lithology of the Shiniulan Formation is mainly composed of a set of bioclastic limestone in an argillaceous limestone and calcareous mudstone interbedding, which is in a conformity relationship with the underlying strata. The thickness of limestone increases from bottom to top, while that of calcareous mudstone diminishes slightly (Figure 2).

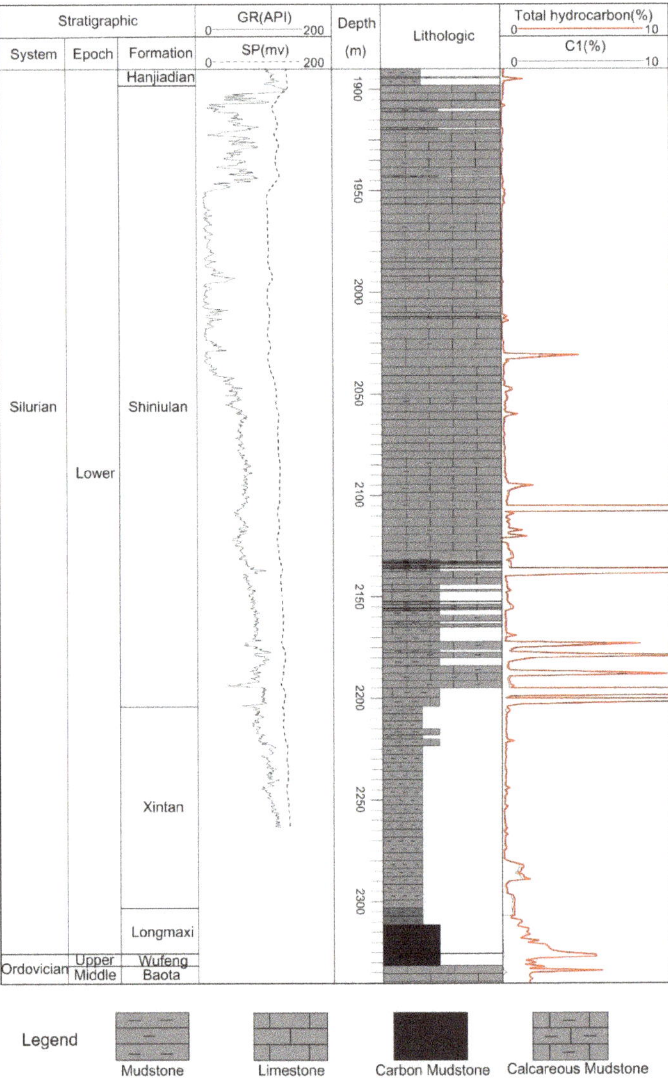

Figure 2. The generalized stratigraphic column and characteristics of logging curves of the upper Ordovician Wufeng–lower Silurian Longmaxi and lower Silurian Shiniulan Formation in the northern Guizhou.

3. Materials and Methods

3.1. Sample Selection

Twenty-four fresh core samples from depths of 2110 to 2348 m of AY-1 well were analyzed, 15 of which were from the Shiniulan Fm., and nine of which were from the Longmaxi Fm. A series of experiments were used for evaluating the characteristics of mudstone samples and the gas potential of the Shiniulan Fm. in the upper Yangtze area.

Eleven gas samples from Shiniulan Fm. in AY-1 were collected from the wellheads or separators for the molecular composition and isotopes of carbon and hydrogen analyses. To collect the gas samples, double-ended stainless-steel bottles equipped with a maximum pressure of 15 MPa and vacuumed before sampling were used. During sampling, the stainless-steel bottles were repeatedly flushed and replaced with a gas stream, and then the exhaust valve was closed. The filling gas pressure was made stable at 4 MPa, the intake valve was closed, and the bottle was covered in water in case of gas leakage.

3.2. Organic Matter Properties

The total organic carbon (TOC) content was determined using a Leco CS230 carbon/sulfur analyzer following the Chinese National Standard GB/T19145-2003. [14]. Vitrinite reflectance values (Ro%), a representative indicator for characterizing the thermal maturity of organic matter, were evaluated by an MPV-SP microphotometer under an oil-immersion objective lens. The photometry of samples was obtained at the temperature of 23 °C according to the Chinese standard SY/T5124-2012 [15]. Maceral composition observations were performed using polished sections under reflected white and fluorescent light with a Leica MPV microscope. According to the type index (TI) following the criteria of the Chinese oil and gas industry standard SY/T5125-1996 [16]. and Cao [17], the abundances of the maceral components could be evaluated for organic matter type identification. All of these experiments and analyses were performed by at the geochemistry laboratory of Yangtze University, Jingzhou, China.

3.3. Parameterization of the Gas Content

3.3.1. Indirect Calculation of the Gas Content

By adding together the adsorbed and free gas contents, the total gas content can be obtained. The content of adsorbed gas in mudstone cannot be obtained directly. At present, methane isothermal adsorption is commonly used to signify the capacity of gas adsorption and to measure the amount of adsorbate (methane); that is, through isothermal adsorption, a simulation experiment is conducted to establish the relationship model between the adsorption gas content and pressure and the temperature, as pointed out in previous publication by Wang et al. [18]. The experimental process involves testing the capacity of the adsorption gas under different pressures at a constant temperature.

The methane adsorption isotherm was formed at a temperature of 30 °C and a humidity level of 2.31% by the isothermal adsorption/desorption analyzer (HPVA-200-4). Through the isothermal adsorption test, eight equilibrium pressure points were measured at 30 °C (about the same as the ground temperature at 500 m depth), as was the equilibrium moisture of pure methane at 0, 0.35, 1.07, 2.25, 4.31, 6.24, 8.72, and 11.13 MPa, respectively. The equilibrium time of each pressure point was approximately 12 h, and then it was pressurized to the next pressure point under the experimental environment with a methane gas concentration greater than 99.999%. The adsorption volume of methane was obtained by the following equation; see Dang [19], Gasparik et al., [20], Ji et al., [21], and Tan et al., [4], Dang [22].

$$\mathrm{m} = m_{total} - \rho_{gas} V_{void} \tag{1}$$

where m is the volume of methane adsorption; m_{total} represents the amount of gas drawn into the sample cell; and ρ_{gas} is the density of gas; the void volume (V_{void}) generally is contingent on helium expansion. The adsorption profile used the Gibbs adsorption calculation which neglects the volume occupied by the adsorbed phase in calculating the amount of unadsorbed gas and V_{void} is viewed as being available to the unadsorbed gas suggested by Sudibandriyo et al. [23].

According to the Langmuir isotherm equation, as pointed out by Langmuir [24], which is related to the adsorption of gas molecules on a solid surface under a fixed gas pressure and temperature, the methane adsorption's experimental results can be obtained:

$$V = V_L P / (P_L + P) \tag{2}$$

P_L is the pressure at which the adsorbed gas content is equivalent to the half of the Langmuir volume; V means the adsorbed gas volume; and P is the gas pressure; V_L indicates the maximum capacity of the adsorbent's adsorption

Free gas is generally stored in the pores and microfractures of mudstone. The main influential factors of the content include porosity, pressure conditions, the temperature system, and gas saturation. The content of free gas was calculated by the following equation proposed by Lewis et al. [25].

$$G_f = \frac{\varphi \times S_g}{B_g \times \rho_b} \tag{3}$$

where G_f is the content of free gas, $m^3 \cdot t^{-1}$; Φ is the rock porosity (%); S_g is the original gas saturation (%); B_g is the factor of gas formation volume; and ρ_b is the formation density, $g \cdot cm^{-3}$.

The factor of gas formation volume is the volume of natural gas per unit volume under formation standard state (20 °C, 0.101325 MPa). The formula of the factor of gas formation volume as follows.

$$B_g = \frac{V_g}{V_{sc}} \tag{4}$$

$$P_{sc} V_{sc} = nRT_{sc} \tag{5}$$

$$PV_g = nZRT \tag{6}$$

where V_g is the volume of n mol gas under formation conditions, m^3; V_{sc} is the volume of n mol gas under formation conditions, m^3; n is the number of moles, k mol; T is the formation temperature, $K(T = (273.15 + t)K)$; P is the formation pressure, MPa; R is the universal gas constant, $MPa \cdot m^3 (k\ mol \cdot K)^{-1}$; t is the formation temperature, °C; t_{sc} is the formation temperature in standard state, $t_{sc} = 20$ °C; and Z is the compressibility factor, which can be obtained by the two-parameter compression factor chart of natural gas. $P_{sc} = 0.101235$ MPa; $T_{sc} = (273.15 + 20)$ °C (under standard state).

$$B_g = \frac{P_{sc}}{(20 + 273.15)} \times \frac{Z(t + 273.15)}{P} = 3.456 \times 10 - 4Z\ \frac{t + 273.15}{P} \tag{7}$$

$$G_f = k\frac{2893.5\varphi S g P}{\rho Z(t + 273.15)} \tag{8}$$

The key to this calculation is to determine the effective porosity and gas saturation of free gas. The reservoir space of free gas consists of matrix pores and fractures. Both of these factors directly affect the content of free gas. The reservoir space can be characterized by effective porosity.

Therefore, the effective porosity of the free gas in the reservoir is the summation of the matrix porosity and fracture porosity, which are obtained through high-precision experiments and log interpretation. The comparative matrix's porosity can be measured by sound waves, neutrons, density, and other well logging data; and accurate fracture porosity is commonly obtained by dual lateral logging data. The gas saturation is obtained from a formula named the total shale model, which was described by Schlumberger (1972) [26]. The calculation of the total shale model is as follows:

$$\frac{1}{R_t} = \frac{\varphi^m \times S_w^n}{aR_w \times (1 - V_{sh})} + \frac{V_{sh} \times S_w}{R_{sh}} \tag{9}$$

where, R_{sh} is the resistivity of mudstones; R_t presents the rt true formation resistivity; and S_w is the water saturation. Then, R_w is the formation water resistivity, which has units of $\Omega \cdot m$; φ is the formation porosity (%); a is the lithology factor; n is the saturation exponent; and m is the formation cementation exponent.

3.3.2. USBM Direct Method

The USBM direct method, as one of the industry standards, is used to evaluate the gas content, as presented by Diamond et al. [27], Yee et al. [28], Diamond and Schatzel [29], and Dang et al. [30]. The core samples lifted from the wellhead are quickly loaded into stainless-steel cans and sealed. The adsorbed gas content can be measured by spherical particles, which are characterized by a surface concentration of zero and a constant initial gas concentration (see Dang et al. [26] for details). The lost gas content is estimated by linear regression analysis of the square root of the adsorbed gas content and desorption time when the desorption of gas is completed.

3.4. Original TOC and Hydrocarbon Potential

An analysis was carried out with 10 g of powdered sample. Each sample was heated to 600 °C by Rock-Eval 6 pyrolysis equipment manufactured by Vinci® Technologies (Nanterre, France). Herein, we focused mainly on the following parameters, which were recorded during the pyrolysis phase: Some parameters can be obtained by experimentation, including the volatile hydrocarbon content, mg HC·g^{-1} Rock (S_1); remaining hydrocarbon generative potential, mg HC·g^{-1} Rock (S_2); and the temperature of maximum pyrolysis yield (T_{max}). The hydrogen index (HI) and the production index (PI), which were proposed by Behar et al. [31], were calculated through the calculation formula. Finally, we used the following equation, according to Jarvie et al. [32], to acquire the original hydrocarbon generative potential:

$$S_{2o} = HI_o \times TOC_o \times 100^{-1} \tag{10}$$

where HI_o is the original hydrogen index; TOC_o is the total organic carbon; and S_{2o} is the original generative potential of hydrocarbon, mg HC·g^{-1} Rock.

3.5. Natural Gas Geochemistry

The compositions of the 11 gas samples in this study were analyzed by a gas chromatograph (GC); i.e., an Agilent 6890 N integrated with a flame ionization detector (FID) and a thermal conductivity detector (TCD). A capillary column (PLOT Al_2O_3 50 m × 0.53 mm) was used for hydrocarbon gas components (C_1–C_3) fractions. The temperature program of the GC oven included an initial heating at 30 °C, which was then increased to the temperature of 180 °C with a final hold time for 20–30 min at 10 °C·min^{-1}.

The carbon isotope compositions of the gas components were obtained using a mass spectrometer (Finnigan Mat Delta Plus). The target gas component was separated on a fused silica capillary column. The temperature program of the GC oven was increased from 35 °C to 80 °C at 8 °C·min^{-1} and reached 260 °C at 5 °C·min^{-1} (held for 10 min). The stable hydrogen isotope analyses of alkane gases were performed on a mass spectrometer (Finnigan GC/TC/IRMS). Helium, as the carrier gas with a rate of 1.5 mL·min^{-1}, was injected into an HP-PLOT Q column (30 m × 0.32 mm × 20 µm) to separate the alkane gas components. The GC oven temperature was initially held at 40 °C for 4 min, increased at a rate of 10 °C·min^{-1} to 80 °C, then to 140 °C at °C·min^{-1}, and finally, turned up to 260 °C at a rate of 3 °C·min^{-1} and held for 10 min. The values of stable carbon and hydrogen isotopes are presented in the δ-notation ($\delta^{13}C$, ‰) as V-PDB (Vienna-Pee Dee Belemnite) and V-SMOW (Vienna-standards mean ocean water), respectively. The analytical error of the carbon and hydrogen isotopic compositions was less than 3‰, as suggested by Ni et al. [33], Dai et al. [34], Dai et al. [35].

3.6. Observation and Statistical Classification Method for Fractures

The observation and statistical analyses of the length, opening, dip, and density of core fractures are fundamental for the study of the fracture development and distribution features of cores.

In the identification and observation of core fractures, the core is generally immersed in water so that fractures can be clearly observed before the core surface is entirely dry. The reason for this is that there are many microfractures in cores. After the water immersion, the water on the surface completely evaporates, but does not completely evaporate in the microfractures; thus, water marks with the same dip size were observed on the core.

(1) Statistical analysis and research into the fracture length of cores:

The length distribution of fractures can be divided into three different types: length (L) < 6.5 cm, length $L \geq 10$ cm, and 6.5 cm $\leq L < 10$ cm.

(2) Statistical analysis and research into the degree of the fracture development of cores:

A fracture's density is a significant parameter to describe the degree of fracture development, which can intuitively reflect the intensity of a fracture. The fracture liner density is one of the most effective parameters indicating the degree of fracture development. It refers to the ratio of the number of fractures intersecting a straight line to the length of the line:

$$D = \frac{N}{H} \tag{11}$$

N is the total number of fractures observed in the core unit; H is the core length in the unit, m; and D is the liner density of a fracture, m^{-1}.

4. Results and Discussion

4.1. Organic Geochemical Bulk Parameters and the Hydrocarbon-Generative Source Rock

The type of organic matter affects the gasification capacity and gas content, both of which were demonstrated in the pioneering work by Boyer, C. et al. [36]. Some information about the depositional environment can be revealed by the compositional characteristics of kerogen, proposed by Romero and Philp [37]. The type of kerogen not only has an influence on the hydrocarbon generation of the rock, but can also affect the adsorption and diffusion rates of natural gas.

The results of the microfractional analysis for samples in the Longmaxi Formation show that the organic matter types of the organic-rich shale are mainly distributed in type I and II_1, and the macerals are mainly composed of sapropelinite and inertinite groups and lack-of-vitrinite groups (Table 1). In the chitin group, the sapropelinite group is mainly composed of the dispersed mineral asphalt matrix. Although the content of organic carbon is fairly low, the analysis of the microstructure shows that the organic type of the calcareous mudstone in the Shiniulan Fm. is type II_1 which indicates good hydrocarbon generation capability (Figure 3).

Table 1. The microscopic examination of organic matter microscopic components of AY-1.

Maceral (%)	Shiniulan Formation					Longmaxi Formation			
	AY-1-1	AY-1-3	AY-1-4	AY-1-5	AY-1-6	AY-1-17	AY-1-18	AY-1-19	AY-1-20
Sapropelinite	82	83	88	84	96	95	87	96	86
Liptinite	-	-	-	-	-	-	-	-	-
Vitrinite	17	16	10	15	3	2	9	3	8
Inertinite	1	1	2	1	1	3	4	1	6
Type index	68	70	78	71	93	91	76	93	74
Type	II_1	II_1	II_1	II_1	I	I	II_1	I	II_1

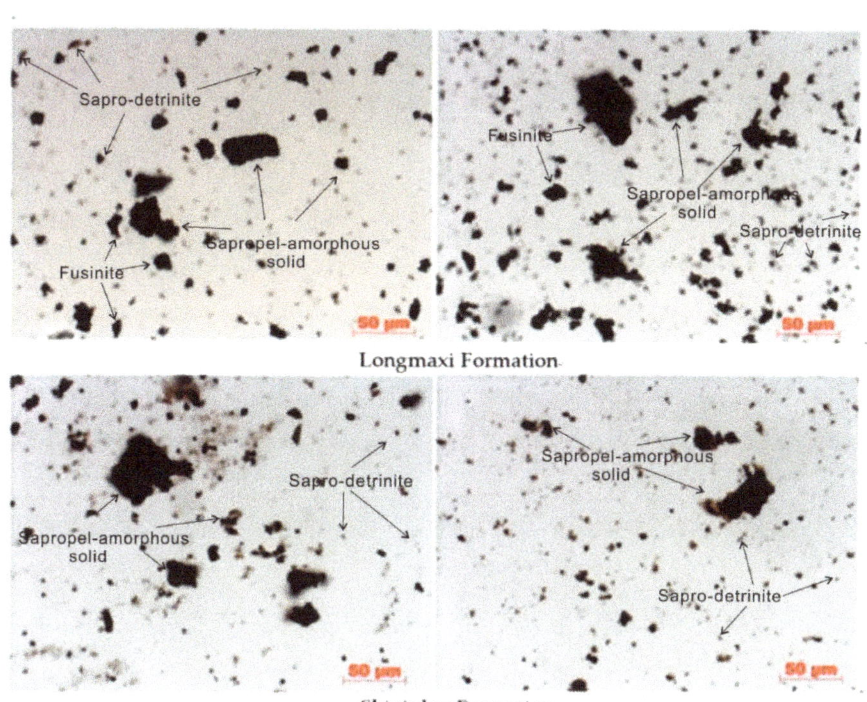

Figure 3. Microscopic compositional characteristics of organic matter in the Shiniulan Formation and Longmaxi Formation of AY (AnYe)-1.

As proposed by Bowker et al. [38], there is a large amount of organic matter in the shale reservoir, and its abundance and maturity have a significant impact on shale gas resources. Boyer C et al. [36] demonstrated that the organic carbon content (TOC) is an important controlling factor for the accumulation of shale gas, dominating not only the physicochemical properties of shale, but more importantly, controlling the gas content of shale.

The geochemical experiment of calcareous mudstone in the Shiniulan Formation shows that the TOC content is generally less than 1% and is distributed mostly at 0.1%, with an average value of 0.16%; the distribution of organic carbon in the Wufeng–Longmaxi Formation ranged from 3.5% to 5.4%, with an average of 4.5%. According to geochemical logging data, the shale gas enrichment section of the Wufeng–Longmaxi Fm. was concentrated at 2311–2331 m; the organic carbon content was between 0.2% and 4.8%, with an average of 2.96%; and the organic carbon content (TOC) increased with the depth (Figure 4). However, the TOC of samples in the lower Shiniulan Fm. were generally low, being less than 1% at most.

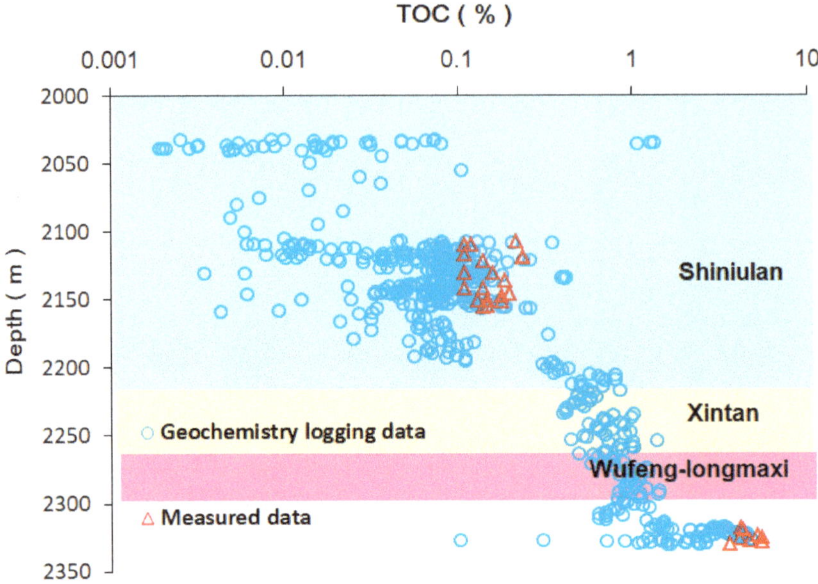

Figure 4. The relationship of the organic matter abundance of samples and the depth of AY-1 well.

The maturity of all samples was greater than 2%. The maturity of the Silurian mudstone samples changed within the range of 2.79% to 3.32%, with an average of 3.11%, both of which are in the over-mature stage. Owing to the rather high thermal evolution in the Shiniulan Fm., the maturity was between 2.75% and 2.92% with an average of 2.95% (Figure 5).

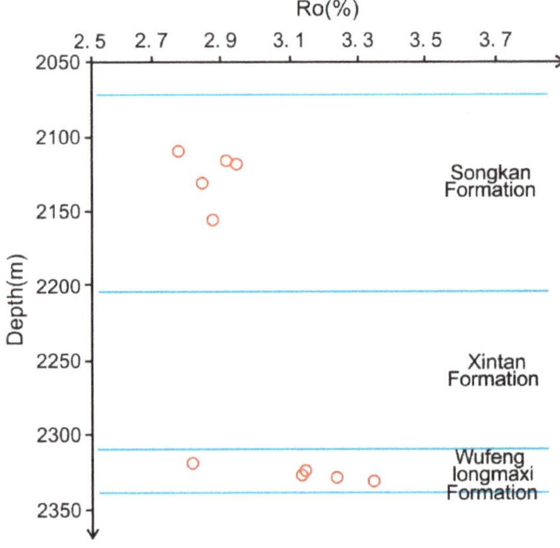

Figure 5. Relationship between maturity of samples and depth.

For hydrocarbon source rocks of the same or similar quality, in general, a higher R_o indicates a greater possibility of producing, fracture development (the greater the relative content of free gas), and a greater production of gas. Thermal maturity has some effect on the hydrocarbon generation potential of organic matter and directly influences the amount of gas generated by source rock and affects the occurrence condition, migration, and aggregation state of natural gas after hydrocarbon generation.

The organic matter content before the threshold of oil generation when hydrocarbons are not produced in large numbers or expelled, is considered to be the original organic matter content. However, the organic matter content of core samples measured at present is usually the residual organic matter content after massive hydrocarbon expulsion. The abundance of organic matter drops significantly during maturation, according to Qin et al. [39]. Thus, the present organic matter cannot show the original geochemical characterization of the mudstone cores accurately, owing to the thermally over-mature organic matter. As a result, we probably underestimated the hydrocarbon generation potential of source rock. Thus, restitution of the original TOC was needed. Therefore, we used a formula proposed by Jarvie et al. [32] to evaluate the hydrocarbon generation potential by arithmetic and parameters, such as HI_o, TRHI (Transformation ratio), TOC_o, and S_{2o} according to Dang et al. [40] (Table 2).

The results obtained by different formulas show that the TRHI is distributed in the range of 92%-100%, with a mean value of 97.6%, which occurs when organic matter is at the post-mature stage. Furthermore, the quite low S_2 indicates that the current organic matter has hydrocarbon generation.

For an amount of generated hydrocarbon equal to S_{2o} minus S_2, the result ranges from 1.18 to 3.39 with an average number of 2.8 mg $HC·g^{-1}$ rock for Shiniulan Fm. mudstones. As the final step, we obtained the Q_{HC}, which was transformed to a volume of CH_4 per unit mass of rock by means of the formulas shown below the table, and then obtained the average result of about 3.67 $m^3·t^{-1}$ rock of Shiniulan mudstone. It was proven that the Shiniulan mudstone has a certain hydrocarbon generation potential. However, it is far from enough to satisfy the gas content in the reservoir.

4.2. The Gas Content Measurement of Source Rock in the Shiniulan Formation

The amount of adsorbed gas with pressure changes can be divided into three stages. When the pressure is low (0.00-2.25 MPa), the amount of adsorbed gas and the pressure are approximately linear, with rapidly increasing values; when the pressure is moderate (2.25-11.13 MPa), the amount of adsorbed gas rises slowly with the increasing pressure. The adsorption reaches monolayer saturation when the pressure is large enough (greater than 11.13 MPa), and the amount of adsorption does not change with pressure, which is consistent with the adsorption process demonstrated by the Langmuir equation theory. Table 3 shows the experimental data of four mudstone samples of the Shiniulan Fm., indicating that the adsorption performance of mudstone in the Shiniulan Fm. is not strong based on the analysis of the result of the isothermal adsorption experiment (Figure 6).

Table 2. The calculated results of parameters for evaluating hydrocarbon generative potential of samples.

Sample	Depth (m)	Formation	Measured Data							Calculated Data				
			S_1 (mgHC·g⁻¹ Rock)	S_2 (mgHC g⁻¹ Rock)	TOC (wt%)	HI_o (mgHC·g⁻¹ Rock)	PI_{pd} (%)	HI (mgHC·g⁻¹Rock)	TR_{HI}	TOC_o (wt%)	S_{2o} (mgHC·g⁻¹ Rock)	Q_{HC} (mgHC·g⁻¹ Rock)	V_{re} (cm³·g⁻¹ Rock)
AY-1-1	2110	Shiniulan	0.007	0.023	0.12	637	0.23	34.1	0.97	0.22	1.39	1.37	2.14
AY-1-2	2116	Shiniulan	0.017	0.037	0.11	643	0.31	98.7	0.92	0.21	1.38	1.34	2.11
AY-1-3	2118	Shiniulan	0.014	0.035	0.24	674	0.29	40.9	0.97	0.49	3.28	3.24	5.02
AY-1-6	2131	Shiniulan	0.012	0.022	0.16	649	0.35	17.4	0.99	0.33	2.14	2.11	3.27
AY-1-7	2156	Shiniulan	0.007	0.034	0.14	680	0.17	24.9	0.98	0.28	1.88	1.85	2.88
AY-1-18	2318	Wufeng	0.021	0.021	4.11	686	0.50	5.4	1.00	10.80	74.10	74.08	113.47
AY-1-23	2328	Wufeng	0.141	0.065	0.80	692	0.68	20.0	0.99	1.86	12.88	12.81	19.72
AY-1-24	2334	Wufeng	0.004	0.004	1.20	705	0.50	22.5	0.99	2.87	20.24	20.24	31.00

(1)　$HI_O = \left(\frac{\%\text{typeI}}{100} \times 750\right) + \left(\frac{\%\text{typeII}}{100} \times 450\right) + \left(\frac{\%\text{typeIII}}{100} \times 125\right) + \left(\frac{\%\text{typeIV}}{100} \times 50\right)$, and the % type is the percentages of kerogen macerals.

(2)　$TR_{HI} = 1 - \frac{HI_{pd}[1200 - HI_O(1 - PI_O)]}{HI_O[1200 - HI_{pd}(1 - PI_{pd})]}$; $TOC_o = \frac{HI_{pd}(TOC_{pd}) \times 83.33}{HI_O(1 - TR_{HI})(83.33 - TOC_{pd}) - HI_{pd}(TOC_{pd})}$ (83.33 is the mean number of hydrocarbons carbon). $PI_O = 0.02/PI_{pd}$; $S_{2o} = HI_o\text{-}TOC_o/100$, mg HC·g⁻¹ Rock; $Q_{HC} = S_{2o} - S_2$, mg HC·g⁻¹ Rock.

(3)　$V_{re} = (n \times Q_{HC})/M_{CH4}$, where n is molar volume of gas under the experimental environment with the temperature of 25 °C and the pressure of 1.01×10^5 Pa, 24.5 L/mol; M_{CH4} is the methane molar mass, 16 g·mol⁻¹.

Table 3. The Langmuir fitting results and thermodynamic parameters of methane adsorption.

	Adsorption Capacity under Different Pressures									Adsorption Constants			Depth (m)
Pressure (Mpa)	0	0.35	1.07	2.25	4.31	5.25	6.24	8.72	11.13	V_L (m³·t⁻¹)	P_L (MPa)	V (m³·t⁻¹)	
Adsorbed gas (m³·t⁻¹)	0	0.16	0.33	0.45	0.47	0.48	0.51	0.55	0.57	0.63	1.10	0.56	2110 (AY-1-1)
	0	0.25	0.5	0.6	0.7	0.72	0.75	0.77	0.79	0.88	1.04	0.78	2120 (AY-1-5)
	0	0.23	0.4	0.52	0.57	0.59	0.62	0.65	0.66	0.72	1.08	0.68	2131 (AY-1-6)
	0	0.2	0.35	0.5	0.51	0.55	0.58	0.6	0.62	0.68	1.11	0.66	2156 (AY-1-7)

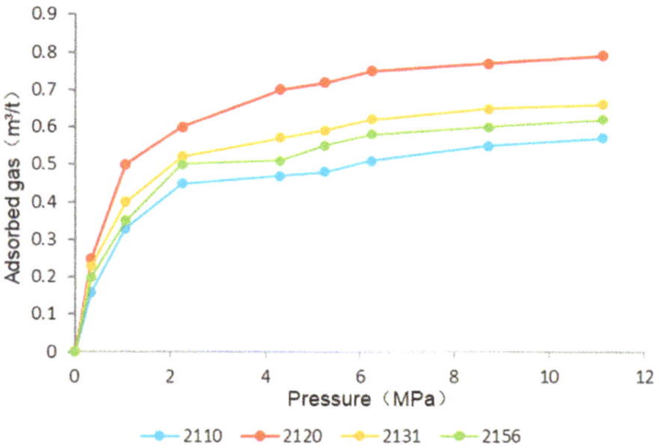

Figure 6. Methane adsorption isotherms of different Shiniulan mudstone samples at 30 °C.

It can be seen from Figure 7 that the TOCs of samples from Shiniulan Fm. are positively correlated with the adsorption capacity of mudstone (Figure 7). The higher the organic matter abundance, the more micropores that develop in mudstone, resulting in a greater adsorption capacity.

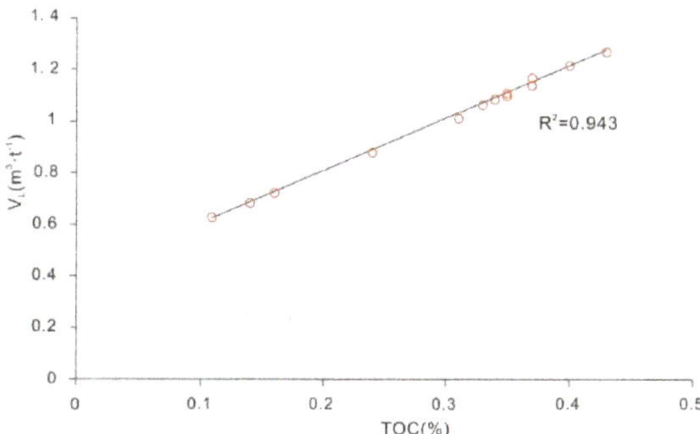

Figure 7. The relationship of total organic carbon (TOC) of samples and adsorption capacity (V_L) of mudstone.

The isothermal adsorption experiments mentioned above also show a relationship in which the Langmuir volume is negatively correlated with the Langmuir pressure (Figure 8). The fitting coefficient (R^2) reaches 0.899, indicating that the Langmuir pressure is one of the influential factors of Langmuir volume.

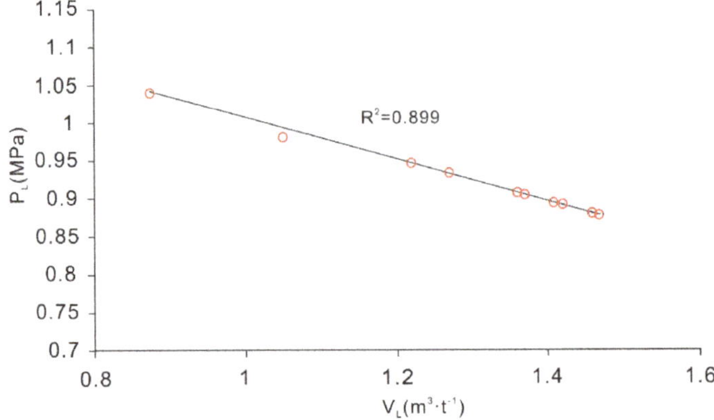

Figure 8. The relationship of Langmuir pressure (P_L) and Langmuir volume (V_L).

Adding up the data points obtained from the isothermal adsorption experiments of the samples implies a good fit for the formulas regarding the relationship between the Langmuir pressure and the Langmuir volume, and the Langmuir volume and the TOC, respectively. Consequently, the formulas for the adsorption gas content of the Shiniulan Fm. samples can be derived.

$$P_L = -0.276\, V_L + 1.2839 \tag{12}$$

$$V_L = 2.0404TOC + 0.393 \tag{13}$$

Substituting the fitting formula into the Langmuir model (using Equation (2)), the formation pressure is converted into an equation between the formation pressure coefficient and depth and other parameters. The adsorption gas content of calcareous mudstones (units of m³/t) was obtained based on the following equations.

$$P = k \times P_h \tag{14}$$

$$P_h = \rho g h \tag{15}$$

$$Q = \frac{(20TOC + 3.8514)kh}{-506TOC + 1173.9 + 9.8kh} \tag{16}$$

where k is the formation pressure coefficient, h is the depth, P_h is the hydrostatic pressure, ρ is the density of water, and g is the gravitational acceleration.

Through the calculated volumes of the adsorption gas and free gas from the Shiniulan Fm., the theoretical gas content of the Shiniulan Fm. gas can be obtained (Table 4).

Table 4. Free gas content of mudstone samples using computational method.

Sample	Effective Porosity (%)	Gas Saturation (%)	Formation Pressure (MPa)	Density (g·cm^{-3})	Compressibility Factor	Formation Temperature (°C)	The Free Volume (m^3·t^{-1})
AY-1-1	3.40	30	28.54	2.7	0.94	70.64	0.96
AY-1-3	3.50	67	28.64	2.7	0.95	70.83	2.12
AY-1-6	3.70	55	37.80	2.7	0.92	71.14	2.31
AY-1-7	3.90	36	41.62	2.7	0.93	71.74	1.95
AY-1-8	5.00	56	44.99	2.7	0.93	72.70	4.19
AY-1-9	5.10	62	44.56	2.7	0.93	72.73	4.70
AY-1-10	4.80	57	44.56	2.7	0.93	72.75	4.10
AY-1-11	4.10	65	44.56	2.7	0.93	72.78	3.96
AY-1-12	2.20	78	44.78	2.7	0.93	72.80	2.56
AY-1-13	2.30	45	44.78	2.7	0.93	72.82	1.50
AY-1-14	2.86	58	44.78	2.7	0.93	72.85	2.47
AY-1-15	4.66	49	44.78	2.7	0.93	72.88	3.40
AY-1-16	5.70	52	42.15	2.7	0.93	72.90	4.17

The porosity data of core samples from the Shiniulan Fm. show a good linear correlation with the content of free gas (Figure 9). Under the condition that the amount of generated gas is definite, the volume of the free gas is mainly affected by the porosity of mudstone. The extent of the porosity is controlled by the development of the pores and the fractures. Thus, the more developed the pores and the fractures, the higher the content of free gas. In addition, a large number of fractures are conducive to the desorption of adsorbed gas, which was converted into more free gas.

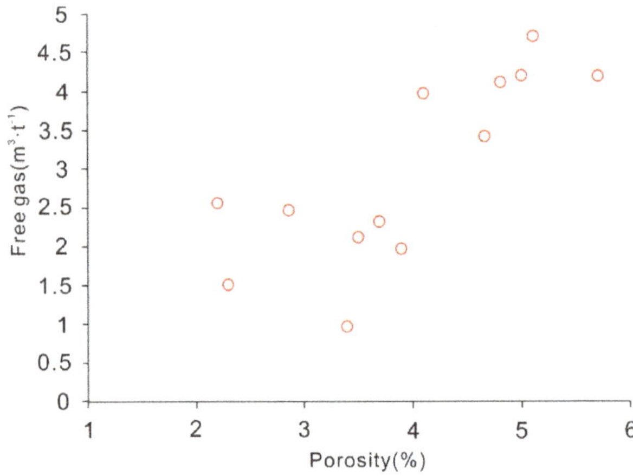

Figure 9. The relationship of porosity of samples and the free gas content of mudstone.

Thus, the theoretical gas content of samples can be obtained by the volumes calculated of the adsorption gas and free gas together, as shown in Table 5.

Table 5. The results of calculated total gas content of samples.

Sample	Depth(m)	TOC (%)	The Content of Free Gas ($m^3 \cdot t^{-1}$)	The Content of Adsorption Gas ($m^3 \cdot t^{-1}$)	Theoretical Gas Content ($m^3 \cdot t^{-1}$)
AY-1-1	2110	0.11	0.96	0.63	1.59
AY-1-4	2119	0.24	2.12	0.88	3.00
AY-1-6	2131	0.16	2.31	0.72	3.03
AY-1-7	2156	0.14	1.95	0.68	2.63
AY-1-8	2196	0.35	4.19	1.10	5.29
AY-1-9	2197	0.34	4.67	1.08	5.75
AY-1-10	2198	0.31	4.10	1.01	5.11
AY-1-11	2199	0.37	3.96	1.14	5.10
AY-1-12	2200	0.33	2.56	1.06	3.62
AY-1-13	2201	0.43	1.50	1.27	2.77
AY-1-14	2202	0.40	2.50	1.21	3.71
AY-1-15	2203	0.37	3.40	1.16	4.56
AY-1-16	2204	0.35	4.17	1.11	5.28

Table 6 provides the results of the gas content (including the desorbed gas content and lost gas content) of the samples. The total gas contents of the mudstone in Shiniulan Fm. range from 0.065 to 0.841 $m^3 \cdot t^{-1}$. The total gas contents of Wufeng (2.3-6.2 $m^3 \cdot t^{-1}$)-Longmaxi (3.1-6.1 $m^3 \cdot t^{-1}$) Fm. are generally greater than 3 $m^3 \cdot t^{-1}$, even reaching 6 $m^3 \cdot t^{-1}$ at the depth of 2323–2326 m. It is apparent that the content of the Shiuniulan mudstone is much lower than that of the Wufeng–Longmaxi Fm. shale (Figure 10).

Table 6. The contents of desorbed gas are measured by USBM direct method.

Sample	Depth (m)	Formation	Sample Weight (g)	Desorbed Gas Content ($m^3 \cdot t^{-1}$)	Lost Gas Content ($m^3 \cdot t^{-1}$)	Total Gas Content ($m^3 \cdot t^{-1}$)
B1-3	2107.99	Shiniulan	3580	0.556	0.053	0.609
1-3 (A3)	2108.88	Shiniulan	3291	0.023	0.042	0.065
B1-2	2110.27	Shiniulan	3528	0.113	0.004	0.117
1-2 (A2)	2111.97	Shiniulan	3778	0.032	0.140	0.172
B1-1	2113.56	Shiniulan	3438	0.104	0.038	0.142
1-1 (A1)	2114.82	Shiniulan	3675	0.046	0.170	0.216
B2-1	2119.55	Shiniulan	3654	0.111	0.029	0.140
2-2 (A2)	2119.55	Shiniulan	3639	0.057	0.114	0.171
B2-2	2121.98	Shiniulan	3428	0.154	0.027	0.181
2-1 (A1)	2121.98	Shiniulan	3654	0.072	0.118	0.190
B3-3	2123.66	Shiniulan	2932	0.153	0.119	0.272
3-4 (A5)	2123.98	Shiniulan	3720	0.070	0.134	0.204
3-3 (A3)	2129.05	Shiniulan	3497	0.074	0.137	0.211
B3-2	2130.26	Shiniulan	3233	0.151	0.078	0.229
3-2 (A2)	2133.45	Shiniulan	3746	0.069	0.178	0.247
B3-1	2135.43	Shiniulan	2586	0.190	0.107	0.297
3-1 (A1)	2139.46	Shiniulan	3597	0.184	0.657	0.841
B4-2	2142.67	Shiniulan	3545	0.140	0.571	0.711
B4-1	2153.33	Shiniulan	2910	0.162	0.580	0.742
B5-3	2317.86	Longmaxi	1626	1.698	2.763	4.461
B5-2	2318.19	Longmaxi	1618	1.616	2.654	4.270
A4-1	2318.41	Longmaxi	1342	1.026	2.095	3.121
B5-1	2319.20	Longmaxi	1609	1.469	2.475	3.944
(A8-1)	2319.70	Longmaxi	1255	1.596	3.512	5.108
B6-4	2320.57	Longmaxi	1548	0.728	3.267	3.995
(A3-1)	2322.10	Longmaxi	1312	1.509	1.984	3.493
B6-3	2323.54	Longmaxi	1461	2.218	3.926	6.144
B6-5	2324.26	Longmaxi	1559	2.331	3.666	5.997
(A2-1)	2325.92	Wufeng	1741	1.959	4.126	6.085
B6-2	2326.10	Wufeng	1364	1.222	5.005	6.227
B6-1	2327.11	Wufeng	1476	1.484	2.385	3.869
(A8-2)	2329.04	Wufeng	1337	1.035	1.247	2.282
B7-2	2329.66	Wufeng	1496	0.996	1.630	2.626
(A7-1)	2331.04	Wufeng	1753	0.800	1.377	2.177
B7-1	2331.36	Wufeng	1471	1.331	1.963	3.294

However, the desorbed gas contents we collected of samples with active oil-gas were especially low, ranging mainly from 0.1 to 0.2 $m^3 \cdot t^{-1}$; the maximum was 0.842 $m^3 \cdot t^{-1}$. The gas of Shiniulan mudstone mainly exists in free states. Thus, the low desorbed gas contents tend to be primarily caused by the escape of free gas from horizontal fractures (Figure 11) during the processes from lifting the drill to canning. What you should note here is that the USBM direct method is usually used to predict the gas contents of shale or coal. These data would be fairly low due to the gas samples extracted from the interbedded rock association, with only appropriate meaning as a reference.

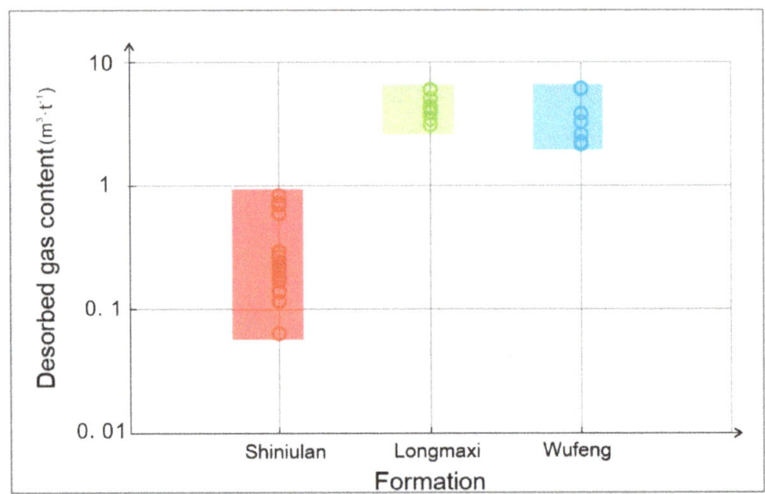

Figure 10. The distribution of total gas contents from different formations in the AY-1 well.

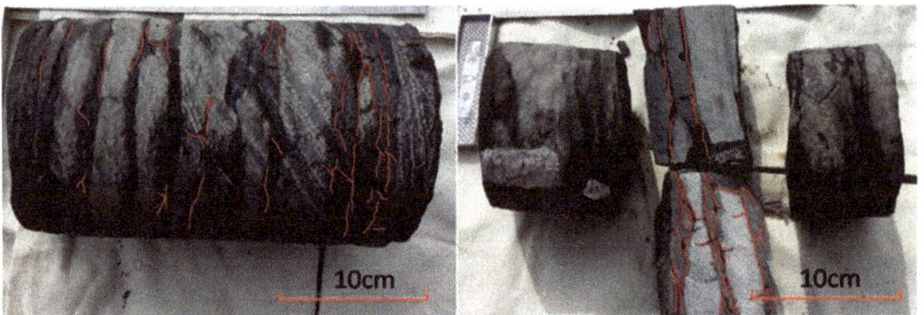

Figure 11. The photos about fracture characteristics of cores from Shiniulan Formation.

4.3. The Origin of Gas from the Siniulan Formation

Alkane gas is one of the most fundamental and major components of natural gases in the Shiniulan Fm. As shown in Table 7, for eleven gas samples from sample AY-1, the CH_4 and C_{2-3} contents were in the range of 95.244%-96.432% and 2.167%-2.2%, respectively. CO_2 is the most fundamental and major part of the non-hydrocarbons in the gas samples, with a content in the range of 0.709%-0.864%.

Table 7. The molecule compositions of gas samples of AY-1 (data quoted from Liu et al. [10]).

Sample	Strata	Gas Composition (%)				$\delta^{13}C(‰)VPDB$			$\delta^2H(‰)VSMOW$
		CH_4	C_2H_6	C_3H_8	CO_2	CH_4	C_2H_6	CO_2	CH_4
1	S₁s	96.289	2.067	0.139	0.795	−33.6	−36.8	−18.6	−149
2	S₁s	95.375	2.041	0.126	0.864	−33.6	−36.9	−18.1	−156.6
3	S₁s	96.432	2.082	0.143	0.709	−33.2	−37.0	−17.6	−145.8
4	S₁s	96.276	2.058	0.132	0.798	−33.5	−36.8	—	—
5	S₁s	95.244	2.042	0.124	0.864	−33.9	−36.5	—	—
6	S₁s	95.682	2.063	0.127	0.823	−33.4	−36.8	—	—
7	S₁s	96.278	2.053	0.134	0.805	−33.4	−36.6	—	—
8	S₁s	96.258	2.049	0.128	0.821	−33.3	−36.5	—	—
9	S₁s	96.274	2.052	0.132	0.812	−33.9	−36.8	−20.8	−150
10	S₁s	95.874	2.065	0.131	0.818	−33.2	−36.2	−20.6	−153.9
11	S₁s	95.946	2.066	0.134	0.806	−33.5	−37.0	−19.4	−146

Footnote: S₁s: the Shiniulan Formation of the Silurian.

The experimental results of the stable carbon and hydrogen isotopic compositions of alkanes indicate that the $\delta^{13}C_1$ values in methane vary from −33.9‰ to −33.2‰ (average −33.5‰), compared with −37.0‰ to −36.2‰ (average −36.7‰) in ethane. Furthermore, the hydrogen isotope (δD) data were obtained, which are mainly distributed from −156.6‰ to −145.8‰ (average −150‰). The $\delta^2HC_2H_6$ values could not be detected due to the low concentration of ethane (Table 7).

The compositions of carbon and hydrogen isotopes in association with the molecular composition can be utilized extensively to discern the gas origin, as suggested by Berner and Faber [41], Chung et al. [42], Schoell [43], Whiticar [44], Strąpoć et al. [45], and Strąpoć et al. [46]. Galimov [47], Colombo et al. [48], Stahl [49], and Schoell [50] showed that the molecular composition of natural gases and the isotope ratios in the hydrocarbons are both in control of processes during the formation of the gases.

The X-identification chart, which was proposed by Zhang et al. [51] is used to determine the natural gas genetic type according to Dai. [52]. As shown in Figure 12, all gas samples from the Shiniulan Fm. could be recognized as a mixed source when deep-seated. According to the carbon isotope of methane and ethane distribution characteristics, the natural gas genetic type identification criteria and principles are used to determine the origin of the natural gas—which is a mature–highly mature oil-based mixture in the Shiniulan Fm.(Figure 13)—as presented by Ding et al. [53].

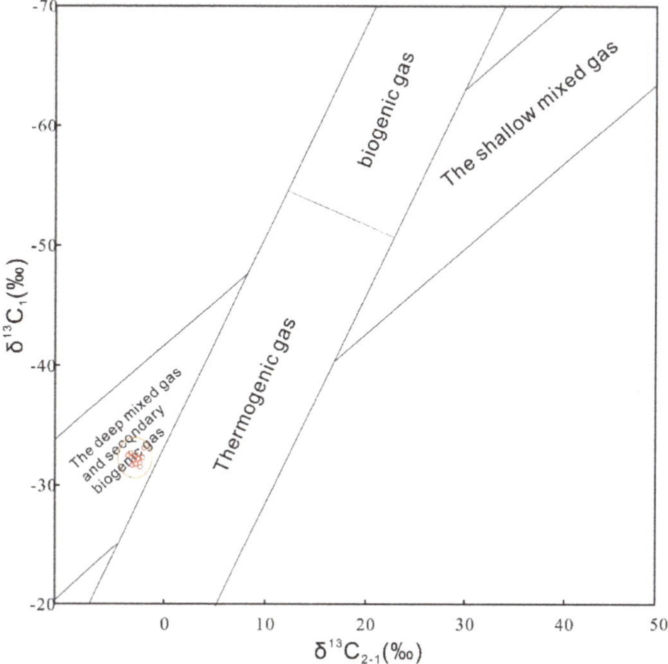

Figure 12. X identification chart of natural gas genetic type in Shiniulan Formation (After Zhang et al. [51]).

That combination of the $\delta^{13}C$ values of ethane and methane is also useful for determining the genetic types of the gases. According to the $\delta^{13}C_2$-$\delta^{13}C_1$ identification chart to identify the gases from Shiniulan Fm., in the study area, they fall into field M in the chart, which might be due to the methane in over-mature zones below the formation possibly migrating through these areas (Figure 14).

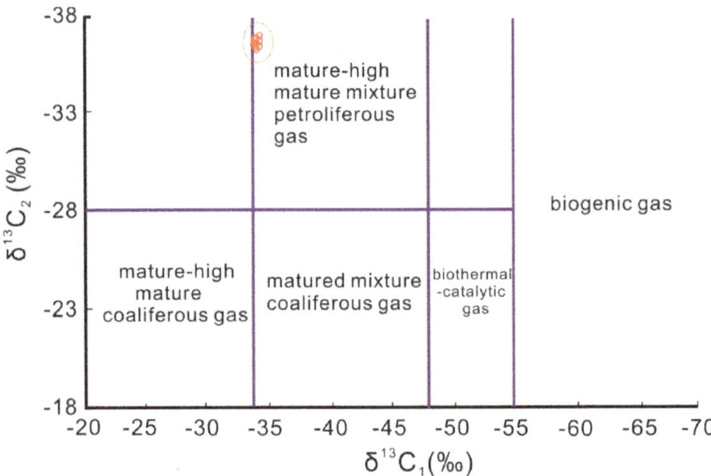

Figure 13. Discrimination of natural gas genetic types of Shiniulan Formation (using Ding et al. [53]).

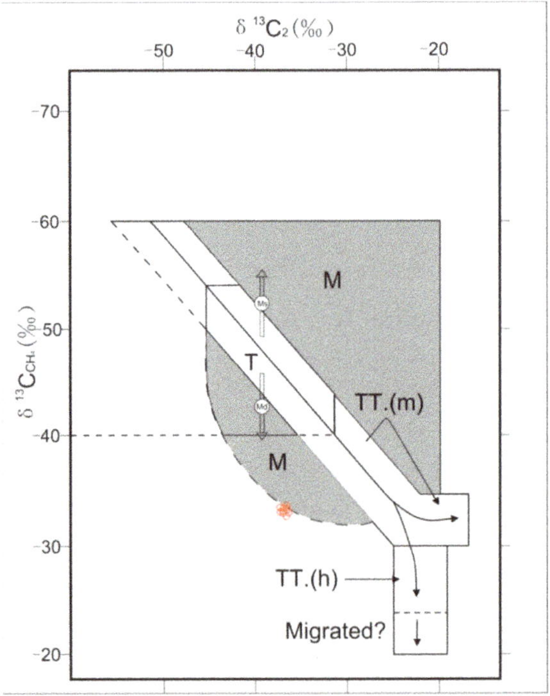

Figure 14. Carbon isotope variations in ethane related to carbon isotope variations in methane (taking after Schoell M. [50]).

In general, the carbon isotope of organic origin primary alkane gas is a series of normal carbon isotopes ($\delta^{13}C_1 < \delta^{13}C_2 < \delta^{13}C_3 < \delta^{13}C_4$) according to Dai et al. [54]. The cause of the reverse of the carbon isotope series of organic alkane hydrocarbons in Shiniulan Fm. is a mixed gas which is derived from the same source rock with different gas generating stages or the same genetic types in different source rocks.

Natural gas in the Shiniulan Fm. is mainly distributed in the carbon isotope inverted series gas zone based on the carbon isotope chart of methane and ethane of natural gas proposed by Dai (Figure 15) [55]. Furthermore, $\delta^{13}C$ values of alkane gases in Shiniulan Fm. are close to those reported previously in the Wufeng–Longmaxi shale in southern Sichuan Basin, indicating that Wufeng–Longmaxi Fm. may have been a potential gas source rock for Shiniulan Fm. The isotopic rollovers of natural gas in Shiniulan Fm. may have resulted from carbon exchange at high temperature (Dai et al. [56]). The potential contribution of Wufeng–Longmaxi Fm. is also consistent with the gas content results. Through the comprehensive comparison of the in-place gas content reconstructed, the amount of gas measured by the indirect methods and direct methods from the Shiniulan calcareous mudstone samples reflect the fact that mudstone samples with lower organic abundance have certain hydrocarbon-generation potential, which is largely insufficient to satisfy the gas content currently stored within its layer. It can be preliminarily concluded that the gas source of the Shiniulan Fm. mudstone is partly derived from the gas source rocks in other formations, probably Wufeng–Longmaxi Fm. However, there are not enough figures and information to determine definite gas sources in Shiniulan Fm., and thus further studies are needed.

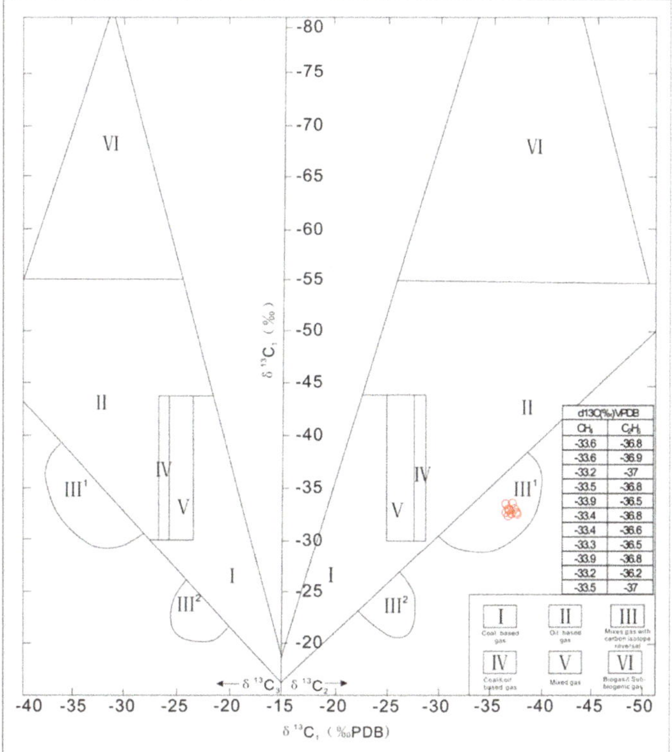

Figure 15. Determination of natural gas genesis based on carbon isotope composition (using Dai. [55]).

Through the comprehensive comparison of the reconstructed in-place gas contents, the amounts of gas measured by the indirect methods and direct methods from the Shiniulan calcareous mudstone samples reflect the fact that mudstone samples with lower organic abundance have certain hydrocarbon-generation potential, which is largely insufficient to satisfy the gas content currently stored within its layer. It can be preliminarily concluded that the gas source of the Shiniulan Fm. mudstone is partly derived from the gas source rocks in other formations. In addition, it can be concluded that there is a mixed-nature gas gathering in the Shiniulan Fm. by the isotope chart. The shale of Wufeng–Longmaxi Fm. is thought to be a potential gas source rock for Shiniulan Fm. However, there are not enough figures and information to determine definite gas sources in Shiniulan Fm., and thus further research is necessary. Through the preliminary judgment of the natural gas types in the Shiniulan Fm., the natural gas accumulation model of the Shiniulan Fm. can be further deduced.

4.4. Accumulation Factors and Natural Gas Accumulation Model of Shiniulan Formation

4.4.1. Accumulation factors of the natural gas from the Shiniulan Formation

It has been confirmed by two-dimensional seismic exploration that the direction of the Anchang slope axis is in the NW–NE direction; the southeast wing is steep, while the northwest wing is gentle. The northwest wing's inclination angle is approximately 24°, while the southeast wing's is approximately 30°. It gradually closes, with the increased buried depth of the strata developing near the fault of the slanting core. In the slanting direction, due to the northwest–south–eastward compression of the Yanshan movement and the north–south direction of the early Himalayan movement, three northeast–southwest strike faults were developed. The length of the fault in the bottom of the Longmaxi Fm. varies from 3 to 11 km, with a distance from 80 to 670 m and an inclination between 30° and 50°.

In addition, through the observation and measurement of the outcrop in the Lower Silurian Shiniulan Fm. (Figure 16), it was found that the structural fractures of the Shiniulan Fm. included tensile fractures and shear fractures. The shear fractures have a deep cutting-bed section, stable occurrence, and further extension, with fracture angles in the range of 45°-90°. Tension fractures often appear in parallel and in groups, with obvious directionality and regular distribution. Structural fractures are generally filled with materials such as calcite and are produced in the areas in which the deformation is severe or the faults are well-developed.

(**a**) The thick-bedded limestone of the Shiniulan Fm. develops shear fractures, which are filled with calcite.

(**b**) Tensile fractures are developed in the interlayer formed by mudstone and the middle-thick limestone of the Shiniulan Fm., which are filled with large amounts of calcite.

Figure 16. Photos about the development of fractures at the outcrop.

The systematic observation (Figure 17) and statistical analyses of the fractures, in which the length of the cores was 51.5 m and the number of fractures was 1176, from cores of Shiniulan Fm., was completed.

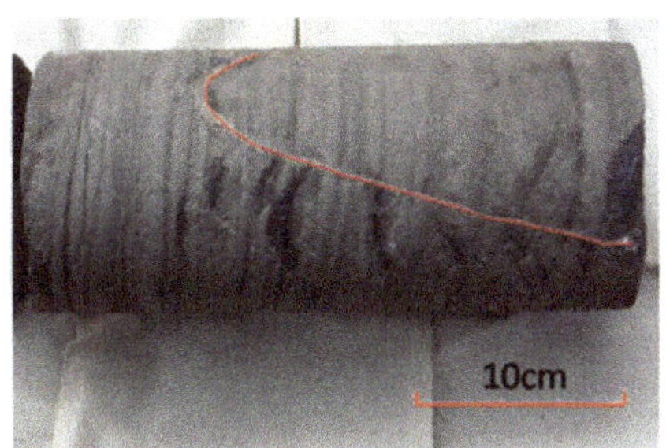

(**a**) Depth. 2107.20–2108.08 m high-angle shear fractures filled with calcite.

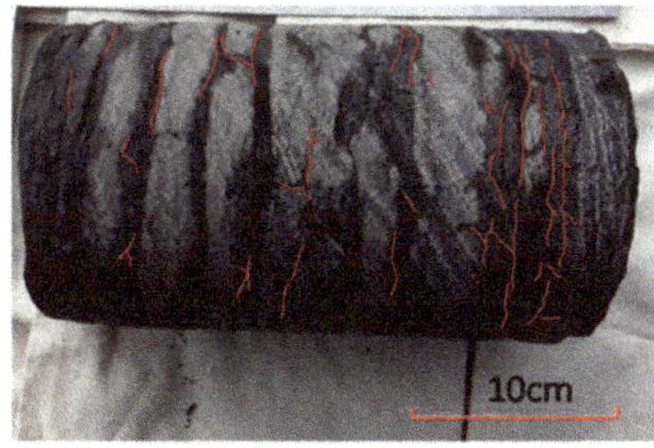

(**b**) The 2039.19–2040.16 m shrinkage fractures resulting from diagenesis.

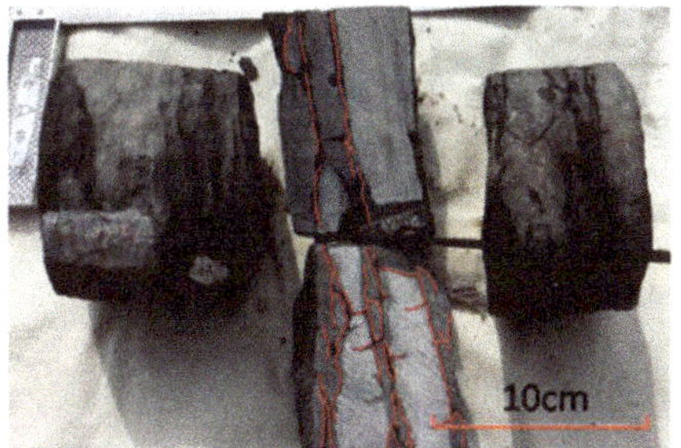

(**c**) Depth: 2111.41–2112.15 m shrinkage fractures resulting from diagenesis.

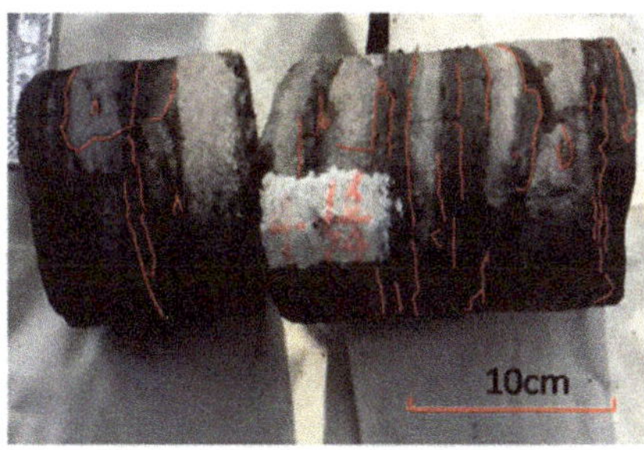

(**d**) Depth: 2125.61–2126.4 m shrinkage fractures resulting from diagenesis.

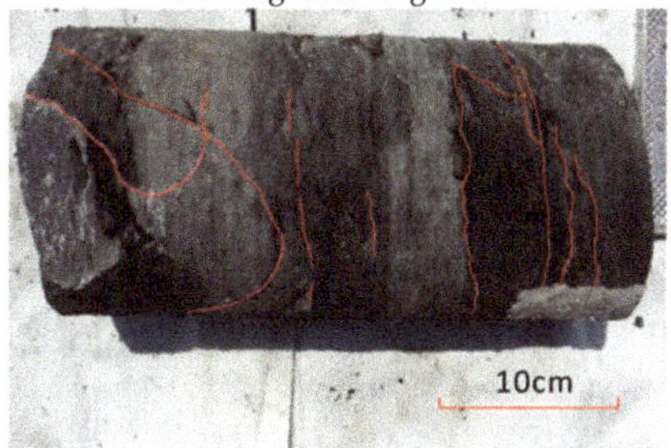

(**e**) Depth: 2137.80–2138.75 m multi-stage structural fractures and shrinkage fractures resulting from diagenesis.

(**f**) Depth: 2151.96–2152.91 m shrinkage fractures resulting from diagenesis.

Figure 17. Photos about the core fractures of Shiniulan Formation.

Through the statistical analysis of core fractures, shrinkage fractures resulting from diagenesis in the Shiniulan Fm. mudstone are well-developed, while the structural fractures formed due to stress are rare (Figure 18). The microfractures that developed in mudstone are mainly caused by mineral phase transformation or shrinkage caused by water loss during diagenesis, and consolidation and had nothing to do with structuring. Moreover, most of the microfractures in the Shiniulan Fm. shale are not filled with minerals, while most of the microfractures developed in the limestone are filled with calcite.

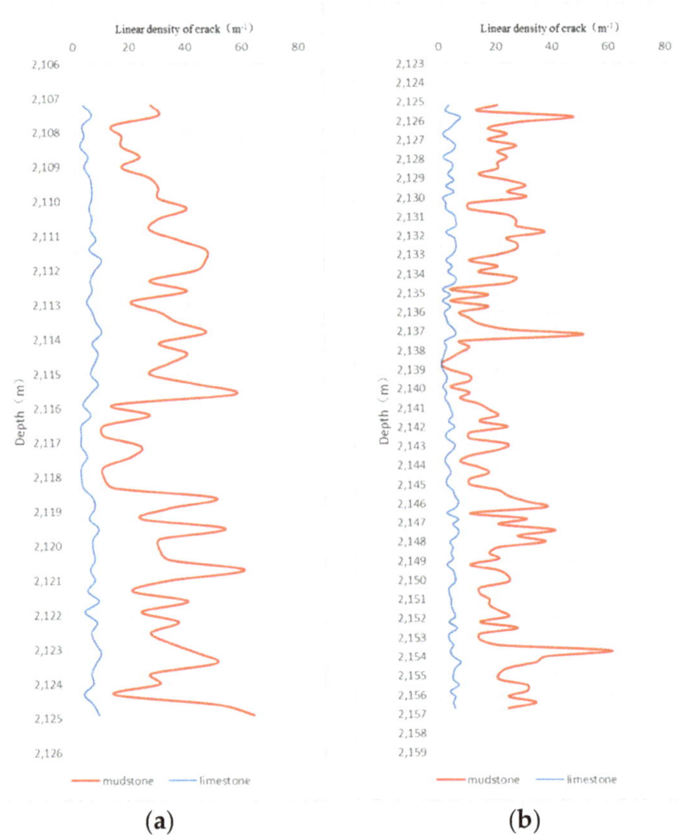

Figure 18. Comparison of mudstone and limestone fractures of Shiniulan Formation. (**a**) The variety trend of liner density of cracks from 2107 to 2125 meters. (**b**) The variety trend of liner density of cracks from 2125 to 2156 meters.

Microfractures, as important channels for natural gas, increase the porosity and permeability, and thus improve the capacity of the reservoir, which is conducive to the increase of the free gas volume and the desorption capability of adsorbed gas. A large number of fractures due to diagenesis developed in the special combination of limestone and mudstone to accumulate free methane gas in the Shiniulan Fm. Fractures are hardly developed in limestone, so the limestone is an available cap rock to prevent the leakage of shale gas (Figure 19).

A series of oil and gas exploration practices have shown that the presence of fracture development zones in reservoirs results in high-yield zones with favorable porosity and permeability. The fractures, on the one hand, provide a storage space for oil and gas; on the other hand, they can significantly improve the matrix permeability and the pore connectivity of tight reservoirs, providing an effective channel for fluid migration.

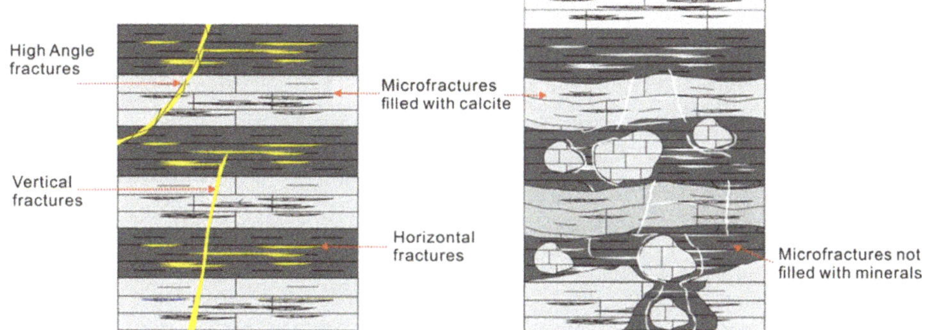

Figure 19. The Fractures' development pattern in the Shiniulan Formation.

4.4.2. Natural Gas Accumulation Model of Shiniulan Formation

The origin of natural gas in Shiniulan Fm. is a special mixed gas derived from different strata. In addition, the pore size distribution of mudstone and limestone in the Shiniulan Fm. shows that the pore size of the limestone is concentrated at approximately 2.88 nm, while the pore size of the mudstone is mainly concentrated at approximately 7.13 nm. The limestone is denser than the mudstone and can effectively seal the natural gas in the mudstone. The combination of lithology as a good preservation condition makes the formation relatively closed; thus, facilitating the preservation of gas reservoirs.

Natural gas is mainly concentrated in a large number of horizontal fractures developed in mudstone. The partly-developed, high-angle fractures provide good channels for horizontal migration, which allows natural gas to migrate upwards at short distances and accumulate, causing abnormally high pressures. When the drilling reaches this layer, a local decompression occurs, which causes the natural gas in the fractures to accumulate in the wellbore. Finally, a high yield is formed where horizontal fractures develop while high-angle fractures are scarce.

The stratum of the Silurian stopped moving during the Triassic, while the three faults that developed in the Anchang syncline core continue to be active under the Yanshan–Hishan movement. Due to the simultaneous extrusion of multi-phase tectonic movements, the fractures in the oblique core are developed, accompanied by a large number of associated microfractures; thus, providing an important channel for the supply of gas to the Shiniulan Fm. from the lower formation. The well-developed fractures, which are dominated by shrinkage fractures that result from diagenesis in the Shiniulan Fm., are more conducive to the formation of a low potential area to accumulate gases. Similarly, the limestone in the Baota Fm. also developed a small amount of fractures due to the effect of fractures, while the shale gas from the Longmaxi Fm. was partially migrated and stored in the limestone of the Baota Fm. through the fractures. Therefore, a high-pressure fractured gas reservoir is predicted to be in the limestone of the Baota Fm.

The high yield in the Shiniulan Fm. is due to the accumulation of mudstone gas and the mixing of the shale gas from underlying strata. Natural gas accumulates along fractures in the Shiniulan Fm. through fault and hydrocarbon self-generation, forming a unique, abundant accumulation model in the Shiniulan Fm. (Figure 20).

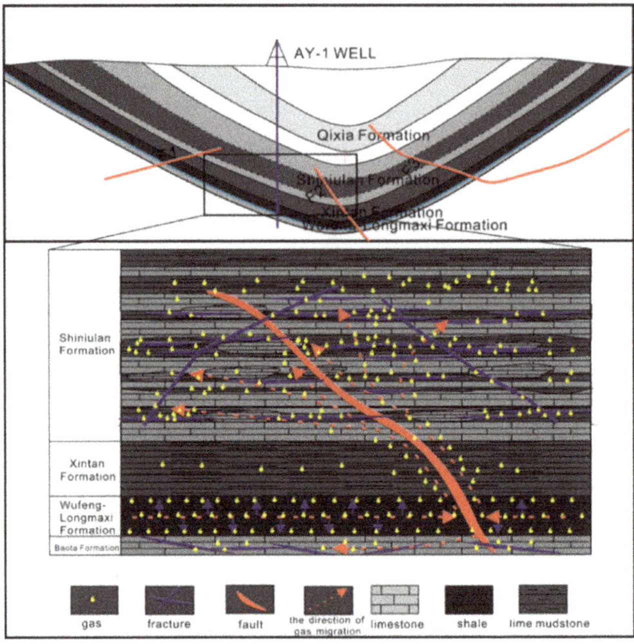

Figure 20. Natural gas accumulation model of Shiniulan Formation in AY-1 well.

5. Conclusions

1. Combined with the gas contents by different methods above, it was indicated that the Shiniulan mudstone has a certain hydrocarbon generation potential; however, it is far from enough to satisfy the gas content in the reservoir.

2. The origin of natural gas from the Shiniulan Fm. is a mixed source type. It can be considered that the gas source of the Shiniulan Fm. is not only derived from its own source rock but also migrated from other gas-source rock, such as the shale of the Wufeng–Longmaxi Fm.

3. The diagenetic microfractures are well-developed, while the tectonic fractures formed by stress are rare. Moreover, the microfractures in the limestone are filled with calcite. This lithological association as a good element facilitates the accumulation of gas reservoirs.

4. Due to the specific lithological association consisting of frequent calcareous mudstone and limestone interbedding, the limestone with low porosity and permeability and poorly developed fractures act as cap layers to restrain the natural gas from escaping to other layers and effectively seal the gas into well-developed fractures of mudstone.

5. The development of local faults and high-angle microfractures provides a channel for the upward and downward migration of shale gas, allowing the abundant accumulation of natural gas in horizontal fractures developed in the mudstone of the Shiniulan Fm.

Author Contributions: Conceptualization, Z.L. and R.G.; Methodology, X.T.; Software, P.Z.; Validation, R.G., P.Z. and Z.L.; Formal Analysis, R.G.; Investigation, P.Z.; Data Curation, Z.L.; Writing-Original Draft Preparation, R.G.; Writing-Review & Editing, J.Z.

Funding: This work was supported by the Research on Shale Gas Resource Potential Evaluation Method and Exploration Technology (2016ZX05034), Shale Gas Resources Evaluation and Factors Optimization in Key Areas of South China (G20171901), the Excellent Supervisor Funded Program of the Ministry of Education (grant number 2-9-2017-317), Gas accumulation model of well AY-1 in Guizhou Province, and Evaluation and parameter optimization of shale gas resources in typical areas of South China.

Acknowledgments: Many thanks to Zhang for his help with experiments in the geochemistry laboratory of Yangtze University, Jingzhou, China.

Conflicts of Interest: The authors declare no conflict of interest.

Abbreviations

Fm   Formation
USBM   United States Bureau of Mine

References

1. Jia, B.; Tsau, J.S.; Barati, R. A review of the current progress of CO_2 injection EOR and carbon storage in shale oil reservoirs. *Fuel* **2019**, *236*, 404–427. [CrossRef]
2. U.S. Crude Oil and Natural Gas Proved Reserves. *Year-End 2017. U.S.*; Energy Information Administration (EIA): Washington, USA, 2017.
3. Zhang, J.C.; Nie, H.K.; Xu, B.; Jiang, S.L.; Zhang, P.X.; Wang, Z.Y. Geological condition of shale gas accumulation in Sichuan Basin. *Nat. Gas Ind.* **2008**, *28*, 151–156.
4. Zou, C.N.; Dong, D.Z.; Wang, S.J.; Li, J.Z.; Li, X.J.; Wang, Y.M.; Li, X.J.; Wang, Y.M.; Li, D.H.; Cheng, K.M. Geological characteristics and resource potential of shale gas in China. *Pet. Exp. Dev.* **2010**, *37*, 641–653. [CrossRef]
5. Wang, S.Q.; Chen, G.S.; Dong, D.Z.; Yang, G.; Lu, Z.G.; Xu, Y.H.; Huang, Y.B. Accumulation conditions and exploitation prospect of shale gas in the lower Paleozoic Sichuan Basin. *Nat. Gas Ind.* **2009**, *29*, 51–58.
6. Tan, J.Q.; Weniger, P.; Krooss, B.; Merkel, A.; Horsfield, B.; Zhang, J.C. Shale gas potential of the major marine shale formations in the upper Yangtze platform, South China, Part II: Methane sorption capacity. *Fuel* **2014**, *129*, 204–218. [CrossRef]
7. Zou, C.N.; Dong, D.Z.; Yang, H.; Wang, Y.M.; Huang, J.L.; Wang, S.F.; Fu, C.X. Conditions of shale gas accumulation and exploration practices in China. *Nat. Gas Ind.* **2011**, *31*, 26–39.
8. Han, S.B.; Zhang, J.C.; Li, Y.X.; Horsfield, B.; Tang, X.; Jiang, W.L.; Chen, Q. Evaluation of Lower Cambrian Shale in Northern Guizhou Province, South China: Implications for Shale Gas Potential. *Energy Fuels* **2013**, *27*, 2933–2941. [CrossRef]
9. Zhao, W.Z.; Li, J.Z.; Yang, T.; Wang, S.F.; Huang, J.L. Geological difference and its significance of marine shale gases in South China. *Pet. Exp. Dev.* **2016**, *43*, 547–559. [CrossRef]
10. Liu, Y.; Zhang, J.C.; Zhang, P.; Liu, Z.Y.; Zhao, P.W.; Huang, H.; Mo, X.X. Origin and enrichment factors of natural gas from the Lower Silurian Songkan Formation in northern Guizhou province, south China. *Int. J. Coal Geol.* **2018**, *187*, 20–29. [CrossRef]
11. Wang, S.J.; Wang, L.S.; Huang, J.L.; Li, X.J.; Li, D.H. Accumulation conditions of shale gas reservoirs in Silurian of the Upper Yangtze region. *Nat. Gas Ind.* **2009**, *29*, 45–50.
12. Liang, C.; Jiang, Z.X.; Zhang, C.M.; Guo, L.; Yang, Y.T.; Li, J. The shale characteristics and shale gas exploration prospects of the Lower Silurian Longmaxi shale, Sichuan Basin, South China. *J. Pet. Sci. Eng.* **2014**, *21*, 636–648. [CrossRef]
13. Wang, Z.H.; Tan, Q.G.; He, L.; Cheng, J.X.; Wang, R.H. Deposition and sequence stratigraphy of the Silurian Shiniulan Formation in southeastern Sichuan-northern Guizhou province. *Oil Gas Geol.* **2013**, *34*, 499–507.
14. Supervision, G.A.O.Q. *Determination of Total Organic Carbon in Sedimentary Rock*; GB/T19145-2003; CHINESE GB Standards: Beijing, China, 2003.
15. SY/T5124-1995. *Method for Determining the Vitrinite Reflectance in the Sedimentary Rocks*; China Petroleum Standardization Committee: Beijing, China, 1995.
16. SY/T5125-1996. *Identification and Classification Method of the Maceral Composition of Organic Matter Using Fluorescence-Fluorescent Light*; China Petroleum Standardization Committee: Beijing, China, 1996.
17. Cao, Q.Y. Identification of microcomponents and types of kerogen under transmitted light. *Pet. Explor. Dev.* **1985**, *5*, 14–23.
18. Wang, X.; Gao, S.; Gao, C. Geological features of Mesozoic continental shale gas in south of Ordos Basin, NW China. *Pet. Exp. Dev.* **2014**, *41*, 326–337. [CrossRef]

19. Dang, W.; Zhang, J.C.; Wei, X.L.; Tang, X.; Chen, Q.; Li, Z.M.; Zhang, M.C.; Liu, J. Geological controls on methane adsorption capacity of Lower Permian transitional black shales in the Southern North China Basin, Central China: Experimental results and geological implications. *J. Pet. Sci. Eng.* **2017**, *152*, 456–470. [CrossRef]

20. Gasparik, M.; Bertier, P.; Gensterblum, Y.; Ghanizadeh, A.; Littke, R. Geological controls on the methane storage capacity in organic-rich shales. *Int. J. Coal Geol.* **2014**, *123*, 34–51. [CrossRef]

21. Ji, L.; Zhang, T.; Milliken, K.; Qu, J.; Zhang, X. Experimental investigation of main controls to methane adsorption in clay-rich rocks. *Appl. Geochem.* **2012**, *27*, 2533–2545. [CrossRef]

22. Dang, W.; Zhang, J.C.; Nie, H.K.; Wang, X.; Tang, X.; Wu, N.; Chen, Q.; Wei, X.L.; Wang, R. Isotherms, thermodynamics and kinetics of methane-shale adsorption pair under supercritical condition: Implications for understanding the nature of shale gas adsorption process. Available online: https://www.sciencedirect.com/science/article/pii/S1385894719326038 (accessed on 23 October 2019).

23. Sudibandriyo, M.; Pan, Z.J.; Fitzgerald, J.E.; Robinson, R.L.; Gasem, K.A.M. Adsorption of methane, nitrogen, carbon dioxide, and their binary mixtures on dry activated carbon at 318.2 k and pressures up to 13.6 mpa. *Langmuir* **2003**, *19*, 5323–5331. [CrossRef]

24. Langmuir, I. The adsorption of gases on plane surfaces of glass, mica and platinum. *J. Chem. Phys.* **2015**, *40*, 1361–1403. [CrossRef]

25. Lewis, R.; Ingraham, D.; Sawyer, W. New Evaluation Techniques for Gas Shale Reservoirs. *Reserv. Symp.* **2004**, *2002*, 1–11.

26. Schlumberger. *Log Interpretation Volume 1: Principles*; Schlumberger Education Services: Huston, TX, USA, 1972.

27. Diamond, W.; LaScola, J.; Hyman, D. *Results of Direct-method Determination of the Gas Content of US Coalbeds*; Bureau of Mines Information Circular: Pittsburgh, PA, USA, 1986.

28. Yee, D.; Seidle, J.; Hanson, W. Gas sorption on coal and measurement of gas content. *AAPG Stud. Geol.* **1993**, *38*, 203–218.

29. Diamond, W.; Schatzel, S. Measuring the gas content of coal: A review. *Int. J. Coal Geol.* **1998**, *35*, 311–331. [CrossRef]

30. Dang, W.; Zhang, J.C.; Tang, X.; Wei, X.L.; Li, Z.M.; Wang, C.H.; Chen, Q.; Liu, C. Investigation of gas content of organic-rich shale: A case study from Lower Permian shale in southern North China Basin, central China. *Geosci. Front.* **2018**, *9*, 559–575. [CrossRef]

31. Behar, F.; Beaumont, V.; Penteado, H.D.B. Rock-Eval 6 technology: Performances and developments. *Oil Gas Sci. Technol.* **2001**, *56*, 111–134. [CrossRef]

32. Jarvie, D.M.; Hill, R.J.; Ruble, T.E.; Pollastro, R.M. Unconventional shale-gas systems: The Mississippian Barnett Shale of north-central Texas as one model for thermogenic shale-gas assessment. *AAPG Bull.* **2007**, *91*, 475–499. [CrossRef]

33. Ni, Y.Y.; Dai, J.X.; Zhu, G.Y.; Zhang, S.C.; Zhang, D.J.; Su, J.; Tao, X.W.; Liao, F.R.; Wu, W.; Gong, D.Y. Stable hydrogen and carbon isotopic ratios of coal-derived and oil-derived gases: A case study in the tarim basin, NW China. *Int. J. Coal Geol.* **2013**, *116*, 302–313. [CrossRef]

34. Dai, J.X.; Zou, C.N.; Liao, S.M.; Dong, D.Z.; Ni, Y.Y.; Huang, J.L.; Wu, W.; Gong, D.Y.; Huang, S.P.; Hu, G.Y. Geochemistry of the extremely high thermal maturity Longmaxi shale gas, southern Sichuan Basin. *Org. Geochem.* **2014**, *74*, 3–12. [CrossRef]

35. Dai, J.X.; Gong, D.Y.; Ni, Y.Y.; Huang, S.P.; Wu, W. Stable carbon isotopes of coal-derived gases sourced from the Mesozoic coal measures in China. *Org. Geochem.* **2014**, *74*, 123–142. [CrossRef]

36. Boyer, C.; Kieschnick, J.; Suarez-Rivera, R.; Lewis, R.E.; Waters, G. Producing gas from its source. *Oilfield Rev.* **2006**, *18*, 36–49.

37. Romero, A.M.; Philp, R.P. Organic geochemistry of the Woodford Shale, southeastern Oklahoma: How variable can shales be? *AAPG Bull.* **2012**, *96*, 493–517. [CrossRef]

38. Bowker, K.A. Barnett Shale gas production, Fort Worth Basin: Issues and discussion. *Aapg Bull.* **2007**, *91*, 523–533. [CrossRef]

39. Qin, J.; Zheng, L. Study on the restitution coefficient of original total organic carbon for high mature marine source rocks. *Front. Earth Sci.-China* **2007**, *1*, 482–490. [CrossRef]

40. Dang, W.; Zhang, J.C.; Tang, X.; Chen, Q.; Han, S.B.; Li, Z.M.; Du, X.R.; Wei, X.L.; Zhang, M.Q.; Liu, J. Shale gas potential of Lower Permian marine-continental transitional black shales in the Southern North China Basin, central China: Characterization of organic geochemistry. *J. Nat. Gas Sci. Eng.* **2016**, *28*, 639–650. [CrossRef]

41. Berner, U.; Faber, E. Maturity related mixing model for methane, ethane and propane, based on carbon isotopes. *Org. Geochem.* **1988**, *13*, 67–72. [CrossRef]

42. Chung, H.M.; Gormly, J.R.; Squires, R.M. Origin of Gaseous Hydrocarbons in Subsurface Environment; Theoretical Considerations of Carbon Isotope Distribution. *Chem. Geol.* **1988**, *71*, 97–104. [CrossRef]

43. Schoell, M. The hydrogen and carbon isotopic composition of methane from natural gases of various origins. *Geochim. Cosmochim. Acta* **1980**, *44*, 649–661. [CrossRef]

44. Whiticar, M.J. Carbon and hydrogen isotope systematic of microbial formation and oxidation of methane. *Chem. Geol.* **1999**, *161*, 291–314. [CrossRef]

45. Strąpoć, D.; Mastalerz, M.; Eble, C.; Schimmelmann, A. Characterization of the origin of coalbed gases in southeastern Illinois basin by compound-specific carbon and hydrogen stable isotope ratios. *Org. Geochem.* **2007**, *38*, 267–287. [CrossRef]

46. Strąpoć, D.; Mastalerz, M.; Schimmelmann, A. Variability of geochemical properties in amicrobially dominated coalbed gas system from the eastern margin of the Illinois Basin, USA. *Int. J. Coal Geol.* **2008**, *76*, 98–110. [CrossRef]

47. Galimov, E.M. 13C enrichment of methane during passage through rocks. *Geochem. Int. USSR* **1967**, *4*, 1180–1181.

48. Colombo, U.; Gazzarrini, F.; Gonfiantini, R.; Tongiorgi, E.; Caflisch, L. Carbon Isotopic Study of Hydrocarbons in Italian Natural Gases. *Adv. Org. Geochem.* **1969**, *43*, 499–516.

49. Stahl, W. Carbon isotope fractionations in natural gases. *Nature* **1974**, *251*, 134–135. [CrossRef]

50. Schoell, M. Genetic Characterization of Natural Gases. *AAPG Bull.* **1983**, *67*, 2225–2238.

51. Zhang, F.K.; Zhang, Y.G. Identification of carbon isotope method for natural gas. In Proceedings of the Organic Geochemistry, Beijing, China, 1 March 1987; Geological Publishing House: Beijing, China, 1987; pp. 1–14.

52. Dai, J.X. Carbon and hydrogen isotopic compositions and origin identification of different types natural gas. *Nat. Gas Geosci.* **1993**, *4*, 1–40.

53. Ding, W.J.; Hou, D.J.; Zhang, W.W.; He, D.S.; Cheng, X. A new genetic type of natural gases and origin analysis in Northern Songnan-Baodao Sag, Qiongdongnan Basin, South China Sea. *J. Pet. Sci. Eng.* **2018**, *50*, 384–398. [CrossRef]

54. Dai, J.X.; Xia, X.Y.; Qin, S.F.; Zhao, J.Z. Causation of partly reversed orders of δ13C in biogenic alkane gas in China. *Oil Gas J.* **2003**, *24*, 1–6.

55. Dai, J.X. *Selected Works of Natural Gas Geology and Geochemistry: Vol. 2*; Petroleum Industry Press: Beijing, China, 2000; p. 201.

56. Dai, J.X.; Zou, C.N.; Dong, D.Z.; Ni, Y.Y.; Wu, W.; Gong, D.Y.; Wang, Y.M.; Huang, S.P.; Huang, J.L.; Fang, C.C.; et al. Geochemical characteristics of marine and terrestrial shale gas in China. *Mar. Pet. Geol.* **2016**, *76*, 444–463. [CrossRef]

Article

Volumetric Measurements of Methane-Coal Adsorption and Desorption Isotherms—Effects of Equations of State and Implication for Initial Gas Reserves

Jamiu M. Ekundayo * and Reza Rezaee

Discipline of Petroleum Engineering, Western Australian School of Mines: Minerals, Energy and Chemical Engineering, Curtin University, 26 Dick Perry Avenue, Kensington, WA 6151, Australia; r.rezaee@curtin.edu.au
* Correspondence: Jamiu.Ekundayo@student.curtin.edu.au

Received: 10 May 2019; Accepted: 23 May 2019; Published: 27 May 2019

Abstract: This study presents the effects of equations of state (EOSs) on methane adsorption capacity, sorption hysteresis and initial gas reserves of a medium volatile bituminous coal. The sorption experiments were performed, at temperatures of 25 °C and 40 °C and up to 7MPa pressure, using a high-pressure volumetric analyzer (HPVA-II). The measured isotherms were parameterized with the modified (three-parameter) Langmuir model. Gas compressibility factors were calculated using six popular equations of state and the results were compared with those obtained using gas compressibility factors from NIST-Refprop® (which implies McCarty and Arp's EOS for Z-factor of helium and Setzmann and Wagner's EOS for that of methane). Significant variations were observed in the resulting isotherms and associated model parameters with EOS. Negligible hysteresis was observed with NIST-refprop at both experimental temperatures, with the desorption isotherm being slightly lower than the adsorption isotherm at 25 °C. Compared to NIST-refprop, it was observed that equations of state that gave lower values of Z-factor for methane resulted in "positive hysteresis", (one in which the desorption isotherm is above the corresponding adsorption curve) and the more negatively deviated the Z-factors are, the bigger the observed hysteresis loop. Conversely, equations of state that gave positively deviated Z-factors of methane relatively produced "negative hysteresis" loops where the desorption isotherms are lower than the corresponding adsorption isotherms. Adsorbed gas accounted for over 90% of the calculated original gas in place (OGIP) and the larger the Langmuir volume, the larger the proportion of OGIP that was adsorbed.

Keywords: adsorption and desorption isotherms; sorption hysteresis; medium volatile bituminous coal; equation of state; NIST-Refprop; gas compressibility factors; original gas in-place

---

1. Introduction

Coal-seams are typically characterized by large proportions of the gas (mostly methane [1–3], with traces of impurities such as CO_2 and N_2) existing predominantly as adsorbed phase in the internal surface areas of the rock matrix while the natural cleats' macropores may contain some water and free gas [1]. The primary storage mechanism in coal-seams is thus gas adsorption in the internal surface areas of the coal matrix [4]. The amounts of gas adsorbed by a coal depend on different factors such as pressure, temperature and moisture contents [5]. The relation of amount of adsorbed gas to pressure at a constant temperature is referred to as adsorption isotherm and has been the primary means of assessing the gas adsorption capacity of coals [2]. Adsorption isotherms are mostly measured experimentally using different techniques such as volumetric, manometric, gravimetric, and some combinations thereof [6] most of which rely on some equation of state to obtain gas compressibility factors for data interpretation [7].

The dependencies of adsorption capacity and gas reserves estimates on gas compressibility factors have been illustrated by different researchers [7–11]. Despite this knowledge, indiscriminate uses of EOSs for Z-factors needed for generating sorption isotherms is very common [10,12–14]. Redlich-Kwong EOS [12], 32-parameter modified Benedict-Webb-Rubin (MBWR) EOS [13] and a combination of McCarty and Arp and Setzmann and Wagner EOSs [12] are some examples found in the literature.

To further demonstrate the influence of equations of state on high-pressure sorption data, we extend our investigation to methane-coal sorption hysteresis and gas in-place volumes. Hysteresis in high-pressure sorption isotherms of supercritical methane on coal has become a common observation and yet, it remains difficult to explain. Some researchers believe it is caused primarily by measurement uncertainties [5,15,16] while others have argued that it could be too significant to be attributed to measurement errors alone [5]. Parameters such as equilibrium time [6,17], (residual) moisture contents of coal [8,17], coal ranks [8,18], particle sizes of coal [19,20], irreversible structural changes to coal due to gas adsorption [21,22], irreversible absorption associated with gas adsorption [16,23], coal's surface heterogeneity [24], gas trapping [8], irreversible chemical bonding of gas molecules with molecules of coal [16,17,21] and selectiveness of adsorption sites during desorption [21] have been offered as explanations for observed high-pressure sorption hysteresis of CO_2-coal and/or CH_4-coal systems.

It is generally believed that, where hysteresis exists, desorption isotherms should lie above their adsorption counterparts. This type of hysteresis is referred to as "positive hysteresis" [25] and was the case with all the examples cited above. Less commonly, cases of "negative hysteresis" and cross-overs (desorption isotherm intercepting adsorption isotherm) have also be reported [6,25–27]. However, it was hitherto believed that such observations could only be caused by pressure non-equilibration, desorption of pre-adsorbed gas in the adsorbent [6] or irreversible structural changes to the solid as a result of gas adsorption [25,26].

From the above discussion, it is obvious that there is no single universal explanation for CO_2-coal and CH_4-coal sorption hysteresis at high-pressure. A similar conclusion was reached by Wang et al. [28] in their review paper on sorption hysteresis of CO_2 and CH_4 on coals. It is noteworthy however, that majority of the articles reviewed were based on volumetric/manometric method. This invariably means that the authors used some equations of state to obtain the fluid data required for experimental data analysis. Interestingly, none of the authors considered that the choice of EOSs (and hence, calculated Z-factors) could be the cause of the observed hysteresis. The primary aim of this paper is therefore to establish that the choice of EOS has a significant effect on high-pressure adsorption and desorption isotherms of methane on coal, associated model parameters, degree and type of sorption hysteresis as well calculated gas reserves from coalbed seams.

2. Materials and Methods

2.1. Sample

The coal sample used in this study is ranked medium volatile bituminous with a mean vitrinite reflectance of 1.43%. The results of the sample's petrographic and proximate analyses are summarized in Table 1. The sample was pulverized and sieved to particle sizes 45–75 μm (325–200 mesh) and then subjected to low-pressure nitrogen adsorption to determine its pore-size distributions (PSDs).

Table 1. Properties of the coal sample.

Components	Composition (%)
A. Petrographic Analysis	
Vitrinite	75.0
Fusinite	3.3
Semi-fusinite	13.0
Macrinite	0.3
Inertodetrinite	4.0
Mineral matter	4.5
Mean maximum reflectance	1.43
B. Proximate Analysis	
Ash content	9.6
Moisture content	1.2
Volatile matter	21.2
Fixed carbon	68.0

2.2. Sample Characterization Using Low-Pressure Nitrogen Adsorption

Pore size distribution was achieved using low-pressure nitrogen ad/desorption tests at a temperature of 77.3 K (≈ -196 °C) and equilibrium relative pressure (P/P$_o$, where the saturation pressure, P$_o$ = 0.1 MPa and P = equilibrium pressure) range of about 0.01 to 1 using a Micromeritics® Tristar II 3020 instrument. The sample was degassed at a temperature of 110 °C for a minimum of 8 h prior to the measurements. The experimental isotherms were parameterized using Brunauer-Emmett-Teller (BET) model [29] and coupled with the non-local density-functional theory (NLDFT) to compute the PSDs.

2.3. Measurement of Methane Sorption Isotherms

Methane adsorption and desorption isotherms were measured for the sample using Particulate System's HPVA-II® 200 unit described in [10,13,30,31]. This equipment uses volumetric technique for gas adsorption and desorption measurements. The volume of free-space was measured using helium expansion method [10,13,30,31]. In addition to the isotherms computed by the equipment (reference data), six other isotherms were calculated, at each temperature, from the raw experimental data, using Z-factors obtained from the EOSs discussed in Ekundayo and Rezaee [10]. The total amount adsorbed at any pressure step r is given by the following equation [10]:

$$\widetilde{V}_r^{ads} = \frac{V_{STP}}{RM_s} \sum_{i=1}^{r} \left[\left(\frac{P_{mA}^i V_{man}}{T_{mA}^i Z_{mA}^i} - \frac{P_{mB}^i V_{man}}{T_{mB}^i Z_{mB}^i} \right) - \left(\frac{P_{sc}^i V_{sc}}{T_{sc}^i Z_{sc}^i} + \frac{P_{sc}^i V_{us}}{T_{us}^i Z_{us}^i} + \frac{P_{sc}^i V_{ls}}{T_{ls}^i Z_{ls}^i} \right) + \left(\frac{P_{sc}^{i-1} V_{sc}}{T_{sc}^{i-1} Z_{sc}^{i-1}} + \frac{P_{sc}^{i-1} V_{us}}{T_{us}^{i-1} Z_{us}^{i-1}} + \frac{P_{sc}^{i-1} V_{ls}}{T_{ls}^{i-1} Z_{ls}^{i-1}} \right) \right]$$

(1)

where V_{STP}, R and M_s are the standard molar volume of gas, the universal gas constant (≈ 8.3145 Jmol^{-1} K^{-1}) and the weight of the adsorbent respectively; P_{mA}, T_{mA} & Z_{mA} are the pressure, temperature and the corresponding methane Z-factor condition of the manifold (of volume, V_{man}) before dosing gas into the sample cell, P_{mB}, T_{mB} & Z_{mB} are the manifold's final condition after dosing. P_{sc}, T_{sc} & Z_{sc} represent the equilibrium condition of the sample cell (volume V_{sc}), T_{us} & Z_{us} are the temperature and Z-factor in the upper stem (volume, V_{us}) of the sample holder while T_{ls} & Z_{ls} are the corresponding values in the lower stem (volume V_{ls}). The parameters V_{sc}, V_{ls} & V_{us}, are related to the volume free space at ambient condition (V_{Amfs}) and can be calculated as discussed in Ekundayo and Rezaee [10]. A plot of $\widetilde{V}_r^{ads}$ against P_{sc} gives an adsorption isotherm.

If the amounts adsorbed at any two consecutive adsorption steps $i-1$ and i are $\widetilde{V}_{i-1}^{ads}$ and $\widetilde{V}_i^{ads}$ respectively, then the differential amount adsorbed at step i can be obtained as:

$$\Delta \widetilde{V}_i^{ads} = \widetilde{V}_i^{ads} - \widetilde{V}_{i-1}^{ads}$$

(2)

Thus, the target amount during the desorption from step i to $i-1$ is $\Delta \widetilde{V}_i^{ads}$. In practice, however, desorption step terminates at a new pressure point j (typically, $P_{i-1} < P_j < P_i$) desorbing an amount $\Delta \widetilde{V}_j^{des}$. Therefore, the amount adsorbed at the pressure point P_j is given as:

$$\widetilde{V}_j^{ads} = \widetilde{V}_i^{ads} + \Delta\widetilde{V}_j^{des} \tag{3}$$

A plot of $\widetilde{V}_j^{ads}$ against P_j on the same axes as the adsorption isotherm gives the desorption isotherm. In a rare case where $P_j = P_{i-1}$, $|\Delta\widetilde{V}_j^{des}| = \Delta\widetilde{V}_i^{ads}$. For this ideal case, the isotherms do not only overlay, they are also equal. In reality, $P_j \neq P_{i-1}$ and $|\Delta\widetilde{V}_j^{des}|$ may be less or greater than $\Delta\widetilde{V}_i^{ads}$ resulting in one of three scenarios:

(1) *Negligible (or Zero) Hysteresis* in which the isotherms overlay even though $|\Delta\widetilde{V}_j^{des}| \neq \Delta\widetilde{V}_i^{ads}$.

(2) *Positive Hysteresis* in which $|\Delta\widetilde{V}_j^{des}| < \Delta\widetilde{V}_i^{ads}$ and as such, desorption isotherm lies above the adsorption curve

(3) *Negative Hysteresis* in which $|\Delta\widetilde{V}_j^{des}| > \Delta\widetilde{V}_i^{ads}$ and as such, desorption isotherm lies below the adsorption curve

2.4. Theory

2.4.1. Isotherm Modeling

The measured excess adsorbed amounts are usually converted to absolute amounts, which are needed for practical applications [32]. In this work, the measured isotherms were parameterized using the 3-parameter Langmuir model [7,13,21,32] which includes a factor to convert the excess amount adsorbed to absolute amount. If the excess amount at any equilibrium pressure, P is V_{exc}, then the model is given as:

$$V_{exc} = \frac{V_L P}{P + P_L}\left(1 - \frac{\rho_{bulk}}{\rho_{ads}}\right) \tag{4}$$

where V_L, P_L = Langmuir volume and pressure (respectively); ρ_{ads} = Adsorbed gas phase density

$$\rho_{bulk} = \frac{PM}{ZRT} = \text{Density of the bulk gas phase} \tag{5}$$

M = Molecular weight of methane.

The model parameters $(V_L, P_L, \text{ and } \rho_{ads})$ were determined by non-linear curve fitting with each parameter bounded to $(0, +\infty)$. The adsorbed gas phase density was assumed to be pressure-independent.

2.4.2. Gas in-Place Calculation

The total volume of gas originally in-place, in a coal-seam, is the sum of the free gas and adsorbed gas and can be calculated using Equation (6) [33]:

$$Gi = Ah\left[\frac{43560\varnothing_i(1 - S_{wi})}{B_{gi}} + 1359\rho_{rock}\frac{V_L^{res}P_i}{P_i + P_L^{res}}\right] \tag{6}$$

In this equation, A (acre) is the area of the coal seam, h (ft) is the thickness, $\varnothing_i$, S_{wi} represent the respective initial porosity and water saturation in the cleat system, ρ_{rock} (g/cc) is the density of the coal seam, P_i (MPa) is initial reservoir pressure, B_{gi} (rcf/scf) represents the initial gas formation volume factor (FVF) calculated from Equation (7), V_L^{res} (scf/ton) is the Langmuir volume for the adsorption isotherm at reservoir temperature, T_{res} (°C) and P_L^{res} (MPa) is the corresponding Langmuir pressure. Table 2 shows the parameters used in calculating the OGIP for each EOS.

$$B_{gi} = \frac{0.000195\, Z_i(1.8T_{res} + 492)}{P_i} \tag{7}$$

Z_i in Equation (7) is the Z-factor of methane at initial coal seam pressure and temperature.

Table 2. Parameters used for initial gas reserves estimates.

Parameters	
Reservoir Temperature, °C	40
Initial Res. Pressure, MPa	10.34
Rock density, g/cc	1.34 *
Initial porosity, fraction	0.1 *
Initial water saturation, fraction	0.8 *

* data from [27].

It is obvious from Equation (7) that the gas FVF is directly related to Z- factor and as such is also EOS-dependent. To isolate the effect of the gas FVF on the calculated volume of gas in-place, two different cases were investigated as follows:

Case 1: Gas FVF equals the value calculated for each EOS.

Case 2: Gas FVF equals constant for all EOS. The constant value was set as the value obtained for the reference EOS.

3. Results and Discussion

3.1. Low-Pressure Nitrogen Sorption Isotherms and Pore-Size Distribution

Figure 1 shows the low-pressure nitrogen sorption isotherms (Figure 1a) and the corresponding pore-size distribution, PSD (Figure 1b) for the coal sample used in the study. The nitrogen adsorption isotherm shows no monolayer coverage but exhibits a significant multilayer coverage [34], up to a relative pressure of about 0.8, followed by a sharp increase in the amount adsorbed (an indication of capillary condensation [35]) thereafter. Negligible hysteresis of type H3 (IUPAC classification, [34]) characterized by a type II adsorption isotherm and a lower closure of the hysteresis loop indicate that the sample is composed of non-rigid aggregates of slit-like particles and possible presence of a network of macropores [34]. The PSD shows that the sample is composed of approximately bi-modal distributed pores with mesopores accounting for more than 90% of the distribution and dominant pores sizes of approximately 4 nm and 25 nm. The PSD also shows negligible micropores thereby confirming the absence of monolayer coverage observed in the nitrogen adsorption isotherm.

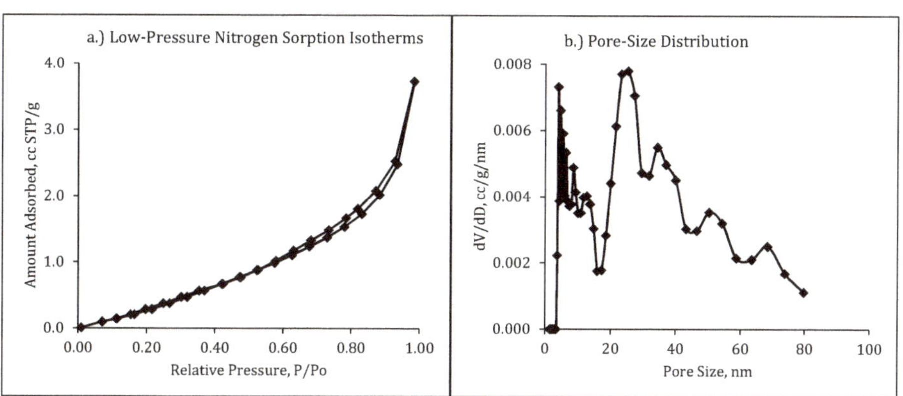

Figure 1. Pore-Size Distribution from Low-Pressure Nitrogen Adsorption Data.

3.2. Methane-Coal Adsorption Isotherms

Figure 2 shows the variations in the calculated Z-factors of helium (Figure 2a), relative to those obtained from McCarty and Arp's EOS [36], at the test temperatures and the corresponding variations in the resulting ambient volumes of free space (Figure 2b). The deviations are minimal for helium

Z-factor (within ±0.25%) for all the EOS. Although the corresponding deviations of free space volume are also small (within±0.5%), they can have significant effects on the amounts adsorbed, especially at high pressure [10,37,38].

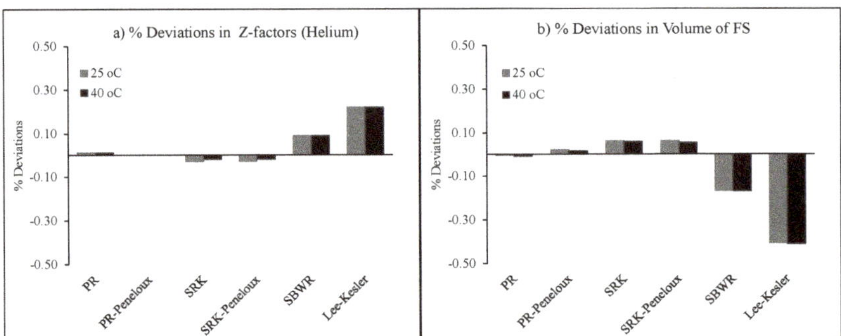

Figure 2. Effect of Equation of State on Free-Space (**a**) Variation in Z-factor of helium gas by EOS (**b**) Corresponding variation in volume of free space.

Figure 3 shows an example of the deviations in Z-factors of methane at 25 °C as a function of pressure for each EOS. For each sample, the deviations increase with pressure with PR-Peneloux showing the largest negative deviation while SBWR-EOS gave the closest Z-factor of methane relative to Setzmann and Wagner's EOS. As shown in Figure 4, significant variations were also seen in the resulting adsorption isotherms for all EOSs. As expected, EOSs that gave negative deviations for Z-factor of methane resulted in adsorption isotherms higher than NIST-refprop's. The size of the resultant isotherms, relative to that of the reference EOS, is determined by the combined effects of volume of free-space and Z-factor of methane. For example, PR-EOS and PR-Peneloux EOS gave nearly the same free space volume as the reference EOS but gave significant larger isotherms due to their lower Z-factors of methane while SBWR-EOS and Lee-Kesler, despite their higher Z-factors of methane, gave higher isotherms than the reference isotherm due their lower free space volumes. It is not surprising that the SRK-EOS and its volume-shifted version gave significantly lower isotherms relative to the reference given that the volumes of free-space and Z-factors of methane are higher for both compared to the reference EOS.

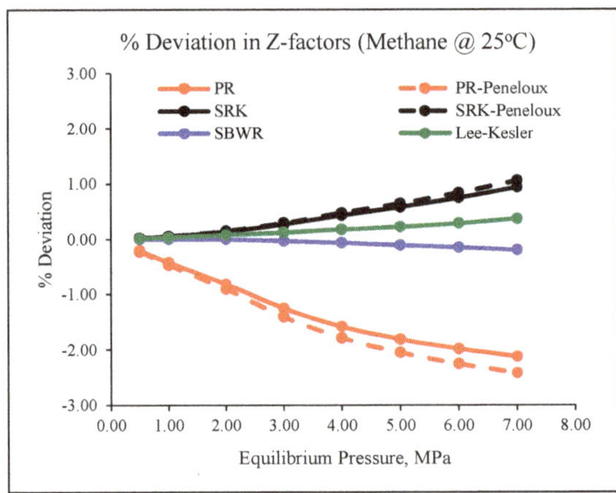

Figure 3. Example of variations in gas compressibility factor of methane at 25 °C.

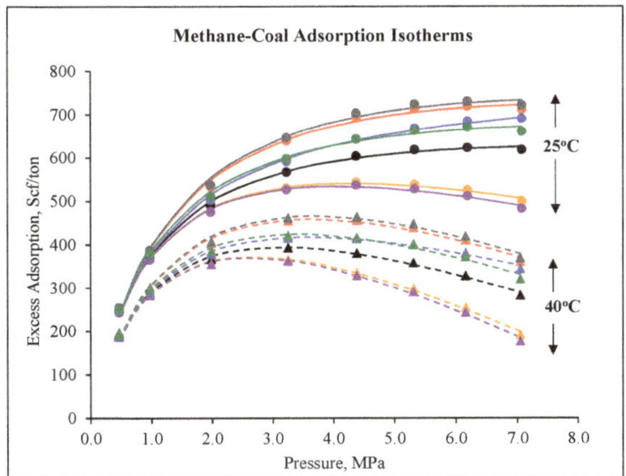

Figure 4. Methane-coal adsorption isotherms by equation of state (legends: markers = Excess adsorbed amount, lines = Langmuir fit; black = NIST-refprop, red = PR-EOS, brown = PR-Peneloux, orange = SRK-EOS, purple = SRK-Peneloux, blue = SBRW-EOS and green = Lee-Kesler).

3.3. Methane-Coal Desorption Isotherms & Sorption Hysteresis

It is clear from Section 3.2 that adsorption isotherms from the same experimental data vary significantly for different EOSs. Similar observations are made with desorption isotherms and hysteresis between the adsorption and desorption isotherms. Figure 5 shows the adsorption and desorption isotherms for different EOSs.

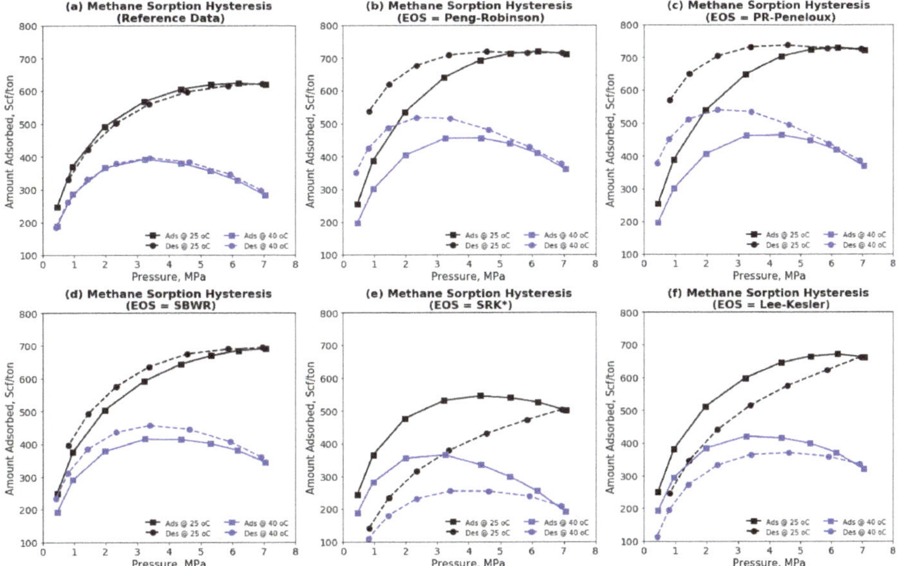

Figure 5. Methane-coal sorption hysteresis by equation of state (* SRK-Peneloux is similar).

Negligible hysteresis can be seen between the adsorption and desorption isotherms for NIST-refprop$^{®}$ at both temperatures. For the measurements at 25 °C, the desorption isotherm is slightly lower than the adsorption isotherm for most test pressures (Figure 5a). Although the reason for this behavior is not clear, it is unlikely due to any significant irreversible structural changes to the coal sample, as postulated by He et al. [26], given the near negligible difference between the isotherms. Also, pressure equilibrium was achieved (as shown in Figure 6) for each desorption step, therefore, pressure non-equilibration was also ruled out as a reason for this anomaly. Thus, it is assumed that this observation may be related to measurement uncertainties especially fluctuations in experimental temperature.

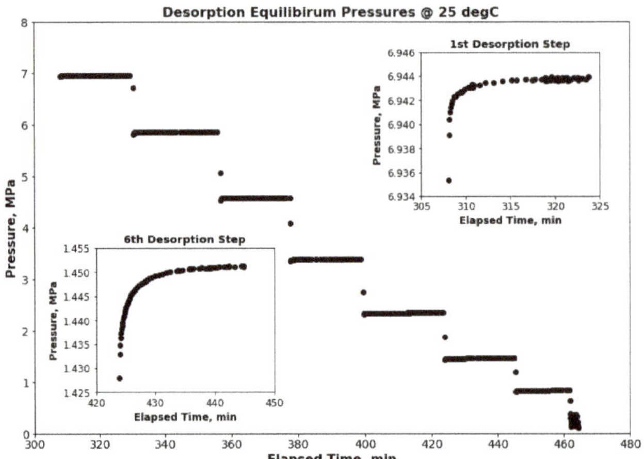

Figure 6. Equilibrium pressures for all desorption steps at 25 °C.

Compared to refprop, each of the comparison EOS gave significant differences between the adsorption and desorption isotherms (Figure 5b–f) leading to two types of observed sorption hysteresis. On one hand, EOSs (that is PR-Peneloux, PR-EOS and SBWR-EOS) that gave lower values of Z-factor for methane resulted in positive hysteresis and the more negatively deviated the Z-factors are, the bigger the observed hysteresis loop. As shown in the example plot in Figure 7a, these EOSs gave lower differential amounts of desorbed gas than the corresponding differential amounts adsorbed. Conversely, EOSs (that is SRK-Peneloux, SRK-EOS and LK-EOS) that gave positive relative deviations for Z-factors of methane [10] produced negative hysteresis loops and the more positively deviated the Z-factors are, the larger the loop. As shown in the example plot in Figure 7b, each of these EOSs gave differential amounts of gas desorbed greater, for most of the pressure steps, than the corresponding differential amounts adsorbed.

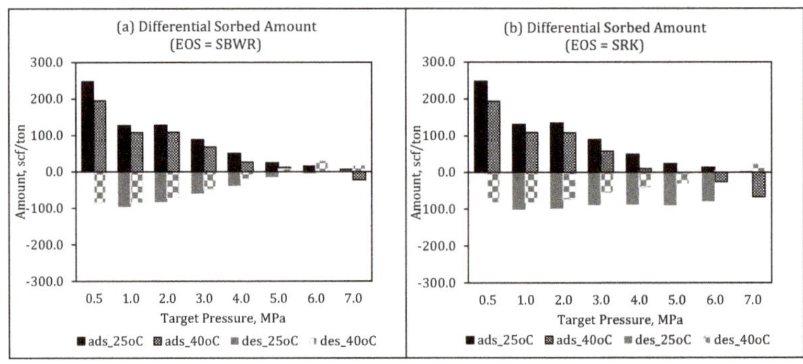

Figure 7. Comparing differential ad/desorbed amounts vs. EOS.

3.4. Langmuir Parameters

Table 3 shows the model parameters obtained by fitting the 3-parameter Langmuir model to each of the adsorption isotherms. The results showed that model achieved over 99% match with each of the isotherms as indicated by the R^2 values. For each EOS, it was also observed that Langmuir volume and adsorbed phase density are lower, while Langmuir pressure is higher, at 40 °C than the corresponding values at 25 °C due to the reducing effect of temperature on gas adsorption [12–14,31,32]. The lower adsorbed phase density at 40 °C is an indication of the temperature-dependent nature of this parameter which is often assumed to be constant [13,21,32,39]. SBWR-EOS gave the closest values of Langmuir volume and pressure compared to NIST-refprop at both temperatures. Conversely, PR-EOS and PR-Peneloux gave the largest deviations for these parameters relative to the reference values. These observations agree with the deviations of Z-factors of methane obtained from these EOSs. Lastly, the adsorbed phase density obtained from SBWR-EOS at 25 °C is large and as such, the ratio of the bulk gas density at each pressure to adsorbed phase density is negligibly small so that excess adsorption approximates absolute adsorption for this EOS at this temperature.

Table 3. Langmuir parameters for different equations of state (adsorption isotherms only).

Temp.	Parameters	EOS						
		Refprop	PR	PR-Pen.	SRK	SRK-Pen.	SBWR	Lee-Kesler
25 °C	V_L, Scf/ton	772.9	943.3	964.6	806.6	815	800	836.8
	P_L, MPa	1.02	1.34	1.38	1.08	1.1	1.09	1.13
	ρ_{ads}, Kg/m^3	689.9	590.9	578.8	183.8	165.4	1.30E+09	742.9
	R^2	0.9985	0.9969	0.9966	0.9978	0.9977	0.9983	0.9973
40 °C	V_L, Scf/ton	705.8	860.9	880.1	740	748.2	699.7	768.4
	P_L, MPa	1.26	1.63	1.68	1.34	1.35	1.23	1.39
	ρ_{ads}, Kg/m^3	92.1	103.7	104.6	69.4	66.9	116.6	98.2
	R^2	0.9953	0.9918	0.9915	0.9946	0.9947	0.9957	0.9911

3.5. Original Gas in-Place

Figure 8 shows the calculated initial Z-factor, initial gas FVF, OGIP and relative differences in the calculated OGIP for each EOS. Similar to previous observations on variations in z-factors of methane, PR-Peneloux, PR and SBWR EOSs gave lower values of Z-factor of methane compared to the Setzmann & Wagner's (shown as refprop in Figure 8) while SRK-Peneloux, SRK and Lee-Kesler EOSs gave higher values. The corresponding values of initial gas FVF are shown in Figure 8b (grey bars). Given the direct relationship between Z-factor and gas FVF (as shown in Equation (7)), the calculated initial gas FVF followed the same trend as the Z-factors. The black bars in Figure 8b represent the constant value of gas FVF for case 2. This constant value is the reference gas FVF obtained from Setzmann and Wagner's EOS. As shown in Figure 8c, the calculated OGIPs for cases 1 and 2 show minimal differences. This implies that gas FVF (by implication z-factor) has negligible effect on OGIP for this coal. Therefore, it can be concluded that the observed variation in calculated OGIPs is primarily caused by the variation in Langmuir parameters obtained with the different EOSs. As shown in Figure 8d, SBWR-EOS gave the lowest deviation in OGIP relative to refprop while PR-Peneloux EOS gave the largest deviation. As shown in Figure 9, these deviations correlate directly with deviations of Langmuir volume (or pressure) relative to the refprop.

The proportion of OGIP that is adsorbed is shown in Figure 10. It is not unsurprising that the adsorbed gas accounted for over 90% of gas in-place (for each EOS) given the initial porosity and initial water saturation in the cleats (see Table 2). This is also supported by the sample's pore-size distribution which contains less than 10% of macropores.

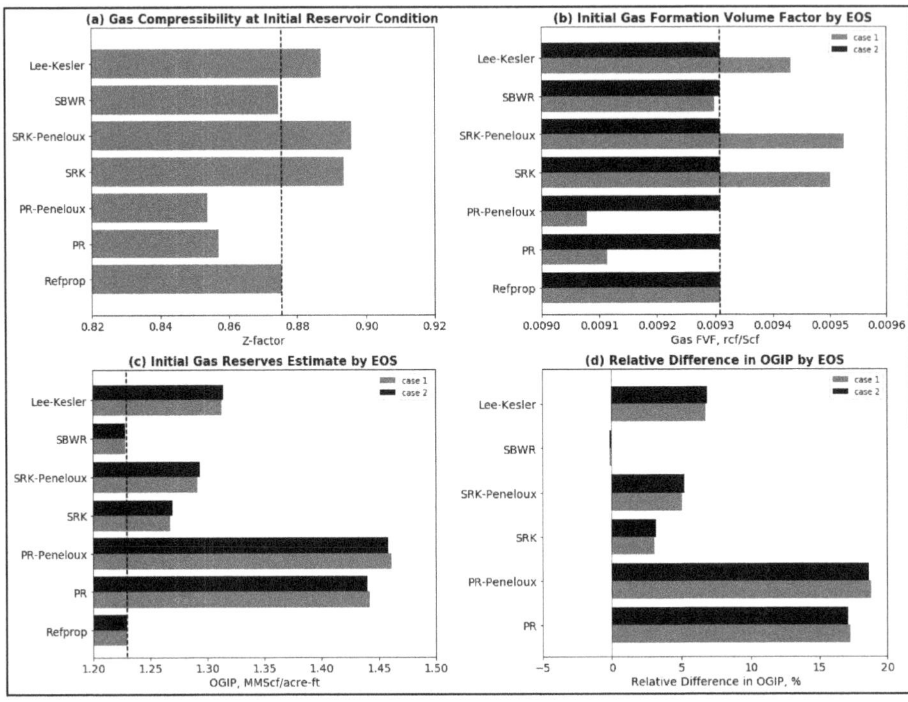

Figure 8. Effects of equation of state on OGIP (**a**) EOS Vs Z-factor at initial reservoir conditions (**b**) EOS Vs Initial FVF (**c**) EOS Vs OGIP (**d**) Relative difference in calculated OGIPs.

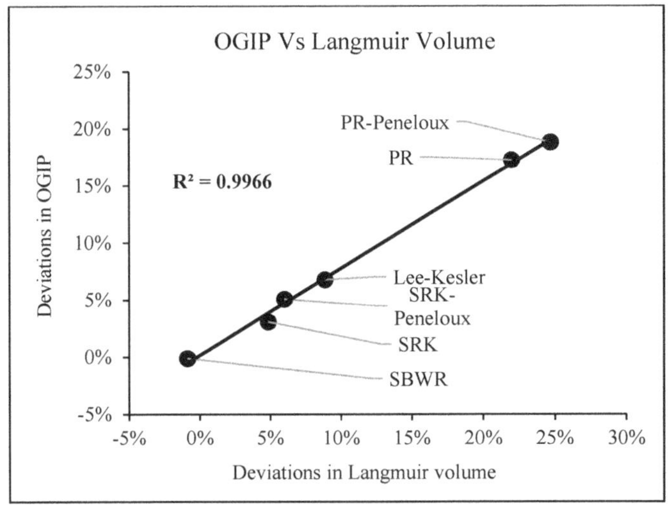

Figure 9. Relationship between calculated gas reserves and Langmuir volume.

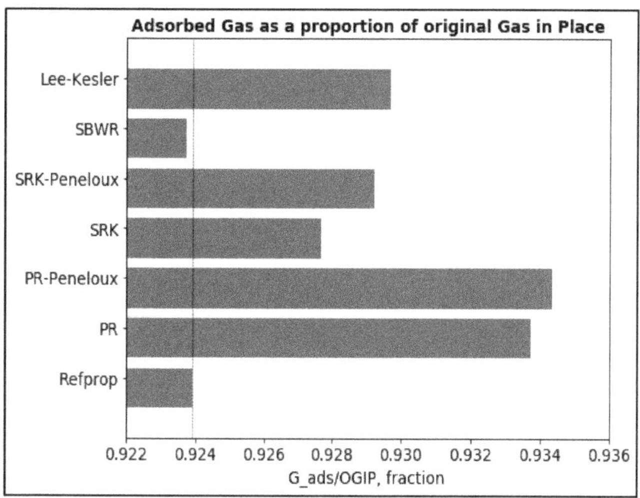

Figure 10. Adsorbed gas as a ratio of the original gas in-place.

4. Conclusions

This paper presents the effects of equations of state on the results of high-pressure volumetric measurements of methane adsorption and desorption isotherms on a coal sample. The results showed that all the equations of state tested gave varied deviations in the measured isotherms, calculated Langmuir parameters, observed type and degree of sorption hysteresis and original gas in-place because of the variations in the calculated Z-factors of both helium and methane relative to NIST-refprop. It can be concluded that:

(1) Gas compressibility factors have significant effects on high-pressure adsorption and desorption isotherms of methane on coal.

(2) Langmuir parameters also varied significantly with the choice of EOS with the SBWR-EOS having the closest values of Langmuir volumes and pressures to NIST-refprop at both test temperatures.

(3) For each EOS, Langmuir volume and adsorbed phase density are lower at 40 °C than the corresponding values at 25 °C because of the negative effect of temperature on adsorption.

(4) High-pressure methane-coal adsorption-desorption hysteresis is not a unique phenomenon; its existence, degree and type depend on the applied EOS.

(5) For both test temperatures, negligible hysteresis was observed for the reference EOS while a relationship was observed between the deviations of methane Z-factor for each EOS (relative to Setzmann and Wagner's) and the type of sorption hysteresis observed. Negatively deviating EOSs produced positive sorption hysteresis while positively deviating ones gave negative sorption hysteresis for the same dataset.

(6) Calculated OGIP is predominantly in adsorbed phase and as such, varies with Langmuir volume.

In summary, it has been demonstrated that, significant uncertainties may be incurred in estimated gas contents and gas reserves of unconventional gas reservoirs due to the choice of EOS for computing the gas compressibility factors needed for experimental data analysis.

Author Contributions: Conceptualization, J.M.E.; Data curation, J.M.E.; Formal analysis, J.M.E.; Investigation, J.M.E.; Methodology, J.M.E.; Resources, R.R.; Supervision, R.R.; Writing–original draft, J.M.E.; Writing–review & editing, R.R.

Funding: This research received no external funding.

Acknowledgments: The authors would like to acknowledge the contributions of Australian Government Research Training Program and Curtin Research Scholarships and the Unconventional Gas Research group at the discipline of Petroleum Engineering, Western Australian School of Mines: Minerals, Energy and Chemical Engineering in supporting this research.

Conflicts of Interest: The authors declare no conflict of interest.

References

1. Rachmat, S.; Pramana, A.A.; Febriana, L. Indonesia's Unconventional Resources, Modified Resource Triangle, and a Typical Example of Stimulation of Coalbed Methane Reservoir. *Mod. Appl. Sci.* **2012**, *6*, 99–111. [CrossRef]
2. Seidle, J. Sorption of Gas on Coals. In *Fundamentals of Coalbed Methane Reservoir Engineering*; PennWell: Tulsa, OK, USA, 2011; pp. 125–153.
3. Penuela, G.; Ordonez, A.; Bejarano, A. A Generalized Material Balance Equation for Coal Seam Gas Reservoirs. In Proceedings of the SPE Annual Technical Conference and Exhibition, New Orleans, LA, USA, 27–30 September 1998.
4. King, G.R. Material-Balance Techniques for Coal-Seam and Devonian Shale Gas Reservoirs With Limited Water Influx. *SPE Reserv. Eng.* **1993**, *8*, 67–72. [CrossRef]
5. Bell, G.J.; Rakop, K.C. Hysteresis of Methane/Coal Sorption Isotherms. In Proceedings of the 61st SPE-ATCE, New Orleans, LA, USA, 5–8 October 1986.
6. Keller, J.U.; Staudt, R. *Gas Adsorption Equilibria: Experimental Methods and Adsorptive Isotherms*; Springer: Berlin, Germany, 2005.
7. Gasparik, M.; Rexer, T.F.T.; Aplin, A.C.; Billemont, P.; Weireld, G.D.; Gensterblum, Y.; Henry, M.; Krooss, B.M.; Liu, S.; Ma, X.; et al. First international inter-laboratory comparison of high-pressure CH4, CO_2 and C2H6 sorption isotherms on carbonaceous shales. *Int. J. Coal Geol.* **2014**, *132*, 131–146. [CrossRef]
8. Goodman, A.L.; Busch, A.; Duffy, G.J.; Fitzgerald, J.E.; Gasem, K.A.M.; Gensterblum, Y.; Krooss, B.M.; Levy, J.; Ozdemir, E.; Pan, Z.; et al. An Inter-laboratory Comparison of CO_2 Isotherms Measured on Argonne Premium Coal Samples. *Energy Fuels* **2004**, *18*, 1175–1182. [CrossRef]
9. Lutynski, M.A.; Battistutta, E.; Bruining, H.; Wolf, K.H.A. Discrepancies inthe assessment of CO_2 storage capacity and methane recovery from coal with selected equation of states Part I. Experimental isotherm calculation. *Physicochem. Probl. Miner. Process.* **2011**, *47*, 159–168.
10. Ekundayo, J.; Rezaee, R. Effect of Equation of States on High Pressure Volumetric Measurements of Methane-Coal Sorption Isotherms—Part 1: Volumes of Free Space and Methane Adsorption Isotherms. *Energy Fuels* **2019**, *33*, 1029–1036. [CrossRef]
11. Al-Fatlawi, O.; Hossain, M.M.; Osborne, J. Determination of best possible correlation for gas compressibility factor to accurately predict the initial gas reserves ingas-hydrocarbon reservoirs. *Int. J. Hydrogen Energy* **2017**, *42*, 25492–25508. [CrossRef]
12. Tang, X.; Wang, Z.; Ripepi, N.; Kang, B.; Yue, G. Adsorption Affinity of Different Types of Coal: Mean Isosteric Heat of Adsorption. *Energy Fuels* **2015**, *29*, 3609–3615. [CrossRef]
13. Zhang, Y.; Xing, W.; Liu, S.; Liu, Y.; Yang, M.; Zhao, J.; Song, Y. Pure methane, carbon dioxide, and nitrogen adsorption on anthracite from China over a wide range of pressures and temperatures: Experiments and modeling. *RSC Adv.* **2015**, *5*, 52612–52623. [CrossRef]
14. Yang, F.; Hu, B.; Xu, S.; Meng, Q.; Krooss, B.M. Thermodynamic Characteristic of Methane Sorption on Shales from Oil, Gas, and Condensate Windows. *Energy Fuels* **2018**, *32*, 10443–10456. [CrossRef]
15. Ozdemir, E. Dynamic nature of supercritical CO_2 adsorption on coals. *Adsorption* **2017**, *23*, 25–36. [CrossRef]
16. Dutta, P.; Bhowmik, S.; Das, S. Methane and carbon dioxide sorption on a set of coals from India. *Int. J. Coal Geol.* **2011**, *85*, 289–299. [CrossRef]
17. Battistutta, E.; van Hemert, P.; Lutynski, M.; Bruining, H.; Wolf, K.H. Swelling and sorption experiments on methane, nitrogen and carbon dioxide on dry Selar Cornish coal. *Int. J. Coal Geol.* **2010**, *84*, 39–48. [CrossRef]
18. Busch, A.; Gensterblum, Y.; Krooss, B.M. Methane and CO_2 sorption and desorption measurements on dry Argonne premium coals: Pure components and mixtures. *Int. J. Coal Geol.* **2003**, *55*, 205–224. [CrossRef]
19. Feng, Y.Y.; Yang, W.; Chu, W. Coalbed methane adsorption and desorption characteristics related to coal particle size. *Chin. Phys. B* **2016**, *25*, 068102. [CrossRef]
20. Zhang, L.; Aziz, N.; Ren, T.; Nemcik, J.; Tu, S. Influence of coal particle size on coal adsorption and desorption characteristics. *Arch. Min. Sci.* **2014**, *59*, 807–820. [CrossRef]

21. Zhang, R.; Liu, S. Experimental and theoretical characterization of methane and CO_2 sorption hysteresis in coals based on Langmiur desorption. *Int. J. Coal Geol.* **2016**, *171*, 49–60. [CrossRef]

22. Liu, Z.; Zhang, Z.; Lu, Y.; Choi, S.K.; Liu, X. Sorption Hysteresis Characterization of CH4 and CO_2 on Anthracite, Bituminous Coal, and Lignite at Low Pressure. *J. Energy Resour. Technol.* **2017**, *140*, 012203. [CrossRef]

23. Weishauptová, Z.; Přibyl, O.; Sýkorová, I.; Machovič, V. Effect of bituminous coal properties on carbon dioxide and methane high pressure sorption. *Fuel* **2015**, *139*, 115–124. [CrossRef]

24. Jessen, K.; Tang, G.Q.; Kovscek, A.R. Laboratory and Simulation Investigation of Enhanced Coalbed Methane Recovery by Gas Injection. *Transp. Porous Media* **2008**, *73*, 141–159. [CrossRef]

25. Kim, H.J.; Shi, Y.; He, J.; Lee, H.H.; Lee, C.H. Adsorption characteristics of CO_2 and CH4 on dry and wet coal from subcritical to supercritical conditions. *Chem. Eng. J.* **2011**, *171*, 45–53. [CrossRef]

26. He, J.; Shi, Y.; Ahn, S.; Kang, J.W.; Lee, C.H. Adsorption and Desorption of CO_2 on Korean Coal under Subcritical to Supercritical Conditions. *J. Phys. Chem. B* **2010**, *114*, 4854–4861. [CrossRef]

27. Rodrigues, C.F.A.; da Silva, J.M.M.; Dinis, M.A.P.; de Sousa, M.J.L. Effect of gas compressibility factor estimation in coal sorption isotherms accuracy. *Int. J. Oil Gas Coal Technol.* **2018**, *19*, 230–247. [CrossRef]

28. Wang, K.; Wang, G.; Ren, T.; Cheng, Y. Methane and CO_2 sorption hysteresis on coal: A critical review. *Int. J. Coal Geol.* **2014**, *132*, 60–80. [CrossRef]

29. Brunauer, S.; Emmett, P.H.; Teller, E. Adsorption of Gases in Multimolecular Layers. *J. Am. Chem. Soc.* **1938**, *60*, 309–319. [CrossRef]

30. Zou, J.; Rezaee, R. Effect of particle size on high-pressure methane adsorption of coal. *Pet. Res.* **2016**, *1*, 53–58. [CrossRef]

31. Zou, J.; Rezaee, R.; Liu, K. Effect of Temperature on Methane Adsorption in Shale Gas Reservoirs. *Energy Fuels* **2017**, *31*, 12081–12092. [CrossRef]

32. Gasparik, M.; Ghanizadeh, A.; Bertier, P.; Gensterblum, Y.; Bouw, S.; Krooss, B.M. High-Pressure Methane Sorption Isotherms of Black Shales from the Netherlands. *Energy Fuels* **2012**, *26*, 4995–5004. [CrossRef]

33. Seidle, J. Gas and Water Mass Balances in Coals. In *Fundamentals of Coalbed Methane Reservoir Engineering*; PennWell: Tulsa, OK, USA, 2011; pp. 217–246.

34. Thommes, M.; Kanero, K.; Neimark, A.V.; Olivier, J.P.; Rodriguez-Reinoso, F.; Rouquerol, J.; Sing, K.S.W. Physisorption of gases, with special reference to the evaluation of surface area and pore size distribution (IUPAC technical Report). *Pure Appl. Chem.* **2015**, *87*, 1051–1069. [CrossRef]

35. Jun-yi, L.; Zheng-song, Q.; Wei-an, H.; Yang, L.; Ding-ding, S. Nano-pore structure characterization of shales using gas adsorption and mercury intrusion techniques. *J. Chem. Pharm. Res.* **2014**, *6*, 850–857.

36. McCarty, R.D.; Arp, V.D. New wide range equation of state for helium. *Adv. Cryog. Eng.* **1990**, *35*, 1465–1475.

37. Rouquerol, J.; Rouquerol, F.; Llewellyn, P.; Denoyel, R. Surface excess amounts in high-pressure gas adsorption: Issues and benefits. *Colloids Surf. A Physicochem. Eng. Asp.* **2016**, *496*, 3–12. [CrossRef]

38. Do, D.D.; Do, H.D. Appropriate volumes for adsorption isotherm studies: The absolute void volume, accessible pore volume and enclosing particle volume. *J. Colloid Interface Sci.* **2007**, *316*, 317–330. [CrossRef]

39. Harpalani, S.; Prusty, B.K.; Dutta, P. Methane/CO_2 Sorption Modeling for Coalbed Methane Production and CO_2 Sequestration. *Energy Fuels* **2006**, *20*, 1591–1599. [CrossRef]

Article

A Prediction Model for Methane Adsorption Capacity in Shale Gas Reservoirs

Jie Zou * and Reza Rezaee *

Curtin WA School of Mines: Minerals, Energy and Chemical Engineering, Perth 6151, Australia
* Correspondence: 18312722@student.curtin.edu.au (J.Z.); r.rezaee@curtin.edu.au (R.R.)

Received: 5 December 2018; Accepted: 13 January 2019; Published: 16 January 2019

Abstract: Estimation of methane adsorption capacity is crucial for the characterization of shale gas reservoirs. The methane adsorption capacity in shales is measured using high-pressure methane adsorption to obtain the adsorption isotherms, which can be fitted by Langmuir model. The determined Langmuir parameters can provide the methane adsorption capacity under actual reservoir conditions. In this study, a prediction model for the methane adsorption in shales was constructed based on 66 samples from 6 basins in China and Western Australia. The model was established in four steps: a model of Langmuir volume at experimental temperature, the temperature dependence of Langmuir volume, a model of Langmuir pressure, the temperature dependence of Langmuir pressure. In the model of Langmuir volume at experimental temperature, total organic carbon (TOC) and clay content (V_{sh}) were considered. A positive relationship was observed between the TOC and the temperature effect on the Langmuir volume. As the Langmuir pressure is sensitive to various factors, the Langmuir pressure at experimental temperature shows no trend with the TOC, clay content and thermal maturity, but a positive trend with the Langmuir volume. The results of this study can help log analysts to quantify adsorbed gas from well-log data since TOC and V_{sh}, which are the measure inputs of the introduced models, can be obtained from well-log data as well.

Keywords: shale gas; methane adsorption capacity; Langmuir volume; Langmuir pressure; total organic carbon; clay content

---

1. Introduction

Shale gas contains not only free gas in pore volume but also a significant amount of adsorbed gas on the surface area of the pore wall [1]. The usual method for assessing the methane adsorption capacity in shale gas is high-pressure methane adsorption experiment to obtain adsorption isotherms. The high-pressure methane adsorption isotherms can be fitted by many models. The most popular model is the Langmuir model because of its simplicity and accuracy [2,3]. Langmuir parameters such as Langmuir volume and Langmuir pressure can be determined using Equation (1) to characterize the methane adsorption isotherms of shale samples [4].

$$V_{ads} = \frac{V_L P}{P + P_L} \tag{1}$$

where V_L is the Langmuir volume, defined as the maximum gas content that can be adsorbed on shale at an infinite pressure; P_L is the Langmuir pressure, defined as the pressure at which one-half of the Langmuir volume can be adsorbed; V_{ads} is the adsorbed gas content and P is the experimental pressure.

With the Langmuir parameters, the methane adsorption capacity of shale sample at certain pressure can be predicted. Thus, it is necessary to have a quantitative model of the Langmuir parameters for evaluating the methane adsorption capacity in shales.

However, the Langmuir parameters are controlled by various factors [5]. It has been reported that the Langmuir volume is related to the compositional properties (total organic content, thermal maturity, and clay minerals content), pore structure properties (specific surface area and micropore volume) and reservoir conditions (pressure, temperature and moisture content) [6–9]. As for the Langmuir pressure, the most considered controlling factor is temperature, but it was also observed that the Langmuir pressure is related to composition and volume of small pores [6,10,11]. Furthermore, a power-law decrease trend was found between the Langmuir pressure and Vitrinite Reflectance [5]. Based on the controlling factors of the Langmuir parameters, the prediction model for the methane adsorption capacity in shale gas has been constructed by many scholars [11–13]. As listed in Table 1, however, the considered factors in the models are different: in terms of the Langmuir volume, TOC is the only considered factor for organic-rich shales by Zhang, Ellis [11]; apart from the TOC, Liu, Chen [13] also considered the clay content for low TOC shales. Li, Tian [12] used other parameters in the model to improve accuracy, such as the amount of residual hydrocarbon and temperature. For the Langmuir pressure, Zhang, Ellis [11] classified the shale samples by the thermal maturity and employed the temperature to model the Langmuir pressure; the temperature is the only considered factor in the model of the Langmuir pressure by Liu, Chen [13], while the content of clay minerals, illite, feldspar, and carbonate was used by Li, Tian [12]. However, some of the considered parameters are hard to obtain from well log, such as the content of residual hydrocarbon, illite, feldspar, and carbonate. Furthermore, the prediction models of methane adsorption capacity in previous studies were established on data from specific formations or basins. To assess the methane adsorption capacity in shales, it is required to establish a prediction model based on representative data and proper factors.

Table 1. Considered factors of the prediction model for Langmuir parameters in previous studies.

References	Considered Factors of Langmuir Volume	Considered Factors of Langmuir Pressure
Zhang, Ellis [11]	TOC	temperature and thermal maturity
Liu, Chen [13]	TOC and clay content	temperature
Li, Tian [12]	TOC, clay content, temperature and residual of hydrocarbon (s_1)	the content of clay minerals, illite, feldspar, and carbonate minerals

In this study, we analyzed the experiment results of high-pressure methane adsorption on shales using available published and our unpublished data. The experimental procedures of our unpublished data have been described in a previous work [14]. Given that the experimental temperatures are not constant among the related studies, the Langmuir parameters were modeled at experimental temperature first and then the temperature dependence of Langmuir parameters was explored. With various shale samples, the results can contribute to the evaluation of the methane adsorption in shales.

2. Materials

As the high-pressure methane adsorption experiments are intrinsically controlled by various factors [15], all the considered data must be obtained under similar experimental conditions (e.g., dry with the particle size of <250 μm) with available compositional and geochemical information. It is worth mentioning that data of wet shale samples were not employed because the moisturization level is not constant and the number of shale samples in wet condition is constrained. Meanwhile, the 60 mesh (<250 μm) was applied in this study, as it has been widely used in the related studies. In addition, 60-80 mesh was also carried out with close particle diameter to 60 mesh to involve as much data as possible. Under these conditions, a total of 66 samples from 6 Basins in China and Western Australia were studied [14,16–19]. The samples have a TOC range of 0.23 to 28.48 wt % and clay content range of 20.1 to 83.5. Thermal maturity of the samples directed measured by rock-eval in T_{max}, ranges from 424 to 589 °C. Vitrinite Reflectance (R_o) is not used for the thermal maturity, because the convention from T_{max} to R_o is not constant for different basins. The wide ranges of compositional and geochemical parameters indicate good representativeness of the studied shale samples. The detailed information about the studied samples is provided in an Appendix A at the end of the paper.

3. Results and Discussion

Since the studied shale samples were measured at different temperatures and the amount of data at each high temperature is limited, a model of the Langmuir volume at experimental temperature (30 °C) was considered first. The experimental temperature in the range of 25–30 °C was regarded as the similar condition due to the little temperature difference. Secondly, the model was updated for the methane adsorption under actual reservoir conditions, at higher reservoir temperature.

3.1. Model of Langmuir Volume at Experimental Temperature

As the collected adsorption data of 10 samples are not available at experimental temperature, a total of 56 samples in 5 basins were studied for the model of Langmuir volume at experimental temperature. Figure 1 shows the positive relationship between the TOC and Langmuir volume at experimental temperature, with a coefficient of determination of 0.87, indicating the critical role of organic matter in methane adsorption in shales. However, more data is still required as the shale samples in the TOC range of 10 to 25 wt % is limited in Figure 1.

Comparing with the TOC, clay content appears to have a much less relationship with the Langmuir volume (Figure 2), demonstrating a limited contribution to the methane adsorption. However, it has been reported that the contribution of clay minerals to the methane adsorption is significant for low TOC shale samples [6,20]. To explore the effect of clay on methane adsorption in shales, the studied samples were classified into three groups of low (0–1.5%), medium (1.5–3%), and high (>3%) based on the TOC content.

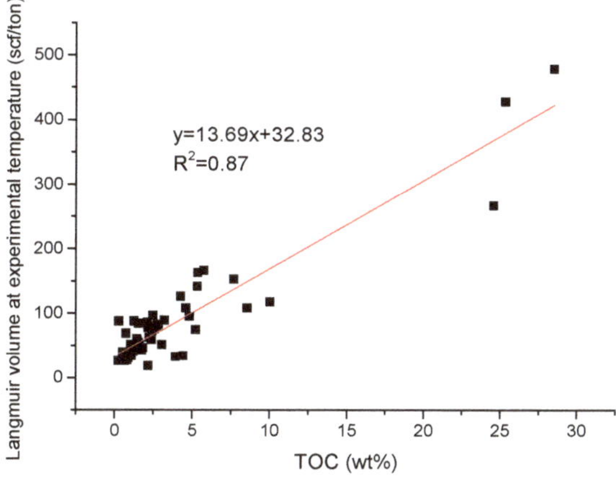

Figure 1. The relationship between the TOC and Langmuir volume at experimental temperature (30 °C) for the studied shale samples.

As for the clay content, Figure 3 displays a good relationship between the clay content and Langmuir volume at experimental temperature for the low TOC samples but not for medium and high TOC samples. The phenomenon demonstrates that the clay content is the critical controlling factor of methane adsorption for low TOC shales, but its effect weakens in higher TOC samples due to the organic matter. This is because both organic matter and clay minerals can adsorb methane. Comparing with low TOC samples, high TOC samples have smaller proportion of methane adsorbed on clay content. Furthermore, the slopes of the relationship between the clay content and Langmuir volume at experimental temperature for the three groups of shales have no big difference, with the range of 1 to 1.73. The similarity of the slopes indicates that there is no remarkable difference between organic-poor and organic-rich shales with respect to the adsorption capacity of clay minerals. Note that the type of clay minerals was not specified for their relationship to Langmuir volume. The reason is that the content of each clay type is hard to obtain directly from well log and illite dominates the clay content for the applied data.

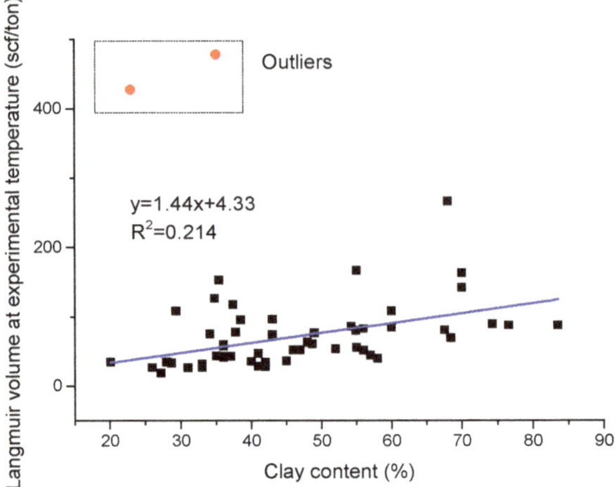

Figure 2. A weak relationship exists between the clay content and Langmuir volume at experimental temperature (30 °C) for the studied shale samples.

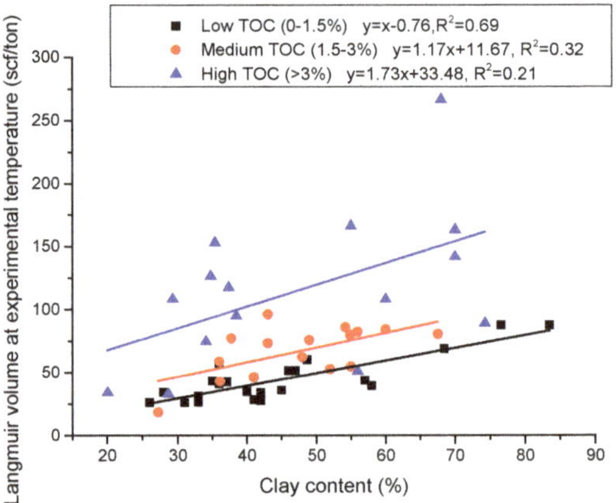

Figure 3. The relationship between the clay content and Langmuir volume at experimental temperature for low, medium and high TOC shale samples.

Apart from the TOC and clay content, thermal maturity is also believed to control the methane adsorption in shales. It has been reported that the relationship between the TOC-normalized maximum methane adsorption capacity (Langmuir volume divided by the TOC) and thermal maturity is positive for mature shales and negative for over mature shales [5,11]. However, the thermal maturity in the form of T_{max} displays no relationship to the TOC-normalized Langmuir volume at experimental temperature for the collected data (Figure 4). The phenomenon can be explained in two aspects: 1. The TOC-normalized Langmuir volume can be influenced by clay content especially for the low TOC shale; 2. other factors such as kerogen type, thermal maturity levels, and depositional environment may have impacts on methane adsorption capacity. Therefore, the thermal maturity is not considered in the model of the Langmuir volume at experimental temperature.

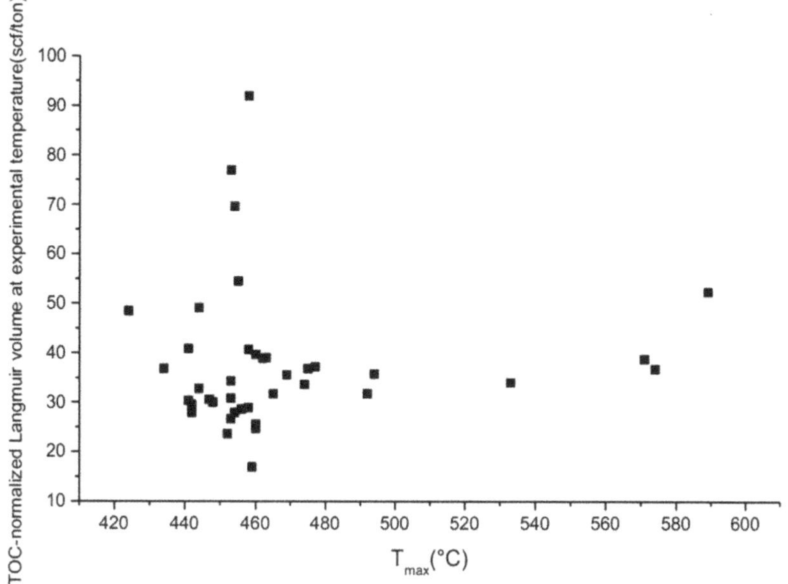

Figure 4. There is no meaningful relationship between the TOC-normalized Langmuir volume at experimental temperature and T_{max} for the studied shale samples.

According to the linear relationship of the Langmuir volume to TOC and clay content, a model for the Langmuir volume at experimental temperature is proposed in the form of the following equation:

$$V_L = a \times TOC + b \times V_{sh} + c \tag{2}$$

where a, b and c are the fitting coefficients, which can be determined using the 56 studied samples in by multiple linear regression. Thus, the prediction model is written as follows:

$$V_L = 13.87TOC + 0.79V_{sh} - 4 \tag{3}$$

where V_L is the Langmuir volume at experimental temperature, scf/ton; TOC is the total organic carbon, wt %; V_{sh} is the total clay content, %.

The predicted Langmuir volume and measured Langmuir volume at experimental temperature are plotted in Figure 5, with R-square 0.88.

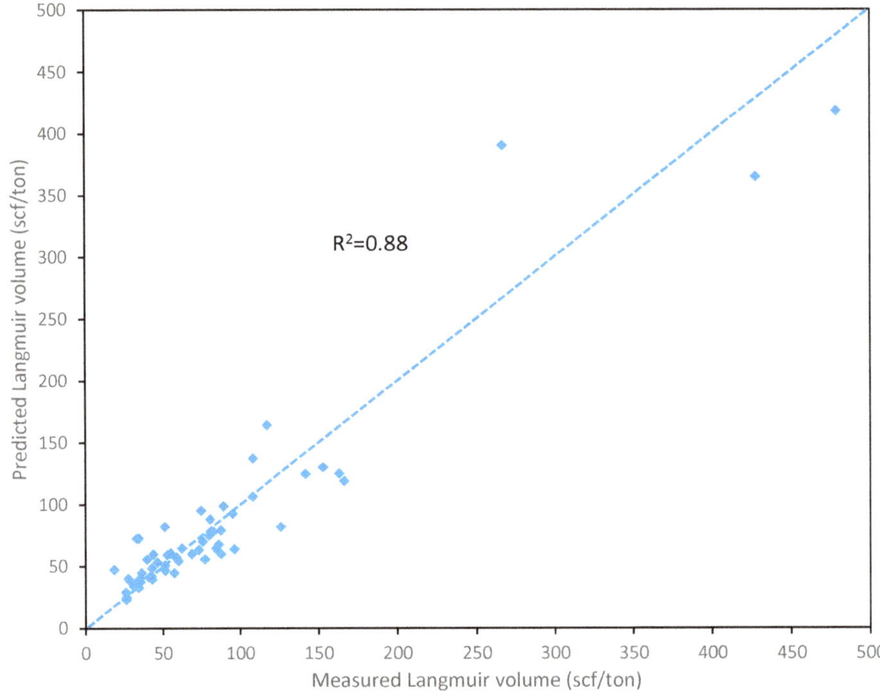

Figure 5. The relationship between the measured and predicted Langmuir volume at experimental temperature.

3.2. Model of Langmuir Volume at Reservoir Temperature

As the methane adsorption in shale is an exothermic process, the methane adsorption capacity is reduced at a higher temperature. It has been observed that the Langmuir volume decreases with increasing temperature [7,17]. A linear negative correlation exists between the Langmuir volume and temperature, which can be written in the following equation:

$$V_L(T) = -dT + e \qquad (4)$$

where $V_L(T)$ is the Langmuir volume at reservoir temperature, scf/ton; T is the reservoir temperature, °C; d and e are the fitting coefficients. The value of the trend-line slope, d, is described as the decrease rate of Langmuir volume with increasing temperature, which quantitatively describes the temperature effect on the Langmuir volume. It has been concluded that the methane adsorbed on the organic matter is more sensitive to the temperature than the methane adsorbed on the clay minerals [6]. The finding is also confirmed in Figure 6, which displays a positive relationship between the TOC and decrease rate of V_L, with the R^2 of 0.58. With this relationship, the decrease rate of V_L (d) can be calculated based on the TOC using Equation (5):

$$d = 0.35TOC - 0.05 \qquad (5)$$

Given that the Langmuir volume at experimental temperature discussed in last section, the Langmuir volume at reservoir temperatures can be estimated using the decrease rate of V_L or the d value from Equation (5). Thus, the Langmuir volume at reservoir temperature can be written as:

$$\frac{V_L(T) - V_L}{T - T_0} = -d \qquad (6)$$

By rearrangement:

$$V_L(T) = [13.87TOC + 0.79V_{sh} - 4] - (T - T_0)(0.35TOC - 0.05) \tag{7}$$

where $V_L(T)$ is the Langmuir volume at reservoir temperature, scf/ton; T is the reservoir temperature, °C; T_0 is the experimental temperature, °C; TOC is the total organic carbon, wt %; V_{sh} is the total clay content, %.

As the available data for the temperature dependence of Langmuir volume has a TOC range of 0.23 to 5.15 wt %, the result here may not be reliable for shale with larger TOC. Moreover, the samples with the TOC range of 3.03 to 5.15 wt % have a larger variation on the relationship than low TOC samples. Therefore, more data is required for the shale samples with TOC larger than 3.03 wt % in terms of the temperature dependence of Langmuir volume.

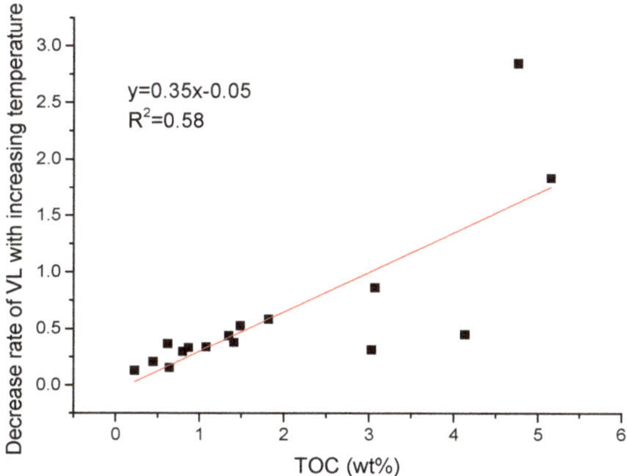

Figure 6. The relationship between the TOC and decrease rate of Langmuir volume with increasing temperature.

3.3. Model of Langmuir Pressure at Experimental Temperature and Reservoir Temperature

Langmuir pressure is also required to calculate the methane adsorption capacity in actual reservoir conditions. The reciprocal of the Langmuir pressure represents the affinity of the gas for sorbent. It has been concluded that the adsorption affinity on the organic matter is stronger than that on clay minerals [6]. Thermal maturity and volume of small pores were also regarded as controlling factors of the Langmuir pressure [5,10]. Herein, the Langmuir pressure shows no trend with the thermal maturity, TOC or clay content, but a logarithmic-law trend exists to the Langmuir volume, with R^2 of 0.31 (Figure 7). The low correlation might result from the sensitive and various controlling factors of the Langmuir pressure. The shale sample with a large Langmuir volume has a high Langmuir pressure, which represents a weak adsorption affinity of methane. As reported, the organic matter and small pore have stronger adsorption affinity of methane comparing to the clay and large pore, respectively. In this case, the adsorption affinity of methane in shale reflects the proportion of adsorbed methane in the small pore and organic matter. Thus, the weak adsorption affinity in the shale sample with a large Langmuir volume might infer that the proportion of adsorbed methane in the small pore and organic matter is low. Since the relationship between the Langmuir volume and Langmuir pressure is obtained from a large amount of data, the relationship can be informative. Therefore, the Langmuir pressure can be predicted using the following equation:

$$P_L = 93.8ln(V_L - 9.3) \tag{8}$$

where P_L and V_L are the Langmuir pressure and Langmuir volume at experimental temperature in psi and scf/ton, respectively.

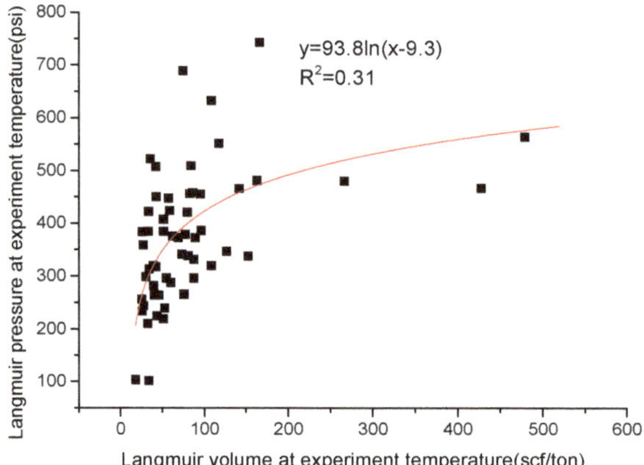

Figure 7. The relationship between the Langmuir pressure and Langmuir volume at experimental temperature for the studied samples.

The temperature dependence of the Langmuir pressure has been described by the following equation [21]:

$$ln\left(\frac{1}{p_L}\right) = \frac{m}{T + 273} + n \tag{9}$$

where m and n is the fitting coefficient, resulting from the thermal dynamic parameters: the heat of adsorption and the standard entropy of adsorption. These parameters have been compared between the organic matter and clay minerals, concluding that the methane adsorbed on the organic matter releases more heat than the methane adsorbed on clay. A linear relationship between the heat of adsorption and the standard entropy of adsorption has been proposed for different types of kerogen, clay and shale samples at different thermal maturity [5]. It might imply that the thermal dynamic parameters are related to the TOC. For each shale sample, the coefficient m and n are determined using linear fitting on $ln\left(\frac{1}{p_L}\right)$ and $\frac{1}{T+273}$. In terms of the studied samples, the plot of m with TOC is listed in Figure 8 as the following equation:

$$m = 1215.3TOC^{0.179} \tag{10}$$

Combined with the prediction model for the Langmuir pressure at experimental temperature, the Langmuir pressure at reservoir temperature ($P_L(T)$) can be obtained using the following equations:

$$ln\left(\frac{P_L(T)}{P_L}\right) = m\left(\frac{1}{T_0 + 273} - \frac{1}{T + 273}\right) \tag{11}$$

$$P_L(T) = P_L \times e^{(1215.3TOC^{0.179}) \times \left(\frac{1}{T_0+273} - \frac{1}{T+273}\right)} \tag{12}$$

$$P_L(T) = 93.8ln(13.87TOC + 0.79V_{sh} - 13.3) \times e^{(1215.3TOC^{0.179}) \times \left(\frac{1}{T_0 + 273} - \frac{1}{T + 273}\right)} \tag{13}$$

where m is the fitting coefficient in Equation 9; T_0 is the experimental temperature, $^\circ$C; T is the reservoir temperature, $^\circ$C.

Therefore, the methane adsorption at certain pressure and temperature can be predicted by the Langmuir model, Equation (14):

$$gc(T) = \frac{V_L(T) \times P}{P_L(T) + P} \tag{14}$$

where $gc(T)$ is the adsorbed gas content at certain temperature and pressure, scf/ton; $V_L(T)$ is the Langmuir volume at reservoir temperature, scf/ton; $P_L(T)$ is the Langmuir pressure at reservoir temperature, psi; P is the reservoir pore pressure, psi.

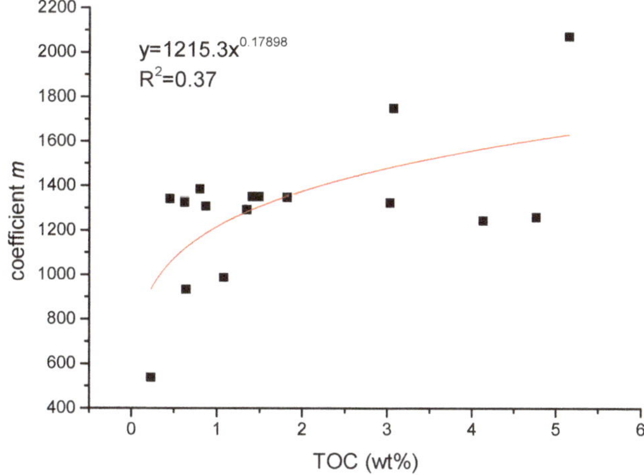

Figure 8. The relationship of the TOC to the fitting coefficient m.

4. Conclusions

In this study, we proposed a prediction model for the methane adsorption capacity in shales based on the high-pressure methane adsorption experiment result. The methane adsorption capacity at certain pressure and temperature can be calculated using the Langmuir model with the Langmuir parameters. Herein, the prediction model for methane adsorption in shales was built in 4 steps: a model of the Langmuir volume at experimental temperature, the temperature dependence of the Langmuir volume, a model of the Langmuir pressure at experimental temperature, the temperature dependence of Langmuir pressure.

The model of the Langmuir volume at experimental temperature considers the TOC and clay content without the thermal maturity, which shows no relationship with the TOC-normalized Langmuir volume. The predicted Langmuir volume at experimental temperature was plotted against the measured results, showing a good R-square. However, more data is still required to improve the model, as the shale samples in the TOC range of 10 to 25 wt % is rarely measured.

For the other three steps, the relationships are informative but not precise enough to provide a reliable prediction. A positive relationship exists between the TOC and decrease rate of Langmuir volume with increasing temperature based on the published data, which requires more data for the shale samples with TOC larger than 3 wt %. As the Langmuir pressure is sensitive to too many factors, it is hard to estimate using the TOC and clay content. However, a logarithmic-law trend is observed between the Langmuir volume and Langmuir pressure at experimental temperature; the temperature dependence of Langmuir pressure is related to the TOC. Furthermore, as the high-pressure methane adsorption experiments on shales were measured under different conditions in the references, the amount of samples for the temperature dependence of the Langmuir parameters is insufficient, which constrains the accuracy of the related models.

Moisture was not considered in this study, which is also regarded as an important controlling factor on methane adsorption in shales [22–24]. The existing moisture in shales occupies pore volume or blocks pore throat to reduce the methane adsorption capacity. However, the moisture content employed in the references are various, which are not available to compare with each other or collect sufficient data. Moreover, it is very uncertain for the moisture content under in-situ conditions and its variation within a shale reservoir.

The major application of this study is that the well log data can calculate TOC and V_{sh} without any problem and therefore this study can help to calculate V_L and P_L at the reservoir condition for volumetric calculation of absorbed gas in shale reservoirs.

Author Contributions: Investigation and writing—original draft preparation, J.Z.; writing—review and editing and supervision, R.R.

Funding: This research was funded by China Scholarship Council, File No.201506440050.

Acknowledgments: The authors would like to thank the Unconventional Gas Research Group in Curtin University for providing support for the experimental work.

Conflicts of Interest: The authors declare no conflict of interest.

Appendix A

Table A1. The detailed information about the studied shale samples in this paper.

Source	TOC (wt %)	Clay (%)	T (°C)	Mesh	VL (scf/ton)	PL(psi)	T_{max} (°C)	Decrease Rate of VL
	7.68	35.4	30	60	152.9	337.9	N/A	N/A
	4.24	34.8	30	60	126.4	346.6	N/A	N/A
	2.18	37.8	30	60	77.3	378.5	N/A	N/A
	1.46	48.7	30	60	60.4	287.2	N/A	N/A
	5.23	34.1	30	60	74.9	688.9	N/A	N/A
Wang, Zhu	4.82	38.5	30	60	95.3	455.4	N/A	N/A
[18]	1.76	36.1	30	60	42.7	507.6	N/A	N/A
	8.54	29.3	30	60	108.4	632.4	N/A	N/A
	10.02	37.4	30	60	117.6	551.1	N/A	N/A
	2.17	27.2	30	60	18.4	103.0	N/A	N/A
	4.4	20.1	30	60	34.3	101.5	N/A	N/A
	3.9	28.7	30	60	32.8	210.3	N/A	N/A
	0.28	76.58	25	60	87.7	296.1	442	N/A
	0.52	N/A	25	60	40.0	281.9	453	N/A
	1.26	83.49	25	60	87.7	331.3	454	N/A
	3.2	74.28	25	60	89.3	372.4	454	N/A
Our	2.82	67.52	25	60	80.6	338.2	456	N/A
unpublished	2.6	54.90	25	60	80.0	420.2	453	N/A
data	2.11	54.22	25	60	86.2	458.3	441	N/A
	1.24	37.10	25	60	42.6	317.1	453	N/A
	2.76	55.97	25	60	82.6	455.6	448	N/A
	0.75	68.44	25	60	69.0	372.5	458	N/A
	0.5	26	30	60-80	26.1	233.5	589	N/A
	0.95	40	30	60-80	35.0	313.3	475	N/A
	0.81	42	30	60-80	27.5	358.2	533	N/A
	0.76	41	30	60-80	28.3	243.7	477	N/A
	1.05	46	30	60-80	51.6	407.6	444	N/A
	0.7	42	30	60-80	33.9	384.4	424	N/A
	0.98	45	30	60-80	36.0	522.1	574	N/A
	1.3	47	30	60-80	51.6	384.4	460	N/A
	5.76	55	30	60-80	166.3	742.6	458	N/A
Guo, Lü [19]	1.55	60	30	60-80	84.4	509.1	455	N/A
	0.87	33	30	60-80	31.1	298.8	494	N/A
	2.24	48	30	60-80	62.5	375.6	442	N/A
	2.57	49	30	60-80	75.9	265.4	442	N/A
	1.22	36	30	60-80	41.0	264.0	474	N/A
	2.42	43	30	60-80	73.5	340.8	441	N/A
	2.47	43	30	60-80	96.4	385.8	463	N/A
	5.35	70	30	60-80	163.1	481.5	447	N/A
	5.33	70	30	60-80	142.0	465.6	453	N/A
	4.59	60	30	60-80	108.4	319.1	452	N/A
	1.55	55	30	60-80	55.1	295.9	469	N/A

Table A1. Cont.

Source	TOC (wt %)	Clay (%)	T (°C)	Mesh	VL (scf/ton)	PL(psi)	T_{max} (°C)	Decrease Rate of VL
Guo, Lü [19]	1.48	36	30	60-80	57.6	446.7	462	N/A
	25.31	23	30	60-80	427.6	467.0	501	N/A
	2.39	36	30	60-80	59.0	423.5	460	N/A
	24.58	68	30	60-80	266.6	480.1	424	N/A
	1.02	58	30	60-80	39.6	319.1	571	N/A
	1.17	35	30	60-80	43.1	449.6	434	N/A
	1.38	57	30	60-80	43.8	224.8	492	N/A
	1.62	52	30	60-80	53.0	239.3	444	N/A
	28.48	35	30	60-80	478.5	564.2	436	N/A
		N/A	20	60	93.9	195.8	N/A	
		N/A	40	60	85.5	220.5	N/A	
	1.41	N/A	60	60	84.0	380.0	N/A	0.3778
		N/A	80	60	73.5	433.7	N/A	
		N/A	100	60	62.2	475.7	N/A	
		N/A	20	60	135.6	207.4	N/A	
		N/A	40	60	128.9	220.5	N/A	
	4.13	N/A	60	60	119.7	382.9	N/A	0.452
		N/A	80	60	109.8	422.1	N/A	
		N/A	100	60	99.9	464.1	N/A	
		N/A	20	60	59.7	192.9	N/A	
		N/A	40	60	54.4	223.4	N/A	
	0.45	N/A	60	60	51.9	381.5	N/A	0.2066
		N/A	80	60	47.0	427.9	N/A	
		N/A	100	60	42.7	469.9	N/A	
		N/A	20	60	77.3	197.3	N/A	
		N/A	40	60	71.7	224.8	N/A	
	0.87	N/A	60	60	68.9	393.1	N/A	0.3337
		N/A	80	60	57.9	420.6	N/A	
		N/A	100	60	50.9	471.4	N/A	
[16,17]		N/A	20	60	74.5	185.7	N/A	
		N/A	40	60	71.0	221.9	N/A	
	0.8	N/A	60	60	68.9	382.9	N/A	0.2966
		N/A	80	60	61.8	435.1	N/A	
		N/A	100	60	49.4	464.1	N/A	
		N/A	20	60	100.6	191.5	N/A	
		N/A	40	60	98.2	227.7	N/A	
	1.49	N/A	60	60	92.5	387.3	N/A	0.5279
		N/A	80	60	74.5	438.0	N/A	
		N/A	100	60	59.7	468.5	N/A	
		N/A	20	60	71.7	198.7	N/A	
		N/A	40	60	66.4	219.0	N/A	
	0.62	N/A	60	60	60.0	384.4	N/A	0.3655
		N/A	80	60	54.0	436.6	N/A	
		N/A	100	60	41.3	468.5	N/A	
		N/A	20	60	85.8	195.8	N/A	
		N/A	40	60	80.5	217.6	N/A	
	1.35	N/A	60	60	74.2	381.5	N/A	0.4379
		N/A	80	60	64.3	427.9	N/A	
		N/A	100	60	50.1	449.6	N/A	
		N/A	30	60	216.8	207.4	N/A	
		N/A	40	60	202.7	216.1	N/A	
	5.15	N/A	50	60	184.7	307.5	N/A	1.8362
		N/A	60	60	159.6	388.7	N/A	
		N/A	70	60	146.6	419.2	N/A	
		N/A	30	60	205.9	206.0	N/A	
		N/A	40	60	184.7	214.7	N/A	
	4.76	N/A	50	60	140.9	248.0	N/A	2.8497
		N/A	60	60	111.6	298.8	N/A	
		N/A	70	60	99.9	320.5	N/A	

Table A1. *Cont.*

Source	TOC (wt %)	Clay (%)	T (°C)	Mesh	VL (scf/ton)	PL(psi)	T_{max} (°C)	Decrease Rate of VL
[14]	3.03	51.3	25	60	51.3	218.9	459	0.3135
			45	60	46.3	277.8		
			60	60	40.2	346.9		
			80	60	37.6	433.8		
	0.64	26	25	60	26.0	256.1	458	0.1538
			45	60	23.1	306.7		
			60	60	20.6	322.5		
			80	60	19.8	429.2		
	1.82	46.5	25	60	46.5	263.8	460	0.5846
			45	60	35.2	352.3		
			60	60	26.0	492.2		
			80	60	N/A	N/A		
	1.08	34.2	25	60	34.2	422.1	465	0.337
			45	60	29.5	544.6		
			60	60	22.2	562.0		
			80	60	20.5	729.7		
	0.23	26.5	25	60	26.5	383.5	N/A	0.1257
			45	60	24.0	398.3		
			60	60	22.1	445.3		
			80	60	20.7	505.4		
	3.07	76	25	60	80.6	338.2	452	0.8646
			45	60	63.7	371.7		
			60	60	50.3	481.4		
			80	60	N/A	N/A		

References

1. Curtis, J.B. Fractured Shale-Gas Systems. *AAPG Bull.* **2002**, *86*, 1921–1938.
2. Chalmers, G.R.L.; Bustin, R.M. The organic matter distribution and methane capacity of the Lower Cretaceous strata of Northeastern British Columbia, Canada. *Int. J. Coal Geol.* **2007**, *70*, 223–239. [CrossRef]
3. Ross, D.J.K.; Bustin, R. The importance of shale composition and pore structure upon gas storage potential of shale gas reservoirs. *Mar. Pet. Geol.* **2009**, *26*, 916–927. [CrossRef]
4. Langmuir, I. The evaporation, condensation and reflection of molecules and the mechanism of adsorption. *J. Frankl. Inst.* **1917**, *183*, 101–102. [CrossRef]
5. Gasparik, M.; Bertier, P.; Gensterblum, Y.; Ghanizadeh, A.; Krooss, B.M.; Littke, R. Geological controls on the methane storage capacity in organic-rich shales. *Int. J. Coal Geol.* **2014**, *123*, 34–51. [CrossRef]
6. Ji, L.; Zhang, T.; Milliken, K.L.; Qu, J.; Zhang, X. Experimental investigation of main controls to methane adsorption in clay-rich rocks. *Appl. Geochem.* **2012**, *27*, 2533–2545. [CrossRef]
7. Guo, S. Experimental study on isothermal adsorption of methane gas on three shale samples from Upper Paleozoic strata of the Ordos Basin. *J. Pet. Sci. Eng.* **2013**, *110*, 132–138. [CrossRef]
8. Zhang, T.; Ellis, G.S.; Ruppel, S.C.; Milliken, K.; Lewan, M.; Sun, X. Effect of Organic Matter Properties, Clay Mineral Type and Thermal Maturity on Gas Adsorption in Organic-Rich Shale Systems. In Proceedings of the Unconventional Resources Technology Conference, Denver, CO, USA, 12–14 August 2013.
9. Dang, W.; Zhang, J.; Wei, X.; Tang, X.; Chen, Q.; Li, Z.; Zhang, M.; Liu, J. Geological controls on methane adsorption capacity of Lower Permian transitional black shales in the Southern North China Basin, Central China: Experimental results and geological implications. *J. Pet. Sci. Eng.* **2017**, *152*, 456–470. [CrossRef]
10. Myers, A.L. Characterization of nanopores by standard enthalpy and entropy of adsorption of probe molecules. *Colloids Surfaces A Physicochem. Eng. Asp.* **2004**, *241*, 9–14. [CrossRef]
11. Zhang, T.; Ellis, G.S.; Ruppel, S.C.; Milliken, K.; Yang, R. Effect of organic-matter type and thermal maturity on methane adsorption in shale-gas systems. *Org. Geochem.* **2012**, *47*, 120–131. [CrossRef]
12. Li, T.; Tian, H.; Xiao, X.; Cheng, P.; Zhou, Q.; Wei, Q. Geochemical characterization and methane adsorption capacity of overmature organic-rich Lower Cambrian shales in northeast Guizhou region, southwest China. *Mar. Pet. Geol.* **2017**, *86*, 858–873. [CrossRef]

13. Liu, Y.C.; Chen, D.X.; Qiu, N.S.; Wang, Y.; Fu, J.; Huyan, Y.; Jia, J.K.; Wu, H. Reservoir characteristics and methane adsorption capacity of the Upper Triassic continental shale in Western Sichuan Depression, China. *Aust. J. Earth Sci.* **2017**, *64*, 807–823. [CrossRef]
14. Zou, J.; Rezaee, R.; Liu, K. The effect of temperature on methane adsorption in shale gas reservoirs. *Energy Fuels* **2017**. [CrossRef]
15. Gasparik, M.; Rexer, T.F.T.; Aplin, A.C.; Billemont, P.; De Weireld, G.; Gensterblum, Y.; Henry, M.; Krooss, B.M.; Liu, S.; Ma, X.; et al. First international inter-laboratory comparison of high-pressure CH4, CO2 and C2H6 sorption isotherms on carbonaceous shales. *Int. J. Coal Geol.* **2014**, *132*, 131–146. [CrossRef]
16. Ji, W.; Song, Y.; Jiang, Z.; Wang, X.; Bai, Y.; Xing, J. Geological controls and estimation algorithms of lacustrine shale gas adsorption capacity: A case study of the Triassic strata in the southeastern Ordos Basin, China. *Int. J. Coal Geol.* **2014**, *134–135*, 61–73. [CrossRef]
17. Ji, W.; Song, Y.; Jiang, Z.; Chen, L.; Li, Z.; Yang, X.; Meng, M. Estimation of marine shale methane adsorption capacity based on experimental investigations of Lower Silurian Longmaxi formation in the Upper Yangtze Platform, south China. *Mar. Pet. Geol.* **2015**, *68, Part A*, 94–106. [CrossRef]
18. Wang, Y.; Zhu, Y.; Liu, S.; Zhang, R. Methane adsorption measurements and modeling for organic-rich marine shale samples. *Fuel* **2016**, *172*, 301–309. [CrossRef]
19. Guo, S.; Lü, X.; Song, X.; Liu, Y. Methane adsorption characteristics and influence factors of Mesozoic shales in the Kuqa Depression, Tarim Basin, China. *J. Pet. Sci. Eng.* **2017**, *157*, 187–195. [CrossRef]
20. Fan, E.; Tang, S.; Zhang, C.; Guo, Q.; Sun, C. Methane sorption capacity of organics and clays in high-over matured shale-gas systems. *Energy Explor. Exploit.* **2015**, *32*, 16. [CrossRef]
21. Xia, X.; Litvinov, S.; Muhler, M. Consistent Approach to Adsorption Thermodynamics on Heterogeneous Surfaces Using Different Empirical Energy Distribution Models. *Langmuir* **2006**, *22*, 8063–8070. [CrossRef]
22. Li, J.; Li, X.; Wang, X.; Li, Y.; Wu, K.; Shi, J.; Yang, L.; Feng, D.; Zhang, T.; Yu, P. Water distribution characteristic and effect on methane adsorption capacity in shale clay. *Int. J. Coal Geol.* **2016**, *159*, 135–154. [CrossRef]
23. Wang, L.; Yu, Q. The effect of moisture on the methane adsorption capacity of shales: A study case in the eastern Qaidam Basin in China. *J. Hydrol.* **2016**, *542*, 487–505. [CrossRef]
24. Zou, J.; Rezaee, R.; Xie, Q.; You, L.; Liu, K.; Saeedi, A. Investigation of moisture effect on methane adsorption capacity of shale samples. *Fuel* **2018**, *232*, 323–332. [CrossRef]

Article

Investigation of Analysis Methods for Pulse Decay Tests Considering Gas Adsorption

Guofeng Han [1], Yang Chen [2] and Xiaoli Liu [3,*

[1] Institute of Mechanics, Chinese Academy of Sciences, Beijing 100190, China
[2] Architectual Design and Research Institute of Tsinghua University, Beijing 100084, China
[3] State Key Laboratory of Hydroscience and Engineering, Tsinghua University, Beijing 100084, China
* Correspondence: xiaoli.liu@tsinghua.edu.cn; Tel.: +86-10-62794910

Received: 10 May 2019; Accepted: 1 July 2019; Published: 3 July 2019

Abstract: The pulse decay test is the main method employed to determine permeability for tight rocks, and is widely used. The testing gas can be strongly adsorbed on the pore surface of unconventional reservoir cores, such as shale and coal rock. However, gas adsorption has not been well considered in analysis pulse decay tests. In this study, the conventional flow model of adsorbed gas in porous media was modified by considering the volume of the adsorbed phase. Then, pulse decay tests of equilibrium sorption, unsteady state and pseudo-steady-state non-equilibrium sorption models, were analyzed by simulations. For equilibrium sorption, it is found that the Cui-correction method is excessive when the adsorbed phase volume is considered. This method is good at very low pressure, and is worse than the non-correction method at high pressure. When the testing pressure and Langmuir volume are large and the vessel volumes are small, a non-negligible error exists when using the Cui-correction method. If the vessel volumes are very large, gas adsorption can be ignored. For non-equilibrium sorption, the pulse decay characteristics of unsteady state and pseudo-steady-state non-equilibrium sorption models are similar to those of unsteady state and pseudo-steady-state dual-porosity models, respectively. When the upstream and downstream pressures become equal, they continue to decay until all of the pressures reach equilibrium. The Langmuir volume and pressure, the testing pressure and the porosity, affect the pseudo-storativity ratio and the pseudo-interporosity flow coefficient. Their impacts on non-equilibrium sorption models are similar to those of the storativity ratio and the interporosity flow coefficient in dual-porosity models. Like dual-porosity models, the pseudo-pressure derivative can be used to identify equilibrium and non-equilibrium sorption models at the early stage, and also the unsteady state and pseudo-steady-state non-equilibrium sorption models at the late stage. To identify models using the pseudo-pressure derivative at the early stage, the suitable vessel volumes should be chosen according to the core adsorption property, porosity and the testing pressure. Finally, experimental data are analyzed using the method proposed in this study, and the results are sufficient.

Keywords: adsorption; unconventional reservoirs; pulse decay test; unsteady state non-equilibrium sorption; pseudo-steady-state non-equilibrium sorption; equilibrium sorption

---

1. Introduction

The pulse decay test is the most used method for determining the permeability of low permeability rocks, and was proposed by Brace et al. in 1968 [1]. Thereafter, analytical solutions of pulse decay tests under a variety of conditions were obtained. Based on these solutions, the asymptotic solutions at early and late time are used to determine the core permeability [2–5].

Modifications are also made to the primary testing method in order to improve the accuracy and the flexibility, and to decrease the testing time [6,7]. All of these analysis methods are based upon the homogeneous model. In recent years, with the development of unconventional oil/gas reservoirs, the pulse decay method is widely used in testing shale, coal rock and tight sandstone cores [8–15]. This situation makes the analysis methods based on the homogeneous model no longer applicable. Usually helium, methane, nitrogen and carbon dioxide are used in these pulse decay tests [16,17]. For shale and coal rock, the absorptivity of helium is weak enough to be neglected, but methane and carbon dioxide are strongly adsorbed substances. When using them for testing, some phenomena other than those found using conventional non-adsorbing gases are found. The experiments of Ghanizadeh et al. and Gensterblum et al. [18–20] indicate that adsorption reduces core permeability. In addition, using reservoir fluid in tests is more realistic, and it can be absorbed by gas shale and coal rock. Therefore, there is a need to investigate the analysis method for adsorbing gas.

Cui et al. [21] are the earliest to investigate the influence of adsorption on the performance of pulse decay tests, and they propose a correction method by defining an effective porosity due to gas adsorption. Based on this, Civan et al. [22] suggest a more elaborate analysis method considering pressure-dependent properties by fitting several pulse decay tests. Although the method of Cui et al. [21] has become the mainstream analysis method for the pulse decay test with gas adsorption, it is only fit for equilibrium sorption, and does not involve the volume of the absorbed phase. In addition to equilibrium sorption, non-equilibrium also exists [23,24]. A few experiments of gas diffusion indicate that the unipore diffusion model is not good enough for representing cores flow characteristics [25–28]. Therefore, the non-equilibrium sorption models need to be investigated.

In this study, pulse decay tests of non-equilibrium sorption and equilibrium sorption models involving adsorbed phase volume are simulated and analyzed. A new correction method is proposed involving adsorbed phase volume, and an identifying method for non-equilibrium sorption models is suggested.

2. Mathematical Models and the Numerical Method

The principle of the pulse decay test is shown in Figure 1. It consists of an upstream vessel, a downstream vessel and a core holder. At the beginning, the fluid pressure in the upstream and downstream vessels and the core is balanced. Then, a pressure pulse is applied in the upstream vessel. The fluid in the upstream vessel flows through the core to the downstream vessel, thereby reducing the upstream pressure and increasing the downstream pressure. By analyzing the changes in the upstream and downstream pressures over time, the permeability of the core can be obtained.

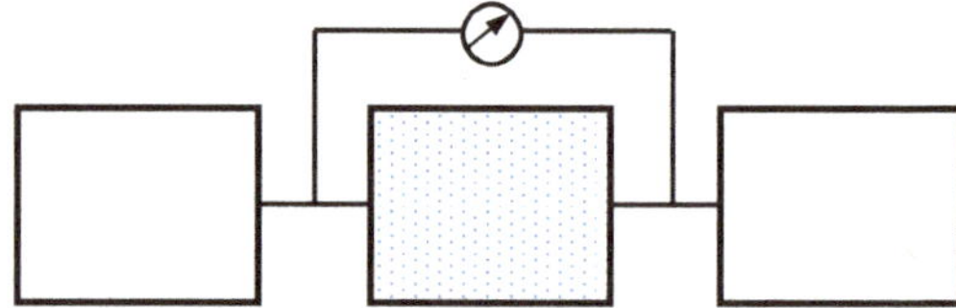

Figure 1. Schematic diagram for the arrangement of pulse decay tests.

Coal and shale are adsorptive to methane. If methane is used for pulse decay tests, it is necessary to use a flow model that considers adsorption for data analysis. The flow models of coal and shale are divided into equilibrium sorption and non-equilibrium sorption models. Non-equilibrium sorption models include pseudo-steady-state and unsteady state models. Adsorbed gas occupies pore volume, but current models do not consider this effect [21,23,24]. In this study, these models are modified to consider this factor. During the test, the device is placed in a constant temperature bath to ensure that the temperature is well controlled.

Therefore, the following assumptions are taken in this study: (1) The test is carried out under constant temperature conditions; (2) the test fluid is a single-phase absorbing gas; (3) the fluid flow conforms to Darcy's law for the equilibrium sorption model, and in fracture for the non-equilibrium sorption model, and conforms to Fick's diffusion law in matrix for the non-equilibrium sorption model, and the secondary pressure gradient term is negligible; (4) the pore compressibility and core permeability are constant; and (5) the upstream and downstream vessels can be regarded as an isobaric body, and the gas leakage can be neglected. Then, equations for pulse decay tests can be written in the following. Since the pulse used in tests generally is small, the permeability change due to adsorption can be ignored. But the results obtained by this method are the apparent permeability. In order to investigating the effect of gas adsorption on permeability, tests under different pressures are required. If the pulse is large, the effect of adsorption upon permeability needs to be considered in the model.

2.1. Equilibrium Sorption Model

Most of the adsorbed gas exists in micropores, and these micropores' conductivity is small. If the adsorbed gas can be desorbed into the flow channel within the characteristic time scale of the study, it is called equilibrium sorption, otherwise it is non-equilibrium sorption. Taking adsorbed phase volume into consideration, the governing equation of the equilibrium sorption model for pulse decay tests can be written as the following [21].

$$\frac{\partial}{\partial t}\left[\rho(\varphi - \varphi_a) + \frac{\rho_{sc}\rho_s V_L p}{p_L + p}\right] = \frac{\partial}{\partial x}\left(\rho\frac{k}{\mu}\frac{\partial p}{\partial x}\right), \tag{1}$$

where p is the pressures in Pa; t is the time in s; x is the coordinate along the sample which takes the upstream reservoir as the origin in m; μ is the viscosity of the fluid in Pa·s; k is the permeability of the sample in m^2; ρ_{sc} and ρ_s are the density of the gas at the standard condition and the apparent density of the sample, respectively, in kg/m^3; V_L is the Langmuir volume in m^3/kg; p_L is the Langmuir pressure in Pa; φ is the porosity in %; the porosity occupied by the absorbed phase φ_a is

$$\varphi_a = \frac{\rho_{sc}\rho_s V_L p}{\rho_a(p_L + p)}, \tag{2}$$

where ρ_a is the density of the adsorbed gas in kg/m^3. It should be pointed out that, unlike Cui et al. [21], ρ_s in Equation (1) is the apparent density of the core, and not the skeleton density, which is the ratio of the mass to the total volume of the skeleton and the pores. The equivalent total compressibility is defined as follows.

$$c_t^e = c_f - \frac{\rho_{sc}\rho_s V_L p_L}{\varphi\rho_a(p_L + p)^2} + c_g\frac{\varphi - \varphi_a}{\varphi} + \frac{\rho_{sc}\rho_s V_L p_L}{\rho\varphi(p_L + p)^2}, \tag{3}$$

where c_t^e is the equivalent total compressibility in Pa^{-1}; c_g is the compressibility of the testing gas in Pa^{-1}; c_f is the pore volume compressibility of the sample in Pa^{-1}. Then, the governing equation of the equilibrium sorption model becomes:

$$\rho\varphi c_t^e\frac{\partial p}{\partial t} = \frac{\partial}{\partial x}\left(\rho\frac{k}{\mu}\frac{\partial p}{\partial x}\right). \tag{4}$$

And the boundary conditions are:

$$p(0,t) = p_u(t) \quad t \geq 0 \ , \ p(L,t) = p_d(t) \quad t \geq 0 \ , \tag{5}$$

$$\frac{dp_u}{dt} = \frac{k}{\left(c_g + c_{V_u}\right)\mu\varphi L}\frac{V_p}{V_u}\frac{\partial p}{\partial x}\bigg|_{x=0} \quad t > 0 \ , \tag{6}$$

$$\frac{dp_d}{dt} = \frac{-k}{\left(c_g + c_{V_d}\right)\mu\varphi L}\frac{V_p}{V_d}\frac{\partial p}{\partial x}\bigg|_{x=L} \quad t > 0 \ , \tag{7}$$

where L is the sample length in m; V_u, V_d and V_p are the volumes of the upstream reservoir, downstream reservoirs and the pores, respectively, in m^3; c_{Vd} and c_{Vu} are the compressibilities of both upstream and downstream reservoirs, respectively, in Pa^{-1}; the subscripts u and d denote these upstream and downstream reservoirs, respectively.

Since at the initial moment the pressure in the core is balanced with the downstream pressure vessel, and the upstream pressure vessel applies a pressure pulse, the initial conditions are:

$$p(x,0) = p_d(0) \quad 0 < x < L \ , \ p(0,0) = p_u(0). \tag{8}$$

2.2. Non-Equilibrium Sorption Model

For non-equilibrium sorption, the pores are divided into macropores/fractures (for convenience, they are collectively referred to as fractures) and micropores in the matrix. If the difference in the conductivity of the pores is up to orders of magnitude, the pores with strong conductivity can constitute preferential flow channels, and the fluids in other pores will converge toward the preferential flow channel. If the convergence time is greater than the characteristic time scale of the study, the pores of the preferential flow channel are called fractures, and the other pores form the matrix. Taking the adsorbed phase volume into consideration, the governing equations of the pseudo-steady state non-equilibrium sorption model are as follows [23,24]:

$$\rho(\varphi c_t)_f \frac{\partial p_f}{\partial t} = \frac{\partial}{\partial x}\left(\frac{\rho k_f}{\mu}\frac{\partial p_f}{\partial x}\right) - \frac{\partial V}{\partial t}, \tag{9}$$

$$\frac{\partial V}{\partial t} = \frac{6 D_m \pi^2}{R_m^2}(C_E - C), \tag{10}$$

where D_m is the gas diffusion coefficient in m^2/s; k_f is the fracture permeability in m^2; R_m is the radius of the spherical matrix in m; p_f is the pressure of the fracture in Pa; $c_t = c_g + c_f$ is the total compressibility of the sample in Pa^{-1}; V is the total gas mass occurred in the matrix in kg/m^3; C and C_E are the gas concentration in the matrix and the equivalent gas concentration in the fracture, respectively, in kg/m^3; φ_f is the fracture porosity in %; and the subscripts f and m denote the fracture and the matrix, respectively.

Equations (9) and (10) represent the flow in the fracture and the matrix, respectively. Thus, the gas concentration and the total content of the gas in the matrix are:

$$C = \frac{pM}{ZRT}, \tag{11}$$

$$V = \frac{\rho_{sc}\rho_s V_L p_f}{p_L + p_f} + \frac{p_f M}{ZRT}(\varphi_m - \varphi_a), \tag{12}$$

where M is the molecular molar mass in kg/mol; Z is the gas deviation factor; R is the Universal Gas Constant in J/(mol·K) ; T is the temperature in K; φ_m is the matrix porosity in %.

Supposing the matrix is spherical, the governing equations of the unsteady state non-equilibrium sorption model are as follows [23,24]:

$$\rho(\varphi c_t)_f \frac{\partial p_f}{\partial t} = \frac{\partial}{\partial x}\left(\frac{\rho k_f}{\mu}\frac{\partial p_f}{\partial x}\right) - \frac{3 D_m}{R_m}\frac{\partial C}{\partial r_m}\Big|_{r_m = R_m}, \tag{13}$$

$$\frac{\partial V}{\partial t} = \frac{1}{r_m^2}\frac{\partial}{\partial r_m}\left(r_m^2 D_m \frac{\partial C}{\partial r_m}\right). \tag{14}$$

where r_m is the coordinate of the spherical matrix with the origin located at the center of the sphere in m.

The pseudo-steady state model assumes that the concentration of the matrix can quickly reach a pseudo-steady state; that is, the concentration in the matrix changes at the same speed. Therefore, a single concentration parameter can be used to characterize the state of the matrix. The mass exchange between the matrix and the fracture is positively correlated to their concentration difference, which is similar to the steady state model. It is described in Equation (10). The unsteady state model considers that the concentration distribution in the matrix is difficult to reach a pseudo-steady state, and is in an unsteady state for a long time. This unsteady flow can be governed by Equation (14).

It is only necessary to replace p in Equations (5)–(7) with p_f to obtain the boundary conditions for the non-equilibrium sorption models. In addition, the gas concentration at the initial time can be obtained by substituting the pressure of the downstream vessel at the initial time into Equation (11). For the unsteady state non-equilibrium sorption model, the gas concentration on the outer boundary of the matrix and the fracture pressure satisfy Equation (11). The following conditions exist on the inner boundary of the matrix.

$$\frac{\partial C}{\partial r_{\mathrm{m}}} = 0. \tag{15}$$

The initial condition of the pseudo-steady-state non-equilibrium sorption model is

$$p_f(x,0) = p_{\mathrm{m}}(x,0) = p_{\mathrm{d}}(0) \quad 0 < x < L \ , \ p_f(0,0) = p_{\mathrm{u}}(0). \tag{16}$$

The initial condition of the unsteady state non-equilibrium sorption model is

$$p_f(x,r_{\mathrm{m}},0) = p_{\mathrm{d}}(0) \quad 0 < x < L \ , \ p_f(0,R_{\mathrm{m}},0) = p_{\mathrm{u}}(0). \tag{17}$$

In the non-equilibrium models, the adsorption in fractures is neglected. The subsequent analysis will indicate that the basic conclusions cannot be affected, even if that is considered.

3. Sensitivity Analysis

To consider the changes of the gas properties, the normalized pseudo-pressure is defined as:

$$p_p = \frac{(\mu Z)_i}{p_i} \int_0^p \frac{p}{\mu Z} dp, \tag{18}$$

where the subscript p and i denote the normalized pressure and the reference status, respectively. The dimensionless variables are defined as:

$$t_{\mathrm{D}} = \frac{k_f t}{\mu \varphi c_{t0} L^2}, \ x_{\mathrm{D}} = \frac{x}{L}, \ p_{\mathrm{D}} = \frac{p_p - p_{pd}(0)}{p_{pu}(0) - p_{pd}(0)},$$
$$A_{\mathrm{u}} = \frac{V_p c_{t0}}{V_{\mathrm{u}}\left(c_{g0} + c_{V\mathrm{u}}\right)}, \ A_{\mathrm{d}} = \frac{V_p c_{t0}}{V_{\mathrm{d}}\left(c_{g0} + c_{V\mathrm{d}}\right)}, \tag{19}$$

where the subscript 0 and D denote the initial status and the dimensionless variable, respectively.

If the newly defined equivalent total compressibility c_t^e (Equation (3)) is used instead of the abovementioned conventionally-defined total compressibility $c_t = c_g + c_f$ in Equation (19), a new dimensionless definition can be obtained, which will be referred to as the new dimensionless definition and the old dimensionless definition, respectively.

In the latter analysis using the numerical methods presented in Appendices A–C, the testing gas is methane, and its physical parameters are computed by the PVT formulae. The absorbed density is $\rho = 374 \ \mathrm{kg/m^3}$ [29], the rock apparent density is $\rho_s = 2600 \ \mathrm{kg/m^3}$, the rock pore compressibility is $c_f = 1.0 \times 10^{-3} \ \mathrm{MPa^{-1}}$ and the vessel compressibility is $c_{V\mathrm{u}} = c_{V\mathrm{d}} = 10^{-5} \ \mathrm{MPa^{-1}}$.

3.1. Equilibrium Sorption Model

It can be seen that the shape of the pressure curve of the equilibrium sorption model is similar to that of the homogenous model (Figures 2–4). If using the old dimensionless definitions, unlike the homogeneous model, the values of the dimensionless pseudo-pressure curves are impacted by the testing pressure and the Langmuir sorption parameters. The lower the testing pressure, the larger the Langmuir volume and pressure are, and the more rapidly the upstream pseudo-pressure decreases; the slower the downstream pseudo-pressure increases, the smaller the balanced pseudo-pressure (Figures 2a, 3a and 4a). When the Langmuir volume is 0, it becomes a homogenous model. Therefore, the gas adsorption speeds up the decrease of the pseudo-pressure of the upstream vessel, and reduces the increase of that of the downstream vessel. When using the new dimensionless definition, for the same A_u and A_d, the pulse decay test curves for different Langmuir pressure, Langmuir volume and test pressure, are almost completely coincident, and they are the same as that of the homogenous model (Figures 2b, 3b and 4b). Therefore, the equivalent total compressibility defined in this study is reasonable.

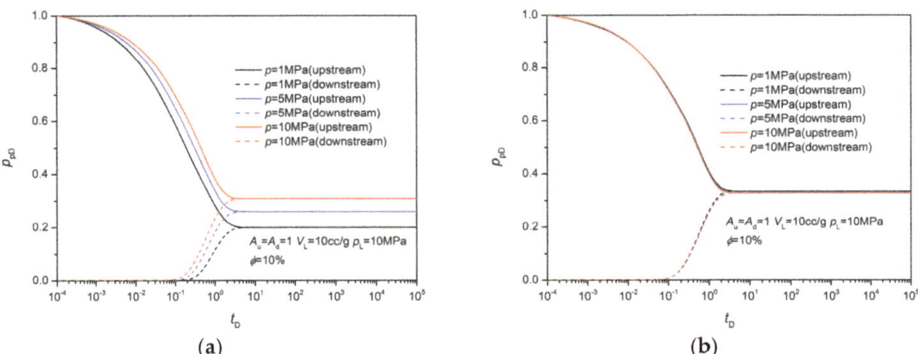

Figure 2. The influence of the testing pressure on the pseudo-pressure curves of the equilibrium sorption model. (**a**) Old dimensionless definition. (**b**) New dimensionless definition.

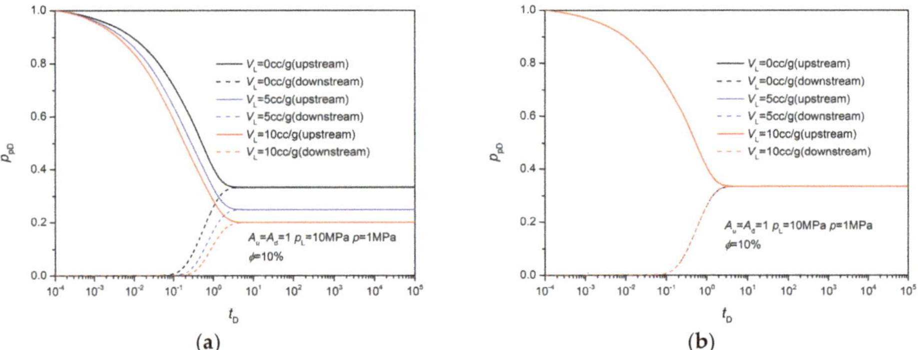

Figure 3. The influence of the Langmuir volume on the pseudo-pressure curves of the equilibrium sorption model. (**a**) Old dimensionless definition. (**b**) New dimensionless definition.

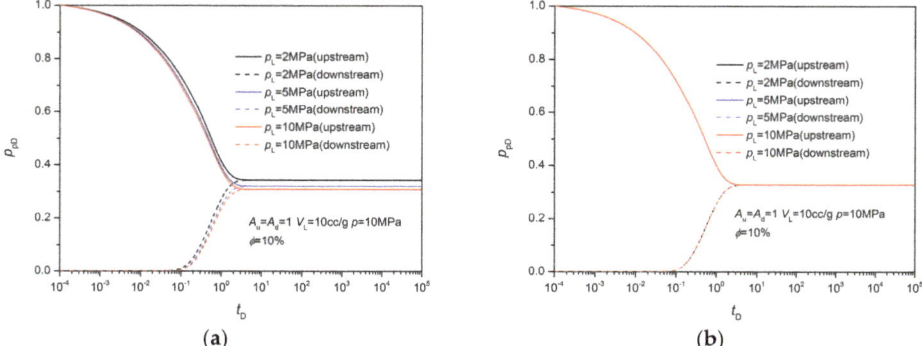

Figure 4. The influence of the Langmuir pressure on the pseudo-pressure curves of the equilibrium sorption model. (**a**) Old dimensionless definition. (**b**) New dimensionless definition.

3.2. Non-Equilibrium Sorption Model

For non-equilibrium models, the dimensionless definitions in Equation (19) can be changed to the follow definitions.

$$t_{\mathrm{D}} = \frac{k_f t}{\mu \wedge L^2}, \ A_{\mathrm{u}} = \frac{V_{\mathrm{p}}\wedge}{\varphi V_{\mathrm{u}}\left(c_{g0} + c_{V_{\mathrm{u}}}\right)}, \ A_{\mathrm{d}} = \frac{V_{\mathrm{p}}\wedge}{\varphi V_{\mathrm{d}}\left(c_{g0} + c_{V_{\mathrm{d}}}\right)}. \tag{20}$$

Meanwhile, the following is defined.

$$\wedge = \frac{V_{\mathrm{ic}}}{\rho_0\left(p_{pu0}-p_{pd0}\right)} + (\varphi c_t)_{f0}, \ V_{\mathrm{D}} = \frac{V}{V_{\mathrm{ic}}}, \ C_{\mathrm{D}} = \frac{C}{V_{\mathrm{ic}}},$$
$$\text{pseudo} - \text{storativity ratio } \omega = \frac{(\varphi c_{tf})_0}{\wedge}, \tag{21}$$

where V_{ic} is the initial total gas mass occurring in the matrix in kg/m^3. For the unsteady state of the non-equilibrium sorption model, the pseudo-interporosity flow coefficient is defined as follows.

$$\lambda_{\mathrm{D}} = \frac{6 D_{\mathrm{m}} \pi^2 \mu_0 \wedge L^2}{R_{\mathrm{m}}^2 k_{f0}}. \tag{22}$$

For the pseudo-steady-state model, it is defined as follows.

$$\lambda_{\mathrm{D}} = \frac{3 D_{\mathrm{m}} \mu_0 \wedge L^2}{R_{\mathrm{m}}^2 k_{f0}}. \tag{23}$$

It should be noted that in order to get the same dimensionless pressure for the different fracture porosity at equilibrium stages, the following definition is used in the following figures.

$$\wedge = \frac{V_{\mathrm{ic}}}{\rho_0 p_{pd0}} + (\varphi c_t)_{f0}. \tag{24}$$

Because of the small change of the testing pressure, the gas compressibility and viscosity can be assumed as a constant in the testing. Therefore, the dimensionless governing equations of the pseudo-steady-state model can be written as:

$$\omega \frac{\partial p_{pfD}}{\partial t_{\mathrm{D}}} = \frac{\partial}{\partial x_{\mathrm{D}}}\left(\frac{\partial p_{pfD}}{\partial x_{\mathrm{D}}}\right) - (1-\omega)\frac{\partial V_{\mathrm{D}}}{\partial t_{\mathrm{D}}}, \tag{25}$$

$$\frac{\partial V_D}{\partial t_D} = \lambda(C_{ED} - C_D).\tag{26}$$

The dimensionless governing equations of the unsteady state model are:

$$\omega\frac{\partial p_{pfD}}{\partial t_D} = \frac{\partial}{\partial x_D}\left(\frac{\partial p_{pfD}}{\partial x_D}\right) - (1-\omega)\lambda_D\left.\frac{\partial C_D}{\partial r_{mD}}\right|_{r_{mD}=1},\tag{27}$$

$$\frac{\partial V_D}{\partial t_D} = \frac{\lambda_D}{3}\frac{1}{r_{mD}^2}\frac{\partial}{\partial r_{mD}}\left(r_{mD}^2\frac{\partial C_D}{\partial r_{mD}}\right).\tag{28}$$

It can be found that the governing equations for the unsteady state and the pseudo-steady-state non-equilibrium sorption models are similar in the forms to the unsteady and the pseudo-steady-state dual-porosity models, respectively, which have similar matrix-fracture exchange terms. The difference is that the dual-porosity model does not consider gas adsorption, and the flow in the matrix conforms to Darcy's law, but the non-equilibrium sorption model considers it to comply with Fick's law of diffusion. In the meantime, their initial and boundary conditions can be easily found to be the same [28]. Hence, their properties are almost the same for pulse decay tests.

Figure 5 shows that the shape of the pseudo-pressure curves of the non-equilibrium sorption model is not similar to that of the equilibrium sorption model. After the upstream and downstream pseudo-pressures of the non-equilibrium adsorption model are balanced, they will continue to fall together until the system pseudo-pressure is at equilibrium. This finding is similar to of the feature of the pressure curves of the dual-porosity models, and is consistent with the previous analysis of the mathematical models [28]. Under the same conditions presented in the figure, the decrease ($t_D < 3$) of the upstream vessel pseudo-pressure of the non-equilibrium sorption model is slower than that of the equilibrium sorption model, but the upstream vessel pseudo-pressure is contrary. In addition, the balanced pseudo-pressure and the equilibrium pseudo-pressure of the non-equilibrium sorption models are higher than those of the equilibrium sorption model (Figure 5). This is due to the different definitions of their dimensionless parameters. As used herein, the balanced pseudo-pressure refers to the pseudo-pressure at which the upstream and downstream vessel pressures initially become equal; the equilibrium pseudo-pressure refers to the pseudo-pressure at which all pressures no longer change.

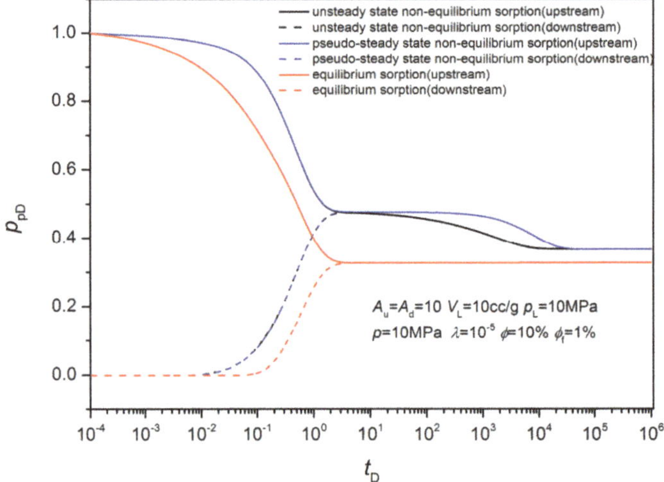

Figure 5. The comparison of the curves of the equilibrium sorption and the non-equilibrium sorption models.

The sensitivity analysis of the fracture porosity, testing pressure, Langmuir volume and pressure and vessel volumes on the pulse decay tests, were simulated by the numerical methods in the Appendices B and C. If the total porosity is constant, the balanced pseudo-pressure of the non-equilibrium sorption model decreases with the increase of the fracture porosity. Meanwhile, the pseudo-pressure of the upstream vessel decreases more rapidly, and the downstream vessel pseudo-pressure increases more slowly with the increase of the fracture porosity. However, the final equilibrium pseudo-pressure was not affected (Figure 6). The final equilibrium pseudo-pressure increased with the increase of the testing pressure and the Langmuir volumes, and decreased with the increase of the Langmuir pressures (Figures 7–9). With the increase of the vessel volumes, the pseudo-pressure curves of the equilibrium and non-equilibrium sorption model approaches that of no adsorption (Figure 10). Therefore, when the vessel volumes are big enough, the influence of gas adsorption can be ignored, and the non-equilibrium and equilibrium models cannot be distinguished.

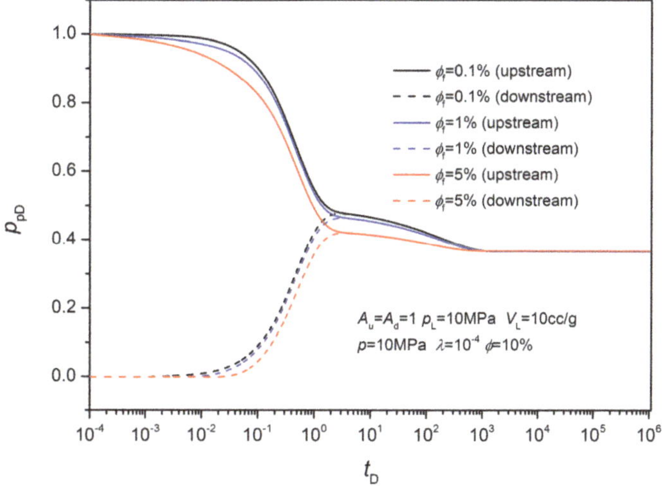

Figure 6. The influence of the fracture porosity on the curve of the non-equilibrium sorption model.

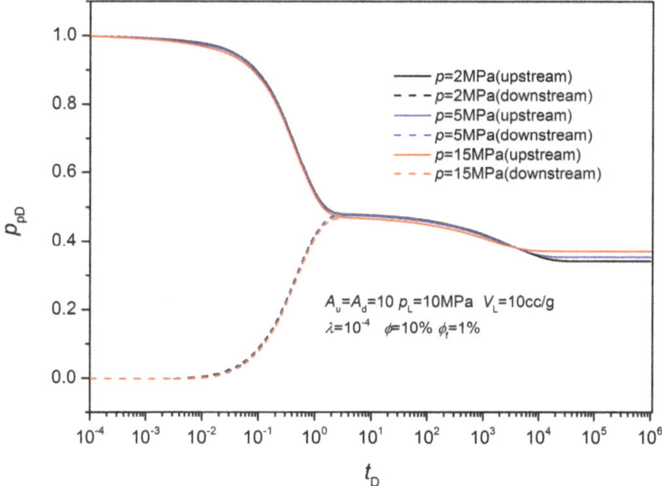

Figure 7. The influence of the testing pressure on the curve of the non-equilibrium sorption model.

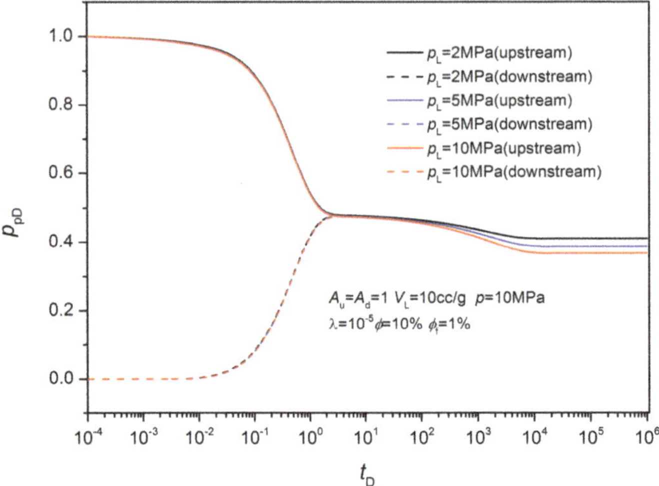

Figure 8. The influence of the Langmuir pressures on the pseudo-pressure the non-equilibrium sorption.

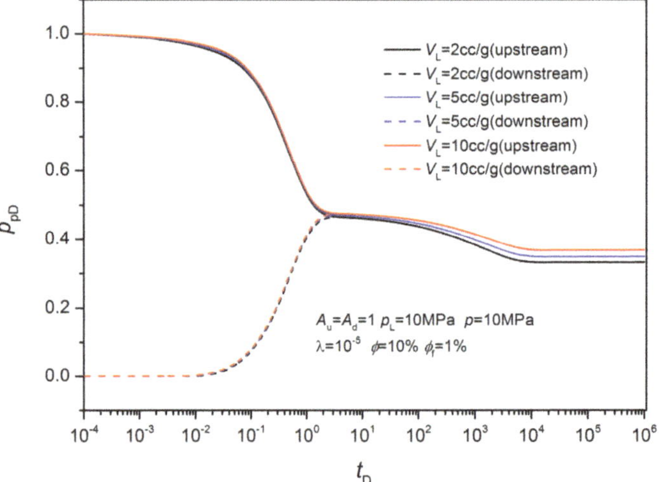

Figure 9. The influence of the Langmuir pressure on the pseudo-pressure curves of the non-equilibrium sorption model.

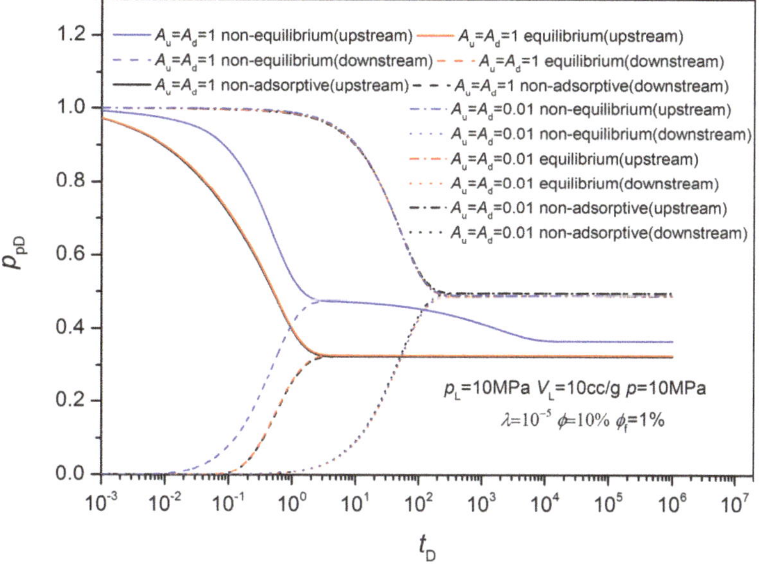

Figure 10. The influences of the vessel volumes on the pseudo-pressure curves.

4. The Analysis Method for Adsorptive Gas

4.1. Equilibrium Sorption Model

According to the analysis in Section 3.1, if the total compressibility c_t of the homogeneous model is replaced by the equivalent total compressibility c_t^e, it can be used to analyze the pulse decay test data of the equilibrium sorption. Therefore, the equilibrium sorption model is essentially a homogenous model. Thus it can be inferred that ignoring adsorption in the fracture in the previous section does not change the characteristics of the non-equilibrium sorption model, and the total compressibility of the fracture can be substituted for by the equivalent total compressibility to account for the adsorption in the fractures. Cui et al. [21] corrected the porosity to involve gas absorption, but their correction does not consider the influence of the absorbed phase volume. The Cui-correction is equivalent to a total compressibility correction, which leads to the same permeability. Here, it is named the Cui-correction as well, and the corrected equivalent total compressibility is as follows:

$$c_t^{\mathrm{Cui}} = c_f + c_g + \frac{\rho_{sc}\rho_s V_L p_L}{\rho\varphi(p_L + p)^2}.$$ (29)

In the following, the equivalent total compressibility and permeability determined by the Cui-correction and our new correction are compared with those of no correction. The errors in the following figures are relative to our new correction.

If the testing pressure is low, the non-correction total compressibility is less than the equivalent total compressibility, which is contrary at high pressure (Figures 11 and 12). Therefore, these two total compressibilities are close at certain pressures. Their difference increases with the increase of V_L, and could be larger than 80%. Figure 11 shows that the influence of p_L is not monotonous. When the testing pressure is very large or very small, the difference decreases with p_L.

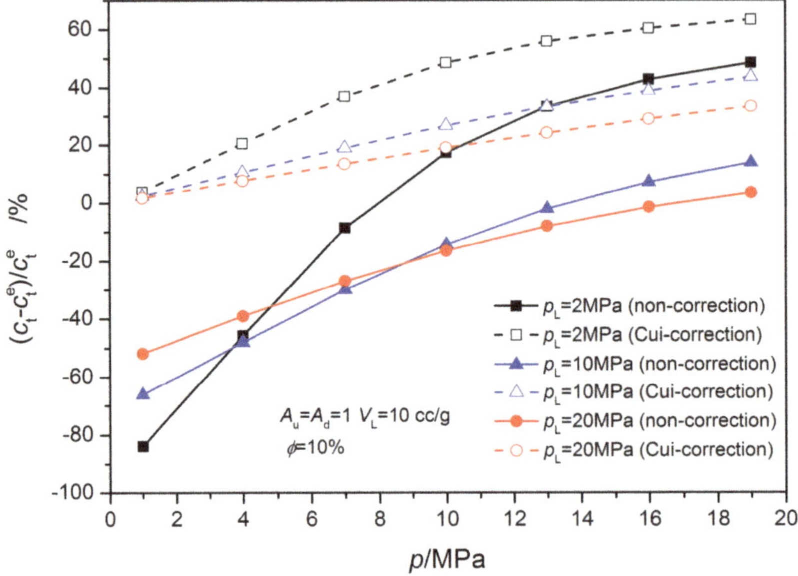

Figure 11. The influence of p_L on the total compressibility.

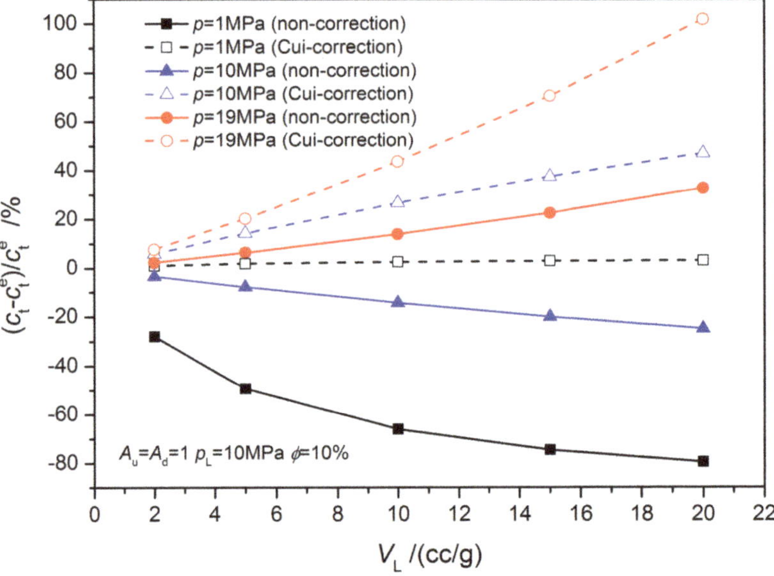

Figure 12. The influence of V_L on the total compressibility.

Figures 11 and 12 show that the Cui-correction total compressibility is larger than the equivalent total compressibility. The difference increases with the testing pressure and V_L, and decreases with p_L, and it can be more than 100% (Figures 11 and 12).

The Cui-correction method is excessive. Only when the testing pressure of V_L is very small, the Cui-correction method works well. It can be found that the influences of the testing pressure,

V_L and p_L, on the permeability, follow the same law for total compressibility shown in Figures 13 and 14. Although, the influence of gas adsorption on permeability is not as intense as on the total compressibility, it still can generate an error of more than 45% (Figures 13 and 14).

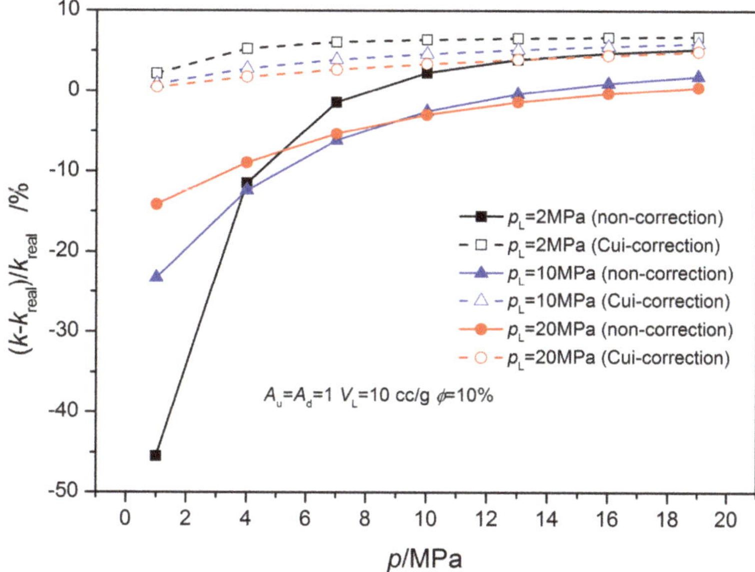

Figure 13. The influence of p_L on the tested permeability.

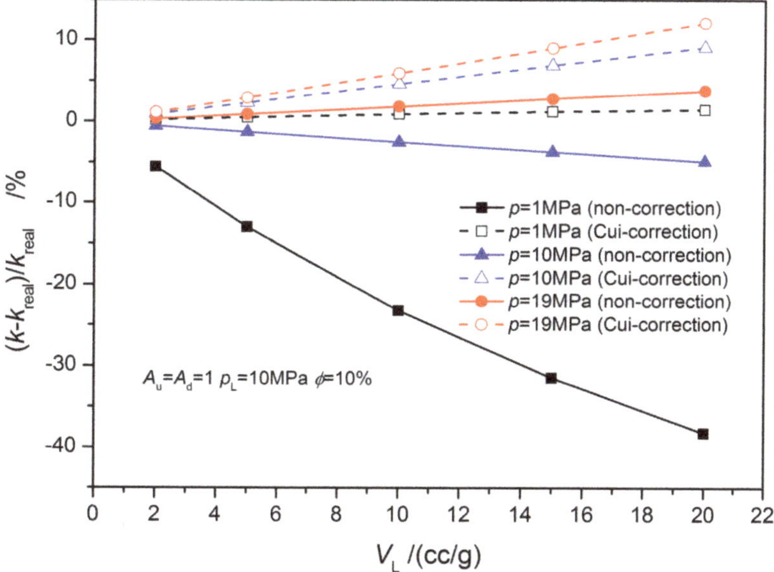

Figure 14. The influence of V_L on the tested permeability.

Figures 15 and 16 show that the bigger the vessel volume is, the larger the error of non-correction permeability is. If $A_u = A_d = 0.1$, the error is less than 1%. Therefore, if the vessel volume is very large, it does not need to be corrected. The permeability error of the Cui-correction method increases with the decrease of the vessel volumes. When $A_u = A_d = 10$, $\varphi = 10\%$, and $p = 19$ MPa, the error is approximately 30%. When $A_u = A_d = 1$, $\varphi = 5\%$, and $p = 19$ MPa, the error is also larger than 14%. When the testing pressure is very high, the result of the Cui-correction method is worse than that of the uncorrected, and it intensifies with the decrease of the vessel volume (Figure 16).

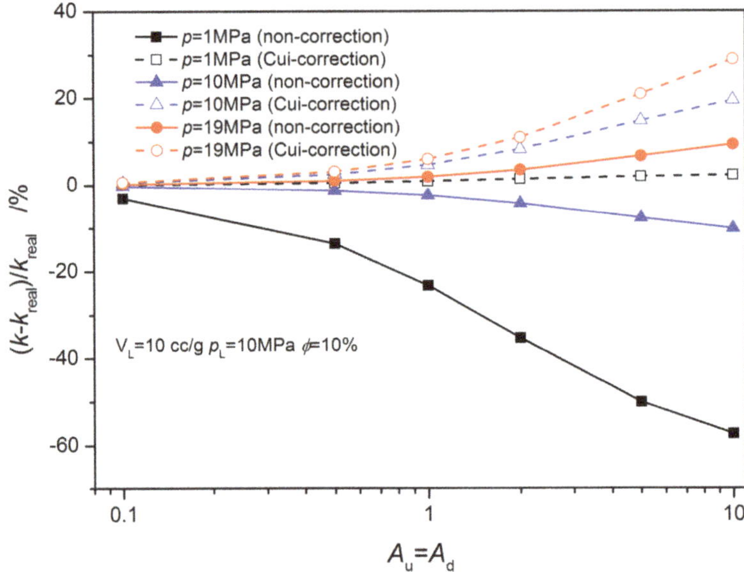

Figure 15. The influence of vessel volumes on the tested permeability.

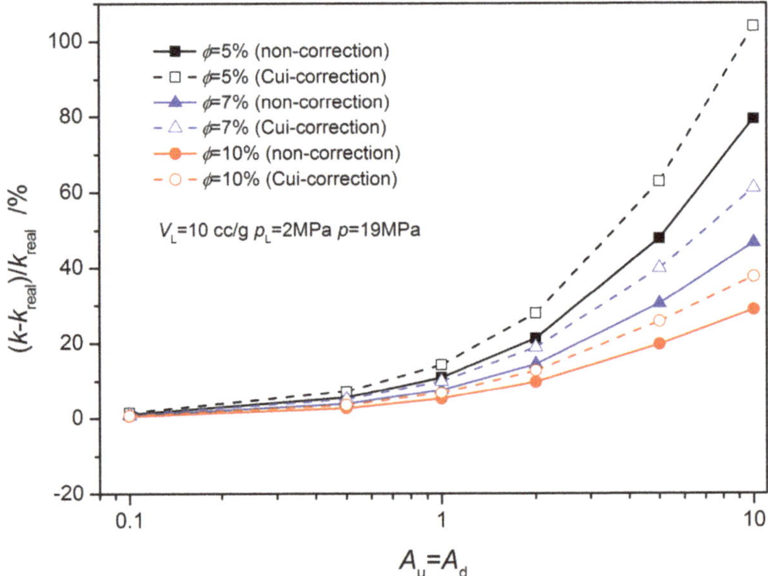

Figure 16. The influence of porosity on the tested permeability.

When the testing pressure is very low, the error of non-correction is large, and the result of the Cui-correction method is appropriate. Therefore, when the testing pressure is very large, the permeability needs to be determined by the equivalent total compressibility involving the adsorbed phase density in Equation (3), instead of by the total compressibility.

4.2. Non-Equilibrium Sorption Model

Figure 17 shows that the pseudo-pressure derivative curve of the non-equilibrium sorption model has a plateau under the proper vessel volume. However, this is not true for the equilibrium sorption model. These patterns are similar with the dual-porosity models [28]. Therefore, the pseudo-pressure derivative at the early stage can be used to differentiate the equilibrium and the non-equilibrium sorption models. If there is a plateau on the pseudo-pressure derivative curve, it means that the non-equilibrium sorption model is suitable, and the test must be continued to determine the diffusion coefficient. When there is gas sorption, besides the vessel volume, the adsorption parameters and the testing pressure also affect the pseudo-pressure derivative plateau, which is different from the dual-porosity media. Therefore, the vessel volumes should be chosen according to the core adsorption property and the testing pressure before the test. The unsteady state and the pseudo-steady state sorption model can also be identified by the pseudo-pressure derivative at the late stage. There is a plateau at the late stage on the pseudo-pressure derivative curve of pseudo-steady-state sorption model, and a lean straight line for the unsteady state sorption model (Figure 18). Since an analysis method similar to dual-porosity cores can be used, it will not be described in more detail herein [28].

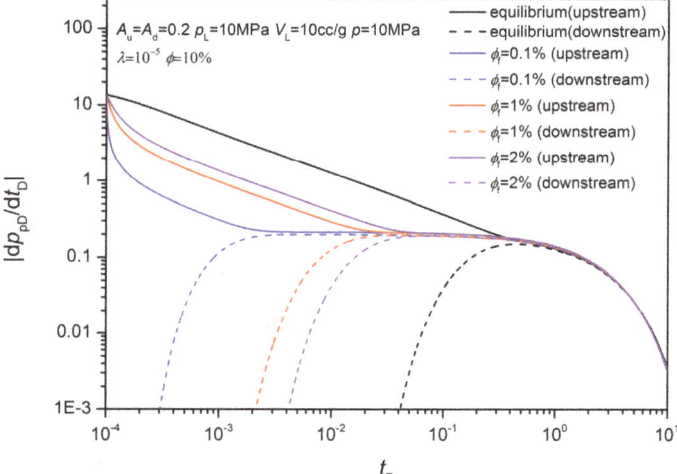

Figure 17. The pseudo-pressure derivative at the early stage.

For the two non-equilibrium sorption models, the pseudo-pressure derivative decreases with the Langmuir volume, which is the opposite observed for the pseudo-pressure (Figure 19). The influences of the Langmuir pressure and the testing pressure have similar impacts upon the pseudo-pressure derivative. These influences will not be detailed here.

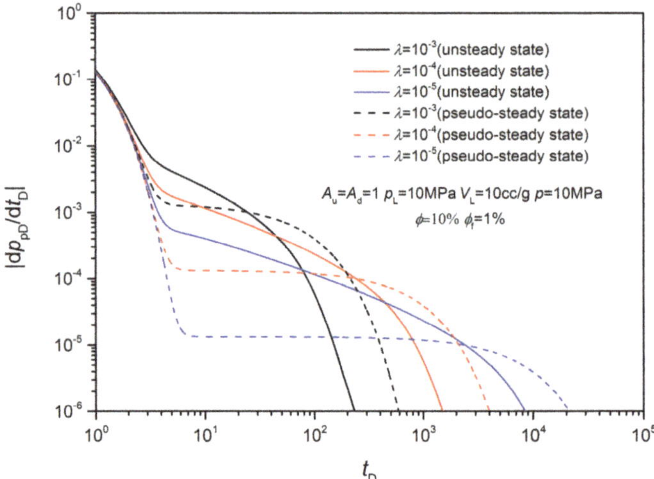

Figure 18. The pseudo-pressure derivative at the late stage.

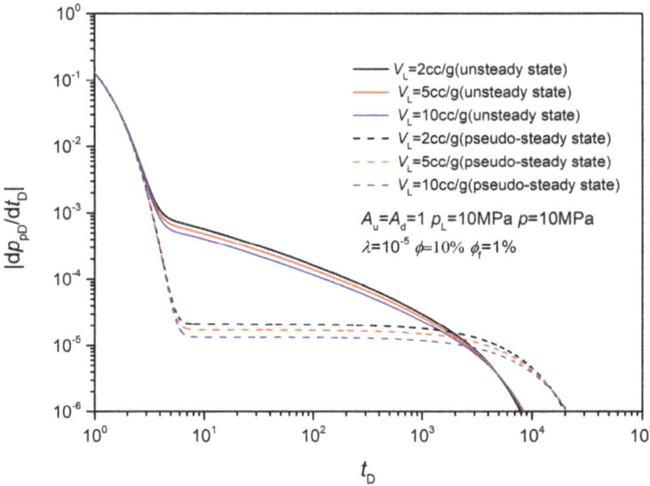

Figure 19. The influence of V_L on the pseudo-pressure derivative curves at the late stage.

5. Case Study

Pulse decay test data of N_2 and CH_4 for shale were extracted from Figure 6 of Aljamaan et al. [16]. Since the adsorption behavior of H_2 and CO_2 do not conform to the Langmuir model, which is used in this study, their data were not extracted. The pressure history and the pressure derivative history of N_2 and CH_4 were fitted using the method proposed in this study. The results are shown in Figure 20. The pressure derivative plots indicate that the unsteady state non-equilibrium sorption model is more suitable for this sample. The fitting quality of the early time is lower. Because the early pressure in the original figures changes almost vertically, it is hard to extract the data exactly in this period.

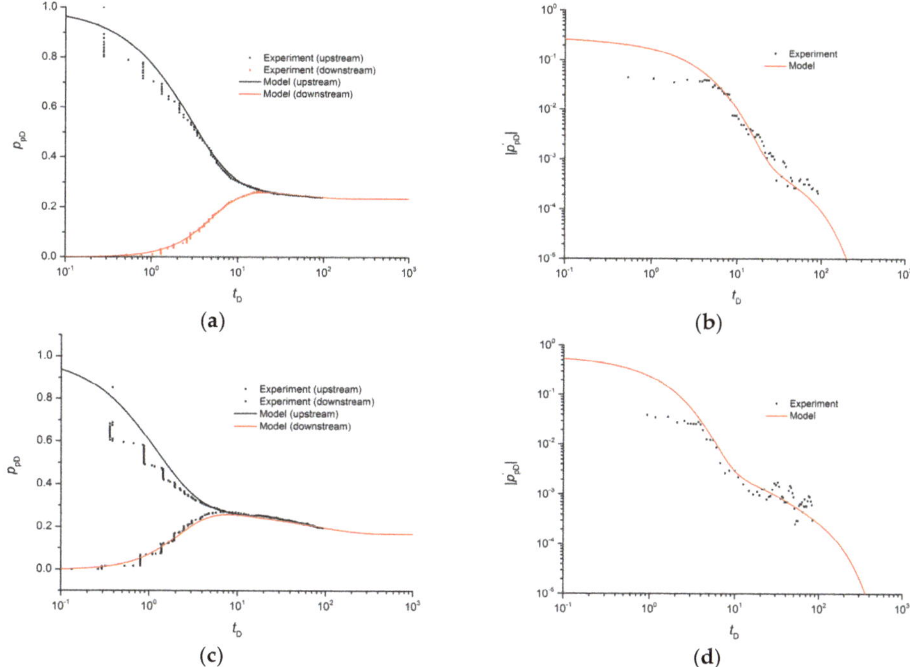

Figure 20. Comparisons between the simulation results and the experiment of Barnett 26-Ha. (**a**) Pressure histories of N_2. (**b**) Pressure derivative histories of N_2. (**c**) Pressure histories of CH_4. (**d**) Pressure derivative histories of CH_4. (Experimental data from Aljamaan et al. [16])

In addition, the time coordinate of Figure 20 is logarithmic, which leads to dispersed data at an early time. However, overall, the fitting results are sufficient. The parameters of the adsorption isotherm for CH_4 and N_2 measured by Aljamaan et al. [16] were used for fitting. Since the size of the sample was not specifically explained by Aljamaan et al. [16], the commonly used size was utilized in the simulation (the diameter is 2.54 cm (one inch) and the length is 5.08 cm (two inches)). The fitting results for N_2 are the fracture permeability $k_f = 11$ μD and $\lambda = 3.8 \times 10^{-4}$. The results for CH_4 are the fracture permeability $k_f = 14.5$ μD and $\lambda = 4.0 \times 10^{-4}$. In the fittings for these two gases, $\varphi_f = 0.3\%$ and $\varphi_m = 4.4\%$ were used. However, near this value, the fitting result is not sensitive to the fracture porosity, so the results of the fracture porosity may have large errors.

According to the parameters of the slippage effect measured by Aljamaan et al. [16], the apparent permeability of N_2 and CH_4 should be 7.98 μD (equilibrium pressure 53.01 psi) and 8.02 μD (equilibrium pressure 47.92 psi), respectively. These values are close to the results of this study, but there are obvious differences. Han et al. [28] points out that using a homogeneous model to analyze the permeability of dual-porosity cores may lead to significant error. A similar problem exists in the results of Aljamaan et al. [16], when using the approach of the Cui et al. [21] model to analyze non-equilibrium sorption. From the view of basic parameters, the sample length assumed in this study may differ from the actual. Furthermore, the adsorption isotherm fitted by Aljamaan et al. [16] deviates significantly from the measured point in the range of 0–100 psi (see Figure 9 in Aljamaan et al. [16]). These factors may lead to the difference between these two results. This example preliminarily validates the analysis method proposed for non-equilibrium sorption models. In order to more rigorously verify the methods proposed in this study, more pulse decay tests with gas adsorption are needed in the future.

6. Conclusions

Absorptive gases are sometimes used in pulse decay tests, and the influence of gas adsorption must be considered in data analysis. In this study, the volume of the adsorbed phase is considered, and the conventional models of adsorbing gas flow in porous media are modified to simulate the performance of pulse decay tests for equilibrium sorption, unsteady state and pseudo-steady-state non-equilibrium sorption.

For a pulse decay test of the equilibrium sorption model, an equivalent total compressibility is defined, resulting in the same form as the homogeneous model. Therefore, the pseudo-pressure curve of an equilibrium sorption model is similar to the homogeneous model. Numerical simulation also proves this conclusion. By using this equivalent total compressibility instead of the total compressibility of the analysis method for the homogeneous model, a correction method involving the absorbed phase volume is proposed. Further error analysis indicates that the model does not need any correction when the vessel volumes are very large. The Cui-correction method is excessive, and it is suitable only when the testing pressure is very low. If the testing pressure is very high, the error of the Cui-correction method cannot be ignored.

By defining appropriate dimensionless quantities, a dimensionless form of the non-equilibrium sorption equation similar to dual-porosity models can be obtained. Numerical simulations also show that the unsteady state and pseudo-steady state non-equilibrium sorption models are similar to the unsteady state and pseudo-steady state dual-porosity models, respectively. Like dual-porosity media, the equilibrium and non-equilibrium sorption models can be identified by the pseudo-pressure derivative at the early stage. When the non-equilibrium sorption models are affirmed to be the suitable model, the test must be continued to determine the model parameters. The unsteady state and the pseudo-steady state sorption models can also be identified by the pseudo-pressure derivative at the late stage. To use the pseudo-pressure derivative successfully, the suitable vessel volumes must be chosen according to the core sorption property, the testing pressure and the porosity. This proposed method was verified by a case study.

Author Contributions: Conceptualization, G.H. and X.L.; validation, Y.C.; formal analysis, G.H.; investigation, Y.C. and X.L.; writing—original draft preparation, G.H.; writing—review and editing, Y.C. and X.L.; funding acquisition, G.H.

Funding: This research was funded by the National Natural Science Foundation of China, grant number 11602276 and the Key Laboratory for Mechanics in Fluid Solid Coupling Systems, CAS.

Conflicts of Interest: The authors declare no conflict of interest.

Nomenclature

p, p_f	pressure and fracture pressure [Pa];
t	time [s];
x	coordinate along the sample, takes the upstream vessel as the origin [m];
L	length of the sample [m];
V_u, V_d, V_p	the volume of the upstream vessel, downstream vessels and the sample pore, respectively [m^3];
V, V_{ic}	total gas mass occurred and initial total gas mass occurred in the matrix [kg/m^3];
C, C_E	gas concentration in the matrix and equivalent gas concentration in the fracture, respectively [kg/m^3];
V_L, p_L	Langmuir volume and pressure respectively [m^3/kg, Pa];
c_g, c_f	compressibility of testing gas and pore volume, respectively [Pa^{-1}];
c_t, c_f^e	sample total compressibility and equivalent total compressibility [Pa^{-1}];
c_{Vd}, c_{Vu}	compressibilities of upstream and downstream vessels respectively [Pa^{-1}];
k, k_f, k_m	permeability, permeabilities of fractures and matrix respectively [m^2];
D_m	gas diffusion coefficient [m^2/s];
R_m, r_m	the radius and coordinate of the spherical matrix respectively [m];
M	molecular molar mass [kg/mol];

Z	gas deviator factor;
R	Universal Gas Constant [J/(mol·K)];
T	temperature [K]

Greeks

$\varphi, \varphi_a, \varphi_f, \varphi_m$	porosity, porosity of the adsorbed phased, fracture and matrix, respectively [%];
μ	viscosity [Pa·s];
$\rho, \rho_a, \rho_{sc}, \rho_s$	density of the testing gas, adsorbed gas, gas at the standard condition and apparent density of samples [kg/m^3];
λ_D	pseudo-interporosity flow coefficient;
ω	pseudo-storativity ratio;

Subscripts

f, m	macropores and micropores respectively;
u, d	upstream and downstream vessels respectively;
a	adsorbed phase;
i	reference status;
0	initial status;
p	normalized pseudo-pressure;
D	dimensionless variable

Appendix A. Numerical Solution for Equilibrium Sorption Model

For the convenience of writing, the following parameters are defined.

$$\theta = \frac{\Delta t}{\Delta x}, \tau = \frac{\Delta t}{\Delta x^2}, E = \frac{k_f}{\left(c_g + c_{V_d}\right)\mu\varphi L}\frac{V_P}{V_d}, F = \frac{k_f}{\left(c_g + c_{V_d}\right)\mu\varphi L}\frac{V_P}{V_d}, H = \rho\varphi c_t^e, G = \frac{\rho k}{\mu} \tag{A1}$$

Therefore, the governing equation Equation (4) and the boundary conditions Equations (5)–(7) of the equilibrium sorption model can be written as follows.

$$H\frac{\partial p}{\partial t} = \frac{\partial}{\partial x}\left(G\frac{\partial p}{\partial x}\right) \tag{A2}$$

$$\frac{dp_u}{dt} = E\frac{\partial p}{\partial x}\bigg|_{x=0}, \quad t > 0 \tag{A3}$$

$$\frac{dp_d}{dt} = -F\frac{\partial p}{\partial x}\bigg|_{x=L}, \quad t > 0 \tag{A4}$$

The sample is divided into N segments with element length Δx and is numbered from 0 to N. The finite difference method was used to discretize the equations. The governing equation Equation (A3) of the equilibrium sorption model can be discretized as follows.

$$H_j^{(s)}\frac{p_j^{(s+1)} - p_j^n}{\Delta t} = \frac{1}{\Delta x}\left(G_{j+1/2}^{(s)}\frac{p_{j+1}^{(s+1)} - p_j^{(s+1)}}{\Delta x} - G_{j-1/2}^n\frac{p_j^{(s+1)} - p_{j-1}^{(s+1)}}{\Delta x}\right) \tag{A5}$$

The subscript n and s represent the nth time step and the sth iteration in one time step, respectively. The subscript j represents the jth node. The boundary conditions of the upstream and downstream vessels can be discretized as follows.

$$\frac{p_0^{(s+1)} - p_0^n}{\Delta t} = E_{1/2}^{(s)}\frac{p_1^{(s+1)} - p_0^{(s+1)}}{\Delta x} \tag{A6}$$

$$\frac{p_N^{(s+1)} - p_N^n}{\Delta t} = -F_{N-1/2}^{(s)}\frac{p_N^{(s+1)} - p_{N-1}^{(s+1)}}{\Delta x} \tag{A7}$$

The above Equations (A5) to (A7) can be simplified to the following Equations (A8) to (A10), respectively.

$$- \tau G^{(s)}_{j+1/2} p^{(s+1)}_{j+1} + \left(\tau G^{(s)}_{j+1/2} + \tau G^{(s)}_{j-1/2} + H^{(s)}_{j} \right) p^{(s+1)}_{j} - \tau G^{(s)}_{j-1/2} p^{(s+1)}_{j-1} = H^{(s)}_{j} p^{n}_{j} \tag{A8}$$

$$- \theta E^{(s)}_{1/2} p^{(s+1)}_{1} + \left(\theta E^{(s)}_{1/2} + 1 \right) p^{(s+1)}_{0} = p^{n}_{0} \tag{A9}$$

$$\left(\theta F^{(s)}_{N-1/2} + 1 \right) p^{(s+1)}_{N} - \theta F^{(s)}_{N-1/2} p^{(s+1)}_{N-1} = p^{n}_{N} \tag{A10}$$

The initial condition Equation (5) can be discretized as:

$$p_{j}(0) = p_{\mathrm{d}}(0) \quad 0 < j \leq N+1 \ , \ p_{0}(0) = p_{\mathrm{u}}(0). \tag{A11}$$

There is an unknown variable p_j on each node for a total of $N + 1$ unknown variables. There are a total of $N + 1$ equations: $N - 1$ equations on $N - 1$ internal nodes with Equation (A8), and Equations (A9) and (A10) on the two boundary nodes. They form a matrix of $N + 1$ rank, which can be iteratively solved in combination with the initial condition Equation (A11).

Appendix B. Numerical Solution for the Pseudo-Steady-State Non-Equilibrium Sorption Model

In addition to the definition of Equation (A1), the following parameters are defined for the pseudo-steady-state non-equilibrium sorption model.

$$H = \rho(\varphi c_{\mathrm{t}})_{\mathrm{f}}, \ G = \frac{\rho k_{\mathrm{f}}}{\mu}, \ W = \frac{6 D_{\mathrm{m}} \pi^{2}}{R^{2}_{\mathrm{m}}}. \tag{A12}$$

Therefore, the governing equations Equations (9) and (10) of the pseudo-steady-state non-equilibrium sorption model can be written as:

$$H \frac{\partial p_{\mathrm{f}}}{\partial t} = \frac{\partial}{\partial x} \left(G \frac{\partial p_{\mathrm{f}}}{\partial x} \right) - \frac{\partial V}{\partial t}, \tag{A13}$$

$$\frac{\partial V}{\partial t} = W(C_{\mathrm{E}} - C). \tag{A14}$$

Like the equilibrium adsorption model, the sample can be divided into N segments with element length Δx, and the node number is from 0 to N. Using the finite difference method, the governing Equations (A13) and (A14) can be discretized into:

$$H^{(s)}_{j} \frac{p^{(s+1)}_{\mathrm{f},j} - p^{n}_{\mathrm{f},j}}{\Delta t} = \frac{1}{\Delta x} \left(G^{(s)}_{j+1/2} \frac{p^{(s+1)}_{\mathrm{f},j+1} - p^{(s+1)}_{\mathrm{f},j}}{\Delta x} - G^{(s)}_{j-1/2} \frac{p^{(s+1)}_{\mathrm{f},j} - p^{(s+1)}_{\mathrm{f},j-1}}{\Delta x} \right) - \frac{V^{(s+1)}_{j} - V^{n}_{j}}{\Delta t}, \tag{A15}$$

$$\frac{V^{s+1}_{j} - V^{n}_{j}}{\Delta t} = W\left(C^{s+1}_{\mathrm{E},j} - C^{s+1}_{j} \right). \tag{A16}$$

The governing Equations (A15) and (A16) can be simplified into the following forms, respectively.

$$-\tau G^{(s)}_{j+1/2} p^{(s+1)}_{\mathrm{f},j+1} + \left(\tau G^{(s)}_{j+1/2} + \tau G^{(s)}_{j-1/2} + H^{(s)}_{j} \right) p^{(s+1)}_{\mathrm{f},j} - \tau G^{(s)}_{j-1/2} p^{(s+1)}_{\mathrm{f},j-1} + W \Delta t C^{(s+1)}_{\mathrm{E},j} - W \Delta t C^{(s+1)}_{j} = H^{(s)}_{j} p^{n}_{\mathrm{f},j}, \tag{A17}$$

$$- W \Delta t C^{(s+1)}_{\mathrm{E},j} + W \Delta t C^{(s+1)}_{j} + V^{(s+1)}_{j} = V^{n}_{j} \tag{A18}$$

The discretized equations for the boundary conditions and the initial conditions can be directly obtained by using pf instead of p in Equations (A9)–(A11). $C_{\mathrm{E},j}$ can be written as a function of $p_{\mathrm{f},j}$, thereby reducing one variable. There are three unknown variables, $p_{f,j}$, V_j and C_j, on each node, and V_j and C_j can be written as a function of the same pressure, thereby reducing one unknown variable on every node. Therefore, there is a total of $2(N+1)$ unknown variables.

There are $2(N-1)$ equations on $N-1$ internal nodes with Equations (A17) and (A18), and four equations on two boundary nodes with Equation (A18) and Equations (A9)–(A11). Hence, a matrix of rank $2(N+1)$ can be formed. A numerical solution can be iteratively obtained by adding the initial conditions Equation (A11).

Appendix C. Numerical Solution for the Unsteady State Non-Equilibrium Sorption Model

In addition to the definitions of Equations (A1) and (A12), the following parameters are defined.

$$\theta_{\mathrm{m}} = \frac{\Delta t}{\Delta r_{\mathrm{m}}}, \; \tau_{\mathrm{m}} = \frac{\Delta t}{\Delta r_{\mathrm{m}}^2}, \; W = \frac{3D_{\mathrm{m}}}{R_{\mathrm{m}}}. \tag{A19}$$

Therefore, the governing equation, that is, Equation (12) for the unsteady state non-equilibrium sorption model, can be written as:

$$H\frac{\partial p_f}{\partial t} = \frac{\partial}{\partial x}\left(G\frac{\partial p_f}{\partial x}\right) - W\frac{\partial C}{\partial r_{\mathrm{m}}}\bigg|_{r_{\mathrm{m}}=R}. \tag{A20}$$

For the unsteady state non-equilibrium sorption model, the sample is divided into N segments with fracture element length Δx, and the node number is from 0 to N. In addition, the matrix needs to be divided into M segments with an element length Δr_{m}, and the node number is from 0 to M. Using the finite difference method, the governing equations, Equations (A20) and (13) for the unsteady state non-equilibrium sorption model, can be discretized into:

$$H_j^{(s)}\frac{p_{f,j}^{(s+1)} - p_{f,j}^n}{\Delta t} = \frac{\partial}{\partial x}\left(G_{j+1/2}^{(s)}\frac{p_{f,j+1}^{(s+1)} - p_{f,j}^{(s+1)}}{\Delta x} - G_{j-1/2}^{(s)}\frac{p_{f,j}^{(s+1)} - p_{f,j-1}^{(s+1)}}{\Delta x}\right) - W\frac{C_M^{(s+1)} - C_{M-1}^n}{\Delta r_{\mathrm{m}}}, \tag{A21}$$

$$\frac{V_{j,i}^{(s+1)} - V_{j,i}^n}{\Delta t} = \frac{1}{r_{\mathrm{m},i}^2}\frac{1}{\Delta r_{\mathrm{m}}}\left(r_{\mathrm{m},i+1/2}^2 D_{\mathrm{m}}\frac{C_{j,i+1}^{(s+1)} - C_{j,i}^{(s+1)}}{\Delta r_{\mathrm{m}}} - r_{\mathrm{m},i-1/2}^2 D_{\mathrm{m}}\frac{C_{j,i}^{(s+1)} - C_{j,i-1}^{(s+1)}}{\Delta r_{\mathrm{m}}}\right). \tag{A22}$$

where subscript i indicates the node number of the matrix. The matrix boundary conditions Equation (14) can be discretized into:

$$\frac{C_{j,1}^n - C_{j,0}^n}{\Delta r_{\mathrm{m}}} = 0. \tag{A23}$$

The above Equations (A21)–(A23) can be simplified to:

$$-\tau G_{j+1/2}^{(s)}p_{f,j+1}^{(s+1)} + \left(\tau G_{j+1/2}^{(s)} + \tau G_{j-1/2}^{(s)} + H_j^{(s)}\right)p_{f,j}^{(s+1)} - \tau G_{j-1/2}^{(s)}p_{f,j-1}^{(s+1)} + W\theta_{\mathrm{m}}\left(C_M^{(s+1)} - C_{M-1}^{(s+1)}\right) = H_j^{(s)}p_{f,j}^n. \tag{A24}$$

$$-D_{\mathrm{m}}\tau_{\mathrm{m}}\frac{r_{\mathrm{m},i+1/2}^2}{r_{\mathrm{m},i}^2}C_{j,i+1}^{(s+1)} + \left(D_{\mathrm{m}}\tau_{\mathrm{m}}\frac{r_{\mathrm{m},i+1/2}^2}{r_{\mathrm{m},i}^2} + D_{\mathrm{m}}\tau_{\mathrm{m}}\frac{r_{\mathrm{m},i-1/2}^2}{r_{\mathrm{m},i}^2}\right)C_{j,i}^{(s+1)} + V_{j,i}^{(s+1)} - D_{\mathrm{m}}\tau_{\mathrm{m}}\frac{r_{\mathrm{m},i-1/2}^2}{r_{\mathrm{m},i}^2}C_{j,i-1}^{(s+1)} = V_{j,i}^n. \tag{A25}$$

$$C_{j,1}^n - C_{j,0}^n = 0. \tag{A26}$$

The discretized equations for the boundary and initial conditions can be directly obtained by replacing p into Equations (A9)–(A11) with p_f. For each fracture node, there are $M+1$ matrix unknown variables, $C_{j,i}$ and $V_{j,i}$, and one fracture unknown variable $p_{f,j}$, totaling $(2M+3)$ unknown variables. If the $(M+1)$th matrix node unknown variable is written as a function of the fracture pressure at that location, an unknown variable can be reduced for every fracture node. Also note that C_j and V_j can be written as a function of the same pressure. Therefore, there are a total of $(M+1) \times (N+1)$ unknown variables. There are $M - 1$ Equation (A25) on matrix internal nodes, and Equation (A26) on the matrix inner boundary node for each fracture node. There are $N - 1$ equations on inner fracture nodes with Equation (A24), and Equations (A9) and (A10) on boundary nodes. They make up a $(M + 1)(N + 1)$ rank equation group. Combined with the initial condition Equation (A11), a matrix of $(N + 1)$ $(M + 1)$ order can be formed for solving.

The following calculation strategy is taken in each time step: For the calculation of the fracture pressure, the gas concentration of the matrix is fixed; for the calculation of the concentration in the matrix, the fracture pressure is fixed. This approach can reduce the rank of the matrix by iteratively solving $N + 1$ M-rank matrices and one $(N + 1)$-rank matrix. In addition, the chasing method can be used to improve the calculation speed.

It should be pointed out that in order to improve the calculation accuracy of fracture-matrix exchange, the MINC modeling techniques and random walks methods were proposed in the literature [30,31].

References

1. Brace, W.F.; Walsh, J.B.; Frangos, W.T. Permeability of granite under high pressure. *J. Geophys. Res.* **1968**, *73*, 2225–2236. [CrossRef]
2. Bourbie, T.; Walls, J. Pulse decay permeability: Analytical solution and experimental test. *SPE J.* **1982**, *22*, 719–921. [CrossRef]
3. Hsieh, P.A.; Tracy, J.V.; Neuzil, C.E.; Bredehoeft, J.D.; Silliman, S.E. A transient laboratory method for determining the hydraulic properties of 'tight' rocks—I. Theory. *Int. J. Rock Mech. Min. Sci. Geomech. Abstr.* **1981**, *18*, 245–252. [CrossRef]
4. Chen, T.; Stagg, P.W. Semilog analysis of the pulse decay technique of permeability measurement. *SPE J.* **1984**, *24*, 639–642. [CrossRef]
5. Jones, S.C. A technique for faster pulse-decay permeability measurements in tight rocks. *SPE Form. Eval.* **1997**, *12*, 19–25. [CrossRef]
6. Pan, Z.; Ma, Y.; Connell, L.D.; Down, D.I.; Camilleri, M. Measuring anisotropic permeability using a cubic shale sample in a triaxial cell. *J. Nat. Gas Sci. Eng.* **2015**, *26*, 336–344. [CrossRef]
7. Jang, H.; Lee, W.; Kim, J.; Lee, J. Novel apparatus to measure the low-permeability and porosity in tight gas reservoir. *J. Pet. Sci. Eng.* **2016**, *142*, 1–12. [CrossRef]
8. Firouzi, M.; Alnoaimi, K.; Kovscek, A.; Wilcox, J. Klinkenberg effect of predicting and measuring helium permeability in gas shales. *Int. J. Coal Geol.* **2014**, *123*, 62–68. [CrossRef]
9. Heller, R.; Vermylen, J.; Zoback, M. Experimental investigation of matrix permeability of gas shales. *AAPG Bull.* **2014**, *98*, 975–995. [CrossRef]
10. Mokhtari, M.; Tutuncu, A.N. Characterization of anisotropy in the permeability of organic-rich shales. *J. Pet. Sci. Eng.* **2015**, *133*, 496–506. [CrossRef]
11. Zhang, R.; Ning, Z.; Yang, F.; Wang, X.; Zhao, H.; Wang, Q. Impacts of nanopore structure and elastic properties on stress-dependent permeability of gas shales. *J. Nat. Gas Sci. Eng.* **2015**, *26*, 1663–1672. [CrossRef]
12. Cronin, M.B.; Flemings, P.B.; Bhandari, A.R. Dual-permeability microstratigraphy in the Barnett Shale. *J. Pet. Sci. Eng.* **2016**, *142*, 119–128. [CrossRef]
13. Cao, C.; Li, T.; Shi, J.; Zhang, L.; Fu, S.; Wang, B.; Wang, H. A new approach for measuring the permeability of shale featuring adsorption and ultra-low permeability. *J. Nat. Gas Sci. Eng.* **2016**, *30*, 548–556. [CrossRef]
14. Zhang, R.; Ning, Z.; Yang, F.; Zhao, H.; Wang, Q. A laboratory study of the porosity-permeability relationships of shale and sandstone under effective stress. *Int. J. Rock Mech. Min. Sci.* **2016**, *81*, 19–27. [CrossRef]
15. Alnoaimi, K.R.; Duchateau, C.; Kovscek, A.R. Characterization and measurement of multiscale gas transport in shale-core samples. *SPE J.* **2016**, *21*, 573–588. [CrossRef]
16. Aljamaan, H.; Al Ismail, M.; Kovscek, A.R. Experimental investigation and Grand Canonical Monte Carlo simulation of gas shale adsorption from the macro to the nano scale. *J. Nat. Gas Sci. Eng.* **2017**, *48*, 119–137. [CrossRef]

17. Alnoaimi, K.R.; Kovscek, A.R. Influence of microcracks on flow and storage capacities of gas shales at core scale. *Transp. Porous Media* **2019**, *127*, 53–84. [CrossRef]

18. Ghanizadeh, A.; Gasparik, M.; Amann-Hildenbrand, A.; Gensterblum, Y.; Krooss, B.M. Experimental study of fluid transport processes in the matrix system of the European organic-rich shales: I. Scandinavian Alum Shale. *Mar. Pet. Geol.* **2014**, *51*, 79–99. [CrossRef]

19. Ghanizadeh, A.; Amann-Hildenbrand, A.; Gasparik, M.; Gensterblum, Y.; Krooss, B.M.; Littke, R. Experimental study of fluid transport processes in the matrix system of the European organic-rich shales: II. Posidonia Shale (Lower Toarcian, northern Germany). *Int. J. Coal Geol.* **2014**, *123*, 20–33. [CrossRef]

20. Gensterblum, Y.; Ghanizadeh, A.; Krooss, B.M. Gas permeability measurements on Australian subbituminous coals: Fluid dynamic and poroelastic aspects. *J. Nat. Gas Sci. Eng.* **2014**, *19*, 202–214. [CrossRef]

21. Cui, X.; Bustion, A.M.M.; Bustion, R.M. Measurements of gas permeability and diffusivity of tight reservoir rocks: Different approaches and their applications. *Geofluids* **2009**, *9*, 208–223. [CrossRef]

22. Civan, F.; Rai, C.S.; Sondergeld, C.H. Shale permeability determined by simultaneous analysis of multiple pressure-pulse measurements obtained under different conditions. In Proceedings of the North American Unconventional Gas Conference and Exhibition, The Woodlands, TX, USA, 14–16 June 2011. [CrossRef]

23. Anbarci, K.; Ertekin, T. A comprehensive study of pressure transient analysis with sorption phenomena for single-phase gas flow in coal seams. In Proceedings of the SPE Annual Technical Conference and Exhibition, New Orleans, LA, USA, 23–26 September 1990. [CrossRef]

24. Kolesar, J.E.; Ertekin, T.; Obut, S.T. The unsteady-state nature of sorption and diffusion phenomena in the micropore structure of coal: Part 1-theroy and mathematical formulation. *SPE Form. Eval.* **1990**, *5*, 81–97. [CrossRef]

25. Yuan, W.; Pan, Z.; Li, X.; Yang, Y.; Zhao, C.; Connell, L.D.; Li, S.; He, J. Experimental study and modelling of methane adsorption and diffusion in shale. *Fuel* **2014**, *117*, 509–519. [CrossRef]

26. Jia, B.; Tsau, J.S.; Barati, R. Evaluation of core heterogeneity effect on pulse-decay experiment. In Proceedings of the International Symposium of the Society of Core Analysts, Vienna, Austria, 27 August–1 September 2017.

27. Jia, B.; Tsau, J.S.; Barati, R. Experimental and numerical investigations of permeability in heterogeneous fractured tight porous media. *J. Nat. Gas Sci. Eng.* **2018**, *58*, 216–233. [CrossRef]

28. Han, G.; Sun, L.; Liu, Y.; Zhou, S. Analysis method of pulse decay tests for dual-porosity cores. *J. Nat. Gas Sci. Eng.* **2018**, *59*, 274–286. [CrossRef]

29. Haydel, J.J.; Kobayashi, R. Adsorption equilibria in the Methane-Propane-Silica gel system at high pressures. *Ind. Eng. Chem. Fundam.* **1967**, *6*, 546–554. [CrossRef]

30. Ding, D.Y.; Farah, N.; Bourbiaux, B.; Wu, Y.S.; Mestiri, I. Simulation of matrix/fracture interaction in low-permeability fractured unconventional reservoirs. *SPE J.* **2018**, *23*, 1–389. [CrossRef]

31. Noetinger, B.; Estebenet, T.; Landereau, P. A direct determination of the transient exchange term of fractured media using a continuous time random walk method. *Transp. Porous Media* **2001**, *44*, 539–557. [CrossRef]

Article

Numerical Simulation of Gas Production from Gas Shale Reservoirs—Influence of Gas Sorption Hysteresis

Jamiu M. Ekundayo * and Reza Rezaee

Discipline of Petroleum Engineering, Western Australian School of Mines: Minerals, Energy and Chemical Engineering, Curtin University, 26 Dick Perry Avenue, Kensington, WA 6151, Australia
* Correspondence: Jamiu.Ekundayo@student.curtin.edu.au

Received: 17 July 2019; Accepted: 2 September 2019; Published: 4 September 2019

Abstract: The true contribution of gas desorption to shale gas production is often overshadowed by the use of adsorption isotherms for desorbed gas calculations on the assumption that both processes are identical under high pressure, high temperature conditions. In this study, three shale samples were used to study the adsorption and desorption isotherms of methane at a temperature of 80 °C, using volumetric method. The resulting isotherms were modeled using the Langmuir model, following the conversion of measured excess amounts to absolute values. All three samples exhibited significant hysteresis between the sorption processes and the desorption isotherms gave lower Langmuir parameters than the corresponding adsorption isotherms. Langmuir volume showed positive correlation with total organic carbon (TOC) content for both sorption processes. A compositional three-dimensional (3D), dual-porosity model was then developed in GEM® (a product of the Computer Modelling Group (CMG) Ltd., Calgary, AB, Canada) to test the effect of the observed hysteresis on shale gas production. For each sample, a base scenario, corresponding to a "no-sorption" case was compared against two other cases; one with adsorption Langmuir parameters (adsorption case) and the other with desorption Langmuir parameters (desorption case). The simulation results showed that while gas production can be significantly under-predicted if gas sorption is not considered, the use of adsorption isotherms in lieu of desorption can lead to over-prediction of gas production performances.

Keywords: organic-rich shale; gas adsorption and desorption; sorption hysteresis; Langmuir model; compositional 3D; dual-porosity system; total organic carbon (TOC); Computer Modelling Group (CMG); GEM®

---

1. Introduction

Gas production from shale rocks has gained attention worldwide due to advances in hydraulic fracturing and multi-lateral drilling technologies [1,2]. Shale rocks have affinities for gas storage in the internal surface areas of their pore structures, particularly their organic matter pores, and their natural fractures [3,4]. Therefore, gas adsorption and desorption mechanisms must be considered during shale gas reserves evaluations and shale gas production predictions respectively [5–7]. Research has found that gas production can be significantly under-predicted if "adsorption" is not accounted for in the calculation [1,2,8,9]. The word "adsorption" in this context signals a traditional practice of using adsorption isotherms to obtain desorbed gas volumes during gas production calculations [1,2,8,9]. This practice is based on the assumption that both sorption processes follow the reversible monolayer adsorption theory underpinning the Langmuir model [10], which is arguably the standard model for shale gas adsorption isotherms [2,11]. This assumption invariably ignores the hysteresis behavior of shale gas adsorption and desorption isotherms at in situ conditions.

Sorption hysteresis is the term used to describe the difference between adsorption and desorption isotherms for a given adsorbate-adsorbent system. As the pressure is lowered to produce gas, adsorbed gas molecules in the pores and surface of the adsorbent begin to desorb. Where the adsorption process is characterized by capillary condensation, the amount of desorbed gas at a given pressure is often different from the amount initially adsorbed at the same pressure [12–14]. Different hysteresis behavior have been reported, at conditions beyond which capillary condensation can occur, mostly through experimental studies, in methane-coal, carbon dioxide-coal and nitrogen-coal systems [7,15–17]. However, very few literatures exist on shale gas desorption isotherms and by implication, limited resources exist on shale gas sorption hysteresis in spite of the large volume of published work on shale gas adsorption isotherms [7,16,18–20]. While it is not exactly clear why shale gas desorption is not so widely studied like adsorption, despite that it is clearly understood that desorption is akin to shale gas production, it can be speculated that the assumption that the Langmuir isotherm is valid for modeling shale gas sorption processes invariably makes researchers assume that the desorption isotherm should be exactly the same as the adsorption isotherm. Hence, it is the norm to use adsorption isotherm data for desorbed gas volume during shale gas production calculations [1,2,8,9].

Although it is well-acknowledged that gas desorption is one of the multiple complex flow mechanisms characterizing gas production from shale reservoirs [1,2,6,8,11], the true contribution of gas desorption is often masked by the assumption of reversible sorption isotherms. Despite the published results on the contributions of "adsorption" to shale gas production, we are not aware of any literature that explicitly discussed or quantified the amount of gas over-predicted by using adsorption isotherms in lieu of desorption isotherms for gas production predictions. Hence, the main objective of this work is to demonstrate that the existence of sorption hysteresis means different model parameters and consequently, different gas production performances for adsorption and adsorption isotherms. Assuming adsorption equates desorption can result in significant over-prediction of gas production performances.

2. Samples and Methods

2.1. Samples

The samples used in this study are clay-rich shale samples from the Canning Basin in Western Australia. Sample 1 has a total organic carbon (TOC) content of 1.26 wt% and total clay content of 83.49%, sample 2 has a TOC content of 3.20 wt% and total clay content of 74.28%, while sample 3 has a TOC content 2.82 wt% and total clay content of 67.52%. The mineralogical and geochemical properties of the samples are summarized in Tables 1 and 2, respectively.

Table 1. Mineralogical composition of study samples.

Sample ID	Depth (m)	Quartz (%)	K-feldspar (%)	Plagioclase (%)	Kaolinite (%)	Illite/Mica (%)	Chlorite (%)	Calcite (%)	Pyrite (%)
Sample 1	1390	12.42	1.24	1.84	1.54	67.45	14.50	0.25	0.77
Sample 2	1473	17.39	1.78	3.54	7.14	61.27	5.87	0.42	2.59
Sample 3	1478	20.11	2.63	4.21	1.26	56.19	10.07	3.93	1.60

Table 2. Geochemical properties of study samples. TOC: total organic carbon.

Sample ID	TOC (wt%)	S1 (mg/g)	S2 (mg/g)	S3 (mg/g)	T_{max} (°C)	Ro (%)
Sample 1	1.26	0.63	2.43	0.28	454	1.01
Sample 2	3.20	2.12	7.55	0.51	454	1.01
Sample 3	2.82	1.57	4.66	0.43	456	1.05

The parameters S1, S2 and S3 in Table 2 represent the amount of hydrocarbon and non-hydrocarbon compounds liberated during Rock-Eval 6® (a product of Vinci Technologies, France) pyrolysis of the shale samples [21–23]. S1 is the amount of free hydrocarbon released from the kerogen at 300 °C prior to thermal cracking [21–23] and it corresponds to the first peak detected by the flame ionizing detector (FID) in the Rock-Eval 6® equipment [22,23]. S2, the second peak detected by the FID at a temperature, T_{max}, is the amount of hydrocarbon generated from the thermal cracking of the kerogen and heavy hydrocarbons present in the rock samples [21–23]. S3, the third peak picked by the FID, is the volume of carbon dioxide generated from thermal cracking of kerogen at about 390 °C [21–23]. Combined with the TOC content, these parameters are useful in assessing the hydrocarbon-generating potential of a rock sample. They also give insights, through several indices such as the vitrinite reflectance (Ro), hydrogen index, oxygen index and productivity index, into the type and level of thermal maturity of the organic matters present in the rock sample [21–23].

2.2. Measurements and Modeling of Methane Adsorption and Desorption Isotherms

A high-pressure volumetric analyzer (HPVA-II 200®) purchased from Particulate Systems (a division of Micromeritics, Norcross, GA, United States) was used to measure the methane adsorption and desorption capacities of the shale samples at reservoir temperature (80 °C). The equipment measures gas adsorption and desorption isotherms on pulverized samples using volumetric method, and using helium expansion technique for measuring void volumes. It is equipped with a Microsoft Excel®-based software that automatically computes the isotherms using fluid data obtained from refprop® software. Details of the formulae used in calculating the isotherms and the dependence of the isotherms on equation of state have been discussed in our previous studies [18,24]. Thus, based on the findings from those studies, the equation of state by Soave-Benedict-Webb-Rubin (SBWR-EOS) [25] was adopted to calculate the gas compressibility factors for both helium (for free-space volume calculation) and methane in this study.

For each sample, the measured excess ad/desorbed amounts (V_{exc}) were converted to absolute amounts (V_{abs}) using Equation (1) below. The resulting absolute adsorption was then described with Langmuir model (Equation (2)) for application in gas production calculations:

$$V_{exc} = V_{abs}\left(1 - \frac{\rho_{bulk}}{\rho_{ads}}\right) \tag{1}$$

$$V_{abs} = \frac{V_L P}{P + P_L} \tag{2}$$

A constant adsorbed phase density (ρ_{ads}), obtained as the value of the bulk gas density (ρ_{bulk}) at which the measured excess adsorption isotherm is zero [26], was applied for the conversion of excess sorption isotherms to absolute isotherms. At zero excess adsorption, the adsorbed gas is identical to and indistinguishable from the free bulk gas, and as such, the adsorbed phase density at saturation can be taken as the density of the bulk gas at this point [26]. Thus, adsorbed phase density at saturation was achieved, in this study, by plotting the excess adsorption isotherm as a function of bulk gas density and then extrapolating the decreasing linear trend of the excess adsorption isotherm to the bulk gas density axis [26–28].

The parameters, V_L and P_L in Equation (2) represent the Langmuir volume and pressure, respectively, while P represents equilibrium pressure.

2.3. Numerical Simulation of Shale Gas Production

To illustrate the influence of adsorption and desorption on shale gas production, a pure component (100% methane), single-phase, three-dimensional, dual-porosity compositional model was developed in GEM® based on the "Fractured Gas Reservoir with DUALPOR" example dataset (GMFRR006.DAT) provided by the Computer Modelling Group Ltd. [29]. Table 3 contains a list of the key reservoir and well parameters used for the base case (no-sorption case). For each sample, an adsorption case (where the Langmuir parameters obtained from the adsorption isotherms were used with the assumption of fully reversible sorption isotherms) and a desorption case (where the Langmuir parameters obtained from the desorption isotherm were used) were compared against the no-sorption case. The "ADGMAXC" and "ADGCSTC" keywords in GEM [29] were used to define the Langmuir volume and the reciprocal of Langmuir pressure for the two comparison cases. Langmuir parameters were taken from Table 4. Natural fractures were assumed to be uniformly spaced and run perpendicular to the I- and J-directions through the entire reservoir thickness.

Table 3. Model parameters used for base case (no-sorption case).

Parameter	Value
Reservoir area, ft^2	1378 by 1378
Reservoir thickness, ft	66
Reservoir pressure, psi	2750
Reservoir temperature, °F	176
Initial gas saturation, %	100
Matrix porosity, fraction	0.04
Fracture porosity, fraction	0.001
Matrix permeability, mD	1×10^{-5}
Fracture permeability, mD	0.001
Fracture spacing, ft	26
Number of layers	1
Number of wells	1
Wellbore radius, ft	0.12
Minimum flowing bottom-hole pressure, psi	350
Compressibility factor, psi^{-1}	1×10^{-6}
Rock density, g/cc	2.65
Duration, year	10

Table 4. Langmuir parameters for methane adsorption and desorption isotherms.

Sample	Adsorbed Phase Density, Kg/m^3	Adsorption		Desorption	
		V_L, Scf/ton	P_L, Psi	V_L, Scf/ton	P_L, Psi
Sample 1	118	75	379	65	83
Sample 2	105	151	924	104	202
Sample 3	86	105	508	98	195

3. Results and Discussion

3.1. Methane Adsorption and Desorption Isotherms

Figure 1 shows the excess adsorption and desorption isotherms of methane on all three samples. At each pressure, samples 2 and 3 adsorbed higher amounts of methane than sample 1 at approximately the same equilibrium pressure. This is expected given that both samples have higher TOC contents and lower clay contents than sample 1. However, samples 2 and 3 adsorbed approximately equal amounts up to about 500 psi, pressure beyond which the amount adsorbed by sample 2 becomes greater than those of sample 3. This could be due to the counteracting effects of TOC contents and clay contents. Sample 2 has higher TOC and should have adsorbed more than sample 3, but its higher clay content seems to have counteracted the expected effect than TOC content at pressures up to 500 psi. Additionally, for each sample, the excess adsorbed amounts increased up to a maximum and then decreased slightly afterwards. This is a typical characteristic of the excess adsorption of supercritical fluids on solids [28]. The maximum excess adsorption occurs at a certain pressure where the rate of densities of the adsorbed and free phases is changing at the same rate with pressure [26,28]. Moreover, significant hysteresis can be seen between the adsorption and desorption isotherms for each sample. It is believed that the primary reason for the observed large hysteresis between the sorption isotherms for each sample is the equation of state used in calculating volume changes during the sorption processes [18]. It has been reported that both the size and type of hysteresis between methane adsorption and desorption isotherms are dependent on the applied equation of state [18].

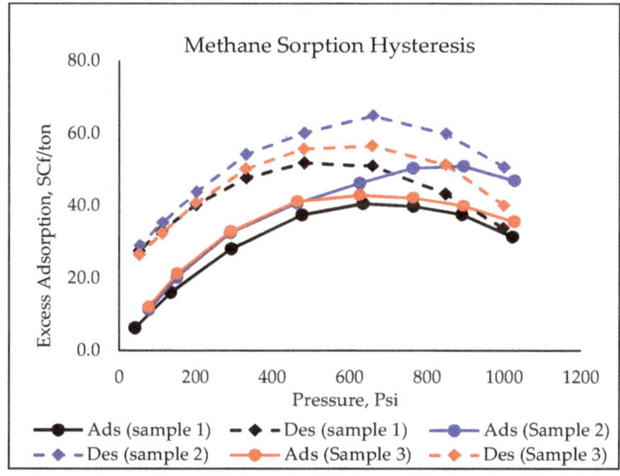

Figure 1. Shale gas adsorption and desorption isotherms measured at 80 °C (176 °F).

Figure 2 shows the excess adsorption isotherm plotted as a function of the bulk gas density for each sample. The dashed lines in the figure indicate the linear "trend-line" fits extrapolated to the bulk-density axis. The density at this point represents the adsorbed phase density at saturation and is equivalent to the ratio of the vertical intercept (of the trend-line) to the slope of the line. While some researchers have argued that this method of estimating the adsorbed phase density can sometimes over-estimate the parameter [30,31], there is no agreement on which method works best. Besides being relatively easier to implement than most of the other known methods, such as the molecular simulation techniques and the adsorbed volume mapping (AVM) method which are generally computationally intensive [32], published data have revealed that the method adopted in this study gives results comparable to other methods such as the three-parameter Langmuir model and Dubinin-Radushkevitch (D-R) equation [30]. While both the three-parameter Langmuir and D-R isotherms have been reported to adequately fit excess adsorption, both have been found to sometimes give unphysical values of the adsorbed phase density [26,30,31]. Moreover, the use of three-parameter Langmuir fit has been reported to be inappropriate at not so high pressures [26], which is the case in this study. As shown in the example (Figure 3) below, while the three-parameter Langmuir model could fit the excess adsorption isotherm, it resulted in low adsorbed phase density leading to high values of Langmuir parameters.

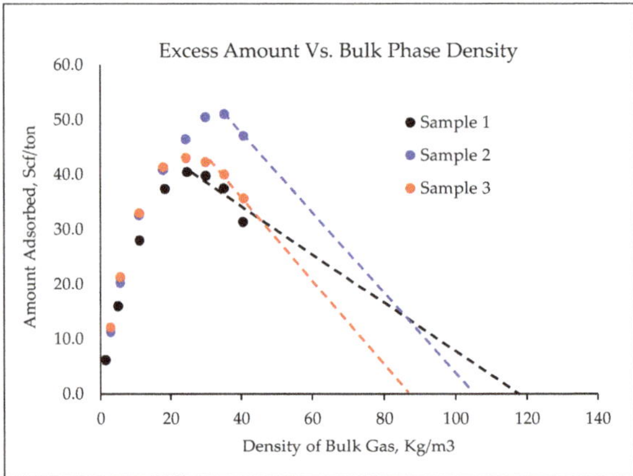

Figure 2. Estimation of adsorbed phase density.

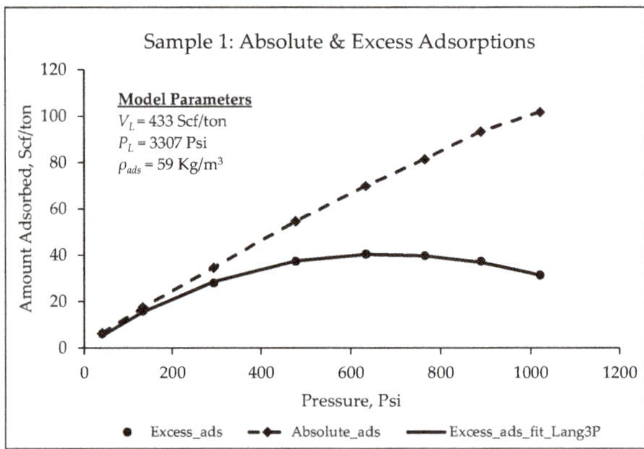

Figure 3. Example of fitting excess adsorption with three-parameter Langmuir model.

The adsorbed phase density obtained from Figure 2 was used to convert each sample's excess adsorption and desorption isotherms to absolute sorption isotherms using Equation (1). The resulting absolute isotherms were then fitted to Langmuir model (Equation (2)) [10]. Table 4 summarizes the results of the Langmuir fit to the isotherms and the values of the adsorbed phase densities used in converting the isotherms from excess to absolute. For all three sample, Langmuir volumes and pressures obtained from the desorption isotherms are lower than the corresponding values obtained from adsorption isotherms. As shown in Figure 4, the Langmuir volumes for both adsorption and desorption processes correlate positively with TOC content. Such a relationship between Langmuir volume and TOC content has also been reported in the literature [33,34].

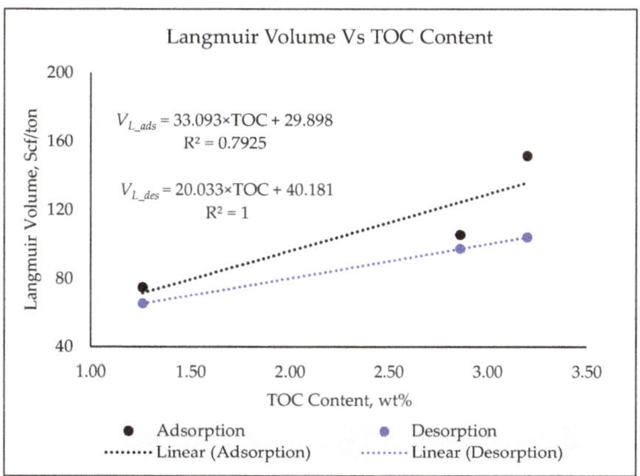

Figure 4. Relationship between Langmuir volumes and TOC content.

3.2. Simulation Results

The results of the numerical simulation with and without sorption parameters are shown in Figure 5 for all three samples. For each sample, the base case gave the lowest gas production rate and 10-year cumulative gas production compared to the other two cases. This confirms the importance of gas sorption parameters on shale gas production calculations. For each sample, the adsorption case gave the highest production rate and cumulative gas production after 10 years resulting in 7%, 18% and 11% additional gas production for samples 1, 2 and 3 respectively (Figure 6). This agrees with existing literatures on the effect of adsorption on shale gas production [1,2,8,9] and also correlates with the TOC contents of the samples. However, the use of adsorption model parameters assumes that the sorption isotherms are fully reversible and no hysteresis exists between them. Such an assumption can lead to a significant over-prediction of gas production depending on the size of the observed hysteresis. Although the desorption case gave higher gas production rate and cumulative gas production than the base case for each sample, its production rate and cumulative gas production are significantly lower than the corresponding values from the adsorption case (Figure 5). The difference between the cumulative gas productions from these two sorption cases was quantified as the effect of sorption hysteresis on gas production. As shown in Figure 6, the assumption of reversible isotherms would result in the over-prediction of the cumulative gas productions, after 10 years, by 5% for each of samples 1 and 3 and by 12% for sample 2.

It follows from the above discussion that while gas sorption generally has a positive effect on gas production from shale reservoirs, the effects are often exaggerated by the assumption of reversible sorption isotherms. Where hysteresis exists between adsorption and desorption isotherms, the (Langmuir) model parameters are usually lower for desorption isotherms than for the adsorption counterparts and the differences will directly translate to lower gas production with the desorption parameters.

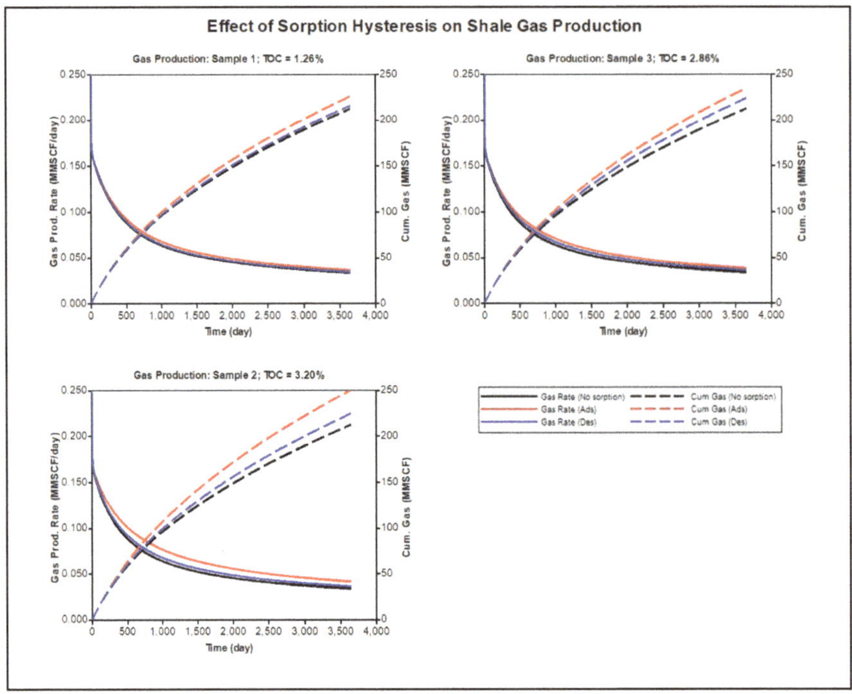

Figure 5. Comparison of shale gas production simulation results.

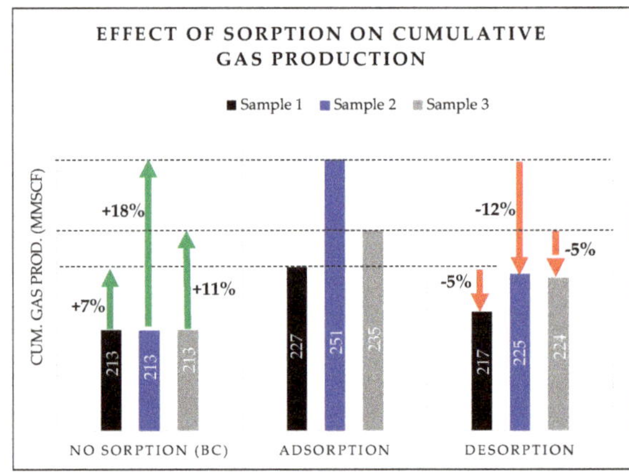

Figure 6. Summary of effects of adsorption and sorption hysteresis on cumulative gas production.

4. Study Limitations

While this study offers insights into the effect of sorption hysteresis on gas production, the following limitations are acknowledged:

1. First, this study included a limited number of samples, which, although revealed the expected trend between Langmuir volume (and hence, gas production) and TOC content, may be too small to make adequate statistical correlation of gas production with TOC content. However, given the positive relationships observed between Langmuir volume and TOC contents (Figure 4)

as well as the positive correlation reported between gas production and Langmuir volume [9], it is envisaged that the findings reported in this study, will be valid irrespective of the number of samples.

2. Additionally, the method employed in this study to calculate the adsorbed phase density uses a linear equation to fit the last few data points, following the maximum excess adsorption, of each isotherm. Thus, the value of the adsorbed phase density obtained depends on the linearity of these data points. For example, as shown in Figure A1 (Appendix A), if the last data point is included for sample 1 the resulting adsorbed phase density becomes lower than the value reported in Table 4. With this lower adsorbed phase density, the calculated Langmuir parameters become higher than they currently are and so are the cumulative gas productions for the two comparison cases. However, these values are still lower than the corresponding values for samples 2 and 3 and as such, the observed trend with TOC content remains unchanged.

3. Finally, the results of the numerical simulation presented in this study are focused only on the effects of sorption processes. In reality, shale gas transport is a multiphysics process and the inclusion of other flow mechanisms may affect the simulation outputs. However, in keeping with the objectives of this study, it was necessary to keep other parameters constant to isolate the effect of sorption parameters and hysteresis on gas production. Several publications exist that adopted similar approach in their studies [1,2,8,9].

5. Conclusions

This study presented the effect of gas adsorption and sorption hysteresis on shale gas production. The results showed that gas desorption might not necessarily follow the same path as gas adsorption, and the sorption isotherms may exhibit significant hysteresis. In such a case, desorption model parameters are much lower than the corresponding adsorption model parameters. Consequently, using adsorption model parameters to calculate desorbed gas volumes during gas production could lead to significant over-prediction of gas production performances. In summary, it can be concluded that:

1. Significant hysteresis was observed between adsorption and desorption isotherms of methane on all three samples under high-pressure, high-temperature conditions.

2. The desorption isotherms gave lower Langmuir parameters compared to the adsorption counterparts.

3. Langmuir volumes for both adsorption and desorption processes showed positive correlation with TOC content.

4. Sorption has a significant positive influence on gas production and as such, neglecting the gas sorption during gas production predictions can lead to under-estimation of gas production performances.

5. The additional gas production due to gas sorption consideration in the gas production calculations increased with TOC content. This is expected given the positive correlation between Langmuir volumes and TOC contents.

6. The use of adsorption Langmuir parameters during gas production calculations can lead to over-prediction of the gas production performances.

Author Contributions: Conceptualization, J.M.E.; Data curation, J.M.E.; Formal analysis, J.M.E.; Investigation, J.M.E.; Methodology, J.M.E.; Resources, R.R.; Supervision, R.R.; Writing—original draft, J.M.E.; Writing—review & editing, R.R.

Funding: This research received no external funding.

Acknowledgments: The authors would like to acknowledge the contributions of Australian Government Research Training Program and Curtin Research Scholarships and the Unconventional Gas Research group at the discipline of Petroleum Engineering, Western Australian School of Mines: Minerals, Energy and Chemical Engineering in supporting this research. The authors would also like to acknowledge the support of Western Australia's Department of Mines, Industry Regulation and Safety and Finder Energy in providing the shale samples used in this study. Pooya Hadian, Lukman M. Johnson and Jie Zou are also acknowledged for their efforts in the initial characterization of the samples.

Conflicts of Interest: The authors declare no conflict of interest.

Appendix A

Figure A1 below illustrates one of the limitations of this study, which is about the method used to determine the adsorbed phase densities for conversion of excess adsorption to absolute amounts.

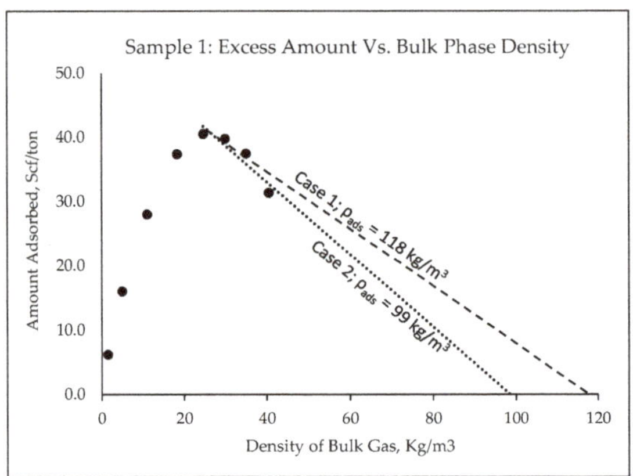

Figure A1. Possible uncertainty in adsorbed phase density determined from excess adsorption versus bulk density plot.

The different between the two cases in the figure lies in the inclusion (or otherwise) of the last data point in the linear fitting of the excess adsorption—bulk density plot post saturation. The Langmuir parameters obtained from these two cases are different, with the case 2 resulting in higher Langmuir parameters than case 1 as shown in Table A1 below:

Table A1. Variations in Langmuir parameters and cumulative gas production between cases 1 and 2.

Case ID	Adsorption			Desorption		
	V_L Scf/ton	P_L Psi	Cum Gas Prod MMScf	V_L Scf/ton	P_L Psi	Cum Gas Prod MMScf
Case 1	75	379	227	65	83	217
Case 2	87	462	231	72	101	218

The 10-year cumulative gas productions obtained from case 2 is higher than case 1 due to its higher Langmuir volumes for both sorption processes. However, these values are still lower than the cumulative gas productions for the other two samples, which is consistent with our conclusions.

References

1. Yu, W.; Huang, S.; Wu, K.; Sepehrnoori, K. Development of A Semi-Analytical Model for Simulation of Gas Production in Shale Gas Reservoirs. In Proceedings of the Unconventional Resources Technology Conference, Society of Exploration Geophysicists, American Association of Petroleum Geologists, Society of Petroleum Engineers, Denver, CO, USA, 25–27 August 2014; pp. 2187–2204.
2. Feast, G.; Wu, K.; Walton, J.; Cheng, Z.F.; Chen, B. Modeling and Simulation of Natural Gas Production from Unconventional Shale Reservoirs. *Int. J. Clean Coal Energy* **2015**, *4*, 23–32. [CrossRef]
3. Curtis, J.B. Fractured shale-gas systems. *AAPG Bull.* **2002**, *86*, 1921–1938.
4. Leahy-Dios, A.; Das, M.; Agarwal, A.; Kaminsky, R.D. Modeling of Transport Phenomena and Multicomponent Sorption for Shale Gas and Coalbed Methane in an Unstructured Grid Simulator. In Proceedings of the SPE ATCE, Denver, CO, USA, 30 October–2 November 2011; pp. 1–9.

5. Wang, J.; Dong, M.; Yang, Z.; Gong, H.; Li, Y. Investigation of Methane Desorption and Its Effect on the Gas Production Process from Shale: Experimental and Mathematical Study. *Energy Fuels* **2016**, *31*, 205–216. [CrossRef]

6. Wang, J.; Luo, H.; Liu, H.; Cao, F.; Li, Z.; Sepehrnoori, K. An Integrative Model To Simulate Gas Transport and Production Coupled With Gas Adsorption, Non-Darcy Flow, Surface Diffusion, and Stress Dependence in Organic-Shale Reservoirs. *SPE J.* **2015**, *22*, 244–264. [CrossRef]

7. Wei, G.; Xiong, W.; Gao, S.; Hu, Z.; Liu, H.; Yu, R. Impact of temperature on the isothermal adsorption/desorption characteristics of shale gas. *Pet. Explor. Dev.* **2013**, *40*, 514–519.

8. Wang, W.; Yu, W.; Hu, X.; Liu, H.; Chen, Y.; Wu, K.; Wu, B. A semianalytical model for simulating real gas transport in nanopores and complex fractures of shale gas reservoirs. *AICHE J.* **2018**, *64*, 326–337. [CrossRef]

9. Zhang, W.; Xu, J.; Jiang, R. Production forecast of fractured shale gas reservoir considering multi-scale gas flow. *J. Pet. Explor. Prod. Technol.* **2017**, *7*, 1071–1083. [CrossRef]

10. Langmuir, I. The adsorption of gases on plane surfaces of glass, mica and platinum. *J. Am. Chem. Soc.* **1918**, *40*, 1361–1403. [CrossRef]

11. Yu, W.; Sepehmoori, K.; Patzek, T.W. Evaluation of Gas Adsorption in Marcellus Shale. In Proceedings of the SPE ATCE, Amsterdam, The Netherlands, 27–29 October 2014; pp. 1–16.

12. Rajniak, P.; Yang, R.T. A Simple Model and Experiments for Adsorption-Desorption Hysteresis: Water Vapor on Silica Gel. *AICHE J.* **1993**, *39*, 774–786. [CrossRef]

13. Kierlik, E.; Monson, P.A.; Rosinberg, M.L.; Tarjus, G. Adsorption hysteresis and capillary condensation in disordered porous solids: A density functional study. *J. Phys. Condens. Matter* **2002**, *14*, 9295–9315. [CrossRef]

14. Monson, G. A model of adsorption-desorption hysteresis in which hysteresis is primarily developed by the interconnections ina network of pores. *Proc. R. Soc. Lond. Ser. A Math. Phys. Sci.* **1983**, *390*, 47–72.

15. Bell, G.J.; Rakop, K.C. Hysteresis of Methane/Coal Sorption Isotherms. In Proceedings of the 61st SPE-ATCE, New Orleans, LA, USA, 5–8 October 1986.

16. Wei, G.; Hu, Z.; Zhang, X.; Yu, R.; Wang, L. Shale Gas Adsorption and Desorption Characteristics and its Effects on Shale Permeability. *Energy Explor. Exploit.* **2017**, *35*, 463–481.

17. Zhang, R.; Liu, S. Experimental and theoretical characterization of methane and CO_2 sorption hysteresis in coals based on Langmiur desorption. *Int. J. Coal Geol.* **2016**, *171*, 49–60. [CrossRef]

18. Ekundayo, M.J.; Rezaee, R. Volumetric Measurements of Methane-Coal Adsorption and Desorption Isotherms—Effects of Equations of State and Implication for Initial Gas Reserves. *Energies* **2019**, *12*, 2022. [CrossRef]

19. Bhowmik, S.; Dutta, P. A study on the effect of gas shale composition and pore structure on methane sorption. *J. Nat. Gas Sci. Eng.* **2019**, *62*, 144–156. [CrossRef]

20. Zhang, Y.; Xing, W.; Liu, S.; Liu, Y.; Yang, M.; Zhao, J.; Song, Y. Pure methane, carbon dioxide, and nitrogen adsorption on anthracite from China over a wide range of pressures and temperatures: Experiments and modeling. *RSC Adv.* **2015**, *5*, 52612–52623. [CrossRef]

21. El Nady, M.M.; Ramadan, F.S.; Hammad, M.M.; Lotfy, N.M. Evaluation of organic matters, hydrocarbon potential and thermal maturity of source rocks based on geochemical and statistical methods: Case study of source rocks in Ras Gharib oilfield, central Gulf of Suez, Egypt. *Egypt. J. Pet.* **2015**, *24*, 203–211. [CrossRef]

22. Peters, K.E. Guidelines for Evaulating Petroleum Source Rock Using Programmed Pyrolysis. In *American Chemical Society, Symposium on Organic Geochemistry of Humic Substances, Kerogen and Coal*; American Association of Petroleum Geologists: Philadelphia, PA, USA, 1985.

23. Lafarge, E.; Marquis, F.; Pillot, D. Tock-Eval 6 Applications in Hydrocarbon Exploration, Production, and Soil Contamination Studies. *Rev. De L'institut Français Du Pétrole* **1998**, *53*, 421–437.

24. Ekundayo, J.; Rezaee, R. Effect of Equation of States on High Pressure Volumetric Measurements of Methane-Coal Sorption Isotherms—Part 1: Volumes of Free Space and Methane Adsorption Isotherms. *Energy Fuels* **2019**, *33*, 1029–1036. [CrossRef]

25. Soave, G.S. An effective modification of the Benedict-Webb-Rubin equation of state. *Fluid Phase Equilibria* **1999**, *164*, 157–172. [CrossRef]

26. Zhang, J.; Clennell, M.B.; Liu, K.; Pervukhina, M.; Chen, G.; Dewhurst, D.N. Methane and Carbon Dioxide Adsorption on Illite. *Energy Fuels* **2016**, *30*, 10643–10652. [CrossRef]

27. Moellmer, J.; Moeller, A.; Dreisbach, F.; Glaeser, R.; Staudt, R. High pressure adsorption of hydrogen, nitrogen, carbon dioxide and methane on the metal–organic framework HKUST-1. *Microporous Mesoporous Mater.* **2011**, *138*, 140–148. [CrossRef]

28. Do, D.D.; Do, H.D. Adsorption of argon from sub- to supercritical conditions on graphitized thermal carbon black and in graphitic slit pores: A grand canonical Monte Carlo simulation study. *J. Chem. Phys.* **2005**, *123*, 084701. [CrossRef] [PubMed]

29. Computer Modelling Group. *GEM Compositional & Unconventional Reservoir Simulator*; USER GUIDE C.M.G. Ltd.: Calgary, AB, Canada, 2016.

30. Zhou, S.; Xue, H.; Ning, Y.; Guo, W.; Zhang, Q. Experimental study of supercritical methane adsorption in Longmaxi shale: Insights into the density of adsorbed methane. *Fuel* **2018**, *211*, 140–148. [CrossRef]

31. Tian, Y.; Yan, C.; Jin, Z. Characterization of Methane Excess and Absolute Adsorption in Various Clay Nanopores from Molecular Simulation. *Sci. Rep.* **2017**, *7*, 12040. [CrossRef] [PubMed]

32. Murata, K.; El-Merraoui, M.; Kaneko, K. A new determination method of absolute adsorption isotherm of supercritical gases under high pressure with a special relevance to density-functional theory study. *J. Chem. Phys.* **2001**, *114*, 4196–4205. [CrossRef]

33. Gasparik, M.; Bertier, P.; Gensterblum, Y.; Ghanizadeh, A.; Krooss, B.M.; Littke, R. Geological controls on the methane storage capacity in organic-rich shales. *Int. J. Coal Geol.* **2014**, *123*, 34–51. [CrossRef]

34. Zou, J.; Rezaee, R.; Liu, K. Effect of Temperature on Methane Adsorption in Shale Gas Reservoirs. *Energy Fuels* **2017**, *31*, 12081–12092. [CrossRef]

Article

Numerical Analysis of Transient Pressure Behaviors with Shale Gas MFHWs Interference

Dapeng Gao [1,2,3], Yuewu Liu [2,3,*], Daigang Wang [4,*] and Guofeng Han [2]

[1] State Key Laboratory of Oil and Gas Reservoir Geology and Exploitation, Southwest Petroleum University, Chengdu 610500, China; gaodapeng@imech.ac.cn
[2] Institute of Mechanics, Chinese Academy of Sciences, Beijing 100190, China; hanguofeng@imech.ac.cn
[3] School of Engineering Science, University of Chinese Academy of Sciences, Beijing 100049, China
[4] Beijing International Center for Gas Hydrate, Peking University, Beijing 100871, China
* Correspondence: lywu@imech.ac.cn (Y.L.); dgwang@pku.edu.cn (D.W.)

Received: 6 November 2018; Accepted: 9 January 2019; Published: 15 January 2019

Abstract: After the large-scale horizontal well pattern development in shale gas fields, the problem of fast pressure drop and gas well abandonment caused by well interference becomes more serious. It is urgent to understand the downhole transient pressure and flow characteristics of multi-stage fracturing horizontal well (MFHW) with interference. Therefore, the reservoir around the MFHW is divided into three regions: fracturing fracture, Stimulated reservoir volume (SRV), and unmodified matrix. Then, multi-region coupled flow model is established according to reservoir physical property and flow mechanism of each part. The model is numerically solved using the perpendicular bisection (PEBI) grids and the finite volume method. The accuracy of the model is verified by analyzing the measured pressure recovery data of one practical shale gas well and fitting the monitoring data of the later production pressure. Finally, this model is used to analyze the effects of factors, such as hydraulic fractures' connectivity, well distance, the number of neighboring wells and well pattern arrangement, on the transient pressure and seepage characteristics of the well. The study shows that the pressure recovery double logarithmic curves fall in later part when the well is disturbed by a neighboring production well. The earlier and more severe the interference, the sooner the curve falls off and the larger the amplitude shows. If the well distance is closer, and if there are more neighboring wells and interconnected corresponding fracturing segments, the more severe interference appears among the wells. Moreover, the well interference may still exist even without interlinked fractures or SRV. Especially, severe interference will affect production when the hydraulic fractures are connected directly, and the interference is weaker when only SRV induced fracture network combined between wells, which is beneficial to production sometimes. When severe well interference occurs, periodic well shut-in is needed to help restore the reservoir pressure and output capacity. In the meanwhile, the daily output should be controlled reasonably to prolong the stable production time. This research will help to understand the impact of well interference to gas production, and to optimize the well spacing and achieve satisfied performance.

Keywords: shale gas; multi-stage fracturing horizontal wells; well interference; transient pressure; numerical analysis

---

1. Introduction

The multi-stage fracturing horizontal well (MFHW) is a crucial technology in shale gas development, and the large-scale horizontal wells pattern haves achieved remarkable performance in many fields in North America and China [1–5]. However, some well groups have shown increasingly dangerous well interference after producing for several years, due to the well pattern infilling and enhancement of hydraulic fracturing. For instance, The Jiaoshiba shale gas play is the most successful shale gas reservoir in China with some wells' cumulative production over 0.1 billion cubic meters in the first year; unfortunately, parent wells crop up jumps in water production during hydraulic fracturing processing of child wells in the later production period. Besides, North American shale plays, such as Arkoma Basin, have also shown an obvious loss of gas production because of well interference [6,7]. Fracture pressure hits and production interference are two main factors influencing the shale gas permanent development and determining the inter-well connectivity [8].

By a combination of hydraulic fracturing process and production data, the existence of well interference from the adjacent wells when well interference happens and the influence it imposes on the target well can be detected. Maintaining a high and stable gas productivity faces many challenges in the future. It is urgent to study in depth the transient pressure behaviors of shale gas MFHWs with neighboring wells, considering the complex fracture network and the multi-flow mechanism, such as the desorption and diffusion of shale gas [9,10], to recognize the well interference in time and analyze its impact on production. Mezghani et al. [11] combined gradual deformation and upscaling techniques for direct conditioning of fine-scale reservoir models to interference test data, as consequence, both fine- and coarse-scale models are updated by dynamic data during the history matching process. Yaich et al. [12] presented a methodology to quantify the impact of well interference and optimize well spacing in the Marcellus shale. Marongiu-Porcu et al. [13] proposed a numerical simulation method for shale gas reservoirs based on geophysical, completion and development data of Eagle Ford shale gas fields, and studied the propagation of hydraulic fractures and their respective network with natural fractures. The magnitude and orientation of in-situ stress were evaluated. Pang et al. [14] studied the effect of well interference on shale gas well SRV interpretation. Compared with the previous literature, we analyzed the interference using the numerical simulation pressure double logarithmic curve method and evaluated the influence of various factors on the interference.

To study the porous media flow and transient pressure behaviors of shale gas MFHWs, many scholars have established kinds of multi-linear flow region coupled models. Bello et al. [15] used the layered double porosity model and the Warren-Root dual-porosity model to analyze the pressure response and production dynamics of multi-stage fracturing horizontal wells in shale reservoirs. Ozkan [16] and Al-Rbeawi [17] divided the stimulated shale reservoirs into hydraulic fractures, stimulated reservoir volume (SRV), and matrix. Moreover, they simplify the flow in these three regions into the one-dimensional linear flow by establishing a three-linear-flow model. Additionally, Stalgorova [18] and Zhang et al. [19] improved the three-linear-flow model by considering the un-stimulated areas between two hydraulic fracturing SRV and proposed a five-region coupled flow model. Zeng et al. [20] further subdivided the five-region coupled flow model and proposed a seven-region coupled flow model. Based on these coupled models, Wang et al. [21] and Kim et al. [22] analyzed the stress-sensitive effects of gas reservoirs and fractures. Overall, the flow around shale MFHWs is mainly characterized by coupled linear flow models with multiple subdivided regions. In these models, hydraulic fracture (HF), SRV, and matrix are commonly applied, and they are separately discussed as follows. The flow in fracturing fractures usually satisfied Darcy law or high-speed non-Darcy law [23]. The SRV can be treated as dual media, or characterized by complex fracture network models [24]. The matrix can be regarded as a homogenous ultra-low-permeability medium. However, some scholars treat it as a dual medium with natural fracture network [25]. The equivalent permeability can be analyzed in the multiple regions coupled model to simplify the effects of desorption and diffusion in the SRV and matrix. From these three main regions, five regions or seven regions are further subdivided, but the physical parameters in the added regions are difficult to obtain. Well interference may be caused by interaction between primary hydraulic fractures and/or secondary natural fractures activated during hydraulic fracturing. Well interference has had a significant influence on the SRV interpretation. Well interference has drawn people's attention in recent years, but its impact on transient behaviors is rarely reported.

As shown in the previous literatures, Laplace transform and Stehfest method are mainly applied to obtain the analytical solution or semi-analytical solution of transient pressure behaviors. It is principally applicable for the analysis of one single shale gas well, but it's really difficult to solve the transient pressure with multiple well interferences. So it's crucial to investigate a numerical method to analyze the well pattern pressure dynamic. Therefore, to fully understand the transient pressure behaviors with well interference, a three-medium coupled numerical model is given in the paper, considering connected and unconnected HFs, SRV, and matrix. Also, we numerically solve the model by PEBI grids and the finite volume method. Different factors' influences on transient pressure behaviors with well interference are studied. This research may help to characterize well interference and related factors and to optimize well spacing.

2. Model Description

Due to the significant difference in porosity and permeability between the fracturing and un-fracturing areas in the reservoir, shale reservoirs around MFHWs are divided into sub-regions, including hydraulic fractures, SRV with abundant inducing micro-fractures, and matrix as shown in Figure 1. Among them, only SRV region is simplified as dual medium due to micro-fracture development; matrix region is single-porosity single-permeability medium, hydraulic fracture is a high permeability medium. Assumptions: (1) Water flow is ignored, and there is only single-phase gas flow existed in each sub-region. (2) Just viscous flow exists in the HF and satisfies the Darcy law [26], neglecting the longitudinal flow. (3) SRV is treated as dual media, and each hydraulic segment's SRV overlaps with each other in one MFHW. (4) Matrix is regarded as homogeneous ultra-low permeability media. (5) Pseudo pressure function (m) is introduced to simplify the gas composition change with temperature and pressure [27]. (6) There are three connection modes between the well and its neighboring wells, including the connection of the inducing micro-fracture clusters in the SRV and the connection of hydraulic fractures as shown in Figure 1.

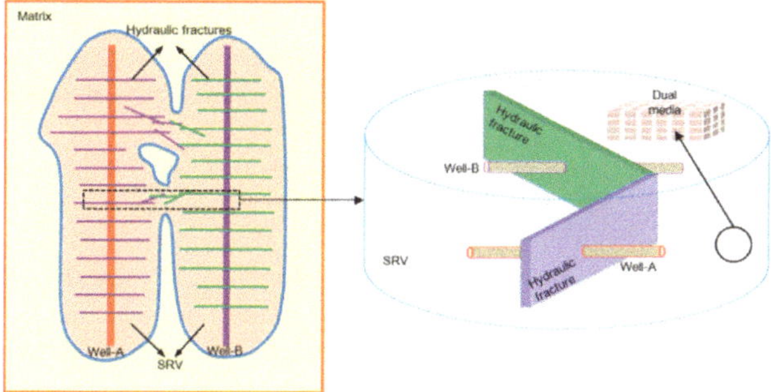

Figure 1. Multi-region coupled shale reservoir physical model (Well-A and well-B are two multi-stage fracturing horizontal wells (MFHWs), the hydraulic fractures connected with the SRV and the SRVs are dual medium.).

3. Mathematical Model

3.1. HF Flow Model

It is assumed that fluid exchanges exist among the fracture, the SRV and the wellbore, the boundary between the well and the HFs is defined as Γ_1, and the boundary between the SRV and the HFs is Γ_2. If the HFs directly connected in one well pair, the fluid exchange between two wells' HFs needs to be considered. Commonly, shale gas wells produce at a given production rate first according to the development scheme, and the gas supply capacity of the reservoir gradually tends to insufficient as the pressure continues to drop. The shale gas wells are converted to produce with constant pressure later. Based on this, the flow equations in the finite conductivity fractures are established as follows:

$$\nabla^2 m_F = -\frac{\phi_F \mu_g c_g}{k_F} \frac{\partial m_F}{\partial t} \tag{1}$$

The initial conditions:

$$m_F|_{t=0} = m_i \tag{1a}$$

Inner boundary condition:
Constant pressure boundary conditions:

$$m_F|_{\Gamma_1} = m_w \tag{1b}$$

Fixed output boundary conditions:

$$\left.\frac{\partial m}{\partial n}\right|_{\Gamma_1} = q_{iw,F} \tag{1c}$$

Outer boundary condition:

$$m_{S,f}|_{\Gamma_2} = m_F|_{\Gamma_2} \tag{1d}$$

3.2. SRV Flow Model

SRV exist around MFHWs in unconventional shale reservoirs, which leads to the flow characteristics of the fluid that are different from those in unstimulated formations. It is necessary to integrate various measurements and surveillance data to build a variable SRV reservoir model. The variable SRV model described here has the following building blocks [28]: (1) Formation evaluation: included all the reservoir characterization data derived from logs and 3D seismic inversions and structural attributes. (2) Surveillance data integration: micro-seismic data are integrated with chemical and radioactive tracer logs. (3) Well performance data integration: Production data is used to determine different flow regimes during the well history and to set bounds for stimulation parameters, such as HF half-length and permeability. (4) Numerical simulation: Micro-seismic attributes (density and magnitude) are converted to a permeability model after being calibrated with tracer logs and production flow regime parameters. pressure, volume, temperature (PVT) data is matched against an Equation of State and input into the model. Due to the abundant micro-fractures induced by hydraulic fracturing in SRV, SRV is regarded as a dual medium containing the matrix and fracture systems.

(1) Matrix system flow model

Assuming that gas desorbed from the SRV matrix system, the desorption gas satisfies the Langmuir isothermal adsorption equation on the surface of the matrix bedrock. The migration of gas includes viscous flow, Knudsen diffusion, and surface diffusion. So, the matrix system flow model is Equation (2):

$$\nabla^2 m_{S,m} - \alpha(m_{S,m} - m_{S,f}) = \frac{\mu_g c_g}{k_{S,m}} \frac{\partial}{\partial t}\left[\phi_{S,m} + (1 - \phi_{S,m})\frac{M_{S,g} V_{S,L} m_{S,L}}{V_{std}(m_{S,L} + m_{S,m})^2}\right] \tag{2}$$

(2) Fracture system flow model

Because the micro-fractures in the SRV region are very developed, how to characterize the fracture network equivalently in the seepage model has been a difficult problem to solve. For this reason, many scholars hypothesize that the development and spread of fracture networks satisfy the fractal characteristics and propose a fractal model that characterizes natural fracture networks [29,30]. However, the critical parameters such as the fractal dimension in the model are difficult to determine.

Also, considering complex networks will greatly increase the complexity of meshing and numerical calculations. Therefore, the equivalent permeability is used to characterize the comprehensive permeability of the fracture system in the SRV region. The gas in the fracture medium in the SRV is mainly in the form of free gas. Therefore, only the viscous gas flow and Knudsen diffusion are considered in the fracture medium, and the apparent permeability $k_{\text{s,f}}$ is used to represent the permeability of the fracture medium [31]:

$$k_{\text{S,f}} = \frac{b^3}{12c}(1 + \beta K_{\text{n}})\left(1 + \frac{4K_{\text{n}}}{1 + K_{\text{n}}}\right) \tag{3}$$

Assume that the wellbore only has fluid exchange with the fracture, neglecting the direct fluid exchange between the fracture system and the wellbore in the SRV, and defining the interface between the fracture and the matrix systems in the SRV as Γ_3. Then the flow model in the fracture system of the SRV is presented as Equation (4):

$$\nabla^2 m_{\text{S,f}} + \alpha(m_{\text{S,m}} - m_{\text{S,f}}) = \frac{\phi_{\text{S,f}}\mu_{\text{g}}c_{\text{g}}}{k_{\text{S,f}}}\frac{\partial m_{\text{S,f}}}{\partial t} \tag{4}$$

The initial conditions:

$$m_{\text{S,m}}\big|_{t=0} = m_{\text{S,f}}\big|_{t=0} = m_{\text{i}} \tag{4a}$$

Inner boundary condition:

$$m_{\text{S,f}}\big|_{\Gamma_2} = m_{\text{F}}\big|_{\Gamma_2} \tag{4b}$$

Outer boundary condition:

$$m_{\text{S,f}}\big|_{\Gamma_3} = m_{\text{M}}\big|_{\Gamma_3} \tag{4c}$$

3.3. Matrix Flow Model

The shale gas reservoir is rich in kerogen organic matter, and the hydrocarbon gas generated in the kerogen satisfies the saturation adsorption and then spreads from the kerogen pores to the inorganic matrix pore space where the hydrocarbon concentration is relatively reduced. The gas in the kerogen occurs in two forms: free gas and adsorbed gas. The pores in the kerogen have the same order of magnitude as the gas molecules in the shale gas. Therefore, the free gas will generate Knudsen Diffusion in the kerogen nanoporous network. At the same time, the kerogen is saturated with a large amount of adsorbed gas, and the adsorbed gas on the surface of the skeleton will produce surface diffusion. Assuming that the shale gas reservoir is isothermally developed, the Langmuir isotherm adsorption equation is used to describe the adsorption and desorption of kerogen.

The apparent permeability of the unmodified Matrix region proposed by Singh et al. [32] and Civan et al. [29,30] is:

$$k_{\text{M,a}} = k_0\left[1 + \frac{128}{15\pi^2}\tan^{-1}\left(4K_{\text{n}}^{0.4}\right)K_{\text{n}}\right]\left[1 + \frac{4K_{\text{n}}}{1 + K_{\text{n}}}\right] \tag{5}$$

Thus, the kerogen-medium continuity equation considering Knudsen diffusion, adsorption-desorption and surface diffusion is obtained as Equation (6):

$$\nabla^2 m_{\text{M}} = \frac{\phi_{\text{M}}\mu_{\text{g}}c_{\text{g}}}{k_{\text{M,a}}}\frac{\partial m_{\text{M}}}{\partial t} \tag{6}$$

The initial conditions:

$$m_{\text{M}}\big|_{t=0} = m_{\text{i}} \tag{6a}$$

Inner boundary condition:

$$m_{\text{S,f}}\big|_{\Gamma_3} = m_{\text{M}}\big|_{\Gamma_3} \tag{6b}$$

Outer boundary condition:

$$\frac{\partial m_{\mathrm{M}}}{\partial n}\bigg|_{\Gamma_4} = 0 \tag{6c}$$

3.4. Model Solution

Accuracy and efficiency of reservoir simulators in complex systems depend highly upon a proper grid selection. Grids based on a cartesian coordinate system have been widely used, but have some disadvantages: (a) Flexibility in the description of faults, pinchouts, hydraulic fractures, horizontal wells and general discontinuities presented in reservoirs; (b) inflexibility in representing well locations; and (c) suffer from grid orientation effects. PEBI grids have been applied to the oil industry for about a decade. On the other hand, generation and construction of PEBI grids are not as easy as cartesian grids. The construction of PEBI grids for a reservoir is feasible only if it is done by a numerical grid generation procedure. These PEBI grids are locally orthogonal. It means the block boundaries are normal to lines joining the nodes on the two sides of each boundary. This allows a reasonable accurate computation of inter-block transmissibility for heterogeneous but isotropic permeability distribution.

The irregular geologic body boundary can be depicted by PEBI grids. In this paper, the unstructured PEBI grid is applied to mesh the solution area and carry on local grid refinement around MFHWs, in which the connection between the center node of each grid and the adjacent grid center node is perpendicular to the interface, as shown in Figure 2.

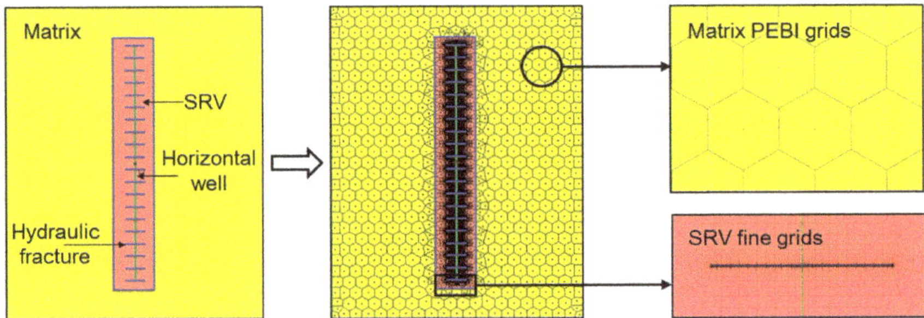

Figure 2. Perpendicular bisection (PEBI) meshing grids of HFs and matrix.

Finding a pressure solution for large-scale reservoirs that takes into account fine-scale heterogeneities can be very computationally intensive. One way of reducing the workload is to employ multi-scale methods that capture local geological variations using a set of reusable basis functions. One of these methods, the multi-scale finite-volume (MsFV) method is well studied for 2D Cartesian grids, but has not been implemented for stratigraphic and unstructured grids with faults in 3D. With reservoirs and other geological structures spanning several kilometers, running simulations on the meter scale can be prohibitively expensive in terms of time and hardware requirements. Multiscale methods are a possible solution to this problem, and extending the MsFV method to realistic grids is a step on the way towards fast and accurate solutions for large-scale reservoirs.

Moyner and Lie [33] presented an MsFV solver along with a coarse partitioning algorithm that can handle stratigraphic grids with faults and wells. The solver is an alternative to traditional upscaling methods, but can also be used for accelerating fine-scale simulations. In this paper, we use MsFV to discretize the reservoir area into non-overlapping control volumes, and each center node has a controlled volume around it. According to the seepage equation and boundary conditions of multiple composite flow models, the partial differential equations to be solved are integrated for each control volume, and a set of discrete equations is obtained.

Based on the flow control Equation (1) in the fracturing fracture, the control volume V of this point mesh is integrated to obtain Equation (7):

$$\int_V \nabla^2 m_F \, dV = -\left(\frac{\phi_F \mu_g c_g}{k_F}\right) \int_V \frac{\partial m_g}{\partial t} \, dV \tag{7}$$

Use the Gaussian formula to convert the volume fraction of the left-side diffusion term of Equation (7) into area fractions, as Equations (8) and (9):

$$\oint_S \nabla m_F \, dA = -\left(\frac{\phi_F \mu_g c_g}{k_F}\right) \frac{\partial}{\partial t} \int_V m_F \, dV \tag{8}$$

$$\sum_i^n (\nabla m_F A)_i = -\left(\frac{\phi_F \mu_g c_g}{k_F}\right) \frac{\partial}{\partial t} \int_V m_F \, dV \tag{9}$$

Assuming that the difference between the two time steps is Δt and the distance between the two neighboring grid center nodes is L, then the discrete Equation (9) can be converted to Equation (10):

$$\sum_i^n \left(A \frac{m_F^a - m_F^i}{L_{a,i}}\right)_i = -\left(\frac{\phi_F \mu_g c_g}{k_F}\right)^{n+1} \frac{m_F^{n+1} - m_F^n}{\Delta t} \Delta V \tag{10}$$

Based on this, the discrete equations for the SRV and matrix area can be further deduced. The discrete flow equation for the SRV region matrix system is shown in Equation (11):

$$\sum_i^n \left(A \frac{m_{S,m}^a - m_{S,m}^i}{L_{a,i}}\right)_i - \alpha(m_{S,m} - m_{S,f})\Delta V$$
$$= -\left(\frac{\phi_F \mu_g c_g}{k_F}\right)^{n+1} \frac{\left[\phi_{S,m} + (1-\phi_{S,m})\frac{M_{S,g} m_{S,L} V_{S,L}}{V_{std}\left(m_{S,L} + m_{S,m}\right)^2}\right]^{n+1} - \left[\phi_{S,m} + (1-\phi_{S,m})\frac{M_{S,g} m_{S,L} V_{S,L}}{V_{std}\left(m_{S,L} + m_{S,m}\right)^2}\right]^n}{\Delta t} \Delta V \tag{11}$$

The discrete flow equation for the fracture system in the SRV region is shown in Equation (12):

$$\sum_i^n \left(A \frac{m_{S,f}^a - m_{S,f}^i}{L_{a,i}}\right)_i - \alpha(m_{S,m} - m_{S,f})\Delta V = -\left(\frac{\phi_F \mu_g c_g}{k_F}\right)^{n+1} \frac{\left(m_{S,f}^{n+1} - m_{S,f}^n\right)}{\Delta t} \Delta V \tag{12}$$

Discrete flow equation in matrix can be expressed as Equation (13):

$$\sum_i^n \left(A \frac{m_M^a - m_M^i}{L_{a,i}}\right)_i = -\left(\frac{\phi_F \mu_g c_g}{k_F}\right)^{n+1} \frac{m_M^{n+1} - m_M^n}{\Delta t} \Delta V \tag{13}$$

4. Results and Discussion

4.1. Model Verification

Taking the A1 well group in Chinese Jiaoshiba shale gas field for instance, we use the multi-region coupled model to analyze MFHWs' transient pressure recovery behaviors during the shut-in process. In this shale gas field, the gas investigated permeability is 200–300 nD, porosity is 3%, and the reservoir thickness is approximately 90 m. Well A1-1 (in A1 well group) has operated 15 segments of hydraulic fracturing and put into production from April 2014. The average daily gas production was 6×10^4 m^3/d up to March 2018. In the same layer, there is an adjacent A1-2 well 300 m away from well A1-1. A1-2 has operated 19 segments of hydraulic fracturing and put into production from May 2014. Up to March 2018, its average daily gas production was 10×10^4 m^3/d. The interpretation of the micro-seismic monitoring results shows that the two wells have partial overlap in the hydraulic fracture network. Therefore, it is preliminarily judged that the well communicates with the adjacent well through the SRV and HFs. In order to understand the reservoir parameters around the A1-1 well and its connectivity with the neighboring well A1-2, a pressure build-up well test was conducted during the production process of well A1-1, and the gas production of the neighboring well was maintained at 10×10^4 m^3/d.

According to the pressure recovery test data of Well A1-1, the logarithm curve of the pseudo pressure and its derivative are plotted. It is found that the derivative curve dropped significantly in the last part, and it was initially judged that the well A1-2 is more severely interfered by well A1-1. Based on the spatial relationship of the two wells and the characteristics of the hydraulic fracture network, a numerical seepage model was established under the mode of 'HF + SRV connection', and the permeability and wellbore storage were obtained by fitting the pseudo pressure and its derivative curve (Figure 3). As shown in the Figure 3, the early time data is related to wellbore storage capacity, the early bulge data is related to skin effect, the middle time data reflects linear flow under fracture interference, this part would control fracture half-length, fracture conductivity and fracture spacing. The equivalent permeability of the SRV fracture system is 0.0507 mD, which is far greater than the permeability of the matrix (200–300 nD), and the ratio of horizontal to vertical permeability is only 0.0252, reflecting the general characteristics of shale reservoirs with low vertical permeability. SRV permeability is calculated by fitting pressure data. The average HF half-length of the two wells has reached about 100 m, and the HFs in the toe connecting section are as long as about 170 m. In order to verify the reliability of the model and the parameters, these parameters explained by the pressure recovery test data were assigned to the well group geological model, and the bottomhole pressure (BHP) changes in the later production process of the well were simulated (Figure 4). The comparison error with the BHP monitoring data is quite small.

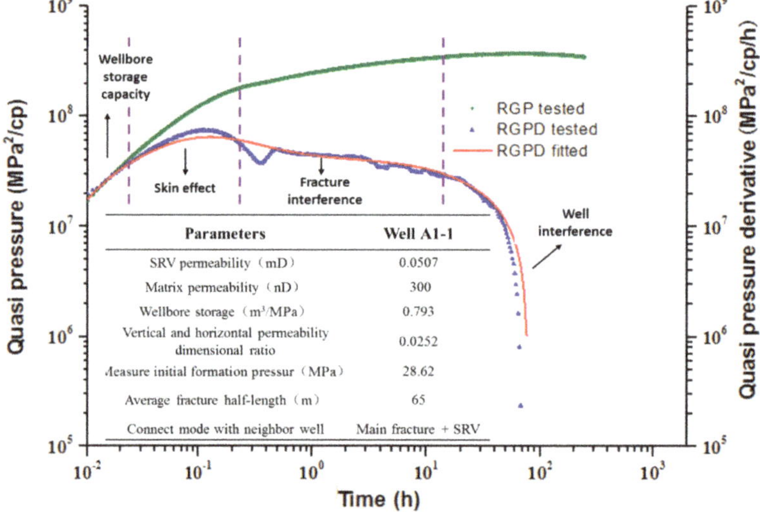

Parameters	Well A1-1
SRV permeability （mD）	0.0507
Matrix permeability （nD）	300
Wellbore storage （m³/MPa）	0.793
Vertical and horizontal permeability dimensional ratio	0.0252
Measure initial formation pressur （MPa）	28.62
Average fracture half-length （m）	65
Connect mode with neighbor well	Main fracture + SRV

Figure 3. Double logarithmic curve fitting results for pressure recovery in well A1-1.

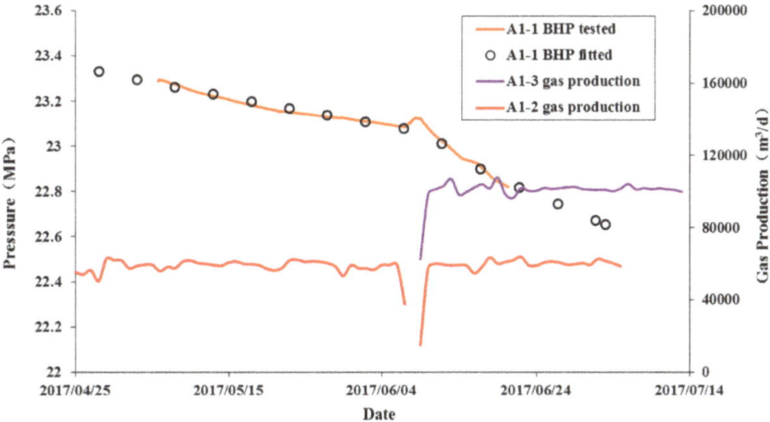

Figure 4. A1-1 bottomhole pressure (BHP) fitting results (A1-3 was a new neighboring well produced at June 2017).

4.2. Mechanism Model

The transient pressure behavior under well interference is influenced by various factors, such as the inter-well communication mode (IWCM), well spacing, number of neighboring wells, and well pattern arrangement [34,35]. The influence of different factors is analyzed by combining the established multi-zone coupled flow model. According to the reservoir and fracture parameters obtained from the well test and laboratory test data of Jiaoshiba shale gas field, it is assumed that the permeability of the matrix is 0.001 mD to 0.003 mD, and the SRV has a permeability of 50 to 200 times that of the matrix. The horizontal well-length is 1500 m. Each well is fractured by 19 sections, and except for connecting HFs, other HFs are of the same length and have the same conductivity; the initial pressure of the reservoir is 34 MPa, the temperature is 96 °C., and the gas component is dominated by methane (98%) with a small amount of nitrogen (0.7%), carbon dioxide (0.6%), ethane (0.4%), and propane (0.3%); the well spacing between the well and neighboring well is 300–600 m. Based on these parameters, a corresponding mechanism model was established. The well pair sketch and meshing grids under the coexistence conditions are shown in Figure 5.

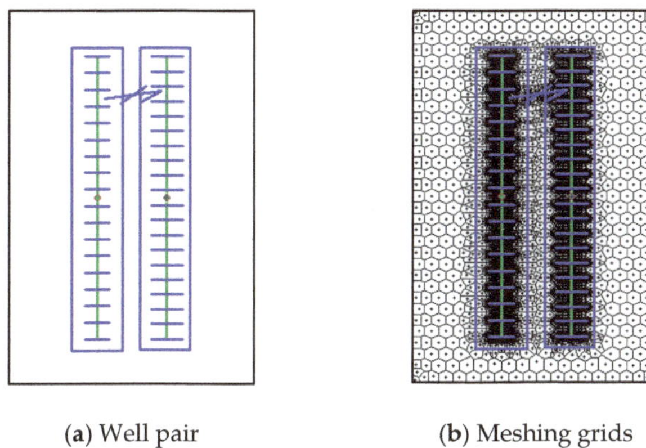

(a) Well pair   **(b)** Meshing grids

Figure 5. Mechanism model of two adjacent shale gas wells.

4.3. Effect of Inter-Well Connection Mode

The different connectivity modes and fracturing effects have considerable influence on the seepage characteristics under the disturbance of adjacent wells. By establishing the mechanism model, the effects of hydraulic fracture connected layers, connectivity modes, and the development degree of micro-fractures in SRV on the transient pressure of the well are discussed. A variety of simulation schemes have been designed based on different inter-well connection modes (IWCMs) and SRV parameters. Table 1 shows the IWCMs and essential physical property of each plan. In each scheme, the investigated well (right L2 well) and its neighboring well (left L1 well) are designed to produce at 6×10^4 m^3/d for 15 months, and the pressure field around these two wells is shown in Figure 6; afterward, the L2 well is shut-in for 20 days with pressure recovery, while L1 well maintains at the 6×10^4 m^3/d gas rate. The BHP of L2 well in each plan is shown in Figure 7, and the double logarithmic curve of the pseudo pressure and its derivative in the shut-in stage is shown in Figure 8.

Table 1. Inter-well communication modes (IWCMs) and essential physical properties of each plan.

Plan	IWCMs	HF Conductivity/mD·m	HF Half-Length/m	SRV Permeability/mD	Matrix Permeability/mD
Plan 1	One HF connected	20	60	0.3	0.0003
Plan 2	Four HFs connected	20	60	0.6	0.0003
Plan 3	One SRV segment connected with induced micro-fractures	20	60	0.15	0.0003
Plan 4	One SRV segment connected with abundant induced micro-fractures	20	60	0.6	0.0003
Plan 5	Two SRV segments and HF connected	20	60	0.6	0.0003
Plan 6	No SRV or HF connected	20	60	0.15	0.0003

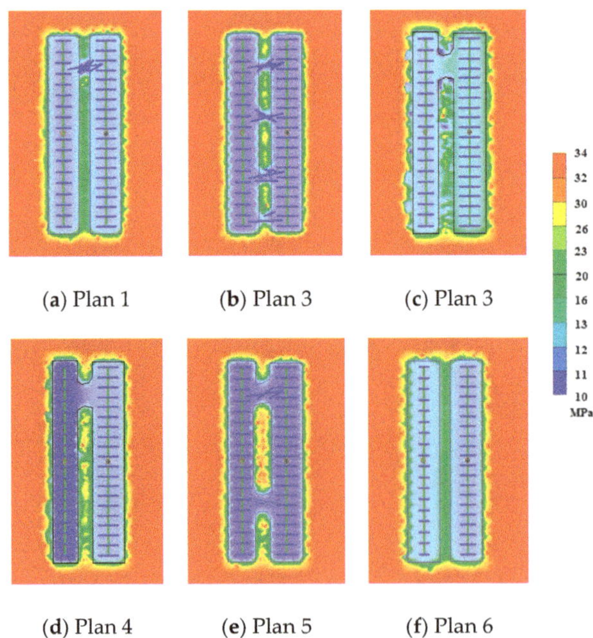

(**a**) Plan 1   (**b**) Plan 3   (**c**) Plan 3

(**d**) Plan 4   (**e**) Plan 5   (**f**) Plan 6

Figure 6. Pressure distribution after 15 months of simulated production of different plan.

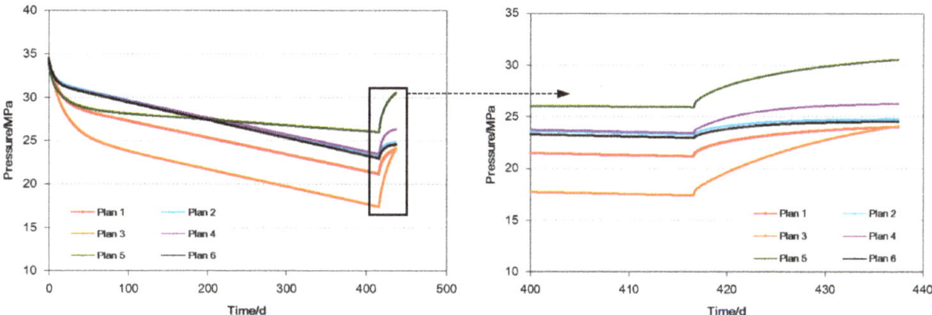

Figure 7. BHP changes during production and shut-in of L2 well.

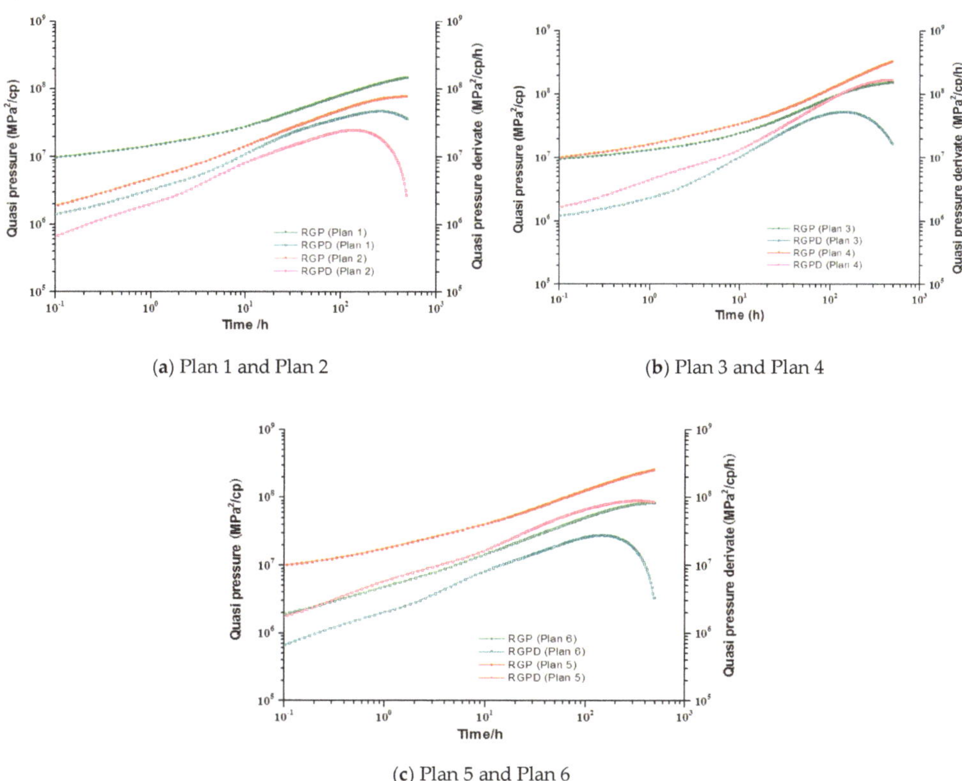

(**a**) Plan 1 and Plan 2

(**b**) Plan 3 and Plan 4

(**c**) Plan 5 and Plan 6

Figure 8. Double logarithmic curve of L2 well recovery pseudo pressure of each plan.

By analysis of the pressure fields (Figure 6) and the BHP of L2 (Figure 7), several points are obtained as follows:

(1)   In Plan 3, induced micro-fractures are not abundant in SRV, and the pressure drop in the SRV is not balanced during the production process, and the pressure near the HFs around the near-well area is severely reduced. However, the overall pressure drop in the SRV is small, so the BHP of L2 well is the largest in the case of the same amount of accumulated gas output. In the latter period, the shut-in pressure recovers the fastest with the greatest extent, but the pressure after recovery is still the lowest.

(2) In Plan 6, the BHP of L2 well is slower and the amplitude is the slowest, and the BHP of L2 shut-in well recoveries faster and higher in the later period, and after the recovery the BHP reaches the highest. In the early stage, due to the poor hydraulic fracturing effect of L2 well, the pressure drops quickly, but in the later period, the pressure drop is not serious due to the interference from the adjacent well. It also shows that even if there is no connection of HF or SRV between the well pairs, there will be some interference.

(3) Comparing Plan 2 and Plan 5 both with strong inter-well connection, the BHP changes of L2 well are very similar. It shows that due to the good effect of fracturing, the SRV area is rich in induced micro-fractures, and the HFs have strong conductivity. Although L2 well gets the interference from neighboring L1 well, the BHP of L2 drops still slowly. After shut-in, disturbed by the production of the neighboring L1 well, the pressure at the later stage of the well shut-in does not recover but decrease, and the inter-well interference is obvious.

(4) In Plan 1, Plan 3, and Plan 6, the two wells have poor fracturing effect and the pressure drop near the HFs is significantly larger than that of the surrounding SRV in each plan. It shows that the pressure drop in the control range of a single well is not balanced enough and the development result is not ideal.

There are two wells in each plan, and L2 on the right is the investigated well, while L1 on the left is the neighboring well.

Through analyzing the pressure recovery curve of each plan in Figure 8, we get the following understanding. (1) the interference degree comes from neighboring wells in each plan is compared based on the fall magnitude in the later part of the pseudo pressure derivative curve (PPDC): Plan 2 > Plan 6 > Plan 4 > Plan1 > Plan 5 > Plan 3; (2) When HFs or SRV's induce micro-fractures is in rich, it first experience a brief stratum-fracture bilinear flow (the slope of the pseudo pressure derivative curve is close to 1/4), and then it converts to formation linear flow (the slope of the pseudo pressure derivative curve is close to 1/4), late-stage well interference causes the pressure curve to drop. When induced micro-cracks are not abundant, they mainly experience formation linear flow, and the characteristics of later well interference are not obvious.

In summary, BHP changes are affected by both the fracturing effect of the well and interference from neighboring wells. When the fracturing effect of the well is good, and it is connected with the neighboring well through the SRV, the BHP in the production process can drop slowly to maintain stable production. When the fracturing effect of the wells is poor, and some parts are connected with the adjacent well, the BHP decline violently during the production, which is unfavorable to the stable high production of gas wells. In the future production process, it is necessary to reduce the steady gas production and timely shut well to recover the BHP.

4.4. Effect of Well Space

Reasonable well spacing is a crucial indicator for the design of horizontal well patterns to develop shale gas fields. If the well spacing is too small, a severe well interference and a decrease in productivity will appear. Large well spacing will lead to a low recovery of the whole shale gas field, and the remaining reserves will be difficult to exploit in the later period. Therefore, interference degree of different well spacing and different IWMDs should be studied necessarily. We design six plans with different IWMDs and well spacing in the mechanism models, whose basic parameters of each plan are shown in Table 2. In each plan, L2 well and its neighboring well (L1 well) are set to produce at 6×10^4 m^3/d for 15 months. The pressure field is shown in Figure 9; after that, the well shut-in and BHP recovers for 20 days. At this time, the neighboring well still maintains 6×10^4 m^3/d gas output. The BHP of L2 well in each plan is shown in Figure 10. The double logarithmic curve of the pseudo pressure and its derivative at the shut-in stage is shown in Figure 11.

Table 2. IWMDs and basic physical properties of six plans.

Plan	IWCMs	HF Conductivity/mD·m	HF Half-Length/m	SRV Permeability/mD	Matrix Permeability/mD
Plan 1	Well spacing 350 m with SRV connected	20	60	0.06	0.0003
Plan 2	Well spacing 450 m with SRV connected	20	60	0.06	0.0003
Plan 3	Well spacing 550 m with SRV connected	20	60	0.06	0.0003
Plan 4	Well spacing 350 m with four HFs connected	200	60	0.06	0.0003
Plan 5	Well spacing 450 m with two HFs connected	200	60	0.06	0.0003
Plan 6	Well spacing 550 m with one HFs connected	200	60	0.06	0.0003

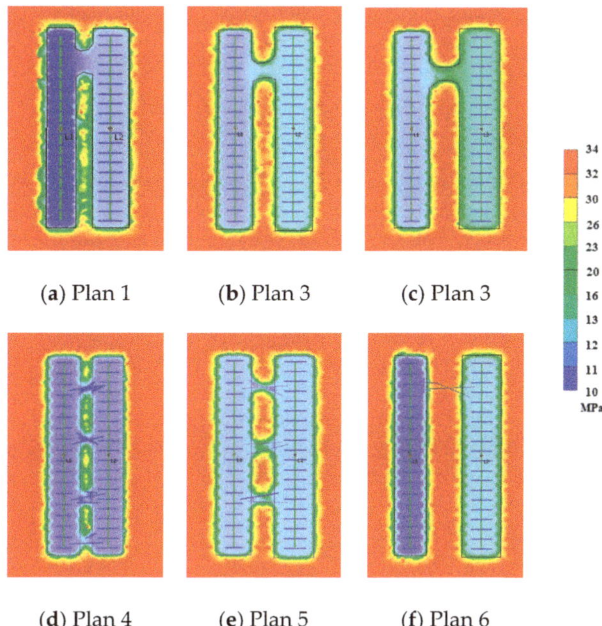

Figure 9. Pressure fields of six plans after producing 15 months under different well spacing.

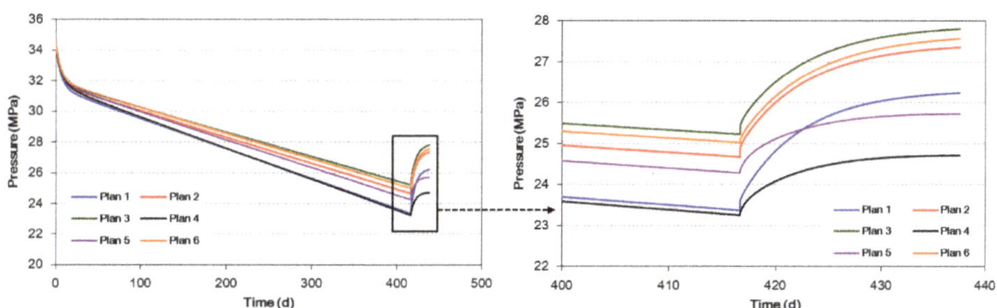

Figure 10. BHP of L2 well after producing 15 months under different well spacing.

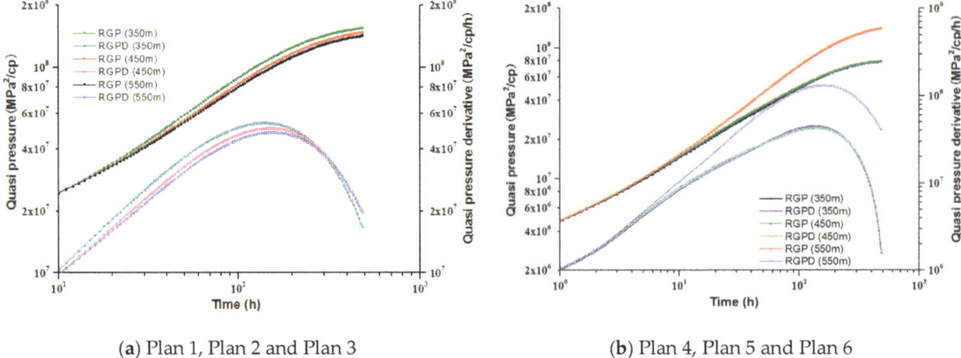

(**a**) Plan 1, Plan 2 and Plan 3   (**b**) Plan 4, Plan 5 and Plan 6

Figure 11. Double logarithmic curves of L2 build-up pressure under different well spacing.

Through analyzing the BHP and pressure fields of each plan in Figures 9 and 10, we get the following understanding:

(1) From the simulation results of the Plan 1 and Plan 4, when the well spacing is small, the pressure drop in the SRV is severe, and the BHP drop of L2 well is relatively large in the production process, and the shut-in pressure recovery ability is relatively weak, as shown in Figure 9a,d, and Figure 10.

(2) From the simulation results of Plans 3 and 6, the pressure decreasing speed in the SRV is not serious when the well spacing is large, and the BHP of L2 wells is smaller in the production process, and the pressure recovery ability is stronger during well shut-in, as shown in Figure 9c,f and Figure 10.

(3) Compared with the HF connection mode, when the inter-well SRVs connected, the pressure drop of L2 well is slow under the same well spacing in the production process, and the pressure recovery after shut-in is obvious, which indicates that the reservoir still maintains strong energy.

Through analyzing the pressure recovery curve of each plan in Figure 11, we obtain some views as follows:

(1) Under close well spacing and multiple HFs connection modes, the interference comes from neighboring wells is more serious, so this kind of multiple HFs connection should be avoided by controlling hydraulic fracturing;

(2) Under the condition of SRV connection mode, the effect of different well spacing is not obvious, and the well interference is also obvious;

(3) For these mechanism models, the well spacing should be controlled above 450 m. For the development of real shale gas fields, the well spacing can be optimized based on the reservoir physical property and the design scale of fracturing, to obtain higher single well productivity and gas field recovery ratio.

4.5. Effect of Well Pattern and Multiple Neighboring Wells

Shale gas field development generally adopts a large-scale horizontal well pattern, and there may be multiple adjacent wells around a well. Based on the mechanism model, the simulation plans for different number of wells and different arrangements are designed, and the designed parameters are shown in Table 3. The wells of each scheme are designed to produce at 6×10^4 m^3/d for 15 months, and the pressure fields are shown in Figure 12. Afterwards, the BHP of the shut-in L2 well recovers for 20 days. At this time, the neighboring wells all maintain at 6×10^4 m^3/d gas output. The BHP of L2 well in each plan is shown in Figure 13. The double logarithm curve of the pseudo pressure and its derivative during the shut-in stage is shown in Figure 14.

Table 3. IWMDs and basic physical properties of six plans.

Plan	IWCMs	HF Conductivity/mD·m	HF Half-Length/m	SRV Permeability/mD	Matrix Permeability/mD
Plan 1	Three wells: L2 well and two adjacent wells (L1 and L3) with seven HFs connected.	100	60	0.03	0.0003
Plan 2	Three wells: L2 well is staggered with two adjacent wells (L1 and L3), and four HFs connected in the corresponding well segments.	100	60	0.03	0.0003
Plan 3	Four wells: L2 well and two adjacent wells (L1 and L3) connected through five HFs, and L1 well connects with L4 well by two HFs.	100	60	0.03	0.0003
Plan 4	Four wells: L2 well is parallel with two adjacent wells (L1 and L3), and head-to-head with L4 well. There are eight HFs connected.	100	60	0.03	0.0003
Plan 5	Five wells: L2 well, and two neighboring wells (L3 and L4) on the left side and two neighboring wells (L1 and L5) on the right side, and total nine HFs connected.	100	60	0.03	0.0003
Plan 6	Six wells: L2 well is parallel with two neighboring wells (L1 and L3), and head-to-head with three wells (L4, L5 and L6)	100	60	0.03	0.0003

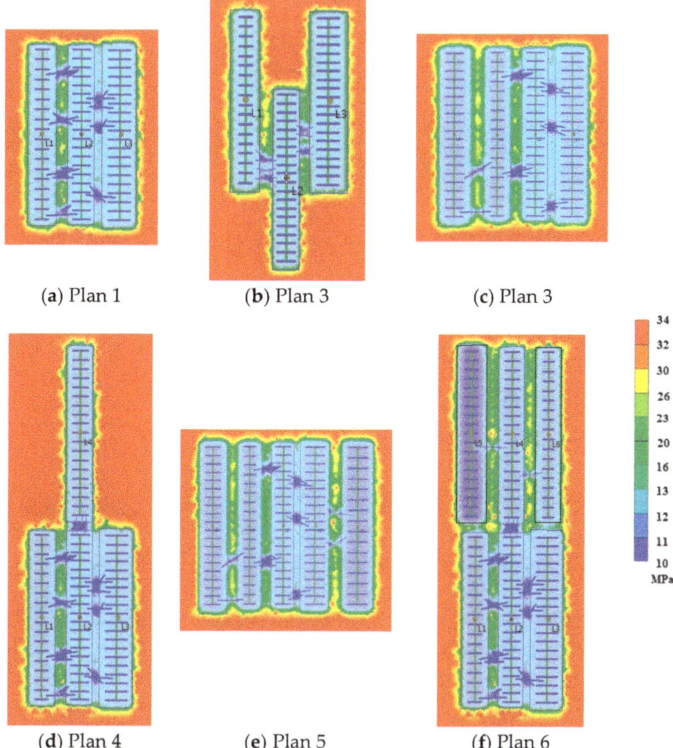

(**a**) Plan 1   (**b**) Plan 3   (**c**) Plan 3

(**d**) Plan 4   (**e**) Plan 5   (**f**) Plan 6

Figure 12. Pressure field under different well pattern and multiple neighboring wells in six designed plans.

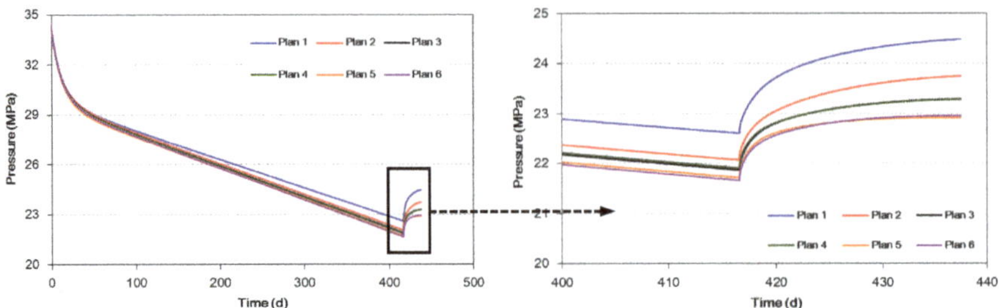

Figure 13. BHP of L2 well with different well pattern and multiple neighboring wells.

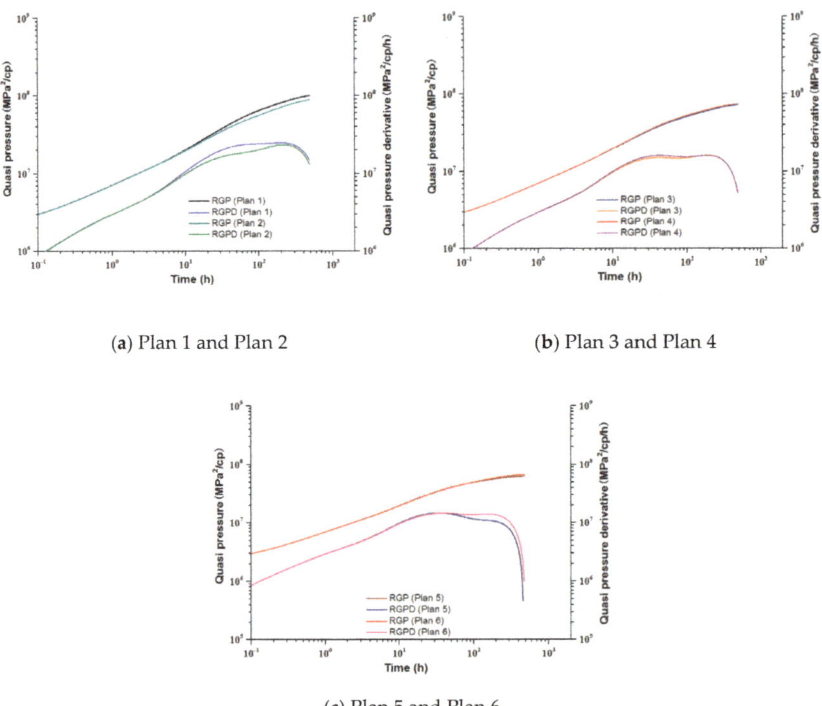

(**a**) Plan 1 and Plan 2   (**b**) Plan 3 and Plan 4

(**c**) Plan 5 and Plan 6

Figure 14. Double logarithmic curve of pressure recovery of shut-in wells under different well pattern and multiple neighboring wells.

By analysis of the pressure fields (Figure 12) and the BHP of L2 (Figure 13), several points are obtained as follows. (1) When well spacing keeps at 350m, increasing number of neighboring wells do not lead to serious pressure drop in the SRV of L2 well. Besides, the pressure drop is not apparent in the outside matrix, but the pressure drop is significant in SRV with HFs connected. (2) Corresponding and connecting with the direct neighboring wells' horizontal segment is the key to generate obvious interference. If the corresponding inter-well horizontal segment is more extended, the interference is more severe (compared Plan 1 with Plan 2). On this basis, the more interconnected wells put into production in the well pattern, the more severe interference is, and even if only the ends of the horizontal well (toe or root) are connected, interference still occurs. (3) Under the condition of the same number of connected neighboring wells, no matter what the well spacing and IWMDs are, the BHP curve is nearly overlapped. It indicates that the number of neighboring wells is crucial to the well interference degree, and the number of HFs connected is relatively secondly. (4) The more severe interference the neighboring wells lead, the more severe the performance is, as well as the fact that BHP in the well is declining faster in the production phase. Even though the neighboring well is not directly adjacent to the well, it is indirectly adjacent to one well after it has been sequestered by one well and still interferes with the well (compared Plan 2 with Plan 3).

Through analyzing the pressure recovery curve of each plan in Figure 14, we obtain some views as follows. (1) Parallel arrangements produce more severe well interference than head-to-head arrangements under a same number of wells. (2) The pseudo pressure derivative curve of each plan shows the characteristics of linear flow, quasi-radial flow segment and pressure drop disturbed by adjacent wells. (3) When the parameters such as HF properties and gas production keep constant, the linear flow in the early stage of each plan is the same, and the difference mainly lies in the late interference stage.

5. Field Application

A2 is a well group (five wells) in the lower gas zone of the Longmaxi section of the Lower Silurian in the Jiaoshiba shale gas field. Log interpretation explains an average porosity of 3.94%, an average permeability of 0.02 mD to 0.06 mD, and an average gas content of 3.47 m^3/t to 3.85 m^3/t. The gas composition is dominated by methane (98.4%) and contains a small amount of ethane, carbon dioxide and nitrogen; the initial formation pressure is 32.3 MPa and the initial temperature is 94.7 °C. During the drilling process, there is no leakage of drilling fluid in the horizontal well section. Each well in the well group put into operation from April 2014. As of March 2018, the cumulative gas production of Well A2-1 was 6272.28 $\times$ 10^4 m^3/d, and the average daily gas production was 6.41 $\times$ 10^4 m^3/d, and the casing pressure dropped from initial 29.4 MPa to 10.7 MPa. The cumulative gas production of A2-2 well is 9633.57 $\times$ 10^4 m^3/d, and the average daily gas production is 8.17 $\times$ 10^4 m^3/d, and the casing pressure reduced from 29.3 MPa to 8.9 MPa. Accumulated gas production in Well A2-3 is 6143.43 $\times$ 10^4 m^3/d, and average daily gas production is 5.21 $\times$ 10^4 m^3/d, and casing pressure reduced from 31.8 MPa to 13.3 MPa. Accumulated gas production in Well A2-4 is 8515.55 $\times$ 10^4 m^3/d, average daily gas production is 8.95 $\times$ 10^4 m^3/d, and casing pressure is reduced from 30MPa to 13.4 MPa. The cumulative gas production of Well A2-5 is 5471.0 $\times$ 10^4 m^3/d, the average daily gas production is 7.00 $\times$ 10^4 m^3/d, and the casing pressure is reduced from 31.4 MPa to 16.4 MPa.

By performing pressure recovery tests on well A2-1 in the center of the well group, inversion of reservoir and fracture property parameters, and analysis of interference from neighboring wells to production. According to the basic parameters of the reservoir, the position relationship of each well, the scale of fracturing construction in each section of the horizontal well, and the tracer test results, a numerical well test model was established to analyze the pressure recovery test data. The simulation results of the simulated pressure field and the pseudo-logarithmic double logarithmic curve are shown in Figure 15. From the simulated pressure field of the numerical well test in Figure 15a, the wells A2-4 and A2-5 arranged opposite to the A2-1 well have weak interference to them. The A2-2 wells arranged in parallel with it and connected to the SRV have strong interference to them, and the A2-3 wells with relatively long distances have weak interference to them. From the plot of the double logarithm curve of the pseudo-pressure in Figure 15b, the lower part of the curve falls, but the amplitude is not large; it shows that there are not many adjacent wells that interfere with it, and there is no multi-segment fracture crack connectivity.

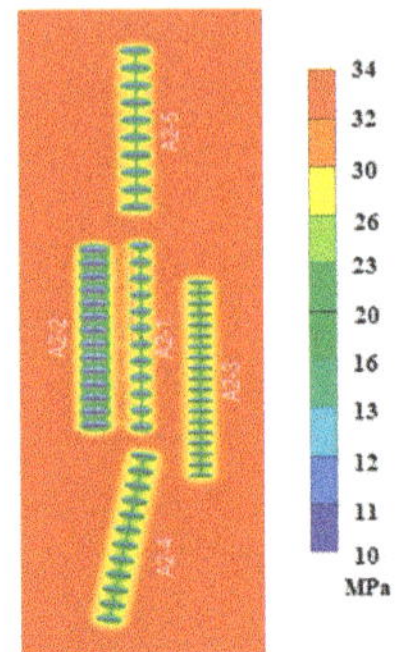

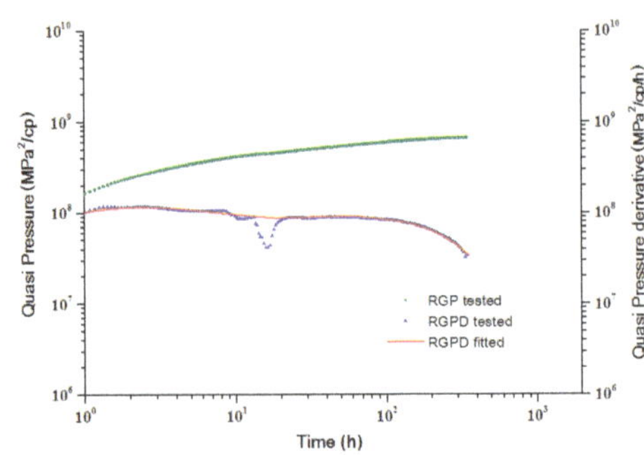

(**a**) A2 well group pressure field   (**b**) A2-1 build-up quasi pressure double logarithmic curves

Figure 15. Numerical well of the simulation results of the simulated pressure field and the pseudo-logarithmic double logarithmic curve.

In order to further verify the interference of 4 neighboring wells to the central well A2-1, the gas production and pressure of the adjacent wells during the pressure recovery of this well were compared (Figure 16). During the recovery of pressure in this well, well A2-2 was normally produced but the pressure decreased slowly or even recovered. The rate of decline of the other three wells did not change substantially during this period. It is confirmed that well A2-2 has a significant influence on well A2-1, which is in agreement with the conclusion of numerical well test analysis. In short, the interference between the two wells is relatively small (A2-1 and A2-4, A2-5), but the interference is smaller when the well spacing is larger (A2-1 and A2-3). Although A2-1 and A2-2 wells interfered with the SRV mode, they did not cause serious pressure drop.

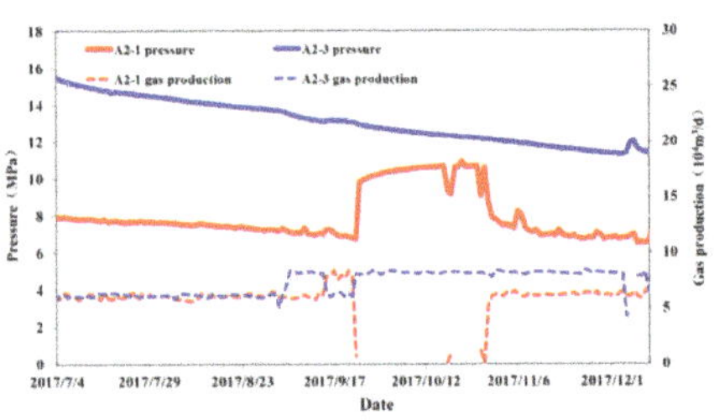

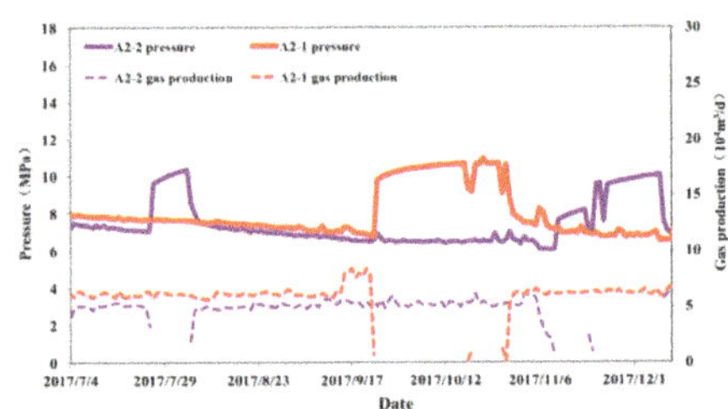

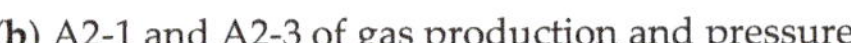

(**a**) A2-1 and A2-2 of gas production and pressure   (**b**) A2-1 and A2-3 of gas production and pressure

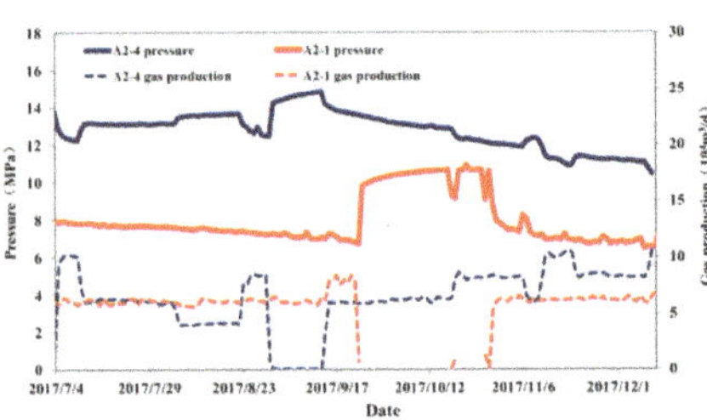

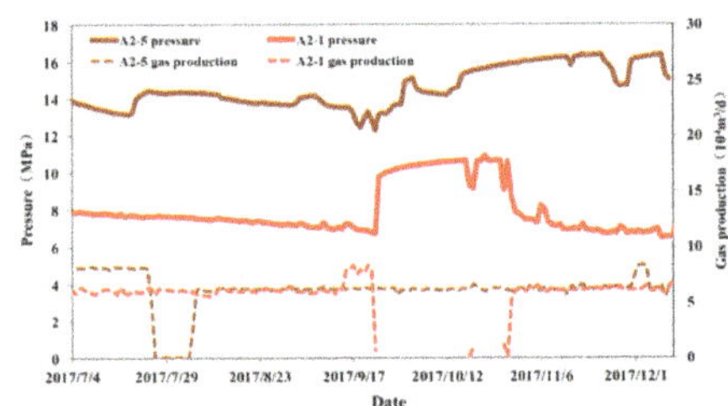

(**c**) A2-1 and A2-4 of gas production and pressure   (**d**) A2-1 and A2-5 of gas production and pressure

Figure 16. Gas production and pressure of A2 well group.

6. Conclusions

In recent years, with the continuous expansion of the shale gas wells pattern's scale and reduction of well spacing, the problem of well interference has become increasingly serious. On the one hand, the shale gas well interference leads BHP drops rapidly, which is harmful to gas production; on the other hand, under the condition of a relatively balanced fracturing scale, it shows that the IWCM is good and can effectively control the resources within the reservoir around the well. In fact, the scale of most fracturing sections of most shale gas wells is not balanced. Interferences are mainly appeared between the wells with good fracturing and close well spacing.

To study transient pressure behaviors of shale gas MFHW under well interference, we establish a multi-region coupled flow model based on the pore fracture structures and flow mechanisms of the different region. Then, PEBI grids and finite volume method are used for the numerical solution. Utilizing the measured shut-in pressure recovery test data of shale gas wells, we explain the property parameters of the reservoir and HFs. Moreover, BHP data monitored in the later stage is fitted to verify the accuracy of the model. Based on the shale gas seepage model, the pressure field and BHP are analyzed under the disturbance of adjacent wells.

The transient pressure behaviors of shale gas wells are affected by fracturing effect, production rules, and well interference. The interference can be observed from the later decline phase of the PPDC during well shut-in. Different IWCM, well spacing, number of neighboring wells, and well pattern arrangement lead different PPDC drops extent in the later phase. A sharp BHP drop can be caused by the direct inter-well connection of HFs, which also has a severe influence on the pressure recovery of the shut-in wells. Sometimes, well interference still exists even without direct HF connection, which reflects the SRV connected with satisfied fracturing effect. It is conducive to higher recovery. Generally, we recommend that it is necessary to optimize the well spacing and the fracturing scale according to the stimulated performance. Especially, the fracturing scale should be controlled for the segments close to the neighboring well; and if serious disturbances have already occurred, it is necessary to timely control the gas production of this well pair, and timely conduct interference test to quantify the degree of interference.

Author Contributions: Conceptualization, D.G. and Y.L.; methodology, D.G.; validation, G.H.; formal analysis, D.G.; investigation, D.G.; resources, D.G.; data curation, D.G.; writing—original draft preparation, D.G.; writing—review and editing, D.W.; visualization, D.G.; supervision, Y.L.; project administration, Y.L.; funding acquisition, D.G.

Funding: This research was funded by Southwest Petroleum University of Open Fund of State Key Laboratory of Oil and Gas Reservoir Geology and Exploitation, grant number PLN201711 and National Natural Science Foundation of China, grant number 11702300 and PetroChina Innovation Fundation, grant number 2017D-5007-0208.

Acknowledgments: The study was supported by Supported By Open Fund (PLN201711) of State Key Laboratory of Oil and Gas Reservoir Geology and Exploitation (Southwest Petroleum University), National Natural Science Foundation of China (11702300), PetroChina Innovation Fundation (2017D-5007-0208) and the Key Laboratory of Fluid-Solid Coupling System Mechanics, Chinese Academy of Sciences.

Nomenclature

m_F is the gas pseudo pressure of HF, MPa^2/(mPa·s); $m_{S,m}$ and $m_{S,f}$ are the gas pseudo pressure of the SRV matrix system and fracture system respectively, MPa^2/(mPa·s); F and S are the marks of HF and SRV respectively, dimensionless; m and f are the marks of SRV matrix system and fracture system respectively, dimensionless; $q_{iw,F}$ is the fluid exchange volume between the HF and wellbore, m^3/d; φ is porosity, dimensionless; k_F is the permeability of HF with finite conductivity, mD; μ_g is gas viscosity, mPa·s; c_g is gas compressibility factor, MPa^{-1}; t is time, d; α is exchange flow coefficient, (m^3·mPa·s)/MPa^2; $V_{S,L}$ is Langmuir gas volume, m^3/kg; V_{std} is gas molar volume at standard temperature and pressure conditions, m^3/kmol; $m_{S,L}$ is Langmuir gas pressure, MPa^2/(mPa·s); $M_{S,g}$ is gas molar weight, kg/kmol; K_n is knudsen number, dimensionless; β is tenuity factor, dimensionless; b is average fracture aperture, m; c is average fracture interval, m.

References

1. Ma, X.; Xie, J. The progress and prospects of shale gas exploration and development in southern Sichuan Basin, SW China. *Pet. Explor. Dev.* **2018**, *45*, 161–169. [CrossRef]
2. Zou, C.; Yang, Z.; He, D.; Wei, Y.; Li, J.; Jia, A.; Chen, J.; Zhao, Q.; Li, Y.; Li, J.; et al. Theory, technology and prospects of conventional and unconventional natural gas. *Pet. Explor. Dev.* **2018**, *45*, 1–13. [CrossRef]
3. Wang, Z. Breakthrough of Fuling shale gas exploration and development and its inspiration. *Oil Gas Geol.* **2015**, *36*, 1–6.
4. Ilkay, U.; Basak, K.; Hossein, K. Multiphase rate-transient analysis in unconventional reservoirs: Theory and application. *SPE Reserv. Eval. Eng.* **2016**, *19*, 1–14.
5. Soeder, D.J. The successful development of gas and oil resources from shales in North America. *J. Pet. Sci. Eng.* **2018**, *163*, 399–420. [CrossRef]
6. Ajani, A.A.; Kelkar, M.G. Interference study in shale plays. SPE 151045. In Proceedings of the SPE Hydraulic Fracturing Technology Conference, Woodlands, TX, USA, 6–8 February 2012.
7. Manchanda, R.; Sharma, M.M.; Holzhauser, S. Time-dependent fracture-interference effects in pad wells. *SPE Prod. Oper.* **2014**, *29*, 1–14. [CrossRef]
8. Sardinha, C.M.; Lehmann, J.; Pyecroft, J.F.; Petr, C.; Merkle, S. Determining interwell connectivity and reservoir complexity through frac pressure hits and production interference analysis. SPE-171628-MS. In Proceedings of the SPE/CSUR Unconventional Resources Conference, Calgary, AB, Canada, 30 September–2 October 2014.
9. Cheng, L.; Jia, P.; Rui, Z.; Huang, S.; Xue, Y. Transient responses of multifractured systems with discrete secondary fractures in unconventional reservoirs. *J. Nat. Gas Sci. Eng.* **2017**, *41*, 49–62. [CrossRef]
10. Zhu, W.; Qi, Q.; Ma, Q.; Deng, J.; Yue, M.; Liu, Y. Unstable seepage modeling and pressure propagation of shale gas reservoirs. *Pet. Explor. Dev.* **2016**, *43*, 261–267. [CrossRef]
11. Mezghani, M.; Roggero, F. Combining gradual deformation and upscaling techniques for direct conditioning of fine scale reservoir models to interference test data. *SPE J.* **2004**, *9*, 79–87. [CrossRef]
12. Yaich, E.; Souza, O.C.D.D.; Foster, R.A.; Abou-Sayed, I.S. A methodology to quantify the impact of well interference and optimize well spacing in the marcellus shale. SPE-171578-MS. In Proceedings of the SPE/CSUR Unconventional Resources Conference, Calgary, AB, Canada, 30 September–2 October 2014.
13. Marongiu-Porcu, M.; Lee, D.; Shan, D.; Morales, A. Advanced modeling of interwell-fracturing interference: An eagle ford shale-oil study. *SPE J.* **2015**, *21*, 1567–1582. [CrossRef]
14. Pang, W.; Ehlig-Economides, C.A.; Du, J.; He, Y.; Zhang, T. Effect of well interference on shale gas well SRV interpretation. SPE 176910. In Proceedings of the SPE Asia Pacific Unconventional Resources Conference and Exhibition, Brisbane, Australia, 9–11 November 2015.
15. Bello, R.O.; Wattenbarger, R.A. Multi-stage hydraulically fractured shale gas rate transient analysis. SPE 126754. In Proceedings of the North Africa Technical Conference and Exhibition, Cairo, Egypt, 14–17 February 2010.
16. Ozkan, E.; Brown, M.L.; Raghavan, R.; Kazemi, H. Comparison of fractured horizontal well performance in tight sand and shale reservoirs. *SPE Reserv. Eval. Eng.* **2011**, *14*, 248–259. [CrossRef]
17. Al-Rbeawi, S. Analysis of pressure behaviors and flow regimes of naturally and hydraulically fractured unconventional gas reservoirs using multi-linear flow regimes approach. *J. Nat. Gas Sci. Eng.* **2017**, *45*, 637–658. [CrossRef]
18. Stalgorova, E.; Mattar, L. Practical analytical model simulate production of horizontal wells with Branch Fractures. SPE 162515. In Proceedings of the SPE Canadian Unconventional Resources Conference, Calgary, AB, Canada, 30 October–1 November 2012.
19. Zhang, L.; Gao, J.; Hu, S.; Guo, J.; Liu, Q. Five-region flow model for MFHWs in dual porous shale gas reservoirs. *J. Nat. Gas Sci. Eng.* **2016**, *33*, 1316–1323. [CrossRef]
20. Zeng, J.; Wang, X.; Guo, J.; Zeng, F. Composite linear flow model for multi-fractured horizontal wells in heterogeneous shale reservoir. *J. Nat. Gas Sci. Eng.* **2017**, *38*, 527–548. [CrossRef]
21. Wang, J.; Jia, A.; Wei, Y.; Qi, Y. Approximate semi-analytical modeling of transient behavior of horizontal well intercepted by multiple pressure-dependent conductivity fractures in pressure-sensitive reservoir. *J. Pet. Sci. Eng.* **2017**, *153*, 157–177. [CrossRef]

22. Kim, T.H.; Lee, J.H.; Lee, K.S. Integrated reservoir flow and geomechanical model to generate type curves for pressure transient responses of a hydraulically-fractured well in shale gas reservoirs. *J. Pet. Sci. Eng.* **2016**, *146*, 457–472. [CrossRef]

23. Xiong, X.; Devegowda, D.; Michel Villazon, G.G.; Sigal, R.F.; Civan, F. A fully-coupled free and adsorptive phase transport model for shale gas reservoirs including non-darcy flow effects. SPE 159758. In Proceedings of the SPE Annual Technical Conference and Exhibition, San Antonio, TX, USA, 8–10 October 2012.

24. Xu, B.; Haghighi, M.; Li, X.; Cooke, D.; Zhang, L. Development of new type curves for production analysis in naturally fractured shale gas/tight gas reservoirs. *J. Pet. Sci. Eng.* **2013**, *105*, 107–115. [CrossRef]

25. Gu, A.; Ding, D.; Gao, Z.; Zhang, A.; Tian, L.; Wu, T. Pressure transient analysis of multiple fractured horizontal wells in naturally fractured unconventional reservoirs based on fractal theory and fractional calculus. *Petroleum* **2017**, *3*, 326–339. [CrossRef]

26. Fan, D.; Yao, J.; Sun, H.; Zeng, H. Numerical simulation of multi-fractured horizontal well in shale gas reservoir considering multiple gas transport mechanisms. *Acta Mech. Sin.* **2015**, *47*, 906–915.

27. Bonyadi, M.; Rahimpour, M.R.; Esmaeilzadeh, F. A new fast technique for calculation of gas condensate well productivity by using pseudopressure method. *J. Nat. Gas Sci. Eng.* **2012**, *4*, 35–43. [CrossRef]

28. Hull, R.; Bello, H.; Richmond, P.L.; Suliman, B.; Portis, D.; Richmond, P. Variable stimulated reservoir volume (SRV) simulation: Eagle ford shale case study. URTEC 1582061. In Proceedings of the Unconventional Resources Technology Conference, Denver, CO, USA, 12–14 August 2013.

29. Civan, F.; Rai, S.C.; Sondergeld, H.C. Shale-gas permeability and diffusivity inferred by improved formulation of relevant retention and transport mechanisms. *Transp. Porous Media* **2010**, *86*, 925–944. [CrossRef]

30. Civan, F. Effective correlation of apparent gas permeability in tight porous media. *Transp. Porous Media* **2010**, *82*, 375–384. [CrossRef]

31. Ye, Z.; Chen, D.; Pan, Z. A unified method to evaluate shale gas flow behaviours in different flow regions. *J. Nat. Gas Sci. Eng.* **2015**, *26*, 205–215. [CrossRef]

32. Singh, H.; Javadpour, F.; Ettehadtavakkol, A.; Darabi, H. Nonempirical apparent permeability of shale. *SPE Reserv. Eval. Eng.* **2014**, *17*, 1–15. [CrossRef]

33. Moyner, O.; Lie, K.A. The multiscale finite volume method on unstructured grids. SPE 163649. In Proceedings of the SPE Reservoir Simulation Symposium, Woodlands, TX, USA, 18–20 February 2013.

34. Dejam, M. Advective-diffusive-reactive solute transport due to non-Newtonian fluid flows in a fracture surrounded by a tight porous medium. *Int. J. Heat Mass Transf.* **2019**, *128*, 1307–1321. [CrossRef]

35. Dejam, M.; Hassanzadeh, H.; Chen, Z. Pre-darcy flow in tight and shale formations. In Proceedings of the 70th Annual Meeting of the APS (American Physical Society) Division of Fluid Dynamics, Denver, CO, USA, 19–21 November 2017.

Article

Performance Evaluation of CO_2 Huff-n-Puff Gas Injection in Shale Gas Condensate Reservoirs

Xingbang Meng [1], Zhan Meng [2,*], Jixiang Ma [3] and Tengfei Wang [1]

[1] Petroleum Engineering, China University of Petroleum (East China), Qingdao 266000, China; 20170049@upc.edu.cn (X.M.); 20170070@upc.edu.cn (T.W.)

[2] Petroleum Systems Engineering, Faculty of Engineering and Applied Science, University of Regina, Regina, SK S4S 042, Canada

[3] PetroChina Huabei Oilfield Company, Renqiu 062552, China; cyy_mjx@petrochina.com.cn

* Correspondence: Zhan.Meng@uregina.ca; Tel.: +1-306-501-7159

Received: 6 November 2018; Accepted: 18 December 2018; Published: 24 December 2018

Abstract: When the reservoir pressure is decreased lower than the dew point pressure in shale gas condensate reservoirs, condensate would be formed in the formation. Condensate accumulation severely reduces the commercial production of shale gas condensate reservoirs. Seeking ways to mitigate condensate in the formation and enhance both condensate and gas recovery in shale reservoirs has important significance. Very few related studies have been done. In this paper, both experimental and numerical studies were conducted to evaluate the performance of CO_2 huff-n-puff to enhance the condensate recovery in shale reservoirs. Experimentally, CO_2 huff-n-puff tests on shale core were conducted. A theoretical field scale simulation model was constructed. The effects of injection pressure, injection time, and soaking time on the efficiency of CO_2 huff-n-puff were examined. Experimental results indicate that condensate recovery was enhanced to 30.36% after 5 cycles of CO_2 huff-n-puff. In addition, simulation results indicate that the injection period and injection pressure should be optimized to ensure that the pressure of the main condensate region remains higher than the dew point pressure. The soaking process should be determined based on the injection pressure. This work may shed light on a better understanding of the CO_2 huff-n-puff-enhanced oil recovery (EOR) strategy in shale gas condensate reservoirs.

Keywords: CO_2 huff-n-puff; condensate recovery; shale gas condensate reservoir

---

1. Introduction

Unconventional resources, especially shale reservoirs, have been widely developed with the techniques of hydraulic fracturing and drilling horizontal wells, and shale gas condensate reservoirs play an important role in regards to unconventional resources. When the reservoir pressure is decreased lower than the dew point pressure in shale gas condensate reservoirs, condensate can be formed near the wellbore or near/in the fracture as shown in Figure 1. This condensate blockage can reduce gas permeability. In addition, the productivity is reduced. Studies indicate that condensate blockage is much more severe when the permeability is low [1,2]. Also, as the condensate is formed by the heavy components of the reservoir fluid, it has a very high economic value. Therefore, it is important to find effective techniques to mitigate condensate blockage. Also, by mitigating condensate blockage in formation, gas permeability can be increased and the productivity can be greatly improved.

Several techniques are used to mitigate the condensate blockage for condensate reservoirs. Drilling horizontal wells and hydraulic fracturing have been widely used. Though the press drop in a horizontal well may be higher, it is distributed over a larger area, and the smaller pressure drop could help to reduce the accumulation of the condensate blockage [3–5]. Hydraulic fracturing can also help to reduce the pressure drop and reduce the formation of condensate blockage around the wellbore [6–10]. Drilling horizontal wells and hydraulic fracturing are two main techniques to enhance commercial production from shale reservoirs with ultra-low permeability. Hence, these two techniques are discussed in the follow discussion in this paper.

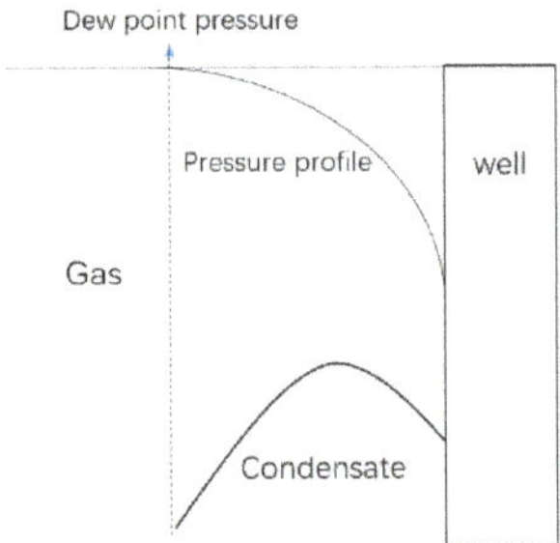

Figure 1. Condensate and pressure profile around the wellbore.

Chemical treatment techniques such as solvent injection and wettability-alteration treatment are also applied to mitigate condensate. By injecting solvent, the interfacial tension between condensate and gas can be reduced, and the solvent could help to dissolve part of the condensate into gas stream. Consequently, the condensate could be mitigated and the productivity of condensate reservoirs could be increased [11–13]. The injection of wettability chemicals can help to change the wettability from liquid wetting to gas wetting and the productivity of condensate reservoirs can be increased [14–17]. However, because of the low permeability of shale reservoirs, chemical treatment is not a suitable technique, as the efficiency of the injection process can be very low.

Gas injection is widely applied to mitigate the condensate recovery. By applying gas injection, pressure could be maintained at a higher rate than the dew point pressure. The accumulation of the condensate can also be prevented. Furthermore, gas injection can revaporize the condensate into a gas state. The accumulated condensate can be produced during the puff process [18,19]. The efficiency of different gas injection modes has been investigated [20–29]. The huff-n-puff process consists of three stages: huff (injection), soaking, and puff (production). The well is operated as both an injection well and a production well. As Figure 2 shows, there is only one well used as both an injection well and a production well in a huff-n-puff well. For wells of this type, the condensate region is located near the injection well. As the function of the well is changed by gas injection, the pressure of near wellbore region can be increased quickly. Consequently, the condensate is revaporized and recovered.

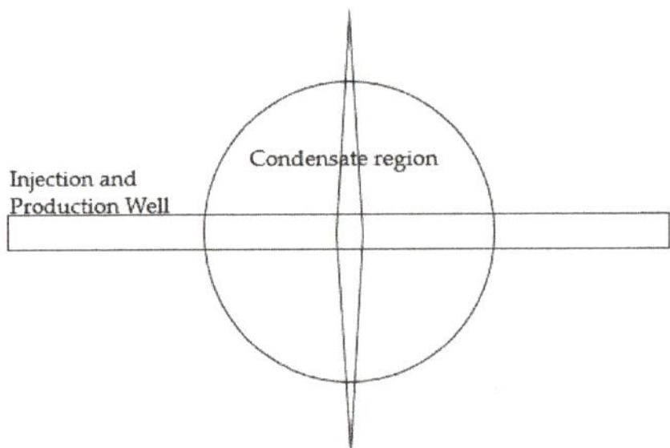

Figure 2. Huff-n-puff gas scenario in shale reservoirs.

The EOR techniques mentioned above are widely investigated in conventional gas condensate reservoirs. However, based on the studies on shale reservoirs [30–35], because of the ultra-low permeability of shale formations, the techniques such as water flooding and chemical flooding are difficult to be applied in shale formations. So far very few studies on EOR in shale gas condensate reservoirs have been conducted. Recently, solvent injection has been investigated to mitigate condensate blockage in shale gas condensate reservoir. Solvent injection could reduce the overall dew point pressure to delay the formation of condensate [36]. However, the efficiency and cost of solvent injection is questionable. Until now, CO_2 huff-n-puff has gained more and more attention in the literature [37–41]. However, the enhanced condensate performance of CO_2 huff-n-puff in shale reservoirs have not been investigated, especially in experimental aspects. The novelty of this study is to evaluate EOR performance of CO_2 huff-n-puff in shale gas condensate reservoirs using experiments and numerical analysis. Experimental work on shale rocks is different because of the ultra-low permeability. It is very difficult to conduct the core-flooding experiments to measure pressure drop or visually observe condensate flow as in sand rocks. In our study, the condensate saturation in shale rock was determined by CT and then the efficiency of huff-n-puff method could be quantified.

In this paper, first, experiments were conducted on shale core. The performance of CO_2 huff-n-puff to enhance condensate recovery in shale gas condensate reservoir was evaluated in core scale study. Then, field scale simulation was performed to investigate the performance of CO_2 huff-n-puff in shale gas condensate reservoir. Finally, the enhanced condensate recovery performance of CO_2 huff-n-puff has been evaluated by analyzing the experimental and simulation results. In addition, the optimization principles of CO_2 huff-n-puff are discussed.

2. Materials and Methods

2.1. Experimental Setup

2.1.1. Experiment Materials

Experiment of CO_2 huff-n-puff was operated on a shale core with 3.8 cm (1.5 in) in diameter and 10.2 cm (4 in) in length. The core was dried first and then the porosity and permeability was measured. Table 1 shows the properties of the core.

Table 1. Core properties.

Diameter (cm)	Length (cm)	Permeability (nD)	Porosity (%)
3.8	10.2	100	6.8

The initial gas mixture used in the experiment was formed of methane and n-butane with a pressure of 2200 psi and a temperature 20 °C (68 °F). Figure 3 shows the phase diagram of the mixture. At 68 °F, this gas mixture has the property of a gas condensate fluid. The liquid drop curve of this gas condensate mixture at 68 °F is shown in Figure 4. As can be seen the methane-butane gas mixture has a wide condensate region, with a dew point of 1860 psi at 20 °C (68 °F).

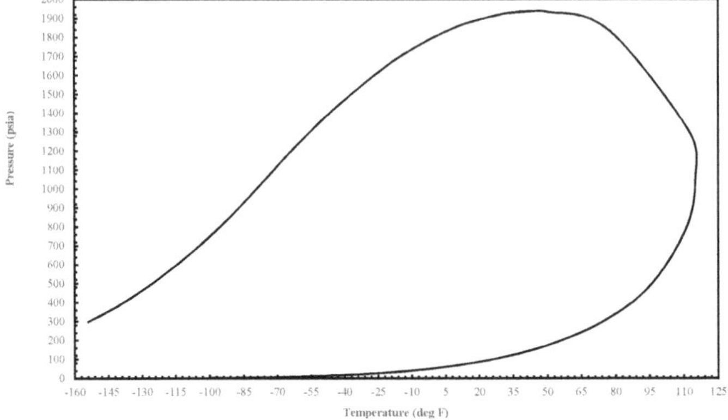

Figure 3. Phase diagram of initial gas mixture.

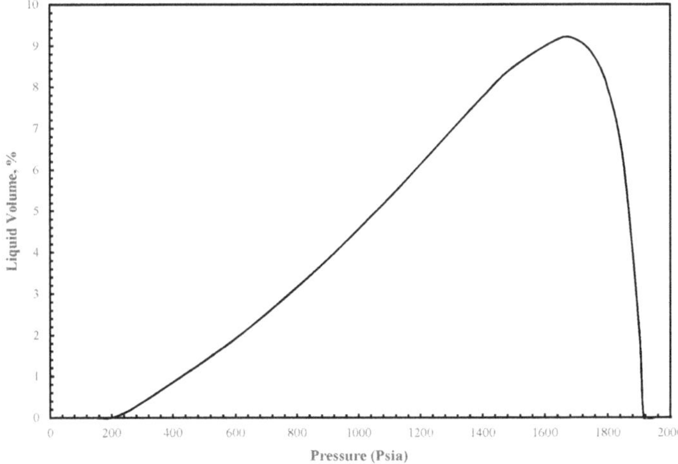

Figure 4. Liquid dropout curve of methane and *n*-butane gas condensate mixture, 20 °C (68 °F).

2.1.2. Experiment Procedure

A schematic of the experiment is shown in Figure 5. Based on the properties of the gas mixture, the gas mixture has a wide condensate region at room temperature. Thus, the experiment was conducted at 20 °C (68 °F). The core holder was placed in the CT scanner during the whole experiment to evaluate the core saturating process and measure the condensate saturation. The general procedure for CO_2 huff-n-puff gas injection experiment is described as follows:

(1) During the experiment, the injection pressure should be higher than 1860 psi (dew point pressure). And the confining pressure should be higher than the injection pressure. The initial gas condensate mixture was injected into the core holder at 2200 psi with a confining pressure of 2500 psi. The CT number was measured during the whole saturating process. When the CT number stopped changing, the core was assumed to be fully saturated with the gas condensate mixture.

(2) After the saturation process, the valve on the left side of core holder was opened and the pressure was decreased to 1460 psi. This step was used to simulate the primary depletion process, with

the CT scanner measuring the condensate saturation in the core. Condensate saturation was calculated by using the following equation [42]:

$$S_c = \frac{CT_{exp} - CT_{gr}}{CT_{cr} - CT_{gr}} \tag{1}$$

CT_{exp} represents the CT number during the experiment. CT_{gr} represents the CT number when the core is full of C_1. CT_{cr} is the CT number when the core is full of nC_4.

(3) Afterwards, the CO_2 huff-n-puff process was applied on the core. Injection pressure was set to 2200 psi and injection time was set to 2 hours. After injection, a soaking time of 1 hour was applied. After the soaking process, depletion process was applied again. The pressure was decreased to 1460 psi. This process represents one cycle of CO_2 huff-n-puff and 5 cycles were operated in total. The condensate saturation in the core was measured after every cycle.

(4) By analyzing the change in condensate saturation after the CO_2 huff-n-puff, the enhanced condensate recovery could be obtained and evaluated in laboratory.

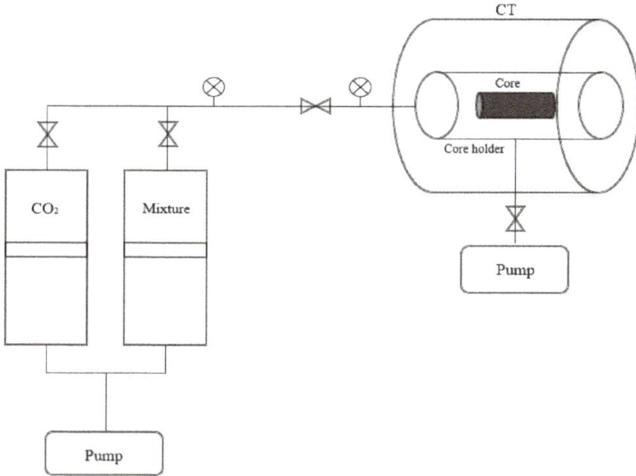

Figure 5. Schematic of CO_2 huff-n-puff.

2.2. Simulation Model Description

A field scale simulation model was built to investigate the enhanced condensate recovery performance of CO_2 huff-n-puff. The simulation work was conducted by using CMG-2015 (Computer Management Group Ltd, Calgary, Canada). Figure 6 shows this simulation model. The reservoir rock properties and gas condensate fluid properties were obtained from published data, as shown in Table 2 [43]. The dimension of the model was 180.44 m (592 ft) × 470.61 m (1544 ft) × 15.24 m (50 ft). In this model, only one fracture was set. Based on the studies [44–47], fracture propagation plays an important role for the development of shale plays. However, the main purpose of this simulation study was to evaluate the enhanced condensate recovery performance of CO_2 huff-n-puff in shale gas condensate reservoirs. In order to make the simulation work more effective, we just set one simple fracture and the fracture propagation was not taken into account. The fracture half-length was 110.34 m (362 ft) and the fracture width was 0.15 m (0.5 ft).

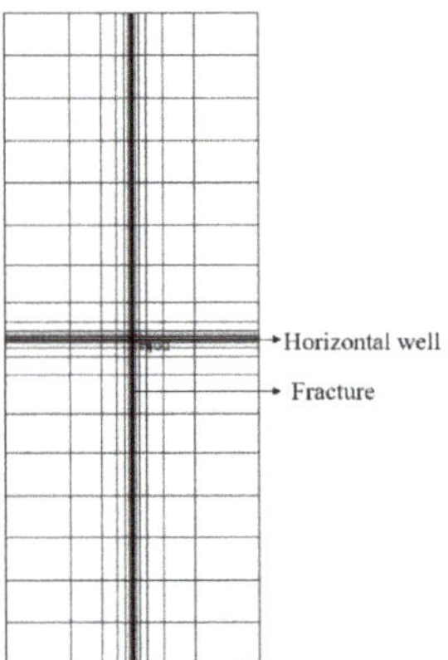

Figure 6. Field scale simulation model in *ij* view.

Table 2. Reservoir and fluid characteristics.

Parameters	Value	Unit
Initial Reservoir pressure	5000	psi
Initial Reservoir Temperature	93.3	°C
Matrix Permeability	0.0001	mD
Matrix Porosity	0.06	-
Fracture Permeability	100	mD

The reservoir condensate composition data is presented in Table 3, with the data obtained from published data [48]. The dew point pressure of the reservoir fluid is 2750 psi as shown in Figure 7, and when the pressure is decreased below the dew point pressure, condensate is formed. As the pressure continues to be decreased to 2460 psi, the liquid volume increases to a maximum value. Following this, the condensate is revaporized as the pressure continues to decrease. Based on the study of the effect of nano-pores on fluid flow, the fluid properties, especially gas condensate fluid properties in nano-pores could be different [49–52]. Whether the condensate saturation could be less or more with the confinement effect is not exactly known. However, it is certain that condensate blockage does exist in shale gas condensate reservoirs. As the objective of this study is to evaluate the enhanced condensate recovery performance of CO_2 huff-n-puff, confinement in our model would not have an impact on our study objective. Confinement is not taken into account.

As Figure 6 shows, only one well was set in the reservoir. The well was used as both an injection well and a production well. During primary depletion, the well was used as a production well. Afterwards, the well was used to inject CO_2. When the injection process was finished, the well was closed to allow for a period of soaking. Following this, the well was opened again as a production well. In our model, the maximum injection pressure was set to 4000 psi when the well was used as an injection well; the minimum bottom-hole pressure was set to 1500 psi when the well was used as a production well.

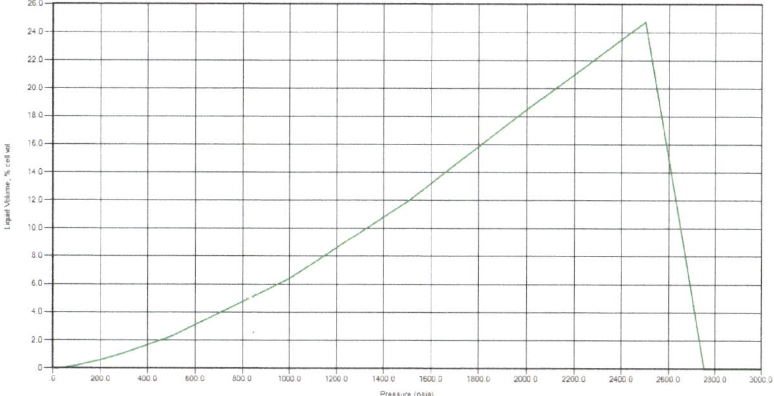

Figure 7. The liquid dropout curve of reservoir fluid at 93.3 °C (200 °F).

Table 3. Reservoir fluid composition.

Name	Composition
CO_2	0.18
N_2	0.13
CH_4	61.92
C_2H_6	14.08
C_3H_8	8.35
IC_4	0.97
NC_4	3.41
IC_5	0.84
NC_5	1.48
NC_6	1.79
NC_7	1.58
NC_8	1.22
NC_9	0.94
$C_{10}{}^+$	3.11

3. Results and Discussion

3.1. Experimental Results

As was mentioned in the previous experiment procedure, five cycles of CO_2 huff-n-puff process were performed on the shale core. After primary depletion, the pressure was decreased to 1460 psi. The accumulated condensate saturation was 10.8% after primary depletion. The condensate saturation was decreased to 7.5% after 5 cycles.

Condensate recovery was obtained by analyzing the condensate saturation decrease as shown in Figures 8 and 9. The condensate recovery was enhanced to 30.36% after 5 cycles of CO_2 huff-n-puff. The experiment results indicate CO_2 huff-n-puff can effectively enhance the condensate recovery from the shale core. In addition, the first cycle of CO_2 huff-n-puff had the highest condensate recovery, at 16.25%. Condensate recovery was reduced significantly after the first cycle, with the 5th cycle only having a 1.2% recovery increment as shown in Figure 10.

Therefore, it is important to set proper cycle numbers during the application of CO_2 huff-n-puff process. Efficiency of CO_2 huff-n-puff can be very low when the number of cycles reaches a critical value.

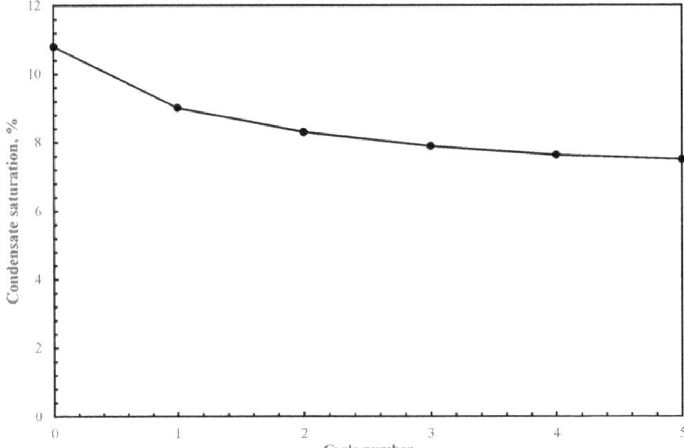

Figure 8. Variation of condensate saturation.

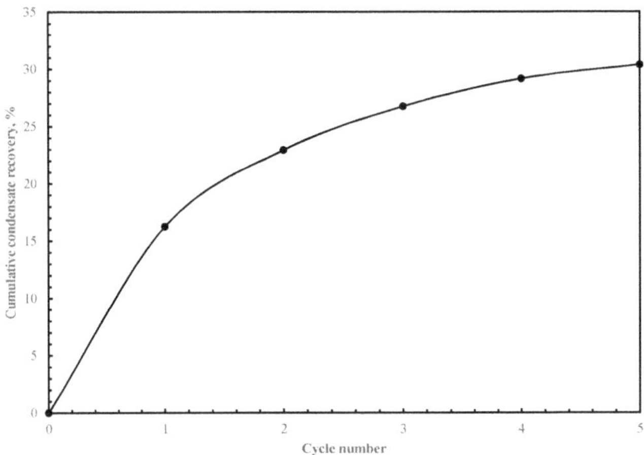

Figure 9. Cumulative condensate recovery.

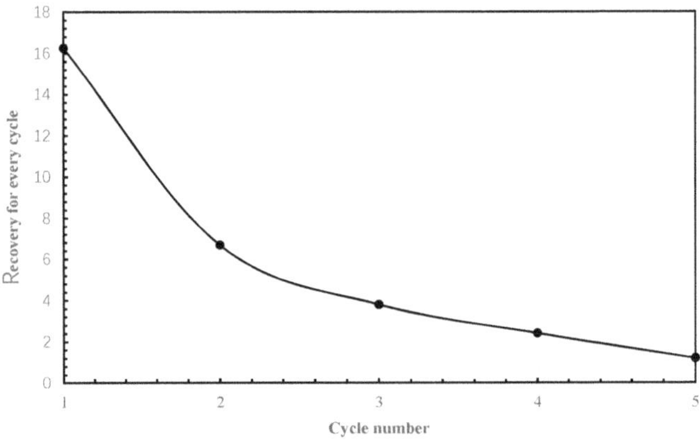

Figure 10. Increment of condensate recovery for every cycle.

3.2. Simulation Results

3.2.1. Base Case

A base CO_2 huff-n-puff case study was conducted with two scenarios and a total exploration time of 8255 days. In the first scenario, the primary depletion period was 8255 days, and the production pressure was 1500 psi. In the second one, the primary depletion time period was 5475 days, after which, CO_2 huff-n-puff was performed. The injection pressure was set to 4000 psi. The production pressure was set to 1500 psi. Four cycles of CO_2 huff-n-puff were performed. The comparison of cumulative condensate recovery is shown in Figure 11. After 8255 days of primary depletion, the condensate recovery was 17.7%. However, after the CO_2 huff-n-puff was applied, the condensate recovery was increased to 24.7%. The condensate recovery was effectively enhanced after the operation of CO_2 huff-n-puff.

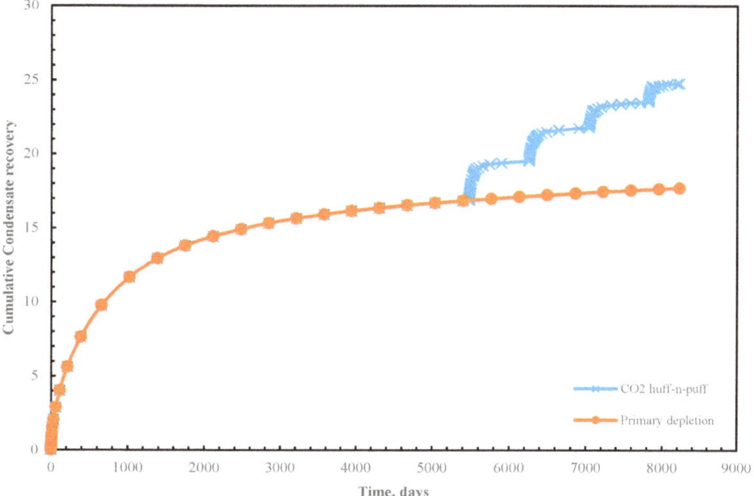

Figure 11. Comparison of cumulative condensate recovery.

3.2.2. Effect of Injection Pressure and Injection Period

Four different cases were conducted in this section, and the effect of injection pressure on the enhanced condensate recovery performance of CO_2 huff-n-puff gas injection is shown in Figure 12. Higher condensate recovery was obtained when the injection pressure was higher, with cumulative condensate recovery factors of 19%, 22%, 24%, and 24.7% corresponding to the injection pressures of 2500 psi, 3000 psi, 3500 psi, and 4000 psi, respectively. The cumulative condensate recovery was increased by 3% when the injection pressure was increased from 2500 psi to 3000 psi. However, the cumulative condensate recovery was only increased by 0.7% when the injection pressure was from 3500 psi to 4000 psi.

As Figure 13 shows, the main condensate region was near the fracture region. After injecting the CO_2, the pressure of condensate region was increased with part of the condensate being revaporized. Thus the condensate could be produced during the puff process. Figure 14 shows pressure distribution of four cases after the huff process of the first cycle. When CO_2 was injected into the formation at 2500 psi, only the minor condensate could be revaporized. Thus, the efficiency of the CO_2 huff-n-puff was low. For case 3 (injection pressure: 3500 psi) and case 4 (injection pressure 4000 psi), it was found that condensate recovery was highly enhanced in both cases, and the condensate recovery of these two cases were similar. As shown in Figure 14, the pressure of the condensate region was increased. Most of the condensate near the fracture could be revaporized and recovered.

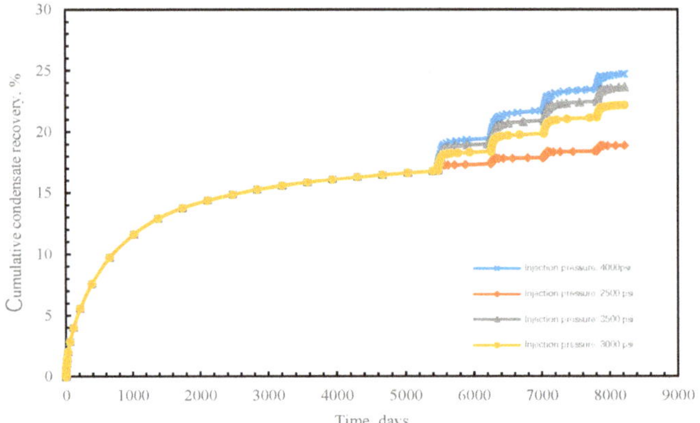

Figure 12. Cumulative condensate recovery with different injection pressure.

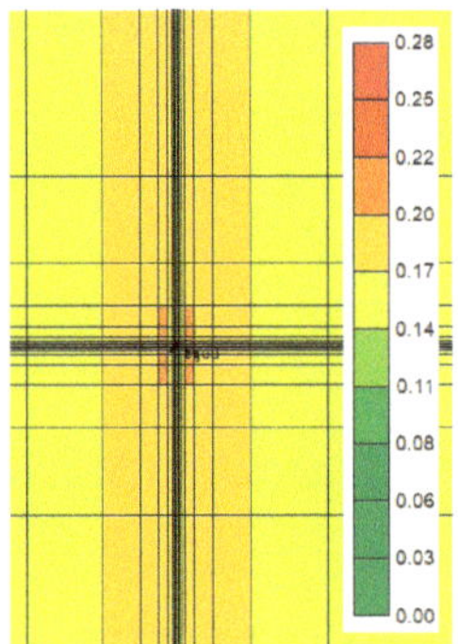

Figure 13. Main condensate region after primary depletion.

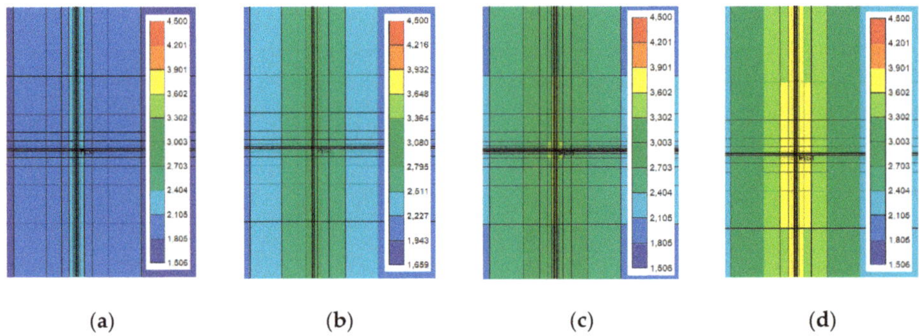

(a) (b) (c) (d)

Figure 14. Pressure distribution for different injection pressure. (**a**) Injection pressure: 2500 psi; (**b**) Injection pressure: 3000 psi; (**c**) Injection pressure: 3500 psi; (**d**) Injection pressure: 4000 psi.

Three cases with different injection time were conducted to investigate the effect of injection time on the performance of CO_2 huff-n-puff as shown in Figure 15. The production was same in three cases. Figure 15 indicates the cumulative condensate recovery for the three cases, being 18.6%, 22.7%, and

24.8%. During the puff process, more condensate could be recovered as the injection period was longer. However, it can be found that during the same reservoir exploitation period, the efficiency of 100 days injection period was similar as the efficiency of 50 days injection period. Figure 16 shows the pressure distribution of the condensate region. After 50 days of injection, the pressure was already higher than 2750 psi. Thus, in this model, a 50 days injection period was long enough to revaporize the condensate into a gas state and increase the condensate recovery.

It can be concluded from the above discussion that the design of the huff process should be based on the pressure variation of the main condensate region. Applying higher injection pressure or a longer injection period did not result in higher condensate recovery. The optimal huff process occurs when the pressure of condensate region is increased higher than the dew point pressure.

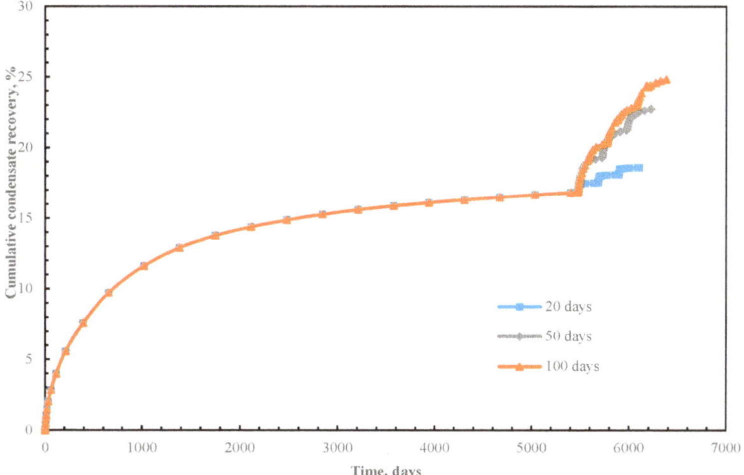

Figure 15. Cumulative condensate recovery for different injection period.

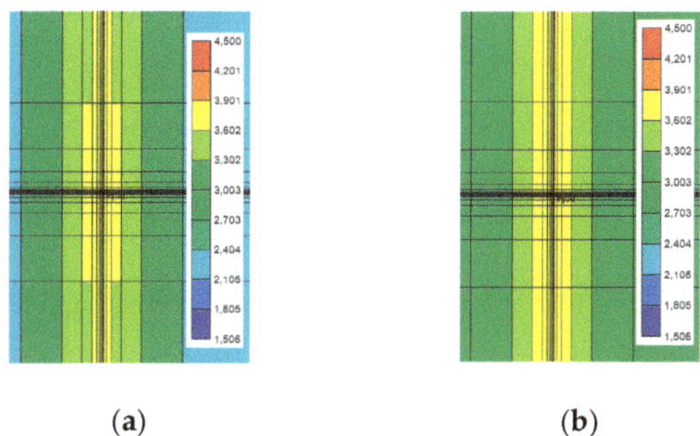

(**a**) (**b**)

Figure 16. Pressure distribution. (**a**) Injection time: 50 days; (**b**) Injection time: 100 days.

3.2.3. Effect of Soaking Period

A series of simulation work was conducted by applying different soaking time at different pressures. Table 4 shows the different scenarios. For all cases, the injection time was 100 days. The production period was 200 days and production pressure was 1460 psi. Three cycles of CO_2 huff-n-puff were operated.

Table 4. Different scenarios used in the study of soaking period effect.

Scenario	Case 1	Case 2	Case 3	Case 4	Case 5	Case 6
Soaking time	0 days	50 days	100 days	0 days	50 days	100 days
Injection pressure	3000 psi	3000 psi	3000 psi	5000 psi	5000 psi	5000 psi

The results show two different trends of cumulative condensate recovery as shown in Figure 17. When the injection pressure was 3000 psi, cumulative condensate recovery was decreased when the soaking time was increased. However, when the injection pressure was 5000 psi, cumulative condensate recovery was increased when the soaking time was increased (Figure 18).

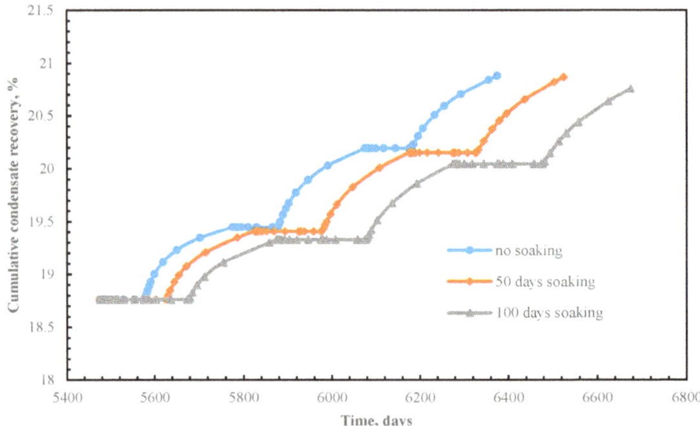

Figure 17. Cumulative condensate recovery for different soaking at 3000 injection pressure.

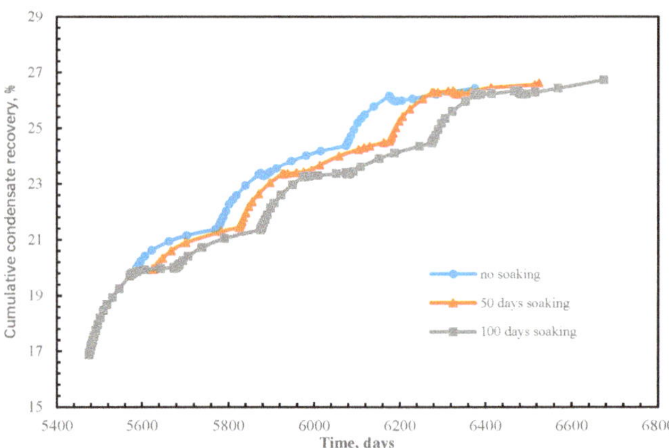

Figure 18. Cumulative condensate recovery for different soaking at 5000 injection pressure.

Figure 19 shows the pressure distribution of condensate region when the injection pressure was 3000 psi. After initial injection, the pressure of the near wellbore region was higher than 2750 psi. However, after a 50 days soaking period, the pressure was decreased and the liquid condensate was accumulated again near the fracture. This is because during the soaking process, the pressure was transferred to the distal region of the reservoir. In this situation, the condensate was still formed near the fracture during the soaking period and it had a negative effect on the efficiency of CO_2 huff-n-puff gas injection.

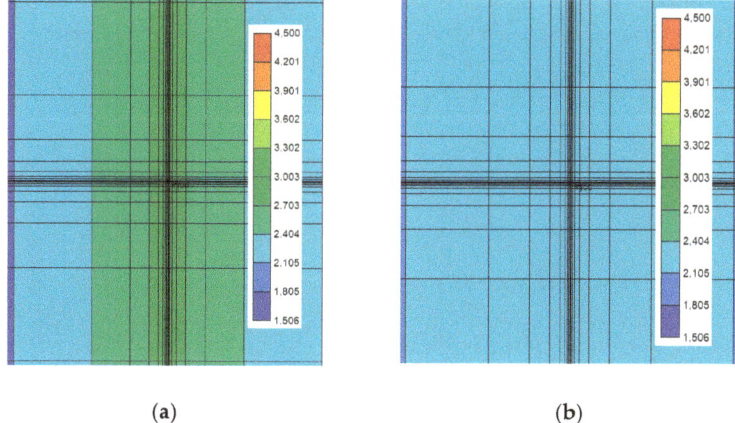

(a)   (b)

Figure 19. Pressure distribution near the fracture, injection pressure: 3000 psi; (**a**) start of soaking time; (**b**) end of soaking time.

However, when the pressure was 5000 psi, the soaking period had a positive effect on the performance of CO_2 huff-n-puff. The pressure was still higher than 2750 psi after 50 days of soaking as shown in Figure 20. More condensate could be revaporized into a gas state. Thus, more condensate could be recovered during the puff process.

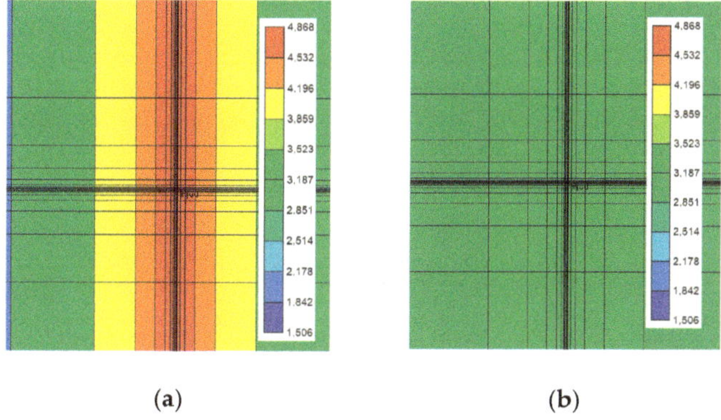

(a)   (b)

Figure 20. Pressure distribution near the fracture, injection pressure: 5000 psi; (**a**) start of soaking time; (**b**) end of soaking time.

It can be concluded that whether a soaking process should be applied or not depends on the injection pressure. When the injection pressure is similar as the dew point pressure, a soaking process could have a negative effect on the efficiency of CO_2 huff-n-puff. However, when the injection pressure is much higher, a soaking process is recommended. The determination of soaking time depends on the area of the condensate region. In general, during the soaking process, the pressure of the main condensate region should be remained higher than the dew point pressure again, otherwise the condensate could be formed again and the efficiency of huff-n-puff would be decreased.

3.2.4. Effect of CO_2 Diffusion

Figure 21 shows the effect of CO_2 diffusion on the performance of CO_2 huff-n-puff. Two cases were conducted in this section. The injection time, soaking time and production time were the same in both cases. Results show that when the CO_2 diffusion coefficient was taken into account in the simulation, the condensate would be lower. When the CO_2 diffusion coefficient was considered in the model, CO_2 could be flowed into the distal region during the 100 days soaking period. The pressure could be decreased and the condensate could be formed again. Thus. CO_2 diffusion plays an important role in enhancing condensate recovery during the application of CO_2 huff-n-puff.

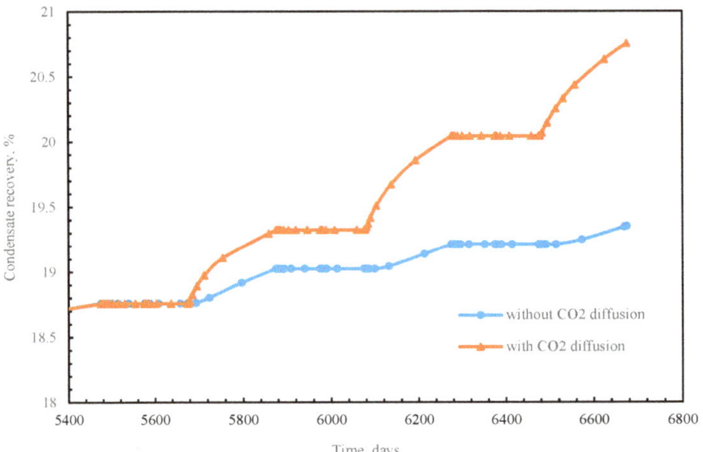

Figure 21. Cumulative condensate recovery with and without CO_2 diffusion.

3.2.5. Effect of Cycle Numbers

Figure 22 indicates the effect of the cycle number on cumulative condensate recovery. As the cycle number of CO_2 huff-n-puff was increased, cumulative condensate recovery increased. The cumulative condensate recovery was enhanced to 27.2% after 10 cycles. Compared with primary condensate recovery, the increment of condensate recovery after 10 cycles was 9%. However, the recovery was increased to 29.3% after 18 cycles. The increment of the recovery was 1.4%. The latter cycles of CO_2 huff-n-puff gas injection resulted in lower efficiency of enhanced condensate recovery. By considering these simulation and experimental results, it can be concluded that the number of cycle numbers of CO_2 huff-n-puff gas injection are important. When the cycle number of CO_2 huff-n-puff gas injection reaches a critical value, the efficiency of CO_2 huff-n-puff could be decreased.

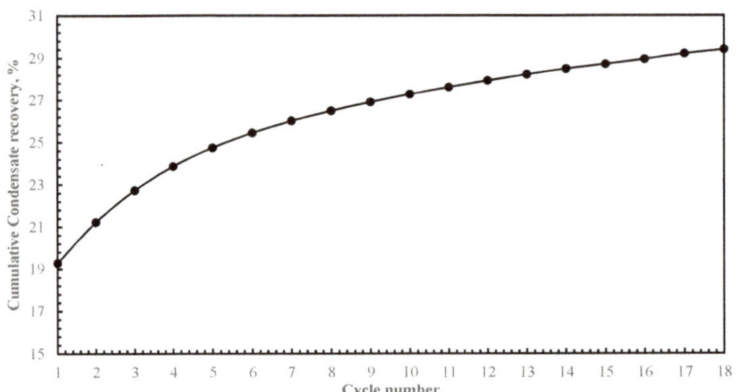

Figure 22. Cumulative condensate recovery at different cycle number.

3.3. Recommended Future Work

In this work, a binary gas mixture was used in the experiment. It was used because this gas condensate mixture has a wide condensate region at the room temperature. Thus, the experiment could be handled in a more convenient way and the accuracy of the experiment could be improved. This study indicates the efficiency of the CO_2 huff-n-puff to enhance condensate recovery in shale gas condensate reservoirs. In future work, gas condensate mixture from a real reservoir is recommended to be used in the experiment. The experiment can be conducted using the reservoir condition.

In addition, the experiment was conducted on the small cores and the results show an impressive high condensate recovery. The core-scale laboratory results cannot be directly applied to predict the field-scale recovery of a practical shale condensate well. Due to the extremely low permeability of shale formation, the injected gas can only penetrate a limited depth of the formation. For the field conditions, if the shale matrix is intersected by high density hydraulic and natural fractures, then the proportion of penetrated matrix by injected gas will be much higher, which yields a higher recovery. However, if only the several main fractures are formed (not forming fracture networks) after the hydraulic fracturing operation, only a near fracture matrix can be invaded by the injection gas, which yields a much lower recovery. Therefore, the size effect plays a significant role for the condensate recovery factor, which is recommended for future work.

4. Conclusions

An experimental study on shale core was operated to evaluate the enhanced condensate recovery efficiency of CO_2 huff-n-puff. Also, a field scale numerical simulation model was built to investigate the performance of CO_2 huff-n-puff. The purpose of this study was to evaluate the enhanced condensate recovery performance of CO_2 huff-n-puff in shale gas condensate reservoirs. The conclusions of this study are drawn as follows:

1. The results indicate that the condensate recovery can be effectively enhanced after the application of CO_2 huff-n-puff. The condensate recovery was increased to 30.36% after 5 cycles of CO_2 huff-n-puff in the experiment. In the simulation work, the condensate was enhanced to 24.7% after 4 cycles.
2. Injection period and injection pressure should be optimized to ensure that the pressure of the condensate region remains higher than the dew point pressure after the huff process.
3. Soaking periods should be based on the injection pressure. During the soaking periods, the pressure of the condensate region should remain higher than the dew point pressure. If this does not occur, condensate can be formed and the efficiency of huff-n-puff is decreased. When the injection pressure is much higher than the dew point pressure, soaking is recommended. Otherwise, the soaking should be neglected.
4. The determination of cycle number should depend on the condensate increment of every cycle. When the cycle number reaches a critical value, condensate recovery decreases as does the efficiency of CO_2 huff-n-puff.

Author Contributions: Conceptualization, X.M.; Methodology, X.M. and Z.M.; Formal analysis, X.M. and Z.M.; Investigation, X.M., Z.M. and J.M.; Writing—original draft preparation, X.M.; Writing—review and editing, X.M., Z.M., and T.W.; Funding Acquisition, X.M.

Funding: This research was funded by China Postdoctoral Science Foundation Funded Project (Grant 2017M622317).

Acknowledgments: The Lab of Unconventional Oil and Gas Resources Development in China University of Petroleum (East China) is highly appreciated.

Conflicts of Interest: The authors declare no conflict of interest.

References

1. Wheaton, R.J.; Zhang, H.R. Condensate banking dynamics in gas condensate Fields: Compositional changes and condensate accumulation around production wells. In Proceedings of the SPE Annual Technical Conference and Exhibition, Dallas, TX, USA, 1–4 October 2000. [CrossRef]
2. Ayyalasomayajula, P.S.; Silpngarmlers, N.; Kamath, J. Well deliverability predictions for a low permeability gas condensate reservoir. In Proceedings of the SPE Annual Technical Conference and Exhibition, Dallas, TX, USA, 9–12 October 2005. [CrossRef]
3. Hinchman, S.B.; Barree, R.D. Productivity loss in gas condensate reservoirs. In Proceedings of the SPE Annual Technical Conference and Exhibition, Las Vegas, NV, USA, 22–26 September 1985. [CrossRef]
4. Muladi, A.; Pinczewski, W.V. Application of horizontal well in heterogeneity gas condensate reservoir. In Proceedings of the SPE Asia Pacific Oil and Gas Conference and Exhibition, Jakarta, Indonesia, 20–22 April 1999. [CrossRef]
5. Miller, N.; Nasrabadi, H.; Zhu, D. On application of horizontal wells to reduce condensate blockage in gas condensate reservoirs. In Proceedings of the SPE International Oil and Gas Conference and Exhibition in China, Beijing, China, 8–10 June 2010. [CrossRef]
6. Carlson, M.R.M.; Myer, J.W.G. The effects of retrograde liquid condensation on single well productivity determined via direct modelling of a hydraulic fracture in a low permeability reservoir. In Proceedings of the Low Permeability Reservoirs Symposium, Denver, Colorado, 19–22 March 1995. [CrossRef]
7. Aly, A.M.; El-Banbi, A.H.; Holditch, S.A.; Wahdan, M.; Salah, N. Optimization of gas condensate reservoir development by coupling reservoir modeling and hydraulic fracturing design. In Proceedings of the SPE Middle East Oil Show, Manama, Bahrain, 17–20 March 2001. [CrossRef]
8. Ignatyev, A.E.; Mukminov, I.; Vikulova, E.A.; Pepelyayev, R.V. Multistage hydraulic fracturing in horizontal wells as a method for the effective development of gas-condensate fields in the arctic region (Russian). In Proceedings of the SPE Arctic and Extreme Environments Conference and Exhibition, Moscow, Russia, 18–20 October 2011. [CrossRef]
9. Zhang, L.; Kou, Z.; Wang, H.; Zhao, Y.; Dejam, M.; Guo, J.; Du, J. Performance analysis for a model of a multi-wing hydraulically fractured vertical well in a coalbed methane gas reservoir. *J. Pet. Sci. Eng.* **2018**, *166*, 104–120. [CrossRef]
10. Dejam, M.; Hassanzadeh, H.; Chen, Z. Semi-analytical solution for pressure transient analysis of a hydraulically fractured vertical well in a bounded dual-porosity reservoir. *J. Hydrol.* **2018**, *565*, 289–301. [CrossRef]
11. Bang, V.S.S.; Pope, G.A.; Sharma, M.M.; Baran, J.R. Development of a successful chemical treatment for gas wells with liquid blocking. In Proceedings of the SPE Annual Technology Conference and Exhibition, New Orleans, LA, USA, 4–7 October 2009. [CrossRef]
12. Al-Anazi, H.A.; Walker, J.G.; Pope, G.A.; Sharma, M.M.; Hackney, D.F. A successful methanol treatment in a gas-condensate reservoir: Field application. In Proceedings of the SPE Production and Operation Symposium, Oklahoma City, OK, USA, 23–26 March 2003. [CrossRef]
13. Al-Anazi, H.A.; Pope, G.A.; Sharma, M.M. Laboratory measurements of condensate blocking and treatment for both low and high permeability rocks. In Proceedings of the SPE Annual Technical Conference and Exhibition, San Antonio, TX, USA, 29 September–2 October 2002. [CrossRef]
14. Li, K.; Firoozabadi, A. Experimental study of wettability alteration to preferential gas-wetting in porous media and its Effects. *SPE Reserv. Eval. Eng.* **2000**, *3*, 139–149. [CrossRef]
15. Ahmadi, M.; Sharma, M.M.; Pope, G.; Torres, D.E.; McCulley, C.A.; Linnemeyer, H. Chemical treatment to mitigate condensate and water blocking in gas wells in carbonate reservoirs. *SPE Prod. Oper.* **2011**, *26*, 67–74. [CrossRef]
16. Zheng, Y.; Rao, D.N. Surfactant-Induced spreading and wettability effects in condensate reservoirs. In Proceedings of the SPE Improved Oil Recovery Symposium, Tulsa, OK, USA, 24–28 April 2010. [CrossRef]
17. Ganjdanesh, R.; Rezaveisi, M.; Pope, G.A.; Sepehrnoori, K. Treatment of condensate and water blocks in hydraulic fractured shale gas-condensate reservoirs. In Proceedings of the SPE Annual Technical Conference and Exhibition, Houston, TX, USA, 28–30 September 2015. [CrossRef]
18. Abel, W.; Jackson, R.F.; Wattenbarger, R.A. Simulation of a partial pressure maintenance gas cycling project with a compositional model, Carson Creek Field, Alberta. *J. Pet. Technol.* **1970**, *22*, 38–46. [CrossRef]

19. Luo, K.; Li, S.; Zheng, X.; Chen, G.; Dai, Z.; Liu, N. Experimental investigation into revaporization of retrograde condensate by lean gas injection. In Proceedings of the SPE Asia Pacific Oil and Gas Conference and Exhibition, Jakarta, Indonesia, 17–19 April 2001. [CrossRef]
20. Aziz, R.M. A 1982 critique on gas cycling operations on gas-condensate reservoirs. In Proceedings of the Middle East Oil Technical Conference and Exhibition, Manama, Bahrain, 14–17 March 1983. [CrossRef]
21. Al-Wadhahi, M.; Boukadi, F.H.; Al-Bemani, A.; Al-Maamari, R. Huff n puff to revaporize liquid dropout in an Omani gas field. *J. Pet. Sci. Eng.* **2007**, *55*, 67–73. [CrossRef]
22. Siregar, S.; Hagoort, J.; Ronde, H. Nitrogen injection vs. gas cycling in rich retrograde condensate-gas reservoirs. In Proceedings of the SPE International Meeting on Petroleum Engineering, Beijing, China, 24–27 March 1992. [CrossRef]
23. Stalkup, F.I. Carbon dioxide miscible flooding: Past, present, and outlook for the Future. *J. Pet. Technol.* **1978**, *30*, 1102–1112. [CrossRef]
24. Gohary, M.E.; Bairaq, A.M.A.; Bradley, D.C.; Saeed, Y. Comparison of condensate recovery by hydrocarbon and nonhydrocarbon injection. In Proceedings of the SPE Abu Dhabi International Petroleum Exhibition and Conference, Abu Dhabi, UAE, 9–12 November 2015. [CrossRef]
25. Odi, U. Analysis and potential of CO_2 huff-n-puff for near wellbore condensate removal and enhanced gas recovery. In Proceedings of the SPE Annual Technical Conference and Exhibition, San Antonio, TX, USA, 8–10 October 2012. [CrossRef]
26. Kurdi, M.; Xiao, J.; Liu, J. Impact of CO_2 injection on condensate banking in gas condensate reservoirs. In Proceedings of the SPE Saudi Arabia section Young Professionals Technical Symposium, Dhahran, Saudi Arabia, 19–21 March 2012. [CrossRef]
27. Gachuz-Muro, H.; Gonzalez Valtierra, B.E.; Luna, E.E.; Aguilar Lopez, B. Laboratory tests with CO_2, N_2 and lean natural gas in a naturally fractured gas-condensate reservoir under HP/HT Conditions. In Proceedings of the SPE Enhanced Oil Recovery Conference, Kuala Lumpur, Malaysia, 19–21 July 2011. [CrossRef]
28. Moradi, B.; Tangsiri Fard, J.; Rasaei, M.R.; Momeni, A.; Bagheri, M.B. Effect of gas recycling on the enhancement of condensate recovery in an Iranian fractured gas/condensate reservoir. In Proceedings of the Trinidad and Tobago Energy Resources Conference, Poet of Spain, Trinidad, 27–30 June 2010. [CrossRef]
29. Sheng, J.J. Increase liquid oil production by huff-n-puff of produced gas in shale gas condensate reservoirs. *J. Unconv. Oil Gas Resour.* **2015**, *11*, 19–26. [CrossRef]
30. Zhao, H.; Firoozabadi, A. Sorption hysteresis of light hydrocarbons and carbon dioxide in shale and kerogen. *Sci. Rep.* **2017**, *7*, 16209. [CrossRef]
31. Dejam, M.; Hassanzadeh, H.; Chen, Z. Pre-Darcy flow in porous media. *Water Resour. Res.* **2017**, *53*, 187–8210. [CrossRef]
32. Jin, Z.; Firoozabadi, A. Flow of methane in shale nanopores at low and high pressure by molecular dynamics simulations. *J. Chem. Phys.* **2015**, *143*, 104315. [CrossRef] [PubMed]
33. Dejam, M. Advective-diffusive-reactive solute transport due to non-Newtonian fluid flows in a fracture surrounded by a tight porous medium. *Int. J. Heat Mass Transf.* **2019**, *128*, 1307–1321. [CrossRef]
34. Dejam, M.; Hassanzadeh, H.; Chen, Z. Shear dispersion in combined pressure-driven and electro-osmotic flows in a capillary tube with a porous wall. *Am. Inst. Chem. Eng. J.* **2015**, *61*, 3981–3995. [CrossRef]
35. Dejam, M.; Hassanzadeh, H.; Chen, Z. Shear dispersion in combined pressure-driven and electro-osmotic flows in a channel with porous walls. *Chem. Eng. Sci.* **2015**, *137*, 205–215. [CrossRef]
36. Sharma, S.; Sheng, J.J. A comparative study of huff-n-puff gas and solvent injection in a shale gas condensate core. *J. Nat. Gas Sci. Eng.* **2017**, *38*, 549–565. [CrossRef]
37. Ren, B.; Ren, S.; Zhang, L.; Chen, G.; Zhang, H. Monitoring on CO_2 migration in a tight oil reservoir during CCS-EOR in Jilin Oilfield China. *Energy* **2016**, *98*, 108–121. [CrossRef]
38. Zuloaga, P.; Yu, W.; Miao, J. Performance evaluation of CO_2 Huff-n-Puff and continuous CO_2 injection in tight oil reservoirs. *Energy* **2017**, *134*, 181–192. [CrossRef]
39. Song, C.; Yang, D. Experimental and numerical evaluation of CO_2 huff-n-puff processes in Bakken formation. *Fuel* **2017**, *190*, 145–162. [CrossRef]
40. Sheng, J.J.; Chen, K. Evaluation of the EOR potential of gas and water injection in shale oil reservoirs. *J. Unconv. Oil Gas Resour.* **2014**, *5*, 1–9. [CrossRef]
41. Abedini, A.; Torabi, F. Oil recovery performance of immiscible and miscible CO_2 huff-and-puff processes. *Energy Fuels* **2014**, *28*, 774–784. [CrossRef]

42. Shi, C.; Horne, R.N. Improved recovery in gas-condensate reservoirs considering compositional variations. In Proceedings of the SPE Annual Technical Conference and Exhibition, Denver, CO, USA, 21–24 September 2008. [CrossRef]

43. Wan, T.; Sheng, J.J.; Soliman, M.Y. Evaluation of EOR potential in fractured shale oil Reservoirs by cyclic gas injection. *Pet. Sci. Technol.* **2015**, *33*, 812–818. [CrossRef]

44. Nezhad, M.M.; Fisher, Q.J.; Gironacci, E.; Rezania, M. Experimental study and numerical modeling of fracture propagation in shale rocks during Brazilian disk test. *Rock Mech. Rock Eng.* **2018**, *51*, 1755–1775. [CrossRef]

45. Josh, M.; Esteban, L.; Delle Piane, C.; Sarout, J.; Dewhurst, D.N.; Clennell, M.B. Laboratory characterisation of shale properties. *J. Pet. Sci. Eng.* **2012**, *88*, 107–124. [CrossRef]

46. Mello, M.R.; Telnaes, N.; Gaglianone, P.C.; Chicarelli, M.I.; Brassell, S.C.; Maxwell, J.R. Organic geochemical characterisation of depositional palaeoenvironments of source rocks and oils in Brazilian marginal basins. *Org. Geochem. Pet. Explor.* **1988**, *13*, 31–45. [CrossRef]

47. Mousavi Nezhad, M.; Gironacci, E.; Rezania, M.; Khalili, N. Stochastic modelling of crack propagation in materials with random properties using isometric mapping for dimensionality reduction of nonlinear data sets. *Int. J. Numer. Methods Eng.* **2018**, *113*, 656–680. [CrossRef]

48. Li, Y.; Pu, H. Modeling study on CO_2 capture and storage in organic-rich shale. In Proceedings of the Carbon Management Technology Conference, Sugar Land, TX, USA, 17–19 November 2015. [CrossRef]

49. Li, B.; Mezzatesta, A. Evaluation of pore size distribution effects on phase behavior of hydrocarbons produced in shale gas condensate reservoirs. In Proceedings of the SPE Middle East Oil & Gas Show and Conference, Manama, Kingdom of Bahrain, 6–9 March 2017. [CrossRef]

50. Wang, X.; Sheng, J.J. Pore network modeling of the non-Darcy flows in shale and tight formations. *J. Pet. Sci. Eng.* **2018**, *163*, 511–518. [CrossRef]

51. Wang, X.; Sheng, J.J. Understanding oil and gas flow mechanisms in shale reservoirs using SLD–PR transport model. *Transp. Porous Media* **2018**, *119*, 337–350. [CrossRef]

52. Altman, R.M.; Fan, L.; Sinha, S.; Stukan, M.; Viswanathan, A. Understanding mechanisms for liquid dropout from horizontal shale gas condensate wells. In Proceedings of the SPE Annual Technical Conference and Exhibition, Amsterdam, The Netherlands, 27–29 October 2014. [CrossRef]

Article

Experimental and Simulation Studies on Adsorption and Diffusion Characteristics of Coalbed Methane

Donghyeon Kim [1,†], Youngjin Seo [1,†], Juhyun Kim [1], Jeongmin Han [2] and Youngsoo Lee [1,*]

[1]   Department of Mineral Resources and Energy Engineering, Chonbuk National University, 567 Baekje-daero, Deokjin-gu, Jeonju-si, Jeollabuk-do 561-756, Korea
[2]   Korea Gas Corporation, 638-1 Sangnok-gu, Ansan, Gyoggi-do 15328, Korea
*   Correspondence: Youngsoo.lee@jbnu.ac.kr; Tel.: +82-63-270-2366
†   These authors equally contribute to this work.

Received: 2 July 2019; Accepted: 3 September 2019; Published: 6 September 2019

Abstract: Coalbed methane (CBM) content is generally estimated using the isotherm theory between pressure and adsorbed amounts of methane. It usually determines the maximum content of adsorbed methane or storage capacity. However, CBM content obtained via laboratory experiment is not consistent with that in the in-situ state because samples are usually ground, which changes the specific surface area. In this study, the effect of the specific surface area relative to CBM content was investigated, and diffusion coefficients were estimated using equilibrium time analysis. The differences in adsorbed content with sample particle size allowed the determination of a specific surface area where gases can adsorb. Also, there was an equilibrium time difference between fine and lump coal, because more time is needed for the gas to diffuse through the coal matrix and adsorb onto the surface in lump coal. Based on this, we constructed a laboratory-scale simulation model, which matched with experimental results. Consequently, the diffusion coefficient, which is usually calculated through canister testing, can be easily obtained. These results stress that lump coal experiments and associated simulations are necessary for more reliable CBM production analysis.

Keywords: coalbed methane; gas content; diffusion coefficient; reservoir simulation

---

1. Introduction

Coalbed gas, coalbed methane (CBM), and coalbed natural gas are terms that refer to the characteristic gases adsorbed in coal [1]. CBM development began in the San Juan basin in the mid-1980s, and CBM production became a new industry in the oil and gas domain [2]. CBM is often extracted from depths lower than conventional natural gas wells, and the methane storage capacity in coal layers is 2–3 times higher than conventional gas reservoirs at the same depth and pressure. Therefore, CBM is usually easier to develop than a traditional natural gas reservoir [3].

CBM is generated during the maturation of organic matter and by microbes residing in coal seams [4]. CBM is composed of methane, with minor amounts of carbon dioxide and nitrogen [5]. Of the produced gas, 98% is adsorbed via van der Waals forces onto the surface of the organic matter. The remaining 2% is compressed in natural cracks or dissolved during water formation [6]. The essential elements in determining gas content are the coal rank and the burial depth. Commonly, a higher coal rank and increased depth denote higher gas content [7].

Typically, the internal surface area of coal, which has a significant impact on its ability to contain methane, differs with coal rank and composition. The internal surface area in the coalbed, which greatly affects the ability to contain CH_4, usually varies according to the rank and composition of the coal; the largest value is observed in bituminous coal, followed by anthracite. Among components, the largest value is observed for vitrinite, and the smallest is observed for inertinite [8]. Also, physical changes cause differences in the amount of gas adsorbed. These primarily occur during the measurement of gas content in laboratories. Specific surface area is one of the main parameters that undergo physical changes. A smaller size of coal particle results in a larger internal surface area that the gas can adsorb onto [9]. Beamish and O'Donnell [10] studied the change in gas content with the increase in surface area by crushing bituminous coal to 250 mm. Zhang [11] suggested that the specific surface area increases as the coal particle size becomes smaller since the amount of ash and moisture interfering with gas adsorption decreases. Thus, a thorough understanding of all factors is essential for the accurate determination of gas content in a CBM reservoir.

Furthermore, the mechanism of gas flow through the tight coal matrix and cleats is highly complex. In particular, diffusional flow is dominant in the matrix due to low permeability. Therefore, the diffusion coefficient is a crucial factor that dramatically affects gas flow. The diffusion coefficient is roughly obtained from the sorption time in the canister test, which is time-consuming and costly.

In this study, adsorption experiments with various size fractions of coal were performed to determine the effect of specific surface area on the amount of gas adsorbed. In this manner, the diffusion flow of gas through coal with different particle sizes could be determined through the analysis of the adsorption equilibrium time. Following this, a history matching technique was applied to calculate the diffusion coefficient by combining the experimental results with the simulation results. This procedure is believed to be more economical in terms of cost and time compared with canister test. It may also serve to address the shortcomings of the indirect method and to provide a more accurate estimation of CBM.

2. Materials and Methods

2.1. Gas Content Measurement

There are two standard ways of measuring gas content, a direct method and an indirect method. The direct method measures the gas content of coal obtained during the drilling operation, and measures the volume of desorbed gas, lost gas, and residual gas via canister desorption [12]. The direct method is widely used in the petroleum industry for gas content estimation because it uses in situ samples, and the diffusion coefficient can be calculated by sorption time analysis. For this reason, it is considered to be a more reliable method than the indirect method. However, the direct method has the following disadvantages: (1) it is difficult to calculate the amount of lost gas that occurs during the core extraction from the underground to the surface; (2) the entire test process usually takes a long time, from two weeks to four months; and (3) the measurement cost is high [13].

In contrast, the indirect method can measure the maximum gas content that can be adsorbed onto the coal surface. It is generally analyzed using the sorption characteristics between coal and gas and a sorption isotherm with pressure [14]. The indirect method usually takes less time, and results are more easily obtained. However, it has the following disadvantages: (1) the physical properties of coal such as ash, moisture, cleat, and pore structure can be altered when a coal sample is crushed to appropriately sized pieces; (2) since the experiment is conducted using single-component gas, it cannot be applied directly where multi-component gases are adsorbed. Therefore, there is an inconsistency between experimental and in situ gas content.

2.2. Sorption Isotherm

Gas adsorption refers to the phenomenon of gas adherence onto a solid surface. The adsorbed volume increases as pressure increases and temperature decreases. The isotherm theory is usually used to explain adsorption characteristics.

An adsorption isotherm shows the relationship between the adsorbed volume and the pressure when the adsorption equilibrium occurs at a specific temperature. Therefore, it is critical to analyze the adsorption characteristics to accurately estimate the storage capacity [15]. Until now, many previous studies applied the adsorption mechanism to different isotherm curves such as the Langmuir, Freundlich, and Brunauer–Emmett–Teller (BET) models. Among them, the Langmuir model is widely accepted in the petroleum industry. In previous CBM studies, various sorption isotherm models were tested using experimental data to find the best-fitted model, and it was found that the Langmuir isotherm model showed the lowest absolute error and a better regression value for all cases [16].

The Langmuir model is derived from the assumptions that gas has single-layer adsorption on a uniform sample surface, and the adsorption amount and the desorption amount are in a dynamic equilibrium state. More precisely, the assumptions are as follows [17]:

One gas molecule is adsorbed onto a single adsorption site;

There is no interaction between the adsorbed molecules;

The adsorbed gas has the properties of an ideal gas.

The Langmuir equation can be written in the following form, described as shown in Figure 1:

$$V_a = \frac{V_L \times P}{P_L + P},\tag{1}$$

where V_L is the maximum content of adsorbed gas, P_L is the pressure at half V_L, and V_a is the total adsorption content.

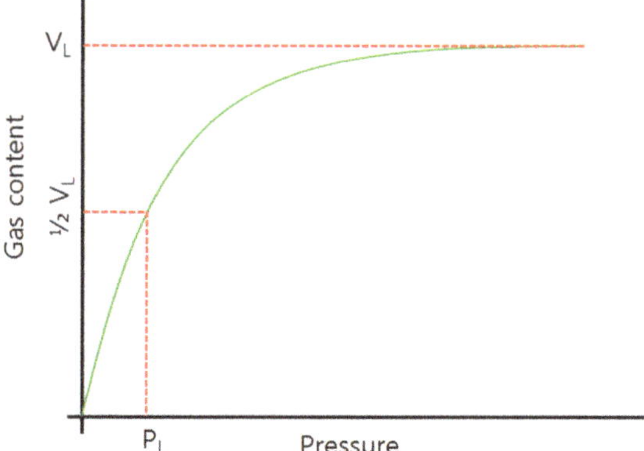

Figure 1. Langmuir isotherm.

2.3. Experimental Method

The indirect methods that measure the adsorbed volume of gas include volumetric measurement and gravimetric measurement. The volumetric analysis calculates the quantities of adsorbed gas with pressure. The gravimetric measurement is a calculation method achieved by directly checking the mass change in a reactor at a constant pressure.

In this study, BELSORP high-pressure adsorption equipment, which applies the volumetric method, was used for the measurement of adsorbed gas. A schematic diagram of the apparatus is drawn in Figure 2. To calculate the adsorbed amount of gas, the gas amount remaining after adsorption was subtracted from the injected amount, since the volumetric method could not directly estimate the amount of adsorbed gas.

The volumetric method calculates the gas content of coal through four steps. With the adsorption calculation equation, the volume of adsorption is estimated according to the equation of state using the temperature and the equilibrium pressure obtained from the measurement.

The total number of moles of adsorbed gas is calculated using Equation (2), and the total volume of adsorbed gas in each step is obtained using Equation (3) [18].

$$V^{(k)} = V^{(k-1)} + \frac{n \times 22,414}{SW},\tag{2}$$

where $V^{(k)}$ is the gas volume at k step, n is the number of molecules of adsorbed gas, k is the adsorption step, and SW is the sample weight.

$$n = n1 - n2 + n3 - n4, \tag{3}$$

where n is the total number of molecules, and $n1$, $n2$, $n3$, and $n4$ represent each molecule of adsorbed gas calculated according to each step.

The value of n for each step can be calculated as described below (Figure 2).

The experimental apparatus used for measuring the adsorption volume is divided into two parts at the location of valve C: a reference volume area Vs, and a dead volume area Vd.

At the $(k-1)^{th}$ point equilibrium, the amount of gas that did not adsorb onto the sample but remained in the gas phase, n1, is expressed as follows:

$$n1 = \frac{Pe^{(k-1)} \times (Vs + Vd)}{R \times Te^{(k-1)} \times Z\left(Pe^{(k-1)}\right)}. \tag{4}$$

At the $(k-1)^{th}$ point measurement, the amount of gas inside Vs after the C valve is closed is expressed in Equation (5), where $Te^{(k-1)}$ is the absolute temperature at $Pe^{(k-1)}$, $Pe^{(k-1)}$ is the measured pressure when the internal pressure in Vs and Vd is at equilibrium at the $(k-1)^{th}$ step, and R is the ideal gas constant (8314.102965 kPa·cm^3·mol^{-1}·K^{-1}.

$$n2 = \frac{Pe2^{(k-1)} \times Vs}{R \times Te2^{(k-1)} \times Z\left(Pe2^{(k-1)}\right)}. \tag{5}$$

The amount of gas inside Vd after the $(k-1)^{th}$ point measurement can be expressed as follows:

$$n1 - n2 \tag{6}$$

At the $(k)^{th}$ point measurement, the amount of gas administered into Vs, n3, is given by

$$n3 = \frac{Pi^{(k)} \times Vs}{R \times Ti^{(k)} \times Z\left(Pi^{(k)}\right)}, \tag{7}$$

where $Ti^{(k)}$ is the absolute temperature at $Pi^{(k)}$, $Pi^{(k)}$ is the measured pressure when the internal pressure in Vs and Vd is at equilibrium at the $(k)^{th}$ step.

At the $(k)^{th}$ point equilibrium, the amount of gas that did not adsorb onto the sample but remained in the gas phase, n4, is expressed by

$$n4 = \frac{Pe^{(k)} \times (Vs + Vd)}{R \times Te^{(k)} \times Z\left(Pe^{(k)}\right)}, \tag{8}$$

where $Te^{(k)}$ is the absolute temperature at $Pe^{(k)}$, and $Pe^{(k)}$ is the measured pressure when the internal pressure in Vs and Vd is at equilibrium at the $(k)^{th}$ point.

Equation (6) is used to calculate the number of gas molecules remaining in Vd at the $(k-1)^{th}$ point before the new gas is injected, while Equations (7) and (8) calculate the number of gas molecules adsorbed at the $(k)^{th}$ point after the new gas is injected.

Commonly, gas compressibility has to be corrected to capture real gas behavior; hence, the virial coefficient (Equation (9)) was applied using the gas property database from the NIST (National Institute of Standards and Technology) chemistry web-book (https://webbook.nist.gov) [19]. Details of the experimental procedure were provided by Kim et al. [20].

$$Z = 1 + B(P)\rho + C(P)\rho^2 + D(P)\rho^3, \tag{9}$$

where Z is a compressibility factor, p is the equilibrium pressure, and $B(P)p$, $C(P)p$, and $D(P)p$ are the second, third, and fourth virial coefficients, respectively.

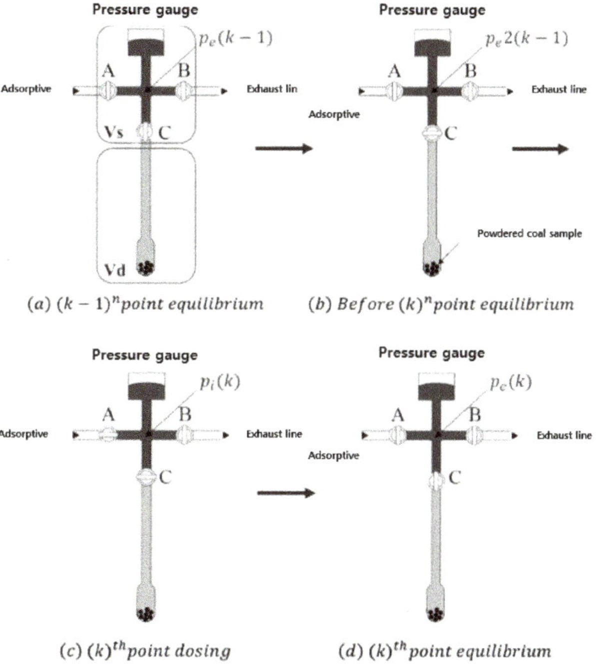

(a) $(k-1)^n$ point equilibrium *(b) Before $(k)^n$ point equilibrium*

(c) $(k)^{th}$ point dosing *(d) $(k)^{th}$ point equilibrium*

Figure 2. The process of the volumetric method.

Since conventional reactor experiments only allow a small volume of powdered coal to be used, a new reactor which can hold a lump coal sample was newly manufactured for the test. This reactor had the following characteristics: (1) a maximum pressure of 7000 kPa and a maximum temperature of 343.14 K are achievable; (2) various sample sizes can be used with a guide to minimize the dead volume. It was possible to measure the gas content of a sample up to 6 cm long on each side (Figure 3). In Figure 3, the left and right reactors have different coal particle sizes. For the left side (a), a coal size of only 1 to 2 cm is possible. For the right side (b), various sizes of coal can be used from 1 to 6 cm. This reactor was also designed to perform not only adsorption and desorption experiments, but also other applications such as CBM production (gas + water).

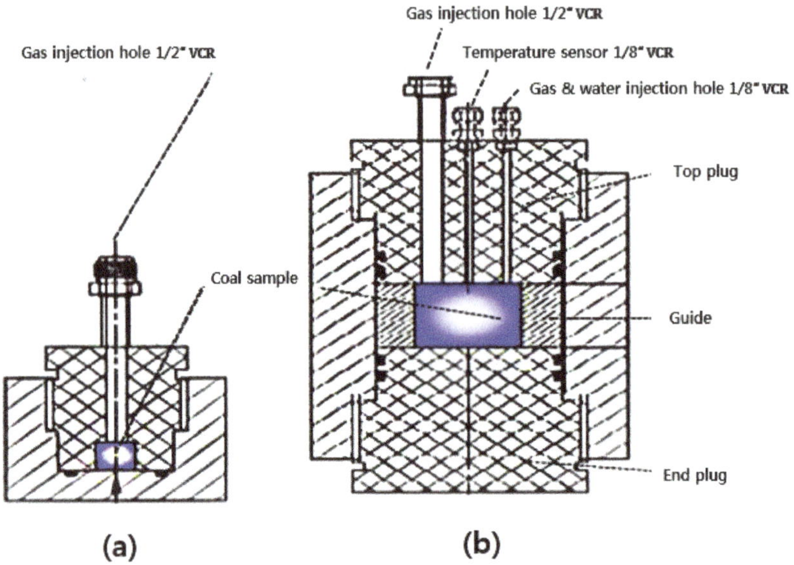

Figure 3. Manufactured reactor for the lump coal experiments. For (**a**), a coal size of only 1 to 2 cm is possible. For (**b**), various sizes of coal can be used from 1 to 6 cm.

3. Results and Discussion

3.1. Sample Preparation

The experiment used two coal samples (coal sample group A and coal sample group B) that were obtained from CBM fields located in Indonesia and Australia. The rank types of coal included lignite, bituminous coal, and anthracite; in some cases, bituminous coal and anthracite were subdivided as sub-bituminous or semi-anthracite by carbonization degree. To check the contents of fixed carbon, volatile matter, and moisture, proximate analysis was performed. Then, the data were converted to ASTM (American Society for Testing Materials) scale using the Parr formula, shown in Equations (11) and (12) (Table 1) [21].

$$Dry,\ Mm-free\ FC\ = \frac{FC-0.15\ S}{100-(M+1.08\ A+0.55\ S)} \times 100\%, \tag{10}$$

$$Dry,\ Mm-free\ VM = 100 - Dry,\ Mm\ FC,\ \%, \tag{11}$$

where FC is fixed carbon, S is sulfur, M is moisture, A is ash, and VM is volatile matter.

Coal sample group A had 2.66% natural moisture and 11.46% ash. On the other hand, coal sample group B had 1.2% inherent moisture and 18.53% ash. This shows that coal sample group B had a lower level of natural moisture and a higher portion of ash compared to coal sample group A. As a result of the analysis, since both samples of coal existed on the ground for a long time, but not underground, it was judged that contents of inherent moisture and ash were low. Therefore, we determined that the two coals were sub-bituminous considering gross calorific value.

Each coal sample was ground into six sizes: 180–252 μm, 253–508 μm, 0.5–1 mm, 2–4 mm, and 2 cm cubes (Figure 4). Then, a mortar was used to crush the small pieces of coal. After cleaning the coal samples with distilled water, a centrifuge was used to remove impurities from the coal samples. As a result of the centrifugation, the soil in the coal and tiny branches were separated. Then, the samples were kept at a temperature of 45 °C for three days.

In general, the adsorption experiments are performed in a dry ash-free (DAF) state. However, in this study, experiments were not carried out in a DAF state because it was impossible to remove ash and inherent moisture from the lump coal. Therefore, all experiments were carried out on an air-dried basis. To prevent the penetration of moisture from the air into the coal, samples were sealed in containers and stored in a temperature chamber at 45 °C prior to experiments for more than one week.

Table 1. Proximate analysis of the samples.

Title	Coal Sample Group A	Coal Sample Group B
Inherent moisture (air-dried basis) (wt.%)	2.66	1.20
Ash (air-dried basis) (wt.%)	11.46	18.53
Volatile matter (air-dried basis) (wt.%)	42.68	25.38
Fixed carbon (air-dried basis) (wt.%)	43.20	54.89
Volatile matter (dried basis) (wt.%)	44.38	25.91
Fixed carbon (dried basis) (wt.%)	44.16	55.56
Volatile matter (dry ash-free, DAF) (wt.%)	49.15	30.33
Fixed carbon (DAF) (wt.%)	50.85	69.67
Gross calorific value (air-dried basis) (MJ/kg)	27.84	27.84
Gross calorific value (dried basis) (MJ/kg)	28.59	27.75

Figure 4. Coal particle sizes: (**a**) 180–254 μm, (**b**) 254–508 μm, (**c**) 0.5–1 mm, (**d**) 1–2 mm, (**e**) 2–4 mm, (**f**) 2-cm cubes.

3.2. Analysis of Adsorption Characteristics According to the Particle Size

The CH_4 adsorption content, dependent on the particle size of coal, was measured with high-pressure sorption measuring equipment. Both coal samples had different V_L values according to the particle size of the samples (Table 2). Coal sample A(b), which was the second smallest coal particle size in coal sample group A, showed the highest V_L of 22,295.85 cm^3/kg. In contrast, coal sample A(f), which had the largest coal particle size, had the lowest V_L of 8670.62 cm^3/kg. Thus, there was a three-fold difference in V_L values between the two samples (Figure 5).

Unlike coal sample group A, coal sample group B had the highest V_L of 20,809.44 cm^3/kg in the smallest coal particle size, sample B(a), and the largest coal particle size sample B(f) had the lowest V_L of 12,000.53 cm^3/kg (Figure 6). These results relate to an increase in the specific surface area and a reduction in ash contents during coal crushing [22]. Coal sample group B had a lower content of volatile matter and a higher content of fixed carbon than coal sample group A. Shan et al. [23] suggested that fixed carbon has a positive effect on the adsorption of CBM, while the volatile matter has an adverse impact. For this reason, when the particle size of coal was identical, more gas was adsorbed onto coal sample group B, because there was more fixed carbon, which had a positive effect on the adsorption.

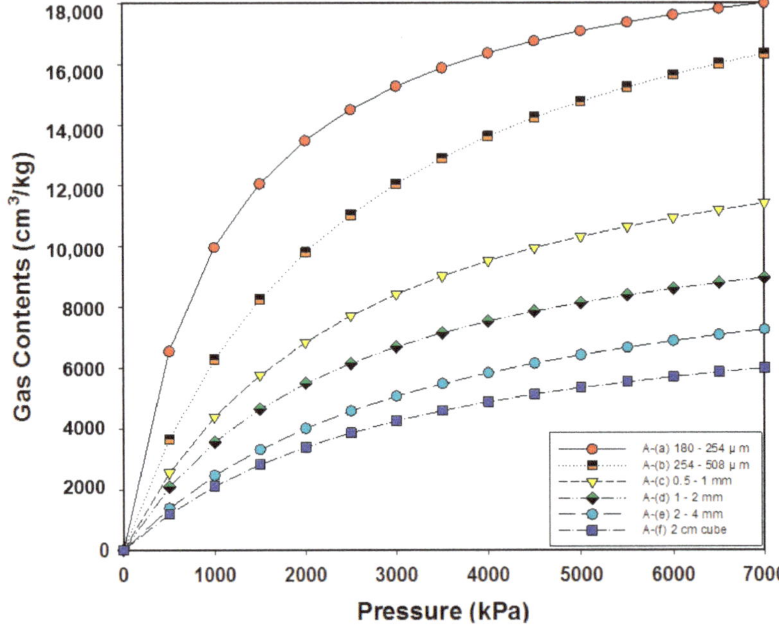

Figure 5. The results of adsorption characteristics with particle size (coal sample group A).

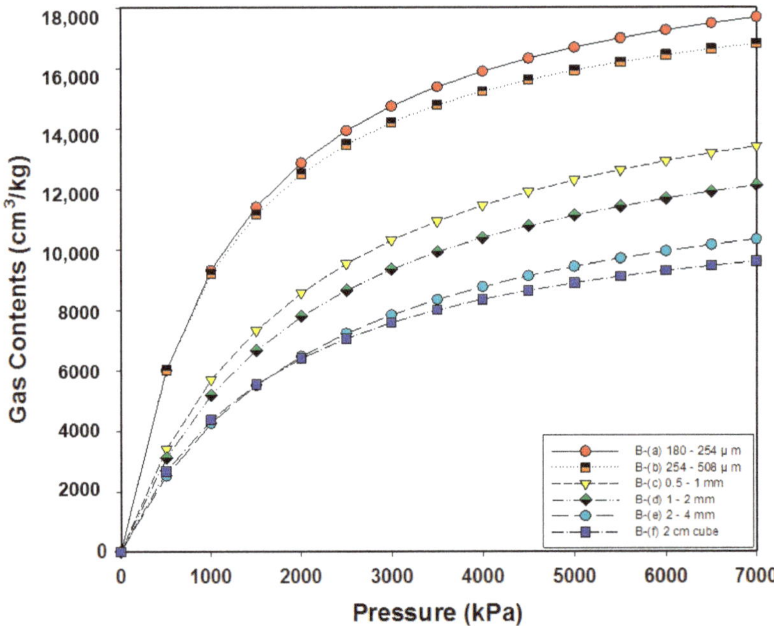

Figure 6. The results of adsorption characteristics with particle size (coal sample group B).

Table 2. Langmuir factors of CH_4 depending on the sample particle size.

Sample Group	Coal Particle Size	(a) 180–254 μm	(b) 254–508 μm	(c) 0.5–1 mm	(d) 1–2 mm	(e) 2–4 mm	(f) 2-cm Cube
A	V_L (cm³/kg)	20,809.44	22,295.85	15,607	12,005.53	10,765.40	8670.62
	P_L (kPa)	1086.76	2545.86	2556.53	2370.44	3371.62	3108.35
B	V_L (cm³/kg)	20,809.44	19,508.75	17,341.25	15,607	13,571.22	12,005.53
	P_L (kPa)	1233.72	1120.04	2047.85	2009.12	2191.64	1749.51

At the same time, the aggregation of coal particles was found to have an impact on V_L. Gas was not able to be sufficiently adsorbed for the powdered samples and caused a V_L reduction because of this effect. This was observed in both powder types of coal sample groups A and B, but especially in coal sample group A. While coal sample A(a) was expected to have the highest adsorption value, coal sample A(b) had the highest V_L value. This result was found to be related to the aggregation of fine particles in the reactor; the crushed particles were clustered together and attached to the wall of the reactor after the test.

Many previous studies described the effect of coal aggregation with an increase in compressive load. Furthermore, this phenomenon intensified with the reduction in coal particle size and its moisture content, as well as the rise in ash content. However, these results varied depending on the type of coal [24]. Consequently, a smaller particle size of coal led to more gas being adsorbed due to the increase in surface area.

3.3. Equilibrium Time Analysis in the Adsorption Process

In the indirect method, gas is artificially adsorbed onto coal in the laboratory. Sufficient time is needed to achieve an equilibrium condition for gas to adsorb onto the coal surface. If the equilibrium time is not enough, the reliability of the experiment diminishes because the test finishes without complete adsorption. Also, crushed coal and lump coal might have different equilibrium times, since diffusion rarely occurs with small particles. Our motivation was based on the fact that, if the equilibrium time varies with coal particle size, it can still be possible to estimate the effects of gas diffusion in an alternative way. Therefore, an accurate equilibrium time measurement was required, and it was measured when the pressure variation was less than 0.1%. Figure 7 shows the adsorption equilibrium time with the particle size of coal sample group A. Through samples A(b) to A(e), the difference was negligible and, thus, they are all represented as A(e). As Figure 7 demonstrates, less equilibrium time was required for adsorption onto a smaller coal sample at a low pressure range (<1400 kPa); the results of sample A(a) can be explained by the difference in equilibrium time of adsorption due to the rearrangement of the aggregated particles as pressure increased. However, these differences were seen only at low pressures. Conversely, when pressure was higher than 1400 kPa, the samples had almost the same equilibrium time, which can be verified by the similar slopes between the samples with smaller particle sizes. These results mean that there was no particular tendency for adsorption, and adsorption time was almost identical from A(a) to A(e).

However, in the case of the lump coal sample A(f), it took a long time to achieve an equilibrium condition. In particular, the slope of the adsorption time was different compared to the powdered samples. More specifically, it took 9000 s, which is about three times longer than for the other powdered samples. This situation occurred because more time was needed for the gas to diffuse through the coal matrix and adsorb onto the surface. Also, we believed that we could obtain diffusion coefficients from the adsorption analysis of lump coal. Therefore, the diffusion coefficient, which is usually calculated through the canister test, can be more accurately obtained using the history matching technique as described in the next section.

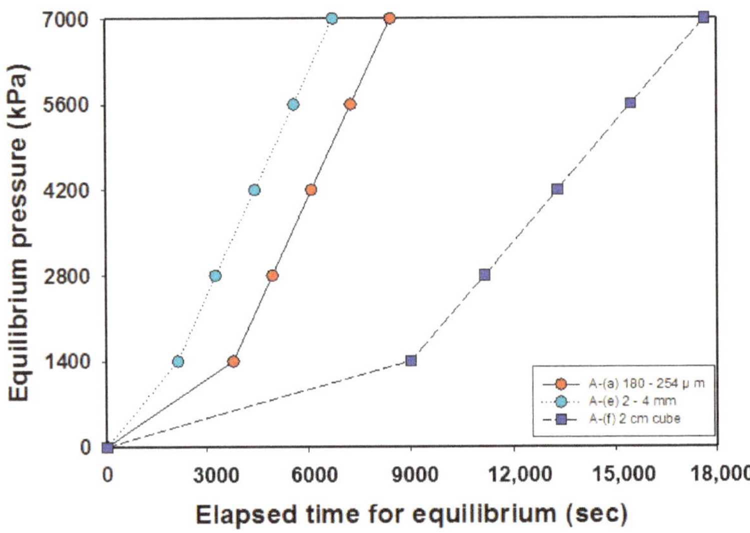

Figure 7. Adsorption equilibrium time with particle size (coal sample group A).

3.4. Model Description

A reservoir simulation was carried out using the results of sample B, using the same size of coal as in the experiment, to determine the diffusion coefficient via the indirect method. The significant assumptions were as follows:

1. The sample was homogeneous and isotropic;
2. The interval of each cleat was equal;
3. Gravity force and the stress of the cleat system were ignored;
4. The gas diffusion coefficient was constant.

CBM is notably different compared to a conventional gas reservoir. The methane is in a near-liquid state and is adsorbed inside pores within the coal matrix. The open fractures in the coal (cleats) can also contain free gas or can be saturated with water. The system is called a "dual-porosity reservoir", which is characterized by a complex interaction between the coal matrix and cleat system coupled with the adsorption/desorption process. Fluid movement in a lump of coal is controlled by diffusion from the coal matrix to the fracture and by Darcy flow in the fracture (cleat) system. The reservoir pressure is decreased upon producing gas from the cleats. This causes the gas to desorb from the coal matrix at the matrix/cleat interfaces, creating a methane concentration gradient across the coal matrix. The gas diffuses through the matrix and is released into the cleat system.

Here, we describe the fracture block governing equations for the dual-porosity approach to modeling naturally fractured reservoirs [25]. These equations are an extension of that derived for single-porosity systems described by Nghiem and Li [26]. In the dual-porosity matrix-diffusion model, the solute cannot move globally through the rock matrix. The matrix simply acts as a one-dimensional source or sink for diffusion between the adjacent fractures. Thus, the diffusion coefficient applies to the flow from the matrix to the fracture.

- Dual porosity formulation—matrix blocks:

$$\varphi_{im} = -\tau_{igmf} - \frac{V}{\Delta t}\left(N_i^{n+1} - N_i^n\right)_m = 0 \qquad i = 1,\ldots,n_c;$$

$$\varphi_{pm} = \sum_{i=1}^{n_c} N_{im}^{n+1} - \varnothing_m^{n+1}\left(\rho_g S_g\right)_m^{n+1} = 0 \qquad (12)$$

- Dual porosity formulation—fracture blocks:

$$\varphi_{if} = \Delta T^s_{gf} y^s_{igf}\left(\Delta p^{n+1} - \gamma^s_g \Delta D\right)_f + q^{n+1}_i + \tau_{igmf} - \frac{V}{\Delta t}\left(N^{n+1}_i - N^n_i\right)_m = 0 \qquad i = 1,\ldots,n_c;$$
$$\varphi_{pf} = \sum_{i=1}^{n_c} N^{n+1}_{if} - \varnothing^{n+1}_f\left(\rho_g S_g\right)^{n+1}_f = 0 \tag{13}$$

In these equations, N_i, $i = 1,\ldots,n_c + 1$ are the moles of i per unit of gridblock volume. The subscript i with $i = 1,\ldots,n_c$ corresponds to the hydrocarbon components. D is the diffusion coefficient from the matrix to the fracture. τ_{igmf} is the matrix–fracture transfer in the gas phase for component i. The subscripts n and $n+1$ denote the old and current time levels, respectively, and the superscript s refers to n for explicit blocks and to $n+1$ for implicit blocks. The subscript f and m correspond to the fracture and matrix, respectively. Many authors performed studies using dual-porosity systems [27–31]. We also utilized this system with two-phase flow to describe gas flow from the matrix to the fracture.

A Cartesian grid system was applied to build a core model. To represent the cleat network with a face and butt cleat, the total core area (2×2 cm^2) was divided into 10×10 grids in the I- and J-directions, and cleat spacing was selected as 0.2 cm in the I-, J-, and K-directions according to the electron microscope images (Figure 8). A detailed description of the model is listed in Table 3.

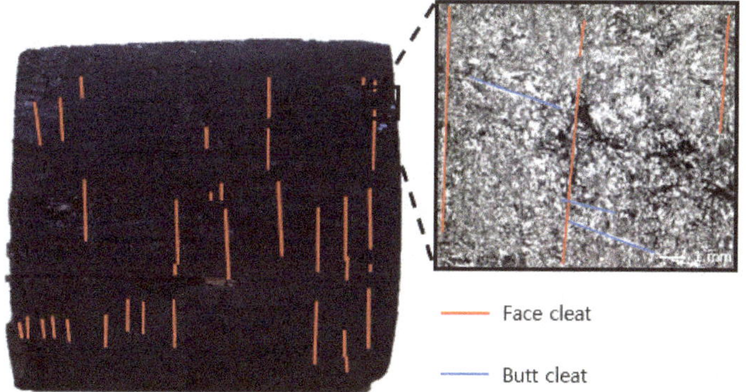

Figure 8. Electron microscope images of cleat structure in sample B.

Table 3. Initial parameter settings in the base model.

Input Parameters	Values
Coal volume (cm^3)	8.00
Grid dimension	$10 \times 10 \times 10$
Matrix porosity	0.017
Fracture porosity	0.005
Matrix permeability (mD)	0.01
Fracture permeability (mD)	4.00
Fracture spacing I (cm)	0.20
Fracture spacing J (cm)	0.20
Fracture spacing K (cm)	0.20
Cleat compressibility (1/kPa)	2.00×10^{-6}
Coal Density (kg/cm^3)	1.43×10^{-3}
Diffusion coefficient (cm^2/s)	3.50×10^{-10}
Langmuir pressure (CH$_4$) (kPa)	3500
Langmuir volume (CH$_4$) (cm^3/kg)	12,920

To simulate the laboratory system accurately, one imaginary injection well was installed, and the recorded experimental data (cell pressure, temperature, gas injection rate) were converted to simulation input data. The measured pressure was selected as an objective function in the history matching. The Langmuir isotherm, which was determined in the lump coal experiment, was used to model adsorption and desorption in coal. In the case of porosity, the dead volume was measured using helium in the experimental process, and the porosity was estimated indirectly. The permeability was measured by core analysis. GEM software (reservoir simulator for compositional, chemical, and unconventional reservoir modeling) from CMG (Computer Modelling Group) was used to conduct the simulation. Table 3 lists the initial parameters of the base model.

3.5. Simulation Results

Firstly, the base simulation was performed to investigate the diffusion coefficient and other reservoir properties. The primary constraint was the flow rate, which was identical to the actual experiment. The injection scenario was designed as an injection fall-off test, and the average period of shut-in was 2500 s after the 300 s of injection. The error between the base model and the actual experiment was around 73%. Many parameters had uncertainty because of the lack of sample properties, and sensitivity analysis was performed to determine the priorities among parameters for the history matching and to understand the interaction effects between each property. RSM (response surface methodology) was used for sensitivity analysis by utilizing the nine parameters listed in Figure 9. In total, 180 cases and a proxy model were also established for more accurate analysis. The quadratic model and the Sobel method were used to calculate the importance (sensitivity) of the parameters affecting the objective function and the interaction between parameters. A large importance means that increasing or decreasing the parameter value leads to a significant change in the objective function (e.g., well bottom-hole pressure). The interaction effect is an index indicating whether or not there is a relationship between parameters; if it has a value of 0, there is no relationship.

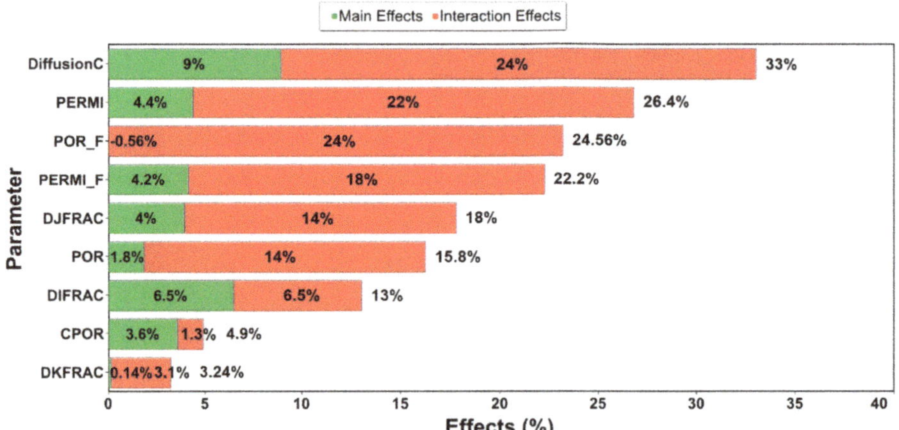

Figure 9. Results of sensitivity analysis.

In Figure 9, the most sensitive parameter was the diffusion coefficient, which had 33% sensitivity and dominantly affected the gas flow in the coal. Next, matrix permeability was shown to have a high sensitivity (26.40%). Also, the sensitivities of fracture porosity and fracture permeability (23.44% and 22.20%, respectively) showed that they dominantly affected the gas flow in the cleat structure. The above parameters also had substantial interaction effects with other parameters. These phenomena, which appeared in the results of the interaction effect analysis, are shown in Figure 10. Notably, the interaction effect between the diffusion coefficient and the fracture porosity was 29.75%. Thus, we determined that the diffusion coefficient is a crucial factor in gas flow through the coal. Meanwhile, the compressibility, the matrix porosity, and the fracture spacing were found to have relatively low sensitivity because of the small size of the core sample. Accordingly, the diffusion coefficient, the matrix permeability, the fracture porosity, and the fracture permeability were selected as the main parameters for the history matching.

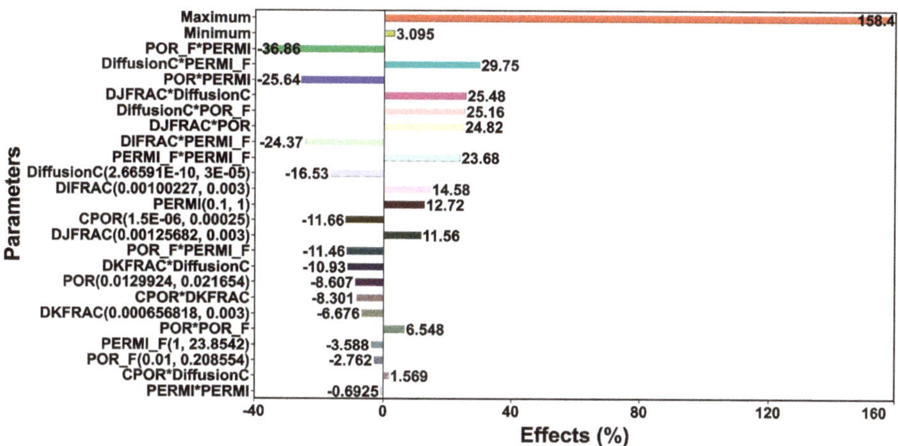

Figure 10. Interaction effect of each parameter in the simulation.

History matching was carried out to achieve reasonable values for each parameter. Of the 300 cases, three different cases were run, and the error rate of the final history-matched case was approximately 2.9% (Figure 11). We determined that this error rate occurred due to the small size of the sample and the assumptions of the model, such as homogeneous medium and isotropic permeability. The diffusion coefficient calculated by the history matching was 2.66×10^{-10} cm^2/s, which is reasonable compared with the typical value calculated from experiment studies [32–34]. The cleat properties, fracture permeability and fracture porosity, were adjusted from 4 mD to 23.85 mD, and from 0.005 to 0.15 because of the large number of cleats compared to the area of coals. The other properties showed small changes. Through this procedure, we found that a relatively accurate diffusion coefficient can be calculated by implementing adsorption experiments and simulation techniques without canister testing. Table 4 lists the parameters of the history-matched model.

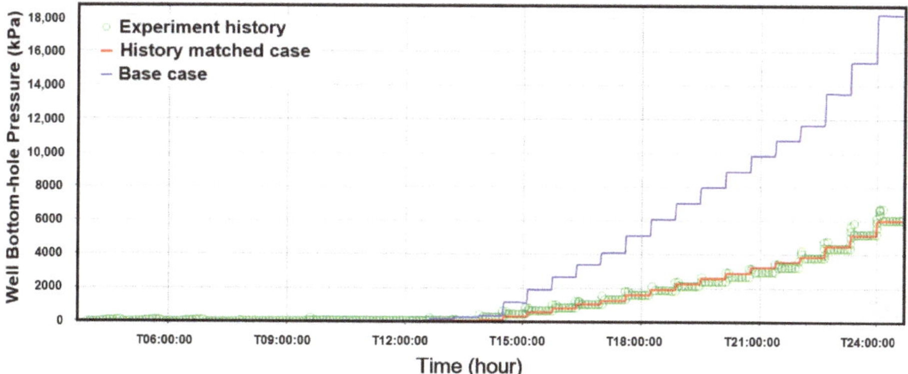

Figure 11. Comparison of the experiment history with the base (blue) and the history-matched (magenta) cases.

Table 4. History-matched parameters.

Matched Parameters	Values
Cleat compressibility (1/kPa)	2.40×10^{-5}
Matrix porosity	0.01
Fracture porosity	0.15
Matrix permeability (mD)	0.11
Fracture permeability (mD)	23.85
Fracture spacing I (cm)	0.24
Fracture spacing J (cm)	0.17
Fracture spacing K (cm)	0.20
Diffusion coefficient (cm^2/s)	2.66×10^{-10}

4. Conclusions

This study presented the adsorption and diffusion characteristics of CBM using the indirect method of measurement and reservoir modeling. The difference in adsorption content of CH_4 depended on the particle size measured with a high-pressure sorption measuring equipment. We used both powdered and lump samples and analyzed factors affecting V_L such as aggregation, ash and moisture content, and specific surface area. As sufficient time is necessary to achieve equilibrium conditions, adsorption time analyses were also conducted.

In conclusion, a smaller coal particle adsorbs more gas due to the increase in surface area. Lump samples had the lowest gas content because they contained ash and moisture. Adsorption isotherms showed different characteristics. Also, when the coal particle size decreased, the content of moisture and ash, which negatively affects adsorption, was reduced, leading to an increase in the amount of adsorbed gas. With regard to adsorption equilibrium time, a smaller coal particle took less time. Furthermore, the lump sample had a different slope because it needed more time to diffuse through the matrix. Therefore, to derive more accurate CBM characteristics that are similar to the actual site, the history-matching technique was applied to optimize coal properties. According to the sensitivity analyses, cleat properties and the diffusion coefficient had a great influence on fluid flow in coal.

Consequently, it is possible to measure the diffusion coefficient, which was previously measurable only using the direct method. However, more lump coal experiments and associated simulations are necessary for a more accurate CBM production analysis. With additional data, it is believed that the processes derived from this study could overcome the shortcomings of the indirect method.

Author Contributions: D.K. wrote the paper, conducted the experimental study, and analyzed the results; Y.S. wrote the paper, contributed to tuning the model, and ran the reservoir simulation; J.K. contributed to processing the raw data and made the initial model; J.H. edited the original draft; Y.L. suggested the main idea and supervised the work, providing continuous feedback.

Funding: This research was funded by the Energy Efficiency and Resources Core Technology Program of the Korea Institute of Energy Technology Evaluation and Planning (KETEP), granted financial resources from the Ministry of Trade, Industry, and Energy, Republic of Korea (No. 20152510101880).

Acknowledgments: This research was supported by the Energy Efficiency and Resources Core Technology Program of the Korea Institute of Energy Technology Evaluation and Planning (KETEP), granted financial resources from the Ministry of Trade, Industry, and Energy, Republic of Korea (No. 20152510101880) and this research was supported by Basic Science Research Program through the National Research Foundation of Korea(NRF) funded by the Ministry of Education(2019R1I1A3A01060375).

Conflicts of Interest: The authors declare no conflict of interest.

References

1. Flores, R.M. *Coal and Coalbed Gas: Fueling the Future*, 1th ed.; Newnes: Waltham, MA, USA, 2013; Volume 1, pp. 210–235.
2. Cho, J.; Kim, J. Global trends of unconventional CBM gas science information. *J. Econ. Environ. Geol.* **2013**, *46*, 351–358. [CrossRef]

3. Yoon, Y.; Oh, S.; Kim, J. An real option application on the feasibility study of an Indonesian CBM project. *J. Korean Soc. Miner. Energy Resour. Eng.* **2014**, *51*, 631–640. [CrossRef]

4. Levine, J.R. Generation, storage and migration of natural gas in coal bed reservoirs. *Alta. Res. Counc. Inf. Ser.* **1990**, *109*, 84–130.

5. Stricker, G.D.; Flores, R.M.; McGarry, D.E.; Stilwell, D.P.; Hoppe, D.J.; Stilwell, K.R.; Ochs, A.M.; Ellis, M.E.; Osvald, K.S.; Taylor, S.L.; et al. Gas desorption isotherm studies in coals in the Powder River Basin and adjoining basins in Wyoming and North Dakota. *US Geol. Surv. Open File Rep.* **2006**, *1174*, 273.

6. Jang, H.; Lee, J.; Shin, C.; Lee, Y.; Kwon, S.; Lee, W. A study on the development status and key technologies of coalbed methane. *J. Korean Soc. Miner. Energy Resour. Eng.* **2012**, *49*, 545–556.

7. Hall, F.E.; Chunhe, Z.; Gasem, K.A.M.; Robinson, R.L.; Dan, Y. Adsorption of pure methane, nitrogen, and carbon dioxide and their binary mixtures on wet Fruitland coal. In Proceedings of the SPE Eastern Regional Conference and Exhibition 1994, Charleston, WV, USA, 8–10 November 1994; pp. 329–344. [CrossRef]

8. Steidl, P.F. Coal as a reservoir. In *A Guide to Coalbed Methane Reservoir Engineering*; Report No. GRI-94/0397; Gas Research Institute: Chicago, IL, USA, 1996; pp. 1–16.

9. He, X.; Zhang, Z. Microscopic pore structural characteristics in coal particles. In Proceedings of the 3rd International Conference on Material, Mechanical and Manufacturing Engineering (IC3ME 2015), Guangzhou, China, 27–28 June 2015. [CrossRef]

10. Beamish, B.B.; O'Donnell, G. Microbalance applications to sorption testing of coal. In Proceedings of the Symposium Coalbed Methane Research and Development, Townsville, Australia, 19–21 November 1992; pp. 31–41.

11. Zhang, L.; Aziz, N.; Ren, T.; Wang, Z. Influence of temperature on the gas content of coal and sorption modeling. In Proceedings of the 11th Underground Coal Operators' Conference, Wollongong, Australia, 10–11 February 2011; pp. 269–276.

12. Owen, L.B.; Sharer, J. Method calculates gas content per foot of coalbed methane pressure core. *Oil Gas J.* **1992**, *2*, 47–49.

13. Diamond, W.P.; Schatzel, S.J. Measuring the gas content of coal: A review. *Int. J. Coal Geol.* **1998**, *35*, 311–331. [CrossRef]

14. Kim, A.G. Estimating methane content of bituminous coalbeds from adsorption data. *US Bur. MinesRep. Investig.* **1997**, *8245*, 22.

15. Lee, S.H.; Kim, C.K.; Park, J.G.; Choi, D.K.; Ahn, J.H. Comparison of steel slag and activated carbon for phosphate removal from aqueous solution by adsorption. *J. Korean Soc. Environ. Eng.* **2017**, *39*, 303–309. [CrossRef]

16. Karmakar, B. Prospective Evaluation and Prediction of Coalbed Methane Production from a Part of Raniganj and Jharia Coalfields in India. Ph.D. Thesis, Indian School of Mines, Dhanbad, India, 2013; pp. 68–116.

17. Langmuir, I. the adsorption of gases on plane surfaces of glass, mica, and platinum. *J. Am. Chem. Soc.* **1918**, *40*, 1403–1461. [CrossRef]

18. BELSORP. *High-Pressure Gas Adsorption Measuring System Operation Manual*; BEL Japan: Osaka, Japan, 2012; Volume 1, pp. 15–20.

19. National Institute of Standards and Technology Chemistry Webbook. Available online: https://webbook.nist.gov/ (accessed on 23 October 2018).

20. Kim, J.; Kim, D.; Lee, W.; Lee, Y.; Kim, H. Impact of total organic carbon and specific surface area on the adsorption capacity in Horn River shale. *J. Pet. Sci. Eng.* **2017**, *149*, 331–339. [CrossRef]

21. Parr, S.W. *The Classification of Coal*; Engineering Experiment Station; College of Engineering; University of Illinois: Champaign, IL, USA, 1928.

22. Zhang, L.; Aziz, N.; Ren, T.; Nemcik, J.; Tu, S. Influence of coal particle size on coal adsorption and desorption characteristics. *Arch. Min. Sci.* **2014**, *59*, 807–820. [CrossRef]

23. Shan, C.; Zhang, T.; Liang, X.; Zhang, Z.; Zhu, H.; Yang, W.; Zhang, K. Influence of chemical properties on CH4 adsorption capacity of anthracite derived from southern Sichuan Basin. *Mar. Pet. Geol. Chin.* **2018**, *89*, 387–401. [CrossRef]

24. King, G.; Rtekin, T.; Schwerer, F.C. Numerical simulation of the transient behavior of coal-seam degasification wells. *SPE Formul. Eval.* **1986**, *1*, 165–183. [CrossRef]

25. CMG. *GEM User Guide: Compositional & Unconventional Simulator*; Computer Modelling Group Ltd.: Calgary, Canada, 2018.

26. Nghiem, L.X.; Li, Y.K. Phase-equilibrium calculations for reservoir engineering and compositional simulation. In Proceedings of the First International Forum on Reservoir Simulation, Alpbach, Austria, 12–16 September 1988.

27. Kim, Y.; Unurbayan, B.; Lee, J. A study on the estimation of CBM resources using probabilistic approach from Mongolian coal basin information. *J. Korean Soc. Miner. Energy Resour. Eng.* **2013**, *50*, 678–691. [CrossRef]

28. Remner, D.J.; Ertekin, T.; Sung, W.; King, G.R. A parametric study of the effects of coal seam properties on gas drainage efficiency. *SPE Reserv. Eng.* **1986**, *1*, 633–646. [CrossRef]

29. Sung, W.; Ertekin, T.; Schwerer, F.C. The development, testing, and application of a comprehensive coal seam degasification model. In Proceedings of the SPE Unconventional Gas Technology Symposium, Louisville, Kentucky, 18–21 May 1986. [CrossRef]

30. Sawyer, W.K.; Paul, G.W.; Schraufnagle, R.A. Development and application of a 3D coalbed simulator. In Proceedings of the International Technical Meeting, Petroleum Society CIM, Society of Petroleum Engineers, Calgary, Alberta, Canada, 10–13 June 1990.

31. Manik, J.; Ertekin, T.; Kohler, T.E. Development and validation of a compositional coalbed simulator. *J. Can. Pet. Technol.* **2002**, *41*, 39–45. [CrossRef]

32. Connell, L.D.; Sander, R.; Pan, Z.; Camilleri, M.; Heryanto, D. History matching of enhanced coal bed methane laboratory core flood tests. *Int. J. Coal Geol.* **2011**, *87*, 128–138. [CrossRef]

33. Dong, J.; Cheng, Y.; Liu, Q.; Zhang, H.; Zhang, K.; Hu, B. Apparent and true diffusion coefficients of methane in coal and their relationships with methane desorption capacity. *Energy Fuels* **2017**, *31*, 2643–2651. [CrossRef]

34. Kumar, H.; Mishra, M.K.; Mishra, S. Laboratory investigation of gas permeability and its impact on CBM potential. *J. Pet. Explor. Prod. Technol.* **2018**, *8*, 1–15. [CrossRef]

Article

Variation of Petrophysical Properties and Adsorption Capacity in Different Rank Coals: An Experimental Study of Coals from the Junggar, Ordos and Qinshui Basins in China

Yingjin Wang [1,2], Dameng Liu [1,2,*], Yidong Cai [1,2] and Xiawei Li [1,2]

[1] School of Energy Resources, China University of Geosciences, Beijing 100083, China; 3006160030@cugb.edu.cn (Y.W.); yidong.cai@cugb.edu.cn (Y.C.); 2006160033@cugb.edu.cn (X.L.)

[2] Coal Reservoir Laboratory of National Engineering Research Center of CBM Development & Utilization, China University of Geosciences, Beijing 100083, China

* Correspondence: dmliu@cugb.edu.cn; Tel.: +86-10-82323971

Received: 29 January 2019; Accepted: 11 March 2019; Published: 13 March 2019

Abstract: The petrophysical properties of coal will vary during coalification, and thus affect the methane adsorption capacity. In order to clarify the variation rule and its controlling effect on methane adsorption, various petrophysical tests including proximate analysis, moisture measurement, methane isothermal adsorption, mercury injection, etc. were carried out on 60 coal samples collected from the Junggar, Ordos and Qinshui basins in China. In this work, the boundary values of maximum vitrinite reflectance ($R_{o,m}$) for dividing low rank, medium rank and high rank coals are set as 0.65% and 2.0%. The results show that vitrinite is the most abundant maceral, but the maceral contents are controlled by sedimentation without any relation to coal rank. Both the moisture content and porosity results show higher values in the low ranks and stabilized with $R_{o,m}$ beyond 1%. $R_{o,m}$ and V_L (*daf*) show quadratic correlation with the peak located in $R_{o,m}$ = 4.5–5%, with the coefficient (R^2) reaching 0.86. P_L decrease rapidly before $R_{o,m}$ = 1.5%, then increase slowly. *DAP* is established to quantify the inhibitory effect of moisture on methane adsorption capacity, which shows periodic relationship with $R_{o,m}$: the inhibitory effect in lignite is the weakest and increases during coalification, then remains constant at $R_{o,m}$ = 1.8% to 3.5%, and finally increases again. In the high metamorphic stage, clay minerals are more moisture-absorbent than coal, and the inherent moisture negatively correlates with the ratio of vitrinite to inertinite (V/I). During coalification, micro gas pores gradually become dominant, fractures tends to be well oriented and extended, and clay filling becomes more common. These findings can help us better understand the variation of petrophysical properties and adsorption capacity in different rank coals.

Keywords: coal rank; petrophysical properties; coalbed methane; adsorption capacity

---

1. Introduction

Coalbed methane (CBM), a form of natural gas extracted from coal seams, has been widely considered a clean alternative energy in fossil fuel development [1–3]. Nevertheless, its accumulation is always deemed a serious safety risk to coal mining [3,4]. Therefore, a deeper understanding of the storage capacity of coal could facilitate safe and effective methane production. Adsorption is the main occurrence state in which CBM exists in most middle and high rank coal reservoirs [3–5], thus investigating the factors of coal affecting adsorption is essential for integrated evaluation of CBM recoverability. A previous study [6] has also confirmed free and soluble gases still account for a certain portion of CBM resources. In addition, the physical characteristics of coal functional groups will evolve during coalification, which also influence the methane adsorption [7–9]. Therefore, we expect to find a regularity in the variation of the inherent physical properties in the entire coal rank, allowing us to ascertain the main controlling factors of coal adsorption. Different from previous studies, this work also attempts to establish some experimental formulas on the basis of numerous experiments. These formulas and conclusions are expected to have practical significance both in CBM production prediction or mining security risk assessment.

Adsorption capacity is determined by the pressure-temperature (P-T) conditions and the nature of coal, the former has been expressed with Langmuir adsorption isotherm [6–10], but the influence of the latter still remains controversial. Many experiments have been carried out to study the controlling factors of coal adsorption, including coal rank, coal macerals, coal quality, moisture content, pore structure and specific area [8–18]. Coal metamorphism dominates the variation of coal adsorption. Su et al. [19] believed that coal rank and Langmuir volume (V_L) present a reverse U relationship. Zhang et al. [20] confirmed this statement and demonstrated that ink bottle pores promote adsorption. In addition, various factors that influence the coal adsorption at different stage of coal metamorphism have been investigated. Weniger et al. [21] concluded that maceral is independent of coal adsorption capacity by comparing CO_2 and CH_4 adsorption experiments. However, other works [9,20,22] demonstrated that both vitrinite and inertinite can enhance the adsorption capacity. Li et al. [23] proved that the specific surface area is irrelevant to adsorption based on coal adsorption barrier theory, whereas An et al. [24] noted that the specific surface area of small micropores decides the CH_4 adsorption in coals.

As an organic rock, coal has a serious heterogeneity, especially regarding the macrolithotype and coal structure, and this heterogeneity has ignored effects on methane adsorption [12–14]. Interlayer inhomogeneity in coal seam results in heterogeneity of physical properties. In order to establish a universal change regularity of the physical properties and adsorption capacity in coals with a wide rank, the influence of irrelevant variables originated from various sublayers should be reduced as much as possible. Therefore, coal macrolithotype and coal structure were constrained in this work, and all selected specimens were semi-bright coals and undeformed. All samples underwent proximate analysis, porosity calculations, maceral identification, equilibrium moisture analysis, and isothermal adsorption experiments.

2. Samples and Experiments

2.1. Sampling Areas

Coal blocks of about 5 kg each were collected using the channel method from the southeastern Junggar basin, and the eastern margin of the Ordos basin and the Qinshui basin (Figure 1a). To prevent sample oxidization and moisture loss, all samples were carefully jacketed with plastic wrap, and then immediately transported to the laboratory for the experiments.

The southeastern Junggar basin (SE-JB) is located between the Junggar basin and the Tianshan orogenic belt (Figure 1b), and the main coal-bearing strata include the Badaowan and Sangonghe formations of the Early Jurassic and the Xishanyao and Toutunhe formations of the Middle Jurassic [25] (Figure 2a). Eleven samples were obtained from two coal mines and a CBM block in SE-JB. The eastern margin of the Ordos basin is a large gentle monoclinal structure of near NS strike and dip NE, and complex secondary folds are developed locally (Figure 1c). The main coal-bearing strata are the Taiyuan formation of the Late Carboniferous and the Shanxi formation of the Early Permian [26] (Figure 2b), and the coal rank increases gradually from north to south. In the Ordos basin, 13 samples were all collected from the east (E-OB) and southeast (SE-OB) (Figure 1c). The remaining 36 samples were all collected from the Qinshui basin, including the Xishan coal field in the northwest (NW-QB) (Figure 1d), the Yangquan-Shouyang coal mine in the northeast (NE-QB), the Jincheng-Qinshui block in the south (S-QB) and the Anze block in the southwest (SW-QB). The Qinshui basin is structurally a large complex monocline with NNE-SSW strike, developing two sets of coal-bearing sedimentary systems in the Shanxi and Taiyuan formations [27–29] (Figure 2c), producing medium-high rank coals.

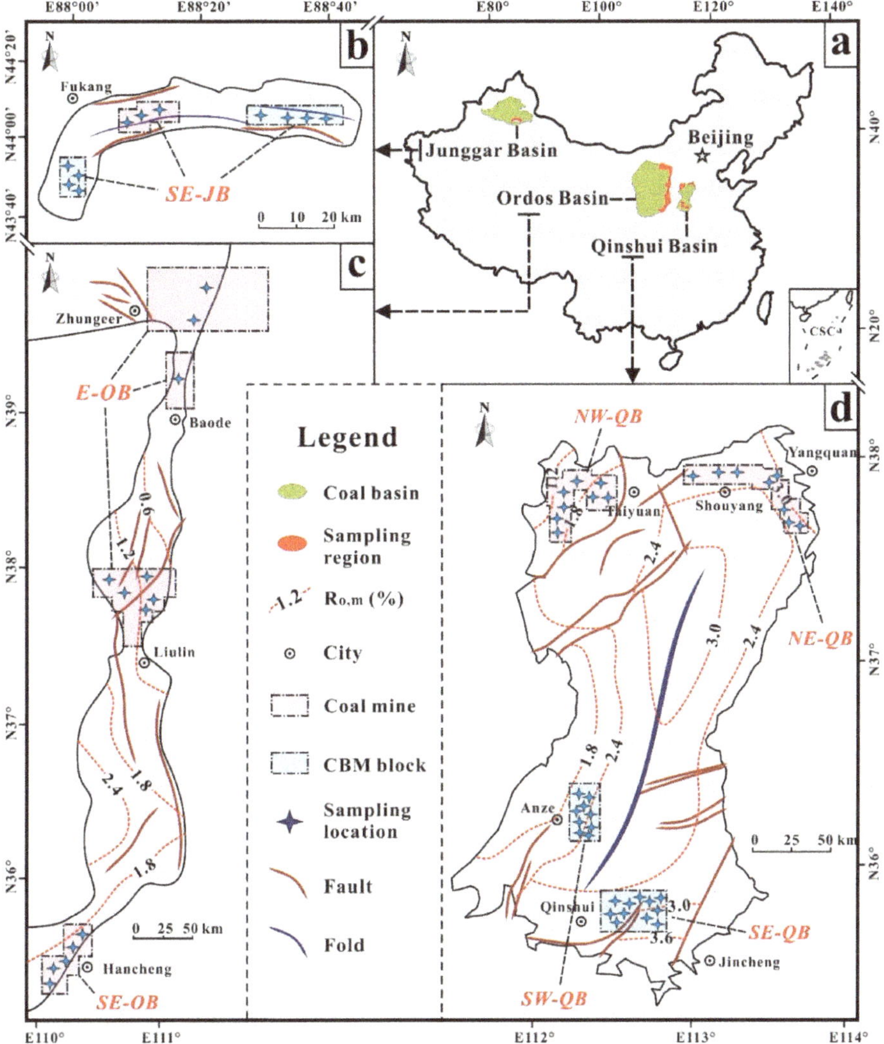

Figure 1. Positions of the Qinshui, Ordos, and Junggar basins in China and the distribution of the sampling locations. SE-JB, southeastern Junggar basin; E-OB, eastern Ordos basin; SE-OB, southeastern Ordos basin; NW-QB, northwest Qinshui basin; NE-QB, northeast Qinshui basin; SW-QB, southwest Qinshui basin; SE-QB, southeast Qinshui basin. (**a**) Locations of the Qinshui, Ordos and Junggar Basins in the China Map; (**b**) Specific locations of coal mines and CBM blocks in the southeastern Junggar Basin; (**c**) Specific locations of coal mines in the eastern margin of the Ordos Basin; (**d**) Specific locations of coal mines and CBM blocks in the Qinshui Basin.

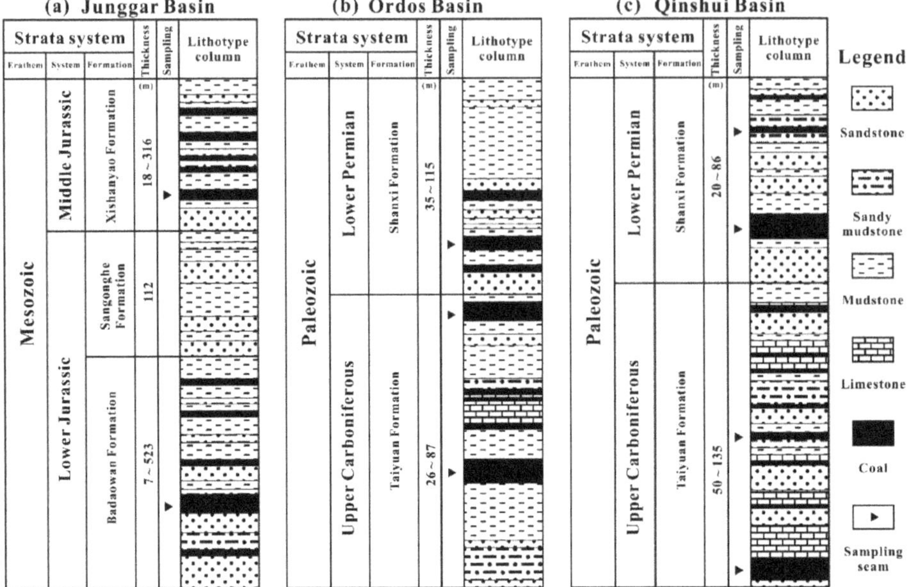

Figure 2. Stratigraphic columns and sampling seams of coal-bearing strata of Junggar basin (**a**), Ordos basin (**b**) and Qinshui basin (**c**).

2.2. Methodology

(1)   Measurement of coal density and porosity

True density (ρ_T) was obtained by applying the pycnometry method, and apparent density (ρ_A) was determined through the waxing method. Porosity (%) of coal can be calculated by Equation (1) with true density (ρ_T) and apparent density (ρ_A):

$$\varphi = (\rho_T - \rho_A)/\rho_T \times 100\% \tag{1}$$

where φ is the porosity (%); ρ_T is true density (g/cm^3); and ρ_A is true density (g/cm^3).

(2)   Vitrinite reflectance ($R_{o,m}$) measurement

Coal rank was described by $R_{o,m}$ (maximum vitrinite reflectance). After the coal slab was polished, petrologic observations (500 points) were taken in oil immersion apparatus under white light using a magnification of 500× using a photometer system, following the Chinese standard (GB/T) 6948-998.

(3)   Coal proximate analysis

Coal proximate analysis, including A_{ad} (ash yield of air dry basis), V_{daf} (volatile matter content of dry ash-free basis), M_{ad} (moisture content of air dry basis), and F_{ad} (fixed carbon of air dry basis), was performed on coal samples with a particle size of <0.2 mm based on the ISO 17246-2010 test standard.

(4)   CH$_4$ isothermal adsorption analysis

CH$_4$ isothermal adsorption analysis were conducted following the GB/T 19560-2004 standard. Coal samples were prepared by sieving to the same particle size fraction of 0.23–0.45 mm, and then placed in a sample cell, at 30 °C and an equilibrium pressure of 8 MPa. Tests for samples under air dry basis and equilibrium moisture conditions were conducted to get $V_L(em)$ and $V_L(ad)$, respectively, $V_L(daf)$ is calculated by:

$$V_L(daf) = V_L(ad) \times (1 - M_{ad} - A_{ad}) \tag{2}$$

Before the adsorption experiment, inherent moisture contents (M_{ad}) and ash yields (A_{ad}) of samples were measured by proximate analysis. To obtain the equilibrium moisture content (M_e), a crushed sample under air dry basis is soaked in water for 24 h, then it is kept in a sealed chamber, where K_2SO_4 solution is present to control the vapour partial pressure within the air in the chamber at a relative humidity of approximately 95% at room temperature. The sample moisture content is allowed to equilibrate with this relative humidity over 48 h. This process ensures the moisture in the coal sample reaches equilibrium. M_e used in this work is defined as:

$$M_e = m_{H_2O} / (m_{coal} + m_{H_2O}) \times 100\% \tag{3}$$

where M_e is the equilibrium moisture content, m_{H_2O} is the total mass of water uptake in coal, m_{coal} is the total mass of the dry coal.

Equilibrium moisture content (M_e) indicates external moisture, which occurs predominantly by adsorption on the walls of the micropore network. Langmuir volume (V_L) and the Langmuir pressure (P_L) were expressed with Equation (4)

$$V / V_L = P / (P_L + P) \tag{4}$$

where V is the gas adsorbed volume (cm^3/g); P is the pressure (MPa); P_L is the Langmuir pressure (MPa); and V_L is the Langmuir volume (cm^3/g).

(5)   Scanning electron microscopy (SEM)

The equipment model was FEI Quanta FEG 450: Acceleration Voltage: 0.2–30 kV; resolution: 1.0 nm at 30 kV, 3.0 nm at 1 kV; Magnification: 1,000,000. Sample preparation: Coal fragments with an area of ~1 cm^2 were cut and fixed on sample mounts using conductive adhesive, which were then sputter coated with gold. The images of the samples were processed with FEI software.

3. Results

Tested parameters of coal samples, including maceral content, proximate analysis, isothermal adsorption, moisture measurements and vitrinite reflectance results, were listed in attached table (Table A1).

3.1. Coal Compositions

The $R_{o,m}$ of the coal samples ranges from 0.35% to 4.26%, including all coal types from lignite to anthracite. The samples were divided into low rank coals (LRC-01~LRC-10), middle rank coals (MRC-01~MRC-25) and high rank coals (HRC-01~HRC-25) with boundary values of 0.65%$R_{o,m}$ and 2.0%$R_{o,m}$. The most abundant maceral is vitrinite, which accounts for more than 40% of the organic constituents (Figure 3a), followed by inertinite (Figure 3b). Liptinite decreases with increasing coal rank and disappears at $R_{o,m}$ of 1.8% (Figure 3c). In fact, liptinite breaks down during coalification and leaves behind materials that greatly resemble either vitrinite or inertinite under reflected light, making it harder to distinguish liptinite against the vitrinite background common at ranks higher than the vitrinite reflectance of about 1.3% R_r [30,31]. However, within some higher rank coals, liptinites can be identified by their morphology [31], for examples, MRCs-12, 13, 15, 18, 20 and 21 in this work. Maceral composition is controlled by peat-forming vegetation type and uncorrelated with coal metamorphism [32]. For example, the vitrinite content in the SE-OB is higher than 80%, that in the NE-QB is higher than 70%, and that in the S-QB is lower than 75%. Minerals generally occupied less than 10%, reaching a minimum at $R_{o,m}$ of 0.5% to 1.5% (Figure 3d).

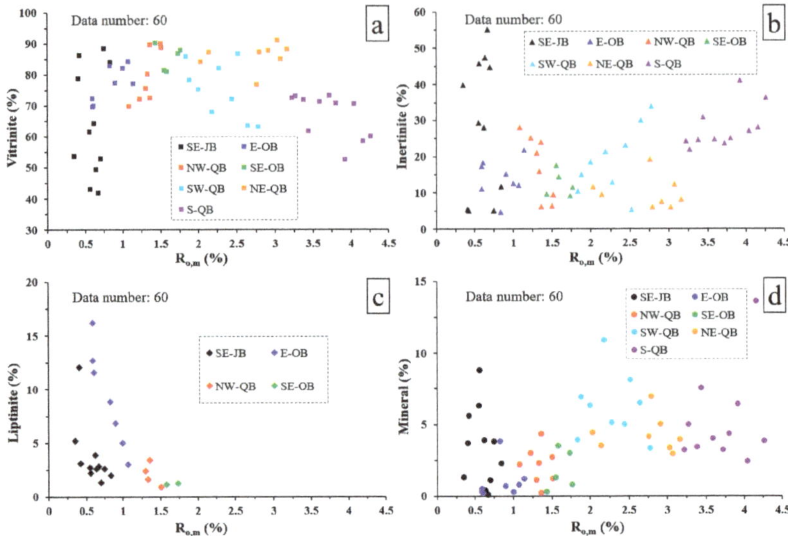

Figure 3. Macerals and mineral contents of all samples, showing data distribution in $R_{o,m}$ scale. Data are shown in Table A1. Colors show different sampling locations (see Figure 1). (**a**) Vitrinite (%) vs. $R_{o,m}$ (%); (**b**) Inertinite (%) vs. $R_{o,m}$ (%); (**c**) Liptinite (%) vs. $R_{o,m}$ (%); (**d**) Mineral (%) vs. $R_{o,m}$ (%).

Volatile matter content (V_{daf}) can also reflect the degree of coal metamorphism, which refers to gas decomposed and escaped from the coal matrix in coal at high temperatures. This composition exhibited a good exponential correlation with coal rank ($R_{o,m}$) (Figure 4a). Due to aliphatic polyester splitting and aromatization proceeding during coalification, fixed carbon (FC_{daf}), the carbon content in the coal, exhibited a high logarithmic correlation with $R_{o,m}$ (Figure 4b). Ash content is the residue after all combustible materials are incinerated and represents shale and mineral impurities [33], thus the A_{ad} can reflect mineral filling during coalification. In this work, both mineral content and ash yield are not well related to coal rank ($R_{o,m}$) (Figures 3d and 4c) and only show a slight increase from 1.5%$R_{o,m}$ to 2.5%$R_{o,m}$, which may indicate that in this stage, partial macropore space can be filled with clays and minerals. M_{ad} is the amount of inherent moisture in coal [34].

Both compaction in diagenesis and coal dehydration during coalification will decrease the moisture content in coal. Besides, the interaction energy between lignite, bituminous, anthracite and water molecules are on order of the combined effect of two strong hydrogen bonds > one hydrogen bond > van der Waals force [34]. Obviously, M_{ad} in low rank coals is much more than that in medium-high rank coals, when $R_{o,m}$ is greater than 1%, the value stabilizes at around 1% (Figure 4d).

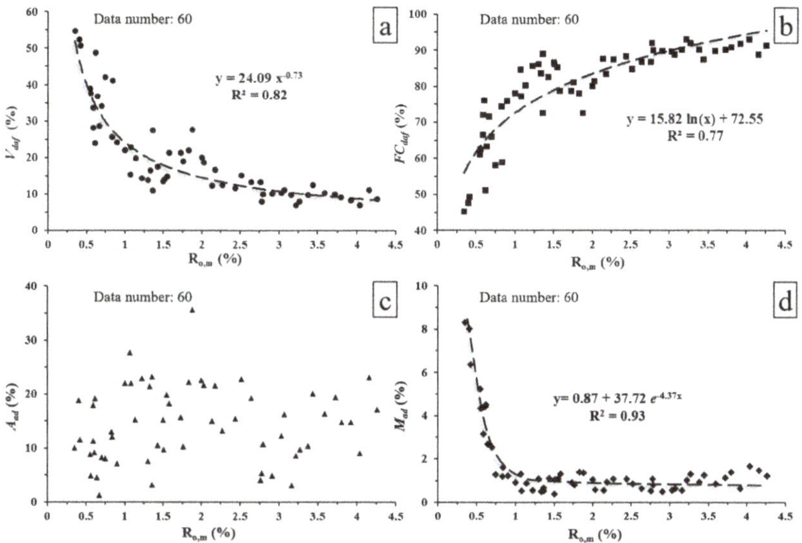

Figure 4. Relationship between the proximate analysis and different rank coals: (**a**) Volatile content (V_{daf}, %) vs. $R_{o,m}$ (%); (**b**) Fixed carbon (FC_{daf}, %) vs. $R_{o,m}$ (%); (**c**) Ash yield (A_{ad}, %) vs. $R_{o,m}$ (%); (**d**) Moisture content (M_{ad}, %) vs. $R_{o,m}$ (%).

3.2. Coal Physical Properties

M_e represents the external moisture that occurs predominantly in the adsorption state, showing a higher value in low rank coals and stabilized with $R_{o,m}$ beyond 1% as inherent moisture does (Figure 5a). Total moisture (M_t), the sum of M_{ad} and M_e, reflects the maximum holding capacity of moisture in coal under a certain relative humidity and temperature. M_{ad}, M_e and M_t of lignites reach amounts of up to 8%, 22% and 30%, respectively (Figures 4d and 5). At a low metamorphic stage, a large number of hydrophilic polar groups are generated in the coal's macromolecular structure [35], causing a high moisture content. In the medium-high metamorphic stage, M_{ad} and M_e are low. The specific surface increases during coalification, but condensed aromatic ring structures in middle-high rank coals contain more hydrogen and are hydrophobic, thus the moisture content is obviously low [34,35]. Therefore, under a same geological environment, the moisture holding capacity of coal is dominated by coal metamorphism.

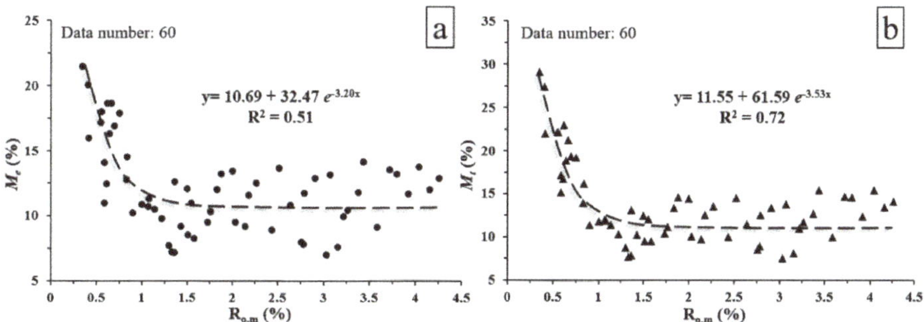

Figure 5. Relationships between (**a**) equilibrium water (M_e) and (**b**) total moisture (M_t) and coal rank ($R_{o,m}$).

Coal density can be subdivided into true density, apparent density and coal matrix density, and the latter can be calculated by removing the influence of minerals and moisture as shown below:

$$\rho_C = (1 - M_{ad} - A_{ad})/(1/\rho_T - M_{ad}/\rho_W - V_{ad}/\rho_M) \tag{5}$$

where: ρ_C, ρ_T, ρ_W and ρ_M represent coal matrix density, true density, moisture density and mineral density respectively, moisture density is 1 g/cm^3, mineral density is approximately 3 g/cm^3; Moisture content and mineral content can be quantitated by M_{ad} and V_{ad} (volatile content in air dry state) respectively.

Coal matrix density (ρ_C, g/cm^3) fluctuates greatly within low-medium rank samples, showing a weak exponential relationship (R^2 = 0.38) with $R_{o,m}$. For low-medium rank coals, structural evolution characterization is closely related to the 1st and 2nd coalification jumps, and the combination of aliphatic cyclization, pyrolysis cracking and aromatization complicates the density change trend [35,36]. During late coalification and graphitization, through Raman spectroscopy, Su et al. [36] indicated that graphite microcrystallite in coal goes through the evolution process from small and disordered to big and ordered. Therefore, the density shows a high linear relationship (R^2 = 0.62) with coal rank beyond 1.6% $R_{o,m}$ (Figure 6a). Density is a property of matter and can be influenced by many factors. For examples, the density of vitrain and clarain is lower than that of durain and fusain [37], mylonitized coal becomes denser with pulverized coal filling in fissures during tectonic movements. Both internal and external issues will lead to density changes.

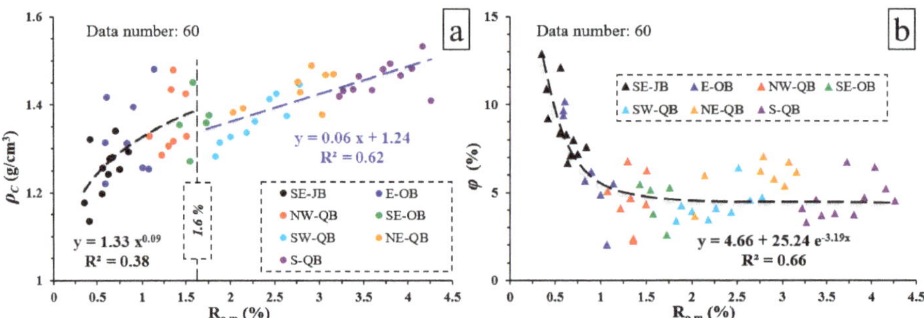

Figure 6. The relationships between coal physical properties and coal rank ($R_{o,m}$): (**a**) Coal matrix density (ρ_C, g/cm^3) vs. $R_{o,m}$; (**b**) Porosity (φ, %) vs. $R_{o,m}$. Colors show different sampling locations (see Figure 1).

Porosity is strong negatively correlated with coal rank when $R_{o,m}$ is less than 1% and stabilizes later (Figure 6b), the same trend can be found in moisture content variation. According to the genetic classification, the pores of coal can be divided into primary pores (plant tissue pores and intergranular pores), organic pores (molecular structure pores and gas pores), exogenous pores and mineral pores [38]. Primary pores are produced in the coal-forming period and preserved more in low-rank coal. At early stage of coal metamorphism, coal experiences the bitumination in the first coalification jump, physical compaction increases due to basin subsidence [29]. Therefore, primary pores are deformed, reduced or even disappear, resulting in a rapid decrease of the porosity. When $R_{o,m}$ is greater than 1.0%, hydrocarbon generation produces a large number of gas pores, while compaction and clay filling lead to a reduction of meso- and macropores. Therefore, in thermal metamorphism, the proportion of pores in each pore diameter fluctuates but tends to be stable as a whole. Exogenous pores and mineral pores are mainly affected by geological structure and have little relation with coal metamorphism, which is the key issue in the research of tectonically deformed coal.

3.3. CH$_4$ Isothermal Adsorption Capacity

The V_L is largest for the dry ash basis, followed by the air dry basis and equilibrium moisture condition for each coal sample, demonstrating that both moisture and mineral filling can reduce the methane adsorption capacity (Figure 7). It can be seen from the Langmuir isothermal adsorption line that the greater the radian of curve is, the higher the Langmuir pressure is, and the easier desorption occurs. Meng et al. [39] divides CBM desorption into three sections, quantifying the proportion of CBM desorption in different pressure drop intervals. Figure 7 shows there is no obvious difference in desorption amount of low-rank coal in any pressure drop interval, while the main desorption interval of high-rank coal is the low-pressure interval.

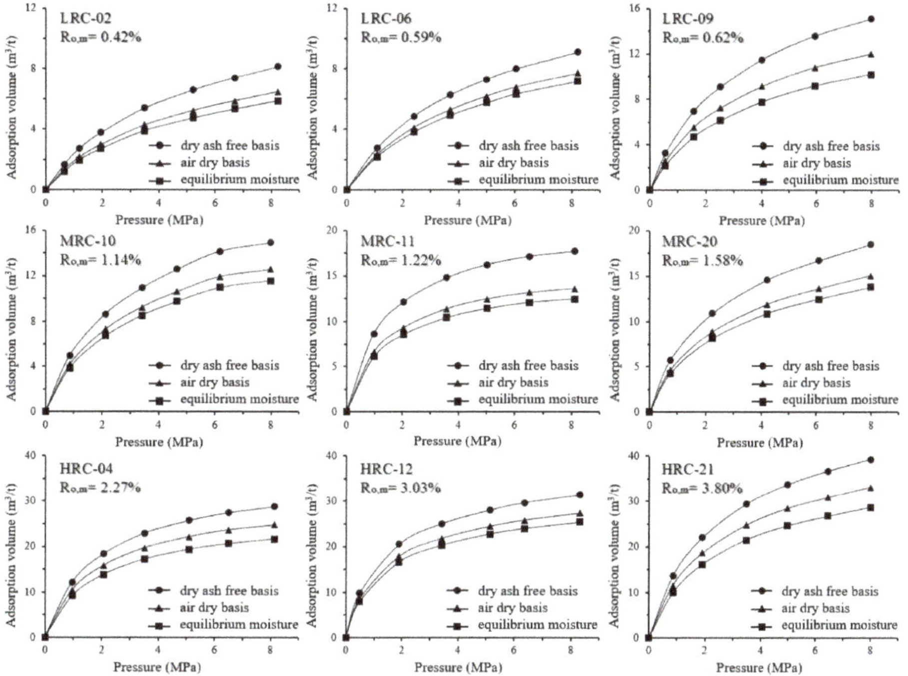

Figure 7. CH$_4$ isothermal adsorption in selected samples.

4. Discussion

4.1. Correlation between Coal Rank ($R_{o,m}$) and Langmuir Constants (V_L and P_L)

Due to different contents of moisture and ash in coal samples, V_L in dry ash-free basis is regarded as the net coal adsorption capacity. In this work, with increasing coal rank, the minimum and maximum appear on LRC-07 (11.7 m^3/t, $R_{o,m}$ = 0.59%) and HRC-24 (52.4 m^3/t, $R_{o,m}$ = 4.16%) respectively. The increase of adsorption capacity gradually slows down (Figure 8a), which has been proven before [19,20]. After coalification, anthracite gradually changes into coal-based graphite or natural coke graphitization, both of which have weak adsorption capacity. Therefore, assuming that the general adsorption capacity of lignite is 5 m^3/t, the results can be speculated by quadratic fitting (Figure 8a) with coefficient (R^2) reaching 0.86, and peak Langmuir volume (V_L) appears at $R_{o,m}$ of 4.5–5%, which agrees with a previous study [19]. However, there are still some data with large residual error, for example, abnormal low values exist at $R_{o,m}$ = 1.0% and $R_{o,m}$ = 2.8% (Figure 8a). Therefore, the factors that can influence the coal adsorption are various, and the change law of coal adsorption capacity during coalification is complex. P_L decrease rapidly before $R_{o,m}$ = 1.5% (Figure 8b), then increase slowly. CBM desorbed rate is the key factor in productivity improvement, which is determined by the absorption capacity and reservoir pressure. The favorable CBM reservoirs should be characterized with high V_L, high P_L and low reservoir pressure. Therefore, insufficient depressurisation in deep burial reservoirs cannot improve the CBM desorbtion rate, especially for high rank CBM zones with low P_L.

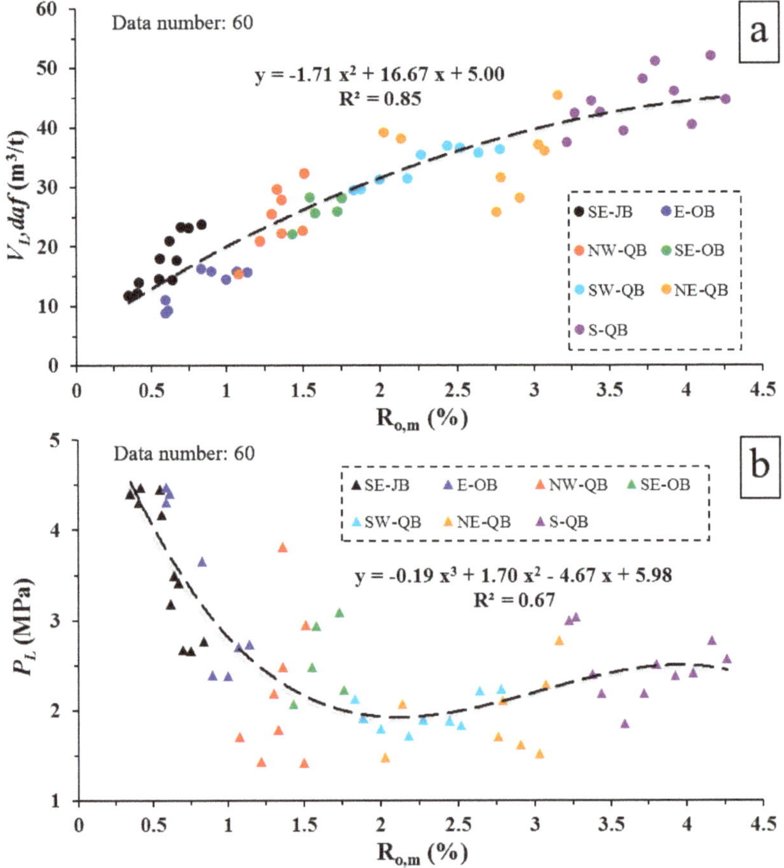

Figure 8. The relationships between adsorption constants (V_L and P_L) and coal rank ($R_{o,m}$), V_L (*daf*) and coal rank showing quadratic correlation (**a**), P_L and coal rank showing cubic correlation (**b**), colors showing different sampling locations (see Figure 1).

4.2. Quantitative Influence of Moisture on CH₄ Adsorption

Copious literature [14,17,31,33] has confirmed that both moisture and mineral have inhibitory effects on the adsorption capacity of coal. Therefore, when calculating V_L on a dry ash-free basis, minerals in coal are often regarded as materials without adsorption, and moisture is deemed as adsorbate equal to CH₄. However, for coals at different stages of metamorphism, the influence of moisture on the adsorption capacity has not been quantified. The adsorption competition between moisture and methane needs to be clarified. Therefore, the quantitative control effect of moisture on coal adsorption can be obtained by calculating the difference between $V_L(daf)$ and $V_L(me)$. Therefore, coal adsorption (air dry basis) in the state of equilibrium moisture should be obtained first, which can be calculated by removing the M_e portion of $V_L(ad)$, and then the decrement of the adsorption percentage of each one percent increase in moisture (DAP) can be calculated by:

$$DAP = 1 - (V_L(me) - V_L(ad) \times (1 - M_e))/M_e \tag{6}$$

where $V_L(me)$ is the Langmuir volume of equilibrium moisture; $V_L(ad)$ is the Langmuir volume of air dry basis; and M_e is the equilibrium moisture content.

Figure 9 shows the controlling mechanism of moisture on the adsorption of coal on different metamorphic grade. The value describes that an increase in the moisture content of a coal reduces the CH_4 adsorption percentage. Two inflection points can be found at $R_{o,m}$ = 1.8% and 3.5%. As the $R_{o,m}$ less than 1.8%, value is less than 1% and increases rapidly. At the range of $R_{o,m}$ from 1.8% to 3.5%, the value beyond 1 stabilizes at 1% to 1.01%. With the increase of coal rank of $R_{o,m}$ over 3.5%, the value increases.

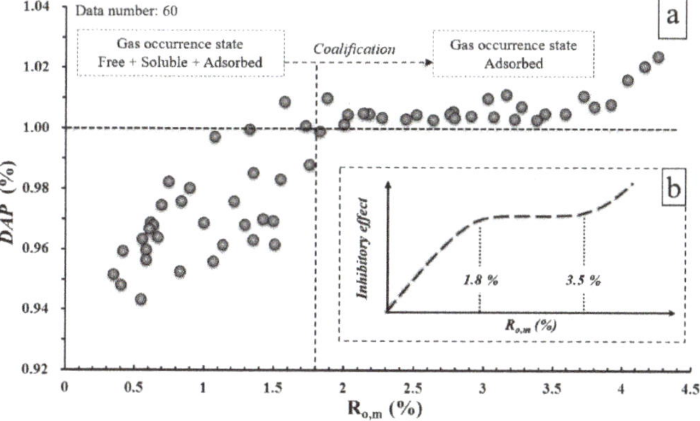

Figure 9. *DAP* distribution in $R_{o,m}$ scale (**a**) and the change trend during coalification (**b**), *DAP* representing the decrement of the adsorption percentage of each one percent increase in moisture.

Therefore, it can be concluded that the increase of moisture content in coal will lead to the decrease of methane adsorption capacity, and moisture has an inhibitory effect on the coal adsorption capacity. However, the effect varies regularly in different stages. Before coal evolves to semi-anthracite ($R_{o,m}$ = 1.8%), the increment of moisture is greater than the decrease of methane. First, in this stage, moisture mainly exists in adsorbed and free form, and free moisture does not compete with methane in adsorption. Especially in the low rank coal stage, large porosity and strong hydrophilicity of coal matrix result in that moisture not only exists in adsorption state [17,40], related research has found that free and water-soluble gas resources in low rank CBM reservoirs are considerable [41,42]. Second, in the process of coalification, the side chains of coal macromolecules are divided, macromolecular arrangement regularity increases, and the matrix polarity changes from hydrophobic to hydrophilic [17,34]. At the same time, the increase of formation pressure leads to the decrease of porosity and free moisture [29]. Therefore, with deepening metamorphism, the inhibitory effect of the moisture increment on the adsorption capacity of methane will become stronger. As $R_{o,m}$ reaches 1.8% (semi-anthracite), methane adsorption decreases more than one percent for every one percent increase in moisture content, but the decrease is small. As can be seen from Figures 5 and 6b, the values of porosity and total moisture decrease to minimum at $R_{o,m}$ from 1.5% to 2.0% and then increases, indicating that the amount of pore collapse caused by compaction is lower than the thermogenic pore increment in metamorphism. Moisture mainly exists in the adsorbed state, and with the decrease of macromolecular layer spacing, the free moisture content is extremely reduced. At the same time, the coalification will lead to the further increase of the polycyclic aromatic hydrocarbons in coal, thus the presence of moisture will hinder the methane adsorption. When the coal rank is beyond 3.5%, further aromatization leads to the decline of porosity and equilibrium moisture quantity, and free fluid almost does not exist. In this stage, the adsorption capacity is mainly controlled by the molecular polarity and specific surface area. Therefore, with the increase of coal rank, the stronger hydrophobic property of the coal matrix leads to an increasing inhibition of moisture on methane adsorption.

Clay minerals are hydrophilic substances and have features of water swelling and softening. Anthracites are hydrophobic, and coal-based graphite has extremely low adsorption. Therefore, for anthracite samples (MRCs 23–25 are collected from anthracite zones), total moisture is positively correlated to ash content (Figure 10). At the late metamorphic stage ($R_{o,m}$ > 1.8%), clay minerals are more moisture-absorbent than coal, which not only leads to further hindering the adsorption of gas content, but their the swelling effect can also plug the pore structure and throat.

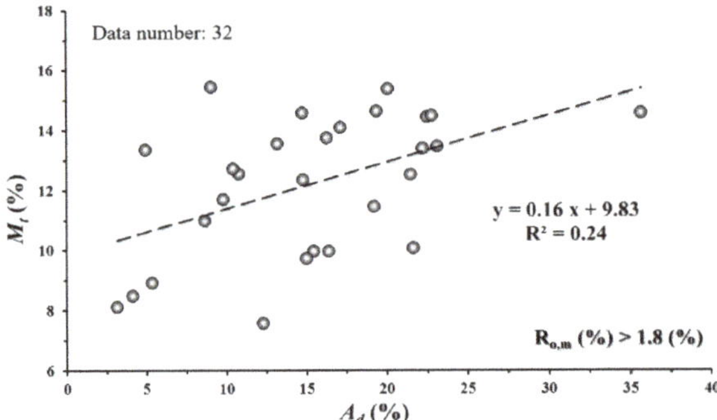

Figure 10. The relationships between ash yield (A_{ad}, %) and equilibrium moisture content (M_e, %) in the range of $R_{o,m}$ > 1.8%.

4.3. Relation between Maceral and CH$_4$ Adsorption

During early coalification, fracturing of macromolecular lipid chains, condensation of aromatic carbons and polarity changes of the material surface are the main processes [14,34,40]. Porosity, moisture content, density and volatiles yield all have obvious gradation laws in this stage, so these physical evolutions fundamentally control the adsorption capacity of coal, and the influence of maceral on the adsorption capacity is very small (Figure 11a). In the middle metamorphic stage, methane adsorption capacity is positively correlated with vitrinite content (Figure 11b), while in the high metamorphic stage, the relationship reverses (Figure 11c). This result is similar to some studies, but the authors suggest that the vitrinite content only slightly improves the adsorption capacity of coal. Figure 11d shows that in medium-high rank coal, the lower the ratio of vitrinite to inertinite is, the more the inherent moisture content is, and the adsorption of methane will be hindered more seriously. Maceral composition is determined by deposition [30,31], so the enrichment of a certain component in a certain area does not change the control of coal rank on methane adsorption capacity, but only indirectly.

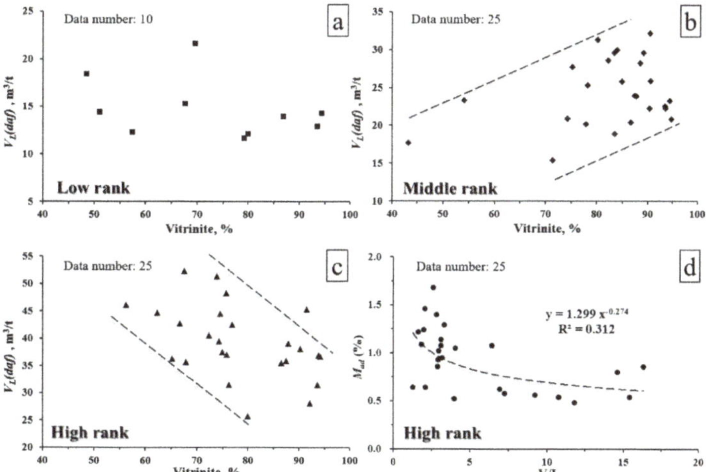

Figure 11. The relationships between vitrinite content and $V_L(daf)$ in Low rank coals (**a**), medium rank coals (**b**) and high rank coals (**c**), respectively; (**d**) The relationship between M_{ad} (%) and the ratio of Vitrinite to inertinite (V/I).

4.4. Relation between Pore Structure and Coal Adsorption Capacity

The pore structure of coal reflects the microstructure features and affects the gas adsorption capacity [43,44]. SEM is used for better analyzing the micromorphology of coal surface and the internal structure of coal internal. The pore-fracture system of lignite (LRC-02, $R_{o,m}$ = 0.41%) can be explained by irregular pores and fractures. Irregularly distributed micro-fractures (Figure 12a), dissolution pores (Figure 12b) and shrinkage-induced pores (Figure 12c) are merged to form pore clusters. Generally, pores and fractures are not filled with minerals or clays. For bituminous coal (MRC-20, $R_{o,m}$ = 1.58%), micro-fractures are not developed and show low extensibility (Figure 12d), wedge-shaped intergranular pores (Figure 12e) and deformed residual plant tissue pores can be identified, followed by some macro gas pores (Figure 12f). Partial pores are filled with or attached to secondary minerals. In the high metamorphic stage, parallel cleats can be observed (Figure 12g), slit-shaped corrosion pores (Figure 12h) and spherical gas pores (Figure 12i) are abundant. In anthracite (HRC-21, $R_{o,m}$ = 3.80%), the pore size distribution tends to be simplified, with dominant micropores, fractures are well oriented and extended, and clay filling is common. Therefore, the pore-fracture system of different rank coals varies a lot, indicating heterogeneity on the coal surface.

In the process of coalification, most of the oxygen-bearing elements in the coal molecules fall off [14], the aromatic ring layers of coal molecules arrange in a fixed orientation, and the high temperature promotes hydrocarbon generation, resulting in the concentration of micropores. Micropores are further generated, and the affinity for methane of coal macromolecules increases, both of which lead to an increase of methane adsorption. After the third coalification jump, the aromatization degree, affinity for methane and micropore ratio of the coal gradually became constant, while the spacing of aromatic rings gradually decreases. Therefore, the increase rate of methane adsorption slows down, and even reverses.

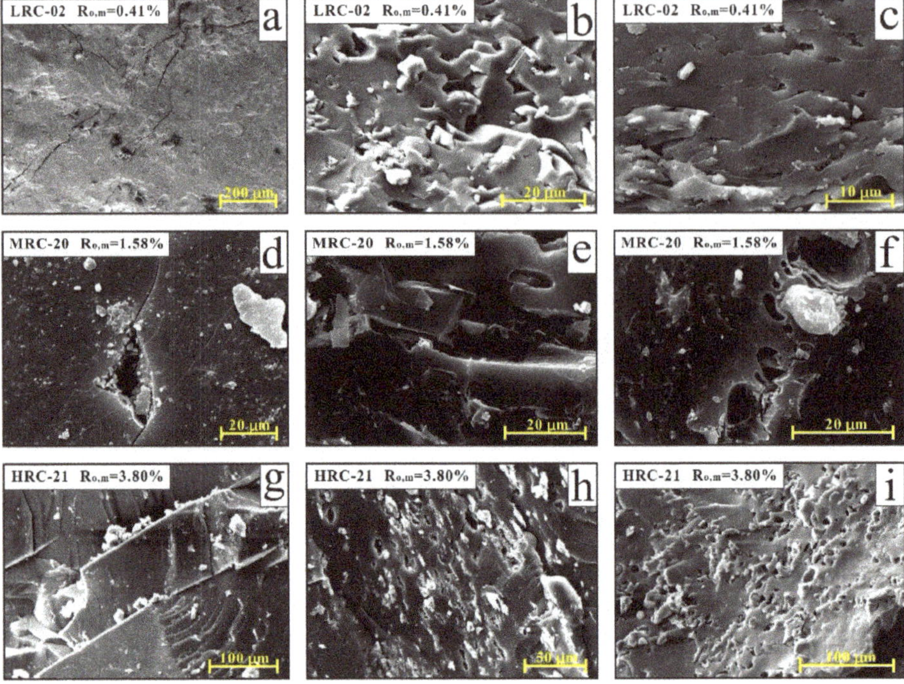

Figure 12. Pore-fracture shapes from SEM result. (**a**) irregular fractures; (**b**) dissolution pores; (**c**) shrinkage-induced pores; (**d**) micro-fractures; (**e**) wedge shaped intergranular pores; (**f**) deformed residual plant tissue pores; (**g**) cleats filled with clay; (**h**) slit-shaped corrosion pores; (**i**) spherical gas pore cluster.

5. Conclusions

Sixty coal samples covering a wide range of coal rank ($R_{o,m}$) were used to explore the variation of petrophysical parameters and their influences on coal adsorption capacity. Tests included CH_4 isothermal adsorption, moisture measurement, proximate analysis, maceral measurement and SEM observation. Maceral composition is controlled by deposition without relation to coal rank, vitrinite is the most abundant maceral, followed by inertinite and liptinite. M_{ad} and M_e decreases sharply to minimum at $R_{o,m}$ of 1.5%, proving that coal metamorphism controls the moisture holding capacity. During coalification, compaction and aromatization can reduce coal porosity, coal matrix density simultaneously increases. $R_{o,m}$ and V_L (*daf*) show a quadratic correlation with the peak located in $R_{o,m} = 5\%$. P_L decrease rapidly before $R_{o,m} = 1.5\%$, then increase slowly.

The inhibitory effect of moisture on CH_4 adsorption varies regularly during coalification. *DAP* is proposed to quantify this effect, which represents the decrement of the adsorption percentage of each one percent increase in moisture. *DAP* first grows linearly as the $R_{o,m}$ increases to 1.8%, then the value beyond 1 stabilizes at 1% to 1.01% with $R_{o,m} = 1.8$–3.5%, finally it increases again when the value of $R_{o,m}$ is over 3.5%,. This indicates that during the early stage of coal metamorphism, large porosity and strong hydrophilicity of coal matrix result in the fact that moisture not only exists in the adsorption state. As coalification progresses, variations of matrix macromolecules and porosity strengthen the inhibitory effect of moisture on adsorption capacity. Vitrinite content only slightly improves the adsorption capacity in medium-high rank coals. The lower the ratio of vitrinite to inertinite is, the more the inherent moisture content is, and the adsorption of methane will be hindered more seriously. During coalification, the pore size distribution tends to be simplified with micropores dominating, fractures tend to be well oriented and extended, and clay filling becomes more common in high rank coals.

This work explored the variation rule of petrophysical properties and adsorption capacity in different rank coals, and then established some experimental formulas. Whether in CBM production prediction or security risk assessment of mining, these formulas and conclusions have practical significance.

Author Contributions: D.L. and Y.C. conceived and designed the experiments; Y.W. performed the experiments and wrote the paper; Y.W. and X.L. analyzed the data; D.L. and Y.C. revised the paper and provided language support; D.L. provided technical support.

Funding: This research was funded by the National Natural Science Foundation of China (grant Nos. 41830427, 41772160 and 41602170), and the National Major Research Program for Science and Technology of China (grant No. 2016ZX05043-001).

Conflicts of Interest: The authors have no conflict of interest to declare.

Appendix A

Table A1. Tested parameters of samples, including maceral content, proximate analysis, isothermal adsorption and vitrinite reflectance results.

Area	Sample No.	$R_{o,m}$ (%)	Maceral Content			Proximate Analysis			Adsorption Capacity		P_L MPa	M_e (%)
			V (%)	I (%)	L (%)	V_{daf} (%)	M_{ad} (%)	A_{ad} (%)	$V_L(me)$ m^3/t	$V_L(ad)$ m^3/t		
SE-JB	LRC-01	0.35	53.63	39.85	5.19	54.81	8.31	9.22	9.01	10.15	4.41	21.47
SE-JB	LRC-02	0.41	78.80	5.40	12.10	52.41	8.01	17.36	8.75	9.64	4.31	20.07
SE-JB	LRC-03	0.42	86.20	5.10	3.10	50.78	6.38	10.87	10.58	11.82	4.48	15.99
SE-JB	LRC-04	0.55	61.61	29.39	2.70	38.96	5.21	8.36	11.95	13.25	4.46	17.18
SE-JB	LRC-05	0.56	43.17	45.84	2.20	37.58	4.32	4.72	14.42	16.78	4.17	18.00
SE-JB	LRC-09	0.62	64.17	28.06	3.86	48.83	4.49	18.32	14.15	16.68	3.18	18.67
SE-JB	LRC-10	0.64	49.50	47.50	2.60	36.71	2.67	4.44	11.72	13.38	3.50	16.30
SE-JB	MRC-01	0.67	42.00	55.10	2.80	28.56	2.68	1.28	14.50	17.00	3.42	18.67
SE-JB	MRC-02	0.70	52.80	44.80	1.30	34.11	2.54	8.13	17.74	20.83	2.67	16.89
SE-JB	MRC-03	0.75	88.40	5.20	2.60	42.01	1.28	7.94	17.66	21.12	2.66	17.89
SE-JB	MRC-05	0.84	84.00	11.70	2.00	41.13	1.62	11.88	18.01	20.66	2.77	14.53
E-OB	LRC-06	0.59	72.30	11.00	16.20	33.57	4.38	10.87	11.00	11.82	4.48	10.99
E-OB	LRC-07	0.59	69.60	17.40	12.70	28.04	3.15	17.36	8.85	9.64	4.31	14.07
E-OB	LRC-08	0.61	69.90	18.30	11.60	23.87	4.41	8.82	9.30	10.15	4.41	12.47
E-OB	MRC-04	0.83	82.80	4.60	8.80	25.60	1.20	12.84	16.21	17.89	3.65	12.77
E-OB	MRC-06	0.90	77.30	15.20	6.80	24.04	1.23	7.12	15.72	17.28	2.39	10.19
E-OB	MRC-07	1.00	82.10	12.60	5.00	21.87	0.91	21.78	14.37	15.74	2.38	10.87
E-OB	MRC-08	1.07	84.10	12.10	3.00	15.31	1.29	27.32	15.76	17.12	2.70	10.70
E-OB	MRC-10	1.14	77.00	21.80		19.70	0.89	15.16	15.58	16.95	2.73	10.47
NW-QB	MRC-12	1.30	75.50	21.00	2.40	13.79	1.07	7.56	21.63	23.17	2.18	7.71
NW-QB	MRC-09	1.08	69.80	28.00		22.70	0.52	21.93	10.63	11.95	1.70	11.33
NW-QB	MRC-11	1.22	72.00	25.00		14.34	0.54	22.75	14.70	16.03	1.43	9.77
NW-QB	MRC-13	1.33	80.20	15.90	1.60	16.41	0.50	21.30	21.48	23.15	1.78	7.22
NW-QB	MRC-14	1.36	89.50	6.20		27.43	0.67	23.09	15.83	16.94	2.48	7.17
NW-QB	MRC-15	1.36	72.50	23.90	3.40	10.96	0.51	3.22	23.82	26.73	3.81	12.63
NW-QB	MRC-17	1.50	90.00	6.30		13.39	0.38	15.25	17.10	19.03	1.42	12.08
NW-QB	MRC-18	1.51	88.50	9.40	0.90	13.92	1.04	9.65	26.64	28.76	2.95	8.51
SE-OB	MRC-16	1.43	90.10	9.60		17.45	1.11	10.43	18.14	19.67	2.06	9.18
SE-OB	MRC-19	1.55	81.20	17.50		14.72	1.09	19.63	20.37	22.67	2.48	10.97
SE-OB	MRC-20	1.58	81.00	14.40	1.10	21.31	1.29	18.09	19.03	20.82	2.94	8.25
SE-OB	MRC-21	1.73	86.70	9.10	1.20	21.30	0.92	15.53	19.56	21.62	3.08	9.49
SE-OB	MRC-22	1.76	87.80	11.40		18.92	0.80	10.27	22.63	25.10	2.22	10.34
SW-QB	MRC-23	1.83	85.63	10.47		21.86	1.38	21.90	19.98	22.70	2.12	12.04

Table A1. *Cont.*

Area	Sample No.	$R_{o,m}$ (%)	Maceral Content			Proximate Analysis			Adsorption Capacity		P_L MPa	M_e (%)
			V (%)	I (%)	L (%)	V_{daf} (%)	M_{ad} (%)	A_{ad} (%)	$V_L(me)$ m^3/t	$V_L(ad)$ m^3/t		
SW-QB	MRC-24	1.88	78.14	14.96		27.53	1.36	35.18	16.38	19.03	1.91	13.23
SW-QB	MRC-25	2.00	75.15	18.55		19.99	1.05	22.27	20.80	24.04	1.79	13.41
SW-QB	HRC-03	2.18	67.89	21.21		16.53	0.95	21.30	21.57	24.46	1.71	11.59
SW-QB	HRC-04	2.27	82.09	12.81		12.43	1.08	13.05	26.65	30.50	1.89	12.48
SW-QB	HRC-05	2.44	72.00	23.00		11.63	1.08	15.31	28.13	30.91	1.88	8.90
SW-QB	HRC-06	2.52	86.58	5.31		15.18	0.86	22.59	24.19	28.08	1.83	13.63
SW-QB	HRC-07	2.64	63.50	30.00		13.21	0.64	19.12	25.56	28.70	2.21	10.83
SW-QB	HRC-09	2.78	63.00	33.70		7.97	1.09	5.33	31.28	33.99	2.23	7.84
NE-QB	HRC-01	2.03	84.03	11.57		18.55	0.58	21.54	27.54	30.48	1.48	9.50
NE-QB	HRC-02	2.14	87.07	9.43		12.26	0.56	14.93	29.19	32.19	2.07	9.16
NE-QB	HRC-08	2.76	76.69	19.21		13.22	0.52	4.13	22.57	24.56	1.70	7.96
NE-QB	HRC-10	2.79	87.14	5.96		9.95	0.80	10.73	24.53	27.85	2.10	11.77
NE-QB	HRC-11	2.91	87.59	7.41		10.16	0.48	4.96	23.06	26.53	1.61	12.88
NE-QB	HRC-12	3.03	90.80	5.90		10.30	0.54	12.26	29.89	32.23	1.52	7.03
NE-QB	HRC-13	3.07	84.87	12.23		11.14	0.62	16.17	25.91	29.89	2.28	13.13
NE-QB	HRC-14	3.16	87.93	8.17		9.83	0.54	3.17	40.22	43.61	2.77	7.58
S-QB	HRC-15	3.22	72.50	24.30		6.94	1.02	8.59	30.47	33.88	3.00	9.97
S-QB	HRC-16	3.27	73.00	22.00		8.01	1.29	9.68	33.62	37.62	3.03	10.43
S-QB	HRC-17	3.38	71.97	24.63		9.82	0.93	10.34	34.81	39.51	2.40	11.81
S-QB	HRC-18	3.44	61.70	30.80		12.49	1.24	19.83	28.93	33.78	2.18	14.15
S-QB	HRC-19	3.59	71.33	24.67		10.23	0.86	16.26	29.73	32.76	1.85	9.11
S-QB	HRC-20	3.72	73.28	23.52		9.88	1.14	19.19	33.17	38.52	2.18	13.51
S-QB	HRC-21	3.80	70.72	24.98		9.12	1.40	14.58	37.38	43.17	2.51	13.19
S-QB	HRC-22	3.92	52.60	41.00		8.21	0.64	14.71	34.39	39.06	2.38	11.71
S-QB	HRC-23	4.04	70.60	27.00		6.94	1.68	8.93	31.03	36.25	2.41	13.78
S-QB	HRC-24	4.16	58.40	28.00		11.17	1.46	22.77	34.68	39.70	2.77	12.02
S-QB	HRC-25	4.26	59.90	36.30		8.58	1.22	16.93	31.61	36.64	2.56	12.88

V: vitrinite (%); I: Inertinite (%); L: Liptinite; M_e: equilibrium moisture content.

References

1. Moore, T.A. Coalbed methane: A review. *Int. J. Coal Geol.* **2012**, *101*, 36–81. [CrossRef]
2. Karacan, C.Ö.; Ruiz, F.A.; Cotè, M.; Sally, P. Coal mine methane: A review of capture and utilization practices with benefits to mining safety and to greenhouse gas reduction. *Int. J. Coal Geol.* **2011**, *86*, 121–156. [CrossRef]
3. Zhou, F.B.; Xia, T.Q.; Wang, X.X.; Zhang, Y.F.; Sun, Y.N.; Liu, J.S. Recent developments in coal mine methane extraction and utilization in China: A review. *J. Nat. Gas Sci. Eng.* **2016**, *31*, 437–458. [CrossRef]
4. Flores, R.M. Coalbed methane: From hazard to resource. *Int. J. Coal Geol.* **1998**, *35*, 3–26. [CrossRef]
5. Crosdale, P.J.; Beamish, B.B.; Valix, M. Coalbed methane sorption related to coal composition. *Int. J. Coal Geol.* **1998**, *35*, 147–158. [CrossRef]
6. Mares, T.E.; Moore, T.A.; Moore, C.R. Uncertainty of gas saturation estimates in a subbituminous coal seam. *Int. J. Coal Geol.* **2009**, *77*, 320–327. [CrossRef]
7. Joubert, J.I.; Grein, C.T.; Bienstock, D. Effect of moisture on methane capacity of American coals. *Fuel* **1974**, *53*, 186–191. [CrossRef]
8. Hao, S.; Wen, J.; Yu, X.; Chu, W. Effect of the surface oxygen groups on methane adsorption on coals. *Appl. Surf. Sci.* **2013**, *264*, 433–442. [CrossRef]
9. Yao, Y.; Liu, D.; Huang, W. Influences of igneous intrusions on coal rank, coal quality and adsorption capacity in Hongyang, Handan and Huaibei coalfields, North China. *Int. J. Coal Geol.* **2011**, *88*, 135–146. [CrossRef]
10. Zhou, F.; Liu, S.; Pang, Y.; Li, J.; Xin, H. Effects of Coal Functional Groups on Adsorption Microheat of Coal Bed Methane. *Energy Fuels* **2015**, *29*, 1550–1557. [CrossRef]
11. Fu, H.J.; Tang, D.Z.; Xu, T.; Xu, H.; Tao, S.; Li, S.; Yin, Z.Y.; Chen, B.L.; Zhang, C.; Wang, L.L. Characteristics of pore structure and fractal dimension of low-rank coal: A case study of Lower Jurassic Xishanyao coal in the southern Junggar Basin, NW China. *Fuel* **2017**, *193*, 254–264. [CrossRef]
12. Levy, J.; Day, S.J.; Killingley, J.S. Methane capacity of Bowen Basin coals related to coal properties. *Fuel* **1997**, *74*, 1–7. [CrossRef]
13. Ritter, U.; Arnt, G. Adsorption of petroleum compounds in vitrinite: Implications for petroleum expulsion from coal. *Int. J. Coal Geol.* **2005**, *62*, 183–191. [CrossRef]
14. Chen, Z.; Jia, C.; Song, Y.; Wang, H.; Wang, Y. Differences and origin of physical properties of low-rank and high-rank coal-bed methanes. *Acta Petrol. Sin.* **2008**, *2*, 179–184, (In Chinese with English abstract). [CrossRef]
15. Yee, D.; Seidle, J.P.; Hanson, W.B. Gas sorption on coal and measurement of gas content. *Hydrocarb. Coal AAPG Stud. Geol.* **1993**, *38*, 203–218.
16. Levine, J.R. Coalification: The evolution of coal as a source rock and reservoir rock for oil and gas. In *SG 38: Hydrocarbons from Coal*; Law, B.E., Rice, D.D., Eds.; AAPG: Tulsa, OK, USA, 1993; Volume 38, pp. 39–77.
17. Laxminarayana, C.; Crosdale, P.J. Controls on methane sorption capacity of Indian coals. *AAPG Bull.* **2002**, *86*, 201–212.
18. Keshavarz, A.; Sakurovs, R.; Grigore, M.; Sayyafzadeh, M. Effect of maceral composition and coal rank on gas diffusion in Australian coals. *Int. J. Coal Geol.* **2017**, *173*, 65–75. [CrossRef]
19. Su, X. Influence of coal rank on coal adsorption capacity. *Nat. Gas Ind.* **2005**, *25*, 19–21, (In Chinese with English abstract).
20. Zhang, K.; Tang, D.; Tao, S.; Liu, Y.; Chen, S. Study on influence factors of adsorption capacity of different metamorphic degree coals. *Coal Sci. Technol.* **2017**, *45*, 192–197, (In Chinese with English abstract).
21. Weniger, P.; Francu, J.; Hemza, P.; Krooss, B.M. Investigations on the methane and carbon dioxide sorption capacity of coals from the SW Upper Silesian Coal Basin, Czech Republic. *Int. J. Coal Geol.* **2010**, *84*, 23–39. [CrossRef]
22. Li, Z.T.; Yao, Y.B.; Zhou, H.P.; Bai, X.H. Study on Coal and Rock Maceral Composition Affected to Methane Adsorption Capacity. *Coal Sci. Technol.* **2012**, *40*, 125–128, (In Chinese with English abstract).
23. Li, J.; Yao, Y.; Cai, Y.; Qiu, Y. Discussion on coal physical properties and formation with different metamorphic degree in North China. *Coal Sci. Technol.* **2012**, *4*, 111–115, (In Chinese with English abstract).
24. An, F.; Cheng, Y.; Wu, D.; Wang, L. The effect of small micropores on methane adsorption of coals from Northern China. *Adsorption* **2013**, *19*, 83–90. [CrossRef]
25. Fu, H.; Tang, D.; Xu, H.; Tao, S.; Xu, T.; Chen, B.; Yin, Z. Abrupt Changes in Reservoir Properties of Low-Rank Coal and Its Control Factors for Methane Adsorbability. *Energy Fuels* **2016**, *30*, 2084–2094. [CrossRef]

26. Li, Y.; Tang, D.; Xu, H.; Elsworth, D.; Meng, Y. Geological and hydrological controls on water co-produced with coalbed methane in Liulin, eastern Ordos Basin, China. *AAPG Bull.* **2015**, *99*, 207–229. [CrossRef]

27. Wang, Y.; Liu, D.; Cai, Y.; Yao, Y.; Zhou, Y. Evaluation of structured coal evolution and distribution by geophysical logging methods in the Gujiao Block, northwest Qinshui basin, China. *J. Nat. Gas. Sci. Eng.* **2018**, *51*, 210–222. [CrossRef]

28. Zhang, J.; Liu, D.; Cai, Y.; Pan, Z.; Yao, Y.; Wang, Y. Geological and hydrological controls on the accumulation of coalbed methane within the No. 3 coal seam of the southern Qinshui Basin. *Int. J. Coal Geol.* **2017**, *182*, 94–111. [CrossRef]

29. Li, Y.; Zhang, C.; Tang, D.; Gan, Q.; Niu, X.; Wang, K. Coal pore size distributions controlled by the coalification process: An experimental study of coals from the Junggar, Ordos and Qinshui basins in China. *Fuel* **2017**, *206*, 352–363. [CrossRef]

30. O'Keefe, J.M.K.; Bechtel, A.; Christanis, K.; Dai, S. On the fundamental difference between coal rank and coal type. *Int. J. Coal Geol.* **2013**, *118*, 58–87. [CrossRef]

31. Pickel, W.; Kus, J.; Flores, D.; Kalaitzidis, S.; Christanis, K.; Cardott, B.J.; Misz-Kennan, M.; Rodrigues, S.; Hentschel, A.; Hamor-Vido, M.; et al. Classification of liptinite—ICCP System 1994. *Int. J. Coal Geol.* **2017**, *169*, 40–61. [CrossRef]

32. Hou, H.; Shao, L.; Li, Y.; Li, Z.; Wang, S.; Zhang, W.; Wang, X. Influence of coal petrology on methane adsorption capacity of the Middle Jurassic coal in the Yuqia coalfield, northern Qaidam Basin, China. *J. Petrol. Sci. Eng.* **2017**, *149*, 218–227. [CrossRef]

33. Gagarin, S.G. Relation between the petrographic content of mineral components in coal and its ash content. *Coke Chem.* **2008**, *51*, 241–246. [CrossRef]

34. Gao, Z.; Yang, W. Adsorption mechanism of water molecule on different rank coals molecular surface. *J. China Coal Soc.* **2017**, *42*, 753–759, (In Chinese with English abstract). [CrossRef]

35. Li, X.; Zeng, F.; Si, J.; Wang, W.; Dong, X.; Cheng, L. High resolution TEM image analysis of coals with different metamorphic degrees. *J. Fuel Chem. Technol.* **2016**, *44*. (In Chinese with English abstract). [CrossRef]

36. Su, X.; Si, Q.; Song, J. Characteristics of coal Raman spectrum. *J. China Coal Soc.* **2016**, *42*, 1197–1202, (In Chinese with English abstract). [CrossRef]

37. Xu, H.; Tang, D.; Mathews, J.P.; Zhao, J.; Li, B.; Tao, S.; Li, S. Evaluation of coal macrolithotypes distribution by geophysical logging data in the Hancheng Block, eastern margin, Ordos Basin, China. *Int. J. Coal Geol.* **2016**, *165*, 265–277. [CrossRef]

38. Cai, Y.; Liu, D.; Pan, Z.; Yao, Y.; Li, J.; Qiu, Y. Pore structure and its impact on CH_4 adsorption capacity and flow capability of bituminous and subbituminous coals from Northeast China. *Fuel* **2013**, *103*, 258–268. [CrossRef]

39. Meng, Y.; Tang, D.; Xu, H.; Qu, Y.; Li, Y.; Zhang, W. Division of coalbed methane desorption stages and its significance. *Petrol. Explor. Dev.* **2014**, *41*, 671–677. [CrossRef]

40. Zhang, L.; Su, X.; Zeng, S. Discussion on the controlling effects of coal properties on coal adsorption capacity. *Acta Geol. Sin.* **2006**, *6*, 910–915, (In Chinese with English abstract). [CrossRef]

41. Pan, Z.; Connell, L.D.; Camilleri, M.; Connelly, L. Effects of matrix moisture on gas diffusion and flow in coal. *Fuel* **2010**, *89*, 3207–3217. [CrossRef]

42. Liu, A.; Fu, X.; Wang, K.; An, H.; Wang, G. Investigation of coalbed methane potential in low-rank coal reservoirs—Free and soluble gas contents. *Fuel* **2013**, *112*, 14–22. [CrossRef]

43. Clarkson, C.R.; Bustin, R.M. The effect of pore structure and gas pressure upon the transport properties of coal: A laboratory and modeling study. 1. Isotherms and pore volume distributions. *Fuel* **1999**, *78*, 1333–1344. [CrossRef]

44. Nie, B.; Liu, X.; Yang, L.; Meng, J.; Li, X. Pore structure characterization of different rank coals using gas adsorption and scanning electron microscopy. *Fuel* **2015**, *158*, 908–917. [CrossRef]

Article

Design and Evaluation of a Surfactant–Mixed Metal Hydroxide-Based Drilling Fluid for Maintaining Wellbore Stability in Coal Measure Strata

Shuya Chen, Yanping Shi, Xianyu Yang, Kunzhi Xie and Jihua Cai *

Faculty of Engineering, China University of Geosciences, Wuhan 430074, China; chenshuya@cug.edu.cn (S.C.); yunus@cug.edu.cn (Y.S.); yxy@cug.edu.cn (X.Y.); kunzhixie@gmail.com (K.X.)
* Correspondence: caijh@cug.edu.cn; Tel.: +86-27-6788-3507

Received: 27 February 2019; Accepted: 13 May 2019; Published: 16 May 2019

Abstract: Co-exploitation of coal measure gases (coalbed gas, shale gas, and tight sandstone gas) puts a higher requirement on drilling fluids. Conventional drilling fluids have disadvantages, such as causing problems of borehole collapse, formation damage, and water blockage. This paper proposes a set of high inhibitive and low-damage drilling fluids that function by electrical inhibition and neutral wetting. Zeta potential results showed that the negative electrical property of Longtan coal in Bijie, Guizhou, can be reversed by organic mixed metal hydroxide (MMH) and the cationic surfactant alkyl trimethylammonium bromide (CS-5) from −3.63 mV to 19.75 mV and 47.25 mV, respectively. Based on the contact angle and Fourier Transform Infrared Spectroscopy (FT-IR) results, it can be concluded that chemical adsorption dominates between the Longmaxi shale and surfactants, and physical adsorption between the Longtan coal and surfactants. A compound surfactant formula (0.001 wt% CS-4 + 0.001 wt% CS-1 + 0.001 wt% CS-3), which could balance the wettability of the Longmaxi shale and the Longtan coal, making them both appear weakly hydrophilic simultaneously, was optimized. After being treated by the compound surfactants, the contact angles of the Longmaxi shale and the Longtan coal were 89° and 86°, respectively. Pressure transmission tests showed that the optimized combination of compound surfactants and inorganic MMH (MMH-1) could effectively reduce permeability of the Longmaxi shale and the Longtan coal, thus retarding pore pressure transmission in coal measure strata. Then, the proposed water-based drilling fluid (WBDF) system (4 wt% sodium bentonite + 1.5 wt% sodium carboxymethyl cellulose + 2 wt% lignite resin + 5 wt% potassium chloride + 3 wt%MMH-1 + 0.001 wt% CS-4 + 0.001 wt% CS-1 + 0.001 wt% CS-3) was evaluated based on parameters including rheology, American Petroleum Institute (API) filtration, electrical property, wettability, inhibition capability, reservoir protection characteristics, and anti-pollution performance. It had an API filtration of 7 mL, reservoir damage rate of 10%, moderate and acceptable viscosity, strong inhibition capability to coal measure strata rocks, good tolerance to inorganic pollutants and drilling cuttings, and environmentally friendly properties. It could meet wellbore stability and reservoir protection requirements in the co-exploitation of coal measure gases.

Keywords: coal measure gases (coalbed gas; shale gas; and tight sand gas); co-exploitation; wellbore stability; wettability; zeta potential; drilling fluid

---

1. Introduction

Coal measure gases (CMG), including coalbed methane (CBM), tight sandstone gas, and shale gas, are important unconventional gas resources. Recent exploration results have revealed the presence of shale gas and tight sandstone gas in the Qinshui basin, located in Shanxi, China, which is an important area of CBM production [1]. Furthermore, in the upper Permian coal bearing strata of the Liupanshui coalfield, the overlaid coal seam and shale have reached the level of high-quality source rock [2].

However, single-reservoir exploitation is always accompanied with problems, such as low production, short drainage periods, and high cost [3]. In order to improve the productivity and utilization efficiency of unconventional gas resources, co-exploitation of CMG is a key factor in the development system [4]. Drilling fluid, known as the blood of drilling engineering, is a vital technique [5]. Besides for balancing formation pressure, cooling and lubricating drilling tools, and carrying drilling cuttings, the main functions of coal measure strata drilling fluid are to maintain wellbore stability and minimize formation damage [6].

Large amounts of natural cleats and tectonic fractures and water-sensitive clay minerals cause borehole collapse and formation damage problems in CMG reservoirs [7,8]. Clay swelling and the water blockage of drilling fluid invasion can reduce the permeability in the near wellbore regions, leading to low productivity of CMG wells [9,10]. Conventional drilling fluids, such as potassium chloride drilling fluid and low solid-phase polymer drilling fluid, may cause leakage problems and borehole collapse due to their low density [11]. In this case, a set of formation-compatible drilling fluids need to be proposed. Several studies were carried out using nanoparticles (NPs) to improve wellbore stability by plugging pore throats, reducing the reservoir permeability and building a thin, tight mud cake [12]. Several types of nanoparticles were investigated, such as cellulose nanocrystals [13], multi-walled carbon nanotubes [14], and different clay particles [15,16]. Accordingly, some researchers proposed rheological models of nanoparticle-based drilling fluid to better estimate the shear stress and viscosity under downhole conditions [17,18]. However, the stability of nanoparticle dispersions remains a huge challenge, along with the agglomerate problem and the high cost, and therefore NP-drilling fluid applications in the industry have not been fully implemented.

Wettability is one of the most important surface properties of rock [19]. It controls the magnitude and direction of capillary force and affects fines migration [20,21]. The wettability of a solid surface can be quantified by contact angle θ via Yang's equation [22]. Generally, when the contact angle is smaller than 90°, the solid surface is considered to be water-wet. If the contact angle is larger than 90°, the solid surface is considered to be oil-wet [23]. Given the pore throat distribution of shale/coal, the invasion of water/filtrate will form a meniscus pointing to the water/filtrate, after which the capillary force is formed [24]. According to the capillary force formula proposed by Liang et al. [25], when the contact angle is equal to 90° (neutral wetting), the capillary force is minimum, which is beneficial to borehole stability and reservoir protection.

Surfactants have been wildly used in production engineering to promote oil recovery rates by changing capillary forces and altering reservoir wettability [26]. They have also been used as emulsifiers in oil-based drilling fluid and as shale-swelling inhibitors to prevent wellbore instability [27]. However, there are not many reports about the application of surfactants in water-based drilling fluid. Yunita et al. [28] discovered that the rheological and filtration properties of water-based drilling fluid were improved with the addition of nonionic and anionic surfactants. Shadizadeh et al. [29] developed a plant-based nonionic surfactant zizyphus spina-christi extract (ZSCE), which showed excellent inhibitive ability to shale. The ZSCE molecules could form a hydrophobic shell on the clay surface through hydrogen bonding between the hydrophilic tail of ZSCE and oxygen atoms available on the silica surface of clay. Our previous study proposed a set of compound surfactants that could increase the contact angle of shale from 34.28° to 64.54° and improve the inhibition of water-based drilling fluid [30]. Compared to the old surfactant formula, this research proposed a set of cationic compound surfactants. After treatment with the proposed surfactants, the contact angles of the Longmaxi shale and the Longtan coal were 89° and 86°, respectively.

Due to the cation exchange capacity of clay minerals, an electrical-double-layer (EDL) can be formed in water systems. Zeta potential (ζ, mV), known as the potential between the slipping plane and the bulk solution, has traditionally been used to measure the stability of clay dispersions and estimate the double-layer thickness [31,32]. The nature of the EDL controls the repulsion/attraction of the system [33]. If the clay zeta potential is smaller than −60 mV, the system is extremely dispersive, if the zeta potential is around −40 mV, the system is strongly dispersive, if the zeta potential is around −20 mV, the system is probably dispersive, and if the zeta potential is close to zero, the system is non-dispersive [34].

Zhang et al. [35] found that the water adsorption rate of clay minerals showed a good correlation with zeta potential, and the negative charge of the formation was the dominant reason of swelling. Therefore, it is feasible to improve drilling fluid inhibition by increasing its zeta potential.

In addition to indicating colloid stability and clay swelling, zeta potential plays a significant role in understanding the adsorption mechanism of inorganic/organic molecules at the solid/solution interface [32]. Previous studies showed that the electric double layer is closely related to wettability [36,37]. Zeta potential measurements can be used to evaluate the stability of the water membrane on a rock surface. If the membrane is instable (zeta potential is between −30 mV and +30 mV), the rock surface is considered to be neutral wetting or oil-wet [38]. Arif et al. [39] measured zeta potentials as a function of temperature and salinity for different ranks of coals. Results showed that zeta potential increased with salinity and coal rank, and contact angle also increased with salinity and coal rank, implying a positive relation between zeta potential and wettability. Song et al. [40] pointed out that coal surfaces appeared to be hydrophobic when zeta potential was close to zero.

Su et al. [34] proposed organic mixed metal hydroxide (MMH) treatment, electrolyte treatment and pH adjustment to control the electrical properties of clay minerals and drilling fluid. Li et al. [41] tested the zeta potential of coal as a function of pH. Results showed that coal particles were negatively charged when the solution was strongly acidic, and then the zeta potential increased when solution became alkaline. Shu et al. [42] developed a modified low-damage mixed metal hydroxide (MMH) drilling fluid system for carbonate reservoirs. Not only can the MMH form a special spatial network structure with bentonite, it can be easily removed by oil or acid, which can further improve/recover reservoir permeability. Furthermore, the structure of MMH–bentonite is larger than the formation pore throat, which limits the invasion depth of the drilling fluid.

Based on previous studies, a set of low-damage, high-inhibitive, water-based drilling fluids based on wettability alteration and electrical inhibition, which could meet wellbore stability and formation protection requirements in the co-exploitation of CMG, was proposed. This paper developed a set of compound surfactant formulas that could effectively reduce the hydrophilicity of the reservoir surface along with the optimized MMH additive, which showed great ability to avoid clay hydration by electrical inhibition. Furthermore, the pressure transmission results confirmed their potential to increase wellbore stability in coal measure strata Among the numerous drilling fluid systems, this paper paves a new way for the design of coal measure strata drilling fluids

2. Materials and Methods

2.1. Materials

The surfactants used in this study were sodium alkylbenzene sulfonate (AS-1), sodium dodecyl sulfate (AS-2), alkyl trimethylammonium chloride (CS-1), alkyl trimethylammonium bromide (CS-2), double-chain quaternary ammonium surfactant (CS-3), benzylmethyl ammonium chloride (CS-4), and alkyl three methyl ammonium bromide (CS-5). Organic MMH (MMH-2) is a light yellow transparent liquid and is easily soluble in water and oil. Inorganic MMH (MMH-1) is a white powder and is soluble in water.

Among drilling fluid additives used, the bentonite used was sodium bentonite, which is widely used in the industry. Sodium carboxymethyl cellulose is a white particulate powder and is soluble in water. Lignite resin is a black powder and is stable at high temperature. Potassium chloride is a white crystalline small particle powder with a purity greater than 99.5%.

Sandstone is barely sensitive to water, and there is almost no appearance of wellbore instability in sandstone formation. This research focused on shale and coal formations, which are troublesome formations in coal measure strata. Shale samples used were collected from the Lower Silurian Longmaxi group, Xiushan, Southwest China (the Longmaxi shale). Coal rock samples used were taken from the Longtan group, Bijie, Southwest China (the Longtan coal). Tables 1 and 2 present mineral compositions of the Longmaxi shale and the Longtan coal using the X-ray diffraction (XRD) method.

Table 1. Mineral composition of the Longmaxi shale samples.

Minerals	Chlorite	Illite	Calcite	Feldspar	Quartz	Dolomite
Contents (%)	10	10	25	5	47	3

Table 2. Mineral composition of the Longtan coal samples.

Mineral Name	Illite	Kaolinite	Albite	Pyrite	Amorphous
Contents (%)	22	7	4.9	1.1	65

2.2. Methodology

2.2.1. X-ray Diffraction (XRD)

Mineral compositions of the Chongqing Longmaxi shale and the Guizhou Longtan coal samples were obtained by an X' Pert PRO Diffractometer (Panalytical B. V., Almelo, The Netherlands). The XRD test was conducted by the State Key Laboratory of Geological Process and Mineral Resources. The quantitative analyses of the samples were performed by XrayRun software. As for the Longtan coal samples, 1 wt% of high-purity silicon (>99.99%) was added into the coal samples as an internal standard. The crystalline minerals were determined using XrayRun, the remains were amorphous mineral (organic matter and minerals with low crystallinity) content. Results are shown in Tables 1 and 2.

Shale is a sedimentary rock and often consists of quartz, clays, and other silicate and carbonate minerals. It usually has low porosity of 2 to 4% [43], and it has very low permeability of 5–100×10^{-5} mD and a compressive strength of 250–300 MPa [44]. Due to its high clay concentration, it tends to swell in water or drilling fluids, causing instability problems of the wellbore. The Longmaxi shale (outcrops) was located in Xiushan County, south of Chongqing City, southwest China. As shown in Table 1, the Longmaxi shale consisted of 47 wt% quartz, 25 wt% calcite, and 20 wt% clay minerals (10 wt% chlorite and 10 wt% illite), which means the Longmaxi shale samples have moderate water sensitivity and brittleness.

The Longtan coal samples were collected from the PQT-1 well, Pu'an county, southwest of Guizhou Province, southwest China. The Longtan coal was black and had weak metallic gloss. Cracks were developed, and with such a clear layered structure, the samples were easily broken. According to Table 2, the Longtan coal had 29.05 wt% clay minerals (22.05 wt% illite and 7 wt% kaolinite), indicating the Longtan coal rock had medium water-sensitivity.

2.2.2. Fourier Transform Infrared Spectroscopy (FT-IR)

The FT-IR spectra of the Longmaxi shale and Longtan coal samples were measured by a Bruker Vertex 70 FTIR spectrophotometer (Bruker Technology Co., Ltd., Madison, WI, USA). Each 2 mg sample was fully mixed with 200 mg potassium bromide pellets. Its resolution was 1.6 cm^{-1} and the measured region extended from 4000 to 400 cm^{-1}.

2.2.3. Zeta Potential Test

Zeta potential of the Longtan coal surface was investigated by a Zetasizer Nano ZS90 nanoparticle size and zeta potential analyzer (Malvern Instruments Ltd., Malvern, UK). Coal powders passing through a 120-mesh sieve were poured into testing liquids. After ultrasonic oscillation treatment for 10 min, the supernatants were removed to check their zeta potentials.

2.2.4. Contact Angle and Surface Tension

Wettability can be used to describe the interaction between oil, water, and reservoir rocks, which can be characterized by the contact angle ($\theta°$). Contact angles of the Longmaxi shale and the Longtan coal samples were measured by a JC2000C contact angle measuring instrument (Shanghai Zhongchen

Digital Technology Equipment Co., Ltd., Shanghai, China). Polished Longtan coal and Longmaxi shale samples were soaked in different surfactant solutions for 24 h, then flushed with water for 10 s to remove residual surfactant, then dried under 100 °C for 2 h and cooled under ambient temperature. Finally, a water droplet was dropped on the polished side of core samples, and photos were recorded by a computer connected to the JC2000C contact angle measuring instrument. The contact angle measurement was conducted using contact angle test software with the angle measuring method. Surface tension of different solutions was tested by a QBZY-2 automatic surface tension meter (Shanghai Fangrui Instrument Co., Ltd., Shanghai, China). Each test was repeated three times to check the confidence of the test.

2.2.5. Pressure Transmission Test

The capability of various fluids to retard pore fluid pressure transmission into rock samples in contact with varied fluids was evaluated with an HKY-3 pressure transmission device developed by Hai'an Petroleum Scientific Research Instrument Co., Ltd., Nantong, China. The basic principle of the pressure transmission test (PTT) is to establish pressure difference on the upper and lower side of the core samples, and the equipment can detect the dynamic change of closed fluid at the upper and the lower end, through a pressure transducer, under the conditions of steady upstream pressure. The experimental setup of the HKY-3 pressure transmission device is shown in Figure 1. Data (including confining pressure, upstream pressure, and downstream pressure) were monitored and recorded by the computer every 60 s. The confining pressure and upstream pressure for the Longmaxi shale and the Longtan coal were 5.5 MPa and 4.5 MPa, and 3 MPa and 2 MPa, respectively. Furthermore, the permeability of the tested core samples was calculated by Equation (1):

$$k = \frac{\mu \beta V L}{A} \frac{\Delta \ln\left[(P_m - P_0) / \left(P_m - P_{(1,t)}\right) \right]}{\Delta t} \tag{1}$$

where k is permeability of the core sample in μm^2, μ is viscosity of downstream reservoir fluid in cP, β is fluid static compression ratio in 10 MPa^{-1}, V is downstream confined volume in cm^3, L is core length in cm, A is core cross section area in cm^2, P_m is upstream pressure in MPa, P_0 is core pore pressure in MPa, $P_{(1,t)}$ is T-time downstream pressure in MPa, Δt is time difference in s.

The core samples used in this study were drilled from rock outcrop collected from the field. The drilled core samples were then cut into slices using a core cutter, flushed with water for 30 s to avoid contaminations, and then dried at 105 °C for 24 h. The Longmaxi shale samples were 5 mm thick and 25 mm in diameter, and the Longtan coal samples were 10 mm thick and 25 mm in diameter.

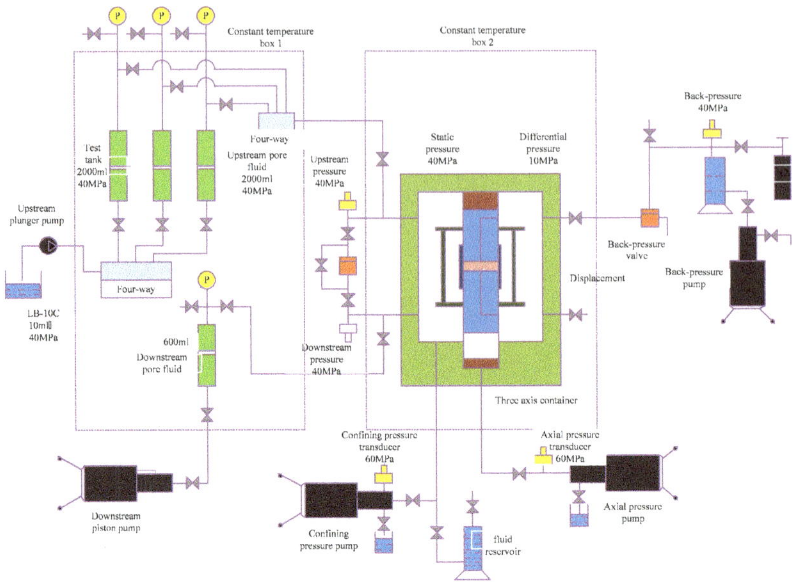

Figure 1. Experimental setup of HKY-3 pressure transmission device.

2.2.6. Low Temperature Nitrogen Adsorption Test

The Longtan coal and the Longmaxi shale samples were crushed and sieved between 60 to 80 meshes. Each 5 g sample was immersed in 1# compound surfactants and 0.8% MMH-2 solutions for 24 h. Then the samples were dried at 100 °C for 2 h. Before the test, all of the samples were outgassed under vacuum at 110 °C for 10 h. The low temperature nitrogen adsorption experiment was performed by a Micromeritics ASAP 2020 surface area and porosity analyzer (Micromeritics, Norcross, GA, USA) at −195.79 °C (liquid nitrogen temperature).

2.2.7. Performance Evaluation of Water-Based Drilling Fluid

(1) Basic performances

An optimized formulation of water-based drilling fluid (WBDF) is shown in Table 3. The order of additives was as follows: after adding bentonite to the water, a sixteen-hour hydration procedure was induced, and then the additives were poured into the mud in order of fluid loss additive 1, fluid loss additive 2, and shale inhibitor. For the water-based drilling fluid proposed in this paper, the 1# compound surfactants (0.001 wt% CS-4 + 0.001 wt% CS-1 + 0.001 wt% CS-3) and the organic MMH-2 needed to be added before the bentonite to prevent inhomogeneous dispersion. The remaining steps were the same.

Table 3. Composition of the water-based drilling fluid.

Additive Type	Additive Name	Concentration (wt%)	Lab Unit (per 360 cm^3)	Field Unit (per bbl)
base fluid	fresh water	87.5	315 cm^3	0.88 bbl
clay	sodium bentonite	4	14.4 g	14.4 lbm
fluid loss additive 1	sodium carboxymethyl cellulose	1.5	5.4 g	5.4 lbm
fluid loss additive 2	lignite resin	2	7.2 g	7.2 lbm
shale inhibitor	potassium chloride	5	18 g	18 lbm

Basic performances including rheological properties, American Petroleum Institute (API) fluid loss, and pH, were measured based on the newest API 13RD standards. The rheological properties were measured with a ZNN-D6 six-speed rotary viscometer. The readings at 600 and 300 rpm were marked as θ_{600} and θ_{300}. The apparent viscosity (AV, cP), plastic viscosity (PV, cP), and yield point (YP, lbf/100 sq ft) of the WBDF were calculated, respectively, according to Equations (2)–(4):

$$AV = 0.5 \times \theta_{600} \tag{2}$$

$$PV = \theta_{600} - \theta_{300} \tag{3}$$

$$YP = \theta_{300} - PV \tag{4}$$

The API fluid loss (FL) of the WBDF was measured with a ZNS-5A moderate pressure filter press at 0.69 MPa (100 psi) at ambient temperature for 30 min. Furthermore, the pH was measured with PHS-3C pH meter.

(2) Inhibition capability

Rolling recovery rate and linear swelling rate were used to evaluate the inhibition capability of drilling fluids to shale or coal samples. Fifty grams of 4 to 10-mesh coal or shale samples (simulated cuttings) were added into three different aging cans, which contained three different drilling fluids (WBDF, WBDF + 3 wt %MMH-1, and WBDF + 0.8 wt% MMH-2). The rolling test was run at 80 °C for 16 h. The cuttings were recovered by a 40-mesh sieve. The recovered cuttings were dried in a hot-roll oven at 100 °C for 4 h and weighed after being cooled under ambient temperature. The rolling recovery rate was calculated according to Equation (5):

$$R = m/50 \tag{5}$$

where m is the weight (g) of the samples after being rolled at 80 °C for 16 h. The higher the shale or coal samples recovery rate was, the better the inhibition performance of the test liquid.

Linear swelling tests were carried out using a ZNP-1 expansion tester. Artificial coal cores were made of 1 mL of sodium silicate, 1 mL of calcium chloride, and 5 g of 80-mesh coal powders. The mixture was suppressed for 15 min under 20 MPa. Then the cores were put into the slot of the expansion tester, where time-dependent swelling increment was measured every 30 min for 8 h.

(3) Formation damage test

Reduction of reservoir permeability can be caused by invasion of solid particles and filtrate in a drilling fluid, which has a negative impact on CMG production at a later stage. Gas permeability of coal samples before and after contamination of various drilling fluids was tested by a JHGP gas permeability tester.

(4) Anti-pollution test

To evaluate pollutants (inorganic salts, drilling cuttings, etc) resistance properties of drilling fluids, a series of anti-pollution tests were carried out. Three wt% NaCl, 1 wt% $CaCl_2$, and 5 wt% attapulgite (simulated drilling cuttings) were used to simulate the possible pollutants in the drilling process of coal measure strata.

(5) Biotoxicity test

Biotoxicity of drilling fluid was determined by the photobacteria method [45]. Luminous intensity of luminescent bacteria is closely related to the toxicity of drilling fluid. Prepared drilling fluids were mixed with 3% NaCl in a ratio of 1 to 9. Then the mixtures' luminosity were tested by a biotoxicity tester after standing for 60 min. Results were compared to the luminosity of 3 wt% NaCl. Then the LC_{50} of drilling fluids were determined when the relative luminosity was decreased by 50%. Furthermore, the toxicity of drilling fluids was assessed following the classification criteria of API shrimp biotoxicity [45].

3. Results and Discussion

3.1. Surface Electrical Property Analysis

Figure 2 shows surface potential variations of the Longtan coal samples with the water solution shifting from acidic to alkaline, which were consistent with the results of Wang et al. [46], which were that the coal fines were usually negatively charged in water system The Longtan coal samples had positive charges under acidic conditions and negative charges under alkaline conditions with an isoelectric point of about 7. The result was consistent with Maršálek et al. [47], where the zeta potential of activated coal fines decreased with the increasing pH of water. That was due to the abundance of free hydroxyl ions in water under acid conditions, which were easy to combine with the coal surface [41].

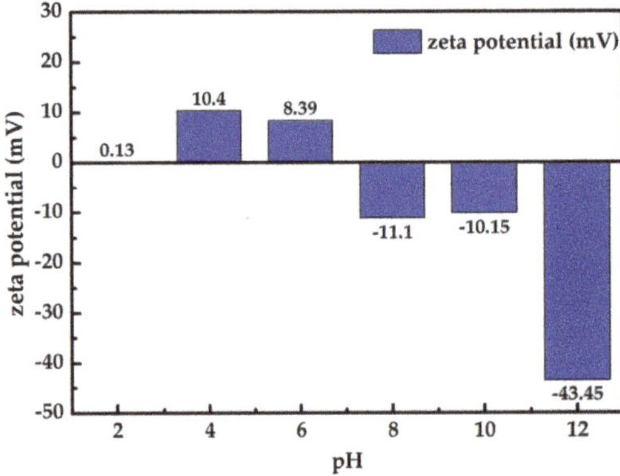

Figure 2. Relationship between zeta potential of the Longtan coal surface and pH.

Drilling fluid is usually weakly alkaline. According to Figure 2, the coal surface is usually negatively charged during the drilling process, which can lead to wellbore instability. Here, the higher the absolute value of the zeta potential, the more stable the drilling fluid is.

As shown in Table 4, with the addition of cationic surfactant CS-5 and organic MMH-2, the zeta potential of the Longtan coal was reversed from negative to positive. Due to the positive hydrophilic group of the cationic surfactant and the positive charge of MMH, these additives were easy to combine with the negative groups on the coal surface, and thus could effectively inhibit coal fines dispersion, and increase the stability of the wellbore.

Table 4. Effect of different dispersions on the surface zeta potential of coal.

Dispersions	Zeta Potential (mV)
fresh water	−3.63
fresh water + 3%MMH-1	−21.75
fresh water + 0.8%MMH-2	19.75
fresh water + 0.2%AS-1	−54.05
fresh water + 0.2%CS-5	47.25

As shown in Table 5, with the addition of MMH additives (MMH-1 and MMH-2), the zeta potential of the WBDF was increased from −56.80 mV to −30.4 mV and −31.8 mV, respectively, while the conventional drilling fluid's zeta potential was −45 mV [41].

Table 5. Effect of different types of MMH on the zeta potential of filtrate.

Formula	Zeta Potential (mV)
WBDF	−56.80
WBDF + 3%MMH-1	−30.40
WBDF + 0.8%MMH-2	−31.80

Due to electrostatic force, MMH can combine with bentonite. Not only can this reaction reduce the negative charge of drilling fluid, but it can also make the drilling fluid has a "solid–liquid" duality, where under static conditions the MMH drilling fluid appears to be in a solid state with a certain elasticity, while under a dynamic situation (stirring) it shows great shear-thinning properties and good flowability [48], which is beneficial to wellbore stability.

3.2. Effect of Surfactants on Wettability of Rock Samples

As shown in Figure 3, the surface tension of water was reduced sharply by a 0.01–0.1 wt% concentration of surfactant. The curve tended to be steady thereafter. It is presumed that critical micelle concentration (CMC concentration) of these surfactants was around 0.01–0.1 wt%. Determination of CMC concentration is critical for the optimization of surfactant concentration. When the concentration is too high, the surfactant molecules will be aggregated to form micelles, which makes it difficult for the molecules to adsorb on the rock surface.

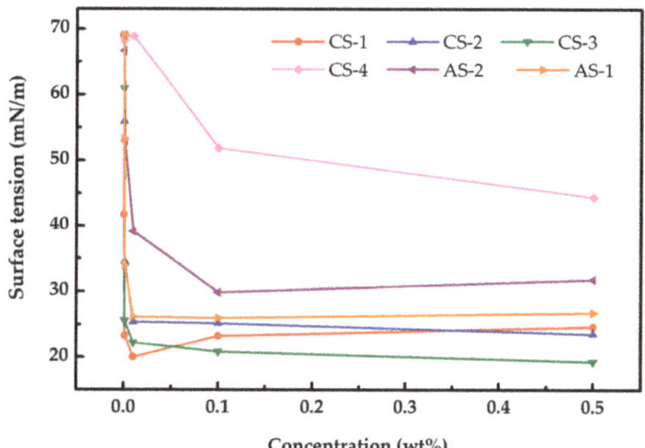

Figure 3. Relationship between surface tension vs surfactant concentration.

Figures 4 and 5 show the effects of different types of surfactants on contact angles of the Longmaxi shale and the Longtan coal samples.

As shown in Figure 4, after being treated with anionic surfactants (AS-1 and AS-2), the surface of the Longmaxi shale remained hydrophilic and all of the contact angles were below 60°, whereas the contact angles of the Longtan coal samples increased. Relatively, Figure 5a reveals that after being treated with cationic surfactants, the hydrophilicity of the Longmaxi shale was decreased, and after being treated with CS-3, the contact angle of the Longmaxi shale reached 98.5°. As for the Longtan coal (Figure 5b), the cationic surfactant CS-3 also increased the contact angle to a certain extent, but CS-4 seemed to be the most effective.

There are two mechanisms to describe the interaction of surfactants and rock samples: chemical adsorption and physical adsorption [49–51]. Taking the Longtan coal samples as an example, XRD results showed a large amorphous content (Table 2), and the contact angle results revealed that the Longtan coal samples were weakly hydrophilic (Figure 5), which indicated a large number of benzene rings and cycloalkyl groups attached to aromatic structures on the coal surface.

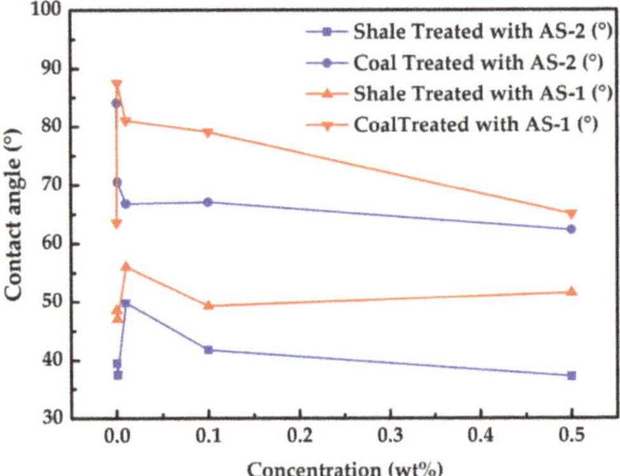

Figure 4. Effect of anionic surfactants to the contact angles of the Longmaxi shale and the Longtan coal samples.

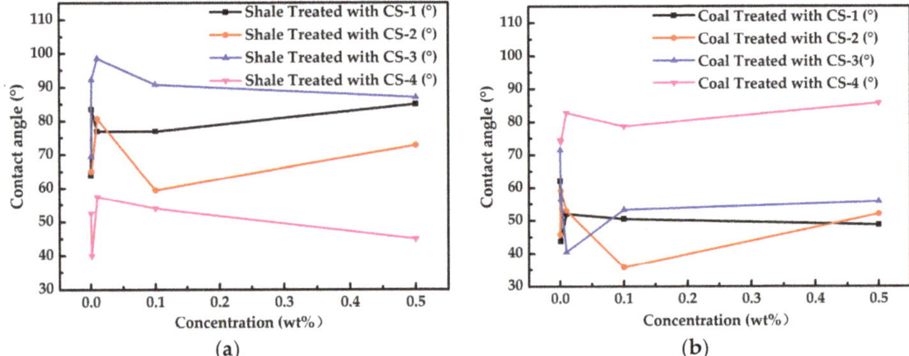

Figure 5. Effect of cationic surfactants to the contact angles of the (**a**) Longmaxi shale and (**b**) the Longtan coal samples.

As shown in Figure 4, the contact angle of coal treated with AS-1 was relatively higher than that of AS-2, probably because AS-1 molecules were easier to adsorb on the coal surface because of the benzene rings. In contact with cationic surfactants (Figure 5b), the contact angle of the Longtan coal samples reduced sharply except CS-4, and the zeta potential increased, which indicated the existence of an electrostatic mechanism (physical adsorption). Alkyl chains of the cationic surfactants were united to the Longtan coal surface, and the polar end of the surfactant pointed to the water, thus making the Longtan coal surfaces more hydrophilic. Among these cationic surfactants, only CS-4 contained benzene rings, the Alkyl chains of the CS-4 were united to the Longtan coal surface, and the benzene rings pointed to the water, forming a hydrophobic film. Therefore the contact angle of the Longtan coal after treatment with CS-4 was higher.

As for the Longmaxi shale samples, due to the large amount of clay mineral content (Table 1), it can be concluded that electrostatic and hydrogen bonding played an important role in the adsorption process [8]. The hydrophilic polar end of the surfactant was attached to the surface of the Longmaxi shale by electrostatic forces and hydrogen bonds. The hydrophobic end of the surfactant pointed to the water, forming a hydrophobic film [51,52]. As surfactant concentration increased and exceeded the CMC concentration (0.01–0.1 wt%), the molecules gradually aggregated to form micelles, and the arrangement on the surface of the Longmaxi shale gradually became disordered, and thus the hydrophobicity was reduced.

As shown in Figures 4 and 5, both the Longtan coal and the Longmaxi shale surfaces were hydrophilic. To reduce the capillary force in the CMG reservoir, the contact angles of the Longmaxi shale and the Longtan coal must ideally remain around 90° simultaneously. According to previous results (Figure 4 to Figure 5), it is difficult to balance the wettability between the Longmaxi shale and the Longtan coal samples using only one kind of surfactant. A set of compound surfactants was optimized via an orthogonal experiment, as listed in Table 6.

Table 6. Experiment design of compound surfactants via orthogonal methods.

Compound Surfactants Combination	CS-3 (%)	CS-1 (%)	CS-4 (%)
1#	0.001	0.001	0.001
2#	0.001	0.005	0.005
3#	0.001	0.01	0.01
4#	0.005	0.001	0.005
5#	0.005	0.005	0.01
6#	0.005	0.01	0.001
7#	0.01	0.001	0.01
8#	0.01	0.005	0.001
9#	0.01	0.01	0.005

Figure 6 shows the effect of compound surfactants on the surface tension and contact angles of the Longmaxi shale and the Longtan coal samples. With the addition of the compound surfactants, the surface tension was clearly reduced from 69.08 mN/m (water surface tension) to less than 30 mN/m. After being treated with the 1# compound surfactant combination (0.001 wt% CS-4 + 0.001 wt% CS-1 + 0.001 wt% CS-3), the contact angles of the Longmaxi shale and the Longtan coal were 89° and 86°, respectively, which indicated that both the Longtan coal and the Longmaxi shale surfaces started to appear weakly hydrophilic. Due to the low concentration (meaning they are quite cost effective) of each surfactant and great wettability adjustment ability, the optimized compound surfactant formula was 0.001 wt% CS-4 + 0.001 wt% CS-1 + 0.001 wt% CS-3.

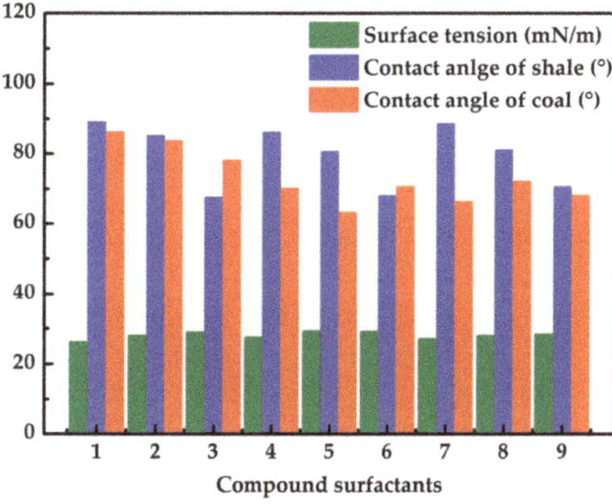

Figure 6. Effect of the optimized compound surfactants on surface tension and contact angle of shale or coal samples.

FT-IR spectroscopy measurements on the Longmaxi shale and the Longtan coal samples were conducted to better illustrate the adsorption mechanism of surfactants (Figure 7). Compared to raw coal (Figure 7a), there was no new adsorption peak after treatment with compound surfactants, indicating that physical adsorption was the dominant factor. As shown in Figure 7b, after being treated with compound surfactants, there was a new absorption peak at 3610.66 cm^{-1}, which was linked to

OH-stretching of structural hydroxyls. It can be speculated that the adsorption mechanism of these compound surfactants is chemical adsorption via hydrogen bonds along with physical adsorption.

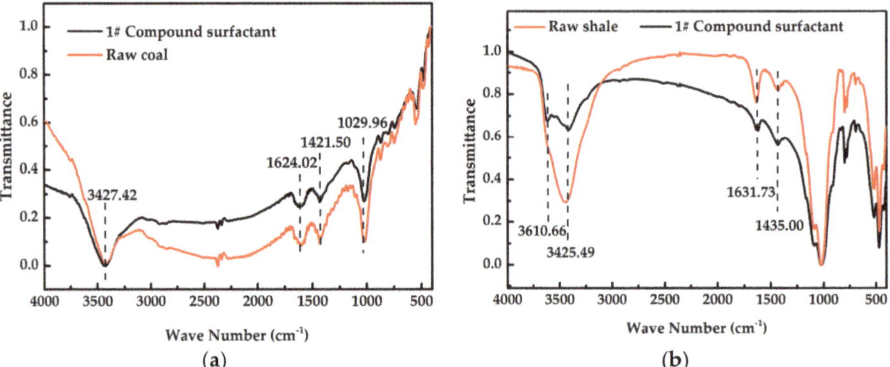

Figure 7. FT-IR spectrum of the Longtan coal (**a**) and the longmaxi shale (**b**) samples before an after treatment with compound surfactants.

As shown in Table 7, with the addition of the 1# compound surfactant combination, the contact angle of the Longmaxi shale increased from 33.5° to 45.5°, while the contact angle of conventional drilling fluid is only 34.5° [51]. Similarly, the contact angle of coal was up to 65°, an increase of 18° compared to the WBDF, while the contact angle of the conventional drilling fluid was 52° [53]. Hydrophilic properties of the Longmaxi shale and the Longtan coal were weakened. However, due to the complex formulation of WBDF, the contact angle of the Longmaxi shale and the Longtan coal were not as high as in the fresh water the presence of the 1# compound surfactant. The stability of the surfactant needs to be improved in future studies.

Table 7. Effect of optimized compound surfactants on contact angle (°) of the Longmaxi shale and the Longtan coal.

Formula	Contact Angle of the Longmaxi Shale (°)	Contact Angle of the Longtan Coal (°)
WBDF	33.5	47
WBDF + 3wt% MMH − 1 + 1# compound surfactants	45.5	65

3.3. Pressure Transmission Test

The primary factor of wellbore instability in shale formations is transmission of hydrostatic pressure of drilling fluid and infiltration of filtrate into wellbores [54]. As shown in Figure 8, pressure transmission test results showed the remarkable effects of the compound surfactants and MMH in retarding pore pressure transmission on the Longtan coal and the Longmaxi shale.

As shown in Figure 8a, when water was used as a medium, the Longmaxi shale started to crack at about 10 h and was broken at about 30 h. When compound surfactant solution was used as the medium, the Longmaxi shale were intact, and the calculated permeability dropped sharply from 2.15×10^{-3} mD to 2.79×10^{-6} mD. When the MMH-1 solution was used as the medium, the calculated permeability was 8.78×10^{-7} mD.

Similarly, when the upstream pressure was 2 MPa and water was used as a medium, the Longtan coal began to crack at about 0.5 h. When the compound surfactant solution was used as the medium, the Longtan coal sample remained intact, and the calculated permeability decreased sharply from 0.61 mD to 6.87×10^{-6} mD, while when MMH-1 was used as the medium, the permeability dropped to 2.134×10^{-4} mD.

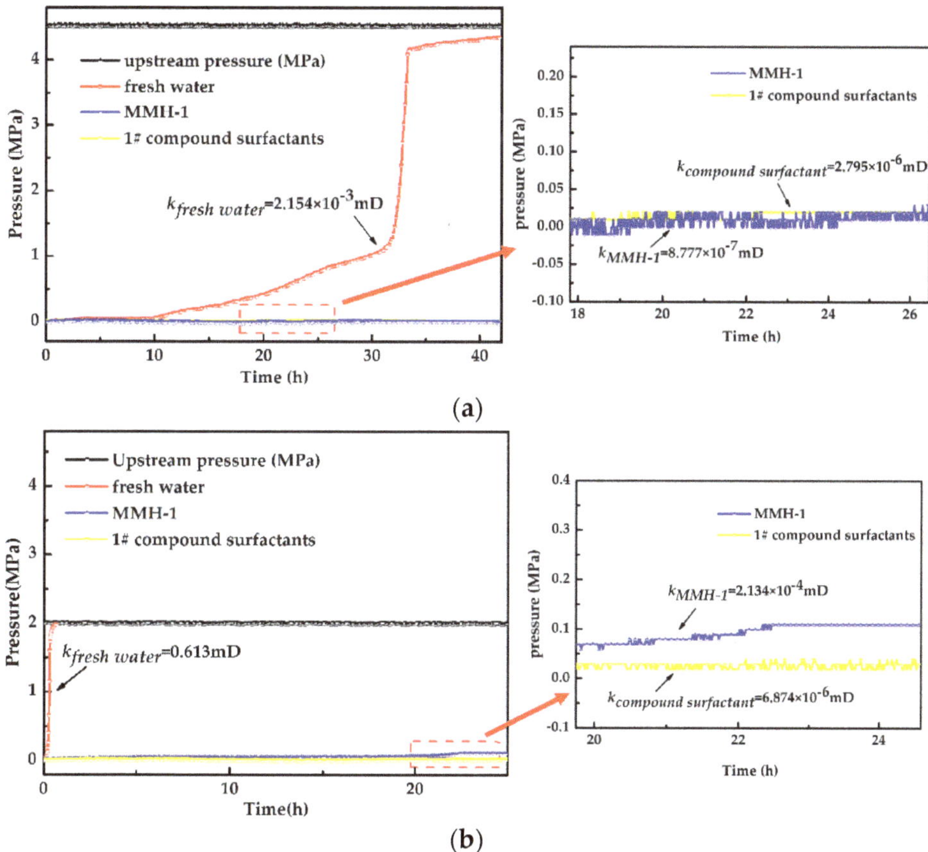

Figure 8. Effect of MMH-1 and 1# compound surfactants (0.001 wt% CS-4 + 0.001 wt% CS-1 + 0.001 wt% CS-3) on retarding pressure transmission of shale and coal. (**a**) Shale sample pressure curve versus time; (**b**) Coal sample pressure curve versus time.

3.4. Pore Structure Comparison Before and After Treatment with 1# Compound Surfactants and MMH-2

FT-IR results (Figure 7) revealed the adsorption mechanisms of different surfactants, and the adsorbed surfactants on the pore structure could not be ignored. A set of low temperature nitrogen adsorption tests were carried out to better understand the adsorption position of surfactants and MMH-2. As shown in Figure 9a, after being treated with 1# compound surfactants, pore volume (diameter larger than 100 nm) decreased significantly compared to the original shale sample, suggesting the adsorption site might have concentrated on the large pore throats. Furthermore, an increase of pore volume (diameter 70–90) was detected, although the reason remains unknown. Possibly, because the surfactant aggregated in large pore throats [55], the space between the surfactant micelles was mistaken for a shale pore. As for coal (Figure 9b), both MMH-2 and 1# compound surfactant minimized the pore volume effectively.

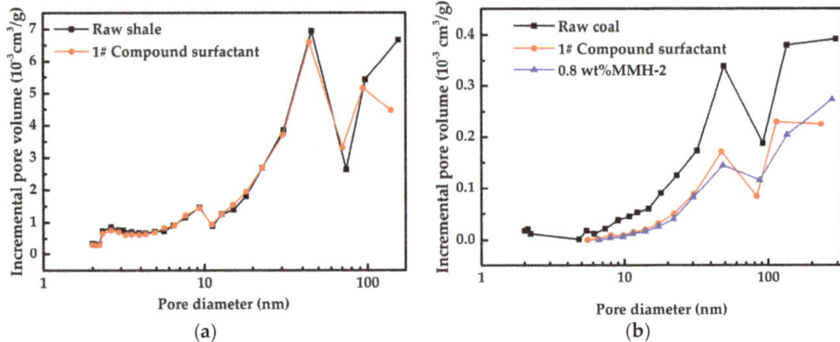

Figure 9. Effect of 0.8 wt%MMH-2 and 1#compound surfactants (0.001 wt% CS-4 + 0.001 wt% CS-1 + 0.001 wt% CS-3) on decreasing pore volume of the Longmaxi shale (**a**) and the Longtan coal (**b**).

3.5. Water-Based Drilling Fluid Performance Test

According to the basic performance tests results shown in Table 8, apparent viscosity of the MMH-1 drilling fluid (WBDF + 3 wt% MMH-1 + 0.001 wt%CS-4 + 0.001 wt% CS-1 + 0.001 wt% CS-3) was slightly lower than that of the MMH-2 drilling fluid (WBDF + 0.8 wt% MMH-2 + 0.001 wt% CS-4 + 0.001 wt% CS-1 + 0.001 wt% CS-3), but its API filter loss was only 3.5 mL, which was much lower than that of the MMH-2 drilling fluid. Thus, the comprehensive performance of the MMH-1 drilling fluid was relatively better.

Table 8. Basic properties of the proposed drilling fluid formula.

Formula	μ_a (cP)	μ_p (cP)	τ_0 (lbf/100 sq ft)	FL$_{API}$ (mL)	pH	Lubrication Factor
WBDF+3wt%MMH-1+1# compound surfactants	33.5	21	23.94	7	10.39	0.18
WBDF+0.8wt%MMH-2+1# compound surfactants	38.5	22	31.6	10.5	8.38	0.17

Note: The lubrication factor of WBDF is 0.46; 1# compound surfactants are 0.001 wt% CS-4 + 0.001 wt% CS-1 + 0.001 wt% CS-3. μ_a is apparent viscosity, μ_p is plastic viscosity, τ_0 is yield point, FL$_{API}$ is American Petroleum Institute fluid loss.

As for the inhibition performance, the MMH-1 drilling fluid achieved a slightly higher rolling recovery rate (Figure 10) and a lower linear expansion rate (Figure 11) compared to the MMH-2 drilling fluid. Overall, the MMH-1 drilling fluid performed better and was more suitable for drilling in coal measure strata.

To better illustrate practicability potential of drilling fluids, formation damage characteristics (Table 9), anti-pollution ability (Table 10), and environmental protection ability of MMH-1 drilling fluid were evaluated. As shown in Table 9, the MMH-1 drilling fluid had a permeability damage rate of only 10% to the coal reservoir, and it could reduce the permeability reduction rate by 3.6% compared to the WBDF.

Table 10 shows the anti-pollution performance of the optimized drilling fluid (WBDF + 3 wt% MMH-1 + 1# compound surfactants). With the addition of inorganic salts and simulated drill cuttings (attapulgite), the viscosity (apparent viscosity, plastic viscosity) and yield point of these drilling fluids showed a slight decrease, which could be attributed to the flocculation of the bentonites [56,57]. The fluid loss appeared to increase when 5 wt% attapulgite was added. However, these changes of the MMH-1 drilling fluid varied within an acceptable range. It showed an excellent capability to withstand common pollutants that may appear in the drilling process of coal measure strata. Another key factor that affects drilling fluids properties is high temperature and high pressure, which need to be studied in the future.

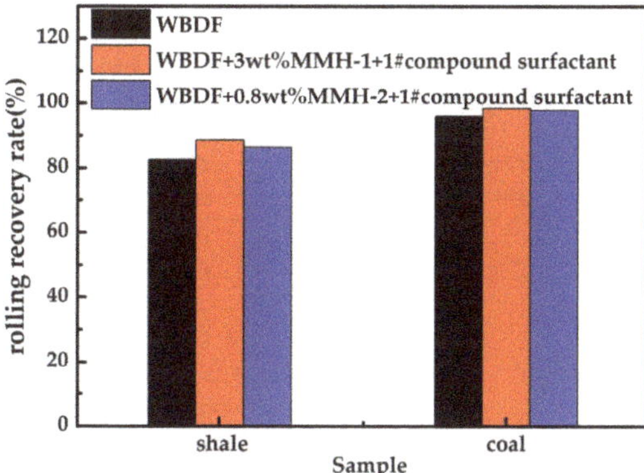

Figure 10. Effect of different drilling fluids on rolling recovery rates of the Longmaxi shale and the Longtan coal. Note: The formula of 1# compound surfactant is 0.001 wt% CS-4 + 0.001 wt% CS-1 + 0.001 wt% CS-3. The same as follows.

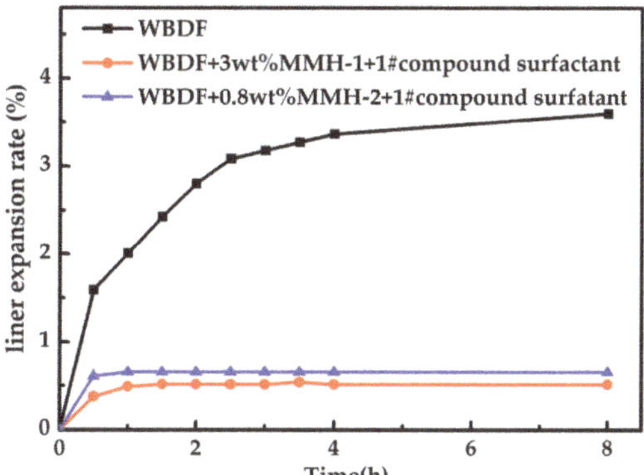

Figure 11. Effect of various drilling fluid on liner expansion rate of the Longtan coal.

Table 9. Effect of drilling fluid on coal sample's permeability.

Formula	k_0 (mD)	k_1 (mD)	Permeability Reduction Rate (%)
WBDF	1.25	1.08	13.6
WBDF + 3 wt% MMH-1 + 1#compound surfactants	0.60	0.54	10

Table 10. Anti-pollution test results.

Formula	μ_a (cP)	μ_p (cP)	τ_0 (lbf/100 sq ft)	FL_{API} (mL)
WBDF + 3 wt% MMH-1 + 1# compound surfactants	33.50	21.00	23.94	7.0
WBDF + 3 wt% MMH-1 + 1# compound surfactants + 3wt %NaCl	26.50	21.00	11.50	6.0
WBDF + 3 wt% MMH-1 + 1# compound surfactant + 1wt% CaCl$_2$	28.00	21.00	13.40	6.0
WBDF + 3 wt% MMH-1 + 1# compound surfactants + 5 wt% attapulgite	31.00	22.00	17.24	9.5

Note: 1# compound surfactants are 0.001 wt% CS-4 + 0.001 wt% CS-1 + 0.001 wt% CS-3. μ_a is apparent viscosity, μ_p is plastic viscosity, τ_0 is yield point, FL_{API} is American Petroleum Institute fluid loss.

3.6. Biotoxicity Evaluation

The LC_{50} (half-life lethal concentration of 96 h luminescent bacteria) of the MMH-1 drilling fluid is 294,000 ppm, as seen in Table 11, and it satisfies the emission standard (30,000 ppm), indicating that it has very low toxicity and is environmentally friendly.

Table 11. Biological toxicity level classification standard.

Level	Poisonous	High Toxicity	Moderate Toxicity	Slight Toxicity	Actually Non-Toxic	Emission Standard
LC_{50} (ppm)	<1	$1 \sim 10^2$	$10^2 \sim 10^3$	$10^3 \sim 10^4$	$>10^4$	$>3 \times 10^4$

3.7. Potential for Field Application

Water-based high inhibitive and low formation damage drilling fluids containing mixed metal hydroxide (MMH) and compound surfactants have been proposed for the first time. Due to its high inhibitive and low formation damage characteristics, this drilling fluid system can not only be used in coal measure strata, but also in single shale and coal formations. While in single coal formations, some additives, such as shale inhibitor (potassium chloride) and fluid loss additive 1 (sodium carboxymethyl cellulose), are not necessary. Furthermore, the proposed 1# compound surfactants showed great performance to increase the contact angle of the Longtan coal and the Longmaxi shale, decreasing their hydrophilicity, which could also be used in fracturing fluid to minimize water-blockage damage and protect the reservoir.

According to Table 10, the proposed drilling fluid system showed great tolerance to $CaCl_2$ and NaCl, indicating the stability when encountering formation water, while the viscosity and filtrate volume showed an increase after attapulgite (simulating drilling cuts) was added. Therefore, a set of four-stage solid control equipment (shale shaker, desander, desilter, centrifuge) was suggested during the drilling process.

4. Conclusions

A systematic study was carried out to characterize the wettability alteration ability and zeta potential adjustment to shale and coal of different types of surfactants and MMH. Based on the results, the following conclusions can be drawn:

(1) When the zeta potential was measured as a function of pH, the results show that the zeta potential of the Longtan coal decreases with increasing pH, the isoelectric point of the Longtan coal is around 7, and the cationic surfactant CS-5 could increase the zeta potential of the Longtan coal up to + 41.25 mV.

(2) According to the contact angle results, a cationic surfactant CS-3 could effectively increase the contact angle of shale up to 98.5°. Furthermore, a set of cationic compound surfactants (0.001 wt% CS-4 + 0.001 wt% CS-1 + 0.001 wt% CS-3) was optimized, which could increase the contact angle of the Longmaxi shale and the Longtan coal up 89° and 86°, respectively.

(3) The dominant adsorption mechanism between the cationic surfactant and the Longmaxi shale is via hydrogen bonds, while that of the Longtan coal was physical adsorption (via electrostatic forces).

(4) Pressure transmission test results show that MMH and the 1# optimized compound surfactants can effectively retard the transmission of pore fluid pressure and decrease the permeability of the core samples, thus increasing wellbore stability.

(5) A set of environmentally friendly water-based drilling fluid systems was proposed (4 wt% sodium bentonite +1.5 wt% sodium carboxymethyl cellulose +2 wt% lignite resin +5 wt% potassium chloride +3 wt% MMH-1 + 0.001 wt% CS-4 + 0.001 wt% CS-1 + 0.001 wt% CS-3). Based on the liner expansion results, the proposed drilling fluid system showed a great inhibitive property. Furthermore, the permeability results confirm its low-damage characteristic to CMG reservoirs.

Author Contributions: S.C. prepared the initial manuscript; Y.S. helped to complete the pressure transmission test of the Longtan coal; X.Y. helped to complete with the pressure transmission test of the Longmaxi shale; K.X. helped with the contact angle test of the Longtan coal; J.C. were responsible for the academic ideas.

Funding: The study is supported by the National Science Foundation of China (No. 41072111), the Key Project of Natural Science Foundation of Hubei Province (No. 2015CFA135), the Technical Innovation Special Project of Hubei Province (No. 2017AHB052), the Youth Chenguang Project of Science and Technology of Wuhan City (No. 2017050304010317) and the National Key Research and Development Program of China (No. 2018YFC1801705).

Acknowledgments: The authors would like to thank Hongbing Zhan at Texas A&M University for kind help in the revision of this article.

Conflicts of Interest: The authors declare no conflict of interest.

References

1. Shao, L.Y.; Yang, Z.Y.; Shang, X.X.; Xiao, Z.H.; Wang, S.; Zhang, W.L.; Lu, J. Lithofacies palaeogeography of the Carboniferous and Permian in the Qinshui Basin, Shanxi Province, China. *J. Palaeogeogr.* **2015**, *4*, 384–412. [CrossRef]

2. Yi, T.S.; Gao, W. Reservoir formation characteristics as well as co-exploration and co-mining orientation of Upper Permian coal-bearing gas in Liupanshui Coalfield. *J. China Coal Soc.* **2018**, *43*, 1553–1564. [CrossRef]

3. Fu, X.; Ge, Y.; Liang, W.; Li, S. Pressure control and fluid effect of progressive drainage of multiple superposed cbm systems. *Nat. Gas Ind.* **2013**, *33*, 35–39. [CrossRef]

4. Qin, Y.; Wu, J.G.; Shen, J.; Yang, Z.B.; Shen, Y.L.; Zhang, B. Frontier research of geological technology for coal measure gas joint-mining. *J. China Coal Soc.* **2018**, *43*, 1504–1516. [CrossRef]

5. Estes, B. Technology Focus: Drilling and completion fluids (November 2009). *J. Pet. Technol.* **2011**, *61*, 62. [CrossRef]

6. Gentzis, T.; Deisman, N.; Chalaturnyk, R.J. Effect of drilling fluids on coal permeability: Impact on horizontal wellbore stability. *Int. J. Coal Geol.* **2009**, *78*, 177–191. [CrossRef]

7. Cai, J.H.; Gu, S.; Wang, F.; Yang, X.Y.; Yue, Y.; Wu, X.M.; Chixotkin, V.F. Decreasing Coalbed Methane Formation Damage Using Microfoamed Drilling Fluid Stabilized by Silica Nanoparticles. *J. Nanomater.* **2016**, 1–11. [CrossRef]

8. Huang, W.; Li, X.; Qiu, Z.; Jia, J.; Wang, Y.; Li, X. Inhibiting the surface hydration of shale formation using preferred surfactant compound of polymine and twelve alkyl two hydroxyethyl amine oxide for drilling. *J. Pet. Sci. Eng.* **2017**, *41*, 37–41. [CrossRef]

9. Yue, Q.S.; Chen, J.; Zou, L.P.; Jiang, G.Z.; Tian, Z.L.; Yang, H.L. Research on coalbed methane drilling fluid for horizontal well based on coal reservoir protection in qinshui basin. *J. China Coal Soc.* **2012**, *37*, 416–419. [CrossRef]

10. Yue, Q.S.; Ma, X.; Chen, J.; Jiang, G.Z. Instability mechanism of borehole and drilling fluid technical countermeasures for coalbed methane horizontal well in qinshui basin. *J. Yangtze Univ.* **2014**, *11*, 73–76. [CrossRef]

11. Zheng, L.H.; Su, G.D.; Li, Z.H.; Peng, R.; Wang, L.; Wei, P.F.; Han, S.X. The wellbore instability control mechanism of fuzzy ball drilling fluids for coal bed methane wells via bonding formation. *J. Nat. Gas Sci. Eng.* **2018**, *56*, 107–120. [CrossRef]

12. Vryzas, Z.; Kelessidis, V.C. Nano-based drilling fluids: A review. *Energies* **2017**, *10*, 540. [CrossRef]

13. Li, M.C.; Wu, Q.L.; Song, K.L.; De Hoop, C.F.; Lee, S.; Qing, Y.; Wu, Y.Q. Cellulose nanocrystals and polyanionic cellulose as additives in bentonite water-based drilling fluids: Rheological modeling and filtration mechanisms. *Ind. Eng. Chem. Res.* **2016**, *55*, 133–143. [CrossRef]

14. Ismail, A.R.; Aftab, A.; Ibupoto, Z.H.; Zolkifile, N. The novel approach for the enhancement of rheological properties of water-based drilling fluids by using multi-walled carbon nanotube, nanosilica and glass beads. *J. Pet. Sci. Eng.* **2016**, *139*, 264–275. [CrossRef]

15. Abdo, J.; Haneef, M.D. Clay nanoparticles modified drilling fluids for drilling of deep hydrocarbon wells. *Appl. Clay Sci.* **2013**, *86*, 76–82. [CrossRef]

16. Barry, M.M.; Jung, Y.; Lee, J.K.; Phuoc, T.X.; Chyu, M.K. Fluid filtration and rheological properties of nanoparticle additive and intercalated clay hybrid bentonite drilling fluids. *J. Pet. Sci. Eng.* **2015**, *127*, 338–346. [CrossRef]

17. Gerogiorgis, D.I.; Clark, C.; Vryzas, Z.; Kelessidis, V.C. Development and parameter estimation for a multivariate Herschel-Bulkley rheological model of a nanoparticle-based smart drilling fluid. In Proceedings of the 12th International Symposium on Process Systems Engineering and 25th European Symposium on Computer Aided Process Engineering, PT C 2015, Copenhagen, Denmark, 31 May–4 June 2015. [CrossRef]

18. Reilly, S.I.; Vryzas, Z.; Kelessidis, V.C.; Gerogiorgis, D.I. First-principle rheological modelling and parameter estimation for nanoparticle-based smart drilling fluids. In Proceedings of the 26th European Symposium on Computer Aided Process Engineering, PT A 2016, Portoroz, Slovenia, 12–15 June 2016. [CrossRef]

19. Zhang, X.; Sun, W.; Ren, D.; Liu, D.; Nan, J. Effects of surfactant on the electrical property of ultra-low permeability sandstone reservoir:the Upper Triassic Chang 8 formation in Longdong area of Ordos basin as an example. *Geol. Sci. Technol. Inf.* **2016**, *35*, 235–241.

20. Adibhatla, B.; Mohanty, K.K. Oil recovery from fractured carbonates by surfactant-aided gravity drainage: Laboratory experiments and mechanistic simulations. *SPE Res. Evaluation Eng.* **2006**, *11*, 119–130. [CrossRef]

21. Hirasaki, G.; Zhang, D.L. Surface chemistry of oil recovery from fractured, oil-wet, carbonate formations. *SPE J.* **2004**, *9*, 151–162. [CrossRef]

22. Young, T. An essay on the cohesion of fluids. *Philos. Trans. Royal Soc. Lond.* **1805**, *95*, 65–87. [CrossRef]

23. Förch, R.; Schönherr, H.; Jenkin, A.T.A. Surface Design: Applications in Bioscience and Nanotechnology. *Mater. Today* **2009**, *12*, 1–54. [CrossRef]

24. Hu, Y.L.; Wu, X.M. Research on coalbed methane reservior water blocking damage mechanism and anti-water blocing. *J. China Coal Soc.* **2014**, *39*, 1107–1111. [CrossRef]

25. Liang, L.; Xiong, J.; Liu, X. Effects of hydration swelling and wettability on propagation mechanism of shale formation crack. *Pet. Geol. Exp.* **2014**, *36*, 780–786. [CrossRef]

26. Alvarez, J.O.; Schechter, D.S. Wettability Alteration and Spontaneous Imbibition in Unconventional Liquid Reservoirs by Surfactant Additives. *SPE Res. Evaluation Eng.* **2017**, *20*, 107–117. [CrossRef]

27. Quintero, L. An Overview of Surfactant Applications in Drilling Fluids for the Petroleum Industry. *J. Dispers. Sci. Technol.* **2002**, *23*, 393–404. [CrossRef]

28. Yunita, P.; Irawan, S.; Kania, D. Optimization of water-based drilling fluid using non-ionic and anionic surfactant additives. *Procedia Eng.* **2016**, *148*, 1184–1190. [CrossRef]

29. Shadizadeh, S.R.; Moslemizadeh, A.; Dezaki, A.S. A novel nonionic surfactant for inhibiting shale hydration. *Appl. Clay Sci.* **2015**, *118*, 74–86. [CrossRef]

30. Cai, J.H.; Ye, Y.; Cao, W.J.; Yang, X.Y.; Wu, X.M. Experimental study on the effect of drilling fluid wettability on shale wellbore stability. *J. China Coal Soc.* **2016**, *41*, 228–233. [CrossRef]

31. Nasralla, R.A.; Nasr-El-Din, H.A. Double-Layer Expansion: Is It a Primary Mechanism of Improved Oil Recovery by Low-Salinity Waterflooding? *SPE Res. Evaluation Eng.* **2014**, *17*, 49–59. [CrossRef]

32. Zadaka, D.; Radian, A.; Mishael, Y.G. Applying zeta potential measurements to characterize the adsorption on montmorillonite of organic cations as monomers, micelles, or polymers. *J. Colloid Interface Sci.* **2010**, *352*, 171–177. [CrossRef]

33. Erbil, H.Y. *Surface Chemistry of Solid and Liquid Interfaces*; Blackwell Pub: Oxford, UK, 2006.

34. Su, C.M.; Fu, J.T. Research on the changing tendency of zeta potential of the clay minerals and drilling fluid. *Driuing Fluid Complet. Fluid* **2002**, *19*, 1–4. [CrossRef]

35. Zhang, C.G.; Hou, W.G. Electrical behavior, inhibition, wall stability and reservoir protection. *Drill. Fluid Complet. Fluid* **2002**, *19*, 5–8. [CrossRef]

36. Alkan, M.; Demirbaş, Ö.; Doğan, M. Electrokinetic properties of kaolinite in mono- and multivalent electrolyte solutions. *Microporous Mesoporous Mater.* **2005**, *83*, 51–59. [CrossRef]

37. Buckley, J.S.; Liu, Y.; Monsterleet, S. Mechanisms of wetting alteration by crude oils. *Spe J.* **1998**, *3*, 54–61. [CrossRef]

38. Alvarez, J.O.; Schechter, D.S. Application of wettability alteration in the exploitation of unconventional liquid resources. *Pet. Explor. Dev.* **2016**, *43*, 832–840. [CrossRef]

39. Arif, M.; Jones, F.; Barifcani, A.; Iglauer, S. Influence of surface chemistry on interfacial properties of low to high rank coal seams. *Fuel* **2017**, *194*, 211–221. [CrossRef]

40. Song, Y. Zeta Potential of the surface of fine coal. *J. Fuel Chem. Technol.* **1998**, *16*, 245–252.

41. Li, X.; Kang, Y.; Yin, Z. Surface electricity and wettability of coal rock under the condition of different chemical environments. *J. China Univ. Min. Technol.* **2014**, *43*, 864–869. [CrossRef]

42. Shu, Y.; Yan, J. Characterization and prevention of formation damage for fractured carbonate reservoir formations with low permeability. *Pet. Sci.* **2008**, *5*, 326–333. [CrossRef]

43. Han, H.; Zhong, N.; Jiao, S. Scanning electron microscope observation of pores in mud stone and shale. *J. Chin. Electron Microsc. Soc.* **2013**, *32*, 325–329. [CrossRef]

44. Yang, J.; Fu, Y.Q.; Chen, H.F.; Zeng, L.X.; Li, J.S. Rock mechanical characteristics of shale reservoirs. *Nat. Gas Ind.* **2012**, *32*, 12–14. [CrossRef]

45. Huang, X. Comparson of biological toxicity assessment methods for drilling fluids. *Environ. Prot. Oil Gas Fields* **2006**, *16*, 25–27. [CrossRef]

46. Wang, B.; Min, L.I.; Zhao, Q.; Qin, Y.; Xie, K. Relationship between surface potential and functional groups of coals. *J. Chem. Ind. Eng.* **2004**, *55*, 1329–1334. [CrossRef]

47. Maršálek, R. The Influence of Surfactants on the Zeta Potential of Coals. *Energy Sources Part A Recovery Util. Environ. Eff.* **2008**, *31*, 66–75. [CrossRef]

48. Su, C.M.; Liu, R.S.; Yu, P.Z.; Wang, C.P. Study and application of electropositive drilling fluid system. *Oil Drill. Prot. Technol.* **2004**, *26*, 17–21. [CrossRef]

49. Liu, S.; Liu, X.; Guo, Z.; Liu, Y.; Guo, J.; Zhang, S. Wettability modification and restraint of moisture re-adsorption of lignite using cationic gemini surfactant. *Colloids Surf. A Physicochem. Eng. Asp.* **2016**, *508*, 286–293. [CrossRef]

50. Chen, S.; Wang, S.; Li, L.; Qu, J.; Tao, X.; He, H. Exploration on the mechanism of enhancing low-rank coal flotation with cationic surfactant in the presence of oily collector. *Fuel* **2018**, *227*, 190–198. [CrossRef]

51. Yue, Y.; Chen, S.; Wang, Z.; Yang, X.; Peng, Y.; Cai, J. Improving wellbore stability of shale by adjusting its wettability. *J. Pet. Sci. Eng.* **2018**, *161*, 692–702. [CrossRef]

52. Zhou, L.; Das, S.; Ellis, B.R. Effect of surfactant adsorption on the wettability alteration of gas-bearing Shales. *Environ. Eng. Sci.* **2016**, *33*, 766–777. [CrossRef]

53. Shi, Y.P.; Chen, S.Y.; Yang, X.Y.; Yu, L. Enhancing wellbore stability of coal measure strata by electricla inhibition and wettability control. *J. Pet. Sci. Eng.* **2019**, *174*, 544–552. [CrossRef]

54. Van Oort, E.; Hale, A.; Mody, F.; Roy, S. Transport in shales and the design of improved water-based shale drilling fluids. *SPE Drill. Complet.* **1996**, *11*, 137–146. [CrossRef]

55. Liu, X.; Liu, S.; Fan, M.; Guo, J.; Li, B. Decrease in hydrophilicity and moisture readsorption of Manglai lignite using lauryl polyoxyethylene ether: Effects of the HLB and coverage on functional groups and pores. *Fuel Process. Technol.* **2018**, *174*, 33–40. [CrossRef]

56. Abdo, J.; Haneef, M.D. Nano-enhanced drilling fluids: Pioneering approach to overcome uncompromising drilling problems. *J. Energy Resour. Technol. Trans. ASME* **2012**, *134*, 014501. [CrossRef]

57. Vryzas, Z.; Kelessidis, V.C.; Nalbantian, L.; Zaspalis, V.; Gerogiorgis, D.I.; Wubulikasimu, Y. Effect of temperature on the rheological properties of neat aqueous Wyoming sodium bentonite dispersions. *Appl. Clay Sci.* **2017**, *136*, 26–36. [CrossRef]

Article

Insight into the Pore Characteristics of a Saudi Arabian Tight Gas Sand Reservoir

Abdulrauf R. Adebayo *, Lamidi Babalola *, Syed R. Hussaini, Abdullah Alqubalee and Rahul S. Babu

Center for Integrative Petroleum Research, King Fahd University of Petroleum & Minerals, Dhahran 31261, Saudi Arabia; rsyed@kfupm.edu.sa (S.R.H.); abdullah.alqubalee@kfupm.edu.sa (A.A.); rahul.babu@kfupm.edu.sa (R.S.B.)

* Correspondence: abdulrauf@kfupm.edu.sa (A.R.A.); lbabalola@kfupm.edu.sa (L.B.);
  Tel.: +966-013-860-7444 (A.R.A.); +966-013-860-2092 (L.B.)

Received: 9 October 2019; Accepted: 29 October 2019; Published: 12 November 2019

Abstract: The petrophysical characterization of tight gas sands can be affected by clay minerals, gas adsorption, microfractures, and the presence of high-density minerals. In this study, we conducted various petrophysical, petrographic, and high-resolution image analyses on Saudi Arabian tight sand in order to understand how a complex pore system responds to measurement tools. About 140 plug samples extracted from six wells were subjected to routine core analyses including cleaning, drying, and porosity–permeability measurements. The porosity–permeability data was used to identify hydraulic flow units (HFU). In order to probe the factors contributing to the heterogeneity of this tight sand, 12 subsamples representing the different HFUs were selected for petrographic study and high-resolution image analysis using SEM, quantitative evaluation of minerals by scanning electron microscope (QEMSCAN), and micro-computed tomography (μCT). Nuclear magnetic resonance (NMR) and electrical resistivity measurements were also conducted on 56 subsamples representing various lithofacies. NMR porosity showed good agreement with other porosity measurements. The agreement was remarkable in specific lithofacies with porosity ranging from 0.1% to 7%. Above this range, significant scatters were seen between the porosity methods. QEMSCAN results revealed that samples with <7% porosity contain a higher proportion of clay than those with porosity >7%, which are either microfractured or contain partially dissolved labile minerals. The NMR T2 profiles also showed that samples with porosity <7% are dominated by micropores while samples with porosity >7% are dominated by macropores. Analysis of the μCT images revealed that pore throat sizes may be responsible for the poor correlation between NMR porosity and other porosity methods. NMR permeability values estimated using the Shlumberger Doll Research (SDR) method are fairly correlated with helium permeability (with an R2 of 0.6). Electrical resistivity measurements showed that the different rock types fall on the same slope of the formation factors versus porosity, with a cementation factor of 1.5.

Keywords: tight gas sand; unconventional; porosity–permeability; hydraulic flow units; electrical resistivity; NMR; micro-CT image; petrophysics; petrography

---

1. Introduction

Population growth and the associated need for sustainable sources of energy have led to an increased interest in the exploration and production from unconventional resources in the recent years. Only four countries—Canada, the USA, China, and Argentina—have fully developed unconventional gas from either shale or tight gas reservoirs [1]. The kingdom of Saudi Arabia is believed to have the fifth largest reserves of unconventional gas [1]. It is also estimated that the kingdom's unconventional gas resources could be as high as 10 times the conventional gas resources [2]. Hence, unconventional gas development is expected to grow exponentially in the kingdom [1]. In Saudi Arabia, tight gas is believed to have more favorable geology than shale because the tight gas deposits are largely found in sandstones with higher permeability and porosity than shale. The bulk mineralogical composition of sandstone, which is dominantly comprised of quartz with or without subordinate amount of feldspars and relatively low clay content, make the tight sand gas reservoirs easier and cheaper to hydraulically fracture than clay-rich oil shale or shale gas reservoirs. However, these tight sands are highly heterogeneous such that an extensive amount of field and laboratory data are required to understand the properties of different reservoir types and to design the appropriate stimulation methods. Improved petrophysical knowledge and methods of low permeability rocks are essential elements toward a successful development of unconventional reservoirs. The most important rock properties that must be known are pay zone thickness, porosity, permeability, water saturation, in-situ stresses, and young modulus [3]. These properties are determined from well logs, well tests, laboratory core measurements, drilling records, etc. Laboratory measurements are the most direct and reliable source of data.

The accurate measurement of porosity and permeability is crucial in all types of reservoir rocks. These measurements are essential for assessing the volume of hydrocarbon in place and the productivity of wells drilled through them. The measurements are also used to calibrate downhole-logging tools. Measurement accuracy is more crucial in very low-porosity and low-permeability rocks such as tight sand or shales [4]. The small unconnected pores present in complex mineralogy and organic matters of unconventional reservoirs (tight sand and shale gas) make their porosity and permeability measurement problematic [5,6]. The presence of accessory minerals, clay types, clay swelling, fine migration, gas slippage effect (or Klinkenberg effect), and turbulence (or Forchiemer's effect) further complicates the accuracy of measuring these reservoir properties. The presence of clay (especially montmorillonite) can be a worrisome obstacle in achieving the desired measurement accuracy [7]. Clays have large surface areas that contain adsorbed water that do not contribute to the accessible storage space of the rock. The accessible storage space is the effective porosity of the rock. Hence, for petrophysical study such as electrical resistivity, permeability, and capillary pressure, it is essential to preserve the clay structure during sample preparation and saturation.

Laboratory analysis in most cases requires that the extracted rock samples be cleaned with solvents, oven dried, and resaturated with synthetic formation fluids as part of a workflow to restore them to their native conditions. Drying can cause the irreversible destruction of the clay fabric and eventually an increase in the porosity and permeability of the samples. To accurately determine porosity types (micropores and mesopores) and their impacts on mud rocks, it is important to follow sample preparation and experimental protocols that would preserve the physical structures of the samples [8]. Contact with incompatible fluids other than their native fluids can cause a variety of problems such as clay swelling, fine migration, and the plugging of pore throats, and ultimately a reduction in the porosity and permeability values. Therefore, it is important to preserve the clay structure during drying such that the clay can be rehydrated to its original state. Morrow et al. [9] showed that clay microstructure (microporosity) can still be partially preserved at temperatures up to 500 °C. However, beyond this temperature the clay structure begins to collapse with a total collapse occurring at over 1000 °C. For a preserved clay structure, the available cations on the clay surface will hydrate when resaturated. The degree of hydration depends on the original cations present at the cation exchange site of the clay. Sodium ion (Na^+) is the most readily exchangeable cation, and in smectite it can cause adsorption of up to 32 layers of water molecules. In samples with large smectitie content, this large expansion after water adsorption can block pore throats and ultimately reduce sample permeability and porosity. Paramagnetic, ferromagnetic, and conductive minerals such as pyrites, hematite, and siderite in tight sandstones are also influencing factors in the rock response to important logging tools such as nuclear magnetic resonance, electrical resistivity, etc. Therefore, it is important that laboratory measurements follow important protocols during the measurements of these sensitive formations. Nonetheless, analyses of tight rocks can further be complicated by their complex pore topology/heterogeneity.

The Saudi Arabia late Ordovician Sarah Formation is a glacial deposit in the northern part of the Rub' al Khali basin. It is a proven gas reservoir in the northwestern and central parts of Saudi Arabia that is being explored as a potential tight gas sandstone reservoir. The formation's paleovalleys cut deeply into the basement and older Paleozoic succession, including Saq formation and all the members of the Qasim Formation including Hanadir, Khafah, Ra'an, and Quwarah members (Figure 1) in northwest and central Saudi Arabia [10,11]. The sub-Sarah unconformity above the underlying Saq and Qasim Formations [10] marks the basal boundary of this formation while the basal occurrence of the Qusaiba Hot Shale delineates its upper contact. The formation is dominantly comprised of pebbly, poorly sorted, fine to coarse-grained, parallel, trough, and planar cross-bedded sandstones. Al-Harbi and Khan [11] described the sandstone as predominantly quartz arenite.

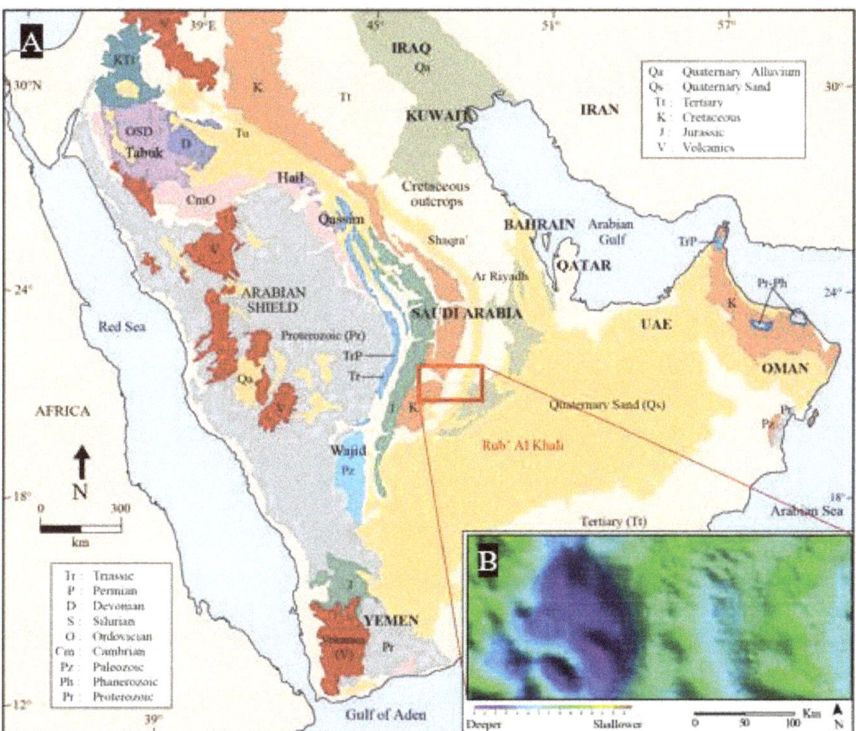

Figure 1. (**A**) Surface geological map of the Arabian Plate, (**B**) gravity map showing the basement topography underneath the study area.

Sedimentological and petrographical analyses previously conducted on core samples from this formation revealed the occurrences of various lithofacies, which were grouped into four facies associations (FAs) (as shown in Figure 2). The facies association are massive to ripple-laminated sandstone (FA1), gray massive sandstone (FA2), diamictites (FA3), and partially deformed, massive-graded sandstone (FA4), which are interpreted as fluvial, glaciofluvial, glaciolacustrine delta, and subglacial outwash deposits, respectively [12]. Alqubalee et al. [13] reported that the clay and cement content in the fluvial facies (well A) and glaciofluvial facies (well E) are relatively low; however, grain compaction, feldspar dissolution, and authigenic illite were observed in these FAs. The pores and pore throats in most of the glaciolacustrine delta (wells B and C) and subglacial facies (wells B, D, and F) were filled by anhydrite, siderite, barite, or detrital illite.

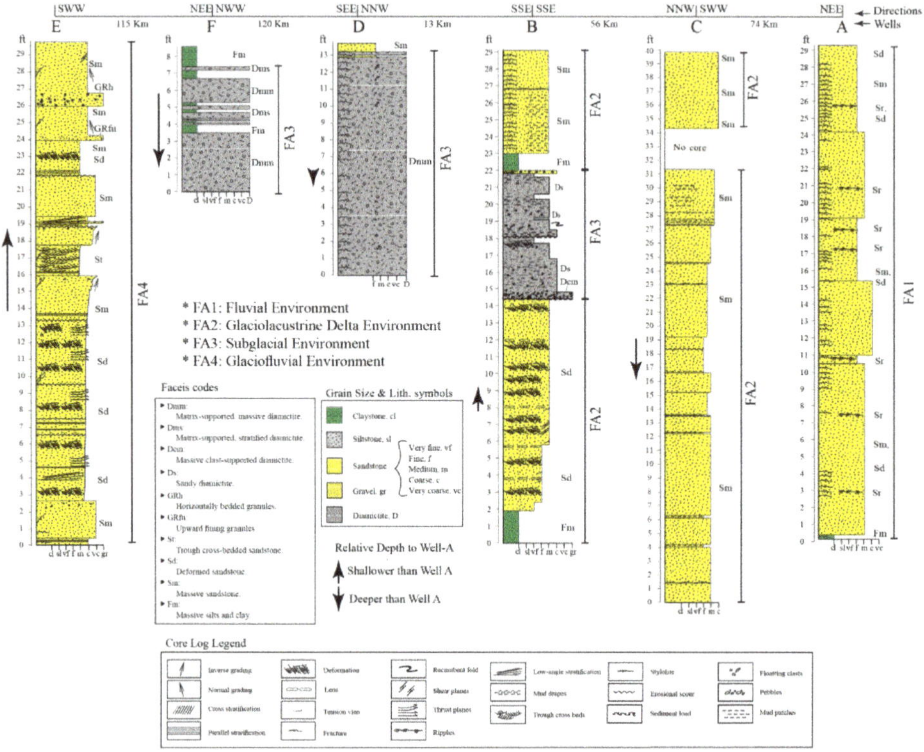

Figure 2. Core stratigraphic columns showing the different facies associations for all wells: FA1: fluvial, FA2: glaciolacustrine delta, FA3: subglacial, and FA4: glaciofluvial outwash facies.

In this paper, we present the results of various petrographic, petrophysical, and high-resolution imaging investigations on the Sarah tight gas reservoir in the Rub' al Khali basin. The measurements were conducted on 140 core plugs extracted from six wells (named well A to well F). First, the porosity and permeability of the samples were measured using pulse decay methods. Then, the flow zone indicator method was used to group the samples into distinct hydraulic flow units. Subsequently, thin-section petrography, scanning electron microscope (SEM), quantitative evaluation of minerals by scanning electron microscope (QEMSCAN), nuclear magnetic resonance (NMR), high-resolution micro-computed tomography (µCT) imaging, and electrical resistivity were conducted on subsamples representing the different flow units in order to understand the different reservoir rock types and the factors controlling their pore behavior. Since petrographic analysis on the samples showed that smectite clay exists in a very insignificant amount, the samples were cleaned and dried using conventional techniques. Nonetheless, the main clay fabrics are believed to be preserved based on the findings of Morrow et al. [9].

2. Methodology

2.1. Rock Sample Preparation

Cylindrical cores of 4-inch diameters were extracted from six wells: namely, well A, well B, well C, well D, well E, and well F. The six wells were drilled through the Sarah reservoir formation in the Rub' al Khali basin. Samples from well A and well B were extracted at a depth of 16xxx feet, while samples from wells C and D were extracted at a depth of 17xxx feet. Samples from well E were extracted at a depth of 14xxx feet and those from well F were extracted at a depth of 18xxx feet. The wells drilled through a very heterogeneous reservoir with significant variation in porosity, permeability, and mineralogy. Then, cylindrical plugs of 1-inch and 1.5-inch diameter with lengths ranging from 1.5–4 inches were extracted from the cores. The plug samples were cleaned to remove hydrocarbon using a Soxhlet-type extractor circulating hot toluene and subsequently with alcohol to remove salt contaminants. The cleaned samples were dried in a vacuum oven. The Soxhlet extractor process involves evaporated toluene circulating through the pores of the rock specimen and thereby cleansing them of any oil present. This cleaning process was continued for several days to ensure a complete removal of residual hydrocarbon. To remove any traces of toluene and salt associated with the formation brine that might be remaining in the sample, the same Soxhlet extractor-cleaning procedure was repeated with isopropanol leach for at least 48 h. After cleaning with isopropanol, the rock specimens were dried under partial vacuum in a vacuum oven at 80 °C for about two days. Some vuggy core plugs or core plugs having surface irregularities, which cannot be used for porosity and permeability measurements under confining pressure, were repaired to remove the surface irregularities using rock powder mixed with epoxy. The average length and diameter of the core plugs were determined from five measurements at different points along the length and diameter, respectively. The averages of the length and diameter were used to calculate the bulk volume of each core plug. Then, the dry weights of the core plugs were measured using a weight balance accurate to 0.001 g.

2.2. Porosity and Permeability

Gas porosity was measured on the cleaned samples using an automated helium permeameter-porosimeter AP-608, which was calibrated using standard samples with known volume and a density of Berea sandstone and two billets of titanium and steel alloy. A high-purity (99.9%) helium gas was used to measure both porosity and permeability at a net confining pressure of 500 psi. Gas permeability, κ_g, was measured by injecting helium at a pore pressure of 200 psi, and the pressure was allowed to decay at a room temperature of 28 °C. Then, liquid permeability κ_l was calculated from the Klinkenberg [14] correlation shown in Equation (1) by plotting gas permeability measurements at different pressures against the reciprocal of the mean pressure $1/(p_m-1)$. The intercept of the best-fit straight line with the gas permeability axis (at $p \approx \infty$) is equal to liquid permeability:

$$\kappa_\infty = \frac{\kappa_g}{1 + \frac{b}{p_m}} \tag{1}$$

where k_∞ is the liquid permeability, k_g is the gas permeability, p_m is the mean pressure of flow, and b is the constant parameter (Klinkenberg's slip factor for a given porous media).

Then, the core samples were saturated with brine solution at room temperature under a pressure of 2000 psi using a high-pressure vessel. Vacuum was used prior to saturation to remove trapped air from the samples. Each sample was left in the saturation vessel for about 48 h after which they were removed and gently wiped on a piece of paper in order to remove excess and surface water from the samples. Then, each sample was quickly weighted on a mass balance accurate to ±1 mg to determine the saturated mass (m_{sat}). Then, the pore volume occupied by brine, V_{pore}, was calculated as $V_{pore} = (M_{sat} - M_{dry})/\rho_{brine}$, while the gravimetric porosity was calculated as $\Phi_g = V_{pore}/V_{bulk}$.

After weighting, the samples were quickly wrapped in thin plastic cling film to avoid drying and to allow NMR porosity experiments to be conducted on them. The composition of the synthetic brine used in saturating the samples fell within the range of the reservoir water samples extracted from the six wells (Table 1).

Table 1. Properties of brine (B: boron, Ba: barium, BCl: chloride, Br: bromide, Ca: calcium, EC: electrical conductivity, Fe: iron, HCO3: bicarbonate, K: potassium, Mg: magnesium, Mn: manganese, Na: sodium, Si: silicon, SO4: sulfate, Sr: stontium, TPH: total petroleum hydrocarbon, Zn: zinc).

Components	Range
pH	4.62–6.6
EC, mmho/cm (25 °C)	96–534
Total Solids (mg/L)	103,250–369,200
Density (g/mL)	1.088–1.264
Cl (mg/L)	46,981–185,596
Br (mg/L)	411–1492
SO_4 (mg/L)	137–1039
HCO_3(mgL)	<1.0
B (mg/L)	25.16–43.56
Ba(mg/L)	311–1275
Ca (mg/L)	16,315–54,500
Fe(mg/L)	695–66,700
K(mg/L)	971–2870
Mg (mg/L)	568–1350
Mn (mg/L)	48.74–221.24
Na (mg/L)	16,012–159,575
Si (mg/L)	10.13–18.46
Sr (mg/L)	272–1336
Zn (mg/L)	92.38–188.63
TPH (mg/L)	1.51–1346.75

2.3. Petrographic and QEMSCAN Analyses

A total number of 140 thin sections selected from the studied core samples were prepared for petrographic study, where grain size, texture, and mineralogical composition (modal composition) were acquired. The concept of QEMSCAN, its applications, and its protocols have been explained and illustrated by several authors [15–17]. This advanced petrographic technique integrates scanning electron microscope (SEM), energy-dispersive spectroscopy, and species identification protocol (SIP)—a predefined spectral mineralogical library—to produce quantitative mineralogical maps from rock samples. We utilized the QEMSCAN technique to evaluate the mineralogical composition of some selected thin sections that represent the different petrophysical rock types in the reservoir. The thin sections were carbon coated using a Quorum EMS 150R ES before the analyses. A standard QEMSCAN setup was conducted, and the system was operated using an X-ray beam voltage of 15 kV and beam current of 10 nA (±0.05). The Field Image Scan mode on an area of 1 cm^2 and 5-µm point spacing was selected. The QEMSCAN measurements took around 7.5 h. After that, the data was processed via iDiscover software, where field stitching, granulator, and boundary phase processors were applied.

2.4. Nuclear Magnetic Resonance Measurements

NMR measurements were performed on the samples using a low magnetic field NMR (0.05 Tesla and 2 MHz) 'Geospec rock core analyzer' from Oxford Instruments. The instrument was used to measure the T_2 relaxation of the samples using optimized scanning parameters as follows: inter-echo spacing (Tau) value of 0.1 milliseconds, signal-to-noise ratio of 200, and a recycle delay of 11,250 milli seconds. T_2 relaxation is a measure of the time that it takes for the transverse magnetization of a hydrogen nuclei to fall to approximately 37% of its initial value, after it was excited by a magnetic field in a direction transverse (perpendicular) to the magnetic field [18]. T_2 is measured in milliseconds (ms), and it is related to the surface-to-volume ratio of the pores and the surface relaxivity (ρ) of the rock minerals coating the pore surface.

Since clay-rich rocks are sensitive to the type of water in contact with them, the synthetic brine used in saturating the rock samples prior to NMR measurements were prepared to simulate the actual reservoir fluid content. Water samples collected from several wells that were drilled through the study reservoir were analyzed for anion and cation compositions using inductive coupled plasma mass spectroscopy (ICP-MS) and inductive coupled plasma optical emission spectroscopy ICP-OES techniques, respectively. Table 1 shows the range of reservoir water properties and components. All TPH (total petroleum hydrocarbon) analysis was conducted on a wet weight basis. An Agilent GC 6890-N equipped with a flame ionization detector (FID) was used for the analysis.

2.5. High-Resolution Images of Samples

A laboratory μCT scanner, Versa XRM-5002D, was used to obtain tomographic images from small rock cuttings (rock cuttings of size approximately 8 mm × 10 mm) selected to represent the varieties of rock types present in the reservoir. X-rays from a microfocused source producing a polychromatic conical X-ray beam were used to generate images of the sample while employing a circular trajectory with 1601 projections. The scan voltage was set to 140 kV and power at 10 W with an exposure time of 1–4 s to achieve the optimal X-ray intensity flux count. The voxel resolution was kept constant around 3 μm for all the samples by adjusting the source and detector distance. Then, the μCT scans for all the samples were processed using PerGeos software.

2.6. Electrical Resistivity Measurements

Electrical resistivity measurements were conducted on the samples after NMR measurements using an electrical resistivity test system that allows the simultaneous measurement of 4-pole and 2-pole resistivity under elevated pressure and temperature. The system utilizes an Agilent 20-MHz waveform generator model 33,220 A to generate electrical currents over a wide range of frequencies, and a North Atlantic LCR meter model 2250 to measure samples' inductance, capacitance, and resistance. Electrical resistivity measurements were conducted on the samples at 100% water saturation at a net confining pressure of 500 psi and at an ambient temperature of 25 °C.

3. Results and Discussions

3.1. Porosity and Permeability

Figures 3 and 4 respectively show the histograms of the porosity and permeability of the studied samples. Figure 3 shows two distinct sample populations based on porosity data alone. The first group of samples has porosities ranging from 0.1%–7% with a modal value of 3%–4%. The samples in the second porosity group have porosities ranging between 7% and 13% with a modal value of 9%–10%. Statistical analysis of the samples' permeabilities alone does not clearly show distinct groups of sample populations as was observed in the porosity data. Nonetheless, two groups of sample populations can be faintly seen based on the permeability values alone (independent of the porosity classification). As shown in Figure 4, one group of samples is very tight with permeability ranging between 0.001

and 0.2 mD (with a modal value of 0.03 mD). The second group of samples has permeability ranging between 0.2 and 1 mD (with a mode of 0.35 mD).

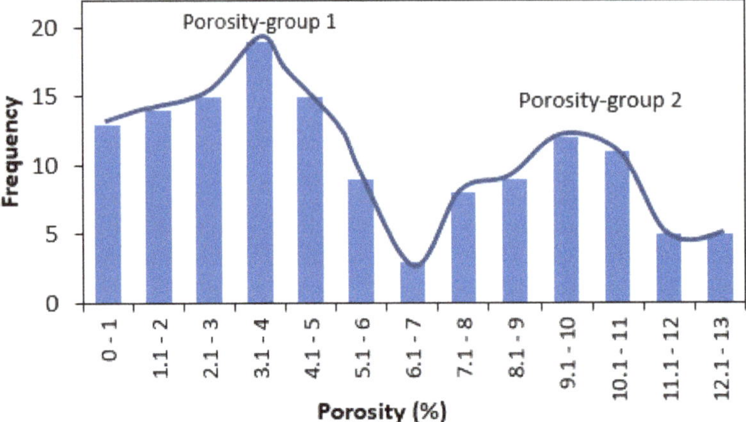

Figure 3. Porosity distribution of 140 rock samples extracted from six wells.

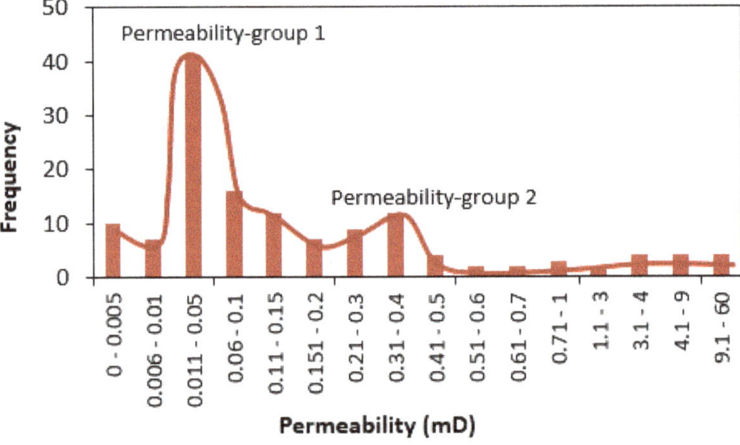

Figure 4. Permeability distribution of 140 rock samples extracted from six wells.

3.2. Porosity–Permeability Relationship

Figure 5 shows the cross-plot of porosity versus log of permeability for all 140 plugs extracted from the six wells (wells A–F). Most of the samples in porosity-group 1 (0.1%–7%) belong to rock samples extracted from wells D, E, and F (at a depth of 14,000–18,000 feet). All the plug samples extracted from well A (16,000 feet) belong to the porosity group 2 (7%–13%). Some of the plug samples from wells B and C belong to porosity group 1, while others belong to porosity group 2. There is no clear correlation between porosity and permeability (Figure 4). The scatter in the porosity–permeability cross-plot can be attributed to the variation in the type of facies, diagenesis, and petrophysical properties of the rocks. In Figure 5, the samples are color-coded based on the wells (and of course the depth) from which they were extracted. It can be seen that the low porosity is not associated with the depth from which the samples were extracted. In the next few subsections, the samples are also classified based on facies and petrophysical rock typing.

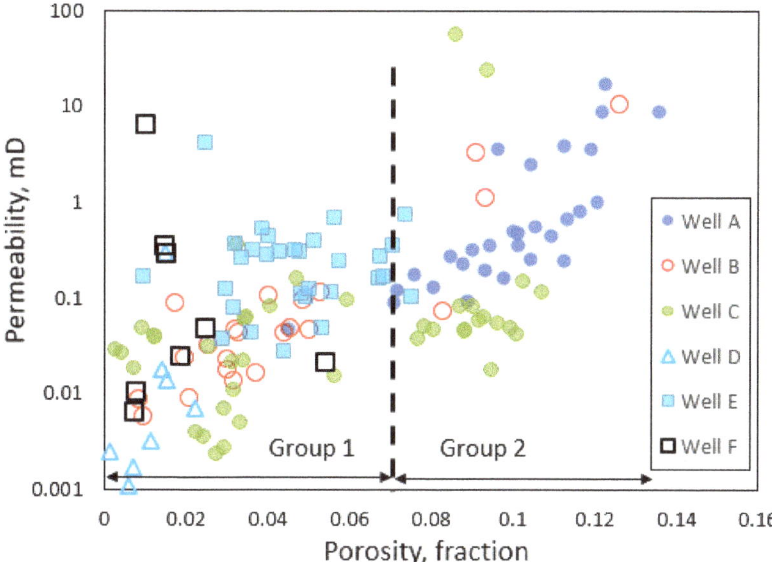

Figure 5. Cross-plot of helium porosity and helium permeability of 140 rock samples classified based on the wells from which they were extracted.

3.3. Facies Classification

As mentioned earlier, sedimentological and petrographical analyses conducted on core samples from the study formation revealed the occurrences of various lithofacies broadly classified into four facies associations (FAs), namely: (1) FA1, massive to ripple-laminated sandstone interpreted as fluvial; (2) FA2, gray massive sandstone (glaciofluvial); (3) FA3, diamictites (glaciolacustrine delta); and (4) FA4, partially deformed massive-graded sandstone (subglacial outwash deposits) [12,13]. The facies associations (FAs) are grossly heterogeneous (Figure 6A) and are also characterized by varied porosity distributions. Samples from FA1 and some samples from FA2 are characterized by primary interparticle porosity (Figure 6B), intraparticle porosity due to grain dissolution (Figure 6C), and microfracture porosity (Figure 6D). Their measured porosity values are greater than 7%. Petrographic examination of the FA1 samples indicates that they are dominantly subarkose arenite with authigenic pore-filling illite (Figure 6E). The low porosity values (<7%) encountered in the samples of the subglacial (FA3) and glaciofluvial (FA4) were probably due to the high matrix content (mainly detrital illite) and poor sorting. The presence of pore-filling anhydrite, barite (Figure 6F), siderite, and differential compaction in some of the FA2 and FA4 samples negatively impacted their reservoir quality.

Figure 7 shows the porosity–permeability plot of all the samples classified based on facies association. It is obvious that facies association FA4 exists exclusively in porosity group 1, while facies association FA1 exists only in porosity group 2. Facie associations FA2 and FA3 occur in both porosity groups. However, a clear correlation is not seen between porosity and permeability in any of the facies associations.

Figure 6. Photomicrographs for microfacies observed in the analyzed samples. (**A**) Heterogenous patterns of grain compactions, (**B**) interparticle porosity associated with grain compaction, (**C**) intraparticle porosity due grains dissolution, (**D**) fractures associated with grain compactions, (**E**) authigenic pore-filling illite, (**F**) pore-filling anhydrite and barite cement.

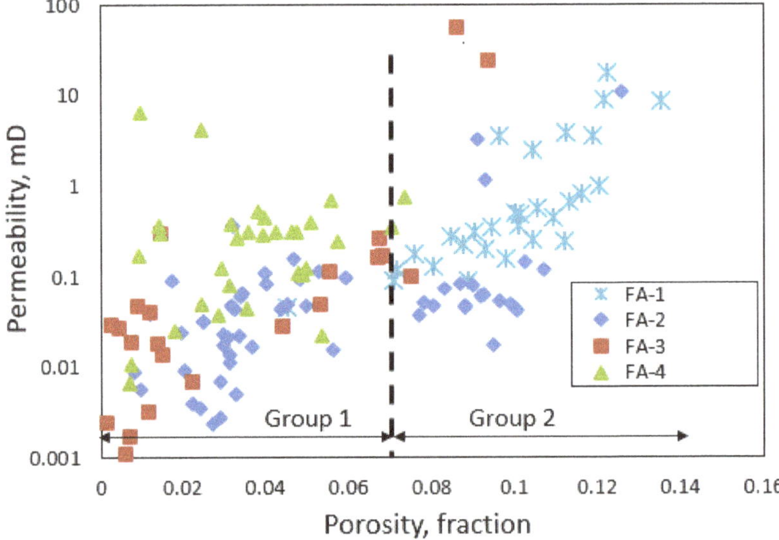

Figure 7. Cross-plot of helium porosity and helium permeability of 140 rock samples classified based on facies associations.

3.4. Petrophysical Rock Typing

In order to identify cores with identical pore and petrophysical characteristics, we used the concept of flow zone indicator (FZI) developed by Amaefule et al. [19]. They developed the FZI by rearranging the Kozeny–Carman equation, as shown in Equation (2):

$$K = \left(\frac{\varnothing_e^3}{(1 - \varnothing_e)^2} \right) \left(\frac{1}{K\tau SV_{gr}^2} \right) \tag{2}$$

where K is permeability (μm^2), $\varnothing_e$ is effective porosity, τ is tortuosity of flow path, and SV_{gr} is the surface area per unit volume of grain.

Rearranging the equation by dividing both sides by porosity and taking the square root yields:

$$\sqrt{\frac{K}{\varnothing}} = \frac{1}{SV_{gr}^2 \sqrt{K\tau}} \left(\frac{\varnothing_e}{1 - \varnothing_e} \right) \tag{3}$$

Amaefule et al. [19] expressed this equation as a reservoir quality index (RQI). If permeability is expressed in millidarcy (mD) and porosity is expressed as a fraction, the left-hand side of the equation becomes:

$$RQI = 0.0314 \sqrt{\frac{K}{\varnothing}} \tag{4}$$

Using Equation (4), the tortuosity and surface area of the rock samples in Equation (3) do not need to be measured, but rather replaced by the left-hand side of the same equation. The FZI is defined from Equation (3) as:

$$FZI = \frac{1}{SV_{gr}^2 \sqrt{K\tau}} \left(\frac{\varnothing_e}{1 - \varnothing_e} \right) \tag{5}$$

Hence, $RQI = FZI \times \varnothing_z$, where $\varnothing_z = \frac{\varnothing_e}{1 - \varnothing_e}$

On a log-log plot of RQI versus $\varnothing_z$ (Figure 8), rock samples with the same hydraulic characteristics and similar petrophysical properties will align around a unit slope. The intercept of that slope in the RQI axis at the $\varnothing_z$ value of 1 defines the flow zone for those samples. Samples with different FZI values will lie on other parallel lines. Hence, rocks having the same FZI tend to obey the same fluid distribution and fluid transport properties. Thus, they can be represented by the same petrophysical measurements such as relative permeability, capillary pressure, and electrical resistivity curves. The rock typing in Figure 8 showed five different hydraulic units. Classification of the rocks based on hydraulic flow units or a flow zones indicator revealed a clear power law correlation between porosity and the log of permeability in each hydraulic flow unit (Figure 9). The coefficients of regression (R^2) range between 0.8 and 0.96. It also apparent from the cross-plot that the 7% cut-off value of porosity marks the deviation of the cross-plots from a linear trend. Further analysis using NMR, QEMSACN, and high-resolution imaging studies were used to probe this phenomenon, as presented in the subsequent subsections.

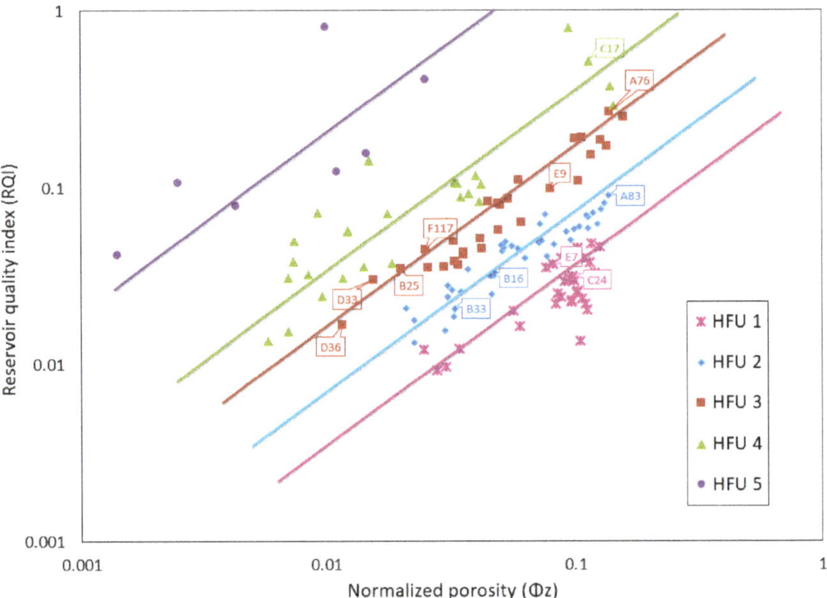

Figure 8. Hydraulic flow units of the reservoir identified from 140 rock samples from six wells.

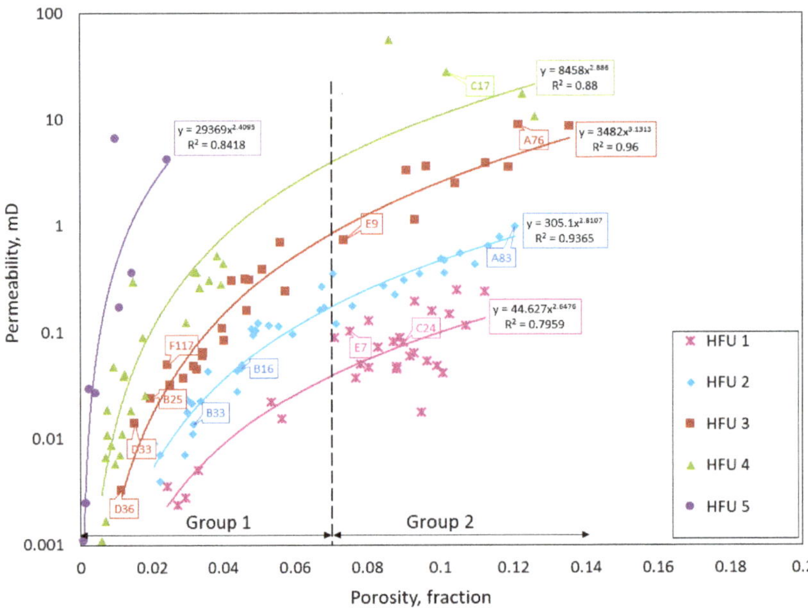

Figure 9. Cross-plot of helium porosity and helium permeability of 140 rock samples classified based on hydraulic flow units.

3.5. QEMSCAN Analysis

Figure 10 shows the petrographic (QEMSCAN) analysis of six thin sections representing the first porosity group (<7% porosity), while Figure 11 shows the analysis of another six sections from porosity group 2 (>7% porosity). All the facies associations are represented in the 12 subsamples. Based on quartz, feldspar, and lithic contents determined from thin-section petrography [12,13], all the samples are comprised of quartz arenite, subarkose, and sublithic arenite. The sediments' grain sizes vary from fine, medium, and coarse, and are poorly to moderately sorted. Quartz was identified as the dominant mineral in all the FAs (Table 2). Subordinate amounts of feldspars, mica, and clay minerals were also identified. The argillaceous sandstones (Figure 10; B33) and diamictites facies (Figure 10; B25, D36, and F117) are rich in matrix content, mainly detrital illite clay mineral. Authigenic illite (Figure 11; A83 and A76), siderite (Figure 10; B25 and D33), and a minor amount of other clay minerals, including chlorite and smectite minerals, are also present in some of the lithofacies. Based on sedimentological background, the reservoir quality in porosity group 2 was generally increased by the preserved and secondary porosity, as well as fractures (e.g., Figure 11; A76), while it was diminished in group 1 by the increase of detrital matrix content (mainly illite), diagenetic siderite cement, and grain compaction (e.g., Figure 10; B16).

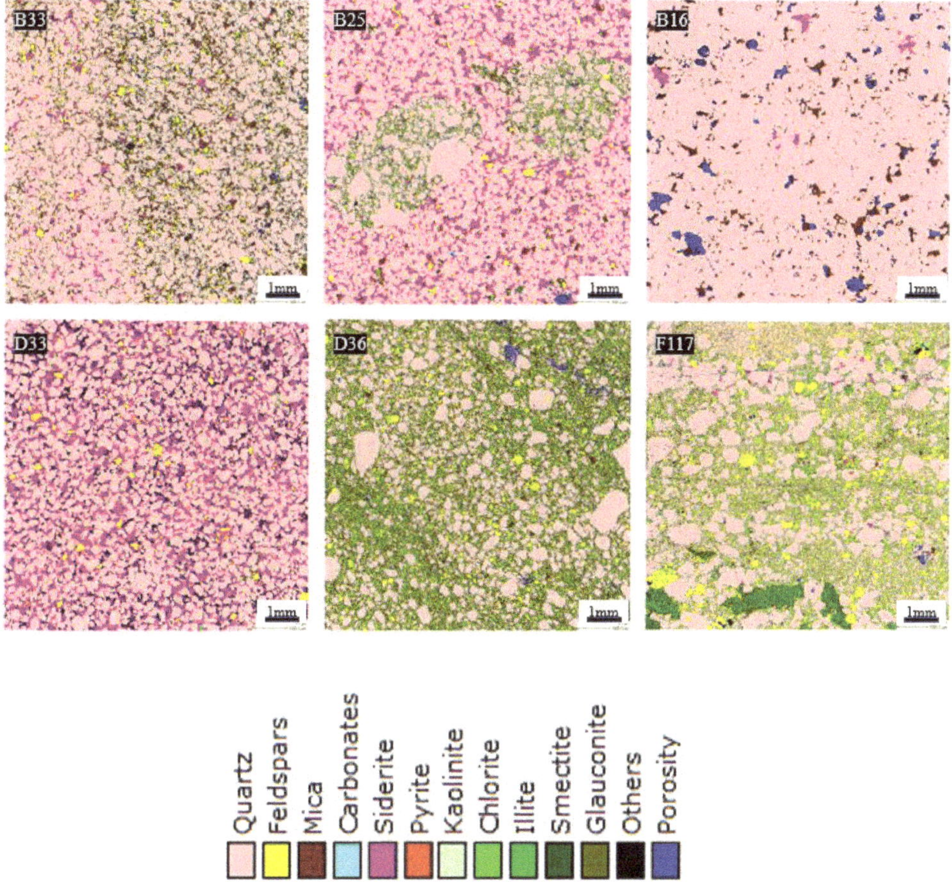

Figure 10. Quantitative evaluation of minerals by scanning electron microscope (QEMSCAN) images of samples in porosity-group 1 (<7% porosity). The color codes for different minerals and pores are shown in the legend.

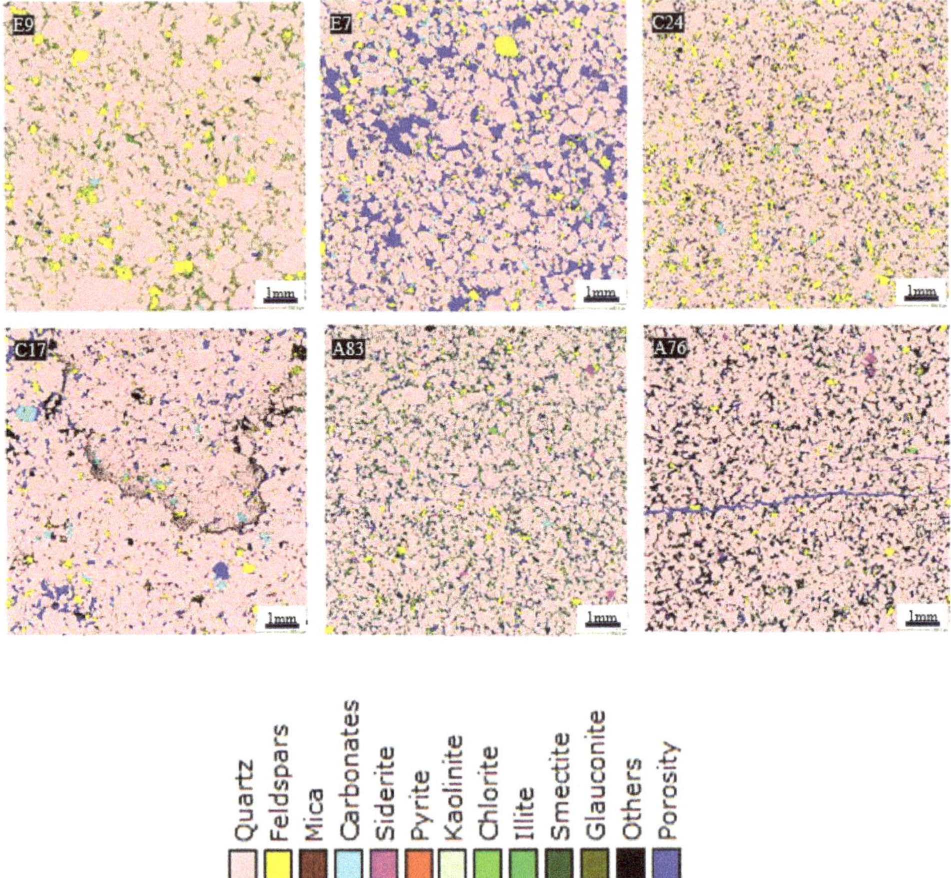

Figure 11. QEMSCAN images of samples in porosity group 2 (>7% porosity). The color codes for different minerals and pores are shown in the legend.

Table 2. Mineralogy compositions of the samples in each of the facies.

Sample ID	Silicates (%)			Carbonates (%)		Pyrite	Clay (%)					Others	FA
	Quartz	Feldspars	Mica	Calcite	Siderite		Kaolinite	Chlorite	Illite	Smectite	Glauconite		
B16	90.66	0.04	3.21	0.01	0.55	0.21	0.01	0.00	0.95	0.00	0.00	1.38	FA2
B25	75.32	1.45	3.06	0.01	13.50	0.06	0.02	0.12	4.07	0.00	0.02	0.39	FA3
B33	75.12	3.10	9.27	0.01	1.76	0.05	0.03	0.11	7.81	0.00	0.01	0.49	FA2
D33	64.45	1.50	1.60	0.03	21.21	0.10	0.01	0.21	1.13	0.00	0.03	8.44	FA4
D36	57.41	5.03	13.16	0.07	0.00	0.17	0.13	0.68	21.43	0.02	0.11	0.70	FA3
F117	67.44	8.91	4.13	0.05	0.26	0.15	0.03	0.48	16.04	0.07	0.04	1.63	FA3
A83	80.89	1.60	2.23	0.05	0.26	0.02	0.23	0.11	7.82	0.02	0.01	1.45	FA1
C24	79.19	7.56	2.19	0.33	0.00	0.03	0.02	0.01	3.84	0.02	0.00	3.09	FA2
E7	73.93	3.65	1.75	0.82	0.00	0.01	0.02	0.22	2.06	0.04	0.06	0.34	FA4
E9	86.15	3.85	3.09	0.14	0.00	0.03	0.07	0.18	5.46	0.03	0.03	0.34	FA4
A76	80.93	1.13	0.83	0.02	0.25	0.03	0.04	0.06	2.38	0.00	0.00	10.38	FA1
C17	85.15	2.05	1.34	0.72	0.00	0.02	0.01	0.00	0.90	0.01	0.00	5.29	FA2

3.6. NMR T_2 Distribution and Permeability

NMR measurements were also conducted on the 12 subsamples representing the two porosity groups, as shown in Figure 12 (for group 1) and Figure 13 (for group 2). The NMR T_2 data show that the pore size distribution in the samples are characterized by unimodal and polymodal pore systems. The samples in group 1 mainly showed polymodal pore distribution. The pore size distribution appears to have been controlled by the proportion of clay content and the presence of fractures in the samples. As shown by their petrographic analysis, they contain a fair amount of clay minerals (Illite: 7.8%–21.44%). They are also predominantly delineated by micropores (with a T_2 range of 0.01–1 ms) and macropores (with a T_2 range of > 1–1000 ms), as shown in the NMR T_2 curves in Figure 12. The micropore-bearing samples are dominantly matrix-rich, and they mainly represent the argillaceous sandstone and mud-rich diamictites facies. On the other hand, the samples in porosity group 2 showed a unimodal pore system (except for sample A83, which appears to be bimodal) characterized by mainly macropores with a T_2 range of > 1–1000 ms (Figure 13). They contain a minor amount of clay minerals (illite: 0.95%–5.56%). The macroporosity in these samples is either due to the preserved primary porosity or secondary porosity created by the digenetic alteration of labile minerals, mainly feldspars grains, in the samples. It was also observed that the pores with a T_2 range of 100–1000 ms in some of the samples may be due to the presence of microfractures (Figure 13).

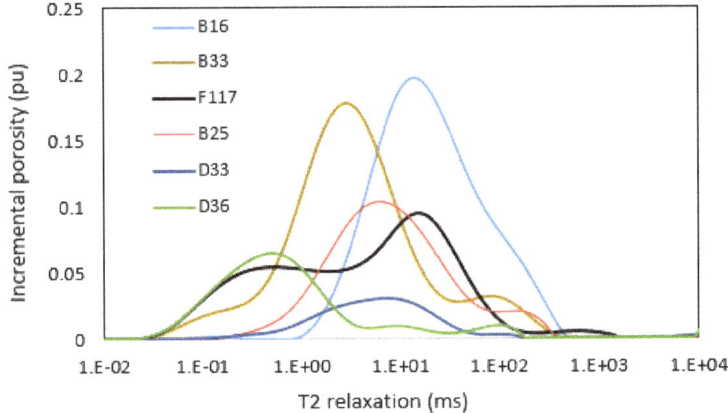

Figure 12. Distribution of NMR T_2 relaxation for subsamples in porosity group 1 (<7%).

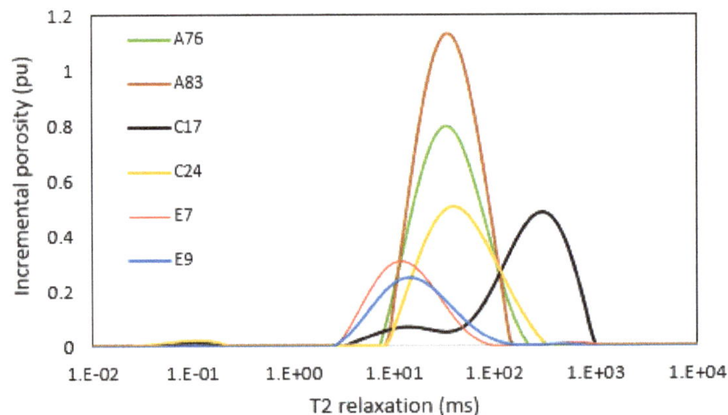

Figure 13. Distribution of NMR T_2 relaxation for subsamples in porosity group 2 (>7%).

3.7. NMR Porosity and Permeability

NMR porosity and pore size distribution were subsequently measured on 56 samples out of the 140 samples. A comparison made between different methods of porosity measurements, namely helium, NMR, and gravimetric porosity, reveals an acceptable match between the helium porosities and those derived from NMR and gravimetric methods (Figures 14 and 15). However, it can be observed that at a porosity value above 7%, there is a scatter between all the porosity measurements. Both NMR and gravimetric porosities were measured under unconfined pressure, while helium porosity was conducted at a net confining pressure of 500 psi. Hence, the fairest comparison is that between NMR and gravimetric porosities (Figure 14). The NMR porosity and gravimetric porosity have a very good match at porosity values of 7% and below. On the other hand, there is a significant disparity between the two measurements at porosity greater than 7%. The reasons for the scatter at the 7% porosity group margin requires a closer interrogation. A closer look at the data in Figure 15 shows a better match between the gravimetric and helium porosities at this porosity range. Consequently, the mismatch could be attributed to errors in NMR measurements, particularly at porosity greater than 7%. The ability to keep the water saturation constant during gravimetric and NMR measurements dictates the level of match between the two measurement methods. Two important factors can affect the ability to sustain a constant water saturation in both measurements. These include (a) the process of wiping off excessive water from the sample surface, and (b) the duration of the porosity measurements. Gravimetric measurements take a few seconds, whereas NMR measurements require multiple scanning in order to achieve a good signal-to-noise ratio, and as such, a considerable amount of time can be expended on a single NMR measurement (an average of 10 min). During this extended time, there is a possibility of losing some of the pore water to evaporation (if the plastic wrap leaks), especially for relatively highly-porosity samples (with internal vugs, microfractures, etc.). Then, this leads to underestimated porosity values. Similarly, for these high-porosity samples, the insufficient wiping of excess water from the surface vugs can cause an overestimation of porosity. This explains why some NMR porosity values are either higher or lower than the gravimetric porosity. The QEMSCAN images of samples E7, E9, A76, A83, C17, and C24 in Figure 11 corroborates our inference. The images showed large pores, microfractures, and well-connected pores compared to the low and not well-connected pores in samples having porosity below 7% (as shown in the QEMSCAN images in Figure 10). For tighter rock samples such as those below the 7% margin, the pore water cannot be easily wiped off during the process of wiping off excessive water. Similarly, the pore water can take a much longer time than the NMR experiment time to start evaporating in the event of leakage in the plastic wrap. This explains why the NMR and gravimetric porosity have a very good match at porosity of 7% and below. It was observed that some samples in the porosity group 2 (>7%) also have a good correlation between the two measurement methods. This indicates that porosity alone may not be responsible for the scatter in this group (>7%) of samples. In order to probe this phenomenon further, high-resolution image analysis was used to investigate the pores and pore throats of samples in both groups. This is discussed in the following subsection.

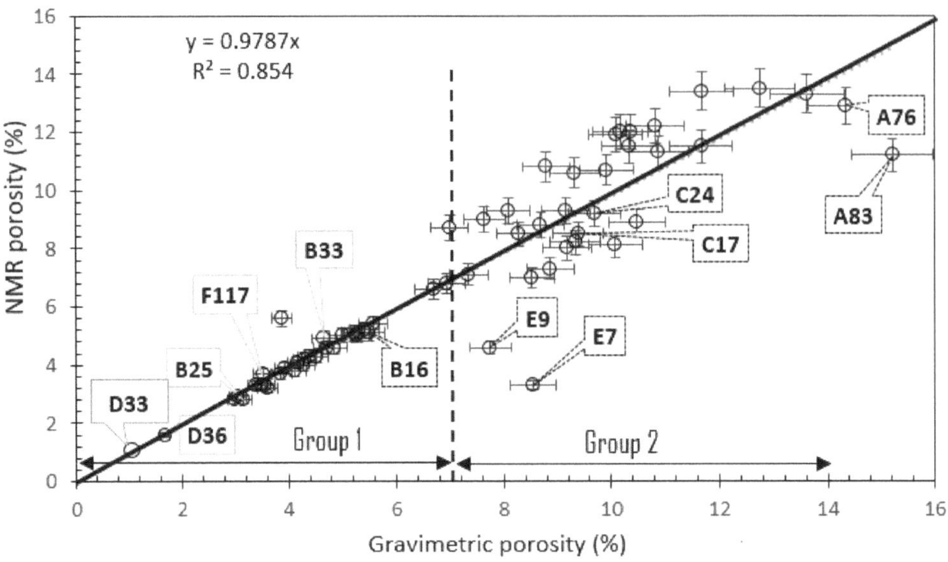

Figure 14. Cross-plot of NMR porosity and gravimetric porosity for 56 subsamples.

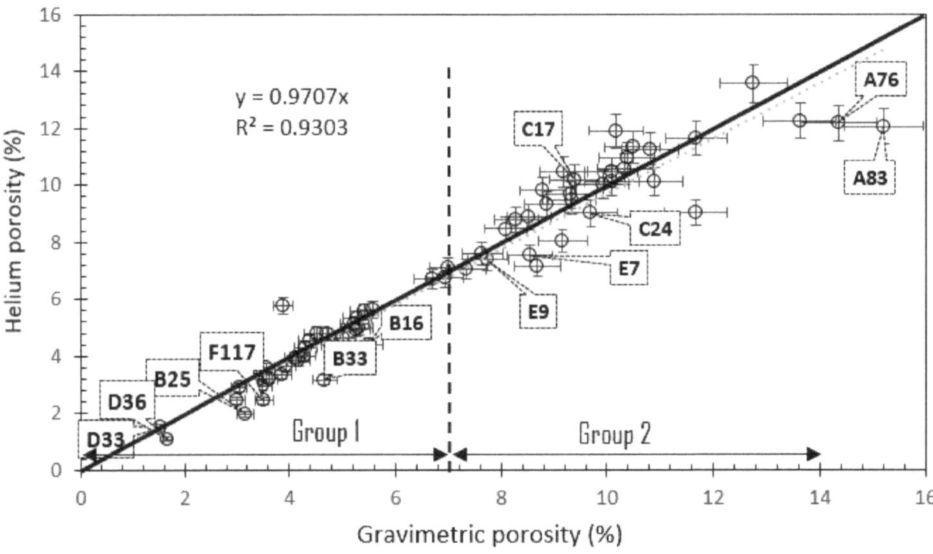

Figure 15. Cross-plot of helium porosity and gravimetric porosity for 56 subsamples.

NMR permeability was also estimated using the SDR model [20] and compared with laboratory measurements. The SDR equation is shown in Equation (6):

$$K = aT_{2LM}^{b}\varnothing^{c} \tag{6}$$

where K is permeability (mD), T_{2LM} is the log mean of relaxation time (milliseconds), and $\varnothing$ is porosity (fraction). The parameters a–c are constants that can be adjusted to have a good match between NMR permeability and measured permeability. Figure 16 shows the cross-plot of the calculated NMR permeability with measured permeability for the 56 samples. The optimized model parameters that gave the best match with measured permeability are a = 14, b = 2, and c = 1, with a root mean square error (RMSE) of 1.83 and R^2 value of 0.6. In Figure 17, a T_2 relaxation time at a 99% sample saturation state was used instead of T_{2LM}. The optimized model parameters for this case are a = 2, b = 4, and c = 2, with an RMSE value of 2.19 and an R^2 value of 0.6.

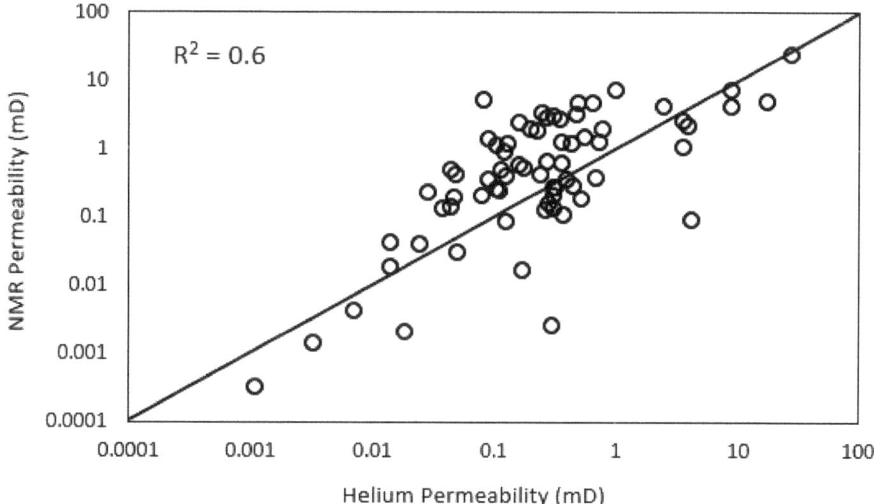

Figure 16. NMR-derived permeability versus core-measured permeability using T_{2LM}.

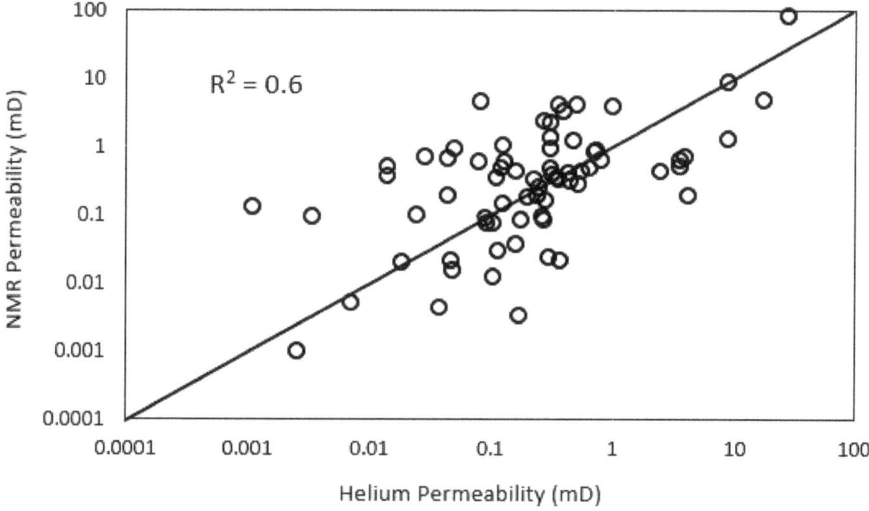

Figure 17. NMR-derived permeability versus core-measured permeability using T_2 at 99% water saturation.

3.8. High-Resolution CT Images

QEMSCAN images do not provide enough details of samples' pores and pore throats. Digital analysis of micro-CT images at high resolutions can elaborate more on the pore structure of these two groups of samples. In order to obtain high-resolution images, seven small rock cuttings selected from the two porosity groups were scanned at about 4-μm voxel resolution and segmented using PerGeos software. A gradient marker-based watershed segmentation was performed, and isolated pores were removed using the Axis Connectivity tool. The connected pore space was used as an input to the pore network extraction module of PerGeos software, and a pore network model was obtained for each sample showing the pores and throats. The pore network model is a spatial graph with branching or endpoints of the network labeled as pores and the connecting lines labeled as throats. The porosities computed using micro-CT analysis were found to be similar to the porosity measured using helium porosimeter (Figure 18).

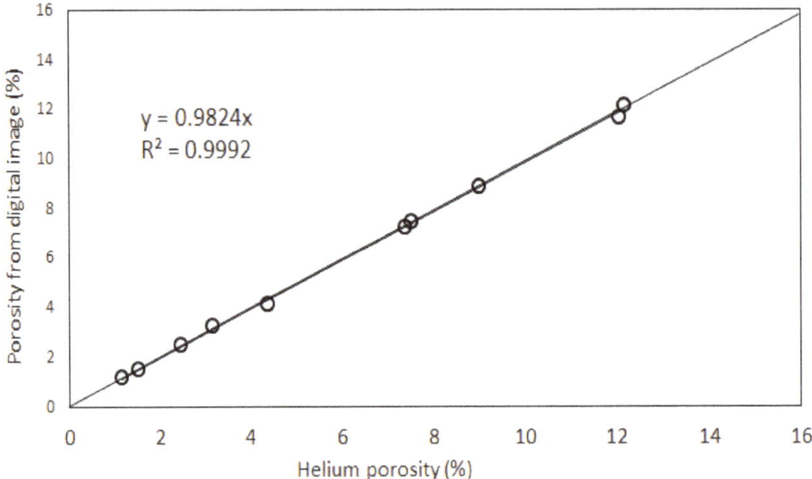

Figure 18. Porosity computed from digital image versus measured porosity.

Figures 19 and 20 show the segmented 2D images for group 1 and group 2, respectively. We could not carry out detailed digital analysis on some of the selected samples (especially the samples with poor connectivity) because of limitations in equipment resolution, which made it impossible to resolve smaller pores. Coincidentally, these poor connectivity samples have an excellent match between the gravimetric and NMR porosity measurements.

The high-resolution images show that the porosity group 1 samples have tight and poorly connected pores compared to the porosity group 2 samples, which have much better connectivity. These CT images conform to the QEMSCAN images and corroborate our inference regarding the disparity in the correlation between gravimetric and NMR porosities. A pore network model of selected samples from both porosity groups revealed the pore throat and pore size distribution from both groups. Interestingly, samples with a good match between gravimetric and NMR porosity have tighter pore throats with a pore throat size in the range of 2–20 μm (Figure 21), while samples that matched poorly between the two porosity methods have bigger pore throats or better connectivity, with pore throat sizes ranging between 2 and 50 μm (Figure 22). Figure 23 shows the pore network model of sample C24 (with a pore throat size range of 2–20 μm) and sample A83 (with pore throat sizes ranging from 2 to 50 μm). One may be tempted to speculate that pore throat sizes in excess of 20 μm may be responsible for the scatter in the porosity cross-plots, since they allow the water in the pores to be drained or lost during NMR measurements. Sample C24 has porosity above the 7% cut-off, but the pore throat size is similar to the samples in group 1 (<7%). This suggests that the scattering observed in the porosity cross-plots is not due to porosity values alone but also due to the connectivity of the pores. This also explains why some samples in group 2 also showed a good match between the two porosity measurements even when their porosity values are above the 7% cut-off.

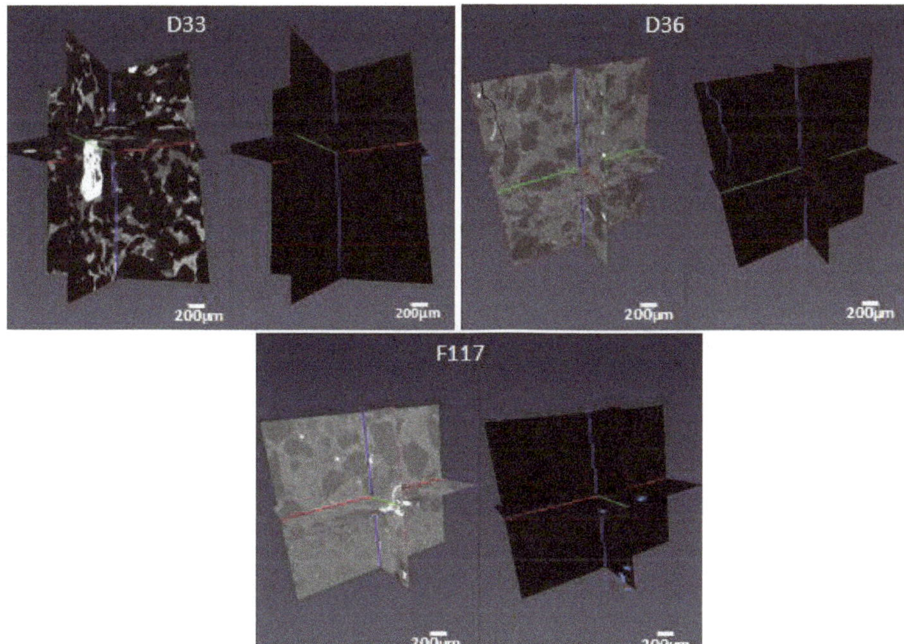

Figure 19. High-resolution μCT images and their segmentations for samples in group 1 (<7% porosity). CT images taken at 3-μm resolution.

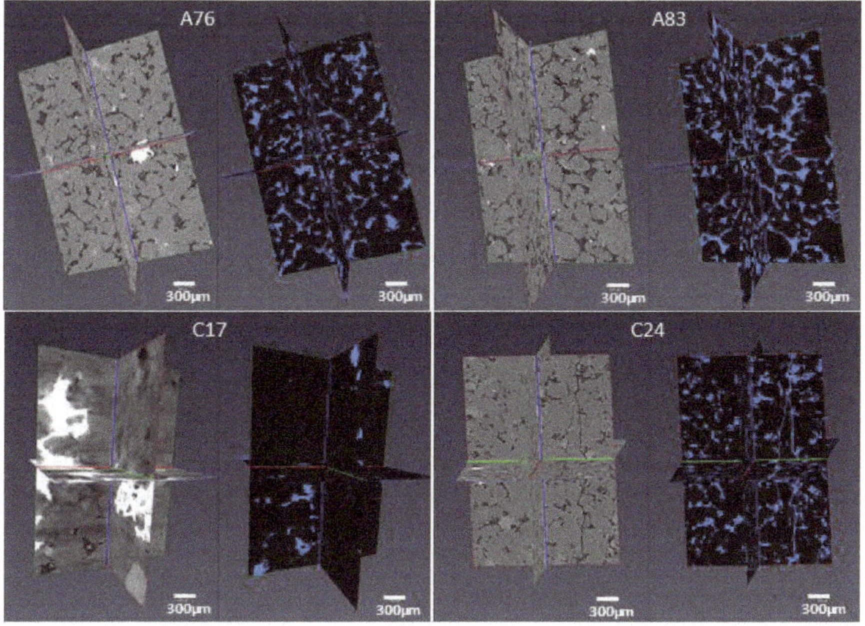

Figure 20. High-resolution μCT images and their segmentation for samples in group 2 (>7% porosity). CT images taken at 4-μm resolution.

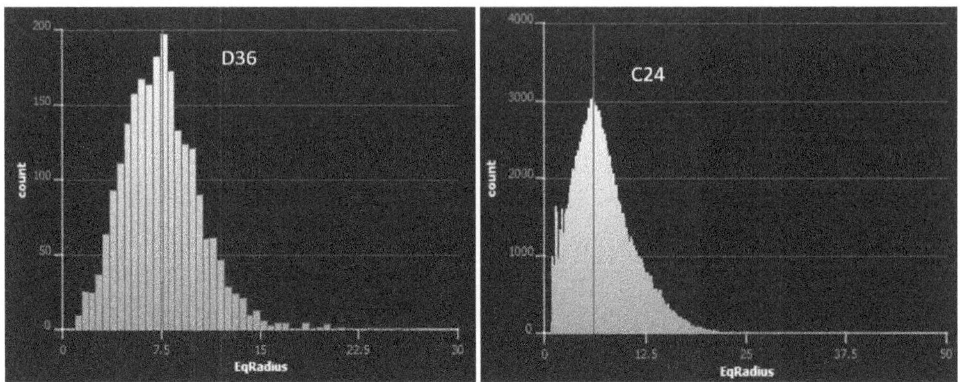

Figure 21. Pore throat size distribution in the range of 1–20 µm. On the y-axis is the count (frequency) and the x-axis is the corresponding equivalent radii (EqRadius in µm) of the pore throats.

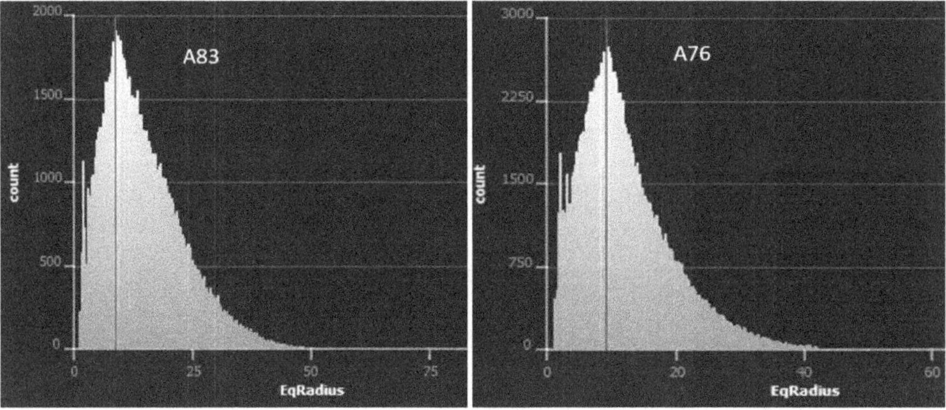

Figure 22. Pore throat size distribution in the range of 1–50 µm. On the y-axis is the count (frequency) and the x-axis is the corresponding equivalent radii (EqRadius in µm) of the pore throats.

3.9. Formation Resistivity and Cementation Factors

Archie defined the ratio of the resistivity of a rock at 100% water saturation (R_o) to the resistivity of the water (R_w) as the formation resistivity factor (or formation factor) given by Equation (7):

$$FF = \frac{R_o}{R_w} \tag{7}$$

The formation factor (FF) is dependent on various properties of the rock such as the porosity, degree of cementation, and clay content. For a clean sand and limestone, where clay is absent, FF remains constant as brine resistivity changes. For a shaly sand (containing clay), FF will decrease as brine resistivity increase and vice versa. This is because the clay content in the rock also serves as a conductor, which does not allow R_o to increase proportionally with R_w. This clay effect depends on the amount, type, and distribution of the clay in the rock. The simplest relationship between FF and porosity ($\varnothing$) is:

$$FF = \frac{1}{\varnothing} \tag{8}$$

Equation (8) was derived with the assumption that the rock matrix is embedded with straight and parallel capillaries. Different researchers developed different forms of Equation (8), as shown in Table 3, Equations (9)–(13). These models to a large extent incorporate a tortuosity factor (τ) in the relationship between FF and porosity. The tortuosity factor is a measure of the departure of a porous medium from the ideal system made of straight and parallel capillaries.

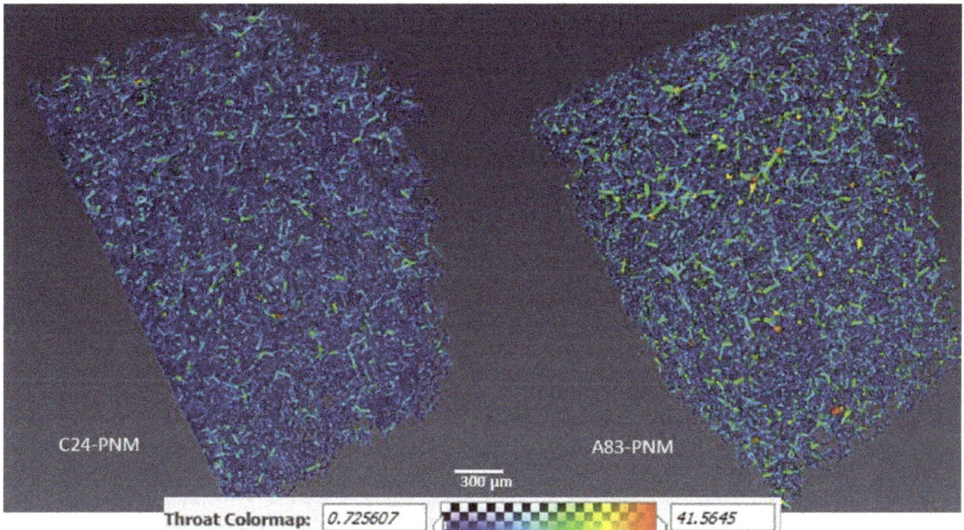

Figure 23. Pore network model for samples C24 and A83. The color legend represents the pore and throats sizes, where dark blue represents small pores and throats, and red represents big pores and throats.

Table 3. Different electrical tortuosity models.

Authors	Tortuosity Model	Equation #
Winsauer et al. [21]	$\tau^2 = (FF \times \varnothing)^{1.2}$	(9)
Wyllie and Spangler [22]	$\tau = (FF \times \varnothing)^2$	(10)
Faris et al. [23]	$\tau^2 = (FF \times \varnothing)^{1.41}$	(11)
Pirson [24]	$\tau^2 = FF \times \varnothing$	(12)
Cornell and Katz [25]	$\tau = FF \times \varnothing$	(13)

FF is also dependent on the degree of cementation of the rock. The degree of cementation depends on the cementing material (i.e., silica, calcite, or clay), amount of cement, and distribution of the cement materials. The higher the cementation, the lower the porosity, and then the higher the FF. Archie derived the relationship between FF, porosity, and cementation factor (m), as shown in Equation (14). The cementation factor or porosity exponent (m) is a function of the large number of factors such as the type, shape, and distribution of pores; packing and sorting of grains; tortuosity; presence of clay; compaction due to overburden stress, etc. Humble modified Equation (14) to include another constant (a), which is also dependent on the rock type. This constant (a) is an empirical constant, and it is sometimes referred to as tortuosity factor or cementation intercept. A value of "a" other than 1 compensates for variation in the FF–$\varnothing$ relationship due to differences in the compaction, pore structure, and grain size [26].

$$FF = \frac{1}{\varnothing^m} \tag{14}$$

$$FF = \frac{a}{\varnothing^m} \tag{15}$$

The cementation factor often ranges between 1.3 for loose unconsolidated sand to as high as 3 in carbonates [15,16].

In Figure 24, the formation factors of 56 samples are plotted against their porosity values. Four pole electrical resistivity measurements were conducted at a 100% water saturation state for 56 subsamples. The samples were saturated with formation brine and subjected to a net-confining pressure of 500 psi and room temperature of 25 °C. The measured formation factor and porosity were fitted with Equation (14) using m values of 1.5. If we allow "a" value other than one (1), the tortuosity factor ("a") estimated as the intercept of the slope at 100% porosity was found to be 5.14 with an estimated cementation factor 0.98 (from the slope of the plot). Tortuosity was also calculated for these samples using the equations provided in Table 3. The tortuosity range based on each equation is given in Table 4. The tortuosity values for the 12 subsamples are listed in Table 5. In addition, Table 5 also contains a list of the different petrophysical measurements conducted on the 12 samples that represent the six wells as well as the different hydraulic flow units.

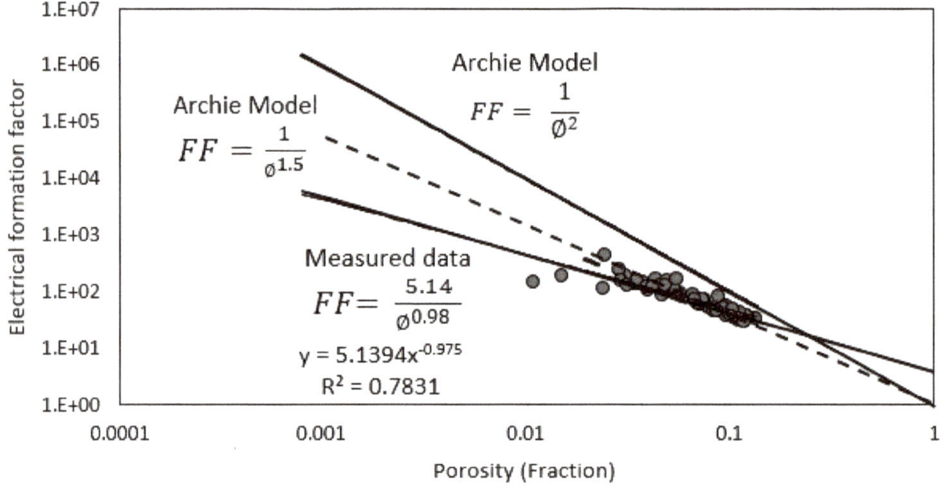

Figure 24. Formation resistivity factor versus porosity for 56 subsamples.

Table 4. Tortuosity range for 56 subsamples using different electrical tortuosity models.

Tortuosity Model	Tortuosity (τ)
Winsauer et al. [21]	1.28–3.89
Wyllie and Spangler [22]	2.28–92.55
Faris et al. [23]	1.34–4.93
Pirson [24]	1.23–3.1
Cornell and Katz [25]	1.51–9.62

Table 5. Summary of measured petrophysical properties on 12 subsamples representing six wells and the different hydraulic flow units. FZI: flow zone indicator, RQI: reservoir quality index.

Sample ID	He-Porosity %	NMR Porosity	Permeability (mD)	RQI (μm)	FZI (μm)	Formation factor (FF)	Tortuosity (τ) *	K_{NMR} (mD)
D36	1.15	1.4	0.0033	0.01682	20	137.2	1.51	0.0967
D33	1.53	0.9	0.014	0.030036	100	238.8	3.65	0.5224
B25	1.97	2.8	0.024	0.034772	100	-	-	0.0976
F117	2.46	3.7	0.05	0.044759	100	376.73	9.25	0.9415
B33	3.17	4.9	0.014	0.020714	4	-	-	0.3646
B16	4.39	5.1	0.044	0.031463	4	-	-	0.6781
E9	7.37	4.6	0.738	0.099349	20	64.05	4.72	0.8955
E7	7.52	3.3	0.103	0.036748	2.5	61.72	4.64	0.075
C24	9	9.2	0.081	0.029844	2.5	-	-	4.5309
C17	10.18	8.5	27.87	0.519591	10,000	-	-	82.9
A83	12.05	11.2	0.99	0.089997	4	-	-	3.9619
A76	12.17	12.9	8.844	0.267724	100	37.29	4.54	8.7779

* Tortuosity based on Cornell and Katz [25] Equation (13). Electrical resistivity was not measured on some samples because of the small sample diameter.

4. Conclusions

An integrated petrophysical and petrological study was conducted on 140 core plugs from a Saudi Arabian tight gas reservoir. A detailed suite of experiments was conducted using a variety of analytical tools such as a helium porosity–permeability meter, NMR, electrical resistivity, micro-CT imaging, thin section, and QEMSCAN. We found that two distinct groups of porosity occur in the studied reservoir and that each group responded differently to the laboratory methods of porosity measurements. In order to probe this observation further, a gradient marker-based watershed segmentation together with digital rock physics computation were used to generate the pore network models, pores, and pore size distribution of representative samples. Then, we speculated how the pore character of the tight formation affected methods of porosity measurements in the laboratory. Then, the following specific conclusions were made from this study.

1. Our analysis identified two major porosity groups of rock samples. One group is characterized by porosity values less than 7% while the second group has porosity values above 7%. There is a generally a good agreement between different methods of porosity measurements in samples with <7% porosity compared to samples with >7% porosity values.
2. Rock typing using the flow zone indicator (FZI) identified five hydraulic flow units (HFUs). Each of the HFUs has reservoir rocks with porosity values that fall within both groups of porosity.
3. Petrographic analysis revealed that the low-porosity samples have a significant amount of clay (mainly illite). However, quartz was identified as the dominant mineral in all the rock samples while subordinate amounts of feldspars, mica, and clay minerals were also identified.
4. NMR analysis showed that the group 1 samples are dominated by micropores, while the group 2 samples are dominated by macropores. Group 1 samples apparently showed a polymodal pore system, while the samples in group 2 generally showed unimodal pore distribution.
5. High-resolution micro-CT images showed that pore throat size plays a very important role in the NMR porosity measurements, and may be responsible for the poor correlation between NMR porosity and gravimetric porosity in the group 2 samples (>7%).
6. NMR-estimated permeability in tight sandstones shows fair correlation with helium permeability with an R^2 of 0.6 and RMSE of 1.8.
7. Electrical resistivity measurements showed that the tight sand samples have a cementation factor of 1.5, when the laboratory measurements are fitted with the Archie model. When a value of tortuosity factor other than 1 is allowed, the cementation factor is about 1 and the tortuosity factor

is 5.14. Electrical tortuosity values computed using various empirical models range between 1.2 and 9.6.

Author Contributions: Conceptualization, A.R.A. and L.B.; Methodology, A.R.A. and L.B.; Software, S.R.H. and A.A.; Formal analysis, A.R.A., L.B., R.S.B., S.R.H., and A.A.; Investigation, A.R.A., L.B., and R.S.B.; Resources, A.R.A. and L.B.; Data curation, R.S.B.; Writing—original draft preparation, A.R.A.; Writing—review and editing, L.B.; Supervision, A.R.A. and L.B.; Project administration, L.B. and A.A.; Funding acquisition, L.B.

Funding: This research was funded by King Abdulaziz City for Science and Technology as part of the National Science, Technology, and Innovation Plan (NSTIP Project # 14-OIL468-04) through the Science and Technology Unit at KFUPM.

Acknowledgments: We duly acknowledge the Ministry of Energy, Industry and Mineral Resources, Saudi Arabia for providing the cores and plug samples used in this study. The contribution by Late Ali Sahin is also highly appreciated. We thank the anonymous reviewers for their critical and constructive review.

Conflicts of Interest: The authors declare no conflict of interest.

References

1. Saudi Aramco Unconventional Gas: A Growing Resource. Available online: https://www.jobsataramco.eu/people-projects/unconventional-gas-growing-resource (accessed on 8 June 2019).
2. Weeden, S. Saudi Arabia Gears Up For Unconventional Gas Exploration. Available online: https://www.hartenergy.com/exclusives/saudi-arabia-gears-unconventional-gas-exploration-19011#p=full (accessed on 8 June 2019).
3. Shehata, A.; Tps, A.U.C.; Aly, A.; Schlumberger, A.U.C.; Ramsey, L.; Tgcoe, S. Overview of Tight Gas Field Development in the Middle East and North Africa Region. In Proceedings of the North Africa Technical Conference and Exhibition, Cairo, Egypt, 14–17 February 2010.
4. Luffel, D.L.; Howard, W.E. Reliability of Laboratory Measurement of Porosity in Tight Gas Sands. In Proceedings of the Low Permeability Reservoirs Symposium, Denver, Colorado, 18–19 May 1987.
5. Sondergeld, C.H.; Newsham, K.E.; Comisky, J.T.; Rice, M.C.; Rai, C.S. Petrophysical considerations in evaluating and producing shale gas resources. In Proceedings of the SPE Unconventional Gas Conference; Society of Petroleum Engineers, Pittsburgh, PA, USA, 23–25 February 2010.
6. Song, L.; Warner, T.; Carr, T. An efficient, consistent, and trackable method to quantify organic matter–hosted porosity from ion-milled scanning electron microscope images of mudrock gas reservoirs. *Am. Assoc. Pet. Geol. Bull.* **2019**, *103*, 1473–1492. [CrossRef]
7. Bush, D.C.; Jenkins, R.E. Proper Hydration of Clays for Rock Property Determinations. *J. Pet. Technol.* **1970**, *22*, 800–804. [CrossRef]
8. Sullivan, M.; Song, L. Briggs colour cubing of spectral gamma ray—A novel technique for easier stratigraphic correlation and rock typing. In Proceedings of the SPWLA 58th Annual Logging Symposium, Oklahoma City, OK, USA, 17–21 June 2017.
9. Morrow, N.R.; Cather, M.E.; Buckley, J.S.; Dandge, V. Effects of Drying on Absolute and Relative Permeabilities of Low-Permeability Gas Sands. In Proceedings of the Low Permeability Reservoirs Symposium, Denver, Colorado, 15–17 April 1991.
10. Laboun, A.A. Regional tectonic and megadepositional cycles of the Paleozoic of northwestern and central Saudi Arabia. *Arab. J. Geosci.* **2013**, *6*, 971–984. [CrossRef]
11. Al-Harbi, O.A.; Khan, M.M. Source and origin of glacial paleovalley-fill sediments (Upper Ordovician) of Sarah Formation in central Saudi Arabia. *Arab. J. Geosci.* **2011**, *4*, 825–835. [CrossRef]
12. Alqubalee, A.; Abdullatif, O.; Babalola, L.; Makkawi, M. Characteristics of Paleozoic tight gas sandstone reservoir: integration of lithofacies, paleoenvironments, and spectral gamma-ray analyses, Rub' al Khali Basin, Saudi Arabia. *Arab. J. Geosci.* **2019**, *12*, 344. [CrossRef]
13. Alqubalee, A.; Babalola, L.; Abdullatif, O.; Makkawi, M. Factors Controlling Reservoir Quality of a Paleozoic Tight Sandstone, Rub' al Khali Basin, Saudi Arabia. *Arab. J. Sci. Eng.* **2019**, 1–19. [CrossRef]
14. Klinkenberg, L.J. The permeability of porous media to liquids and gases. In Proceedings of the Drilling and Production Practice, New York, NY, USA, 1 January 1941.

15. Ayling, B.; Rose, P.; Petty, S.; Zemach, E.; Drakos, P. QEMSCAN®Quantitative evaluation of minerals by scanning electron microscopy: capability and application to fracture characterization in geothermal systems. In Proceedings of the Thirty-Eighth Workshop on Geothermal Reservoir Engineering, Stanford University, Stanford, CA, USA, 30 January–1 February 2012; p. 11.

16. Gottlieb, P.; Wilkie, G.; Sutherland, D.; Ho-Tun, E.; Suthers, S.; Perera, K.; Jenkins, B.; Spencer, S.; Butcher, A.; Rayner, J. Using quantitative electron microscopy for process mineralogy applications. *JOM* **2000**, *52*, 24–25. [CrossRef]

17. Qian, G.; Li, Y.; Gerson, A.R. Applications of surface analytical techniques in Earth Sciences. *Surf. Sci. Rep.* **2015**, *70*, 86–133. [CrossRef]

18. Elster, A. MRIQuestion: T2 Relaxation Definition. Available online: http://mriquestions.com/what-is-t2.html (accessed on 8 October 2019).

19. Amaefule, J.O.; Kersey, D.G.; Marschall, D.M.; Powell, J.D.; Valencia, L.E.; Keelan, D.K. Reservoir Description: A Practical Synergistic Engineering and Geological Approach Based on Analysis of Core Data. In Proceedings of the SPE Annual Technical Conference and Exhibition, Houston, TX, USA, 2–5 October 1988.

20. Al-Ajmi, F.; Holditch, S. NMR Permeability Calibration using a Non-Parametric Algorithm and Data from a Formation in Central Arabia. In Proceedings of the SPE Middle East Oil Show, Manama, Bahrain, 17–20 March 2001.

21. Winsauer, W.O.; Shearin, H.M.; Winsauer, W.O.; Shearin, H.M., Jr.; Masson, P.H.; Williams, M. Resistivity of Brine-Saturated Sands in Relation to Pore Geometry. *Am. Assoc. Pet. Geol. Bull.* **1952**, *36*, 253–277.

22. Wyllie, M.R.J.; Spangler, M.B. Application of Electrical Resistivity Measurements to Problem of Fluid Flow in Porous Media. *Am. Assoc. Pet. Geol. Bull.* **1952**, *36*, 359–403.

23. Faris, S.R.; Gournay, L.S.; Lipson, L.B.; Webb, T.S. Verification of Tortuosity Equations: GEOLOGICAL NOTES. *Am. Assoc. Pet. Geol. Bull.* **1954**, *38*, 2226–2232.

24. Pirson, S.J. *Geological Well Log Analysis*; Gulf Publishing Co.: Houston, TX, USA, 1983.

25. Cornell, D.; Katz, D.L. Flow of Gases through Consolidated Porous Media. *Ind. Eng. Chem.* **1953**, *45*, 2145–2152. [CrossRef]

26. McPhee, C.; Reed, J.; Zubizarreta, I. *Core Analysis: A Best Practice Guide*; Elsevier: Amsterdam, The Netherlands, 2015.

Article

The Characteristics of Oil Migration due to Water Imbibition in Tight Oil Reservoirs

Liu Yang [1], Shuo Wang [1], Zhigang Tao [1,*], Ruixi Leng [2] and Jun Yang [1]

[1] State Key Laboratory for Geomechanics and Deep Underground Engineering, China University of Mining and Technology (Beijing), Beijing 100083, China; shidayangliu@cumtb.edu.cn (L.Y.); xiaohengcupb@163.com (S.W.); caowenk@hotmail.com (J.Y.)

[2] China Research Institute of Exploration and Development, Xinjiang Oilfield Company, PetroChina, Karamay 834000, China; eagle.1983@163.com

* Correspondence: taozhigang1981@163.com

Received: 23 September 2019; Accepted: 30 October 2019; Published: 4 November 2019

Abstract: In tight oil reservoirs, water imbibition is the key mechanism to improve oil production during shut-in operations. However, the complex microstructure and composition of minerals complicate the interpretation of oil migration during water imbibition. In this study, nuclear magnetic resonance (NMR) T_2 spectra was used to monitor the oil migration dynamics in tight oil reservoirs. The factors influencing pore size distribution, micro-fractures, and clay minerals were systematically investigated. The results show that the small pores corresponded to a larger capillary pressure and a stronger imbibition capacity, expelling the oil into the large pores. The small pores had a more effective oil recovery than the large pores. As the soaking time increases, the water preferentially entered the natural micro-fractures, expelling the oil in the micro-fractures. Subsequently, the oil in the small pores was slowly expelled. Compared with the matrix pores, natural micro-fractures had a smaller flow resistance and were more conducive to water and oil flow. Clay minerals may have induced micro-fracture propagation, which can act as the oil migration channels during water imbibition. In contrary to the inhibitory effect of natural micro-fractures, the new micro-fractures could contribute to the oil migration from small pores into large pores. This study characterized the oil migration characteristics and provides new insight into tight oil production.

Keywords: water imbibition; oil migration; tight oil reservoirs; nuclear magnetic resonance

---

1. Introduction

As a key unconventional energy, tight oil has received increased attention by many scholars and has gradually become another research topic following shale gas [1]. The commercial exploitation of tight oil has greatly affected the energy structure of the world, especially North America [2,3]. A tight oil reservoir is characterized by low permeability (<0.1 mD), low porosity (<10%), and micro/nano pores. Currently, multistage hydraulic fracturing is the key technology for commercial exploitation. This technology can induce complex fracture networks by injecting large-scale fracturing fluid into the formation, thus increasing the reservoir's exposed area and improving oil production [4]. After hydraulic fracturing stimulation, shut in operations for a period are beneficial to the production increase of tight oil. Extended shut in time can promote fracturing fluid imbibition into reservoirs to displace crude oil and has been proven to be effective for enhanced oil recovery [1].

In tight oil reservoirs, complex imbibition characteristics result in the uncertainties of appropriate shut-in operations. Lan et al. (2014) [5] conducted a series of spontaneous imbibition experiments to establish dimensionless time model and calculate shut-in time. However, a dimensionless time model cannot involve the effects of pore structure and clay minerals and is not suitable for tight oil reservoirs. Jiang et al. (2018) [6] considered that the water imbibition is generated under confining pressure and carries out a large number of forced imbibition experiments by means of nuclear magnetic resonance (NMR). A modified model of dimensionless time was established by considering the relationship between pore radius and confining pressure [7,8]. Field studies found that not all tight oil wells are suitable for shut-in operation and extending the shut-in time may intensify the damage of the fracturing fluid to the reservoirs, which was not conducive to oil production [9]. Understanding the imbibition characteristics and relevant influencing factors (e.g., complex microstructure and minerals) contributes to conduct-appropriate shut-in operations.

Understanding water imbibition into matrix pores is the key for clarifying microscopic physical mechanism during the shut-in periods. Many published studies have focused on shale, tight sandstone and volcanic rock reservoirs [10,11]. The capillary pressure caused by interfacial tension and the osmotic pressure induced by water activity difference are the main forces that imbibe water into matrix pores. Capillary pressure is inversely proportional to pore radius. Micro/nano pores may result in an ultra-high capillary pressure to imbibe the fracturing fluid [12,13]. The capillary pressure also increases pore pressure within the matrix pores, triggering the propagation of tensile fractures [14,15]. In addition, tight reservoirs have a more complex pore structure than conventional reservoirs, leading to complex imbibition behaviors. Hu et al. (2012) [16] performed imbibition experiments on Barnett shale and found the imbibition behavior to present low pore connectivity. The existence of micro-fractures and bedding planes can enhance the imbibition rate of fracturing fluid. For clay-rich rock, the water imbibition of osmotic pressure is much stronger than that of the capillary pressure, which may result in micro-fracture propagation to increase the imbibition rate [17]. Xu et al. (2018) [1] found that the current imbibition tests were carried out under the atmospheric pressure without considering the effect of forced pressure. Based on low-field nuclear magnetic resonance, a large number of forced imbibition experiments have been conducted on tight sandstone to clarify the imbibition characteristics and influencing factors.

At present, the fracturing fluid imbibition involves physical and chemical process that are not well understood in the tight reservoirs. Oil migration among the multiscale pore structure and the effects of clay minerals on oil migration still need further study. In this study, a large number of imbibition experiments were conducted and the NMR T_2 spectra was used to monitor oil migration dynamics. The imbibition oil recovery of different diameters pores was studied and the effects of pore size distribution, micro-fractures, and clay minerals were analyzed.

2. Experimental Materials and Methods

2.1. Rock Samples and Fluids

The tight reservoirs samples were taken from Ordos Basin, Songliao Basin and Junggar Basin, which are the greatest potential reservoirs for tight oil production in China. Considering that these formations are characterized by a different pore structure and mineral composition, this study can present the oil migration characteristics influenced by microstructure and minerals. The core plugs of 2–5 cm in length and 2.5 cm in diameter were drilled from larger core materials. The end faces of the plugs were cut by a line cutting machine. The X-ray diffraction test for mineral composition was performed on the rock fragments collected from the drilling process. After removing residual oil from the core plugs, the porosity and permeability measurement was conducted on the core plugs.

Table 1 presents the basic properties of tight reservoirs samples that involve source, size, porosity and permeability. The porosity of the samples was measured by a helium porosimeter and ranged from 2.7% to 9.2%. It showed that tight oil reservoirs have the low porosity properties. According to the measurement method of Brace et al. (1968) [18], the pulse-decay permeability was determined under the condition of confining pressure (8 MPa), pore pressure (5 MPa) and room temperature (25 °C). The permeability of samples is not corrected by Klinbenberg effects. In contrary to the porosity, the permeability of samples was significantly different, ranging from 0.0053 to 0.24 mD.

The different permeability characteristics may result in the existence of micro-fractures that can connect the matrix pores and increase the permeability.

Table 1. The physical properties of tight reservoirs samples.

Label	Formation	Diameter, cm	Length, cm	Porosity, %	Permeability, mD
UC7-1	Upper Chang-7	2.49	5.1	3.2	0.24
UC7-2		2.51	3.8	2.5	0.15
LC7-1	Lower Chang-7	2.50	3.6	4.8	0.0062
LC7-2		2.50	4.3	5.3	0.0053
QT-1	Quantou	2.50	3.5	8.1	0.014
QT-2	formation	2.52	5.2	9.8	0.008
WEH-1	Wuerhe	2.51	3.8	6.5	0.098
WEH-2	formation	2.50	3.2	8.9	0.12

The results of the mineral composition are presented in Table 2. The formation of UC7 and LC7 was mainly composed of quartz and feldspar. The content of clay minerals was lower than 16%. Therefore, clay expansion was not considered during water imbibition. The UC7 and LC7 formation samples were used to study the effects of pore size distribution and micro-fractures on oil migration. The content of the clay minerals was larger than 20% in the formation of QT and WEH. The water absorption of clay minerals may change the pore structure. A moderate content of clay minerals (20.5–30.2%) may induce micro-fracture propagation to relieve the reservoir damage and the high content of clay minerals (36.8%) may destroy matrix pores to aggravate the reservoir damage. The QT and WEH formations were used to study for the pore structure change due to clay minerals and their effects on oil migration. In addition, the samples are not organic-rich and the TOC measurement are not conducted.

Table 2. The results of the mineral composition analysis.

Label	Total Mineral Composition, wt.%					Relative Clay Abundance, wt.%				
	Quartz	Feldspar	Calcite	Dolomite	Clay	Smectite	Illite	I/S[a]	Chlorite	Kaolinite
UC7	39.6	35.8	9.3	0	15.3	0	57.2	28.6	6.8	7.4
LC7	45.2	29.4	12.8	0	12.6	5.7	23.6	42.8	8.3	19.6
QT	25.6	33.8	16.9	0	23.7	13.2	17.8	32.3	25.4	11.3
WEH	26.4	11.4	15.7	6.1	40.2	21.9	17.9	43.5	0	16.7

Note: [a] I/S represents the Illite/smectite mixed-layer.

The experimental fluids included deionized water, $MnCl_2$ solution and kerosene. The basic properties of density, viscosity and surface tension are shown in Table 3. The deionized water was prepared for chemical solutions of $MnCl_2$. The NMR signal of hydrogen atom decreased significantly with the $MnCl_2$ solution concentration, and the high concentration of $MnCl_2$ solution did not have the NMR signal [11]. During the imbibition experiments, an $MnCl_2$ solution of 15 wt.% was used as the wetting fluid to displace the non-wetting fluid. The non-wetting fluid was kerosene instead of crude oil.

Table 3. The basic properties of the experimental fluid (25 °C).

Fluid	Density, g/cm³	Viscosity, cP	Surface Tension, mN/m
15 wt% $MnCl_2$	1.2	1.1	74.2
Kerosene	0.81	1.32	29

2.2. Experimental Apparatus

The experimental device for sample mass determination was a Mettler Toledo balance (ME204E, Mettler Toledo, Shanghai, China) with an accuracy of 0.0001 g, as shown in Figure 1a. The volume of the imbibition water and the expelled oil was estimated by sample mass change and the density difference between oil and water. The NMR device (MinNMR, Niumag Analytical Instrument Corporation, Suzhou, China) is shown in Figure 1b. NMR is a non-destructive test method, and the T_2 spectra can well reflect the pore structure and fluid distribution characteristics. The higher the T_2 value, the larger the pore size of the fluid concentration. The T_2 spectra amplitude is positively related to fluid volume in the certain aperture pores. By measuring the changes of T_2 spectra amplitude during the imbibition process, it is possible to understand the oil and water migration characteristics due to capillary pressure imbibition.

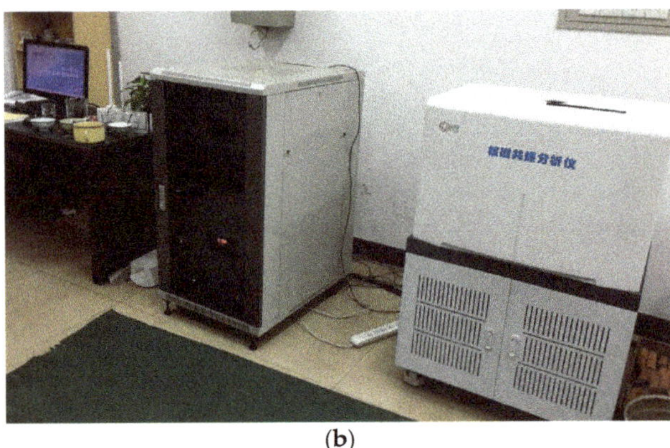

(**a**)   (**b**)

Figure 1. The apparatus for water imbibition experiment: (**a**) analytical balance, (**b**) NMR device (MiniMR-VTP).

The setting parameters during the NMR test had a large influence on the test results. For the different reservoir rocks, the test coefficients needed to be determined. In general, the NMR test had four test coefficients: waiting time (RD), echo time interval (T_E), signal superposition times (SCANS), and echo numbers (NECH). If the RD is too short, the large aperture signals would be lost. On the other hand, if the RD is too long, the measurement time significantly increases. In conventional sandstone, RD > 3000 ms is appropriate. As for tight sandstones, RD > 8000 ms is suitable. Similarly, the large values of NECH and SCANS can improve the test accuracy, but they also can increase the test time. In this study, the values of NECH and SCANS were set to 2048 and 64 respectively. In addition, the T_E was set to 0.3 ms, which is the minimum value of low-field NMR equipment, which contributes to the recognition of micro-pores. The intensity of the magnetic field was about 0.3 ± 0.05 T.

2.3. Experimental Procedure

During the imbibition experiments, the NMR T_2 spectra as a function of time was measured and oil recovery of different-scale pores was comparatively analyzed. The experimental process is presented as follows:

(1)   Before the experiments, the tight reservoirs samples should be cleaned to remove the residual oil. The cleaning solvent are mixture of toluene and ethyl ether, and the process lasts about one month.

(2)   The samples were placed in a saturating device and evacuated for 2 to 3 h to remove air. Then, the kerosene was injected under a 20 MPa pressure. The saturation process lasted for 72 h to fill the samples fully with kerosene.

(3) The samples were taken out and the mass and size of the samples were measured. They were immersed in the MnCl$_2$ solution. After a period of time, the masses of the samples were measured and the T$_2$ spectra was tested using NMR apparatus.

(4) Step (3) was repeated, and the T$_2$ spectra variation was drawn over soaking time.

The schematic diagram of the experimental procedure is shown in Figure 2.

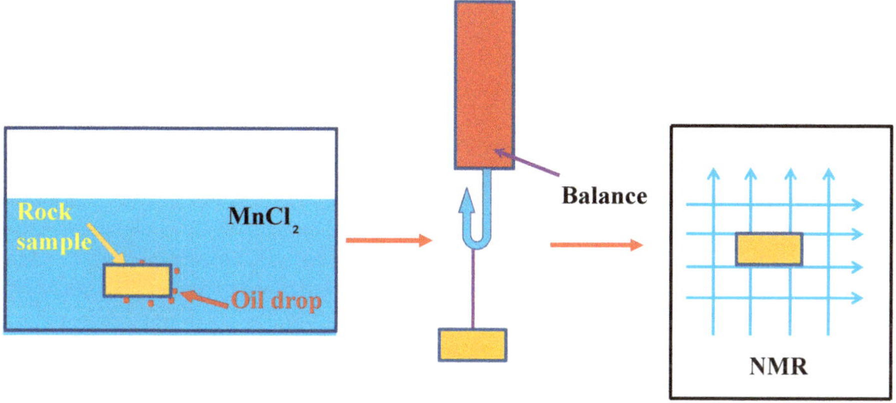

Figure 2. The schematic diagram of the experimental procedure.

3. Experimental Results and Discussions

3.1. Effects of Pore Size Distribution on Oil Migration

The water entered into matrix pores due to the capillary pressure and the oil droplet was gradually expelled to adhere to the surface of the sample. As the volume of the oil droplets was relatively small, the droplets tended to be more evenly distributed on the sample surface, as shown in Figure 3. As for the tight reservoirs, the capillary imbibition can spontaneously displace the oil in the pores, which contributes to the tight oil production. Figure 4 shows the T$_2$ spectra changes of UC7-1 and LC7-1 over time. As Figure 4 shows, the UC7-1 and LC7-1 had the same T$_2$ spectra characteristics. In the initial stage of imbibition, the water preferentially enters into the small pores to displace the oil, causing a rapid decline of oil saturation.

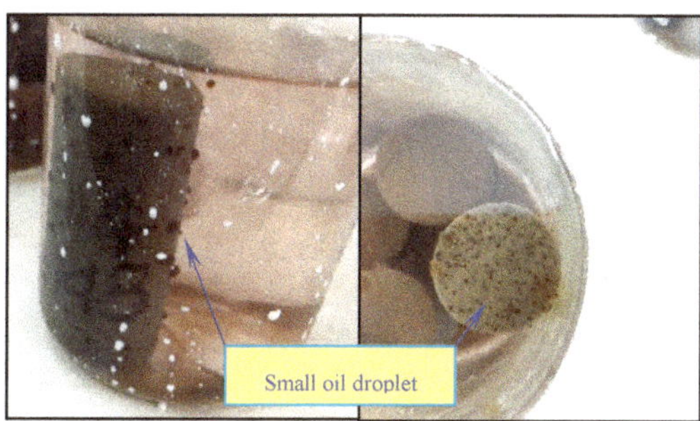

Figure 3. The small oil droplet displaced by water in UC7-1 samples.

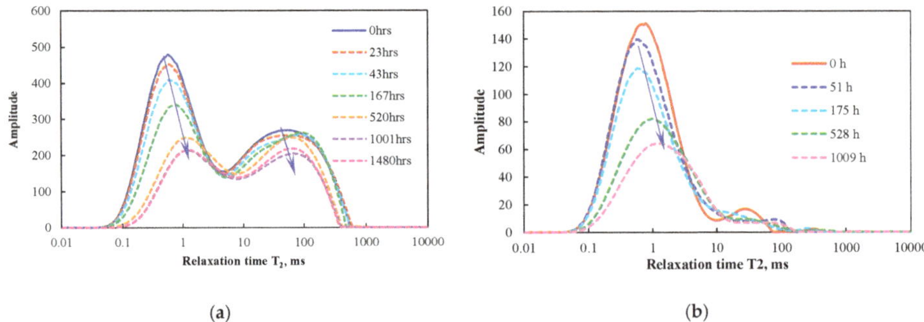

Figure 4. The T_2 spectra vs soaking time in UC7-1 and LC7-1 samples: (**a**) UC7-1 and (**b**) LC7-1.

When the soaking time exceeded 520 h, the T_2 spectra area did not change substantially, indicating that the residual oil in the pores could not be expelled only by capillary pressure (Figure 4). The oil saturation was no longer reduced and the imbibition process ended to some extent. In addition, the samples UC7-1 and LC7-1 have different pore size distribution characteristics. The sample of UC7-1 has a dual-peak distribution with a range of 0.06 to 5 ms for the left peak and 5 to 600 ms for the right peak. The sample LC7-1 has a single peak distribution with a range of 0.06–10 ms. Thus, the permeability (0.0062 mD) of LC7-1 is much smaller than that of UC7-1 (0.24 mD). At a soaking time of 520 h, the imbibition oil recovery of UC7-1 was about 34.2%, and that of LC7-1 was about 35.7%. The sample LC7-1 had a slightly larger imbibition oil recovery than UC7-1.

According to the T_2 relaxation time of Meng et al. (2016) [19], the tight rock pores were divided into micro-pores (<1 ms), small pores (1–10 ms), largish pores (10–100 ms) and large pores (>100 ms). The largish pores and large pores included micro-fractures and matrix pores. Figure 5 presents oil recovery in different diameter pores. The oil recovery of the micro-pores was larger than 70%, that of small pores and largish pores was much lower than 30% and that of largish pores was about 40%. Unexpectedly, the smaller pores did not have a much larger imbibition oil recovery.

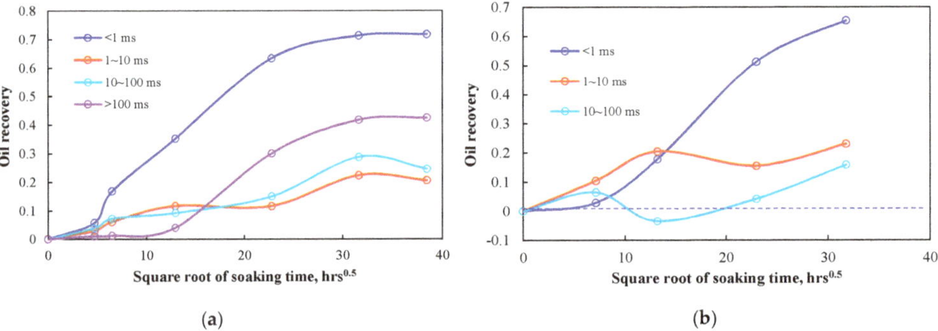

Figure 5. Oil recovery in different diameter pores: (**a**) UC7-1 and (**b**) LC7-1.

In Figure 5a, the oil in micro-pores began to decrease at 0 h, but the oil in large pores began to decrease at 100 h. The water preferentially enters the micro-pores under the capillary pressure and the oil drop displaced by water enters the large pores. Figure 6 presents the schematic diagram of oil migration among the different pores. The smaller pores correspond to a larger capillary pressure and a stronger imbibition capacity. Therefore, the oil of micro-pores tended to migrate into small and largish pores and that of small and largish pores tends to migrate into large pores. The small and largish pores act as bridges to connect the micro-pores and large pores.

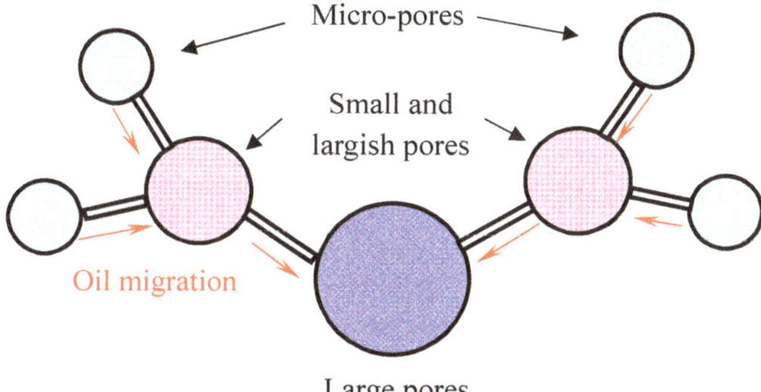

Figure 6. The schematic diagram of oil migration among the different pores.

3.2. Effects of Natural Fractures on Oil Migration

The UC7-2 samples contained many micro-fractures that were visible to the naked eye. They were in the direction of bedding planes. Microscopic observation shows that the crack width was about 100–350 μm (Figure 7b). Figure 8 shows the NMR imaging at a soaking time of 0 h and 59 h. The water imbibition experiments on UC7-2 can help understand the effects of natural fractures on oil migration.

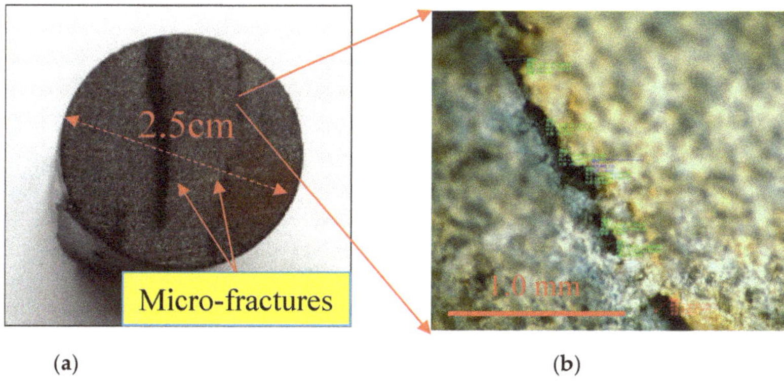

Figure 7. The micro-fractures on the sample surface: (**a**) naked eye observation and (**b**) microscopic observation.

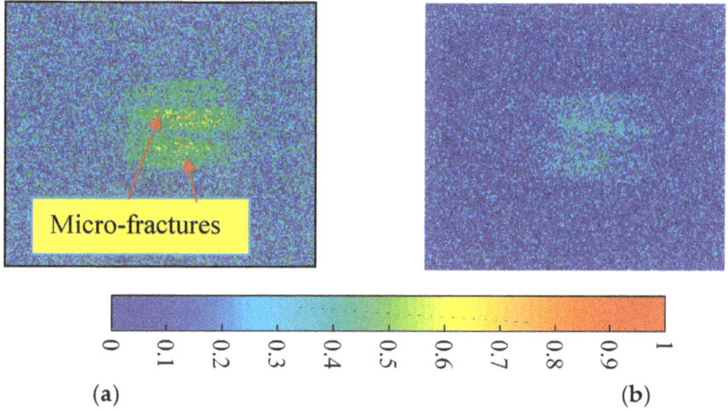

Figure 8. NMR imaging: (**a**) 0 h and (**b**) 59 h. The color bar can reflect oil saturation.

Figure 9 shows the relationship between T_2 spectra and soaking time. The distribution range of fractures corresponds to >10 ms, which involves largish pores and large pores. The sample LC7-2 contained a small amount of micro-fractures that were speculated by T_2 spectra. The samples of UC7-2 and LC7-2 were embedded by micro-fractures and characterized by different oil migration features. The amplitude decline velocity of largish pores and large pores was larger than that of micro-pores. Compared with the micro-pores, the micro-fractures were the dominant channels for oil migration. According to the NMR imaging shown in Figure 8, a large amount of oil exited in the micro-fractures. As the soaking time increased, the water preferentially entered the natural micro-fractures, expelling the oil in the micro-fractures. Subsequently, the oil in the small pores was slowly expelled. When the soaking time was 500 h, the imbibition oil recovery of sample UC7-2 and LC7-2 were about 36.6% and 34.1% respectively. Sample UC7-2 had more micro-fractures that were beneficial to imbibition oil recovery to some extent.

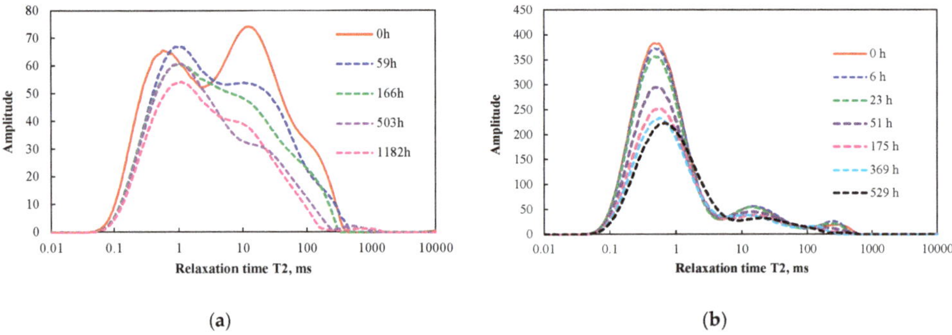

Figure 9. The T_2 spectra vs soaking time in UC7-2 and LC7-2 samples: (**a**) UC7-2 and (**b**) LC7-2.

Figure 10 presents the oil recovery in different diameter pores of UC7-2 and LC7-2. In the UC7-2, the oil recovery of micro-pores, small pores, largish pores and large pores was 32.1, 23.5, 55.2 and 82.7% respectively. In the LC7-2, the oil recovery of micro-pores, small pores, largish pores and large pores was 45.6, 1.5, 30.5 and 66.8% respectively. The oil recovery of large pores was much larger than that of micro-pores. Water preferentially enters the micro-fractures to expel the oil. The micro-pores oil recovery of UC7-2 and LC7-2 was much smaller than that of UC7-1 and LC7-1. Therefore, the existence of micro-fractures may hinder oil migration from micro-pores into large pores. This results in higher residual oil saturation in the micro-pores.

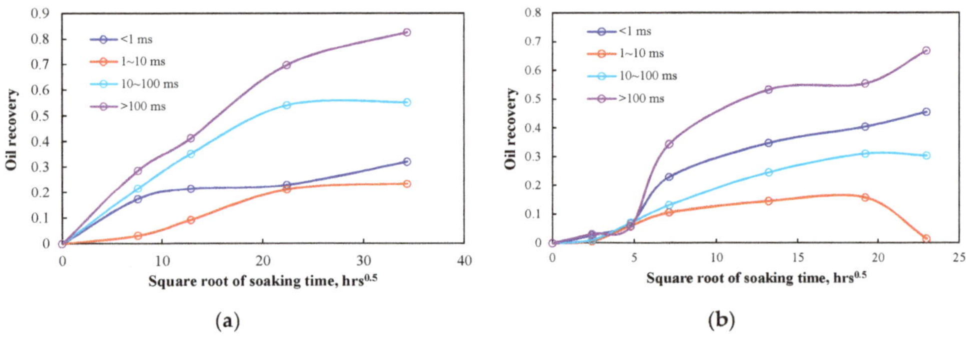

Figure 10. Oil recovery in different diameter pores: (**a**) UC7-2 and (**b**) LC7-2.

3.3. Effects of Fracture Propagation on Oil Migration

The Figure 11 shows the observations of QT-1 before and after experiments. It can be seen that a large amount of oil droplets was precipitated, and the volume of the oil droplets significantly exceeded that of the sample UC7-1. Figure 12 presents the NMR images at different soaking times. Evidently, there was no fractures in the sample at the beginning of the experiment. When the soaking time exceeded 43 h, a large number of micro-fractures were generated on the surface of sample. The micro-fractures were not natural fractures and they were induced during the experiments. It should be noted that nuclear magnetic signals can only be detected when oil is contained in the micro-fractures. This suggests that a large amount of oil migrates into the micro-fractures during the process of micro-fracture propagation. The new fractures were in the direction of the formation bedding planes, which may result from the opening of bedding planes. The QT reservoir had a clay mineral content of about 23.7%, which formed a strong expansion stress to induce the fracture propagation after encountering water [11].

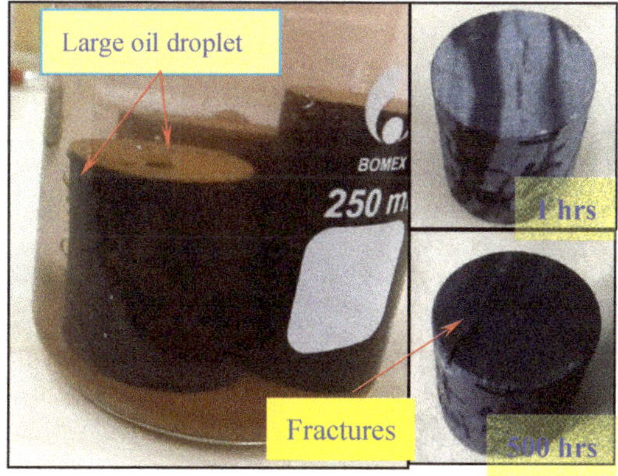

Figure 11. The observations of QT-1 before and after experiments.

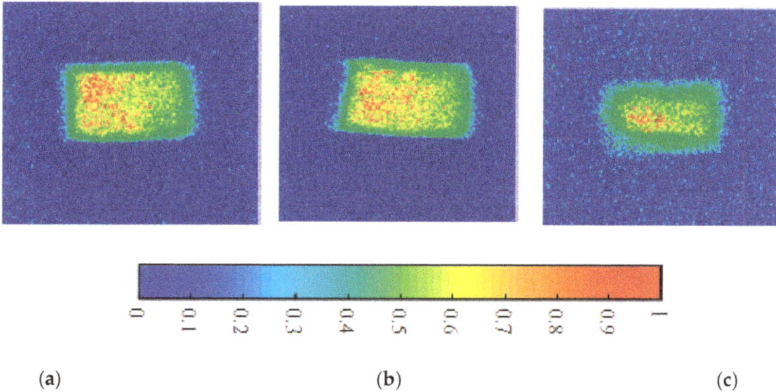

Figure 12. The magnetic resonance imaging of QT-1 sample: (**a**) 0 h, (**b**) 6 h and (**c**) 23 h.

Figure 13 presents the curves of T_2 spectra in the QT-1 and QT-2 samples. The T_2 spectra of QT formation had single peak features and the T_2 value ranged from 0.1 ms to 800 ms. It contained micro-pores, small pores, largish pores and large pores. As the soaking time increased, the total area of T_2 spectra shows a downward trend, suggesting that the water was imbibed into matrix pores to expel the oil. The amplitude of pores corresponding to 5–50 ms decreased at the beginning, rose in the medium term, and fell in the later period. At the beginning, the oil in the pores was gradually displaced, resulting in a T_2 amplitude drop. In the medium term, the clay mineral expansion induced a lot of micro-fractures and the oil gradually migrated into new micro-fractures, leading to a T_2 amplitude rise. In the later period, the oil in the micro-fractures was expelled and the velocity of oil entry was larger than that of oil departure, causing the T_2 amplitude to drop. The imbibition oil recovery of QT-1 and QT-2 was 44.3 and 32%, respectively.

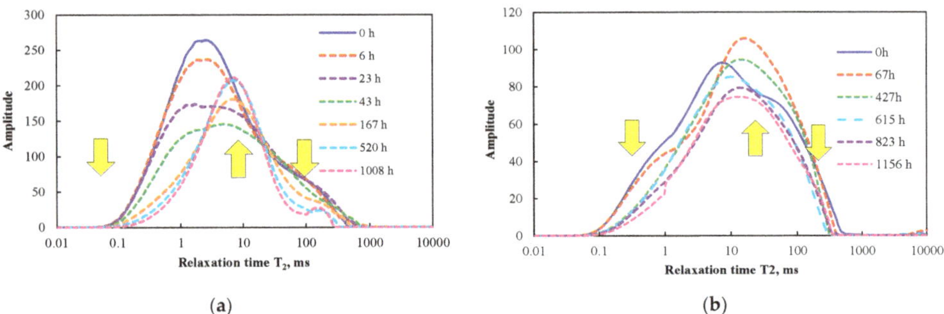

(a)   (b)

Figure 13. The T_2 spectra vs soaking time in QT-1 and QT-2 samples: (**a**) QT-1 and (**b**) QT-2.

Figure 14 presents the oil recovery in different diameter pores of the QT-1 and QT-2 samples. In the QT-1, the oil recovery of micro-pores, small pores, largish pores and large pores was 81.5, 35.4, 21.9 and 63.3% respectively. In the QT-2, the oil recovery of micro-pores, small pores, largish pores and large pores was 57, 27.3, 10.2 and 37.0% respectively. The oil recovery order of different diameters pores was micro-pores > large pores > small pores > largish pores. In addition, the oil recovery of largish pores and large pores descended below the zero at first and then increased above zero. The negative value of oil recovery suggests that the fractures propagated to form new space during water imbibition and oil gradually entered the new fractures to decrease the oil recovery factor. The new micro-fractures correspond to largish pores and large pores. Therefore, part of the oil in micro-pores and small pores may also have migrated into the largish pores and large pores.

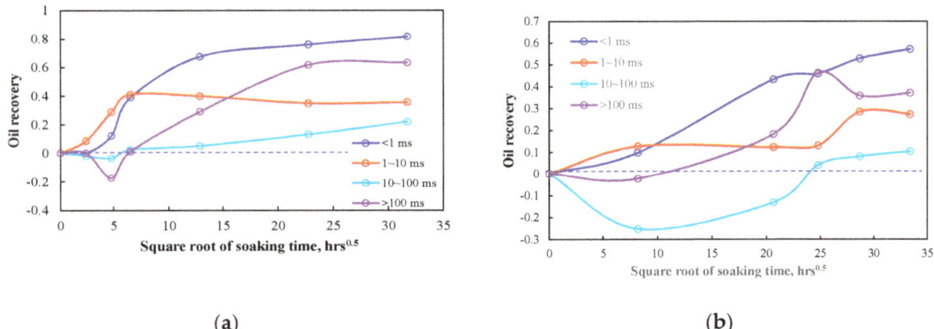

(a)   (b)

Figure 14. Oil recovery in different diameter pores: (**a**) QT-1 and (**b**) QT-2.

3.4. Effects of Clay Mineral on Oil Migration

Figure 15 presents pictures of the samples before and after the imbibition experiments. At 1 h, the samples began to expand and the consolidated strength of samples decreased. At 60 h, the samples had broken into grains, resulting in the interruption of the imbibition experiments. This can be explained by an abundance of clay minerals in WEH formation.

Figure 15. The surface change of the WEH-1 and WEH-2 samples with the soaking time.

Figure 16 presents the change of T_2 spectra during water imbibition. The T_2 spectra of WEH-1 had single peak features and that of WEH-2 had dual-peak features. The T_2 spectra of WEH-1 and WEH-2 decreased gradually over the soaking time. When the soaking time exceeded 43 h, the samples turned into grains and imbibition experiments ends. The imbibition oil recovery of WEH-1 and WEH-2 was 37.5 and 47% respectively. The WEH-2 had larger oil recovery because of more micro-pores and small pores. To an extent, the imbibition oil recovery of WEH-1 and WEH-2 may be close to 100%. The pore structure of the samples completely disintegrated and all the oil trapped by matrix pores was released.

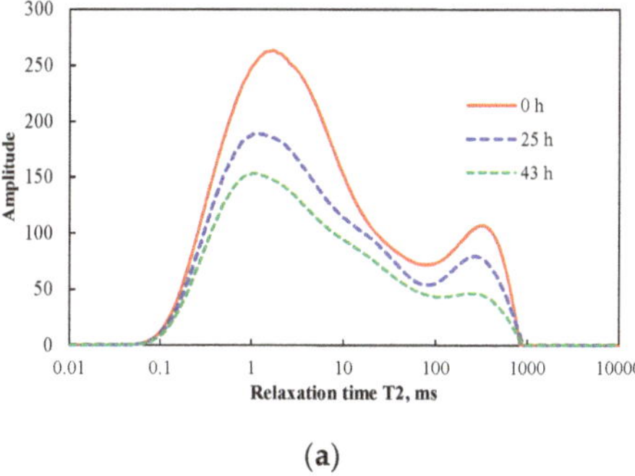

(a)

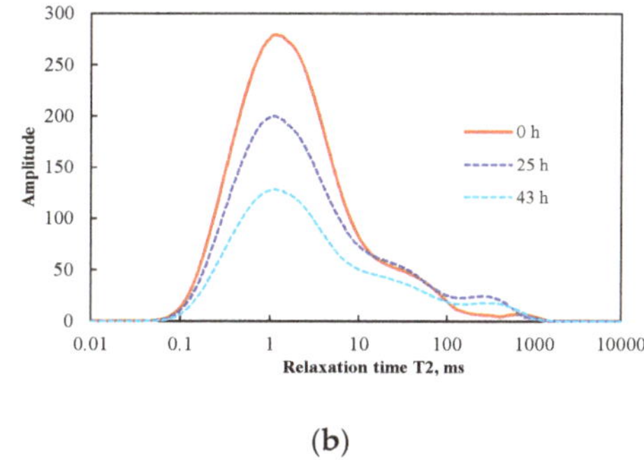

(b)

Figure 16. The T_2 spectra vs soaking time in WEH-1 and WEH-2 samples: (**a**) WEH-1 and (**b**) WEH-2.

Figure 17 shows the oil recovery in different diameter pores of the WEH-1 and WEH-2 samples. In the WEH-1, the oil recovery of micro-pores, small pores, largish pores and large pores was 33.3, 44.4, 31.4 and 54.0% respectively. The two peak features of WEH-1 suggest that it contained a large amount of micro-fractures. The WEH formation are tight conglomerate formation containing a large amount of gravels. These micro-fractures are well developed along the edge of gravels, which are gravel-edge fractures. The water was imbibed into gravel-edge fractures to displace the oil, resulting in larger oil recovery. In the WEH-2, the oil recovery of micro-pores, small pores, largish pores and large pores was 52.7, 53.1, 26.1, and −0.89% respectively. It did not contain the gravel-edge fractures, and the pores were the main channels for water imbibition. Therefore, the smaller pores had a larger oil recovery. The negative value of oil recovery in >100 ms pores suggests that the micro-fracture propagation and oil migration into new micro-fractures decreased the oil recovery during water imbibition.

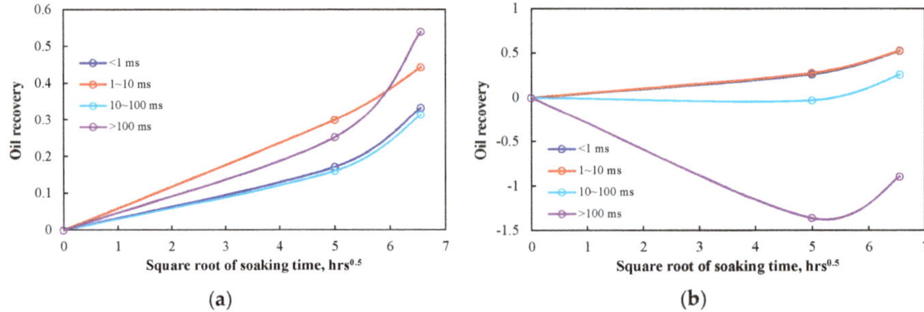

Figure 17. Oil recovery in different diameter pores: (**a**) WEH-1 and (**b**) WEH-2.

3.5. Scaling the Imbibition Results of Different Reservoirs

The ratio of final spectra area to original spectra area is regarded as imbibition oil recovery. In order to study the influencing factors of imbibition oil recovery, the authors used the dimensionless time t_D to scale the experimental results. According to Ma et al. (1997) [20], the dimensionless time t_D is given by

$$t_D = t \sqrt{\frac{k}{\phi}} \frac{\sigma}{\sqrt{\mu_w \mu_o} L_S^2} \tag{1}$$

where k is the rock permeability, ϕ is the fractional porosity of rock, t is the soaking time, σ is the interfacial tension, μ_o is the oil viscosity, μ_w is the water viscosity, and L_s is the characteristic length defined by Ma et al. (1997) [20] that involves the effects of sample shapes and boundary conditions.

During the imbibition experiments, the experimental fluids were $MnCl_2$ solution and kerosene. Considering that all the experiments used the same fluids, the surface tension could be set to 45 mN/m, which is the surface tension of kerosene–water. This may not have had significant effects on the analytical results. Figure 18 shows the results of scaling the imbibtion results using the Ma's model. The Ma's dimensionless time t_D did not function satisfactorily for all the different samples, indicating that it was not suitable for tight oil reservoirs. The mineral composition and pore structure were very different, and they were not scaled by this method. According to Akin et al. (2000) [21], the effects of these influncing factors on the imbibition rate can be studied based on the plots of imbibition oil recovery vs dimensionless time $t_D^{0.5}$ (Figure 18b). In Figure 18b, the different lines represent the different experimens of tight reservoirs samples. Despite of the marked differences in physical property, pore structure and mineral composition, similar trends are found in these curves. The relationship between oil recovery and dimensionless $t_D^{0.5}$ is close to a straight line. The slope of curves represents the imbibition rate that was affected by mineral composition and pore structure. According to Yang et al. (2016) [22], the slope of curves could be defined as a dimensionless imbibition rate, which suggests a stong imbibition potential for water. The imbibition process of tight reservoirs was very slow and need a long time for water front to arrive at the end or center of sample. Therefore, it was difficult to obtain the final oil recovery with the imbibition experiments. However, it can be speculated that the final imbibiton oil recovery was about 35–45% in tight oil reservoirs.

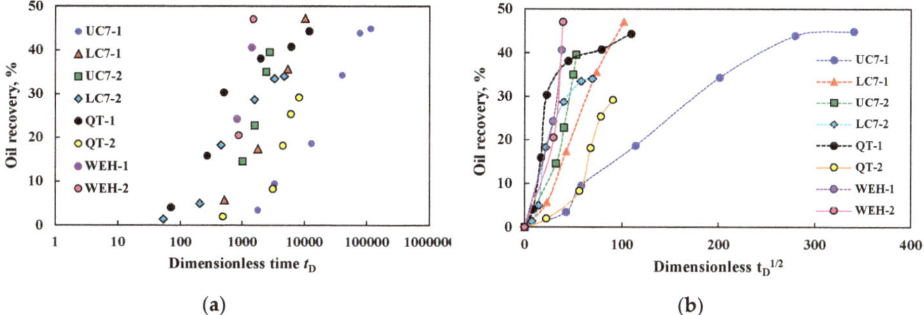

Figure 18. The oil recovery vs dimensionless time t_D and $t_D^{0.5}$: (**a**) oil recovery vs t_D and (**b**) oil recovery vs $t_D^{0.5}$.

The dimensionless imbibition rate of different reservoirs samples is presented in Figure 19. The pore (P) type rock represents the tight reservoirs that only developed matrix pores for the oil migration channel. The natural fracture (NF) type corresponds to the tight reservoirs that developed both matrix pores and micro-fractures. The fracture propagation (FP) type means that the clay expansion could induce fracture propagation during water imbibition. The effects of in situ stress on fracture propagation were not taken into account. The clay (C) type refers to the tight reservoirs that are characterized by a high content of clay minerals (>40%). These different types of tight reservoirs were comparatively studied (Figure 19).

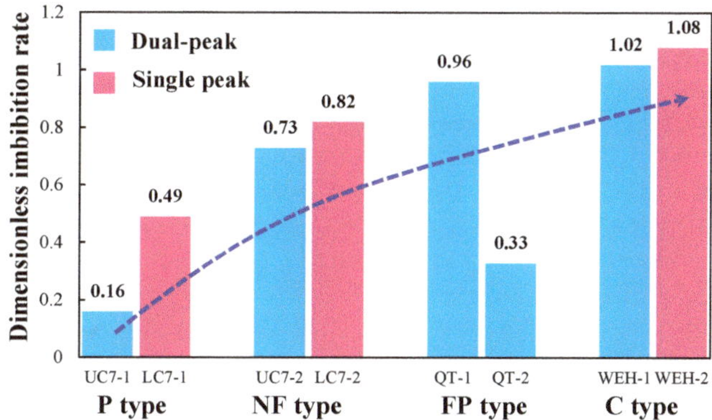

Figure 19. The dimensionless imbibition rate of different reservoirs samples.

For the same type reservoirs, the pore size distribution of single peak tended to have a larger capillary pressure, resulting in a stronger imbibition potential than that of dual-peak. As for the tight reservoirs with the pore size distribution of a single peak, the dimensionless imbibition rates of the P type, NF type, FP type and C type were 0.16, 0.73, 0.33–0.96, and 1.02, respectively. Evidently, the dimensionless imbibition rates of the NF type, FP type and C type rival surpassed those of P type due to clay minerals and micro-fractures. The imbibition rate of the NF type was 4.5 times that of the tight rock P type. Compared with the matrix pores, the micro-fractures had smaller flow resistance and were more conducive to water and oil flow. The imbibition rates of the FP type and C type were six times those of the P type. The more the clay mineral, the larger the imbibition rate. This can be explained by driving force. The imbibition driving force was only capillary pressure in low clay content sample, but the imbibition driving forces were both capillary pressure and osmotic pressure in high clay content sample. However, when the clay mineral content exceeded 23%, the increase in clay mineral content had little effect on the imbibition rate [11]. Excessive clay minerals can also give reservoirs a strong water sensitivity and seriously disperse in water (Figure 15). Under the reservoir condition, solid particle may plug the pores and decrease the in-place permeability, which is not conducive to the production of tight oil. In this studies, the spontaneous imbibition experiments were mainly carried out at atmospheric pressure, which does not reflect the actual situation. In the future, it will be necessary to perform the imbibition experiments under the reservoir conditions.

4. Conclusions

In this study, a series of imbibition experiments were conducted on tight reservoirs samples and the NMR T_2 spectra was used to monitor oil migration dynamics. The oil recovery in different diameters pores was comparatively analyzed. The effects of pore size distribution, micro-fractures, and clay minerals were studied by scaling the imbibition results. The conclusions are as follows:

(1) Concerning the tight reservoirs without clay minerals and micro-fractures, the oil migration due to water imbibition was mainly determined by the pore size. The smaller pores corresponded to a larger capillary pressure and a stronger imbibition capacity. Therefore, the oil of micro-pores tended to migrate into small and largish pores and that of the small and largish pores tended to migrate into large pores. The small and largish pores acted as bridges to connect the micro-pores and large pores. Compared with small and largish pores and large pores, the micro-pores had the largest oil recovery.

(2) As the soaking time increased, the water preferentially entered the natural micro-fractures, expelling the oil in the micro-fractures. Subsequently, the oil in the small pores was slowly expelled. Compared with the matrix pores, micro-fractures have smaller flow resistance and are more conducive to water and oil flow.

(3) The clay minerals of middle content may have induced the fracture propagation. A large amount of oil migrated into the new micro-fractures during the process of micro-fracture propagation. In contrary to the inhibitory effect of natural micro-fractures, the new micro-fractures could have contributed to the oil migration from micro-pores into large pores. The clay minerals of high content can completely decentralize pore structure and significantly increase the imbibition oil recovery at atmospheric pressure. Under the reservoir condition, the effects of excessive clay minerals need to be studied in the future. More clay minerals may result in water sensitivity damage and do not contribute to oil production.

Author Contributions: Methodology, L.Y.; Data curation, S.W.; Writing—review & editing, Z.T.; Visualization, R.L.; Writing—review & editing and Investigation, J.Y.

Funding: This research was funded by the National Natural Science Foundation of China, grant number 11702296 and National Key Research and Development Program, grant number 2016YFC0600906.

Acknowledgments: The financial support of our shale research program is from the Foundation of the National Natural Science Foundation of China (No. 11702296), the National Key Research and Development Program (2016YFC0600906), the Fundamental Research Funds for the Central Universities, the Major National Science, and Technology Projects of China (No. 2017ZX05049003-005).

Conflicts of Interest: The authors declare no conflict of interest. The funders had no role in the design of the study; in the collection, analyses, or interpretation of data; in the writing of the manuscript, or in the decision to publish the results.

References

1. Xu, G.; Shi, Y.; Jiang, Y.; Jia, C.; Gao, Y.; Han, X.; Zeng, X. Characteristics and Influencing Factors for Forced Imbibition in Tight Sandstone Based on Low-Field Nuclear Magnetic Resonance Measurements. *Energy Fuels* **2018**, *32*, 8230–8240. [CrossRef]
2. Liu, P.; Zhang, X.; Wu, Y.; Li, X. Enhanced oil recovery by air-foam flooding system in tight oil reservoirs: Study on the profile-controlling mechanisms. *J. Pet. Sci. Eng.* **2017**, *150*, 208–216. [CrossRef]

3. Gu, X.Y.; Pu, C.; Huang, H.; Huang, F.; Li, Y.; Liu, Y.; Liu, H. Micro-influencing mechanism of permeability on spontaneous imbibition recovery for tight sandstone reservoirs. *Pet. Explor. Dev.* **2017**, *44*, 1003–1009. [CrossRef]

4. Todd, H.B.; Evans, J.G. Improved Oil Recovery IOR Pilot Projects in the Bakken Formation. In Proceedings of the SPE Low Perm Symposium, Denver, CO, USA, 5–6 May 2016.

5. Lan, Q.; Ghanbari, E.; Dehghanpour, H. Water Loss Versus Soaking Time: Spontaneous Imbibition in Tight Rocks. In Proceedings of the SPE/EAGE European Unconventional Resources Conference and Exhibition, Vienna, Austria, 25–27 February 2014.

6. Jiang, Y.; Shi, Y.; Xu, G.; Jia, C.; Meng, Z.; Yang, X.; Zhu, H.; Ding, B. Experimental Study on Spontaneous Imbibition under Confining Pressure in Tight Sandstone Cores Based on Low-Field Nuclear Magnetic Resonance Measurements. *Energy Fuels* **2018**, *4*, 56–78. [CrossRef]

7. Xu, J.; Sun, B.; Chen, B. A hybrid embedded discrete fracture model for simulating tight porous media with complex fracture systems. *J. Pet. Sci. Eng.* **2019**, *174*, 131–143. [CrossRef]

8. Shen, W.; Li, X.; Lu, X.; Guo, W.; Zhou, S.; Wan, Y. Experimental study and isotherm models of water vapor adsorption in shale rocks. *J. Nat. Gas Sci. Eng.* **2018**, *52*, 484–491. [CrossRef]

9. Ezulike, O.D.; Ghanbari, E.; Siddiqui, S.; Dehghanpour, H.; Sidduqui, S. Pseudo-steady state analysis in fractured tight oil reservoirs. *J. Pet. Sci. Eng.* **2015**, *129*, 40–47. [CrossRef]

10. Ge, H.; Yang, L.; Shen, Y.; Ren, K.; Meng, F.; Ji, W.; Wu, S. Experimental investigation of shale imbibition capacity and the factors influencing loss of hydraulic fracturing fluids. *Pet. Sci.* **2015**, *12*, 636–650. [CrossRef]

11. Yang, L.; Zhang, X.; Zhou, T.; Lu, X.; Zhang, C.; Zhang, K. The effects of ion diffusion on imbibition oil recovery in salt-rich shale oil reservoirs. *J. Geophys. Eng.* **2019**, *16*, 525–540. [CrossRef]

12. Dehghanpour, H.; Lan, Q.; Saeed, Y.; Fei, H.; Qi, Z. Spontaneous imbibition of brine and oil in gas shales: Effect of water adsorption and resulting micro fractures. *Energy Fuels* **2013**, *27*, 3039–3049. [CrossRef]

13. Ghaderi, S.M.; Clarkson, C.R.; Ghanizadeh, A.; Barry, K.; Fiorentino, R. Improved Oil Recovery in Tight Oil Formations: Results of Water Injection Operations and Gas Injection Sensitivities in the Bakken Formation of Southeast Saskatchewan. In Proceedings of the SPE Unconventional Resources Conference, Calgary, AB, Canada, 15–16 February 2017.

14. Yuan, J.; Jiang, R.; Zhang, W. The workflow to analyze hydraulic fracture effect on hydraulic fractured horizontal well production in composite formation system. *Adv. Geo-Energy Res.* **2018**, *2*, 319–342. [CrossRef]

15. Habibi, A.; Xu, M.; Dehghanpour, H.; Bryan, D.; Uswak, G. Understanding Rock-Fluid Interactions in the Montney Tight Oil Play. In Proceedings of the SPE/CSUR Unconventional Resources Conference, Calgary, AB, Canada, 20–22 October 2015.

16. Hu, Q.H.; Ewing, P.R.; Dultz, S. Low pore connectivity in natural rock. *J. Contam. Hydrol.* **2012**, *133*, 76–83. [CrossRef] [PubMed]

17. Tian, X.; Cheng, L.; Cao, R.; Wang, Y.; Zhao, W.; Yan, Y.; Liu, H.; Mao, W.; Zhang, M.; Guo, Q. A new approach to calculate permeability stress sensitivity in tight sandstone oil reservoirs considering micro-pore-throat structure. *J. Pet. Sci. Eng.* **2015**, *133*, 576–588. [CrossRef]

18. Brace, W.F.; Walsh, J.B.; Frangos, W.T. Permeability of granite under high pressure. *J. Geophys. Res. Space Phys.* **1968**, *73*, 2225–2236. [CrossRef]

19. Meng, M.; Ge, H.; Ji, W.; Wang, X. Research on the auto-removal mechanism of shale aqueous phase trapping using low field nuclear magnetic resonance technique. *J. Pet. Sci. Eng.* **2016**, *137*, 63–73. [CrossRef]

20. Shouxiang, M.; Morrow, N.R.; Zhang, X. Generalized scaling of spontaneous imbibition data for strongly water-wet systems. *J. Pet. Sci. Eng.* **1997**, *18*, 165–178. [CrossRef]

21. Akin, S.; Schembre, J.; Bhat, S.; Kovscek, A. Spontaneous imbibition characteristics of diatomite. *J. Pet. Sci. Eng.* **2000**, *25*, 149–165. [CrossRef]

22. Yang, L.; Ge, H.; Shi, X.; Cheng, Y.; Zhang, K.; Chen, H.; Shen, Y.; Zhang, J.; Qu, X. The effect of microstructure and rock mineralogy on water imbibition characteristics in tight reservoirs. *J. Nat. Gas Sci. Eng.* **2016**, *34*, 1461–1471. [CrossRef]

Article

Stress-Dependent Permeability of Fractures in Tight Reservoirs

Nai Cao [1], Gang Lei [2,*], Pingchuan Dong [1,*], Hong Li [1], Zisen Wu [1,3] and Yudan Li [1]

[1] College of Petroleum Engineering, China University of Petroleum Beijing, Beijing 102249, China; caonai99@163.com (N.C.); lihong660@163.com (H.L.); 1533141753@ku.edu (Y.L.)
[2] College Petroleum of Engineering & Geoscience, King Fahd University of Petroleum and Minerals, Dhahran 31261, Saudi Arabia; gang.lei@kfupm.edu.sa
[3] Beijing Technical Market Management Office, Beijing 100032, China; wuzisen@163.com
* Correspondence: gang.lei@kfupm.edu.sa (G.L.); dpcfem@163.com (P.D.); Tel.: +86-186-0120-0950 (P.D.)

Received: 4 December 2018; Accepted: 29 December 2018; Published: 29 December 2018

Abstract: Permeability is one of the key factors involved in the optimization of oil and gas production in fractured porous media. Understanding the loss in permeability influenced by the fracture system due to the increasing effective stress aids to improve recovery in tight reservoirs. Specifically, the impacts on permeability loss caused by different fracture parameters are not yet clearly understood. The principal aim of this paper is to develop a reasonable and meaningful quantitative model that manifests the controls on the permeability of fracture systems with different extents of fracture penetration. The stress-dependent permeability of a fracture system was studied through physical tests and numerical simulation with the finite element method (FEM). In addition, to extend capability beyond the existing model, a theoretical stress-dependent permeability model is proposed with fracture penetration extent as an influencing factor. The results presented include (1) a friendly agreement between the predicted permeability reduction under different stress conditions and the practical experimental data; (2) rock permeability of cores with fractures first reduces dramatically due to the closure of the fractures, then the permeability decreases gradually with the increase in effective stress; and (3) fracture penetration extent is one of the main factors in permeability stress sensitivity. The sensitivity is more influenced by fracture systems with a larger fracture penetration extent, whereas matrix compaction is the leading influencing factor in permeability stress sensitivity for fracture systems with smaller fracture penetration extents.

Keywords: fractured tight reservoir; stress-dependent permeability; fracture penetration extent; theoretical model

---

1. Introduction

Fluid flow under stress through fractured media has drawn considerable attention in many engineering fields, including physics [1,2], hydraulics [3,4], chemistry [5], petroleum, and engineering [6–10]. The permeability of a reservoir decreases as effective stress increases during reservoir development [11]. This permeability reduction is more severe for fractured reservoirs, which poses difficulties for oil production in tight reservoirs due to seepage losses [12]. Since permeability is one of the key factors in reservoir production, understanding the mechanism of permeability reduction assists with reservoir dynamic analysis and production optimization.

Many studies have been conducted on the law of reservoir stress sensitivity, with many of them focused on experimentation. Buchsteiner et al. [13] found that the closure of fractures occurs due to rock pore structure deformation under increasing stress. The permeability damage rate under changing stress was defined in the explanation of the stress sensitivity of low permeability gas formation [14]. Stress sensitivity experiments have been conducted under a wide range of stresses [15], and variation patterns have been directly measured [16]. Fractured reservoirs, in the low permeability stage, were found to have greater stress sensitivity, which is weakened when the effective stress is higher beyond a certain criteria [17]. Besides the above experimental works, theoretical works requiring complicated study methods [18] and microimage analysis systems [19] have also been conducted. The physical structure model of the fracture surface [20] and the mechanical structure model of the surface contact of fractures [21] have been established. The change between fracture volume loss and effective stress was determined [22–24]. McKee et al. [25] derived a function for porosity considering effective stress, and substituted the equation for porosity into the Carman–Kozeny equation to calculate the permeability of fractures considering the fracture aperture and fracture height. However, not much attention has been paid to the impact caused by the extent of fracture penetration on stress-dependent permeability, and the fracture has rarely been separated as a distinct object. We still do not fully understand the characteristic behavior of the stress-dependent permeability of a fractured reservoir or the effects of fracture parameters [26].

Numerical methods are effective at simulating the flow behavior in fractured porous media [27–30]. A two-dimensional (2D) fracture model proposed by Perkins and Kern and developed by Nordren (the PKN model) [27], which assumes that fractures have an elliptical cross-section and constant height, has been used to simulate the settling velocity correlation of proppant in foam fracturing. A full three-dimensional (3D) displacement discontinuity method (DDM) has been used to investigate proppant transportation in growing fracture networks [28]. A hybrid Eulerian–Lagrangian model, assuming that foam is a single-phase non-Newtonian fluid, has been used to conduct the simulation of media with hydraulic fractures [29]. A hydraulic fracturing simulator that implicitly couples fluid flow with 3D discrete fracture networks (DFNs) has been established [30]. Due to the advantage of better precision and higher computational efficiency to simulate coupled flow deformation behavior in porous media, FEM simulation was adopted in this work to simulate the stress-dependent permeability change in tight media with fractures.

The major goal of this work is to take the fracture system as a distinct object and explore the impact on stress-dependent permeability caused by a fracture system with different fracture parameters. The specific objectives are to (1) analyze the stress-dependent permeability variation in the fracture system caused by the extent of the fracture penetration, (2) develop a reasonable model to quantify the stress-dependent permeability loss of the fracture system, (3) compare the predicted results with experimental results to support the model results, and (4) to quantitatively determine the influences of different fracture penetration extents on permeability in fractured porous media. A technology roadmap of this work is presented in Figure 1.

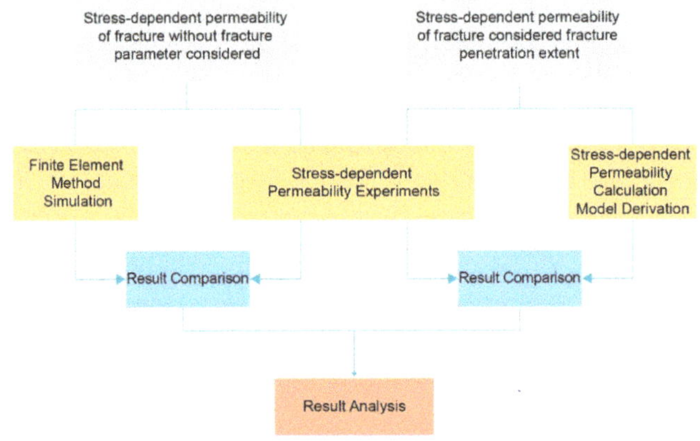

Figure 1. The technology roadmap of the work.

2. Experiment

2.1. Samples

The preparation of samples involved core drilling, eluting, drying, and fracture forming. Cores, 2.5 cm in diameter and 15 cm in length, were drilled along the horizontal direction of full-size core samples from a tight sandstone reservoir. The elution of drilled cores was conducted in an extractor with different solutions. They were first eluted using benzene, and then by a mixed solution of alcohol and toluene at a ratio of 1:3. After the elution, the cores were placed in a vacuum drying oven, vacuumed and dried to a constant weight for 24 h at a temperature range of 60–65 °C, with a relative humidity of 40% to 45%, and then moved to a dryer to cool down to room temperature.

Since the drilled cores potentially had tiny distributed fractures, confining pressure was applied to close those fractures in order to better control the variation. The fractures with different penetration extents in the cores were formed through different methods. For a fully penetrated core, the Brazilian split technique [31,32] was adopted. For the cores with different penetration extents, a specific control method was needed to set the fracture [33]. Alloy structure steel wires were used to form a concentrated stress in order to induce a radial tension crack. As shown in Figure 2, a pressure p (MPa) was exerted on the steel wires and a fracture formed along the direction of the exerted pressure. The extent of the fractures was controlled by monitoring the differential stress.

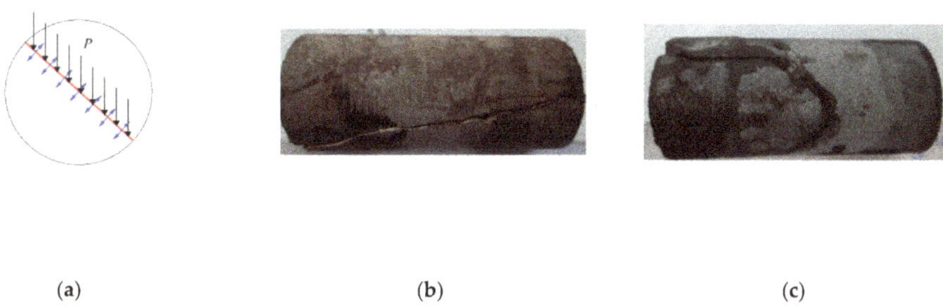

(a)   (b)   (c)

Figure 2. (a) A sketch of the fracture-forming method in cores with different penetration extents. The red line is the alloy structure steel wire, the pressure represented by the black arrows is exerted directly on the wire in the normal direction of the upper plate of the core, and the crack occurs along the direction of the blue arrows. (b) A fully penetrated core and (c) partly penetrated core.

In the stress sensitivity experiments on the matrix and fractured cores, the porosity and permeability were measured in the laboratory (Table 1). Three sets of tight cores with different permeability levels were selected, and each set contained 2 matrixes with the same permeability level. Fractures were formed using the Brazilian split technique [34,35] in one of the two cores in each set, with the fracture surface remaining uncontaminated. The parameters of the cores are summarized in Table 1.

Table 1. Physical parameters of the matrix and fractured cores.

No.	Core Type	Core Diameter (cm)	Core Length (cm)	Width of Fracture (cm)	Porosity (%)	Permeability (10^{-3} μm^2)
1	Matrix	2.46	7.24	0.00	14.7	0.52
	Fractured core	2.46	7.24	2.41	14.7	1.08
2	Matrix	2.53	6.37	0.00	15.2	0.72
	Fractured core	2.53	6.37	2.42	15.2	1.27
3	Matrix	2.48	8.35	0.00	15.4	0.90
	Fractured core	2.48	8.35	2.44	15.4	1.43

In the experiment, the permeability stress sensitivity experiments were conducted on tight cores with five different fracture penetration extents including a matrix without fractures and those with 25, 50, 75, and 100% penetration of the core. The parameters are shown in Table 2.

Table 2. Physical parameters of the cores with different penetration extents.

Core Type	Core Diameter (cm)	Core Length (cm)	Width of Fracture (cm)	Permeability (10^{-3} μm^2)	Permeability Increase Multiple
Matrix	24.62	31.30	0.00	0.052	1.00
Fracture penetration 25%	24.62	31.30	24.16	0.058	1.12
Fracture penetration 50%	24.62	31.30	24.16	0.065	1.25
Fracture penetration 75%	24.62	31.30	24.16	0.074	1.42
Fracture penetration 100%	24.62	31.30	24.16	0.086	1.65

2.2. Experimental Apparatus

The schematic of the apparatus used for the stress-dependent permeability experiment is shown in Figure 3. The driving pressure was provided by a high-pressure injection pump, equipped with a high-pressure container. The recording range of the flow rate of this pumping was between 0.01 μL/min and 50 mL/min, with an accuracy of 0.01 μL/min. Its pressure ranged from 0.068 MPa to 68 MPa, with an accuracy of 7 kPa. A hand pump was connected to the cylindrical core holder to provide confining pressure. The inlet and outlet pressure difference was recorded by a differential pressure transducer.

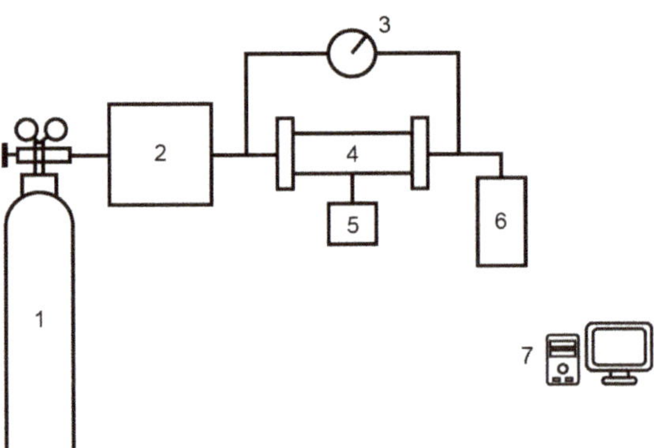

Figure 3. Schematic of the apparatus used for the stress-dependent permeability experiment: 1, High-pressure gas source; 2, Pressure control panel; 3, Pressure sensor; 4, Core holder; 5, Hand pump; 6, Measuring cylinder; and 7, Data acquisition system.

2.3. Experimental Procedure

Determining the relationship between effective stress and permeability is the key objective of stress sensitivity experiments. In this work, stress sensitivity experiments were conducted under variable confining pressure in an unsteady state, and the gas used here was N_2. Primary effective stress was used to evaluate stress sensitivity of the tight core [25,34,35]:

$$\sigma = \sigma_0 - \eta p \tag{1}$$

where σ is the primary effective stress (MPa), σ_0 is the confining pressure (MPa), p is the fluid pressure (MPa), and η is the effective stress coefficient, which can be taken as rock porosity of the matrix system and the unity of the fracture system.

The porosity of the experimental core was obtained using the gas pore volume measurement method. (1) Place the standard sample in the standard chamber under a pressure P_1, (2) place the core in the rock chamber, (3) connect the two chambers until the pressure stabilizes at P_2, and (4) calculate the core porosity by the volume and pressure [36]. The primary permeability experiment can be summarized as follows: (1) put the sample into the core holder, and maintain it under a constant pressure drop; (2) apply confining pressure and gradually increase it to 30 MPa in 5 MPa increments. After the gas flow is stable, record the pressure at the core outlet and inlet and the flow rate of outlet. (3) Calculate the permeability and permeability loss ratio based on the records.

3. Numerical Simulation

3.1. Finite Element Method

Based on computed tomography (CT) scanned images of the tested cores, the pores and the skeleton of the porous media were classified, and then the permeability of the porous media was determined using the FEM using ANSYS software (ANSYS, Inc., Pittsburgh, PA, US) [37–41].

The stress-dependent permeability of the porous media was simulated using a two-way Fluid Structure Interaction (FSI) approach involving both Computational Fluid Dynamics (CFD) and structural mechanics analysis. The pressure/loads of the single-phase water flow in porous media were calculated using ANSYS. The FEM calculation steps with ANSYS software can be divided into three basic steps: mesh division, unit analysis, and overall analysis [42,43]. The mesh division of the matrix and fractured cores are presented in Figure 4a,b, respectively, in which the matrix is presented by grey and the fracture and cores are represented by brown. Along with the ANSYS software, TetraMesh (mesh division method with tetrahedron as mesh unit) was used. The size of the core was $200 \times 200 \times 200$ pixels, with a pixel size of 0.4 μm, porosity 11.24%, with 24,3520 matrix elements, and 336,224 pore and fracture elements. After the fluid pressure on the fluid–solid interface was transferred to the ANSYS structural solver, the meshes on the fluid–solid interface of both the rock matrix and pore model were remeshed according to the deformation data.

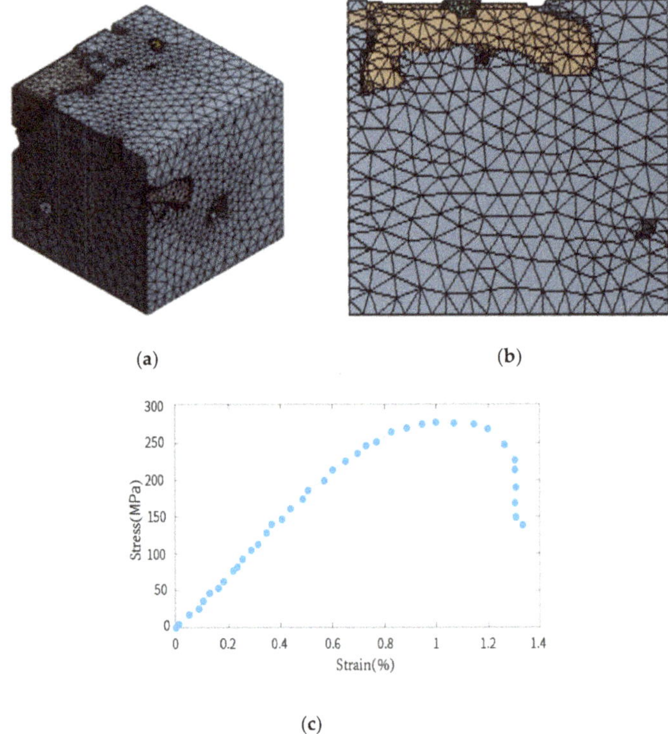

(a)   (b)

(c)

Figure 4. The mesh division and stress–strain curve of the simulation of tested cores: (**a**) mesh division of the core without fractures, (**b**) mesh division for the core with fractures, and (**c**) stress–strain curve.

The specific input and assumptions were required in order to use FEM in engineering materials [44–46]. The simulation is based on the assumption that fluid is incompressible, the density is constant, and the flow is laminar. Since the fracture width was much larger than the fracture opening, the flow in the fracture was simplified as a two-dimensional flow. To present the differences in the fractures and the matrix, different physical material properties were adopted. The pores and fractures were filled with units, the skeleton was built as nonlinear elastic material, and the input model material parameters shown in Table 3 were set based on the FEM work conducted by Zheng [47], which were close to those of natural sandstone and artificial sandstone. The stress–strain curve of the tested cores, which was used as the material property for the structural analysis in the simulation, is shown in Figure 4c.

Table 3. The main input parameters for FEM simulation.

Calculation Unit	Elastic Modulus (GPa)	Poisson's Ratio	Internal Friction Angle (°)	Cohesion Force (MPa)
Matrix	35.6	0.2	15	20
Fracture and Pores	3.56	0.3	20	2

In the simulation, fixed constraints were applied to the upper and lower boundaries of the model, the other two sides were loaded with lateral pressure to simulate confining pressure in the stress sensitivity experiment, and the lateral pressure was gradually increased to 30 MPa in 5 MPa increments. The boundary conditions of the matrix model and pore model were set separately in the ANSYS workbench. The contact surfaces of solid and fluid inside the rock were defined as the fluid–solid interface through which pressure is transmitted between the solid and fluid. The geometric nonlinearity of the system elements was considered. At the end of each load step, the system searched for the approximate balanced configuration of the system structure according to the displacement of the node, considering the change in the stiffness of the model structure, and calculated the model flexibility matrix. The permeability of the porous media under stress conditions can be determined by combining the Navier–Stokes and stress equations. All simulations converged after different numbers of iterations under the condition that the absolute convergence criteria was set to 10^{-5} for all equations, for which default relaxation factors were used.

3.2. Stress-Dependent Permeability Model

Primary effective stress was chosen to evaluate the stress sensitivity of the tight core, which is described as [41]:

$$K = K_0 e^{-\alpha(\sigma_0 - \eta p)} \tag{2}$$

where K_0 is the initial permeability of the rock (10^{-3} μm²), K is the rock permeability under stress (10^{-3} μm²), α is the permeability stress sensitivity coefficient (MPa^{-1}), σ_0 is the overlying rock stress (MPa), p is fluid pressure (MPa), and η is the effective stress coefficient, which can be taken as the rock porosity in the matrix system and 1 in the fracture system [34,35].

Under the assumption that seepage of gas is steady in fractured cores and based on Darcy's equation, the permeability of the matrix system can be described as:

$$K_m = -\frac{Q_m p_0 \mu}{(A - \omega_f l)p} \frac{dx}{dp} \tag{3}$$

where K_m is the matrix system permeability (10^{-3} μm²), Q_m is the gas flow volume of the matrix system under atmospheric pressure (cm³/s), p_0 is the atmospheric pressure (MPa), A is the core section area (cm²), μ is the gas viscosity (mPa·s), l is the width of the fracture (cm), and ω_f is the fracture opening (cm).

A fracture does not always fully penetrate the reservoir, so the fracture penetration extent is the factor discussed here. We assumed fluid seepage only occurs in the horizontal direction. In the equivalent resistance method, the equivalent length of the oblique fracture length can be taken as its projection in the horizontal direction. Therefore, the direction of the fracture can be equated as the horizontal direction, as shown in Figure 5.

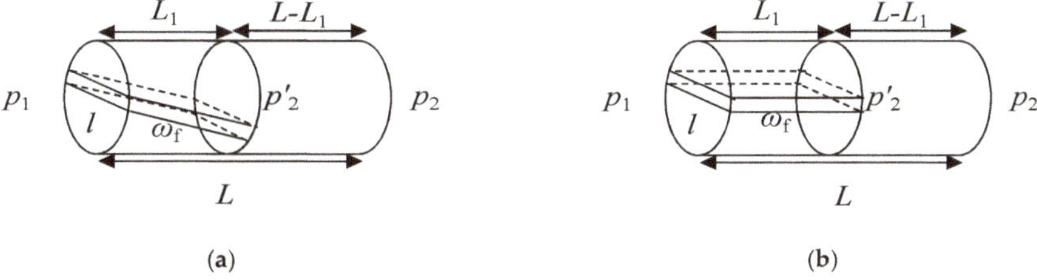

Figure 5. The sketch of a fracture-penetrated core (**a**) with an oblique fracture and (**b**) after an equivalent fracture is made. l is the width of the fracture (equal to the diameter of the section surface) (cm); ω_f is the fracture opening equal to the distance of the two surfaces of the fracture (cm); L_1 is the equivalent length of the fracture (cm); L is core length (cm); p_1 is the absolute pressure on the inlet section (MPa); p_2 is the absolute pressure on the outlet section (MPa); p'_2 is the absolute pressure on the outlet section of the penetrated part (MPa); and the fracture penetration extent is L_1/L.

The gas flow volume through the matrix of the penetrated part is:

$$Q_m = \frac{(A - \omega_{f0}l)\left(p_1^2 - p_2'^2\right)K_{m0}}{2\mu p_0 L_1}$$ (4)

where Q_m is the gas flow volume through the matrix of the penetrated part (cm^3/s), K_{m0} is the initial core matrix permeability (10^{-3} μm^2), p_1 is the absolute pressure on the inlet section (MPa), p'_2 is the absolute pressure on the outlet section of the penetrated part (MPa), L_1 is the equivalent length of the fracture (cm), and ω_{f0} is the fracture system opening before stress is exerted (cm).

The gas flow volume through the fracture system is calculated as:

$$Q_f = \frac{\omega_{f0}l\left(p_1^2 - p_2'^2\right)K_{f0}}{2\mu p_0 L_1}$$ (5)

The gas flow volume through the part without a fracture is calculated as:

$$Q_{s1} = \frac{A\left(p_2'^2 - p_2^2\right)K_{m0}}{2\mu p_0 (L - L_1)}$$ (6)

The total gas flow volume through the core is:

$$Q_{s2} = \frac{A\left(p_1^2 - p_2^2\right)K_{s0}}{2\mu p_0 L}$$ (7)

where Q_f is the gas flow volume through the fracture system under atmospheric pressure (cm^3/s), Q_{s1} is the gas flow volume through the part without a fracture under atmospheric pressure (cm^3/s), Q_{s2} is the total gas flow volume through the fractured core under atmospheric pressure (cm^3/s), p_2 is the absolute pressure on the outlet section (MPa), L is the core length (cm), K_{s0} is the initial total permeability of the fractured core (10^{-3} μm^2), and K_{f0} is the initial fracture permeability (10^{-3} μm^2).

Considering the continuity of the fluid, the gas flow volume through the part without fractures should be equal to the total gas flow volume that flows through the fractured part of the core, which is the sum of the gas flow volume through the fracture system and the matrix. The following equations are obtained:

$$p_2'^2 = p_2^2 + \frac{L - L_1}{L}\frac{K_{s0}}{K_{m0}}\left(p_1^2 - p_2^2\right)$$ (8)

$$AK_{s0} = \frac{K_{m0}L - K_{s0}(L - L_1)}{K_{m0}L_1}\left[(A - \omega_{f0}l)K_{m0} + \omega_{f0}lK_{f0}\right]$$ (9)

Based on cube law, the relationship between the fracture opening and fracture permeability satisfies:

$$K_{f0} = \frac{\omega_{f0}^2}{12}$$ (10)

Then, the following equation can be derived, based on which initial fracture opening and the initial permeability K_{f0} of the fracture system:

$$\omega_{f0}^3 - 12K_{m0}\omega_{f0} + \frac{12AK_{m0}}{l} - \frac{12AK_{s0}}{l}\frac{K_{m0}L_1}{K_{m0}L - K_{s0}(L - L_1)} = 0$$ (11)

With the confining pressure σ_0, the inlet pressure p_1, and the outlet pressure p_2 on the sample, the effective stress (MPa) exerted on the core is:

$$\sigma = \sigma_0 - \eta\left(\frac{p_1 + p_2}{2}\right)$$ (12)

The gas flow volume through the matrix of the penetrated part under effective stress is:

$$Q_{\mathrm{m}} = \frac{K_{\mathrm{m0}}e^{-\alpha_{\mathrm{m}}\sigma}}{2p_0\mu L_1}\left(A - 2\sqrt{3}K_{\mathrm{f}}^{\frac{1}{2}}l\right)\left(p_1^2 - p'^2_2\right) \tag{13}$$

The gas flow volume through the fracture system and the part without a fracture are, respectively:

$$Q_{\mathrm{f}} = \frac{2\sqrt{3}\left(p_1^2 - p'^2_2\right)l}{2p_0\mu L_1}K_{\mathrm{f}}^{\frac{3}{2}} \tag{14}$$

$$Q_{\mathrm{s1}} = \frac{A\left(p'^2_2 - p_2^2\right)K_{\mathrm{m0}}e^{-\alpha_{\mathrm{m}}\sigma}}{2\mu p_0(L - L_1)} \tag{15}$$

The total gas flow volume through the core is calculated as:

$$Q_{\mathrm{s2}} = \frac{A\left(p_1^2 - p_2^2\right)K_{\mathrm{s0}}e^{-\alpha_{\mathrm{s}}\sigma}}{2\mu p_0 L} \tag{16}$$

where α_{m} is the matrix permeability stress sensitivity coefficient (MPa^{-1}), α_{s} is the permeability stress sensitivity coefficient of the fractured core (MPa^{-1}), and K_{f} is the fracture permeability after stress is exerted (10^{-3} μm^2).

Based on the continuity, the following equations are obtained:

$$p'^2_2 = p_2^2 + \frac{L - L_1}{L}\frac{K_{\mathrm{s0}}e^{-\alpha_{\mathrm{s}}\sigma}}{K_{\mathrm{m0}}e^{-\alpha_{\mathrm{m}}\sigma}}\left(p_1^2 - p_2^2\right) \tag{17}$$

$$K_{\mathrm{f}}^{\frac{3}{2}} - K_{\mathrm{m0}}e^{-\alpha_{\mathrm{m}}\sigma}K_{\mathrm{f}}^{\frac{1}{2}} - \frac{\sqrt{3}A}{6l}\left(\frac{K_{\mathrm{s0}}e^{-\alpha_{\mathrm{s}}\sigma}K_{\mathrm{m0}}e^{-\alpha_{\mathrm{m}}\sigma}L_1}{K_{\mathrm{m0}}e^{-\alpha_{\mathrm{m}}\sigma}L - K_{\mathrm{s0}}e^{-\alpha_{\mathrm{s}}\sigma}(L - L_1)} - K_{\mathrm{m0}}e^{-\alpha_{\mathrm{m}}\sigma}\right) = 0 \tag{18}$$

Based on the equations above, the permeability loss ratio of the fracture system γ (%) can be determined as:

$$\gamma = \frac{K_{\mathrm{f}}}{K_{\mathrm{f0}}} \tag{19}$$

4. Results and Discussion

4.1. Experimental Results

4.1.1. Matrix and Fractured Cores

Stress sensitivity experiments under different effective stresses were conducted on all cores according to the Standard SY/T6385—2016 [48], and the permeability stress sensitivity curves are shown in Figure 6a–c. The permeability of matrix and fractured cores was damaged to some degree due to the effective stress increase. Compared with the fractured cores, the permeability stress sensitivity of the matrix was weaker, and the damage range of permeability was relatively smaller. Rock compaction is the dominant factor in the stress-dependent permeability of the matrix. The permeability variation trends are almost linear when compared with those of the fractured cores. A possible reason for this is that the stress-dependent permeability of the matrix is purely determined by rock compaction, whereas those of cores with fractures are a combination of rock compaction and fracture closure. Then, for the fractured cores, a larger damage range and faster permeability rate of decrease were observed. The rate of decrease slows down when the permeability of the fractured cores decreases to a point equal to that of the matrix. The stress-dependent permeability could be divided into two stages. In the first stage, fracture closure is the leading factor with the rock compaction working together; in the second stage, the rock compaction is the dominant factor.

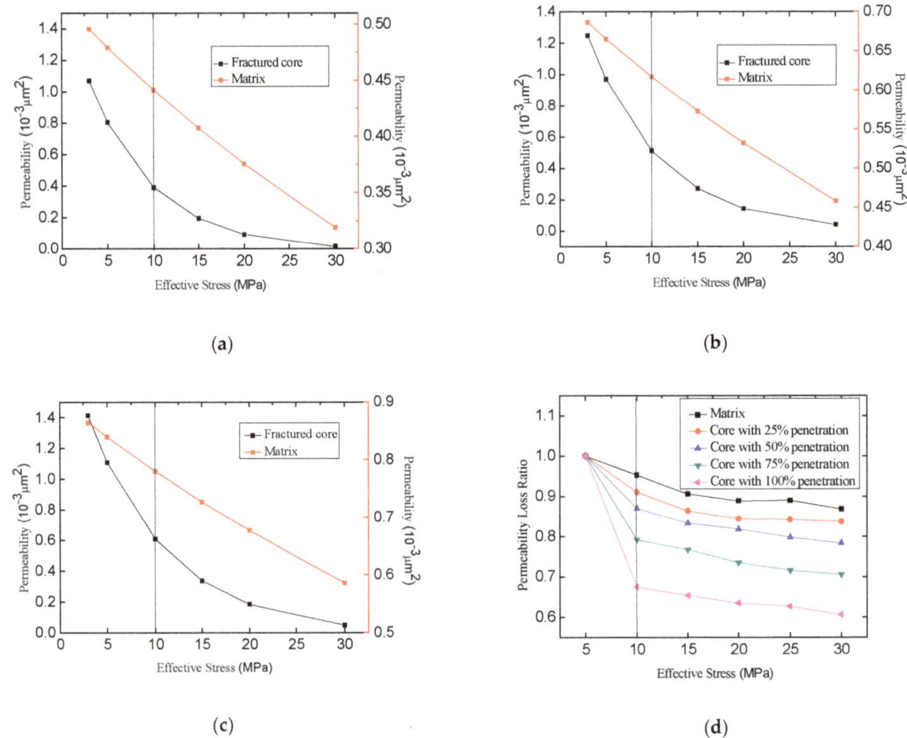

Figure 6. The experimental results of stress-dependent permeability of cores in (**a**) set 1, (**b**) set 2, and (**c**) set 3, and (**d**) fractured cores with different penetration extents.

4.1.2. Fractured Cores with Different Fracture Penetration Extents

The permeability stress sensitivity curve shown in Figure 6d suggests that permeability loss occurred in all tested cores, and stress sensitivity increased with the increasing fracture penetration extent. The permeability stress sensitivity curves for the fractured cores were similar to that for the matrix when the penetration extent was small, suggesting that permeability is mainly dominated by the matrix in this case. The rock stress response process can be divided into the fracture closure stage and the rock compaction stage, with a larger extent of fracture penetration. During the first stage, rock permeability dramatically reduces due to the closure of the fractures, and in the second stage, the permeability decreases slowly with the increase in the effective stress due to matrix compaction.

4.2. FEM Simulation Results

To verify the FEM simulation, the measured permeability and effective stress were compared with the predicted values from FEM simulation (Figure 7). A definitive negative correlation between the permeability and effective stress is demonstrated in Figure 7. As for the matrix (Figure 7a), the permeability decreased at a relatively constant rate, whereas the permeability of the fractured core (Figure 7b) first dramatically decreased and then slowed down with increasing effective stress. The results (Figure 7) suggest that the simulated changing trend of the stress-dependent permeability is the same as that of the experimental results. However, the permeability predicted from FEM simulation is larger than measured during the experiment under a given effective stress. The main reason for this is that the effective resolution of CT is limited by the voxel size: many narrow pores with smaller pore radii cannot be easily detected, and only a limited range of pore throat radii can be detected. Therefore, the permeability predicted by FEM simulation is larger than the experimental findings under a given effective stress.

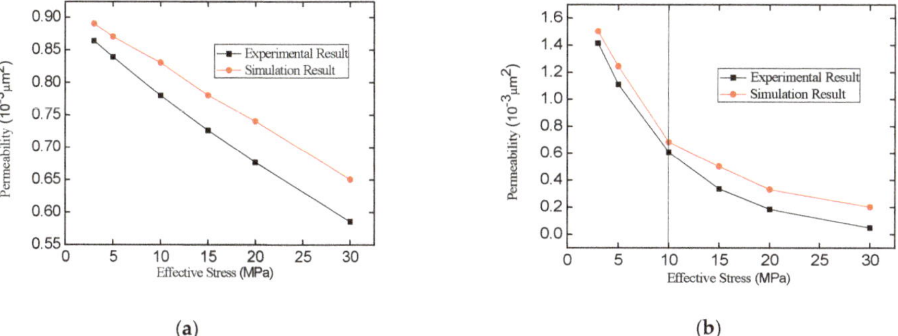

Figure 7. Simulation and experimental results of stress-dependent permeability: (**a**) matrix and (**b**) fractured core.

4.3. Comparison of Model and Experimental Results

To evaluate the performance of the proposed model, Equation (19) and experimental data were used for comparison. Based on the parameters summarized in Tables 1 and 2, the permeability loss ratio was measured, calculated, and plotted (Figure 8). The results suggest that this calculation model can fairly accurately predict the permeability loss ratio; the average relative errors between the measured and predicted results were all within 2.5%.

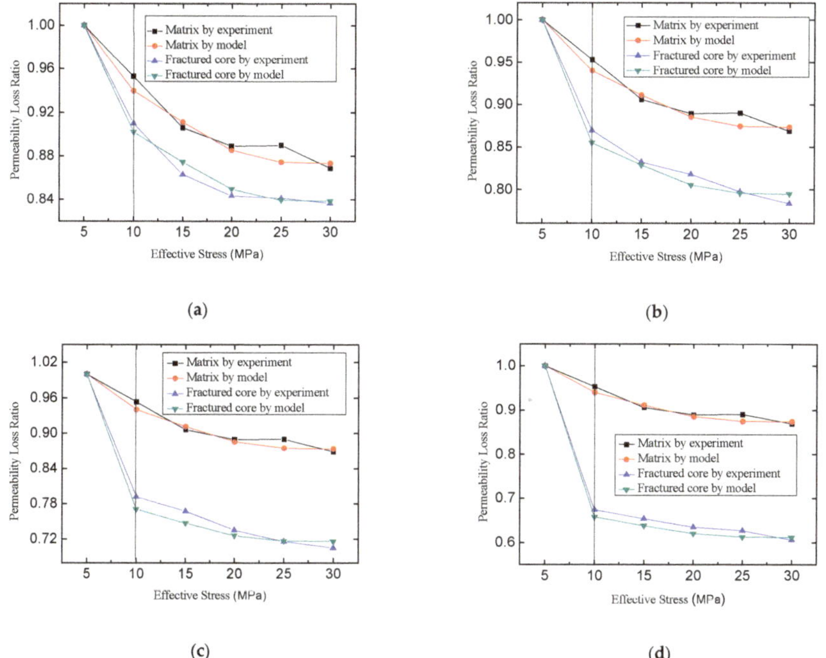

Figure 8. Comparison curve for the experimental data and model results. Core with (**a**) 25%, (**b**) 50%, (**c**) 75%, and (**d**) 100% penetration.

4.4. Influencing Factor Analysis

4.4.1. Fracture Penetration Extent

The resulting permeability loss ratio variation versus effective stress under different fracture penetration extents is shown in Figure 9a. The result illustrates that the permeability of the fracture quickly decreased before reaching 10 MPa and then smoothed (Figure 9a). This suggests that the deformation process of the fractured cores can be divided into two stages: the fracture closure stage and matrix compaction stage. In the first stage, with a lower effective stress, the permeability of the fractured cores quickly decreases. When the effective stress increases to 10 MPa (stage 2), the fracture system closes, and the matrix system is compressed. The permeability stress sensitivity has a positive relationship with the penetration extent, as shown in Figure 9a. The permeability loss ratio increases with increasing fracture penetration extent. The permeability loss ratios of the cores with different fracture penetration extents are compared in Figure 9b–e. This suggests that, under the same effective stress, the fracture has the highest stress sensitivity, followed by the fractured cores, and the matrix has the lowest stress sensitivity of the three. These figures demonstrate that the permeability loss ratio increases with increasing fracture penetration extent. With an increasing penetration extent, the permeability sensitivity curve shows a trend approaching that of the fracture. Thus, during the production process, a reasonable pressure drop and effective stress should be determined according to the fracture penetration extent in order to preclude damage to the permeability of the reservoir due to stress sensitivity.

4.4.2. Permeability Increasing Multiples

The change in the deviation extent versus the increasing multiples of the initial permeability of the fractured core is shown in Figure 9f. This shows that with increasing permeability multiples (i.e., the ratio of fractured core permeability to permeability of matrix core), the deviation extent of the permeability loss ratio of the fractured core to the fracture decreases, whereas that of the fractured core to the matrix system increases. The permeability sensitivity is more heavily influenced by the fracture system with a larger penetration extent and increasing multiples of the initial permeability, and the permeability stress sensitivity curve approaches that of the fracture system, as shown in Figure 9f.

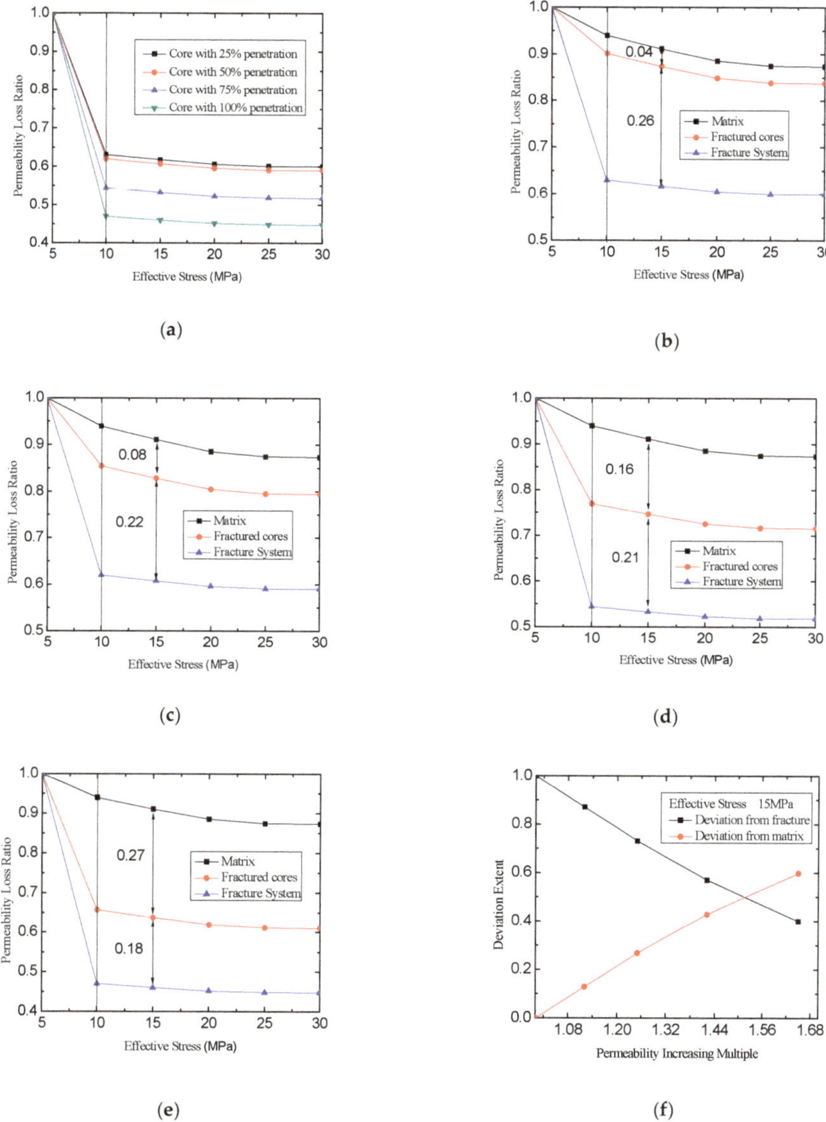

Figure 9. Sensitivity analysis results. Permeability loss ratio curves for (**a**) cores with different penetration extents, (**b**) the core with 25% penetration, (**c**) the core with 50% penetration, (**d**) the core with 75% penetration, and (**e**) the core with 100% penetration, and (**f**) variation curve of the deviation extent and increasing multiples.

5. Conclusions

Through FEM simulation and experiments conducted on the stress-dependent permeability of fractures in cores, we proposed a novel theoretical model for the determination of stress-dependent permeability in tight fractured reservoirs. The model allowed us to analyze the influences caused by a fracture system with different fracture parameters on the stress-dependent permeability reduction in tight fractured reservoirs. Predictions produced by the model presented similar variation trends to the experimental results. The feature of this model is that every parameter in the model has a specific physical significance, while it lacks empirical constants. The novelty of this work is that the fracture penetration extent is newly introduced as an influencing factor and the fracture system is separated as a distinct objective to be analyzed. We drew the following conclusions:

The deformation of fractured cores can be divided into two stages: the fracture closure stage and the matrix compression stage. During the first stage, rock permeability is dramatically reduced due to the closure of the fracture; in the second stage, the permeability decreases slowly with the increase in the effective stress that continues to compress the rocks. The calculation model, considering the fractured penetration extent, can predict the permeability loss of the matrix, fractured core, and fracture system fairly accurately at pressures up to 20 MPa.

Permeability stress sensitivity is more strongly influenced by the fracture system with a larger penetration extent. In the systems with lesser penetration extent, the matrix compression is the leading factor influencing permeability stress sensitivity.

The stress sensitivity experiments and our corresponding model, considering the fracture penetration extent, could be applied to more accurately predict the production of fractured reservoirs and the coupled flow deformation behavior in fractured porous media, including tight carbonates, tight sandstone, and shale rock. Considering the fracture parameters in the production model or seepage model is crucial. As the extent of the fracture penetration directly affects the reservoir permeability and the mechanism of seepage flow, considering the fracture penetration extent in the proposed stress-dependent permeability model makes our work relevant in product development in tight reservoirs. However, it should be noted that the proposed model ignores the crossflow of fractures, and the stress-dependent permeability could be influenced by the composition, microstructures, mineral disintegration interpretation, mechanical properties, and community of reservoirs. The fluid flow in porous media with fracture systems under effective stress is an interesting and challenging topic, and this work is currently underway.

Author Contributions: This work is completed by all of the authors: N.C. and G.L. produced the main contribution, proposed the main ideas employed in the research, and wrote the original paper; P.D. offered meaningful suggestions and helped to improve the work; Z.W. conducted the visualization of the results; and H.L. and Y.L. provided assistance in the process of this research.

Acknowledgments: The authors are grateful for the financial support from the National Science and Technology Major Projects of China (Grant No.2016ZX05037-003; Grant No.2017ZX05049-003). Special thanks to Y.L. for her support and encouragement in the whole process.

References

1. Neto, L.B.; Kotousov, A.; Bedrikovetsky, P. Elastic properties of porous media in the vicinity of the percolation limit. *J. Petrol. Sci. Eng.* **2011**, *78*, 328–333. [CrossRef]
2. Gajo, A.; Loret, B.; Hueckel, T. Electro-chemo-mechanical couplings in saturated porous media: Elastic-plastic behaviour of heteroionic expansive clays. *Int. J. Solids Struct.* **2002**, *39*, 4327–4362. [CrossRef]
3. Zhang, H.J.; Jeng, D.S.; Barry, D.A.; Seymour, B.R.; Li, L. Solute transport in nearly saturated porous media under landfill clay liners: A finite deformation approach. *Int. J. Solids Struct.* **2013**, *479*, 189–199. [CrossRef]
4. Wang, S.J.; Hsu, K.C. The application of the first-order second-moment method to analyze poroelastic problems in heterogeneous porous media. *Int. J. Solids Struct.* **2009**, *369*, 209–221. [CrossRef]
5. Mokni, N.; Olivella, S.; Li, X.L.; Smets, S.; Valcke, E. Deformation induced by dissolution of salts in porous media. *Phys. Chem. Earth* **2008**, *3*, S436–S443. [CrossRef]
6. Cao, N.; Lei, G. Stress sensitivity of tight reservoirs during pressure loading and unloading process. *Petrol. Explor. Dev.* **2019**, *46*, 1–7. [CrossRef]
7. Wang, W.D.; Shahvali, M.; Su, Y.L. A semi-analytical model for production from tight oil reservoirs with hydraulically fractured horizontal wells. *Fuel* **2015**, *158*, 612–618. [CrossRef]
8. Zhang, Q.; Su, Y.L.; Wang, W.D.; Lu, M.J.; Sheng, G.L. Gas transport behaviors in shale nanopores based on multiple mechanisms and macroscale modeling. *Int. J. Heat Mass Transf.* **2018**, *125*, 845–857. [CrossRef]

9. Zhang, Q.; Su, Y.L.; Wang, W.D.; Lu, M.J.; Sheng, G.L. Apparent permeability for liquid transport in nanopores of shale reservoirs: Coupling flow enhancement and near wall flow. *Int. J. Heat Mass Transf.* **2017**, *115*, 224–234. [CrossRef]

10. Zhang, C.; Yu, L.; Feng, R.M.; Zhang, Y.; Zhang, G.J. Numerical Study of Stress Distribution and Fracture Development above a Protective Coal Seam in Longwall Mining. *Processes* **2018**, *6*, 146. [CrossRef]

11. Jiang, H.J.; Yan, J.N. Experimental study on stress sensibility of fractured reservoir. *Spec. Oil Gas Reserv.* **2000**, *3*, 39–41. (In Chinese)

12. Fatt, I. Reduction in permeability with overburden pressure. *J. Petrol. Technol.* **1952**, *4*, 16. [CrossRef]

13. Buchsteiner, H.; Warpinski, N.R.; Economides, M.J. Stress-Induced Permeability Reduction in Fissured Reservoirs. In Proceedings of the SPE Annual Technical Conference and Exhibition, Houston, TX, USA, 3–6 October 1993. SPE26513-MS. [CrossRef]

14. Zhang, Y.; Cui, Y. Experimental study of pressure sensitivity on the gravel low permeability layer. *Oil Drill. Prod. Technol.* **1999**, *21*, 1–6. (In Chinese)

15. Jones, F.O.; Owens, W.W. A laboratory study of low-permeability gas sands. *J. Petrol. Technol.* **1980**, *32*, 1631–1640. [CrossRef]

16. Nelson, R.A. An experimental study of fracture permeability in porous rock. In Proceedings of the 17th US Symposium on Rock Mechanics, Snow Bird, UT, USA, 25–27 August 1976. ARMA-76-0127.

17. Archer, R.A. Impact of Stress Sensitive Permeability on Production Data Analysis. In Proceedings of the SPE Unconventional Reservoirs Conference, Keystone, CO, USA, 10–12 February 2008. SPE-114166-MS. [CrossRef]

18. Zimmerman, R.W.; Kumar, S.; Bodvarsson, G.S. Lubrication theory analysis of the permeability of rough-walled fractures. *Int. J. Rock Mech. Min.* **1991**, *28*, 325–331. [CrossRef]

19. Dobrynin, V.M. Effect of overburden pressure on some properties of sandstones. *Soc. Petrol. Eng.* **1962**, *4*, 360–366. [CrossRef]

20. Lan, L. Research on stress sensitivity and prediction of fracture width in naturally fractured sandstone reservoirs. Master's Thesis, Southwest Petroleum University, Chengdu, China, 2005. (In Chinese)

21. Gangi, A.F. Variation of whole and fractured porous rock permeability with confining pressure. *Int. J. Rock Mech. Min.* **1978**, *15*, 249–257. [CrossRef]

22. Li, D.Q.; Kang, Y.L.; Zhang, H. New evaluation method of permeability stress sensitivity based on visual fracture aperture measurement. *Nat. Gas Geosci.* **2011**, *2*, 494–500. (In Chinese)

23. Lei, G.; Dong, P.C.; Yang, S.; Wang, B.; Wu, Z.S. Study of stress sensitivity of low-permeability reservoir based on arrangement of particles. *Rock Soil Mech.* **2014**, *35*, 209–214. (In Chinese)

24. Advani, S.H.; Lee, T.S.; Avasthi, J.M. Parametric sensitivity investigations for hydraulic fracture configuration optimization: Rock Mechanics as a Guide for Efficient Utilization of Natural Resources. In Proceedings of the 30th U.S. Symposium, Morgantown, WV, USA, 19–22 June 1989. [CrossRef]

25. McKee, C.R.; Bumb, A.C.; Koenig, R.A. Stress-dependent permeability and porosity of coal and other geologic formations. *SPE Form. Eval.* **1988**, *3*, 81–91. [CrossRef]

26. David, C.; Wong, T.F.; Zhu, W.; Zhang, J. Laboratory measurement of compaction-induced permeability change in porous rocks: Implications for the generation and maintenance of pore pressure excess in the crust. *Pure Appl. Geophys.* **1994**, *143*, 425–456. [CrossRef]

27. Tong, S.Y.; Gu, M.; Singh, R.; Mohanty, K.K. Simulation of proppant transport in foam fracturing fluid based on experimental results. In Proceedings of the SPE/AAPG/SEG Unconventional Resources Technology Conference, Houston, TX, USA, 23–25 July 2018. URTEC-2901054-MS. [CrossRef]

28. Shrivastava, K.; Sharma, M.M. Proppant transport in complex fracture networks. In Proceedings of the SPE Hydraulic Fracturing Technology Conference and Exhibition, The Woodlands, TX, USA, 23–25 January 2018. SPE-189895-MS. [CrossRef]

29. Tong, S.Y.; Singh, R.; Mohanty, K.K. Proppant transport in fractures with foam-based fracturing fluids. In Proceedings of the SPE Annual Technical Conference and Exhibition, San Antonio, TX, USA, 9–11 October 2017. SPE-187376-MS. [CrossRef]

30. McClure, M.W.; Babazadeh, M.; Shiozawa, S.; Huang, J. Fully coupled hydromechanical simulation of hydraulic fracturing in 3D discrete-fracture networks. *SPE J.* **2016**. [CrossRef]

31. Wang, C.Y.; Du, X.Y. Experimental study on the splitting characteristics of sandstone in Brazil based on size effect. *Min. Res. Dev.* **2018**, *6*, 44–48. (In Chinese)

32. Peng, C.J.; Chen, C.C.; Xu, J.; Zhang, H.L.; Tang, Y.; Nie, W.; Zhao, K. Loading rate dependency of rock stress-strain curve based on Brazil splitting test. *J. Rock Mech. Eng.* **2018**, *A1*, 3247–3252. (In Chinese)
33. Zhan, Y.P.; Fu, C.L.; Li, S.Y. The artificial cracks preparation method for the tight sandstone fracture reservoir. *Res. Expl. Lab.* **2017**, *1*, 10–12. (In Chinese)
34. Li, C. Double effective stress of porous media. *Nat. Mag.* **1999**, *5*, 288–292. (In Chinese)
35. Qiao, L.P.; Wang, Z.C.; Li, S.C. Effective stress law for permeability of tight gas reservoir sandstone. *J. Rock Mech. Eng.* **2011**, *7*, 1422–1427. (In Chinese)
36. Duan, X.G.; An, W.G.; Hu, Z.M.; Gao, S.S.; Ye, L.Y.; Chang, J. Experimental study on fracture stress sensitivity of Silurian Longmaxi shale formation, Sichuan Basin. *Nat. Gas Geosci.* **2017**, *28*, 1416–1424. (In Chinese)
37. Ahammad, M.J.; Rahman, M.A.; Zheng, L.; Alam, J.M.; Butt, S.D. Numerical investigation of two-phase fluid flow in a perforation tunnel. *J. Nat. Gas. Sci. Eng.* **2017**, *55*, 606–611. [CrossRef]
38. Jiang, C.B.; Xu, Z.M.; Li, X.L. Fractional step finite element formulation for solving incompressible flows. *J. Tsinghua Univ. (Sci. Technol.)* **2002**, *42*, 278–280. (In Chinese)
39. Abdoulkadri, C.; Stiaan, G.; Wang, R.J. Newton–Raphson Solver for Finite Element Methods Featuring Nonlinear Hysteresis Models. *IEEE. Trans. Magn.* **2018**, *54*, 1–8. [CrossRef]
40. Johari, A.; Heydari, A. Reliability analysis of seepage using an applicable procedure based on stochastic scaled boundary finite element method. *Eng. Anal. Bound. Elem.* **2018**, *94*, 44–59. [CrossRef]
41. Walsh, J.B. Effect of pore and confining pressure on fracture permeability. *Int. J. Rock Mech. Min.* **1981**, *18*, 429–435. [CrossRef]
42. Vahab, M.; Akhondzadeh, S.; Khoei, A.R.; Khalili, N. An X-FEM investigation of hydro-fracture evolution in naturally-layered domains. *Eng. Fract. Mech.* **2018**, *191*. [CrossRef]
43. Liu, X.D.; Morita, N. Collapse and bending analysis of slotted liners by 3D FEM under various reservoir Conditions. In Proceedings of the SPE Annual Technical Conference and Exhibition, Dallas, TX, USA, 24–26 September 2018. SPE-191441-MS. [CrossRef]
44. Tong, S.Y.; Mohanty, K.K. Proppant transport study in fractures with intersections. *Fuel* **2016**, *181*, 463–477. [CrossRef]
45. Wang, H.Y.; Sharma, M.M. Modeling of hydraulic fracture closure on proppants with proppant settling. *J. Petrol. Sci. Eng.* **2018**, *171*, 636–645. [CrossRef]
46. Tong, S.Y.; Singh, R.; Mohanty, K.K. A visualization study of proppant transport in foam fracturing fluids. *J. Nat. Gas Sci. Eng.* **2018**, *52*, 235–247. [CrossRef]
47. Zheng, J.T. Study on the stress-dependent properties of low-permeability rock and 3D rock porous structure reconstruction. Ph.D. Thesis, China University of Mining & Technology, Beijing, China, 2016.
48. Min, L.Y.; Lv, C.Y.; Zhang, B.W. SY/T6385—2016 Porosity and permeability measurement under overburden pressure. *Petrol. Indus. Press.* **2016**, *1*, 1–7. (In Chinese)

Article

Applicability Analysis of Klinkenberg Slip Theory in the Measurement of Tight Core Permeability

Jirui Zou [1,2]**, Xiangan Yue** [1,2,]*****, Weiqing An** [1,2]**, Jun Gu** [1,2] **and Liqi Wang** [3]

[1]   State Key Laboratory of Petroleum Resources and Prospecting, China University of Petroleum (Beijing), Beijing 102249, China; 2014312053@student.cup.edu.cn (J.Z.); 2016212184@student.cup.edu.cn (W.A.); 2018212191@student.cup.edu.cn (J.G.)

[2]   Key Laboratory of Petroleum Engineering Ministry of Education, China University of Petroleum (Beijing), Beijing 102249, China

[3]   School of Geoscience, University of Aberdeen, Aberdeen AB24 3FX, UK; l.wang1.18@aberdeen.ac.uk

*   Correspondence: yxa@cup.edu.cn; Tel.: +86-010-8973-3960

Received: 20 May 2019; Accepted: 17 June 2019; Published: 19 June 2019

Abstract: The Klinkenberg slippage theory has widely been used to obtain gas permeability in low-permeability porous media. However, recent research shows that there is a deviation from the Klinkenberg slippage theory for tight reservoir cores under low-pressure conditions. In this research, a new experimental device was designed to carry out the steady-state gas permeability test with high pressure and low flowrate. The results show that, unlike regular low-permeability cores, the permeability of tight cores is not a constant value, but a variate related to a fluid-dynamic parameter (flowrate). Under high-pressure conditions, the relationship between flowrate and apparent permeability of cores with low permeability is consistent with Klinkenberg slippage theory, while the relationship between flowrate and apparent permeability of tight cores is contrary to Klinkenberg slip theory. The apparent permeability of tight core increases with increasing flowrate under high-pressure conditions, and it is significantly lower than the Klinkenberg permeability predicted by Klinkenberg slippage theory. The difference gets larger when the flowrate becomes lower (back pressure increases and pressure difference decreases). Therefore, the Klinkenberg permeability which is obtained by the Klinkenberg slippage theory by using low-pressure experimental data will cause significant overestimation of the actual gas seepage capacity in the tight reservoir. In order to evaluate the gas seepage capacity in a tight reservoir precisely, it is necessary to test the permeability of the tight cores directly at high pressure and low flowrate.

Keywords: tight gas reservoirs; Klinkenberg slippage theory; high pressure and low flowrate; gas permeability measurement

---

1. Introduction

With the continuous depletion of conventional oil and gas reserves, unconventional oil and gas resources play an increasingly important role. Tight sandstone oil and gas, with its huge resources, has become a hotspot in the unconventionals sector [1–5]. In recent years, many scholars have conducted research on enhanced oil recovery in tight sandstone reservoirs [6–9]. The permeability of tight reservoir permeability is one of the most important reservoir physical parameters in the gas reservoir development process [10,11]. The matrix of a tight sandstone reservoir has very low permeability and ultra-small pore structure [12–15]. Gas slip [16] is a phenomenon that occurs when gas flows through low-permeability porous medium. During this process, the velocity of the gas layer near the solid wall of the porous medium is not zero, resulting in a large gas flowrate in the porous medium. Because the effect of gas slip plays an important role in low-permeability core, the acquisition of core permeability in tight reservoirs is more complicated [17–19]. The Knudsen number [20] is a dimensionless parameter that determines whether there is a slippage effect of the gas flow at different scale flow channels. Moreover, it represents the relationship between the mean free path of the molecule and the pore size. It is an important parameter when identifying different gas flow states [21,22]. The mathematical expression of the Knudsen number is:

$$K_n = \frac{\lambda}{r} \tag{1}$$

where, λ represents the mean free path of the gas and r is the average pore radius. K_n varies with permeability and pressure under isothermal conditions. At large Knudsen values ($K_n > 0.001$), the slip effect becomes significant as the mean free path is close to the average pore throat [23].

In 1941, Klinkenberg [24] believed that permeability is a property of porous media and is a constant, and he obtained the mathematical relationship between apparent permeability (K_a) and absolute permeability (K_∞) using first-order slip boundary conditions [25]:

$$K_a = K_\infty \left(1 + \frac{b_K}{\overline{p}}\right) \tag{2}$$

where K_a is the apparent gas permeability observed at the mean pressure; K_∞ is Klinkenberg permeability; $\overline{p}$ is the average pressure at the inlet and outlet of the core; b_K is the slip factor, b_K can be calculated by:

$$b_K = \frac{4c\lambda\overline{p}}{r} \tag{3}$$

where c is the proportionality factor. It can be seen from the Equation (3) that for a particular gas and average pressure, the slip factor b_K increases as the r increases. The pore radius r of a conventional reservoir core is large enough and b_K approaches zero, so that the gas slippage effect is negligible. While the pore radius r of a tight reservoir core is small, the slippage effect has important influence on the gas flow.

It is known from the derivation of the Klinkenberg formula that the nature of the Klinkenberg effect is the gas slippage in the pore throat. The theoretical basis for the Klinkenberg equation deduction is the slip theory of Kundt and Warburg [25], which is only applicable when $K_n < 0.1$. The first-order slip boundary theory is proposed based on the ideal gas state equation, which is used in the Klinkenberg equation deduction process; the calculation of the mean free path of gas molecules is based on the ideal gas assumption. It can be seen that when the pressure increases to a certain extent and the gas must be treated as a real gas, whether the Klinkenberg formula is applicable to the permeability measurement of tight cores is worthy of discussion.

Numerical simulations and experimental studies are two common methods of studying gas flow in micro/nanoscale [26]. Gas flow in micro/nanoscale pores can be described by the molecular modeling (MD) or Lattice–Boltzmann (LB) methods. However, the MD or LB method requires a large amount of computational resources and time, limiting its practical application in shale gas transmission simulation. In recent years, Mohammad et al. [27] solved the Boltzmann equation by reducing its order while using the regularized 13 moment method to ensure its accuracy. An analytical R13-AP model for predicting apparent permeability of shale was established. Density functional theory [28,29] is a theoretical method based on statistical mechanics. It has higher computational efficiency than molecular simulation under the same computational accuracy. Density functional theory [28] is used to simulate ionic fluids in slit-like nanoholes in double-layer capacitors. Therefore, it can be used to simulate gas flow in micron/nanometer porous media. Another way to simulate gas flow in micro/nanopores is to include sliding boundary conditions in the continuum model. Javadpour [30] simply linearly superposes the two transport mechanisms of Knudsen diffusion and slip flow to obtain the apparent permeability. Apparent permeability is a function of pressure, temperature, and gas properties. The results showed that the apparent permeability was much higher than the Darcy permeability in the nanoscale mudstone system. Singh et al. [31] proposed a new LSP permeability model based on Langmuir sliding conditions. The model overcomes the shortcomings of Maxwell's sliding conditions and uses Langmuir adsorption data to determine the slip coefficient of the gas flow. The experimental results showed that the permeability calculated by the LSP permeability model was greater than the permeability calculated by the Klinkenberg model. Based on the analytical model of rare gas flow, Singh et al. [32] proposed an apparent permeability model with no empirical coefficients. The model is based on convective mass transfer and Fick diffusion, and is obtained by simple linear superposition. Their results indicate that the contribution of Knudsen's diffusion to total flow is important in shale and it must be included in the gas flow model. Wang et al. [33] proposed a unified shale matrix apparent permeability model that combines the effects of non-Darcy flow/gas slip, geomechanics (ground compaction), and adsorbed gas layer release into one coherent equation. Cai et al. [34] established an ideal gas fractal transport model based on fractal tortuous capillary bundles in three-dimensional space while considering viscous flow, molecular diffusion, and gas adsorption. The apparent permeability model of three-dimensional fractal shale media was obtained by adding fluxes caused by three flow mechanisms (viscous flow, molecular diffusion, and surface diffusion). Singh and Cai et al. [35,36] proposed a new approach of predicting shale permeability by discretizing fractured shale at the scale of interest into its permeable features, including inorganic matter, organic matter, and fractures. Singh and Cai et al. [36] implemented this approach to predict field-scale permeability of shale by history-matching the production data. Lopez et al. [37] developed a numerical method for calculating the shale adsorption-related permeability. Calculations include Darcy flow and diffusion flow, as well as changes of organic pore radius caused by solid kerogen adsorption/desorption and diffusion. The results show that ignoring viscous flow, Knudsen diffusion and pore radius changes lead to erroneous permeability. Gas diffusion in solid kerogen plays an important role in shale gas reservoirs and should be considered. Chen et al. [38] determined the apparent permeability based on the dust-containing gas model (DGM) while regarding the total flow as a combined result of viscous flow and Knudsen diffusion. This study is the first numerical study of the effective Knudsen diffusivity and apparent permeability based on the actual pore structure of shale. Civan et al. [39] proposed a gas flow model in a tight porous medium, which uses a simplified second-order sliding model combined with several empirical parameters to calculate the sliding flow. Civan et al. [40] used the Knudsen number criterion to determine the slip coefficient and derived an apparent permeability expression in the form of the Knudsen number. Based on the Javadpour model, Darabi et al. [23] considered the effect of nanopore wall surface roughness on Knudsen diffusion. Darabi results show that Knudsen diffusion contributes up to 20% of cumulative shale gas production. Therefore, in addition to slippage flow, the shale gas also includes Knudsen diffusion. In addition to extensive theoretical research, many researchers have also conducted experiments on permeability of compact cores. In recent years, many scholars acquired a non-linear relationship between tight core permeability and the reciprocal of the average pressure from the experiments with back pressure at the end, instead of a linear one described by Klinkenberg. Jones and Owens [41] conducted a series of experiments on low-permeability cores. Although they obtained a reasonable straight line, they noted that the overestimation of absolute permeability could be as high as 25%. Rushing et al. [42] found the apparent permeability follows the Klinkenberg linear theory under low back pressure; however, the value of apparent permeability becomes smaller than the value predicted by Klinkenberg slip theory when the average pressure increases to a certain degree. Besides, the difference was getting larger with the increasing back pressure. Zhu et al. [43] believed that the permeability of the low-permeability core and the reciprocal of the mean pressure did not conform to the first-order Klinkenberg slippage theory equation. The higher-order equations and experimental data under low pressure are recommended to predict the absolute permeability of tight cores based on their research.

Li et al. [44] carried out a single-phase gas flow experiment in tight cores. The experimental results showed that as the average pressure approached infinity, the apparent permeability deviated completely from the linear relationship of the classical Klinkenberg slip theory. They believed this is because the gas slip coefficient is not a fixed value but a variable one, and they found that for extremely low permeability cores, the absolute permeability obtained using the Klinkenberg formula correlation may still include slip effects under low back pressure conditions. Dong et al. [45] measured the permeability of tight cores using C_2H_6. Their experimental results showed that the measured gas permeability (0.026 mD) was slightly smaller than the absolute permeability (0.032 mD) predicted by Klinkenberg slip theory.

Fathi et al. [46] used the Lattice–Boltzmann method to simulate the motion of gases in nanotubes. They found that the permeability predicted by Klinkenberg slip theory was less than the apparent permeability of gas in nanotubes. They explained the deviation from the straight line based on the Lattice–Boltzmann gas dynamics simulation, and an empirical quadratic equation was proposed to fit their experimental results. Moreover, a study conducted by You et al. [47] also indicated that the apparent permeability decreased with the increase of back pressure; they found that when the pressure was 1.42 MPa, the Klinkenberg permeability was 1.23 times of the apparent permeability. Similar conclusions were obtained by Tanikawa [48] and Dion Salam [49]. Additionally, Yue et al. [50] measured the permeability of tight cores under the average pressure of 0–1 MPa, and they found that the measured apparent permeability was bigger than the value predicted by Klinkenberg slip theory, unlike other scholars. Liu et al. [51] carried out gas–water two-phase seepage experiments on compact cores. They believe that the relative permeability of gas phase may be overestimated if slippage effect is not dealt with in the process of determining effective gas permeability.

However, the end pressure of all the experiments above did not exceed 10 MPa, which is rather smaller than the high pore pressure in tight reservoirs. Alireza et al. [52] proposed new theories based on these experimental results to predict and explain the gas slip effect and extended the Klinkenberg slip theory to make it suitable for tight porous media. Its expression is as follows:

$$K_{apparent} = K_{absolute}\left(1 + \frac{b}{P} - \frac{a}{P^2}\right) \tag{4}$$

The same second-order equation was also proposed by Tang et al. [53] and the slip-off quadratic term of the equation could explain why the curve deviated from the straight line. However, whether the model based on slippage theory can predict the gas flow in a tight core at high pressure and low flowrate is still doubtful.

The study of gas permeability under high pressure in reservoirs is done in order to analyze the flow characteristics of gas in the core at very low flowrate and high average pressures. However, the backpressure regulator method is not suitable for the permeability experiment of tight cores at high pressure due to the working principle of the backpressure regulator [54].

In recent years, some scholars have studied the flow pattern in tight cores under high pressure by changing the methods of adding back pressure. Sinha et al. [55] used a pump to add back pressure to measure the gas permeability in tight cores at different pore pressures (0–5000 psi). The experimental results showed that the measured gas permeability at high pore pressure obviously deviates from the classical Klinkenberg slippage theory, which is much smaller than the predicted value. Due to the low gas flowrate through the tight core, the authors' group [54,56] used a cylinder to pressurize the back pressure and measured the permeability of these tight cores under the high back pressure. They found that the permeability under high backpressure was lower than the Klinkenberg permeability and the non-Klinkenberg flow pattern under high pressure.

Since the steady flow in tight core samples requires a long time and accurate measurement of very low flow rates, most laboratories use unsteady state techniques to measure the permeability of tight cores. In the unsteady state method, the pulse attenuation method is the most commonly used. It is a transient technique that generates a pressure pulse through the core [57]. The permeability of the core is then indirectly derived from changes in pressure over time. Pulse attenuation techniques require accurate pore volume and pore compressibility measurements as well as adsorption correction [58]. Deriving permeability from transient pressure responses requires further assumptions and parameters that add uncertainty when dealing with tight sandstones. In tight sandstones, due to the insufficient understanding of the flow laws, the number of hypotheses should be minimized. A study by Rushing et al. [42] showed that the steady-state method can accurately determine the permeability of tight cores. The permeability obtained by an unsteady-state method is larger than that obtained by the steady-state method. The difference is not caused by a simple random measurement error, but by a system phenomenon caused by the fundamental problem of the non-steady-state method. Additionally, in the pulse-decay method, the pressure across the sample changes with time. This causes the flow rate through the sample to change, and therefore, the potential rate dependency of permeability is not captured. The steady-state method is measured when gas flows through the core at a constant pressure differential. When equilibrium is reached, the apparent permeability is calculated by the Darcy equation from the differential pressure and flow rate of the sample. Fundamentally, permeability refers to the steady flow through a porous medium. Therefore, the results of the steady-state method come directly from the definition and do not require assumptions about the flow state [58]. Experiments can be performed at different average pore pressures, confining stresses, and flow rates. Compared to other test techniques, the steady-state method takes a longer time to measure permeability of tight sandstone, but it provides greater flexibility and more fundamentally accurate results.

The permeability of tight reservoir cores has always been measured under low back pressure due to the inaccurate measurement of micro differential pressure and micro flowrate under high pressure. Thus, studies on the permeability of tight cores at high pore pressure and low flowrate were rare. In order to overcome the shortcomings of the original experimental device, a differential pressure meter, a flow meter, and a back pressure regulating device was developed to measure the micro pressure drop and flowrate under high-pressure conditions. This newly designed experimental device was used to study the steady-state [57] gas flow in a tight core under different pore pressures (0–30 MPa) and different differential pressures (0.1–1 MPa). By using this device in experiments, we obtained a relationship between the flowrate and the apparent permeability of low-permeability cores and tight cores, respectively; we then applied both the comparison method and analysis method to clarify the difference between the apparent permeability and predicted permeability by classical Klinkenberg slippage theory at different flowrate. The theoretical velocity and the measured velocity under different backpressure and differential pressure were compared. The result indicated that absolute permeability obtained by the Klinkenberg slippage theory will significantly overestimate the gas seepage capacity in a tight reservoir. This study can provide a theoretical basis and an experimental dataset for the establishment of a tight core permeability test model.

2. Experiments

2.1. Gas Permeability Calculation

The calculation formula [47] of gas permeability is:

$$K_g = \frac{200 Q_1 P_1 \mu L}{A \left(P_1^2 - P_2^2 \right)} \tag{5}$$

where, A is the area of core end (cm^2); μ is the gas viscosity (mPa·s); L is the core length (cm); P_1, P_2 is the absolute pressure at the inlet and outlet, respectively (MPa). Q_1 is the gas flow at the outlet end (mL/s); the PVT simulator CMG WinProp (CMG, 2003) was used to obtain the gas viscosity in this study. Previous studies have shown that the calculation error was less than 1.5% [59,60]. The relationship between pressure and viscosity of nitrogen at 60 °C is shown in Figure 1.

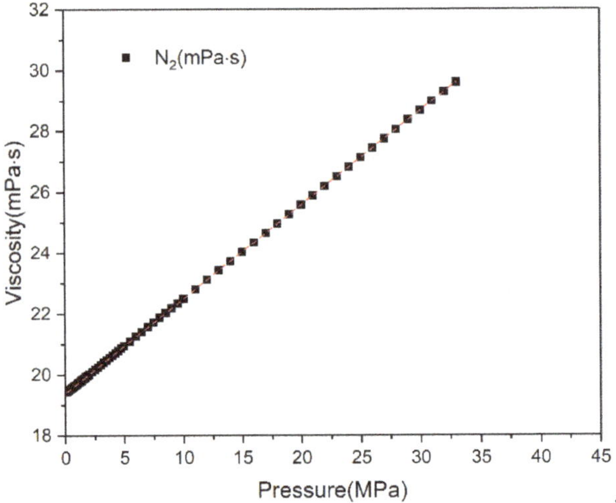

Figure 1. Viscosity of nitrogen as a function of pressure.

2.2. Experimental Apparatus

Figure 2 is the schematic diagram of the experimental device, which mainly includes an oven, ISCO pump, a constant pressure pump, a core holder, two high-pressure resistant intermediate containers, a high-pressure micro differential pressure gauge, a high-pressure micro flowmeter, and a back-pressure regulator. The ISCO pump was linked to a high-pressure resistant intermediate container to provide a stable gas source. The constant pressure pump was connected to the core holder to provide a constant confining pressure. The back-pressure regulator was used to control the outlet pressure. The inlet and outlet pressures were measured using the high-pressure micro differential pressure gauge. The high-pressure micro flowmeter was used to measure gas volume velocity. The high-pressure micro flowmeter (Chinese Invention Patent No: 201510164634.5), high-pressure micro differential pressure gauge (China Invention Patent No: CN104748908A), and back-pressure regulator (China Invention Patent No: CN106647842A) were all novel experimental equipment developed by authors' group.

The HPMR flowmeter can measure extremely low flow (4 nL/min–1 mL/min) at high pressure (0.1 MPa–50 MPa) with high accuracy (error < 1%) [56]. Generally, the working principle of the device is based on the displacement method, which can be used to measure the speed of fluid displacement in a pressure-resistant capillary tube. Prior to this study, the authors' group demonstrated that the effect of high pressure on the diameter of glass tube is negligible [56].

Under high pressure, the high-pressure micro differential pressure gauge can continuously record the dynamic pressure difference from 10^{-6} MPa up to 80 MPa with the error controlled within 0.1% [56,61]. The operating principle rests on the micro differential pressure measurement under a high-pressure system through digitized reading of the fluid level in a pressure resistant U-shaped tube (pressure resistant manometer).

Our group [54,56] found that the traditional back-pressure regulator is not suitable for measuring the extremely low flowrate at high pressure due to its working principle. In this experiment, the self-designed back-pressure regulator was used to limit the pressure fluctuation of the whole experimental facility below 0.002 MPa under the pressure of 40 MPa for 120 min. The working principle of the back-pressure regulator was to stabilize the pressure by using gas–energy storage and adjust the pressure with microtubule and pump assistance.

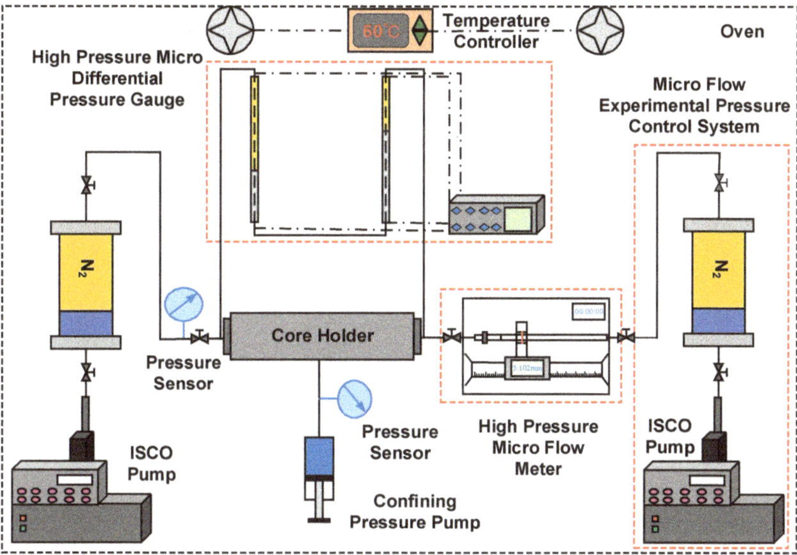

Figure 2. Schematic diagram of the experimental device.

2.3. Core Samples

The experiment used four natural sandstone cores from a certain area of Shanxi Province, China. The lithology of these cores was sandstone. The cores porosities were measured by a KXD-II porosity tester. The particle volume method was used to measure the porosity of the core. Core analysis method SY/T 5336-2006 was used as the analysis basis. The pore throat size distribution curve was obtained from high-pressure mercury injection experiments. The SY/T 5346-2005 rock capillary pressure curve method was used as the basis for analysis. The dominated throat radius ranging from 0.1224 μm to 1.0382 μm, with the porosity distribution ranging from 9.8% to 20.57%. The distribution of their pore throats is shown in Figure 3, and the base data of experiment cores are shown in Table 1. The core plugs were cleaned with toluene to remove hydrocarbons and dried in an oven at 95 °C for a period of 48 h before the measurements. Nitrogen was regarded as displacing medium.

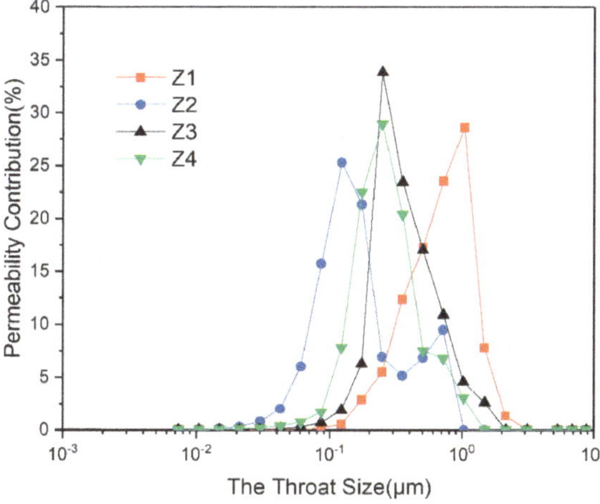

Figure 3. Relationships between throat size and permeability contribution.

Table 1. The base data of experiment cores.

Core Number	Length (cm)	Diameter (cm)	Porosity (%)	Dominated Throat Radius (μm)	Rock Type
Z1	10.78	2.50	20.57	1.0382	Sandstone
Z2	7.46	2.50	10.50	0.1224	Sandstone
Z3	4.09	2.51	11.30	0.2480	Sandstone
Z4	6.26	2.52	9.80	0.2477	Sandstone

2.4. Experimental Procedure

Before the experiment, we adjusted the temperature of the incubator to 60 °C and then placed the experimental core into a core holder to test the sealing of the holder with nitrogen at a pressure of 35 MPa. After the requirement of airtightness was achieved, we modified the inlet and outlet pressure and confining pressure of the intermediate vessel to reach the pre-setting pressure. Core samples were maintained at constant differential pressure for at least 3 h to achieve flow and stress equilibrium. When the pressure and flow state reached the equilibrium, the inlet pressure and the pressure difference (between inlet and outlet) were recorded. The criterion for gas reaching a steady flow was that the flow rate remained constant for a period of time when all other conditions remained the same (effective stress, upstream pressure, downstream pressure, temperature). If the gas flow did not reached a stable state, it would be judged after a period of time until it reached a stable state. The specific steps are to measure the flow rate firs and then measure the flow rate again after 1 h under the same conditions. If the flow rate is the same, it is considered to have reached a steady flow. Additionally, to make the result more accurate, we measured the flow three times and used the average number. The measurement interval was 30 min to detect the gas flow state and air tightness of the experimental system. After the measurement was completed, the inlet and outlet pressure were adjusted for the next measurement and the above steps were repeated. By changing the input cases (differential pressure varied from 0.1 to 1 MPa, the back pressure varied from 0 to 30 MPa and the net confining pressure was kept as 4 MPa), different sets of results can be acquired.

3. Results and Discussion

3.1. Relationships between Permeability and Flowrate

According to the Klinkenberg slippage theory, the permeability decreases with the increase of the average pressure. With a fixed back pressure, the average pressure increases as the flowrate increases. So regardless of the value of back pressure, the permeability predicted by Klinkenberg slippage theory always decreases with increasing flowrate and becomes closer to Klinkenberg permeability. Figures 4–7 shows the permeability versus flowrate at different back pressures by the lines with different colors. The blue line with square markers shows the flowrate–permeability relationship, which is described by classical slippage theory: when the back pressure equals the atmospheric pressure, the apparent permeability decreases as the flowrate increases. The red, dashed line is the Klinkenberg permeability calculated by Klinkenberg slippage theory, which does not change with the flowrate.

It can be seen from Figure 4 that the apparent permeability of the core dropped as the flowrate increased under different back pressures (0.2 MPa, 0.6 MPa, 2 MPa, 8 MPa, 16 MPa, 24 MPa). Moreover, the apparent permeability approached the Klinkenberg permeability when the flowrate became larger. The following experimental data can be used as the proof of this finding: when the back pressure was 0.2 MPa, the apparent permeability at the flowrate of 3.77×10^{-3} cm/s was 1.27 times larger than the Klinkenberg permeability and 0.145 times more than that at the flowrate of 3.37×10^{-2} cm/s; when the back pressure increased to 16 MPa, the apparent permeability at the flowrate of 2.24×10^{-3} cm/s was 1.01 times larger than the Klinkenberg permeability and 0.01 times greater than that at the flowrate of 1.34×10^{-2} cm/s.

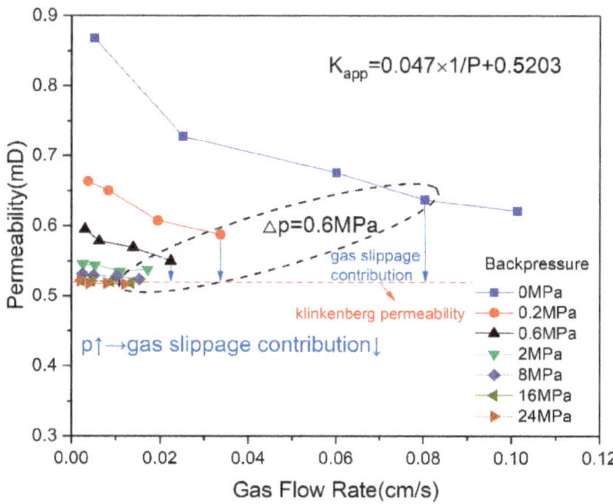

Figure 4. Permeability versus flowrate at diverse backpressures for core Z1.

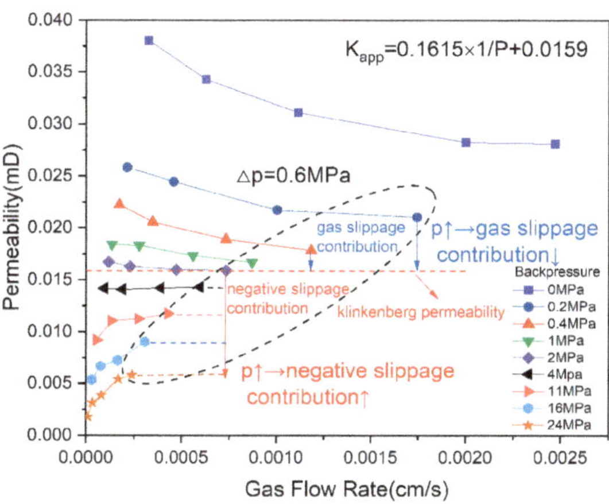

Figure 5. Permeability versus flowrate at diverse backpressures for core Z2.

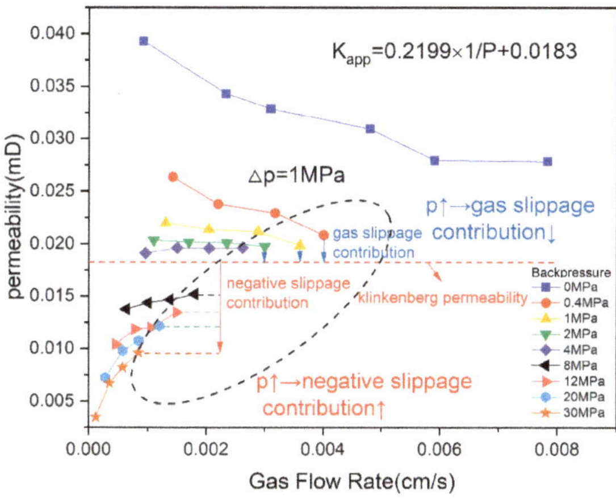

Figure 6. Permeability versus flowrate at diverse backpressures for core Z3.

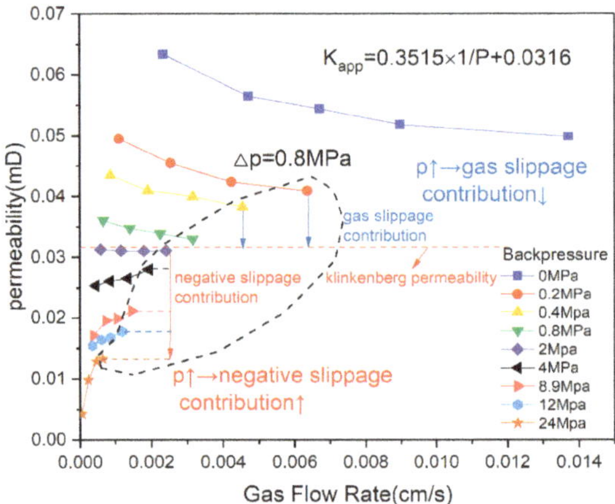

Figure 7. Permeability versus flowrate at diverse backpressures for core Z4.

As shown in the black oval dotted curve in Figure 4, under the same differential pressure (0.6 MPa), the apparent permeability decreased monotonously and approached the Klinkenberg permeability with the increase of the back pressure (the decrease of flowrate). In addition, as the back pressure increased, the contribution of the slippage effect to the apparent permeability decreased: the greater the back pressure and the corresponding apparent permeability got closer to the Klinkenberg permeability. By analyzing the data above, it can be found that the relationship between the flowrate and apparent permeability of the low-permeability core is in accordance with Klinkenberg slippage theory.

The permeability curves of tight core Z2 (Figure 5) to Z4 (Figure 7) have similar characteristics. The results for core Z2 (Figure 5) were used as an example to analyze the change in gas permeability. It can be seen from Figure 5 that under low back pressure (see the results at backpressures of 0.2 MPa, 0.4 MPa, 1 M Pa, 2 MPa), the apparent permeability of the core decreased with the increase of the flowrate, approaching the Klinkenberg permeability predicted by Klinkenberg slippage theory. The pattern shows that the apparent permeability became closer to the Klinkenberg permeability as the flowrate increased, which follows the Klinkenberg slippage theory. When the back pressure was relatively high (referring to the experimental results of 11 MPa, 16 MPa, 24 MPa), the apparent permeability increased with the increase of the flowrate and presented the "negative slippage" law, which is opposite to the Klinkenberg slippage theory. It can also be seen from Figure 5 that the lower the flowrate (the larger the back pressure, the lower the pressure difference), the greater the apparent permeability deviated from Klinkenberg permeability predicted by the Klinkenberg slippage theory. The data from Figure 5 shows that when the back pressure was 0.4 MPa, the apparent permeability decreased with an increasing flowrate. As the flowrate equaled 1.72×10^{-4} cm/s, the apparent permeability was 1.25 times larger than the apparent permeability at the flowrate of 1.18×10^{-3} cm/s. When the back pressure was 24 MPa, the apparent permeability increased with the increase of the flowrate. At the flowrate of 9.28×10^{-6} cm/s, the apparent permeability was 0.312 times larger than the apparent permeability at the flowrate of 2.38×10^{-4} cm/s.

As shown in the black, oval, dotted curve in Figure 5, when the differential pressure was the same and the back pressure was relatively low (0.2 MPa, 0.4 MPa, 1 MPa, 2 MPa), the slippage effect contribution and the apparent permeability decreased as the flowrate decreased (back pressure increased), and the permeability value approached the Klinkenberg value gradually. This regular pattern is in line with the Klinkenberg slippage theory. However, under high back pressure (4 MPa, 11 MPa, 16 MPa, 24 MPa), the relationship between apparent permeability and the flowrate was not consist with the low back pressure. Instead, as the back pressure increased, the flowrate declined, the permeability decreased rapidly, and the phenomenon of "negative slippage" occurred: the apparent permeability was much smaller than the Klinkenberg permeability. Moreover, the deviation degree became greater as the flowrate got lower (the back pressure increased), and the negative slippage contribution increased with the increase of back pressure. Moghadam [58] also drew the conclusion that the apparent permeability increased with the increase of flowrate and reached a stable value at last, which is consistent with our results.

Under low pressure, the permeability gradually decreased with the increase of back pressure and the increase of flow rate to approach the Klinkenberg permeability. This is because at low pressure, the slippage effect will enhance the gas migration capacity in the core [24]. The increase in back pressure and the increase in flow rate can both increase the average pressure in the core. When the average pressure increases, the mean free path of the molecule is reduced and the slippage effect is reduced, so the permeability decreases with increasing back pressure and flow rate. Therefore, increasing test pressure is the most straightforward method to reduce the effect of gas slip, which is the underlying principle of Klinkenberg correlation.

At high pressure, the permeability decreased with increasing back pressure and decreasing flow rate and was much smaller than the Klinkenberg permeability. This is because under high pressure, the mean free path of the gas is significantly smaller, much smaller than the radius of the average pore throat. Thus, the slippage effect can be neglected [44,47] and the micro-scale flow effect of high-pressure gas appears in the tight core [61]. The interaction of high-pressure and microscale effects will result in changes in the physical properties of the gas. As the outlet pressure increases, the gas gradually changes from lean gas to dense gas and further approaches the liquid state. At higher outlet pressures, the smaller the pressure difference, the greater the flow resistance, and the gas flow under different pressure differences in the microtubes exhibits non-linear characteristics. With the increase of outlet pressure and the decrease of tube diameter, the micro-scale flow effect of high-pressure gas becomes more obvious [61]. Therefore, the permeability decreases as the back pressure increases and the flow rate decreases.

Recall from the findings in Section 3.1. that the relationship between the flowrate and the apparent permeability of the low-permeability core is consistent with the Klinkenberg slippage theory. However, the apparent permeability of high-pressure gas in tight cores is not as predicted by Klinkenberg slippage theory; it gets closer to the Klinkenberg permeability as the flowrate decreases (average pressure increases), but is significantly lower than the Klinkenberg permeability. In addition, the apparent permeability increases with the increasing flowrate under high back-pressure condition.

3.2. Measured Flowrate versus Theoretical Flowrate

Figures 8–11 present the measured flowrate versus theoretical flowrate at different back pressures with cores Z1–Z4, respectively. As can be seen from these figures, the points of different colors represent the relationship between the theoretical flowrate and the measured one under different back pressures, and the red, dashed line represents the case where the measured flowrate was the same as the theoretical flowrate. In Figure 8, the flowrate of the core with a mainstream pore radius of 1.0382 µm calculated by the Klinkenberg slippage theory was equal to the measured one (referring to the experimental results of the back pressure from 0 to 24 MPa), which showed that the Klinkenberg slippage theory is capable of properly estimating the flow capacity of a low-permeability core under both high and low pressure.

As can be seen from Figures 9–11, the flowrate calculated by Klinkenberg slippage theory is the same as the measured one under low pressure (see the results in Figures 6 and 7 under back pressure from 0 to 4 MPa and Figure 8 under back pressure from 0 to 1.98 MPa). However, with the increasing back pressure (flowrate declines), the predicted value gradually becomes larger than the actual measured value, and the deviation increases (referring to the results of Figure 6 under back pressure from 11 to 24 MPa, Figure 7 under back pressure from 8 to 30 MPa, and Figure 8 under back pressure from 4 to 24 MPa). The differential pressure within the red, dotted circle is larger than in the black, dotted circle, which indicates that when the back pressure remains same, the smaller the differential pressure (the smaller the flowrate), the larger difference between theory value and actual measured value. It shows that the actual seepage capacity of tight cores under high pressure and low velocity, represented by permeability obtained by Klinkenberg slippage theory, is overestimated. Thus, it is inappropriate to evaluate the actual flow capacity of gas in a tight reservoir using the Klinkenberg permeability, which is calculated by the Klinkenberg equation with low-pressure experimental data.

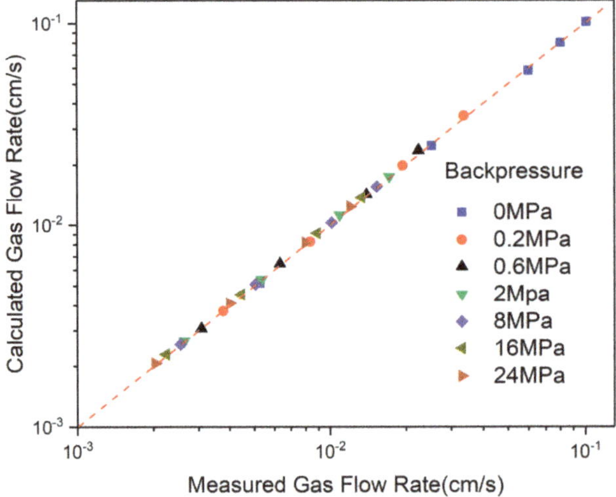

Figure 8. Measured flowrate versus theoretical flowrate at diverse backpressures for core Z1.

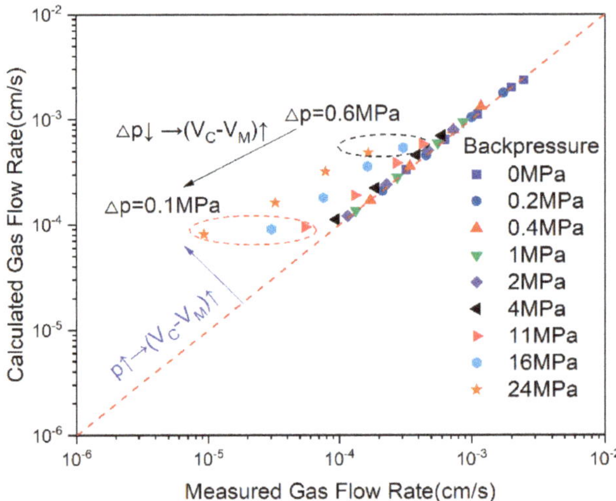

Figure 9. Measured flowrate versus theoretical flowrate at diverse backpressures for core Z2.

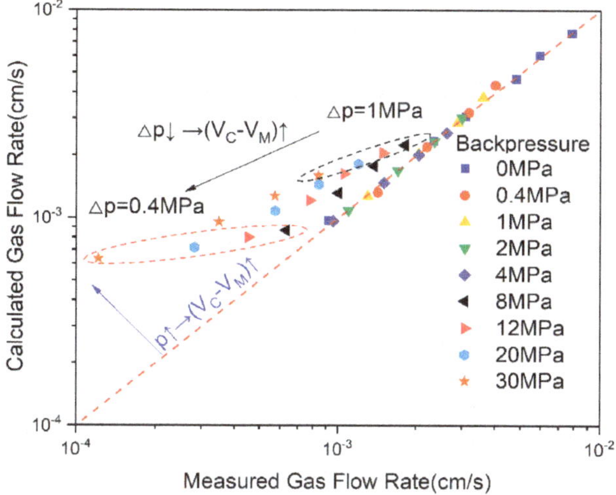

Figure 10. Measured flowrate versus theoretical flowrate at diverse backpressures for core Z3.

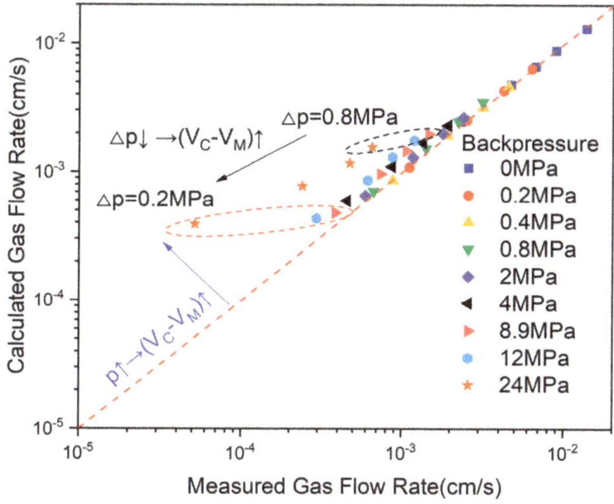

Figure 11. Measured flowrate versus theoretical flowrate at diverse backpressures for core Z4.

4. Conclusions and Suggestions

In this study, new high-pressure gas permeability measuring equipment was applied to measure the nitrogen apparent permeability of a low-permeability core and three tight cores under large back-pressure range (0.1–30 MPa) and low differential pressure (0.1–1 MPa).

Unlike the low-permeability core, the permeability of a tight core is not a constant but a function related to the flowrate. The relationship between flowrate and apparent permeability of low-permeability cores is consistent with Klinkenberg slippage theory, nevertheless, the actual gas seepage law in tight cores under high-pressure conflicts with Klinkenberg slippage theory. At high back pressure, the gas flow in the tight core exhibits a "negative slippage" phenomenon. In terms of the apparent permeability, it is falling away from the Klinkenberg permeability significantly in the real situation, instead of getting closer to the Klinkenberg permeability with declining flowrate (increasing back pressure). Besides, the difference becomes larger when the flowrate gets lower (the back pressure gets larger).

Due to the negative slippage of nitrogen in the tight core under high pressure, the apparent permeability of the tight core at high back pressure and low flowrate is much lower than the Klinkenberg permeability obtained by Klinkenberg slippage theory. Therefore, Klinkenberg slippage theory is only applicable to evaluate the seepage capacity within low-permeability reservoirs or tight reservoirs under low pressure. Which means, it is not suitable to use the Klinkenberg permeability obtained by Klinkenberg slippage theory to evaluate the actual seepage capacity of high-pressure gas in tight reservoirs. Hence, in order to accurately evaluate the actual seepage capacity of the gas in a tight reservoir, it is necessary to directly test the permeability of the tight core at a high back pressure and low flowrate.

Author Contributions: Conceptualization, J.Z. and X.Y.; methodology, J.Z.; validation, W.A.; formal analysis, J.Z.; investigation, J.G. and L.W.; writing—original draft preparation, J.Z.; writing—review and editing, X.Y., L.W. and J.G.; project administration, X.Y.; funding acquisition, X.Y.

Funding: This research was funded by the National Natural Science Foundation of China (51334007), the National Science and Technology Major Projects (2017ZX05009-004), and National Science and Technology Major Projects (2016ZX05050012).

Conflicts of Interest: The authors declare no conflict of interest.

References

1. Yuko, K.; Kazuo, F. Some Predictions of Possible Unconventional Hydrocarbons Availability Until 2100. In Proceedings of the SPE Asia Pacific Oil and Gas Conference and Exhibition, Jakarta, Indonesia, 17–19 April 2001; pp. 1–10. [CrossRef]
2. Jia, C.Z.; Zou, C.N.; Li, J.Z.; Li, D.H.; Zheng, M. China's tight oil evaluation criteria, main types, basic characteristics and resource prospects. *J. Pet.* **2012**, *33*, 343–350.
3. Wang, L.; Tian, Y.; Yu, X.Y.; Wang, C.; Yao, B.W.; Wang, S.H.; Winterfeld, P.H.; Wang, X.; Yang, Z.Z.; Wang, Y.H.; et al. Advances in improved/enhanced oil recovery technologies for tight and shale reservoirs. *Fuel* **2017**, *210*, 425–445. [CrossRef]
4. Velasco, R.; Panja, P.; Pathak, M.; Deo, M. Analysis of North-American Tight oil production. *AICHE J.* **2018**, *64*, 1479–1484. [CrossRef]
5. Hu, S.Y.; Zhu, R.K.; Wu, S.T.; Bai, B.; Yang, Z.; Cui, J.W. Exploration and development of the benefits of China's terrestrial tight oil under the background of low oil prices. *Exp. Dev.* **2018**, *45*, 1–12. [CrossRef]
6. Yu, W.; Lashgari, H.R.; Wu, K.; Sepehrnoori, K. CO_2 injection for enhanced oil recovery in bakken tight oil reservoirs. *Fuel* **2015**, *159*, 354–363. [CrossRef]
7. Wang, X.Z.; Peng, X.L.; Zhang, S.J.; Du, Z.W.; Zeng, F.H. Characteristics of oil distributions in forced and spontaneous imbibition of tight oil reservoir. *Fuel* **2018**, *224*, 280–288. [CrossRef]
8. Kathel, P.; Mohanty, K.K. Wettability Alteration in a Tight Oil Reservoir. *Energy Fuels* **2013**, *27*, 6460–6468. [CrossRef]
9. Huang, S.; Yao, Y.D.; Zhang, S.; Ji, J.H.; Ma, R.Y. A Fractal Model for Oil Transport in Tight Porous Media. *Transp. Porous Med.* **2018**, *121*, 725–739. [CrossRef]
10. Wan, T.; Yang, S.L.; Wang, L.; Sun, L.T. Experimental investigation of two-phase relative permeability of gas and water for tight gas carbonate under different test conditions. *Oil Gas Sci. Technol.* **2019**, *74*, 1–9. [CrossRef]
11. Ghanbarian, B.; Torres-Verdín, C.W.; Lake, L.; Marder, M. Gas permeability in unconventional tight sandstones: Scaling up from pore to core. *J. Pet. Sci. Eng.* **2018**, *173*, 1163–1172. [CrossRef]
12. Nelson, P. Pore-throat sizes in sandstones, tight sandstones, and shales. *AAPG* **2009**, *93*, 329–340. [CrossRef]
13. Liu, Y.M.; Shen, B.; Yang, Z.Q.; Zhao, P.Q. Pore Structure Characterization and the Controlling Factors of the Bakken Formation. *Energies* **2018**, *11*, 2879. [CrossRef]
14. Yao, J.L.; Deng, X.Q.; Zhao, Y.D.; Han, T.Y.; Chu, M.J.; Pang, J.L. Characteristics of tight oil in the Yanchang Formation of the Ordos Basin. *Pet. Explor. Dev.* **2013**, *40*, 150–158. [CrossRef]

15. Gao, H.; Li, H.A. Pore structure characterization, permeability evaluation and enhanced gas recovery techniques of tight gas sandstones. *J. Nat. Gas Sci. Eng.* **2016**, *28*, 536–547. [CrossRef]

16. Maxwell, J.C. On Stresses in Rarified Gases Arising from Inequalities of Temperature. *Philos. Trans. R. Soc. Lond.* **1879**, *170*, 231–256. [CrossRef]

17. Nazari Moghaddam, R.; Jamiolahmady, M. Slip flow in porous media. *Fuel* **2016**, *173*, 298–310. [CrossRef]

18. Wang, H.L.; Xu, W.Y.; Cai, M.; Zuo, J. An Experimental Study on the Slippage Effect of Gas Flow in a Compact Rock. *Transp. Porous Med.* **2016**, *112*, 117–137. [CrossRef]

19. Rubin, C.; Zamirian, M.; Takbiri-Borujeni, A.; Gu, M. Investigation of gas slippage effect and matrix compaction effect on shale gas production evaluation and hydraulic fracturing design based on experiment and reservoir simulation. *Fuel* **2019**, *241*, 12–24. [CrossRef]

20. Ziarani, A.S.; Aguilera, R. Knudsen's Permeability Correction for Tight Porous Media. *Transp. Porous Med.* **2012**, *91*, 239–260. [CrossRef]

21. Zhang, W.M.; Meng, G.; Wei, X.Y. A review on slip models for gas microflows. *Microfluid Nanofluid* **2012**, *13*, 845–882. [CrossRef]

22. Sun, F.R.; Yao, Y.D.; Li, G.Z.; Dong, M.D. Transport behaviors of real gas mixture through nanopores of shale reservoir. *J. Pet. Sci. Eng.* **2019**, *177*, 1134–1141. [CrossRef]

23. Darabi, H.; Ettehad, A.; Javadpour, F.; Sepehrnoori, K. Gas flow in ultra-tight shale strata. *J. Fluid Mech.* **2012**, *710*, 641–658. [CrossRef]

24. Klinkenberg, L.J. The permeability of porous media to liquids and gases. *Am. Pet. Inst.* **1941**, *2*, 200–213. [CrossRef]

25. Kundt, A.; Warburg, E. Poggendorfs. *Annu. Rev. Plant Physiol. Plant Mol.* **1875**, *155*, 337–525.

26. Roy, S.; Raju, R.; Chuang, H.F.; Cruden, B.A.; Meyyappan, M. Modeling gas flow through microchannels and nanopores. *J. Appl. Phys.* **2003**, *93*, 4870–4879. [CrossRef]

27. Kazemi, M.; Takbiri-Borujeni, A. An analytical model for shale gas permeability. *Int. J. Coal Geol.* **2015**, *146*, 188–197. [CrossRef]

28. Pizio, O.; Sokołowski, S.; Sokołowska, Z. Electric double layer capacitance of restricted primitive model for an ionic fluid in slit-like nanopores: A density functional approach. *J. Chem. Phys.* **2012**, *137*, 175002–175124. [CrossRef]

29. Manby, F.R.; Stella, M.; Goodpaster, J.D.; Miller, T.F. A Simple, Exact Density-Functional-Theory Embedding Scheme. *J. Chem. Theory Comput.* **2012**, *8*, 2564–2568. [CrossRef]

30. Javadpour, F. Nanopores and Apparent Permeability of Gas Flow in Mudrocks (Shales and Siltstone). *J. Can. Pet. Technol.* **2009**, *48*, 16–21. [CrossRef]

31. Javadpour, F.; Singh, H. Langmuir slip-Langmuir sorption permeability model of shale. *Fuel* **2016**, *164*, 28–37. [CrossRef]

32. Singh, H.; Javadpour, F. Nonempirical Apparent Permeability of Shale. In Proceedings of the Unconventional Resources Technology Conference, Denver, CO, USA, 12–14 August 2013.

33. Wang, H.Y.; Matteo, M.P. A Unified Model of Matrix Permeability in Shale Gas Formations. In Proceedings of the SPE Reservoir Simulation Symposium, Houston, TX, USA, 23–25 February 2015.

34. Cai, J.; Lin, D.; Singh, H.; Wei, W.; Zhou, S. Shale gas transport model in 3D fractal porous media with variable pore sizes. *Mar. Pet. Geol.* **2018**, *98*, 437–447. [CrossRef]

35. Singh, H.; Cai, J. A feature-based stochastic permeability of shale: Part 1—Validation and two-phase permeability in a utica shale sample. *Transp. Porous Med.* **2018**, *126*, 527–560. [CrossRef]

36. Singh, H.; Cai, J. A feature-based stochastic permeability of shale: Part 2—Predicting Field-Scale Permeability. *Transp. Porous Med.* **2019**, *126*, 561–578. [CrossRef]

37. Lopez, B.; Aguilera, R. Sorption-Dependent Permeability of Shales. In Proceedings of the SPE/C SUR Unconventional Resources Conference, Calgary, AB, Canada, 20–22 October 2015.

38. Chen, L.; Zhang, L.; Kang, Q.; Viswanathan, H.S.; Yao, J.; Tao, W. Nanoscale simulation of shale transport properties using the lattice Boltzmann method: Permeability and diffusivity. *Sci. Rep.* **2015**, *5*, 8089. [CrossRef]

39. Civan, F. Effective Correlation of Apparent Gas Permeability in Tight Porous Media. *Transp. Porous Med.* **2010**, *82*, 375–384. [CrossRef]

40. Civan, F.; Rai, C.; Sondergeld, C. Shale-Gas Permeability and Diffusivity Inferred by Improved Formulation of Relevant Retention and Transport Mechanisms. *Transp. Porous Med.* **2011**, *86*, 925–944. [CrossRef]

41. Jones, F.O.; Owens, W.W. A Laboratory Study of Low-Permeability Gas Sands. *Soc. Pet. Eng.* **1980**, *32*, 1631–1640. [CrossRef]
42. Rushing, J.A.; Newsham, K.E.; Lasswell, P.M.; Cox, J.C.; Balsingame, T.A. Klinkenberg-Corrected Permeability Measurements in Tight Gas Sands: Steady-State versus Unsteady-State Techniques. In Proceedings of the SPE Annual Technical Conference and Exhibition, Houston, TX, USA, 26–29 September 2004. [CrossRef]
43. Zhu, G.Y.; Liu, L.; Yang, Z.M.; Liu, X.G.; Guo, Y.G.; Cui, Y.T. Experiment and Mathematical Model of Gas Flow in Low Permeability Porous Media. In Proceedings of the Fifth International Conference on Fluid Mechanics, Shanghai, China, 15–19 August 2007; pp. 534–537.
44. Li, S.; Dong, M.; Li, Z. Measurement and revised interpretation of gas flow behavior in tight reservoir cores. *J. Pet. Sci. Eng.* **2009**, *65*, 81–88. [CrossRef]
45. Dong, M.Z.; Li, Z.W.; Li, S.L.; Yao, J. Permeabilities of tight reservoir cores determined for gaseous and liquid CO_2 and C_2H_6 using minimum backpressure method. *J. Nat. Gas Sci. Eng.* **2012**, *5*, 1–5. [CrossRef]
46. Fathi, E.; Tinni, A.; Akkutlu, I.Y. Correction to Klinkenberg slip theory for gas flow in nano-capillaries. *Int. J. Coal Geol.* **2012**, *103*, 51–59. [CrossRef]
47. You, L.J.; Xue, K.L.; Kang, Y.L.; Liao, Y.; Kong, L. Pore Structure and Limit Pressure of Gas Slippage Effect in Tight Sandstone. *Sci. World J.* **2013**, *2013*, 1–7. [CrossRef]
48. Tanikawa, W.; Shimamoto, T. Comparison of Klinkenberg-corrected gas permeability and water permeability in sedimentary rocks. *Int. J. Rock Mech. Min. Sci.* **2009**, *46*, 229–238. [CrossRef]
49. Salam, D.D. Novel Analysis to Determine Gas Permeability. In Proceedings of the SPE Annual Technical Conference and Exhibition, Houston, TX, USA, 28–30 September 2015. [CrossRef]
50. Yue, X.A.; Wei, H.G.; Zhang, L.J.; Zhao, R.B.; Zhao, Y.P. Low Pressure Gas Percolation Characteristic in Ultra-low Permeability Porous Media. *Transp. Porous Med.* **2010**, *85*, 333–345. [CrossRef]
51. Liu, G.F.; Fan, Z.Q.; Lu, Y.; Li, S.Y.; Feng, B.; Xia, Y.; Zhao, Q.M. Experimental Determination of Gas Relative Permeability Considering Slippage Effect in a Tight Formation. *Energies* **2018**, *11*, 467. [CrossRef]
52. Moghadam, A.A.; Chalaturnyk, R. Expansion of the Klinkenberg's slippage equation to low permeability porous media. *Int. J. Coal Geol.* **2014**, *123*, 2–9. [CrossRef]
53. Tang, G.H.; Tao, W.Q.; He, Y.L. Gas slippage effect on microscale porous flow using the lattice Boltzmann method. *Phys. Rev. E Stat. Nonlin. Soft Matter. Phys.* **2005**, *72*, 56301. [CrossRef]
54. An, W.Q.; Yue, X.A.; Feng, X.G.; Fang, X. The deviation of gas permeability and classical theory in tight reservoir cores with high pressure. *J. Nat. Gas Sci. Eng.* **2016**, *30*, 331–337. [CrossRef]
55. Sinha, S.; Braun, E.M.; Determan, M.D.; Passey, Q.R.; Leonardi, S.A.; Boros, J.A.; Wood, A.C.; Zirkle, T.; Kudva, R.A. Steady-State Permeability Measurements on Intact Shale Samples at Reservoir Conditions—Effect of Stress, Temperature, Pressure, and Type of Gas. In Proceedings of the SPE Middle East Oil and Gas Show and Conference, Manama, Bahrain, 10–13 March 2013. [CrossRef]
56. An, W.Q.; Yue, X.A.; Feng, X.G.; Fu, J.; Fang, X.; Zou, J.R.; Fang, W. Non-Klinkenberg slippage phenomenon at high pressure for tight core floods using a novel high pressure gas permeability measurement system. *J. Pet. Sci. Eng.* **2017**, *156*, 62–66. [CrossRef]
57. Sander, R.; Pan, Z.; Connell, L.D. Laboratory measurement of low permeability unconventional gas reservoir rocks: A review of experimental methods. *J. Nat. Gas Sci. Eng.* **2017**, *37*, 248–279. [CrossRef]
58. Moghadam, A.A.; Chalaturnyk, R. Rate dependency of permeability in tight rocks. *J. Nat. Gas Sci. Eng.* **2017**, *40*, 208–225. [CrossRef]
59. Cole, W.A.; Wakeham, W.A. The Viscosity of Nitrogen, Oxygen, and Their Binary Mixtures in the Limit of Zero Density. *J. Phys. Chem. Ref. Data* **1985**, *14*, 209–226. [CrossRef]
60. Latto, B.; Saunders, M.W. Viscosity of Nitrogen Gas at Low Temperatures Up to High Pressures: A New Appraisal. *Can. Ioicrnal Chem. Eng.* **1972**, *50*, 765–770. [CrossRef]
61. Fang, X.; Yue, X.A.; An, W.Q.; Feng, X.G. Experimental study of gas flow characteristics in micro-/nano-pores in tight and shale reservoirs using microtubes under high pressure and low pressure gradients. *Microfluid Nanofluid* **2019**, *23*, 5. [CrossRef]

Article

Pressure Transient Performance for a Horizontal Well Intercepted by Multiple Reorientation Fractures in a Tight Reservoir

Guoqiang Xing [1,2], Shuhong Wu [1,2], Jiahang Wang [3], Mingxian Wang [1,4,*], Baohua Wang [1,2] and Jinjian Cao [1]

[1] Research Institute of Petroleum Exploration and Development, China National Petroleum Corporation, Beijing 100083, China; m18600835372@163.com (G.X.); wush@petrochina.com.cn (S.W.); baohuawang@petrochina.com.cn (B.W.); caojinjian@petrochina.com.cn (J.C.)
[2] State Key Laboratory of Enhanced Oil Recovery, Ministry of Science and Technology, Beijing 100083, China
[3] Engineer Institute of SINOPEC Shanghai Offshore Oil & Gas Company, Shanghai 200120, China; wangjiahang.shhy@sinopec.com
[4] School of Earth Science and Engineering, Xi'an Shiyou University, Xi'an 710065, China
* Correspondence: wmx1012683002@163.com

Received: 29 September 2019; Accepted: 5 November 2019; Published: 6 November 2019

Abstract: A fractured horizontal well is an effective technology to obtain hydrocarbons from tight reservoirs. In this study, a new semi-analytical model for a horizontal well intercepted by multiple finite-conductivity reorientation fractures was developed in an anisotropic rectangular tight reservoir. Firstly, to establish the flow equation of the reorientation fracture, all reorientation fractures were discretized by combining the nodal analysis technique and the fracture-wing method. Secondly, through coupling the reservoir solution and reorientation fracture solution, a semi-analytical solution for multiple reorientation fractures along a horizontal well was derived in the Laplace domain, and its accuracy was also verified. Thirdly, typical flow regimes were identified on the transient-pressure curves. Finally, dimensionless pressure and pressure derivative curves were obtained to analyze the effect of key parameters on the flow behavior, including fracture angle, permeability anisotropy, fracture conductivity, fracture spacing, fracture number, and fracture configuration. Results show that, for an anisotropic rectangular tight reservoir, horizontal wells should be deployed parallel to the direction of principal permeability and fracture reorientation should be controlled to extend along the direction of minimum permeability. Meanwhile, the optimal fracture number should be considered for economic production and the fracture spacing should be optimized to reduce the flow interferences between reorientation fractures.

Keywords: semi-analytical model; reorientation fractures; horizontal well; tight reservoir; flow behavior

---

1. Introduction

As one of the most effective stimulation techniques to increase the recovery of hydrocarbons, fractured horizontal wells have been widely adopted to develop tight reservoirs. In some tight reservoirs, reorientation fractures may be formed when an increase in pore pressure in a hydraulically-fractured area or the depletion of the reservoir pressure occurs [1,2]. For reorientation fractures, previous studies mainly focused on field case studies and fracture propagation [2,3]. Compared with hydraulically-fractured vertical wells, fractured horizontal wells can significantly enhance well productivity by increasing the reservoir contact area. For reservoir engineers, pressure transient analysis is one of the most effective techniques to diagnose the flow behavior of hydraulically-fractured horizontal wells and evaluate the influence of key parameters on wellbore pressure, such as fracture spacing, fracture number and fracture configuration, which are intuitively important for efficient and economical development of tight reservoirs.

Research on the performance of horizontal wells with multiple hydraulic fractures has mainly focused on analytical and semi-analytical models [4–10]. Larsen and Herge presented the pressure transient behavior of horizontal wells with single or multiple finite-conductivity fractures in a three-dimensional unbounded reservoir using the Laplace transform method [6,7]. However, they only demonstrated the early-and middle-stage characteristics of multiple fractured horizontal wells without considering the outer boundary. Chen and Raghavan (1997) presented the solution of a multiply-fractured horizontal well in a rectangular drainage region by reforming the point-source solution of Ozkan and Raghavan [5], which laid a solid foundation for this study [8]. Zerzar and Bettam further extended our ability to understand the pressure behavior of a horizontal well with several fractures in closed anisotropic reservoirs [10]. However, most previous models are introduced based on the assumption that the hydraulic fracture is an ideal planar fracture. The direct observation on fractured cores and micro-seismic fracture monitoring demonstrated that the hydraulic fracturing technology can form complex fracture patterns [11–14], such as the non-planar fracture, the reorientation fracture. Thus, the planar fracture solution may not be accurate to analyze the pressure transient of horizontal wells with multiple reorientation fractures.

Recently, in order to analyze the flow behavior of non-planar fractures in vertical wells or horizontal wells, Luo et al. proposed that each fracture-wing should has its own coordinate system and fracture discrete equation to couple with the analytical reservoir solution, which was later called the fracture-wing method [15–18]. However, the fracture-wing method is based on a harsh condition that the exchange of the fluid only occurs in the junction of two fracture wings, which suggests that the fracture-wing model is also not applicable to horizontal wells with multiple reorientation fractures. Lately, by combining an analytical reservoir solution with a discretized fracture panel solution, the nodal analysis technique, which can calculate the pressure and flow rate of each discrete fracture node, was proposed to simulate the flow behavior for the well with complex fracture networks [19–21]. Although this technique may be suitable to analyze the flow behavior of fractured horizontal wells, this semi-analytical approach forms a large sparse matrix which may reduce the computational speed. In terms of the flow behavior of the reorientation fracture on a vertical well, Wu et al. [22] further employed the nodal analysis technique to establish a semi-analytical solution. Similarly, many discrete nodes will be definitely created when the nodal analysis technique is applied to the multiple fractured horizontal well model, which can increase the bandwidth of the related matrix and, thus, is not conductive to the rapid simulation of the multiple reorientation fractures.

To the best of our knowledge, most fractured horizontal well models are established based on the ideal planar fracture model. However, with the rapid development of hydraulic fracturing, reorientation fractures can be formed during the fracturing process, particularly in soft and shallow tight reservoirs [1,11,12]. For these complex multi-stage horizontal wells, the traditional fractured horizontal well model cannot easily describe the geometry of fracture patterns and analyze the transient flow behavior for fractured horizontal wells.

In this study, based on the combination of the nodal analysis technique and the fracture-wing method, we present a novel equation to describe the flow inside the reorientation fracture. Compared with the work of Wu et al. [22], the new fracture equation in this work can help to avoid calculating the pressure and flow rate at all fracture segments and, thus, can greatly improve the computational speed. On the basis of the new semi-analytical approach, we obtain a semi-analytical pressure solution for multiple reorientation fractures along a horizontal well in an anisotropic tight reservoir.

The remainder of this paper is organized as follows: In Section 2, we present the basic physical description of the model. The new semi-analytical solution based on the new fracture equation will be generated in Section 3. In order to verify the accuracy of the semi-analytical solution, we will use two special cases presented by the previous literature to compare with our semi-analytical solution in Section 4. In Section 5, the flow characteristics and production rate distribution along each reorientation fracture will be discussed in detail. In addition, the influence of key parameters on pressure transient will also be investigated in Section 5, such as fracture number, fracture spacing and fracture configuration.

2. Physical Model

In this section, the physical model is introduced for pressure transient analysis of a horizontal well with multiple reorientation fractures in an anisotropic rectangular tight reservoir. Figure 1 presents the schematic of a horizontal well with two reorientation fractures, and every hydraulic fracture reoriented twice, which is significantly different from the physical model of Wu et al. [22]. Three parts, including principle fracture, irregular curve fracture and reoriented fracture, constitute the reorientation fracture. The basic assumptions are made as follows:

(1) The tight reservoir is rectangular in shape. Single-phase Darcy flow occurs in the anisotropic tight reservoir. The permeability of the reservoir in the x- and y-direction is k_x and k_y, respectively.

(2) The horizontal well is fully intercepted by arbitrary reorientation fractures with constant height and width. The horizontal wellbore is deployed parallel to the x-axis, and all fractures' height is equal to the reservoir thickness.

(3) Fluid in the horizontal wellbore and reservoir is constant viscosity and slightly compressible, and flow rates from each reorientation fracture contribute to the horizontal well total rate, although they may change over time. The horizontal wellbore is infinite-conductivity and no pressure loss occurs along the wellbore.

(4) This reservoir is fully penetrated by all reorientation fractures. For the i-th reorientation fracture, its principal fracture angle is $\theta_{i,1}$, and its reoriented fractures angles are $\theta_{i,2}$ and $\theta_{i,3}$, respectively. All reorientation fractures have a finite conductivity and their tips are assumed to be impermeable boundaries.

(5) Gravity effect is negligible, as well as the influence of the temperature on different reservoir parameters.

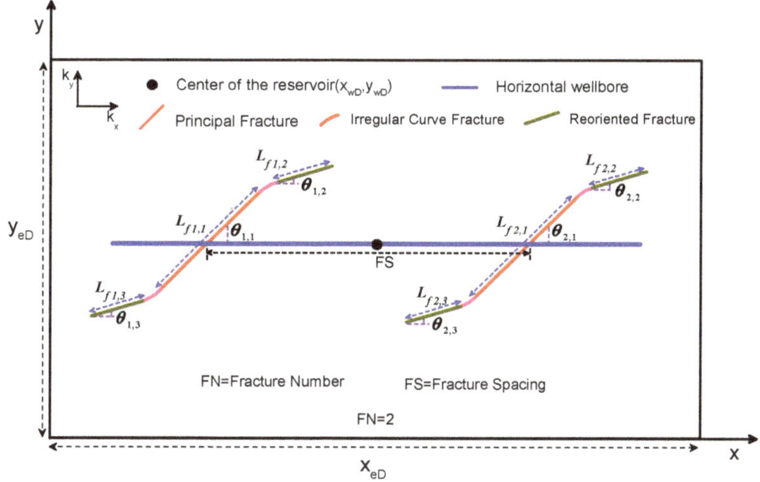

Figure 1. Schematic of two reorientation fractures along a horizontal well.

3. Mathematical Models

3.1. Dimensionless Definitions

To simplify the governing equation and definite conditions, the following dimensionless parameters are defined for the mathematical model.

The dimensionless pressure in the tight reservoir and all reorientation fractures, and the dimensionless time, are expressed as follows:

$$p_D = \frac{2\pi\bar{k}h(p_i - p)}{q_{sc}\mu B} \tag{1}$$

$$p_{fD} = \frac{2\pi\bar{k}h(p_i - p_f)}{q_{sc}\mu B} \tag{2}$$

$$t_D = \frac{\bar{k}t}{\mu(\phi c_t)_m L_R^2} \tag{3}$$

In the fracture model, the dimensionless fracture conductivity, F_{cD}, the dimensionless flow rate, q_{fD}, and the flow rate at the center of each fracture segment, q_{fwD}, are defined as follows:

$$F_{cD} = \frac{k_f w_f}{\bar{k} L_R}, \ q_{fD} = \frac{q_f L_R}{q_{sc}}, \ q_{fwD} = \frac{q_{fw}}{q_{sc}} \tag{4}$$

Other dimensionless definitions in the reservoir model or the reorientation fracture model are given as follows:

$$\beta = \sqrt{\frac{k_x}{k_y}}, \ x_D = \frac{x}{L_R}, \ y_D = \frac{y}{L_R}, \ l_D = \frac{l}{L_R}, \ \bar{k} = \sqrt{k_x k_y}, \ L_R = \left[\sum_{i=1}^{N}\left(L_{fi,1} + L_{fi,2} + L_{fi,3}\right) + \sum_{i=1}^{m_i} L_i\right]/2N \tag{5}$$

where β is the anisotropic factor and L_R is the reference length and $\bar{k}$ is the equivalent system permeability, m_i is the total discrete number of all irregular curve fractures and L_i is the length of the equivalent i-th planar fracture for the irregular curve fracture.

3.2. Reservoir Flow Model

On the basis of the Laplace transform, a fracture segment solution can be obtained for an anisotropic rectangular tight reservoir as follows [22]:

$$\tilde{p}_D = \frac{\pi\beta}{x_{eD}}\int_{l_D} \tilde{q}_{Di}\frac{\cosh\psi_0\left(y_{eD}-|y_D \pm y_{Dmi}|\right)}{\psi_0 \sinh\psi_0 y_{eD}}dl_D$$
$$+\frac{2\pi\beta}{x_{eD}}\int_{l_D} \tilde{q}_{Di}\sum_{n=1}^{\infty}\frac{\cosh\psi_n\left(y_{eD}-|y_D \pm y_{Dmi}|\right)}{\psi_n \sinh\psi_n y_{eD}}\cos\left(\frac{n\pi x_D}{x_{eD}}\right)\cos\left(\frac{n\pi x_{Dmi}}{x_{eD}}\right)dl_D \tag{6}$$
$$\psi_n = \sqrt{\left(s\beta + \left(\beta\frac{n\pi}{x_{eD}}\right)^2\right)} \ n = 0,1,2\cdots$$

where (x_{Dmi}, y_{Dmi}) is a movable infinitesimal unit along the fracture segment and (x_D, y_D) is any point in the fracture segment.

However, Equation (6) has an implicit assumption that the minute fracture segment should be parallel to the coordinate axis. In order to get the pressure solution of a minute fracture segment with a reorientation angle θ (Figure 2) between the fracture and the horizontal wellbore, the following equations are used:

$$x_{Dmi} = x_{Di} + \chi\cos\theta_i \ \ y_{Dmi} = y_{Di} + \chi\sin\theta_i \tag{7}$$

$$\tilde{p}_{Di}(x_D, y_D, x_{Di}, y_{Di}, s) = \frac{\pi\beta}{x_{eD}} \sum_{i=1}^{N_I} \tilde{q}_{D,i} \int_{-\frac{\Delta l_{Di}}{2}}^{\frac{\Delta l_{Di}}{2}} \frac{\cosh\psi_0\left(y_{eD}-|y_D \pm (y_{Di}+\chi\sin\theta_i)|\right)}{\psi_0\sinh\psi_0 y_{eD}}\, d\chi$$

$$\frac{2\pi\beta}{x_{eD}} \sum_{i=1}^{N_I} \tilde{q}_{D,i} \int_{-\frac{\Delta l_{Di}}{2}}^{\frac{\Delta l_{Di}}{2}} \sum_{n=1}^{\infty} \frac{\cosh\psi_n\left(y_{eD}-|y_D \pm (y_{Di}+\chi\sin\theta_i)|\right)}{\psi_n\sinh\psi_n y_{eD}} \cos\left(\frac{n\pi x_D}{x_{eD}}\right)\cos\left(\frac{n\pi(x_{Di}+\chi\cos\theta_i)}{x_{eD}}\right) d\chi \tag{8}$$

$$\psi_n = \sqrt{s\beta + \left(\beta\frac{n\pi}{x_{eD}}\right)^2}\quad n = 0,1,2\cdots;\ i = 1,2\cdots N_I$$

where (x_{Di}, y_{Di}) is the midpoint of the fracture segment, θ_i represents the angle of i-th fracture segment, and Δl_{Di} represents the dimensionless length of i-th reorientation fracture segment.

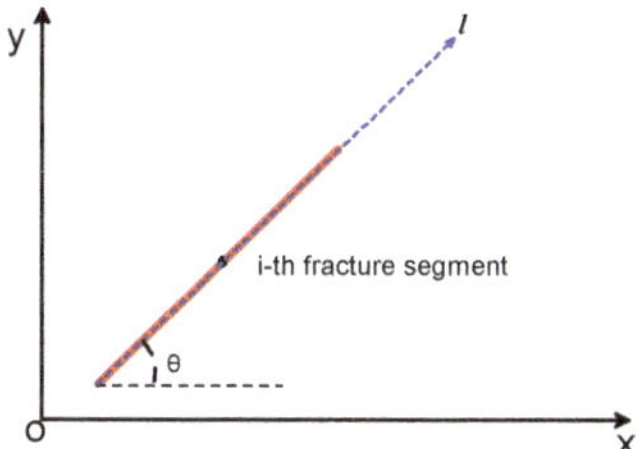

Figure 2. Fracture segment with an azimuth angle of θ.

As stated previously, the reorientation fracture cannot be described by the ideal planar fracture. In order to stimulate the flow characteristics of a horizontal well intercepted by multiple reorientation fractures, we used the approximate approach similar to the boundary element method to handle the irregular shaped boundaries [23]. As illustrated in Figure 3, the irregular curve fracture is replaced by a series of planar fracture segments. Thus, the reorientation fracture system is considered to be composed of several planar fracture segments. The number of planar fracture segments used to describe the irregular curve fracture largely depends on the seepage of the irregular curve fracture.

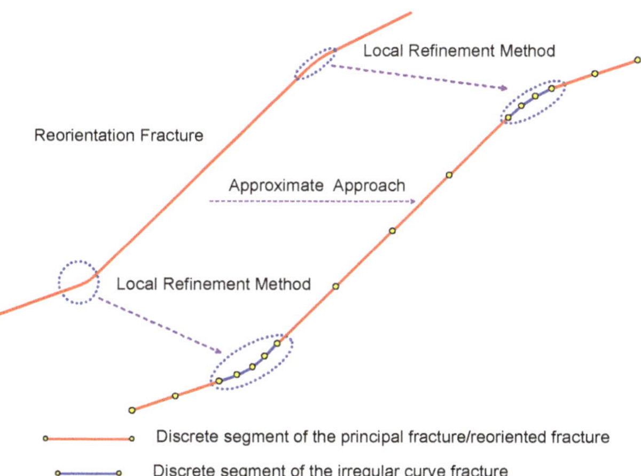

Figure 3. Approximate approach to handle the reorientation fracture.

3.3. Reorientation Fracture Flow Model

In this work, we assumed that the flow in the reorientation fracture is one-dimensional. Taking the i-th fracture segment as an example, we assume that the fluid flows from the ($i+1$)-th fracture segment to the *i*-th fracture segment (Figure 4). Based on the results of Zhou et al. [19], the pressure difference between the *i*-th fracture segment and (*i*+1)-th fracture segment can be written as:

$$p_{fi+1} - p_{fi} = \int_{l_i}^{l_{i+0.5}} \frac{\mu B}{k_f w_f h} q_{fwi}(l,t)dl + \int_{l_{i+0.5}}^{l_{i+1}} \frac{\mu B}{k_f w_f h} q_{fwi+1}(l,t)dl \tag{9}$$

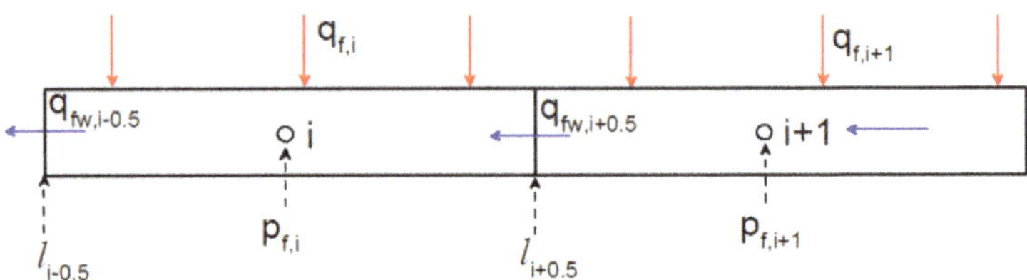

Figure 4. Schematic of fracture discretization.

In order to obtain the wellbore pressure, we take the integration method from the first fracture segment near the horizontal wellbore to the *i*-th fracture segment and derive Equation (10):

$$p_{fi} - p_w = \int_{l_{1-0.5}}^{l_{1+0.5}} \left(\frac{\mu B}{k_f w_f h}\right)_1 q_{fw1}(l,t)dl + \cdots + \int_{l_{(i-1)-0.5}}^{l_{(i-1)+0.5}} \left(\frac{\mu B}{k_f w_f h}\right)_{i-1} q_{fw(i-1)}(l,t)dl$$
$$+ \int_{l_{(i-1)+0.5}}^{l_i} \left(\frac{\mu B}{k_f w_f h}\right)_i q_{fwi}(l,t)dl \tag{10}$$

In addition, according to the principle of mass balance, we can obtain the following equations for each fracture segment:

$$q_{fwi} = q_{fw(i-0.5)} - q_{fi}(l - l_{i-0.5}) \qquad l_{i-0.5} \le l \le l_{i+0.5} \tag{11}$$

The inflow and outflow of each fracture node should satisfy the mass balance:

$$q_{fw,i-0.5} = \sum_{j=1}^{N_I}\left(q_{fj}\Delta l_j\right) - \sum_{j=1}^{i-1}\left(q_{fj}\Delta l_j\right) \tag{12}$$

where N_I is total fracture segments of a wing of the *i*-th fracture.

Substituting Equation (11) and Equation (12) into Equation (10) yields:

$$p_{fi} - p_w = \int_{l_{1-0.5}}^{l_{1+0.5}} \left(\frac{\mu B}{k_f w_f h}\right)_1 \left[\sum_{j=1}^{N_I}\left(q_{fj}\Delta l_j\right) - q_{f1}(l - l_{1-0.5})\right]dl + \cdots +$$
$$\int_{l_{(i-1)-0.5}}^{l_{(i-1)+0.5}} \left(\frac{\mu B}{k_f w_f h}\right)_{i-1} \left[\sum_{j=1}^{N_I}\left(q_{fj}\Delta l_j\right) - \sum_{j=1}^{i-2}\left(q_{fj}\Delta l_j\right) - q_{fi-1}(l - l_{i-1-0.5})\right]dl +$$
$$\int_{l_{(i-1)+0.5}}^{l_i} \left(\frac{\mu B}{k_f w_f h}\right)_i \left[\sum_{j=1}^{N_I}\left(q_{fj}\Delta l_j\right) - \sum_{j=1}^{i-1}\left(q_{fj}\Delta l_j\right) - q_{fi}(l - l_{i-0.5})\right]dl \tag{13}$$

Based on above dimensionless definitions, Equation (13) can be rewritten as follows:

$$p_{wD} - p_{fDi} = \frac{2\pi}{F_{cD}}\left[l_{Di}\sum_{j=1}^{N_I}\left(q_{fDj}\Delta l_{Dj}\right) - \frac{\Delta l_{Di}^2}{8}q_{fDi} - \sum_{j=1}^{i-1}\left(\frac{\Delta l_{Dj}}{2} + l_{Di} - \sum_{n=1}^{j}\Delta l_{Dn}\right)\Delta l_{Dj}q_{fDj}\right] \tag{14}$$

To the best of our knowledge, Equation (14) is a new fracture equation and can be employed to describe the flow inside the reorientation fracture.

In order to obtain the fracture solution in Laplace domain, we should apply the Laplace transform to Equation (14) based on t_D and thus we can have:

$$\tilde{p}_{wD} - \tilde{p}_{fDi} = \frac{2\pi}{F_{cD}}\left[l_{Di}\sum_{j=1}^{N_I}\left(\tilde{q}_{fDj}\Delta l_{Dj}\right) - \frac{\Delta l_{Di}^2}{8}\tilde{q}_{fDi} - \sum_{j=1}^{i-1}\left(\frac{\Delta l_{Dj}}{2} + l_{Di} - \sum_{n=1}^{j}\Delta l_{Dn}\right)\Delta l_{Dj}\tilde{q}_{fDj}\right] \tag{15}$$

For the uniform dimensionless length of Δl_{Di} for each fracture segment, Equation (15) will have the same form as that reported by Cinco-Ley et al., [24]. The simplification of Equation (15) is presented in Appendix A.

The flow equation of the reorientation fracture (Equation (15)) can be given in the matrix form:

$$\mathbf{C\tilde{q}_{fD} + D\tilde{p}_{fD} + E\tilde{p}_{wD} = 0} \tag{16}$$

where $\tilde{\mathbf{q}}_{fD}$ represents the vector of the flow-rate in the reorientation fracture, $\tilde{\mathbf{p}}_{wD}$ is the vector of the horizontal wellbore pressure and $\tilde{\mathbf{p}}_{fD}$ is the vector of the pressure in the reorientation fracture, C, D, and E are the corresponding coefficient matrices obtained from Equation (15).

3.4. Semi-Analytical Solutions

In this paper, we get the semi-analytical pressure solution by discretizing each reorientation fracture into a large number of planar fracture segments, similar to the method illustrated in Figure 4. On the basis of the superposition principle, we can easily get the dimensionless pressure solution for the i-th fracture segment in the Laplace domain, which is presented as a discretized form:

$$\tilde{p}_{Dk,i}\left(x_{Dk},y_{Dk},x_{Di},y_{Di},s\right) = \sum_{k=1}^{N}\sum_{i=1}^{N_I}\tilde{q}_{Dk,i}R_{Dk,i}\left(x_{Dk},y_{Dk},x_{Di},y_{Di},s\right) \tag{17}$$

where N is the reorientation fracture number and:

$$R_{Dk,i}\left(x_{Dk},y_{Dk},x_{Di},y_{Di},s\right) =$$
$$\frac{\pi\beta}{x_{eD}}\int_{-\frac{\Delta l_{Di}}{2}}^{\frac{\Delta l_{Di}}{2}}\frac{\cosh\psi_0\left(y_{eD}-|y_{Dk}\pm(y_{Di}+\chi\sin\theta_i)|\right)}{\psi_0\sinh\psi_0 y_{eD}}d\chi + \frac{2\pi\beta}{x_{eD}}\times$$
$$\int_{-\frac{\Delta l_{Di}}{2}}^{\frac{\Delta l_{Di}}{2}}\sum_{n=1}^{\infty}\frac{\cosh\psi_n\left(y_{eD}-|y_{Dk}\pm(y_{Di}+\chi\sin\theta_i)|\right)}{\psi_n\sinh\psi_n y_{eD}}\cos\left(\frac{n\pi x_{Dk}}{x_{eD}}\right)\cos\left(\frac{n\pi(x_{Di}+\chi\cos\theta_i)}{x_{eD}}\right)d\chi \tag{18}$$
$$\psi_n = \sqrt{s\beta + \left(\beta\frac{n\pi}{x_{eD}}\right)^2}\ n = 0,1,2\cdots i = 1,2\cdots N_I$$

Considering the continuity conditions of pressure and flux along the fracture surface, the following equations can be obtained in the Laplace domain:

$$\tilde{p}_D(x_D,y_D) = \tilde{p}_{fD}(x_D,y_D) \tag{19}$$

$$\tilde{q}_D(x_D,y_D) = \tilde{q}_{fD}(x_D,y_D) \tag{20}$$

Equations (17), (19), and (20) can be further written in the matrix form as follows:

$$\mathbf{A\tilde{q}_{fD} + B\tilde{p}_{fD} + O\tilde{p}_{wD} = 0} \tag{21}$$

where A and B are the corresponding coefficient matrix obtained from Equation (17).

Combining Equations (16) and (21) reveals the following simplified matrix:

$$\mathbf{G}\tilde{\mathbf{q}}_{fD}+\mathbf{H}\tilde{\mathbf{p}}_{wD}=0 \tag{22}$$

where G and H are the corresponding coefficient matrices obtained from Equation (15) and Equation (17).

The sum of the flow rates of the reorientation fracture segments contributes to the total flow rate in the horizontal wellbore:

$$\sum_{k=1}^{N}\sum_{i=1}^{N_I}\left(\Delta l_{Dk,i}\tilde{q}_{fDk,i}\right) = \frac{1}{s} \tag{23}$$

Combining Equations (22) and (23), the following equation in the matrix form is obtained:

$$\begin{bmatrix} G & H \\ M^T & 0 \end{bmatrix}\begin{bmatrix} \tilde{q}_{fD} \\ \tilde{p}_{wD} \end{bmatrix}=\begin{bmatrix} 0 \\ \frac{1}{s} \end{bmatrix} \tag{24}$$

where M^T is the vector of dimensionless length for each reorientation fracture segment.

$$M = \begin{bmatrix} \Delta l_{D1} & \Delta l_{D2} & \cdots & \Delta l_{Dsum} \end{bmatrix}^T \tag{25}$$

Equation (24) is the new semi-analytical solution for multiple reorientation fractures along a horizontal well in a tight reservoir in Laplace domain. According to Equation (24), the reference pressure of all fracture segments except the pressure of the horizontal wellbore does not need to be calculated and, thus, the calculation speed can be significantly increased, which is different from the work of Wu et al. [22]. By solving Equation (25) with the Gaussian elimination method, the pressure of the horizontal wellbore and the flow rate of each reorientation fracture segment can be obtained in Laplace space, and then can be further inverted into the real space by using the Stehfest numerical inverse method [25].

4. Verification and Comparison

The major advantage of the novel semi-analytical model is that we take the fracture reorientation and permeability anisotropy into consideration. To the best of our knowledge, there are no reports on a horizontal well perforated by a single reorientation fracture or multiple reorientation fractures in anisotropic rectangular reservoirs. Therefore, we conduct two special cases presented by Chen and Raghavan to verify the accuracy of our semi-analytical solution [8]. They calculated the pressure transient of a multiple fractured horizontal well in an isotropic rectangular reservoir using the reformulation of the point-source solution presented by Ozkan and Raghavan [4]. The comparison results are shown in Figure 5, where the semi-analytical solution shows a good fit to the results of Chen and Raghavan at every production time [8].

5. Results and Discussion

5.1. Transient Flow Characteristics

The transient flow behavior of this model is illustrated by transient-pressure type curves, which can be employed for pressure and rate transient analysis. In addition, some reservoir parameters can be obtained by type curves matching, such as the average reservoir permeability, fracture number, and fracture reorientation.

For a given set of parameters, the dimensionless pressure and its derivative of two reorientation fractures along the horizontal well in an anisotropic tight reservoir are displayed in Figure 6 to analyze its typical flow regimes. Figure 7 illustrates the fluid flow in the reservoir and reorientation fractures with different flow regimes. Without considering the transition flow regimes, we can easily observe six typical flow regimes in Figure 6: bilinear flow regime (BF), first linear flow regime (FLF), first radial flow regime (FRF), second radial flow regime (SRF), pseudo-radial flow regime (PRF), and pseudo-steady-state flow regime (PSSF).

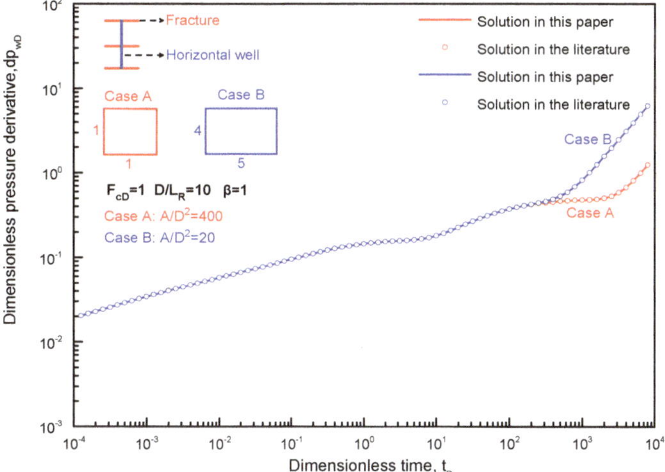

Figure 5. Comparison of pressure derivative for a horizontal well with three vertical fractures with the results of previous literature [8].

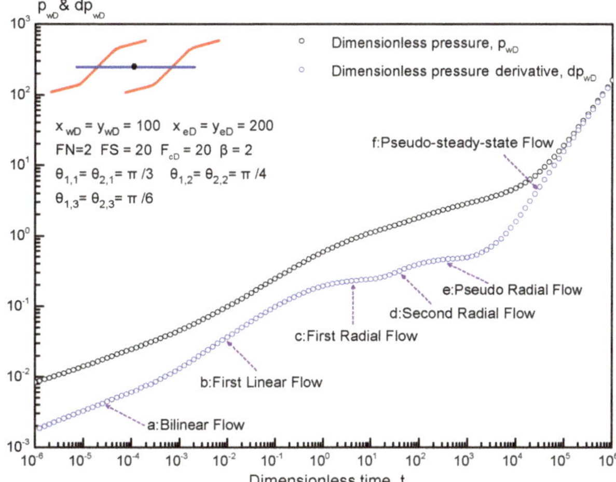

Figure 6. Typical curves of the pressure and pressure derivative for a horizontal well interpreted by two reorientation fractures.

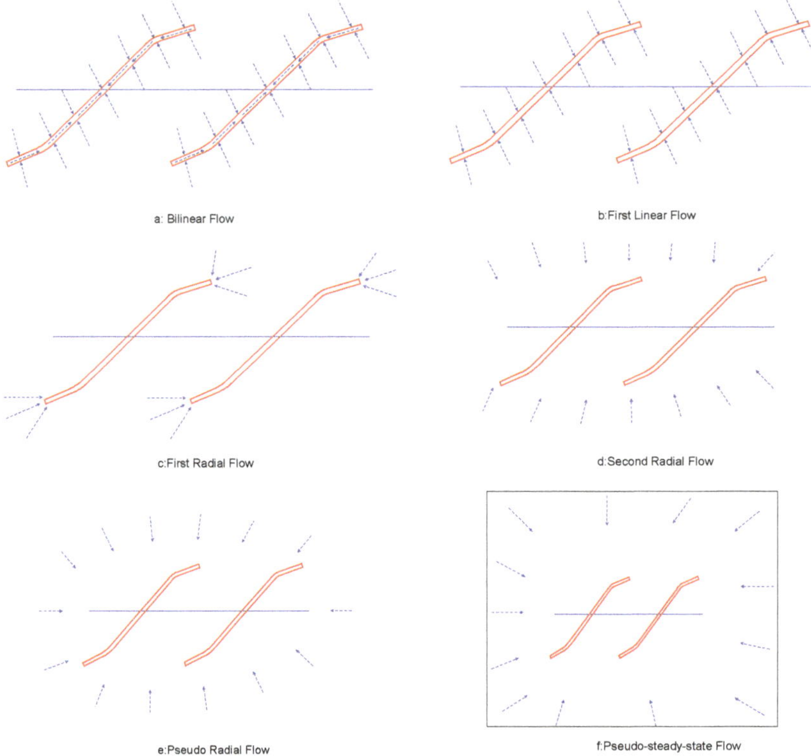

Figure 7. Schematic of the flow regimes for a horizontal well with two reorientation fractures: (**a**) Bilinear flow regime; (**b**) First linear flow regime; (**c**) First radial flow regime; (**d**) Second radial flow regime; (**e**) Pseudo radial flow regime; (**f**) Pseudo-steady-state flow regime.

Bilinear flow: In this regime, two types of linear flow, including the linear flow from the reservoir to reorientation fractures and from reorientation fractures to the horizontal wellbore, occur at the same time (Figure 7a). This flow behavior makes the slope of the dimensionless pressure derivative curve constant. In contrast to conventional multi-stage fractured horizontal wells, the streamline from the reservoir to the reorientation fracture may be not perpendicular to the fracture due to permeability anisotropy and fracture reorientation existing simultaneously. Consequently, the slope of the straight line at this regime is characterized by 1/4–1/3 in the log-log plot, not 1/4 as presented by Cinco-Ley et al. [24]. In addition, this flow behavior only occurs when the reorientation fracture conductivity is considered.

First linear flow: For the conventional finite-conductivity fractured horizontal well model, the slope of the pressure derivative curve is constant to 1/2 in this regime. Due to the fracture reorientation and permeability anisotropy, the slope in this regime is also larger than that of the conventional finite-conductivity fractured horizontal wells. During this regime, the fluid from the reservoir will linearly flow into the individual reorientation fracture and each reorientation fracture works independently on other reorientation fractures.

First radial flow: In the first radial flow regime, one of the most important transient flow characteristics is that the pressure derivative is expressed as a fixed value of 1/(2N). Consequently, the slope of the pressure derivative curve for various fractured horizontal well model may be different, which mainly depends on the fracture number. In this regime, the fluid around reorientation fractures will individually flow into the reorientation fractures. The duration of this regime is highly related to adjacent fracture spacing and reorientation fracture length. Generally, a large reorientation fracture length or a wide fracture spacing can definitely lead to a long duration of the first radial flow regime.

Second radial flow: After the first radial flow, a second radial flow occurs in the tight reservoir. As demonstrated in Figure 6, the dimensionless pressure and its derivative curves are parallel to each other and the slope of these curves is a constant value of 0.36 in this regime. During this period, the interference of reorientation fractures starts to affect the flow between reorientation fractures and reservoir. After this regime, the transient behavior in the reservoir surrounded by the outer fractures reaches a pseudo-radial flow regime.

Pseudo radial flow: In this regime the dimensionless pressure derivative has a constant value of 0.5, and the derivative curve is parallel to the time axis (Figure 6). In the meantime, the fluid flow among the reservoir, the reorientation fractures, and the horizontal wellbore reach a dynamic balance and, thus, the dimensionless pressure derivative stabilizes at 0.5.

Pseudo-steady-state flow: A pseudo-steady-state flow regime occurs in the late-time period for all closed reservoirs. The typical characteristic of this regime is that the imensionless pressure and corresponding derivative curves increase rapidly and finally normalize to a straight line with a constant slope of 1. Generally, the seepage area in this flow regime always presents a circle (Figure 7f).

5.2. Production Rate Distribution

Figures 8 and 9 present the production rate distribution in each reorientation fracture for a horizontal well with three reorientation fractures and the data used in Figures 8 and 9 are presented in Table 1. In this section, we assumed that the inner reorientation fracture is fixed in the center of the tight reservoirs and the distance between the inner reorientation fracture and the outer reorientation fracture is assumed to be constant at FS = 20. Meanwhile, the horizontal wellbore extends parallel to the x-axis.

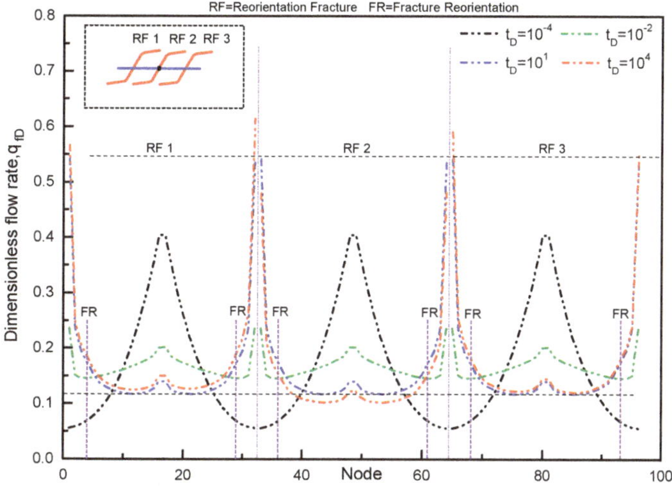

Figure 8. Rate distribution of a horizontal well with three reorientation fractures ($k_x = k_y$).

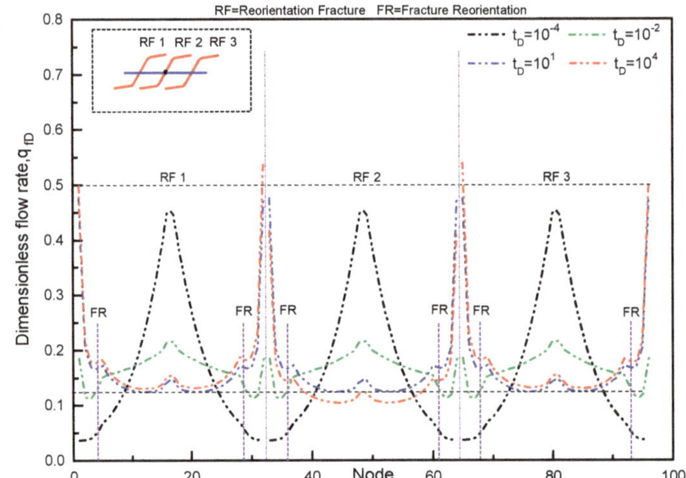

Figure 9. Rate distribution of a horizontal well with three reorientation fractures ($k_x = 4k_y$).

Table 1. Parameters used in Figures 8 and 9 for a horizontal well with three reorientation fractures.

Basic Model Parameters	Value
Drainage area (dimensionless)	200×200
Fracture conductivity (dimensionless)	20
Principal fracture angle (deg)	$\theta_{1x} = \theta_{1y} = \theta_{1z} = 60°$
Reoriented fracture angle (deg)	$\theta_{2x} = \theta_{2y} = \theta_{2z} = 30°$ $\theta_{3x} = \theta_{3y} = \theta_{3z} = 30°$
Fracture spacing (dimensionless)	20
Inner fracture position (dimensionless)	$x_D = y_D = 100$

The production rate distribution of each reorientation fracture along the horizontal well in an isotropic tight reservoir is illustrated in Figure 8 when $k_x = k_y$. In the early flow regime (the black line in Figure 8), the production rate of each fracture segment is symmetrically distributed with respect to the wellbore node. As the pressure wave expands, the production rate in the reoriented fracture significantly increases along the extension of the reoriented fracture (the green line in Figure 8). For a long production time (Figure 8, $t_D = 10^4$), the production rate distribution changes significantly. Since the outermost fractures (RF1 and RF3) have a larger drainage area, they have a higher production rate. In addition, owing to the flow interference, the inner fracture (RF2) has a lower production rate. Furthermore, the production rate distribution of the outermost fractures (RF1 and RF3) (the red line in Figure 8) is asymmetrical. The reoriented fractures away from the horizontal wellbore have a larger production rate due to the larger seepage area caused by fracture reorientation.

Figure 9 presents the production rate distribution for each reorientation fracture in a horizontal well in an anisotropic tight reservoir ($k_x = 4k_y$, FN = 3). The production rate distribution within each reorientation fracture is similar to that in Figure 8. However, for an anisotropic reservoir, fracture reorientation has a more significant effect on the production rate in each reorientation fracture. In the reorientation section, the production rate distribution curves vary sharply (Figure 9), which suggests that the effect of fracture reorientation in anisotropic tight reservoir on type curves is significant and, thus, cannot be neglected.

5.3. Parameter Influence on Transient Pressure Behavior

In this section, we analyze the sensitivity of the transient pressure of multiple reorientation fractures along a horizontal well in an anisotropic tight reservoir. We consider the case of a horizontal well with two reorientation fractures to analyze the effects of some key parameters, such as principal fracture angle (PFA), reoriented fracture angle (RFA), permeability anisotropic factor (PAF), and adjacent reorientation fractures spacing (FS), on the dimensionless pressure and its derivative curves. We also analyze the effect of reorientation fracture number and complex fracture configuration on type curves.

5.3.1. Effect of Principal Fracture Angle (PFA) on Type Curves

Figure 10 presents the effect of PFA on type curves for a horizontal well with two reorientation fractures. The dimensionless fracture spacing is set at 20. As displayed in Figure 10, PFA has a weak impact on type curves in the early-time period, i.e., the bilinear flow regime, first linear flow regime, and first radial flow regime. For an anisotropic tight reservoir, the dimensionless pressure and its derivative decrease as PFA increases, indicating that a large PFA is beneficial to improve the productivity of fractured horizontal wells. In addition, as PFA increases, the first radial flow regime occurs later. The effect of PFA on other flow regimes can be neglected.

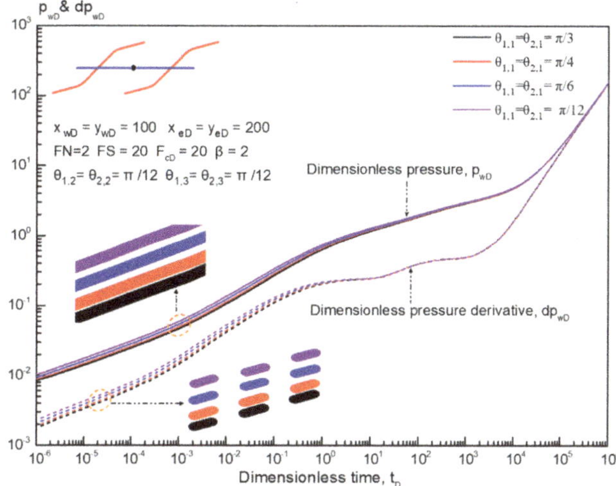

Figure 10. Effect of the principal fracture angle on pressure and the pressure derivative.

5.3.2. The effect of reoriented fracture angle (RFA) on type curves

Figure 11 shows the effect of RFA on the dimensionless pressure and pressure derivative curves. In order to completely reveal the effect of RFA on type curves, we compare the dimensionless pressure (PFA = RFA = $\pi/3$) with the dimensionless pressure (PFA $\neq$ RFA) and then find the point 'MRPC' (maximum relative pressure change) (Figure 12). The relative pressure change (RPC) is defined as follows:

$$RPC = \left(p_{wD|PFA \neq RFA} - p_{wD|PFA=RFA}\right)/p_{wD|PFA=RFA} \times 100\% \tag{26}$$

In Figure 12, t_{DMRPC} represents the corresponding dimensionless time for the MRPC. Before t_{DMRPC}, as the RFA increases, the RPC decreases which means that RFA should be larger to maintain the constant rate under the same pressure depletion rate. However, after t_{DMRPC}, the effect of RFA on RPC weakens, which indicates that the effect of RFA on type curves tends to disappear. When RFA > PFA, RPC gets smaller and even becomes negative, which suggests that the larger the RFA (when RFA > PFA), the higher the well productivity.

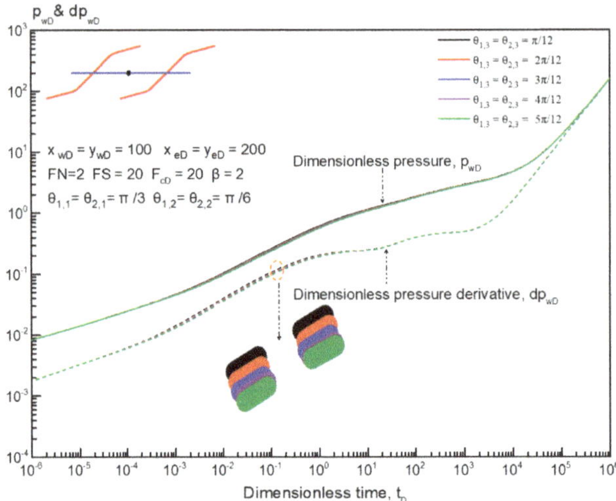

Figure 11. Effect of the reoriented fracture angle on pressure and the pressure derivative.

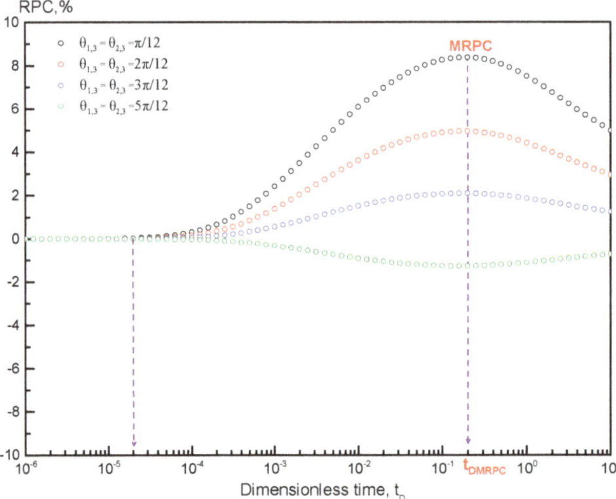

Figure 12. Reoriented fracture angle on relative pressure change.

5.3.3. Effect of Permeability Anisotropy on Type Curves

The impact of permeability anisotropy on dimensionless pressure and pressure derivative is illustrated in Figure 13 ($\beta > 1$) and Figure 14 ($\beta < 1$). In Figures 13 and 14, permeability anisotropy affects all the flow regimes except the pseudo-steady-state flow regime.

The influence of permeability anisotropy ($\beta \geq 1$) on type curves is displayed in Figure 13. Here, the horizontal wellbore is parallel to the principal permeability axis. As permeability anisotropic factor (β) increases, the dimensionless pressure and its derivative decrease in the early linear flow regimes, indicating that the pressure difference of the horizontal wellbore can be smaller to maintain the constant rate. However, before the pseudo-steady-state flow regime, the pressure and its derivative increase when the permeability anisotropic factor (β) increases, for the reason that the first radial flow regime tends to disappear when permeability anisotropic factor (PAF) increases to some extent.

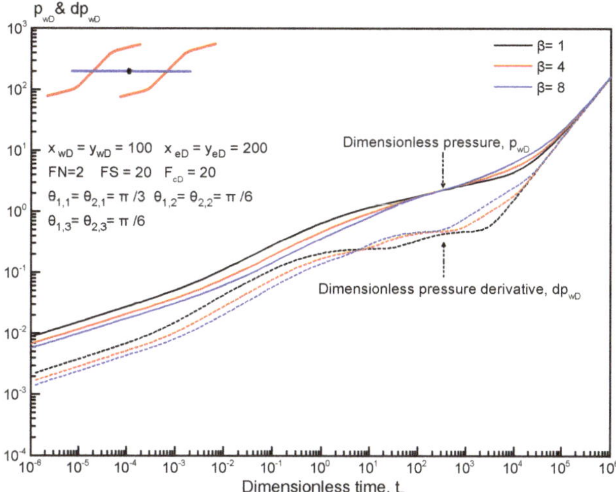

Figure 13. Effect of permeability anisotropy on pressure and the pressure derivative ($\beta \geq 1$).

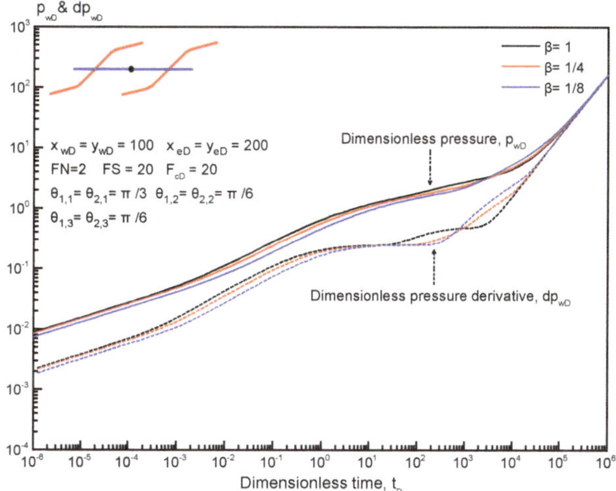

Figure 14. Effect of permeability anisotropy on pressure and the pressure derivative ($\beta \leq 1$).

Similarly, the effect of permeability anisotropic factor ($\beta \leq 1$) on type curves is shown in Figure 14. At this time, the horizontal wellbore is perpendicular to the principal permeability axis. Compared with Figure 13, we can find that the second radial flow regime tends to disappear when the horizontal wellbore is perpendicular to the principal permeability axis. Meanwhile, the compound linear flow regime tends to appear when the first radial flow regime ends, for the reason that the flow in the principal permeability direction dominates and, thus, the second radial flow regime is converted into the compound linear flow regime.

5.3.4. Effect of Fracture Conductivity on Type Curves

Figure 15 depicts the pressure and pressure derivative curves for a horizontal well intercepted by two reorientation fractures with various fracture conductivities: π, 10π, 100π, and 1000, respectively.

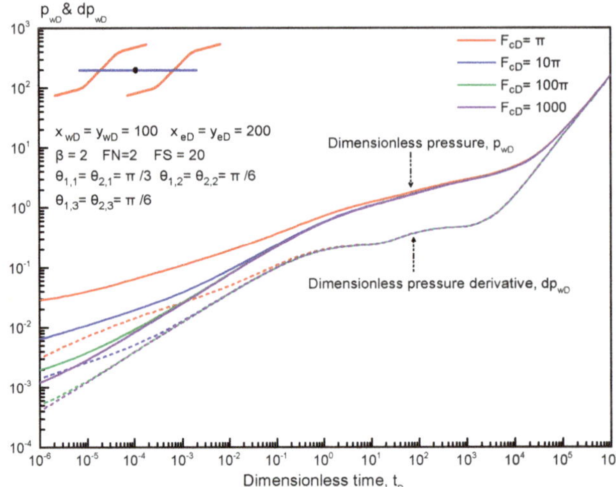

Figure 15. Effect of fracture conductivity on pressure and the pressure derivative.

Here PFA and RFA are kept constant and equal to 60° and 30°, respectively. As seen in Figure 15, the fracture conductivity affects the transient pressure significantly at the early-time period, mainly focusing on bilinear flow regime and first linear flow regime. The lower the fracture conductivity is, the earlier the bilinear flow regime appears. Moreover, when the fracture conductivity is low enough, the first linear flow regime becomes shorter and even tends to disappear. Meanwhile, the bilinear flow regime disappears and consequently the duration of the first linear flow lasts longer when fracture conductivity tends to be infinite ($F_{cD} > 100\pi$).

5.3.5. Effect of Dimensionless Fracture Spacing (FS) on Type Curves

Fracture spacing (FS) determines the intensity of flow interferences happened between the adjacent fractures [10], and can mask some typical flow behaviors. For example, the first radial flow regime disappears when FS is smaller than the fracture length. When FS is large enough, the flow interference occurs later, or each fracture flows independently and does not interfere with each other.

The influence of FS on type curves is presented in Figure 16 and it can be concluded that the difference among the various type curves is mainly concentrated on the duration of first radial flow regime and pseudo-radial flow regime. Smaller fracture spacing generates larger pressure depletion to maintain the constant production rate due to the strong flow interference between the adjacent fractures. Moreover, the larger the fracture spacing is, the later the pseudo radial flow regime appears and also the shorter this regime's duration is. For multiple fractured horizontal wells, the optimal fracture spacing mainly depends on the reorientation fracture number and the drainage area.

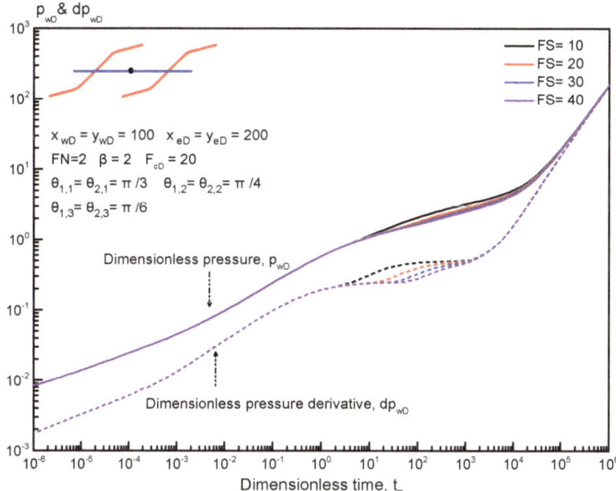

Figure 16. Effect of dimensionless fracture spacing on pressure and the pressure derivative.

5.3.6. Effect of Fracture Number (FN) on Type Curves

Fracture spacing between the adjacent fractures is assumed to be constant (FS = 20). The fracture number can control the area between fractures and the reservoir. Figure 17 presents the influence of fracture number (FN) on pressure transient. The increasing FN means an increasing fractured area when the adjacent fracture spacing is fixed. In general, the pressure drop decreases with the increase of the FN.

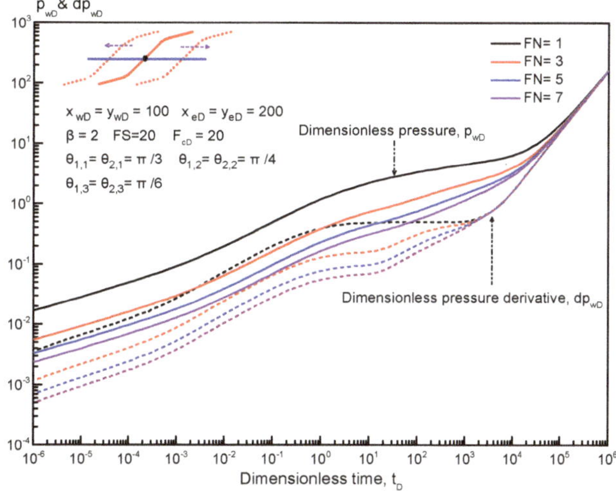

Figure 17. Effect of fracture number on pressure and the pressure derivative.

As seen in Figure 17, bilinear flow, first linear flow, first radial flow and second radial flow are all affected by reorientation fracture number. A large fracture number corresponds to a small pressure and derivative, suggesting that a smaller pressure drop is required to remain the same production rate. Consequently, the production is definitely enhanced by the increase in the fracture number, reflecting the advantage of multi-stage fractured horizontal wells. However, the degree of the decrease of the pressure drop weakens when FN increases to some degree, indicating that the optimal reorientation fracture number needs to be demonstrated to maintain economic production.

5.3.7. Effect of Complex Fracture Configuration on Type Curves

Figure 18 investigates the influence of complex fracture configuration under large fracture spacing (FS = 20) on type curves.

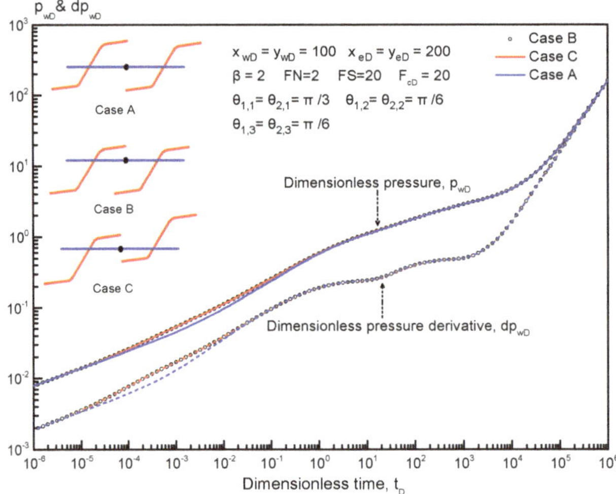

Figure 18. Effect of complex fracture pattern under large dimensionless fracture spacing on pressure and the pressure derivative (FS = 20).

For large fracture spacing, the influence of fracture configuration on type curves is largely focused on the transition flow regime ($10^{-5} \le t_D \le 10^{-1}$). Due to large fracture spacing, the difference between the effects of the staggered form of the reorientation fractures and the asymmetrical reorientation fractures on type curves can be neglected because each reorientation fracture behaves individually and does not interfere with each other. However, symmetrical reorientation fractures have higher production rates than the asymmetrical reorientation fractures when FS is large.

To determine the impact of fracture configuration on type curves under small fracture spacing (FS = 4), three typical cases are considered (Figure 19). Compared with the cases under large fracture spacing (Figure 18), high production rates are obtained when the reorientation fractures are staggered because the mutual dislocation of the reorientation fractures can reduce the flow interference between the adjacent reorientation fractures.

Further, the influence of different fracture configurations under the same fracture spacing (FS = 20) and fracture number (FN = 3) on type curves is illustrated in Figure 20. Case A depicts that the length of the outermost fractures are larger than that of the inner fractures, Case B represents that the length of the outermost fractures are smaller than that of the inner fractures, and Case C shows that each fracture in the system has the same length. Therefore, the equal fracture length model has a low pressure depletion under the same conditions, which suggests that during the hydraulic fracturing treatment the length of each fracture should be equal to maintain economic production (Figure 20).

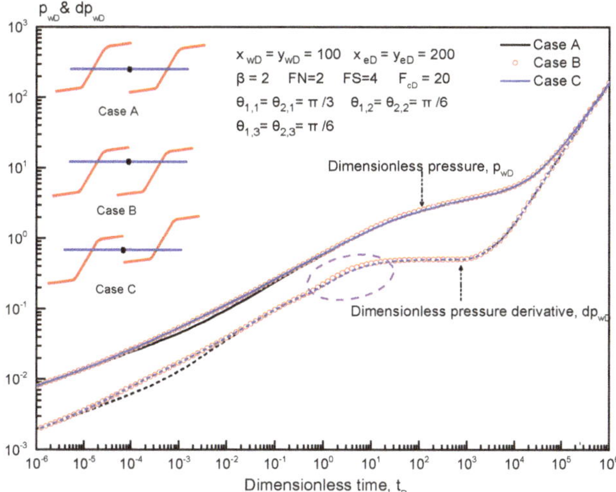

Figure 19. Effect of complex fracture configuration under small dimensionless fracture spacing on pressure and the pressure derivative (FS = 4).

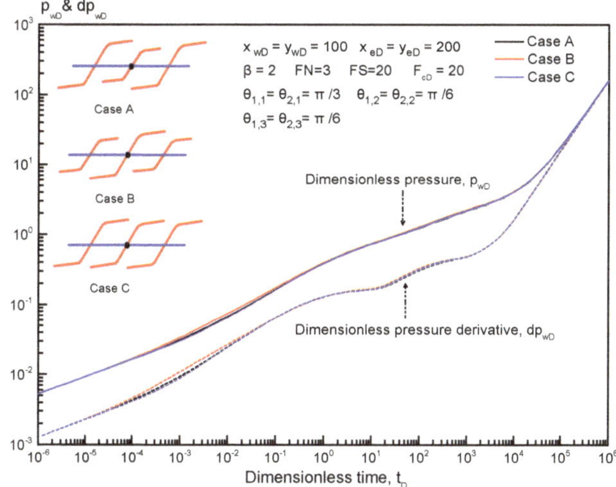

Figure 20. Effect of complex fracture configuration under different fracture distributions on pressure and pressure derivative (FN = 3).

6. Conclusions

On the basis of the new fracture equation, we present a semi-analytical solution to investigate the pressure behavior of multiple reorientation fractures along the horizontal well in an anisotropic tight reservoir. The main conclusions include the following:

(1)   For a horizontal well perforated by multiple reorientation fractures with finite-conductivity in an anisotropic rectangular reservoir, six typical flow regimes can be found on type curves, including bilinear flow regime, first linear flow regime, first radial flow regime, second radial flow regime, pseudo-radial flow regime, and pseudo-steady-state flow regime. Meanwhile, the occurrence and duration of these typical regimes are determined by some significant parameters, such as

the permeability anisotropic factor, fracture reorientation, fracture spacing, fracture number, and fracture configuration.

(2)   Fracture reorientation has an obvious effect on the production rate distribution along the fracture extension, particularly in the anisotropic tight reservoir. The production rate distribution curves for a horizontal well with three reorientation fractures in an anisotropic tight reservoir oscillate, and the outermost reorientation fracture has a larger production rate than the inner reorientation fracture due to a larger contact area with the reservoir.

(3)   Fracture reorientation is one of the highlights of this work. For an anisotropic tight reservoir, the horizontal well should be deployed parallel to the direction of the principal permeability axis, and all reoriented fractures should be perpendicular to the direction of the horizontal well.

(4)   The influence of permeability anisotropy on type curves is significant in all typical flow regimes except the late-time period. When horizontal wellbore is parallel to the principal permeability axis, first radial flow regime tends to disappear. However, when the horizontal wellbore is perpendicular to the principal permeability axis, the pseudo-radial flow regime changes into a compound linear flow regime.

(5)   During the fracturing treatment, it is necessary to make the adjacent fractures stagger over each other and make the fracture spacing large to reduce the flow interference, and the length of each fracture in multiple fractured horizontal wells should remain the same to maintain economic production.

Author Contributions: All authors have contributed to this work. G.X. proposed the mathematical model and wrote the main manuscript text. As the corresponding author, M.W. made substantial contributions to the conception/design of the work. S.W., H.W., B.W., and J.C. contributed to the modeling, programming, results analysis, and discussion. J.W. mainly focuses on the modeling, programming of the manuscript.

Funding: This research was funded by the PetroChina Scientific Study and Technical Development Project (2017A-0906) and National Science and Technology Major Project of China (no. 2016ZX05046-003).

Acknowledgments: I am greatly indebted to my supervisor, Shuhong Wu, for her valuable instructions and suggestions on this work as well as her careful reading of the manuscript. In addition, I deeply appreciate the contribution to this work made in various ways by other co-authors.

Conflicts of Interest: The authors declare that they have no competing interests.

Nomenclature

x = Distance in the x-axis, m
y = Distance in the y-axis, m
x_e = Reservoir length in the x-axis, m
y_e = Reservoir length in the y-axis, m
h = Reservoir thickness, m
A = Drainage area, m^2
D = Distance between outermost fractures, m
m_i = Total discrete number of all irregular curve fractures
L_i = The length of the equivalent i-th planar fracture for the irregular curve fracture
$L_{fi,1}$ = Principal fracture length of the i-th reorientation fracture, m
$L_{fi,2}$, $L_{fi,3}$ = Reoriented fracture length of the i-th reorientation fracture, m
Δl_i = length of the i-th fracture segment, $i = 1, 2 \ldots N_I$, m
l = Fracture segment length, m
L_R = Reference length, m
w_f = Fracture width, m
k_x = Permeability in the x-axis, $10^{-3}\,\mu m^2$
k_y = Permeability in the y-axis, $10^{-3}\,\mu m^2$
k_f = Fracture permeability, $10^{-3}\,\mu m^2$
$\theta_{i,1}$ = Principal fracture angle of the i-th reorientation fracture, rad

$\theta_{i,2}$, $\theta_{i,3}$ = Reoriented fracture angle of the i-th reorientation fracture, rad

p_i = Initial reservoir pressure, MPa

ϕ = Reservoir porosity, fraction

u = Fluid viscosity, mPa•s

c_t = Total compressibility, 1/MPa

q_{sc} = Wellbore flow rate, m^3/d

q_{fw} = Flow rate of a reorientation fracture in the wellbore, m^3/d

q_f = Rate of per unit fracture length from reservoir, m^2/d

p_D = Dimensionless pressure in real time domain

$\tilde{p}_D$ = Dimensionless pressure in Laplace domain

$\tilde{p}_{fD}$ = Dimensionless reoriented fracture pressure in Laplace domain

$\tilde{q}_{fwD}$ = Dimensionless flow rate of a reorientation fracture in the wellbore in the Laplace domain

$\tilde{q}_{fD}$ = Dimensionless flow rate of per unit fracture length from reservoir in the Laplace domain

q_{fD} = Dimensionless flow rate of per unit fracture length from reservoir in the real time domain

q_D = Dimensionless point source flux

F_{cD} = Dimensionless fracture conductivity

l_D = Dimensionless fracture segment length

t_D = Dimensionless time

x_D = Dimensionless distance in the x-axis

y_D = Dimensionless distance in the y-axis

dp_{wD} = Dimensionless pressure derivative

x_{wD} = Dimensionless distance in the x-axis of the center of the reservoir

y_{wD} = Dimensionless distance in the y-axis of the center of the reservoir

RF = Reorientation fracture

FR = The location where the fracture reoriented

x_{eD} = Dimensionless reservoir length in the x-axis

y_{eD} = Dimensionless reservoir width in the y-axis

R_{Dk} = Dimensionless coefficient

s = Dimensionless time variable in Laplace domain

N_I = Total fracture segment

N = Total fracture number

Δl_{Di} = Dimensionless length of the i-th fracture segment, i = 1, 2 … N_I

x_{Dmi} = Dimensionless coordinate along the fracture extension

y_{Dk} = Dimensionless coordinate along the fracture extension

cosh = Hyperbolic cosine function

sinh = Hyperbolic sine function

Appendix A Simplified Algorithm for Discrete Equations of a Reorientation Fracture

The discretized form of the reorientation fracture is:

$$\tilde{p}_{wD} - \tilde{p}_{fDi} = \frac{2\pi}{F_{cD}}\left[l_{Di}\sum_{j=1}^{N_I}\left(\tilde{q}_{fDj}\Delta l_{Dj}\right) - \frac{\Delta l_{Di}^2}{8}\tilde{q}_{fDi} - \sum_{j=1}^{i-1}\left(\frac{\Delta l_{Dj}}{2} + l_{Di} - \sum_{n=1}^{j}\Delta l_{Dn}\right)\Delta l_{Dj}\tilde{q}_{fDj}\right] \tag{A1}$$

For $i > j$, Equation (A1) can be rewritten as:

$$\tilde{p}_{wD} - \tilde{p}_{fDi} = \frac{2\pi}{F_{cD}}\left[\sum_{n=1}^{j}\Delta l_{Dn} - \frac{\Delta l_{Dj}}{2}\right]\Delta l_{Dj}\tilde{q}_{fDj} \tag{A2}$$

For $i = j$, Equation (A1) is given as:

$$\tilde{p}_{wD} - \tilde{p}_{fDi} = \frac{2\pi}{F_{cD}}\left[l_{Di}\Delta l_{Dj} - \frac{\Delta l_{Di}^2}{8}\right]\tilde{q}_{fDj} \tag{A3}$$

Otherwise, Equation (A1) can be expressed as:

$$\tilde{p}_{wD} - \tilde{p}_{fDi} = \frac{2\pi}{F_{cD}} l_{Di} \Delta l_{Dj} \tilde{q}_{fDj} \tag{A4}$$

References

1. Siebrits, E.; Elbel, J.L.; Hoover, R.S.; Diyashev, I.R.; Griffin, L.G.; Demetrius, S.L.; Wright, C.A.; Davidson, B.M.; Steinsberger, N.P.; Hill, D.G. Refracture reorientation enhances gas production in Barnett shale tight gas wells. In Proceedings of the SPE Annual Technical Conference and Exhibition, Dallas, TX, USA, 1–4 October 2000.
2. Benedict, D.S.; Miskimins, J.L. Analysis of reserve recovery potential from hydraulic fracture reorientation in tight gas Lenticular reservoirs. In Proceedings of the SPE Hydraulic Fracturing Technology Conference, The Woodlands, TX, USA, 19–21 January 2009.
3. Tang, S.K.; Li, M.Z.; Qi, M.H.; Han, R.; Li, G. Study of fracture reorientation caused by induced stress before re-fracturing. *Fault-Block Oil Gas Field* **2017**, *24*, 557–560.
4. Soliman, M.Y.; Hunt, J.L.; El Rabaa, A.M. Fracturing aspects of horizontal wells. *J. Pet. Technol.* **1990**, *42*, 966–973. [CrossRef]
5. Ozkan, E. Performance of Horizontal Wells. Ph.D. Thesis, Tulsa University, Tulsa, OK, USA, 1988.
6. Larsen, L.; Hegre, T.M. Pressure-transient behavior of horizontal wells with finite-conductivity vertical fractures. In Proceedings of the International Arctic Technology Conference, Anchorage, AK, USA, 29–31 May 1991.
7. Larsen, L.; Hegre, T.M. Pressure transient analysis of multi-fractured horizontal wells. In Proceedings of the SPE Annual Technical Conference and Exhibition, New Orleans, LA, USA, 25–28 September 1994.
8. Chen, C.C.; Raghavan, R. A multiply-fractured horizontal well in a rectangular drainage region. *SPE J.* **1997**, *2*, 455–465. [CrossRef]
9. Zerzar, A.; Bettam, Y. Interpretation of multiple hydraulically-fractured horizontal wells in closed systems. In Proceedings of the SPE International Improved Oil Recovery Conference in Asia Pacific, Kuala Lumpur, Malaysia, 20–21 October 2003.
10. Wang, J.; Jia, A.; Wei, Y.; Qi, Y. Approximate semi-analytical modeling of transient behavior of horizontal well intercepted by multiple pressure-dependent conductivity fractures in pressure-sensitive reservoir. *J. Petrol. Sci. Eng.* **2017**, *153*, 157–177. [CrossRef]
11. Wright, C.A. Reorientation of propped refracture treatments in the Lost Hills Field. In Proceedings of the SPE Western Regional Meeting, Long Beach, CA, USA, 23–25 March 1994.
12. Wright, C.A.; Conant, R.A. Hydraulic fracture reorientation in primary and secondary recovery from low-permeability reservoirs. In Proceedings of the SPE Annual Technical Conference and Exhibition, Dallas, TX, USA, 22–25 October 1995.
13. Fisher, M.K.; Wright, C.A.; Davison, B.M.; Goodwin, A.K.; Fielder, E.O.; Buckler, W.S.; Steinsberger, N.P. Integrating fracture mapping technologies to optimize stimulations in the Barnett shale. In Proceedings of the SPE Annual Technical Conference and Exhibition, San Antonio, TX, USA, 29 September–2 October 2002.
14. Fisher, M.K.; Heinze, J.R.; Harris, C.D.; Davidson, B.M.; Wright, C.A.; Dunn, K.P. Optimizing horizontal completion techniques in the Barnett shale using miscoseismic fracture mapping. In Proceedings of the SPE Annual Technical Conference and Exhibition, Houston, TX, USA, 26–29 September 2004.
15. Restrepo, D.P. Pressure Behavior of a System Containing Multiple Vertical Fractures. Ph.D. Thesis, The University of Oklahoma, Norman, OK, USA, 2008.
16. Luo, W.J.; Tang, C.F. Pressure-transient analysis of multiwing fractures connected to a vertical wellbore. *SPE J.* **2015**, *20*, 360–367. [CrossRef]
17. Luo, W.J.; Wang, X.; Tang, C.; Feng, C.; Shi, E. Productivity of multiple fractures in a closed rectangular reservoir. *J. Petrol. Sci. Eng.* **2017**, *157*, 232–247. [CrossRef]
18. Tian, Q.; Liu, P.; Jiao, Y.; Bie, A.; Xia, J.; Li, B.; Liu, Y. Pressure transient analysis of non-planar asymmetric fractures connected to vertical wellbores in hydrocarbon reservoirs. *Int. J. Hydrogen Energy* **2017**, *42*, 18146–18155. [CrossRef]
19. Zhou, W.; Banerjee, R.; Poe, B.D.; Spath, J.; Thambynayagam, M. Semianalytical production simulation of complex hydraulic-fracture networks. *SPE J.* **2012**, *19*, 6–18. [CrossRef]

20. Yu, W.; Wu, K.; Sepernoori, K. A semianalytical model for production simulation from nonplanar hydraulic-fracture geometry in tight oil reservoirs. *SPE J.* **2016**, *21*, 1028–1040. [CrossRef]

21. Chen, Z.; Liao, X.; Sepehrnoori, K.; Yu, W. A semi-analytical model for pressure transient analysis of fractured wells in unconventional plays with arbitrarily distributed fracture networks. In Proceedings of the SPE Annual Technical Conference and Exhibition, San Antonio, TX, USA, 9–11 October 2017.

22. Wu, S.; Xing, G.; Cui, Y.; Wang, B.; Shi, M.; Wang, M. A semi-analytical model for pressure transient analysis of hydraulic reorientation fracture in an anisotropic reservoir. *J. Petrol. Sci. Eng.* **2019**, *153*, 157–177. [CrossRef]

23. Wang, L.; Xue, L. A Laplace-transform boundary element model for pumping tests in irregularly shaped double-porosity aquifers. *J. Hydrol.* **2018**, *567*, 712–720. [CrossRef]

24. Cinco, L.H.; Samaniego, V.F.; Dominguez, A.N. Transient pressure behavior for a well with a finite-conductivity vertical fracture. *SPE J.* **1978**, *18*, 253–264. [CrossRef]

25. Stehfest, H. Algorithm 368: Numerical Inversion of Laplace Transform. *Commun. ACM* **1970**, *13*, 47–49. [CrossRef]

Article

Catalytic Effect of Cobalt Additive on the Low Temperature Oxidation Characteristics of Changqing Tight Oil and Its SARA Fractions

Tengfei Wang [1,2,*] and Jiexiang Wang [1]

[1] School of Petroleum Engineering, China University of Petroleum (East China), Qingdao 266580, China
[2] Petroleum Systems Engineering, Faculty of Engineering and Applied Science, University of Regina, Regina, SK S4S 0A2, Canada
* Correspondence: wangtengfei@upc.edu.cn; Tel.: +86-159-6691-6346

Received: 27 June 2019; Accepted: 22 July 2019; Published: 24 July 2019

Abstract: Air flooding is a potential enhanced oil recovery (EOR) method to economically and efficiently develop a tight oil reservoir due to its sufficient gas source and low operational costs, during which low temperature oxidation (LTO) is the key to ensuring the success of air flooding. In addition to inefficiency of conventional LTO, air flooding has seen its limited applications due to the prolonged reaction time and safety constraints. In this paper, a novel air injection technique based on the catalyst-activated low temperature oxidation (CLTO) is developed to improve the operational safety together with its oil recovery in tight oil reservoirs. Experimentally, static oxidation experiments are conducted to examine the influence of the catalyst on the LTO reaction kinetics of Changqing tight oil and its fractions. The catalytic oxidation characteristics are identified by applying a thermogravimetric analyzer coupled with Fourier transform infrared spectrometer (TG-FTIR) with respect to tight oil and its SARA (i.e., saturates, aromatics, resins, and asphaltenes) fractions. Accordingly, the catalyst can obviously decrease the LTO reaction activation energy of the Changqing tight oil and its SARA fraction. Cobalt additive can change the LTO reaction pathways of the SARA fractions, i.e., promoting the formation of hydroxyl-containing oxides and CO_2 from the oxidation of saturates, aromatics and resins, while inhibiting the formation of ethers from the oxidation of aromatics and resins. The LTO of each SARA fraction contains both oxygen addition reaction and bond scission reaction that can be effectively promoted with the cobalt additive. The catalytic effect on the bond scission reaction is continuously enhanced and becomes gradually stronger than that on the oxygen addition reaction as the reaction proceeds.

Keywords: air flooding; catalyst-activated low temperature oxidation; oxidation reaction pathway; catalytic oxidation characteristics; Changqing tight oil

---

1. Introduction

Air flooding is a cost-effective technique for exploiting oil reservoirs, which is remarkably effective for tight oil reservoirs and conventional reservoirs in the late stage of waterflooding [1–4]. Recently, there have seen numerous researches and field pilots for air flooding in the North Sea [5,6], Indonesia [7,8], Argentina [9], Buffalo Red River Unit (BRRU) [10], West Hackberry [11,12], Coral Creek [13], Ekofisk [6,14], Buffalo [15,16], and Horse Creek oilfields [17] with good performance. After air is injected into a hydrocarbon reservoir, the oxygen of air reacts with hydrocarbons in various ways, generating a lot of flue gas and heat, and thus enhancing the oil recovery. To enhance the safety of air flooding, the injected oxygen should be completely consumed before produced. To improve the safety of air flooding without changing flooding pattern in a depleted reservoir, it is a good method to increase the oxygen consumption rate of crude oil by adding catalyst to the LTO process.

Recently, numerous efforts have been made to quantify high-temperature catalytic oxidation during in-situ combustion (ISC) and the high-temperature catalytic viscosity reduction of heavy oil with air injection. As shown in Table 1, effects of metallic additives such as iron, tin, copper, cobalt, and et al. have been examined on the efficiency of ISC, acidity of heavy oil, viscosity reduction rate, and kinetic parameters through combustion tube tests, viscosity reduction tests, and simultaneous thermogravimetry and differential scanning calorimetry (TG-DSC) tests [18–31].

So far, few attempts have been made to quantify catalyst-activated low temperature oxidation (CLTO) and identify the associated catalytic mechanisms. Zhao et al. [32] examined the effect of reservoir rock and clay minerals on the LTO reaction characteristics of light crude oil. By examining the influence of clay-mineral type on the oxidation of crude oil at condition of 120 °C and 30 MPa, they found that clay minerals could catalyze the oxidation of light oil, while the catalytic effect varied for different types of clay minerals due mainly to the different compositions of metal salts in the clay minerals. Jia et al. [33] examined the catalytic effect of clay and rock minerals on the oxidation of light crude oil using the thermogravimetry and differential thermal analysis (TG-DTA) during high-pressure air injection processes. Their results showed that rock cuttings could promote the deposition of fuel (i.e., coke) and that clay minerals can enhance the oxidation degree of light oil. By treating the fracturing proppant with a catalyst and subsequently injecting the supported catalyst into the formation by sand fracturing, Shi et al. [34] found that the effect of the supported catalyst on the LTO of crude oil greatly improved the oxygen consumption rate, and thereby enhanced the safety of air flooding. Such research on the CLTO reaction, however, is still at the early stage, while the underlying mechanisms for the CLTO have not been fully identified.

In our previous work on the CLTO technique [35,36], numerous efforts have been done on the optimization and evaluation of LTO catalysts. Cobalt additive is found to be a proper catalyst that can promote the LTO reaction rate and enhance the safety of air flooding obviously by delaying the oxygen breakthrough time. A preliminary study on the catalytic mechanism has been conducted in our previous work [37] by quantifying the influence of catalyst on the oxidation behavior of the crude oil applying the thermogravimetry coupled with Fourier transform infrared spectroscopy (TG-FTIR) tests. However, the catalytic effect of cobalt additive on different fractions has not been researched before, and the relationships between the catalytic oxidation characteristics of different oil fractions are still not clear. This study tends to investigate the effects of the cobalt additive on the LTO reaction process of the individual SARA fraction of Changiqng tight oil by applying static oxidation experiments and TG-FTIR tests. Then, the experimental measurements are analyzed and discussed to identify the CLTO characteristics of SARA fractions.

Table 1. Summary of previous studies on crude oil catalytic oxidation technologies.

Source	Experimental Type	Crude Oil (API Gravity)	Experimental Conditions (Temperature/Pressure)	Additives	Findings
Bagci et al. [18]	Combustion reaction kinetics experiments	Karakus (29° API) and Beykan (32° API) crude oils from Turkish oilfields	The cell was continuously heated to 500–600 °C with a 1 °C/min heating rate at pressures of 172 or 345 kPa	$CuCl_2$, $FeCl_3$, and $MgCl_2$	The catalyst type and concentration influenced the kinetic parameters (i.e., reaction order and activation energy) of high-temperature oxidation.
Shallcross et al. [19]	Combustion cell experiments	Californian (18.5° API) and Venezuelan (10.5° API) oils	Temperature was increased to 450 °C at a rate of 50 °C/h at pressures of 280 or 550 kPa	$FeCl_2$, $SnCl_2$, $CuSO_4$, $ZnCl_2$, $MgCl_2$, $K_2Cr_2O_7$, Al_2Cl_3, $MnCl_2$, $Ni(NO_3)_2$, and $CdSO_4$	Iron and tin salts could enhance fuel formation, whereas copper, nickel, and cadmium salts have minor effects.
Gerritsen et al. [20]	Combustion tube experiments	Cymric light oil (34° API)	Temperature: 400 °C; Pressure: 68.9 kPa	iron nitrate	Metallic additive improves combustion of light oil. The Cymric light oil cannot sustain combustion without additive.
Ramirez et al. [21,22]	Combustion tube experiments	Heavy oil from Gulf of Mexico (12.5° API)	Ignition temperature: 300 °C; Pressures: 2070 kPa	Additives included Mo, Co, Ni, and Fe, and the anionic part of the salt was acetylacetonate or alkylhexanoate.	The addition of metallic additives could accelerate the propagation velocity of the combustion front, improve combustion efficiency, and increase oil production.
Fassihi et al. [23,24]	Combustion cell experiments	San Ardo oil (11.2° API), Venezuela oil (9.5° API), Huntington Beach oil (18.5° API), and Lynch Canyon oil (10° API)	Temperature was increased to 450 °C at a rate of 55 °C/h at pressures of 690, 550 or 138 kPa	Additives included Cu, Ni, Va, and Fe, but the anionic part of the salt was not reported.	The additives can lower both the activation energy and the occurrence temperature of the combustion reaction under the identical reservoir conditions.
Castanier et al. [25]	Combustion tube experiments	Huntington Beach oil (22° API), Hamaca oil (10° API), Cymric heavy oil (12° API), and Cymric light oil (34° API)	Ignition temperature: 315 °C; Pressure: 690 kPa	Additives included Fe, Sn, and Zn, but the anionic part of the salt was not reported.	The combustion efficiency and front velocities can be improved by the metallic additives. The H/C ratio of the fuel, heat of combustion, air requirements, and density of the produced crude are changed after the addition of metallic salts. In addition, the combustion of light oil can be improved by metallic additives.
Drici et al. [26]	Differential scanning calorimetry (DSC) and thermnogravimetric analysis (TGA)	Crude oil from the Iola field (19.8° API)	Temperature was increased to 600 °C at a rate of 10 °C/h at atmospheric pressure	Heavy metal oxides such as titanium, ferric, nickel, cupric, vanadium, and chromium oxides	The effect of titanium oxide is similar to that of silica and alumina. Vanadium, nickel, and ferric oxides behaves similarly in enhancing the endothermic reactions. The effect of small amount of metal oxide is weak in the presence of a large surface area such as silica.
He et al. [27]	Combustion tube experiments and ramped temperature oxidation (RTO) tests	Heavy (12° API) and light (34° API) oil from Cymric	Tube experiments: Ignition temperature 400 °C; Pressure 690 kPa; RTO tests: Temperature was increased to 470 °C at a rate of 60 °C/h at pressures of 690, 550 or 310 kPa	$Fe(NO_3)_3$	The combustion reaction catalytic mechanism is cation exchange of metallic salts with clay to create activated sites that enhance the combustion reactions between oil and oxygen. The combustion of light oil can be improved by metallic additives.

Table 1. *Cont.*

Source	Experimental Type	Crude Oil (API Gravity)	Experimental Conditions (Temperature/Pressure)	Additives	Findings
Wang et al. [28]	Catalytic oxidation experiment; Viscosity reduction experiment	SZ36-1 crude oil (30° API)	Temperature: 100 °C; Pressure: 2000 kPa	Iron naphthenate, copper naphthenate, zinc naphthenate, manganese naphthenate, cobalt naphthenate	The preferred catalyst is copper naphthenate in an amount of 0.2% by mass of the crude oil. After the catalytic oxidation, the acid value and viscosity of the heavy oil increases. After adding the auxiliary solution to the oxidized oil system, the resulting surfactant can greatly reduce the viscosity of the aqueous heavy oil system by the emulsified viscosity reduction mechanism.
Tang et al. [29]	Catalytic oxidation experiment; Viscosity reduction experiment	Crude oil from Liaohe oilfield in China, API gravity is not given, oil viscosity is 382.6 Pa·s.	Temperature: 200 °C; Pressure: 500 kPa	Water soluble catalyst SP-1, light blue powder, MeL type compound	The catalyst promotes oxidation and cracking of heavy oil and increases the acid value of heavy oil. The surface-active materials formed in the reaction can greatly reduce the viscosity of the aqueous heavy oil system by the emulsified viscosity reduction mechanism.
Pu et al. [30]	Thermogravimetric testing	Tahe heavy crude oil (18.9° API)	Temperature was increased to 800 °C at a rate of 10 °C/min at atmospheric pressure	$ZnSO_4$, $CuCl_2$, $FeCl_2$, and $AlCl_3 \cdot 6H_2O$	Metallic additives exhibit varied catalytic effects on heavy oil oxidation. $CuCl_2$ is found to be an excellent catalyst for improving oil production through positively influencing the oxidation reactions of Tahe heavy crude oil.
Amanam et al. [31]	Ramped temperature oxidation experiments; Isoconversional analysis; Thermogravimetric analysis	Zuata crude oil (9.8° API)	Temperature was increased to 600 °C at a rate of 1–3 °C/min at atmospheric pressure	Copper nanoparticles	The apparent activation energy of the high-temperature oxidation region is decreased. The presence of Cu-NP helps maintain a greater front temperature. Cu-NP changes the type of oil produced and decreases the amount of water generated during the reactions.

2. Experimental

2.1. Materials

In this study, crude oil produced from Dongzhi tight oil reservoir in the Changqing Oilfield in China is collected and used to conduct the experiments. The tight oil physical properties are shown in Table 2. It is worthwhile noting that the oxidation of asphaltenes is not included in this research due to the fact that the state of asphaltenes in a crude oil is completely different from that of the asphaltenes separated from the crude oil [38]. The air used is supplied by the Qingdao Tianyuan Gas Company which is composed of 21.0 mol % oxygen and 79.0 mol % nitrogen. The neutral alumina, reagent-grade n-pentane, HPLC-grade toluene, HPLC-grade methanol, and HPLC-grade tetrahydrofuran used in the SARA fraction separation are all provided by the Sinopharm Chemical Reagent (Co., Ltd., Shanghai, China). The cobalt additives (cobalt naphthenate and cobalt chloride) used as the catalyst are also provided by the Sinopharm Chemical Reagent (Co., Ltd.).

Table 2. Physical properties of the crude oil sample.

Properties		Value	Appearance
Density (g/cm^3)		0.850	N/A
Viscosity at 70 °C (mPa·s)		2.14	N/A
SARA composition (wt %)	Saturates	70.91	Colourless liquid
	Aromatics	16.07	Yellow or red sticky liquid
	Resins	9.78	Brown viscous liquid
	Asphaltenes	3.24	Black fragile powder solid

2.2. Experimental Setup

In this study, a vacuum oven (YZF-6032, Shanghai Yaoshi Instrument Equipment Factory, Shanghai, China), an analytical balance (AB105, Shanghai Precision Instrument Company, Shanghai, China), and an ultrasonic disperser (Scientz-2400F, SCIENTZ, China) are used to separate the tight oil into SARA fractions. The operating temperatures of the vacuum oven is from the room temperature to 250 °C with its temperature accuracy of 0.1 °C and the ultimate vacuum less than 133 Pa. The maximum scale of the balance is 105 g with its weighting accuracy of 0.01 mg, while the frequency of ultrasonic disperser is 19.5–20.5 kHz.

The static oxidation experimental setup (Hai'an Petroleum Research Instrument Co., Ltd., Hai'an, China) used in this research is the same as that of the previous study [35,39]. The schematic of the static oxidation experiment setup is shown in Figure 1.

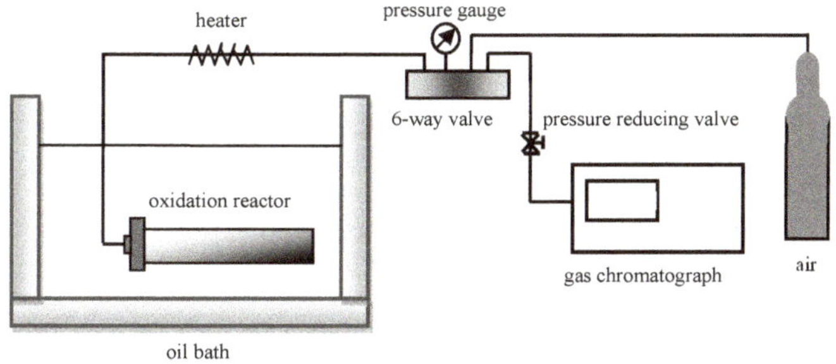

Figure 1. Schematic of the static oxidation experiment setup.

The experimental setup used in thermal analysis consists of a TG-DTG test system (STA6000, PerkinElmer, Waltham, MA, USA) and a FTIR spectrometer (Fourier Infrared Spectrometer, PerkinElmer, Waltham, MA, USA), which is the same as that used in the previous study [37].

2.3. Experimental Procedures

2.3.1. Separation of SARA Fractions

The Changqing tight oil was separated into SARA fractions according to a modified analytical procedure used by Freitag et al. [40]. The separation of Asphaltenes was based on the properties of asphaltenes insoluble in n-pentane. The other three fractions were subsequently separated applying a modified liquid chromatography procedure. Saturates fraction was eluted from the alumina column using n-pentane, aromatics fraction was separated using toluene, and resins fraction was eluted out with methanol and tetrahydrofuran. The vacuum oven was used to remove the remaining solvents from the separated fractions, during which nitrogen was used to protect the SARA fractions from oxidation.

2.3.2. Static Oxidation Experiments

The static oxidation experiment method applied in this study was similar as that used in our previous research [41]. The experimental samples were oil or the SARA fraction with or without catalyst added. The experimental pressure was 16 MPa. The catalyst used in this research is cobalt naphthenate which had good oil solubility, and the catalyst dosage was 0.08 mol/L. The LTO reaction rate is determined applying the material balance method [39,42].

2.3.3. TG-FTIR Tests

The TG-FTIR tests were conducted to quantify the catalytic effect of cobalt additive on the oxidation characteristics of the SARA fractions separated from a tight oil. 12 mg sample (oil or its fraction with or without catalyst added) was placed in the alumina crucible of the thermogravimetric analyzer to conduct the tests. The sample was first heated to 180 °C at a heating rate of 50 °C/min, and then kept at 180 °C for 120 min. By using air as a carrier gas (30 mL/min), the oxidized volatiles were directly introduced to the IR gas cell of the FTIR spectrometer for on-line analysis. The transfer line used in the setup was kept at 180 °C to prevent any condensation of the released gaseous products.

To exclude the influence of the organic groups of catalyst, the anhydrous cobalt chloride ($CoCl_2$) was used as the catalyst in TG-FTIR test. The catalyst dispersion method used is the same as that of previous research [37]. The catalyst dosage was 0.08 mol/L in the tests. The catalytic effect of additive is controlled by the metallic portion of the catalyst and is minimally affected by the anions [39], and the change of cobalt additive has a little effect on the reaction [37].

3. Results and Discussion

3.1. Catalytic Oxidation Kinetics

The LTO reaction rates of Changqing tight oil and SARA fractions with and without catalyst at 16 MPa are tabulated in Table 3. To identify the catalytic mechanisms, the changes in kinetic parameters of LTO before and after adding cobalt additive are studied applying the simplified Arrhenius model [3].

$$\frac{dc(O_2)}{dt} = f \cdot e^{-E/RT} \tag{1}$$

$$ln\frac{dc(O_2)}{dt} = lnf - \frac{E}{RT} \tag{2}$$

The relationship between oxygen consumption rate and absolute temperature at 16 MPa is shown in Figure 2. The LTO activation energies of crude oil and its fractions with or without catalyst added are summarized in Table 4. The cobalt additive can decrease the LTO reaction activation energy of oil and SARA fraction obviously. The activation energy of crude oil is reduced to 21,864 J/mol from 39,571 J/mol due to the catalyst. The activation energy of aromatics is decreased by 44.3% from 35,606 J/mol to 19,849 J/mol, while the activation energy of saturates and resins is decreased to 32,088 and 27,700 J/mol from 37,469 and 30,970 J/mol, respectively. As such, the reduction in activation energy is found to be the main mechanism for the cobalt additive catalyzing the LTO of Changqing tight oil. The reduction in LTO reaction activation energy of SARA fractions is the intrinsic reason for the catalysis of crude oil oxidation.

Table 3. Oxidation rate of oil and its fractions at 16 MPa.

Test Sample	Temperature, °C	Oxidation Rate, (mol O_2/d·m^3 [Sample])	Test Sample	Temperature, °C	Oxidation Rate, (mol O_2/d·m^3 [Sample])
Crude oil	50	26.14	Crude oil + catalyst	50	241.16
	70	53.37		70	271.69
	90	86.62		90	390.46
	110	295.37		110	907.23
Saturates	50	16.97	Saturates + catalyst	50	101.78
	70	35.02		70	185.08
	90	54.64		90	297.07
	110	168.81		110	704.01
Aromatics	50	26.43	Aromatics + catalyst	50	202.17
	70	47.25		70	311.08
	90	65.11		90	377.06
	110	245.96		110	691.65
Resins	50	75.34	Resins + catalyst	50	139.75
	70	137.75		70	313.27
	90	210.43		90	405.37
	110	492.36		110	768.33

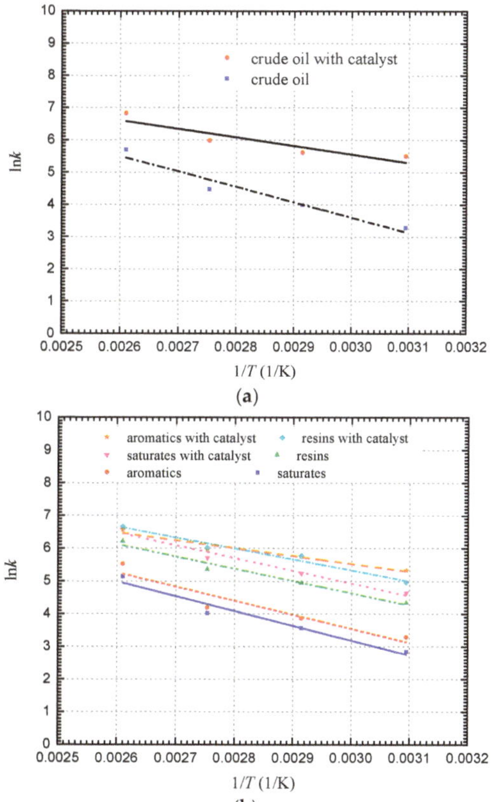

Figure 2. The relationship between oxygen consumption rate and absolute temperature at 16 MPa, (**a**) crude oil, (**b**) SARA fractions.

Table 4. The LTO activation energy of crude oil and its fractions with or without catalyst added.

Crude oil and Fractions		Activation Energy, (J/mol)	Activation Energy with Catalyst Added, (J/mol)
Crude oil		39,571	21,864
SARA fraction	Saturates	37,469	32,088
	Aromatics	35,606	19,849
	Resins	30,970	27,700

3.2. CLTO Reaction Process of Changqing Tight Oil Fractions

The catalytic effect of the cobalt additive on the LTO reaction pathway of SARA fractions is examined by analyzing the composition changes of evolved gas during the LTO and CLTO process by performing TG-FTIR tests. The catalytic oxidation characteristics of Changqing tight oil was researched before, as described in the reference [37].

3.2.1. Saturates Fraction

The mass losses of the saturates with and without catalyst are shown in Figure 3a. At the initial stage of the test, the saturates mass loss is fast. For t < 20.0 min, the mass loss during the CLTO of the saturates is lower than that during the LTO test. Afterwards, the mass loss rate during the LTO of the saturates is reduced, becoming lower than that during the CLTO of the saturates. When the tests end at t = 120.0 min, the residual mass after the CLTO of the saturates is 37.89%, which is 2.32% lower than that of the LTO test. The rapid mass loss of saturates at the initial stage of the test is mainly due to the distillation of low-boiling-point hydrocarbons and some LTO products. At the middle and late stages of the test, the continuous mass loss of the saturates is resulted from the volatilization of the oxidation products.

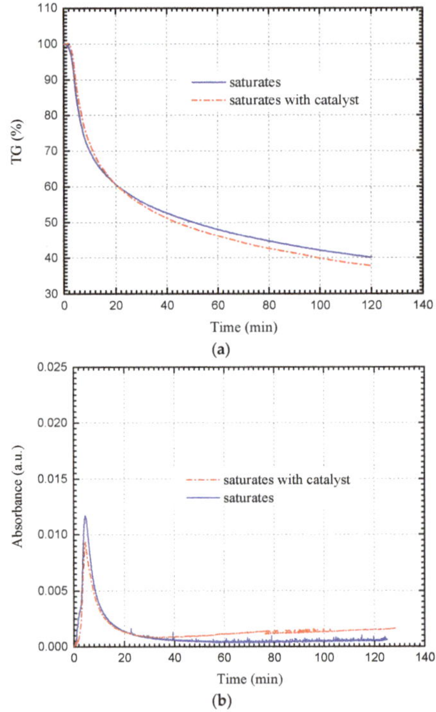

Figure 3. (a) TG, (b) Gram-Schmidt curves of the saturates before and after the catalyst addition.

Figure 4 plots the 3D infrared spectra of the gaseous products during the test. The characteristic peak intensities of the functional groups of the gas products all change significantly with the catalyst, especially in the case of CO_2 (characteristic peaks for the asymmetric stretching (2360 cm^{-1}) and bending vibrations (670 cm^{-1})) [43,44]. Figure 3b is the GS curves of saturates fraction. When t < 20.0 min, the gas release rate during the CLTO of the saturates is lower than that during the LTO of the saturates, whereas, when t > 20.0 min, the gas release rate during the CLTO of the saturates is higher. The trend of the GS curves is consistent with that of the TG curves.

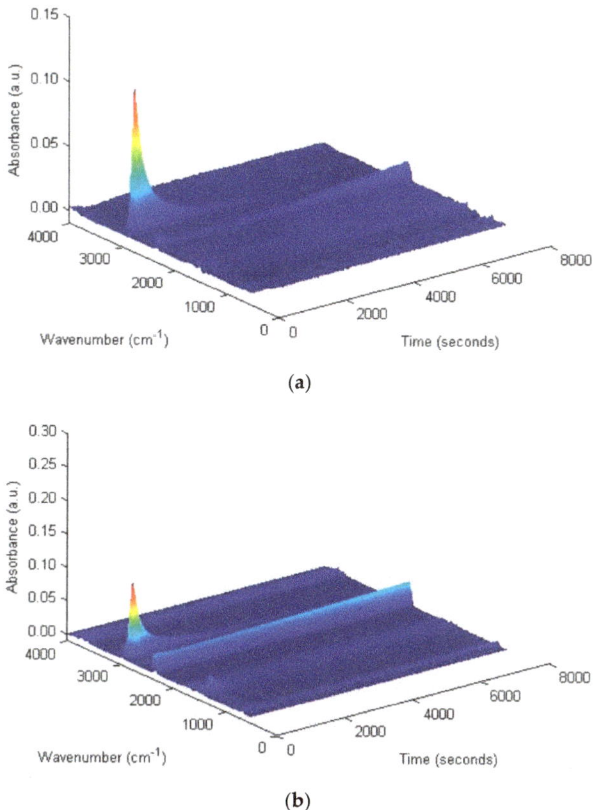

Figure 4. 3D infrared spectra of (**a**) saturates, (**b**) saturates with catalyst.

Figure 5 displays the infrared spectra of the released gas by the LTO of the saturates with and without the catalyst. With the catalyst added, no new peaks appear in the spectra, but the intensities of the original absorption peaks change dramatically. The most remarkable increases are observed for the characteristic peaks of hydroxyl groups between 3400 cm^{-1} and 4000 cm^{-1} [45] and for the characteristic peaks of CO_2 at 2360 cm^{-1} and 670 cm^{-1}, indicating that the catalyst greatly promotes the oxygen addition reaction and bond scission reaction during the saturates LTO process. Additionally, the promotion effect on the bond scission reaction becomes more pronounced as the reaction proceeds. On one hand, the catalytic effect on the oxygen addition reaction causes the oxidation of more hydrocarbons to oxygenated hydrocarbons with higher boiling points, which reduces the mass loss of the saturates during the initial stage of the test. On the other hand, the catalytic effect on the bond scission reaction facilitates the formation of volatile products such as CO_2 and H_2O, which increase the mass loss during the middle and late stages of the test.

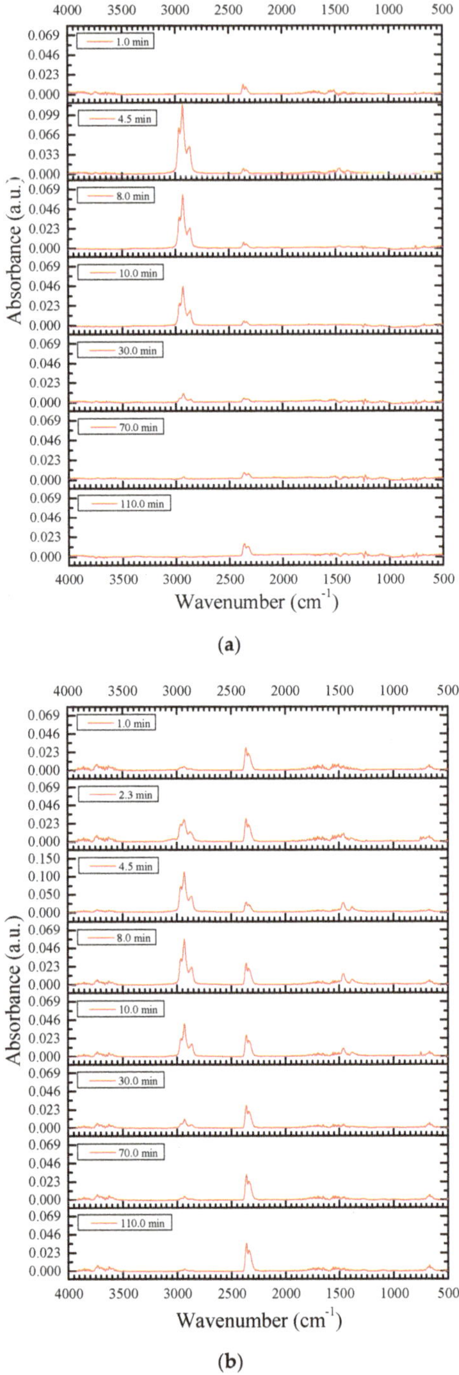

Figure 5. Infrared spectra of gas produced by the LTO of (**a**) saturates, (**b**) saturates with catalyst at 180 °C at different times.

3.2.2. Aromatics Fraction

As shown in Figures 6 and 7, the catalyst imposes no apparent influence on the mass loss trend during the LTO of the aromatics, but it does increase the mass loss rate. At the end of the test, the mass loss during the CLTO of the aromatics is 22.81%, which is 2.97% higher than that during the LTO of the aromatics. Physically, the aromatics has a high molecular weight and a low content of low-boiling-point components, and thus the mass loss due to direct distillation during the test is limited. With the catalyst added, the production of carbon oxides during the LTO is increased, resulting in the rapid increment in gas production, and thereby, the larger mass loss.

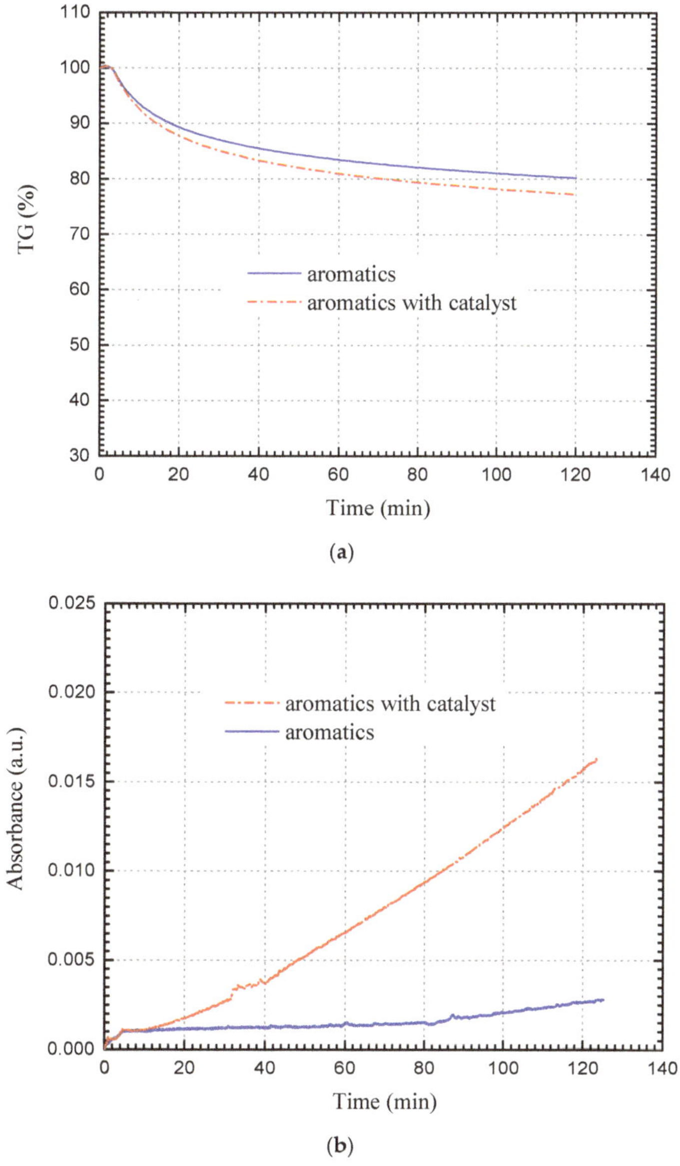

Figure 6. (a) TG, (b) Gram-Schmidt curves of the aromatics with and without the catalyst.

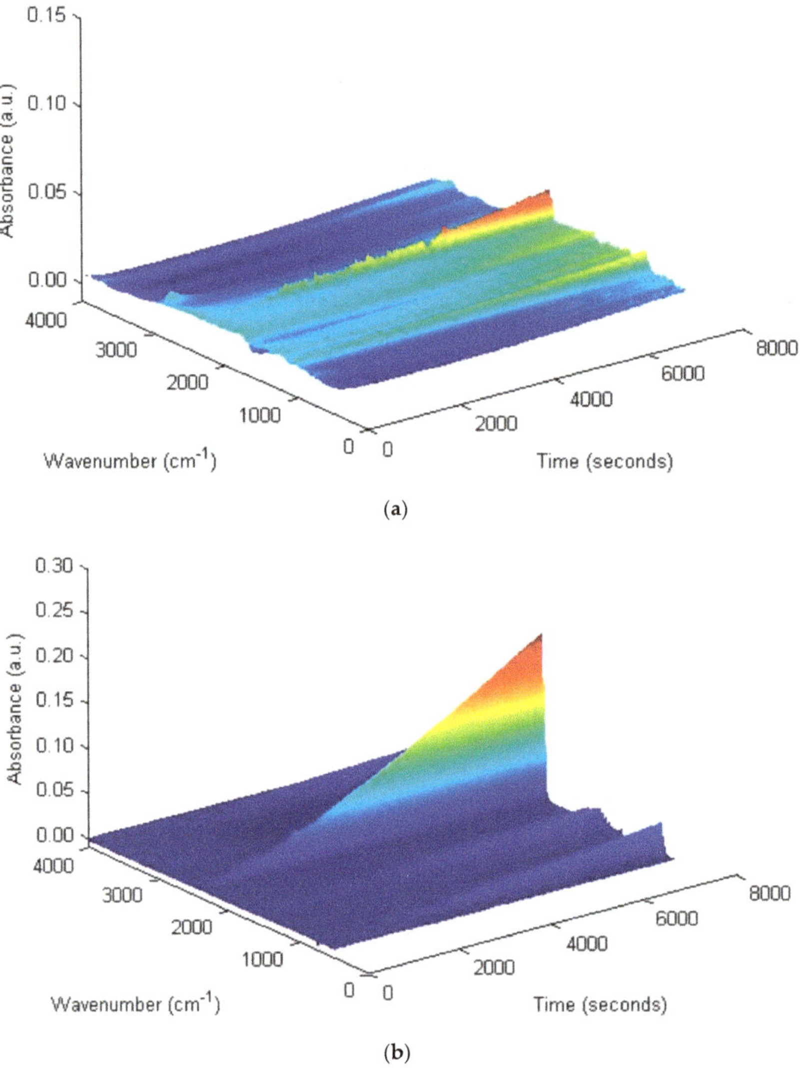

Figure 7. 3D infrared spectra of (**a**) aromatics, (**b**) aromatics with catalyst.

Figure 8 depicts the infrared spectra of gas produced by the LTO of the aromatics as a function of time at 180 °C. The absorption peaks change dramatically with the catalyst. Signals corresponding to ethers (absorption peaks between 1000 cm^{-1} and 1300 cm^{-1}) are greatly weakened or even absent, whereas the peaks corresponding to hydroxyl groups and CO_2 were greatly enhanced. The result suggests that the catalyst substantially promotes the oxygen addition reaction and bond scission reaction during the LTO of aromatics. However, it is worthwhile noting that the catalyst behaves selectively, significantly favoring the formation of hydroxyl-containing oxides while inhibiting the formation of ethers. As the reaction proceeds, the catalytic effect on the bond scission reaction becomes more pronounced, leading to a linear increase in the gas production in the GS curve. This phenomenon is the main reason for the growing difference in mass loss between catalytic and noncatalytic oxidations of aromatics at the middle and late stages of the test.

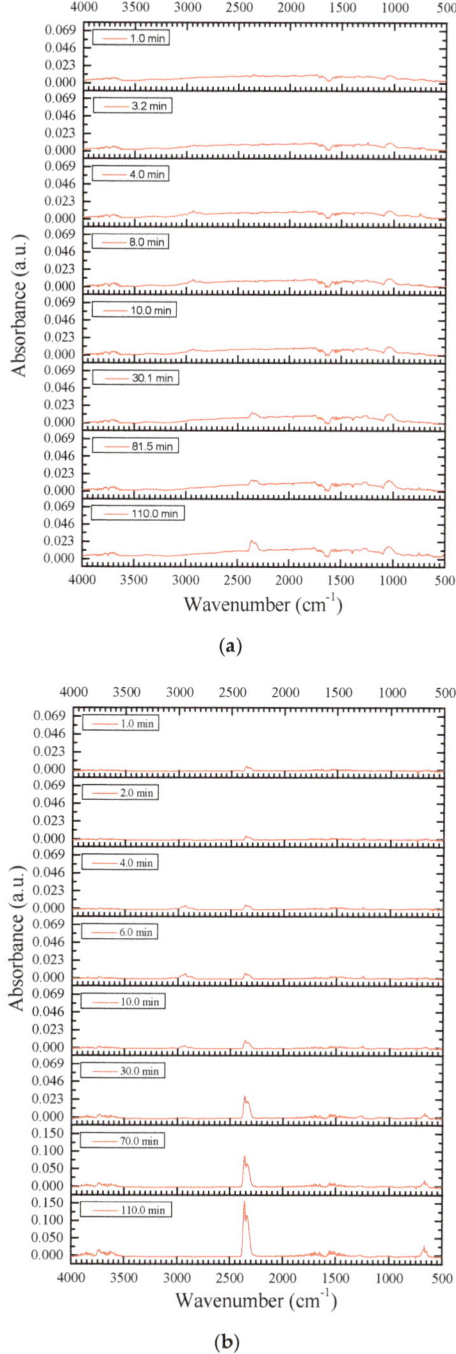

(**a**)

(**b**)

Figure 8. Infrared spectra of gas produced by the LTO of (**a**) aromatics, (**b**) aromatics with catalyst at 180 °C at different times.

3.2.3. Resins Fraction

Resins is a heavy fraction of crude oil, which exhibits complex molecular structures and a high boiling point [44]. The mass loss caused by direct distillation during the test is therefore limited. Similar to the observations on aromatics, the catalyst does not affect the trend of mass loss during the LTO of resins but increases the mass loss rate (see Figure 9a). At the end of the test, the mass loss of the CLTO of the resins is 9.0%, which is 2.07% higher than that of the noncatalytic LTO. With the catalyst added, the intensities of the CO_2 absorption peaks in the 3D infrared spectra of the resins increase significantly as the reaction proceeds (see Figure 10); on the GS curves, the gas production increases linearly (see Figure 9b). These changes are consistent with the changes of the TG curve.

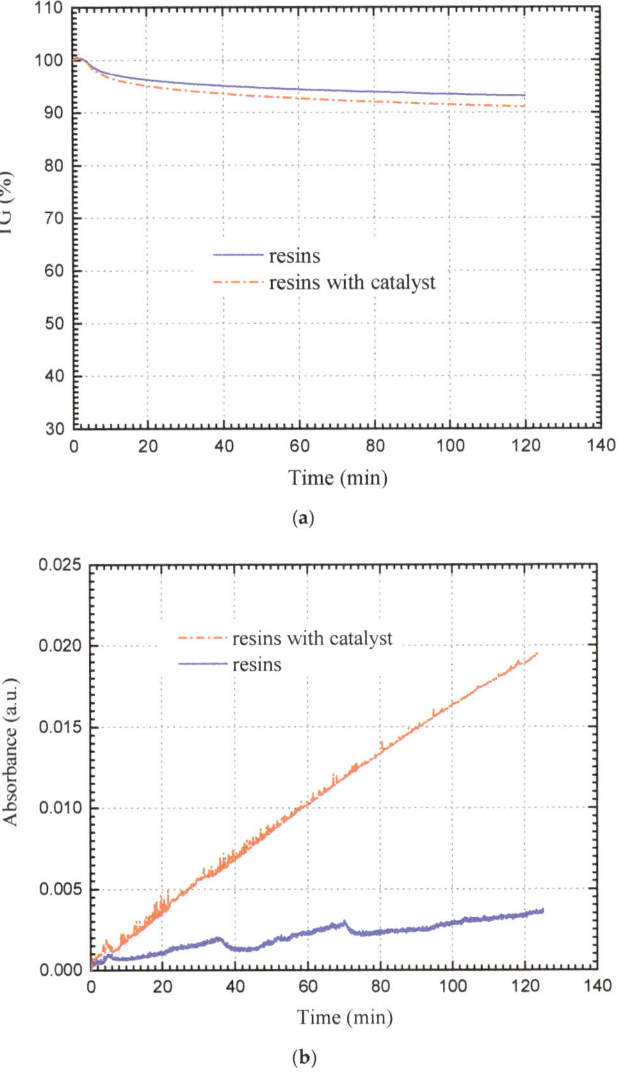

(a)

(b)

Figure 9. (a) TG, (b) Gram-Schmidt curves of the resin fraction with and without the catalyst.

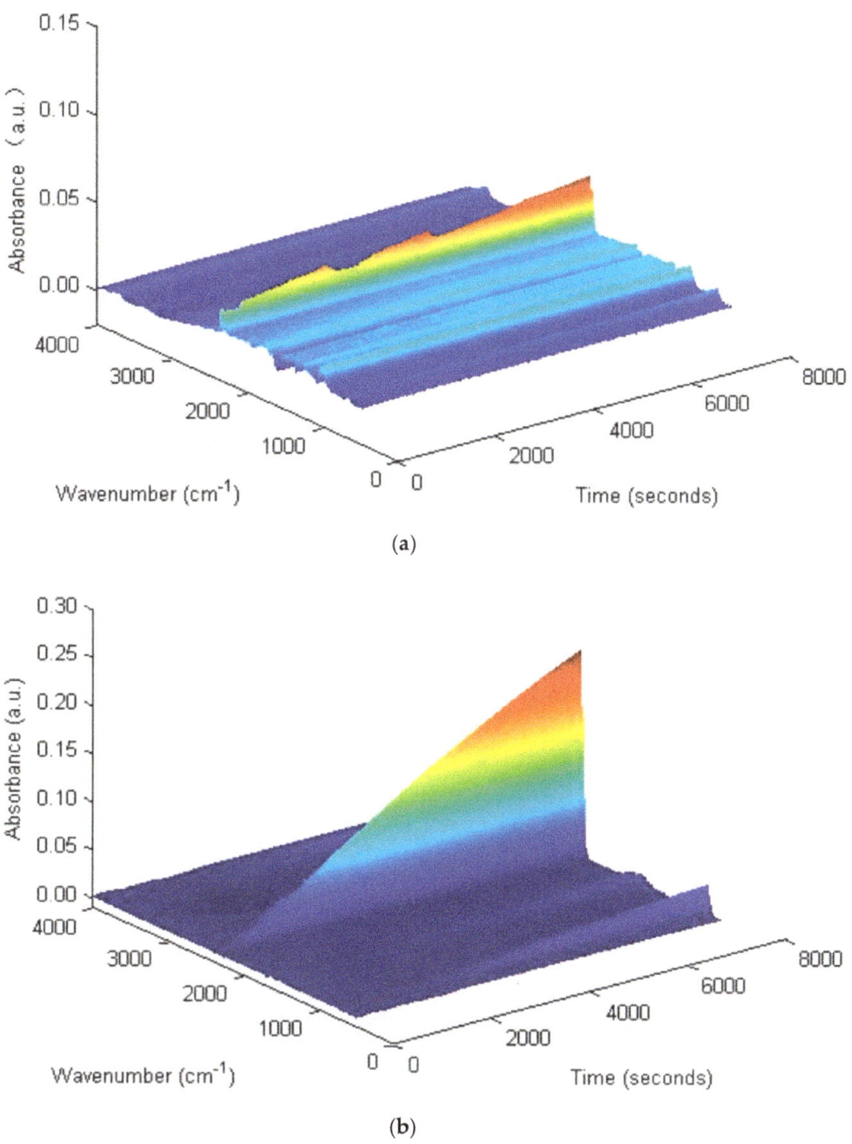

Figure 10. 3D infrared spectra of (**a**) resins, (**b**) resins with catalyst.

With the catalyst, the change in the infrared spectra of the resins is similar to that of the aromatics. The absorption peaks corresponding to ethers are basically absent, whereas the absorption peaks for hydroxyl groups and CO_2 are enhanced significantly (see Figure 11). The result suggests that the catalyst significantly promotes the oxygen addition reaction and bond scission reaction. As the reaction proceeds, the promotion effect on the bond scission reaction becomes more pronounced, and the production of CO_2 increases sharply.

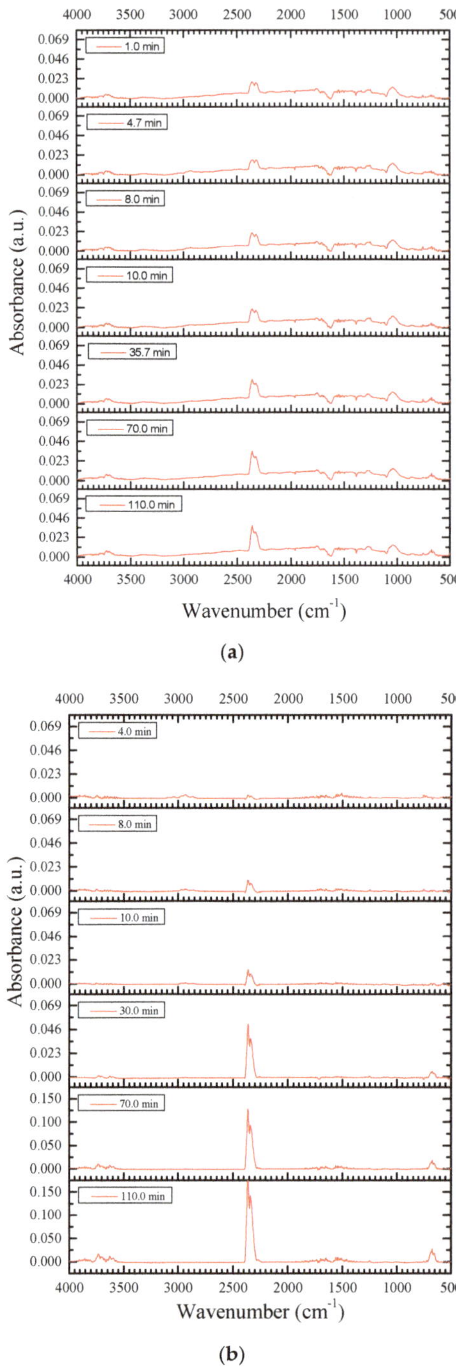

Figure 11. Infrared spectra of gas produced by the LTO of (**a**) resins, (**b**) resins with catalyst at 180 °C at different times.

3.3. Catalytic Oxidation Characteristics of Changqing Tight Oil and SARA Fraction

As can be seen from the above TG-FTIR tests, the cobalt salt imposes an obvious catalytic effect on the LTO of SARA fractions. The catalyst significantly changes the reaction pathways of the LTO of SARA fractions, which is the intrinsic reason for its catalytic effect on Changqing tight oil oxidation.

As shown in Figure 12, during the initial stage of the CLTO of crude oil, the gas is mainly produced from the saturates, and the contributions of aromatics and resins are small. The findings are similar to those observed for the noncatalytic LTO of crude oil; however, the gas production from resin and aromatic fractions increases rapidly at the middle and late stages of the CLTO and exceeds that from the saturate fraction after t = 20.0 min. As evidenced by the TG-FTIR studies, the main component of the gas after 20 min is CO_2 generated by the bond scission reaction. Thus, the intensities of the bond scission reaction of SARA fractions during the middle and late stages of the CLTO reaction follow the order of resins > aromatics > saturates. Considering that the reaction intensities of the three fractions are close without the catalyst, it can be inferred that the promotion effect on the bond scission reaction is the most pronounced for the resins, followed by the aromatics, and the weakest for the saturates. The difference in the catalytic activities can be attributed to the difference in the molecular structures of the SARA fractions. Resins is a heavy fraction in crude oil, and its molecules contain a large number of aromatic rings, alicyclic rings, various long and short branches, and heteroatoms [44]. Resins possesses a very strong polarity and is easier to react with metal ions to achieve a better catalytic effect than with the aromatic and saturate fractions.

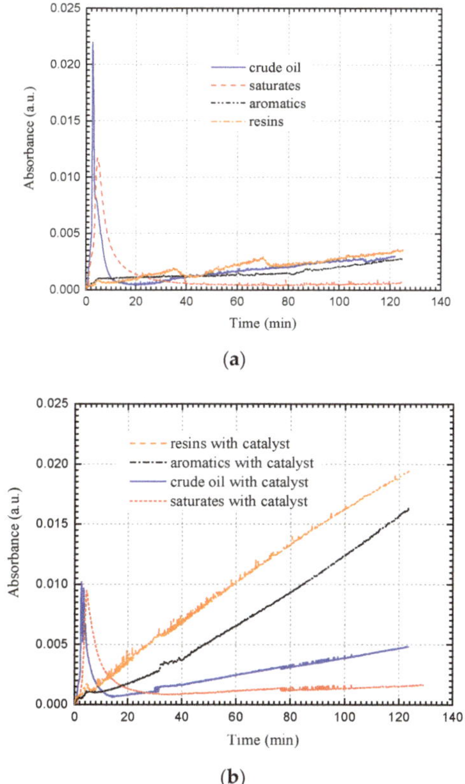

Figure 12. Gram-Schmidt curves of crude oil and its SARA fractions (**a**) without catalyst, (**b**) with catalyst.

Figure 13 illustrates the effect of the catalyst on CO_2 production (i.e., the absorbance of peak corresponding to the asymmetric stretching (2360 cm^{-1}) of CO_2) during the LTO of Changqing tight oil and its SARA fractions. At initial stage of the CLTO, a large amount of CO_2 is produced from the saturate fraction, but the growth rate in CO_2 production from saturates is lower than those from the aromatics and resins. As a result, the amount of CO_2 produced from the saturate fraction is the lowest at the middle and late stages of the CLTO. This result indicates that the saturates in the SARA fractions is most susceptible to the promotion effect on the bond scission reaction, but the catalytic effect is not the strongest. Under the same conditions, the strongest promotion effect on the bond scission reaction occurs with that of the resin fraction.

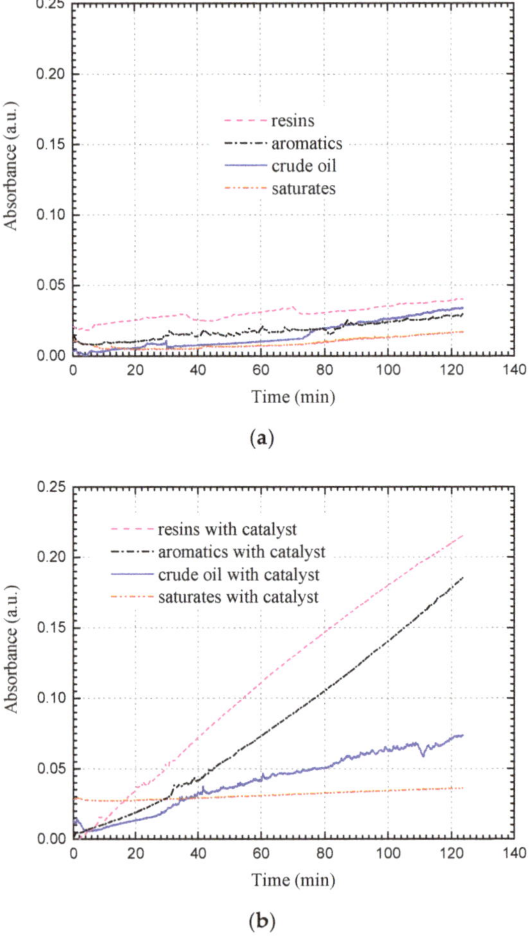

Figure 13. Changes in the CO_2 absorbance of crude oil and its SARA fractions in the TG-FTIR test (**a**) without catalyst, (**b**) with catalyst.

The catalytic oxidation mechanisms of Changqing tight oil were identified by the free radical chain reaction theory and the coordination catalysis theory, as described in our previous research [37]. The catalyst can increase the formation rate of organic peroxides. As a result, the degenerate-branching chain reaction is promoted, and the formation of CO_2 and oxygenated hydrocarbons, such as alcohol and carboxylic acid, is increased. The dispersed cobalt metal ions interact with the aromatic rings or heteroatoms in the crude oil to form coordination complexes between the metal ions and crude oil fractions. Aromatics and resins are heavy fractions of crude oil, which contain more aromatic rings or heteroatoms than saturates [44], so the catalytic effect on the oxidation reaction is more pronounced for aromatics and resins fractions.

4. Conclusions

The cobalt additive can significantly enhance the oxygen consumption capacity of Changqing tight oil and promote the formation of carbon oxides, which are beneficial to the successful application of the air injection technology for tight oil reservoir development.

- The cobalt-salt catalyst can reduce the LTO reaction activation energies of Changqing tight oil and its SARA fractions. The catalytic effect on LTO of SARA fractions is the intrinsic reason for the significant catalysis of Changqing tight oil oxidation.
- The catalyst greatly promotes the oxygen addition reaction and bond scission reaction of saturates, aromatics and resins, and the promotion effect on the bond scission reaction is stronger. The promotion effect on the bond scission reaction is the most pronounced for the resins, followed by the aromatics, and the weakest for the saturates.
- The catalyst significantly changes the reaction pathways of the LTO of SARA fractions. The formation of hydroxyl-containing oxides and CO_2 from the oxidation of saturates, aromatics and resins is promoted, while the formation of ethers from the oxidation of aromatics and resins is inhibited.

Author Contributions: T.W. completed the investigation; T.W. and J.W. designed the research; T.W. completed the experiments; T.W. and J.W. in charge of writing.

Funding: The authors gratefully acknowledge the financial supports from CSC Scholarship (Grant No. 201806455023), Shandong Provincial Natural Science Foundation, China (Grant No. ZR2017BEE023), and The Fundamental Research Funds for the Central Universities (Grant No. 24720182162A).

Conflicts of Interest: The authors declare no conflict of interest.

References

1. Jia, H.; Sheng, J.J. Numerical modeling on air injection in a light oil reservoir: Recovery mechanism and scheme optimization. *Fuel* **2016**, *172*, 70–80. [CrossRef]
2. Huang, S.; Sheng, J.J. Discussion of thermal experiments' capability to screen the feasibility of air injection. *Fuel* **2017**, *195*, 151–164. [CrossRef]
3. Ren, S.R.; Greaves, M.; Rathbone, R.R. Air injection LTO process: An IOR technique for light-oil reservoirs. *SPE J.* **2002**, *7*, 90–99. [CrossRef]
4. Gutierrez, D.; Taylor, A.R.; Kumar, V.; Ursenbach, M.G.; Moore, R.G.; Mehta, S.A. Recovery factors in high-pressure air injection projects revisited. *SPE Reserv. Eval. Eng.* **2008**, *11*, 1097–1106. [CrossRef]
5. Fraim, M.L.; Moffitt, P.D.; Yannimaras, D.V. Laboratory testing and simulation results for high pressure air injection in a waterflooded North Sea oil reservoir. In Proceedings of the SPE Annual Technical Conference and Exhibition, San Antonio, TX, USA, 5–8 October 1997.
6. Adetunji, L.A.; Teigland, R. Light-oil air-injection performance: Sensitivity to critical parameters. In Proceedings of the SPE Annual Technical Conference and Exhibition, Dallas, TX, USA, 9–12 October 2005.
7. Clara, C.; Durandeau, M.; Quenault, G.; Nguyen, T.H. Laboratory studies for light-oil air injection projects: Potential application in Handil field. *SPE Reserv. Eval. Eng.* **2000**, *3*, 239–248. [CrossRef]
8. Duiveman, M.W.; Herwin, H.; Grivot, P.G. Integrated management of water, lean gas and air injection: The successful ingredients to IOR projects on the mature Handil field. In Proceedings of the SPE Asia Pacific Oil and Gas Conference and Exhibition, Jakarta, Indonesia, 5–7 April 2005.
9. Pascual, M.; Crosta, D.; Lacentre, P.; Coombe, D. Air injection into a mature waterflooded light oil reservoir-laboratory and simulation results for Barrancas field, Argentina. In Proceedings of the 67th EAGE Conference Exhibition, Madrid, Spain, 13–16 June 2005.

10. Fassihi, M.R.; Yannimaras, D.V.; Kumar, V.K. Estimation of recovery factor in light-oil air-injection projects. *SPE Reserv. Eng.* **1997**, *12*, 173–178. [CrossRef]

11. Fassihi, M.R.; Gillham, T.H. The use of air injection to improve the double displacement processes. In Proceedings of the SPE Annual Technical Conference and Exhibition, Houston, TX, USA, 3–6 October 1993.

12. Gillham, T.H.; Cerveny, B.W.; Turek, E.A.; Yannimaras, D.V. Keys to increasing production via air injection in Gulf Coast light oil reservoirs. In Proceedings of the SPE Annual Technical Conference and Exhibition, San Antonio, TX, USA, 5–8 October 1997.

13. Glandt, C.A.; Pieterson, R.; Dombrowski, A.; Balzarini, M.A. Coral creek field study: A comprehensive assessment of the potential of high-pressure air injection in a mature waterflood project. In Proceedings of the SPE Mid-continent operations symposium, Oklahoma, OK, USA, 28–31 March 1999.

14. Stokka, S.; Oesthus, A.; Frangeul, J. Evaluation of air injection as an IOR method for the Giant Ekofisk Chalk Field. In Proceedings of the SPE International Improved Oil Recovery Conference in Asia Pacific, Kuala Lumpur, Malaysia, 5–6 December 2005.

15. Gutierrez, D.; Kumar, V.; Moore, R.G.; Mehta, S.A. Air injection and waterflood performance comparison of two adjacent units in the Buffalo Field. *SPE Reserv. Eval. Eng.* **2008**, *11*, 848–857. [CrossRef]

16. Denney, D. 30 years of successful high-pressure air injection: Performance evaluation of Buffalo field, South Dakota. *J. Pet. Technol.* **2011**, *63*, 50–53. [CrossRef]

17. Germain, P.; Geyelin, J.L. Air injection into a light oil reservoir: The Horse Creek project. In Proceedings of the SPE Middle East Oil Show and Conference, Awali, Bahrain, 15–18 March 1997.

18. Bagci, S.; Celebioglu, D. Light oil combustion with metallic additives in limestone medium. In Proceedings of the Canadian International Petroleum Conference, Calgary, AB, Canada, 8–10 June 2004.

19. Shallcross, D.C.; De los Rios, C.F.; Castanier, L.M.; Brigham, W.E. Modifying in-situ combustion performance by the use of water-soluble additives. *SPE Reserv. Eng.* **1991**, *6*, 287–294. [CrossRef]

20. Gerritsen, M.; Kovscek, A.; Castanier, L.; Nilsson, J.; Younis, R.; He, B. Experimental investigation and high resolution simulator of in-situ combustion processes; 1. Simulator design and improved combustion with metallic additives. In Proceedings of the SPE International Thermal Operations and Heavy Oil Symposium and Western Regional Meeting, Bakersfield, CA, USA, 16–18 March 2004.

21. Ramirez-Garnica, M.A.; Mamora, D.D.; Nares, R.; Schacht-Hernandez, P.; Mohammad, A.A.; Cabrera, M. Increase heavy-oil production in combustion tube experiments through the use of catalyst. In Proceedings of the SPE Latin American and Caribbean Petroleum Engineering Conference, Buenos Aires, Argentina, 15–18 April 2007.

22. Ramirez-Garnica, M.A.; Hernandez Perez, J.R.; Cabrera-Reyes, M.D.C.; Schacht-Hernandez, P.; Mamora, D.D. Increase oil recovery of heavy oil in combustion tube using a new catalyst based on nickel ionic solution. In Proceedings of the SPE International Thermal Operations and Heavy Oil Symposium, Calgary, AB, Canada, 20–23 October 2008.

23. Fassihi, M.R.; Brigham, W.E.; Ramey, H.J., Jr. Reaction kinetics of in-situ combustion: Part 1—Observations. *SPE J.* **1984**, *24*, 399–407. [CrossRef]

24. Fassihi, M.R.; Brigham, W.E.; Ramey, H.J., Jr. Reaction kinetics of in-Situ combustion: Part 2—Modeling. *SPE J.* **1984**, *24*, 408–416. [CrossRef]

25. Castanier, L.M.; Baena, C.J.; Holt, R.J.; Brigham, W.E.; Tavares, C. In situ combustion with metallic additives. In Proceedings of the SPE Latin America Petroleum Engineering Conference, Caracas, Venezuela, 8–11 March 1992.

26. Drici, O.; Vossoughi, S. Catalytic effect of heavy metal oxides on crude oil combustion. *SPE Reserv. Eng.* **1987**, *2*, 591–595. [CrossRef]

27. He, B.; Chen, Q.; Castanier, L.M.; Kovscek, A.R. Improved in-situ combustion performance with metallic salt additives. In Proceedings of the SPE Western Regional Meeting, Irvine, CA, USA, 30 March–1 April 2005.

28. Wang, H.; Tang, X.; Meng, K.; Cui, Y.; Wang, Z.; Wang, F. Preparation and evaluation of catalyst for heavy oil injected air catalytic oxidation production. *Fine Chem.* **2009**, *33*, 179–188.

29. Tang, X.; Cui, Y.; He, B.; Liu, Q.; Meng, K. Experimental study on the viscosity reduction of Liaohe heavy oil by catalytic air oxidation. *Oilfield Chem.* **2008**, *25*, 316–319.

30. Pu, W.; Liu, P.; Li, Y.; Jin, F.; Liu, Z. Thermal characteristics and combustion kinetics analysis of heavy crude oil catalyzed by metallic additives. *Ind. Eng. Chem. Res.* **2015**, *54*, 11525–11533. [CrossRef]

31. Amanam, U.U.; Kovscek, A.R. Analysis of the effects of copper nanoparticles on in-situ combustion of extra heavy-crude oil. *J. Pet. Sci. Eng.* **2017**, *152*, 406–415. [CrossRef]

32. Zhao, Y. Effect of Clay Minerals on Oxidation Characteristics of Air Flooding. Ph.D. Dissertation, Southwest Petroleum University, Chengdu, China, 2014.

33. Jia, H.; Zhao, J.Z.; Pu, W.F.; Zhao, J.; Kuang, X.Y. Thermal study on light crude oil for application of high-pressure air injection (HPAI) process by TG/DTG and DTA tests. *Energy Fuels* **2012**, *26*, 1575–1584. [CrossRef]

34. Shi, G. Experimental study on low-temperature oxidation in air injection with supported-catalyst. *Spec. Oil Gas. Reserv.* **2012**, *19*, 126–129.

35. Wang, J.; Wang, T.; Feng, C.; Chen, Z.; Kong, D.; Niu, Z. The catalytic effect of organometallic additives on low-temperature oxidation of light crude oil during air flooding. *Pet. Sci. Technol.* **2015**, *33*, 1118–1125. [CrossRef]

36. Wang, T.; Wang, J.; Yang, W.; Kalitaani, S.; Deng, Z. A novel air flooding technology for light crude oil reservoirs applied under reservoir conditions. *Energy Fuels* **2018**, *32*, 4942–4950. [CrossRef]

37. Wang, T.; Yang, W.; Wang, J.; Kalitaani, S.; Deng, Z. Low temperature oxidation of crude oil: Reaction progress and catalytic mechanism of metallic salts. *Fuel* **2018**, *225*, 336–342. [CrossRef]

38. Groenzin, H.; Mullins, O.C. Asphaltene molecular size and structure. *J. Phys. Chem. A* **1999**, *103*, 11237–11245. [CrossRef]

39. Wang, J.; Wang, T.; Feng, C.; Yang, C.; Chen, Z.; Lu, G. Catalytic effect of transition metallic additives on the light oil low-temperature oxidation reaction. *Energy Fuels* **2015**, *29*, 3545–3555. [CrossRef]

40. Freitag, N.P.; Exelby, D.R. A SARA-based model for simulating the pyrolysis reactions that occur in high-temperature EOR processes. *J. Can. Pet. Technol.* **2006**, *45*, 38–44. [CrossRef]

41. Wang, T.; Wang, J.; Meng, X.; Chu, G.; Liu, C. The low temperature oxidation characteristics of SARA fractions in air flooding process. *Pet. Sci. Technol.* **2018**, *36*, 1125–1130. [CrossRef]

42. Yang, S.; Jin, L.; Kong, Q.; Li, L. *Advanced Engineering Mechanics*; Higher Education Press: Beijing, China, 1988; pp. 156–157, 222–223.

43. Fraust, B. Infrared spectroscopy. In *Modern Chemical Techniques*; The Royal Society of Chemistry: London, UK, 1997; pp. 62–91.

44. Niu, B.; Ren, S.; Liu, Y.; Wang, D.; Tang, L.; Chen, B. Low-temperature oxidation of oil components in an air injection process for improved oil recovery. *Energy Fuels* **2011**, *25*, 4299–4304. [CrossRef]

45. Shokrlu, Y.H.; Maham, Y.; Tan, X.; Babadagli, T.; Gray, M. Enhancement of the efficiency of in situ combustion technique for heavy-oil recovery by application of nickel ions. *Fuel* **2013**, *105*, 397–407. [CrossRef]

Article

Experimental Investigation on Injection and Production Pattern in Fractured-Vuggy Carbonate Reservoirs

Juan He [1,2,3,4], Aowei Li [4,†], Shanshan Wu [4,†], Ruixue Tang [4,*], Dongliang Lv [4], Yongren Li [4] and Xiaobo Li [5]

[1]   Key Laboratory of Renewable Energy and Gas Hydrate, Guangzhou Institute of Energy Conversion, CAS, Guangzhou 510640, China; hejuan@ms.giec.ac.cn
[2]   Guangdong Provincial Key Laboratory of New and Renewable Energy Research and Development, Guangzhou 510640, China
[3]   University of Chinese Academy of Sciences, Beijing 100049, China
[4]   State Key Laboratory of Oil and Gas Reservoir Geology and Exploitation, Southwest Petroleum University, Chengdu 610500, China; lawhhd@163.com (A.L.); hermione533wss@163.com (S.W.); Lvdongliang@swpu.edu.cn (D.L.); liyongren123@163.com (Y.L.)
[5]   Research Institute of Petroleum Engineering and Technology, Northwest Oilfield Company, Sinopec, Urumqi 830011, China; Lxb969@163.com
*   Correspondence: ruixue.tang1@ucalgary.ca
†   The author contributed equally to this work and should be considered co-first authors.

Received: 11 December 2019; Accepted: 21 January 2020; Published: 29 January 2020

Abstract: To constitute and adjust the injection and production pattern in fractured-vuggy reservoirs, we extracted twelve fractured-cave structures, fabricated them into physical models with acrylic plates, and performed experiments via these models. The results show that utilizing oil/water gravity segregation sufficiently and forming valid flow channels should be emphasized. Preferentially exploiting the reservoir body containing intermediate-scaled or large-scaled caves, arranging injection wells in fractures or small-scaled caves while placing production wells in large-scaled caves, and separately putting injection wells and production wells in low/high parts of an intermediate-scaled or large-scaled cave, were found to benefit oil/water gravity segregation and thus gain a better water flooding effect in these experiments. Experiments with combined models also figured out that, after adjusting the injection and production pattern, the valid flow channel newly formed must pass through caves containing enough residual oil to improve the water flooding effect and could be obtained by shutting down the old production well while adding a new production well, adding a new production well, or switching the production well into an injection well while adding a new production well. In the actual field, adjusting the well location and altering the flow channel were proposed to conduct together. This study may provide references on the production management of analogous reservoirs.

Keywords: fractured-vuggy reservoirs; physical model; water flooding effect; injection and production pattern; gravity differentiation; flow channel

---

1. Introduction

Carbonate reservoirs take almost 60 percent of global hydrocarbon resources, so it is crucial to the world's energy supply. In China, this proportion exceeds 25%. The Ordovician reservoir of the Tahe oilfield, Tarim Basin, is the largest fractured-vuggy carbonate reservoir in China. Since the matrix of the carbonate rock cannot store oil and has extremely low permeability [1,2], caves and fractures are reckoned as the main storage spaces, easily causing severe mud loss in drilling like other karst formations in the world [3]. Meanwhile, undergoing tectonic movements and karst processes, these caves and fractures possess various types, scales, and distributions. Therefore, unlike porous and fractured-porous carbonate reservoirs in China, the distinguishing features of the Tahe oilfield suffer from severe heterogeneity, complex space structure, and intricate flow mechanisms [4–6]. These traits cause the difficulty of constituting injection and production patterns in this reservoir. In clastic reservoirs, the well pattern is to enlarge sweep areas, so the water flooding effect can be improved by adjusting the injection layers directly. Nevertheless, using the same method is invalid for the injection and production pattern in this reservoir because of the random distributions and the poor connectivity of storage spaces [7].

Recently, various physical fractured-vuggy models were utilized to investigate the mechanism of water or gas displacement. In some of these models, practical irregular caves were simplified into circles [8–10], spheres [11], or cylinders [12], while original space structures of caves were preserved in other models [13–22]. Although the latter did display the remaining oil distributions much better, their well placements were predetermined in most of these experiments. Both Zhang [23] and Wang et al. [24] verified that a nine-spot well network behaved better in performances of water drive in fractured-vuggy reservoirs by using specific distributions of fractures and caves. However, this conventional well network is difficult to use in fractures and caves with variable spatial distributions. In fact, the phrase 'well net' is not suitable in fractured-vuggy reservoirs because the well placement is governed by reservoir bodies instead of layers. From the practice perspective, Li and Rong [25] proposed that well placement should be considered from the connectivity of wells, the reservoir body type, the position for production and injection, and the fluid distribution. Lu et al. [26] further developed the injection/production well pattern based on reservoir spatial structures, but there are currently few practical effects to support this idea. In addition, because of extreme heterogeneity weakening the function of the Darcy law [27,28], the present numerical simulations about the well pattern design in traditional oil and gas reservoirs [29–38] cannot solve problems effectively in this reservoir. With the lasting exploitation, the exposed disadvantages in the original injection and production pattern are harming the production in this oilfield. However, there are presently few systematic understandings and sustainably valid measures on the injection and production pattern to face up with these problems. Hence, it is urgent to perform relevant surveys.

In this work, physical simulation was employed to explore how to build and adjust the injection and production patterns in fractured-vuggy reservoirs. On the basis of designed and fabricated physical models, experimental schemes were designed and carried out from reservoir body types, injection/production relationships, high/low positions of a reservoir body, and injection and production adjustments, respectively. Through comparing the water flooding effect in four aspects, we analyzed some key points and achieved corresponding adjusting measures on the injection and production patterns in fractured-vuggy reservoirs.

2. Experiment

2.1. Model Design and Fabrication

According to the cognition about the spatial distribution, the geometrical morphology, and the connectivity in reservoir bodies [31,39–44], twelve typical fractured-cave structures were extracted from Block S74 in the Tahe oilfield (Figure 1a). The simplification of a geological explanation was presented in previous work [45]. These structures were divided into the large scale, the intermediate scale, and the small scale. In actual reservoirs, their sizes ranged from 34.5–39, 16.5–24, and 0.6–6 m, respectively. These fractured-cave structures were further fabricated into 3D physical models with a proportion of 300:1 (Figure 1b).

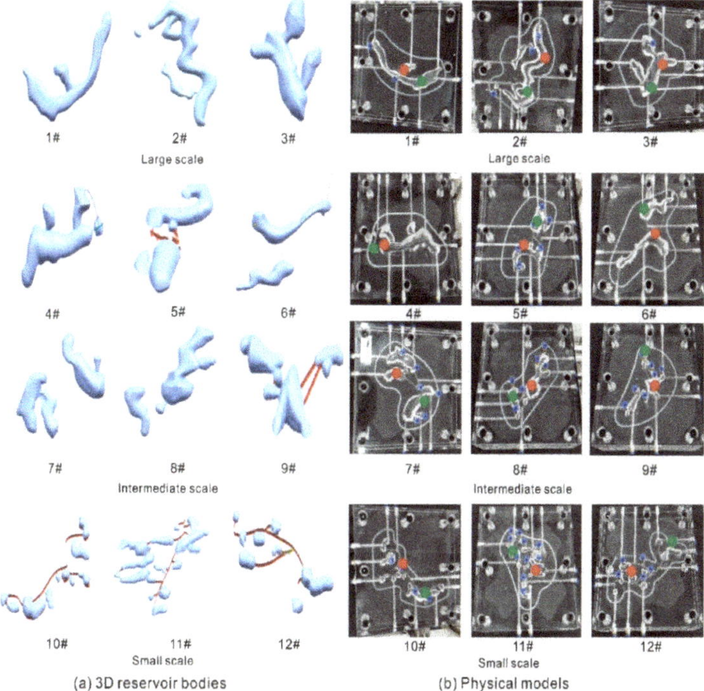

Figure 1. The typical fractured-cave structures (**a**) and the physical models (**b**). Red filled sources represent production wells. Green filled sources are injection wells.

A single model was composed of two weak oil-wet acrylic plates with a scale of $300 \times 300 \times 50$ mm (length × width × height). After the morphology of fractures and caves was engraved in two plates, a packing ring around the caves and fractures was fixed in a groove to avoid fluid leakage (Figure 2). Inlets and outlets were drilled on the surface of a plate with holes with a diameter of 6 mm. Their locations resembled actual well placements. Finally, two plates were combined together with eight screw bolts as a body.

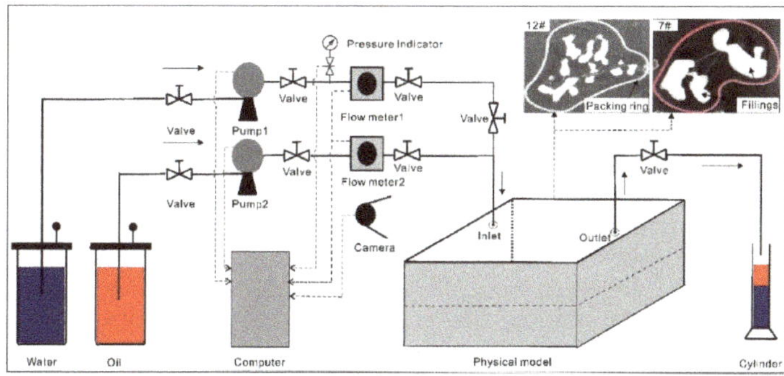

Figure 2. Experimental apparatus.

In these physical models, the size ranges of vugs are 115–130, 55–80, and 2–20 mm, respectively. Fractures were controlled with a width of 0.5–2 mm and crack density of 1–5 m^{-1}. The matrix was treated as an ineffective reservoir [1,2], so it was ignored in these physical models.

2.2. Experimental Materials and Setups

Figure 2 shows the main devices.

Two containers of about 1.2 L of artificial store water and white oil, respectively. The artificial water with a density of 1200 kg/m^3 simulated the underground water. The white oil with a density of 855 kg/m^3 and a dynamic viscosity of 26 mPa·s substituted for the crude oil. The pump (BT100LG) with the power of 22 W supplied the drive to inject artificial water or white oil into the physical model, and its flow rate ranged from 0.0015 mL/min to 380 mL/min. The fluid volume from the pump was measured and corrected by a flow meter (LZB-3WB) in which the range of flow rate varied from 4 to 40 mL/min. Two pumps and two flow meters were prepared to avoid ongoing experiment interruptions from device failures. A pressure gauge (0–500 KPa) with an accuracy of 0.5% measured pressure changes of the corresponding pump.

In the single model, the void consists of fractures and caves. Its bulk ranged from 78.1 to 331.9 mL. Before injecting white oil into the void, the filling was replaced by 10–20 mesh quartz sand with about half of the void bulk. At the same time, a balance (ZH-20H7) of max-capacity of 20 Kg and an accuracy of 0.2 g measured the mass changes of the model. A cylinder with 500 mL reserved displaced fluid from the model during water flooding and metered its volume. Four cameras (JVS-N73-G1 3.6 mm) videoed the whole experimental process. A computer recorded all relevant data.

2.3. Experimental Design

The whole investigation was divided into constituting the injection and production pattern (Table 1) and the adjusting injection and production pattern (Table 2).

Table 1. The experimental contents for well placement design.

Type	Subtype	Models	Amounts
Types of reservoir bodies	Large scale	1#–3#	3
	Intermediate scale	4#–9#	6
	Small scale	10#–12#	3
Injection/production relationships	FIIVO vs. FOIVI	4#, 7#, 8#	6
	FISVO vs. FOSVI	10#–12#	6
	SVIIIV vs. SVOLIVI	2#, 4#, 9#	6
High/low position of the reservoir body	HIHO	2#, 6#, 11#	3
	LIHO	2#, 6#, 11#	3
	HILO	2#, 6#, 11#	3
	LILO	2#, 6#, 11#	3

Note: Fractures (inlets) and intermediate-scaled vugs (outlets) vs. fractures (outlets) and intermediate-scaled vugs (inlets) are abbreviated as FIIVO vs. FOIVI. Fractures (inlets) and small-scaled vugs (outlets) vs. fractures (outlets) and small-scaled vugs (inlets) are simplified into FISVO vs. FOSVI. Small-scaled vugs (inlets) and large-scaled or intermediate-scaled vugs (outlets) vs. small-scaled vugs (outlets) and large-scaled or intermediate-scaled vugs (inlets) are abridged into SVIIIV vs. SVOLIVI.

The contents in Table 1 were designed from types of reservoir bodies, injection/production relationships, and high/low positions of the reservoir body.

The sizes of caves and the connectivity properties are the main contributors of reservoir bodies. In various storage spaces, larger caves may store more oil. In Figure 3, since the sizes of caves in Models 2# and 7# are larger than those in Model 12#, by placing the wells in caves of these two models, it is possible to gain higher outputs. On the other hand, the means of the connectivity among caves encompasses the dissolved caves (yellow rectangles in Figure 3) and fractures (yellow circles in Figure 3). Theoretically, the flow resistance is smaller in the dissolved caves, so it is much easier to flow in them. This may further affect the development effect for different reservoir bodies. Therefore, for reservoir bodies, the sizes and connectivity properties determine which type should be selected preferentially to develop. Both factors were also used to classify these models into three types (Table 1).

Under the fixed injection and production pattern (Figure 1b), the valuable reservoir bodies would be chosen by the comparison of the water flooding effect.

Table 2. The experimental schemes of well placement adjustments.

Adjusting Measures	Combined Models	Before 3.0 PV		After 3.0 PV		Note
		Injection Well	Production Well	Injection Well	Production Well	
Scheme 1	⑦③	S74	TK629	TK629	S74	/
	⑦⑧	S74	TK651CH	TK651CH	S74	/
Scheme 2	⑥④	TK617CH	TK608	TK617CH	TK608	Increasing amounts of connected channels from one to two
	⑦⑪	S74	TK652	S74	TK652	Increasing amounts of connected channels from one to three
	⑪④	TK652	TK608	TK652	TK608	Increasing amounts of connected channels from one to three
Scheme 3	③④	TK629	TK608	TK629	X	Shutting down TK608
	⑥⑦	TK617CH	TK612	TK617CH	TK6104X	Shutting down TK612
	⑥⑪	TK617CH	TK605CH	TK617CH	Z	Shutting down TK605CH
Scheme 4	③⑦	TK629	TK612	TK629	TK612/Z	/
	⑧⑦	TK651CH	TK612	TK651CH	TK612/S74	/
	⑧⑩	TK651CH	TK659	TK651CH	TK659/A	/
Scheme 5	⑦⑪	S74	TK605CH	S74	TK605CH/TK6104X	/
	⑪⑦	TK651CH	TK605CH	TK651CH/B	TK605CH	/
	⑧⑪	TK652	TK612	TK652/TK6104X	TK612	/
Scheme 6	⑥③	TK629	TK617CH	TK629/TK617CH	TK617	/
	⑩⑦	TK659	TK612	TK659/TK612	Z	/
	⑪③	TK652	TK629	TK652/TK629	C	/

Note: Scheme 1 stands for the measure of exchanging locations of the production well and injection well. Scheme 2 means the measure of adding numbers of connected channels. Scheme 3 refers to the measure of closing the old production well while opening the new production well. Scheme 4 represents the measure of opening the old production well while opening the new production well. Scheme 5 is the measure of adding the injection well. Scheme 6 is defined as the measure of switching the old production well into the new injection well while adding the production well.

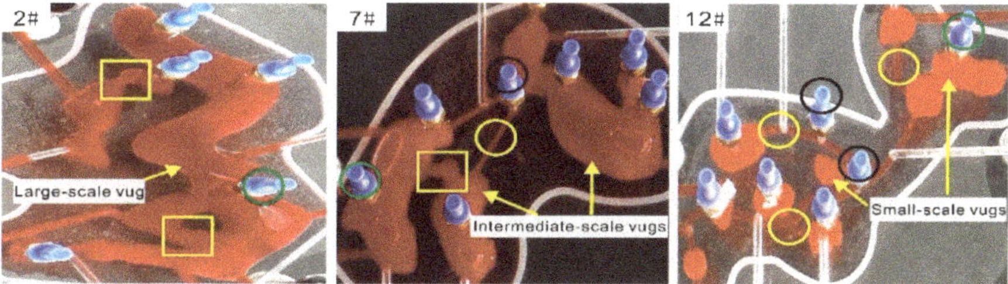

Figure 3. The connectivity features in three kinds of reservoir bodies.

The injection/production relationships describe how to arrange wells in selected reservoir bodies to achieve a better water flooding effect. Since either the injection well or production well may be located in the caves (green circles in Figure 3) or fractures (black circles in Figure 3), there are three pairs of injection/production relationships (Table 1). Their well placements are presented in Figure 4. The suitable well placements will be gained through evaluating water flooding.

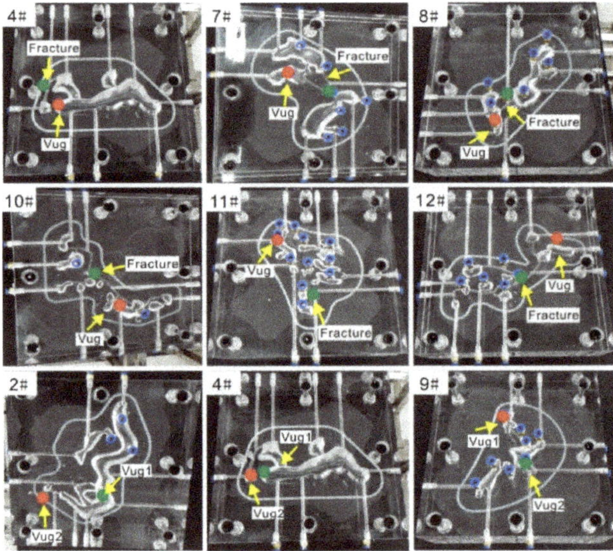

Figure 4. The well locations of FIIVO, FISVO, and SVIIIV. Exchanging orders of red and green filled sources corresponds to FOIVI, FOSVI, and SVOLIVI in the same models.

The high/low position of a reservoir body has impacts on the water flooding effect. Possible well placements include four combinations (Table 1) in which the corresponding locations in the vertical direction are shown in Figure 5a–d, respectively. In every model, the best injection and production pattern was attained by comparing the water flooding effect in four combinations.

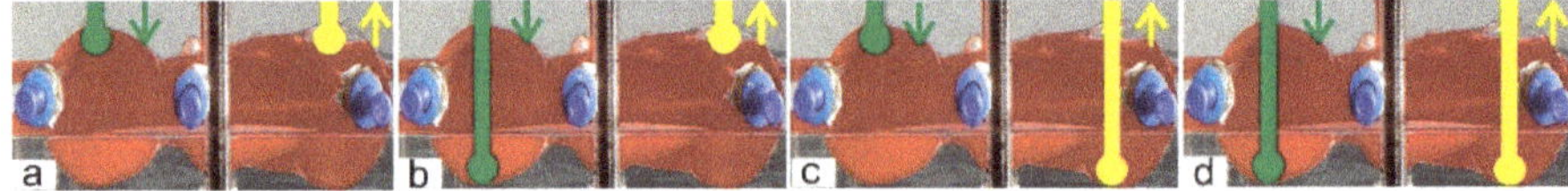

Figure 5. The schematic diagrams of (**a**) HIHO, (**b**) LIHO, (**c**) HILO, and (**d**) LILO.

In Table 2, the combined models were used to investigate the effect of different adjusting measures (Schemes 1–6), respectively. The combined models were composed of two single models. The theory of combination is as follows. Firstly, the connectivity between two wells was judged with dynamic tracer monitoring. Secondly, if two wells were supposed to connect to each other, the corresponding two physical models were combined together. Thirdly, the styles and amounts of connected channels resembled actual connected channels, including large fractures, fractures, and three parallel fractures in Block S74. In these experiments, large fractures and fractures were replaced by a single plastic pipeline with diameters of 6 and 3 mm, respectively. Three parallel fractures were simulated with three plastic pipelines with the same diameter of 3 mm. The above specific messages were presented in our published work [45].

HIHO refers to the high position (inlet) and high position (outlet). LIHO is the low position (inlet) and high position (outlet). HILO represents the high position (inlet) and low position (outlet). LILO means the low position (inlet) and low position (outlet).

2.4. Experimental Procedure

The following contexts are the main procedures of experiments. On the basis of these steps, the injection and production pattern was adjusted after 3.0 PV, and then water injection was stopped at 5.0 PV. The other steps are similar in two kinds of experiments.

At first, some preparations were required. (1) The artificial water was colored with methyl blue, and the white oil was dyed with oil soluble. (2) The mass of quartz sand is the difference value of the mass between the model containing fillings and the model without fillings. (3) The model with fillings was saturated sufficiently with the simulated oil, and then it was weighted. The mass of white oil is equal to the difference value of the mass between the model encompassing oil and the model without oil. (4) Operation (3) was repeated three times to reduce the disturbance of air on the measurement of the saturated oil mass. The main parameters of the quartz sand and the simulated oil are listed in Table 3.

Table 3. Basic parameters of the fillings and the saturated oil.

Number	Mass of the Fillings/g			Mass of the Saturated Oil/g					Simulated Oil Volume/mL	Pore Volume/mL
	Model without Fillings	Model with Fillings	Difference	Run 1	Run 2	Run 3	Average	Difference		
1#	10,482.6	10,537.6	55.0	10,611.5	10,609.9	10,609.8	10,610.4	72.8	85.1	105.9
2#	10,787.8	10,951.6	163.8	11,135.3	11,135.9	11,135.6	11,135.6	184.0	215.2	277.0
3#	10,274.2	10,530.4	256.2	10,681.20	10,682.1	10,682.1	10,681.8	151.4	177.1	273.8
4#	10,756.6	11,085.0	328.4	11,262.30	11,263.2	11,262.9	11,262.8	177.8	208.0	331.9
5#	10,697	10,764.0	67.0	10,842.30	10,843.2	10,842.9	10,842.8	78.8	92.2	117.4
6#	11,164	11,250.6	86.6	11,358.40	11,359.7	11,359.5	11,359.2	108.6	127.0	159.7
7#	10,615.6	10,685.8	70.2	10,804.20	10,805.9	10,805.5	10,805.2	119.4	139.6	166.1
8#	10,699.4	10,792.2	92.8	10,900.80	10,903.2	10,904.4	10,902.8	110.6	129.4	164.4
9#	10,824.6	10,914.4	89.8	10,965.30	10,966.8	10,964.7	10,965.6	51.2	59.9	93.8
10#	10,530	10,564.6	34.6	10,619.90	10,621.4	10,619.3	10,620.2	55.6	65.0	78.1
11#	10,493.6	10,642.0	148.4	10,683.70	10,686.2	10,682.7	10,684.2	42.2	49.4	105.4
12#	10,522.8	10,552.6	29.8	10,676.80	10,673.9	10,671.3	10,674.0	121.4	142.0	153.2

We further carried out the experiments via maintaining the constant flow rate of 20 mL/min and the total injection volume of 3.0 PV. The volume of the measuring cylinder was recorded with an interval of 0.5 PV. Meanwhile, the experimental phenomena were recorded by video.

In addition, the following details in the experiments should be emphasized. Firstly, the mass of the fillings was maintained approximately the same in the identical model. Secondly, exchanging the orders of the inlet and the outlet (Figure 4) was to ensure the unchanged flow path during water flooding. Thirdly, the use of a short pipeline with a diameter of 3 mm was used to model the low position of the reservoir body.

Finally, the water displacement efficiency and the water cut were calculated with Equations (1) and (2), separately.

$$ED = \frac{V_o}{V_{total}} \times 100\% \tag{1}$$

ED represents the water displacement efficiency, %. V_o stands for the output of oil, ml. V_{total} means the total saturated oil volume, mL.

$$f_w = \frac{V_{wi} - V_{wj}}{V_{ti} - V_{tj}} \times 100\% \tag{2}$$

V_{wi} and V_{wj} are the output of the water at different data logging times ($i = j + 1, j = 0, 1, 2, 3,$ and 4), mL. V_{ti} and V_{tj} represent the corresponding total output of the water and the oil, mL.

3. Results and Discussion

3.1. Constituting Injection and Production Pattern

3.1.1. Reservoir Body Types

Figure 6 reveals the relationships between the water displacement efficiency, the water cut, and the injection volume.

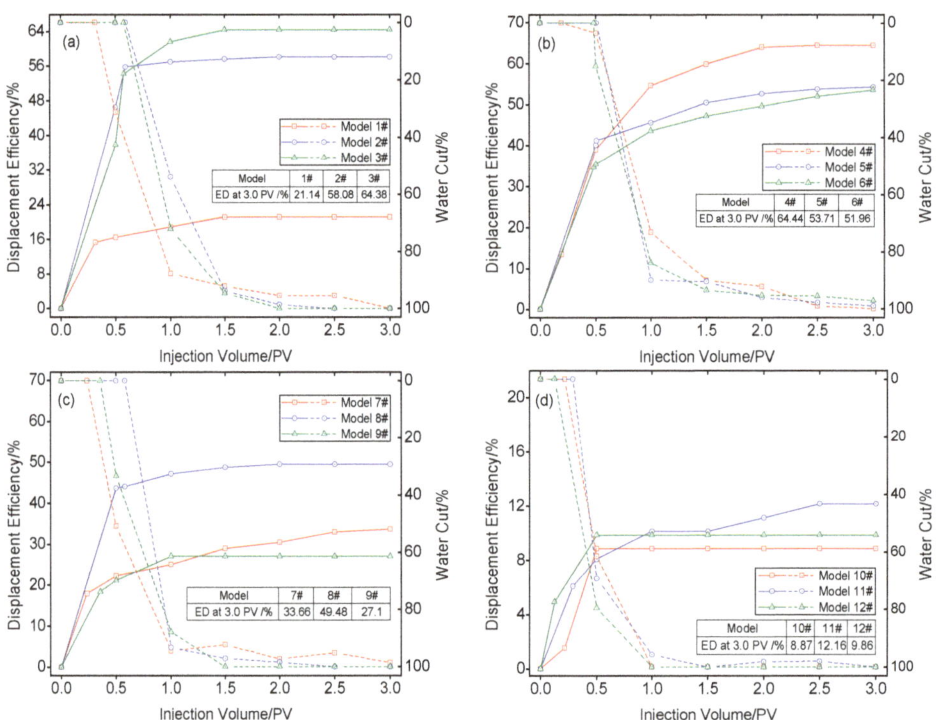

Figure 6. Displacement efficiency (solid lines) and water cut (dash lines) changes of reservoir bodies with large scales (**a**), intermediate scales (**b**,**c**), and small scales (**d**), respectively.

For large scale reservoir bodies (Figure 6a), the averaged final efficiency is 47.87%, and the averaged time of their water-free oil production is 0.49 PV. The water cut arrived at 100% after 2.0 PV in them.

For intermediate scale reservoir bodies (Figure 6b,c), the averaged displacement efficiency of 47.08% at 3.0 PV approached this in the large scale reservoir bodies, but their averaged water breakthrough time was 0.1 PV earlier than the large scale reservoir bodies. The oil still was flushed out at 3.0 PV in Models 5#–7#. Their corresponding water cuts are 98.97% (Figure 6b), 97.08% (Figure 6b), and 98.61% (Figure 6c), respectively.

For small scale reservoir bodies (Figure 6d), their final displacement efficiency is averaged as 10.3%, which is far less than the other two types. Meanwhile, the water broke through the front easily in the three models in which the averaged time of the water-free oil production is merely 0.21 PV. At 1.5 PV, the water cut had reached 100% in them, though about 1 mL of oil was still driven out between 1.5 PV and 2.5 PV in Model 11#.

Figure 7 is an example exhibiting the process of water flooding and the formation of the remaining oil in three sorts of reservoir bodies with Models 1#, 7#, and 12#.

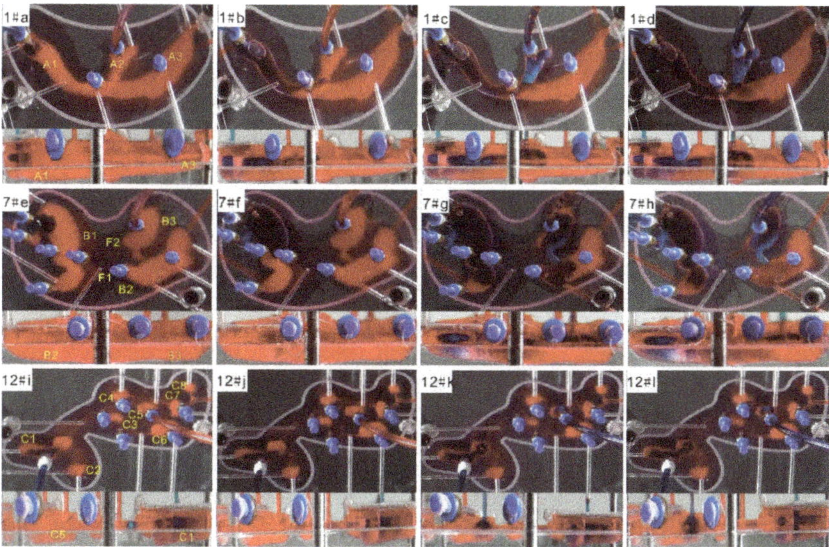

Figure 7. The processes of water flooding in Models 1# (**a–d**), 7# (**e–h**) and 12# (**i–l**).

At the initial stage, depending on the gravity differentiation, the water gathered on the upper surface of the filled areas in the caves (Figure 7a,e,i). Since the flow resistance in fillings is bigger than in unfilled areas, the water flowed predominantly along with the boundaries between two areas in cave A1, B1, and C1. Meanwhile, the water also immersed the fillings slowly.

Then, the oil/water interface moved horizontally towards the outlets (Figure 7b,f,j). However, the means of the connectivity caused the different processes of water drive in three models. In Model 1# (Figure 7a), cave A1, A2, and A3 connect to each other by the dissolved caves. Thus, the oil/water interface moved towards cave A2 rapidly (Figure 7b,c). Because cave A3 was away from the exit, the oil/water interface was not pushed to it. This lead to the residual oil (Figure 7c,d). In Model 7# (Figure 7e), the fractures (F1 and F2) and the dissolved cave link cave B1, B2, and B3. In Model 12# (Figure 7i), the fractures connect cave C1 to C7. The oil/water interfaces in cave B1 and C1 did go up vertically until they came to the locations of fractures (Figure 7f,j). The oil above both oil/water interfaces resided in them. After overcoming the additional force caused by the wettability in the fracture and flowing through fractures, the water entered cave B2 and C5 (Figure 7g,k). In the meantime, the main flow channel was transferred from F1 to F2, causing the discontinuous effect of the water drive on the oil in cave B2, though the water continuously was pushed into cave B2 (Figure 7g). Hence, there existed abundant residual oil in cave B2 (Figure 7h). During this stage, much oil was also replaced in the fillings of three models.

Afterward, the oil/water interface arrived in cave A2, B3, and C5. The elevating oil/water interface made the oil above them displaced. The period of water-free oil production ended up with their oil/water interfaces rising to the locations of outlets, and the water started to be seen in the outlets (Figure 7c,g,k). The oil in the fillings was displaced during this period. The remaining oil belonged to the attic oil in cave B3 and C5.

Finally, the water occupied the dominant flow path, so the exits only produced pure water in cave A2, B3, and C5 (Figure 7d,h,l). During this period, due to the deviation from the main flow channel, the oil could not be washed out in cave C2, C3, C4, C6, C7, and C8 (Figure 7l). Nevertheless, much oil was also observed in the fillings of the caves affected by water.

Because the sizes of caves in Models 2# and 7# are larger than in Model 12#, it was clearer to watch the rising oil/water interfaces in them. Hence, the gravity differentiation was utilized much better in them. Meanwhile, since the fractures connected small caves in Model 12#, the main flow path was easier to form and further caused invalid water displacement for some caves in it. Thus, combined with water displacement efficiency and water breakthrough time, the intermediate and large scale reservoir bodies should be preferentially arranged with wells to exploit.

3.1.2. Injection/Production Relationships

Wells Designed on Fractures or Intermediate-Sized Vugs

For the averaged water displacement efficiency (Figure 8a), FIIVO of 36.22% is higher than FOIVI of 25.64%. The water breakthrough time of FIIVO is 0.17 PV longer than FOIVI (Figure 8b). Meanwhile, the residual oil area of FIIVO is less than FOIVI in Models 4# and 7# (Figure 9). Hence, placing the water well on the fracture and setting the oil well in the intermediate cave can gain a better effect of water flooding.

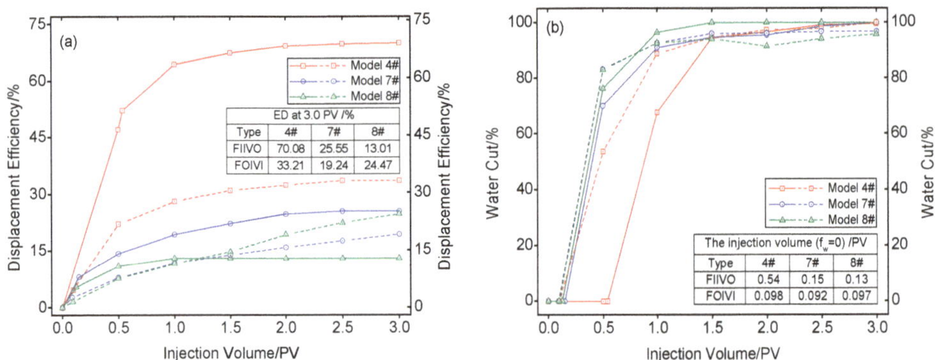

Figure 8. The curves of displacement efficiency (**a**) and the water cut (**b**) for Models 4#, 7#, and 8#. The solid and dash lines are FIIVO and FOIVI, respectively.

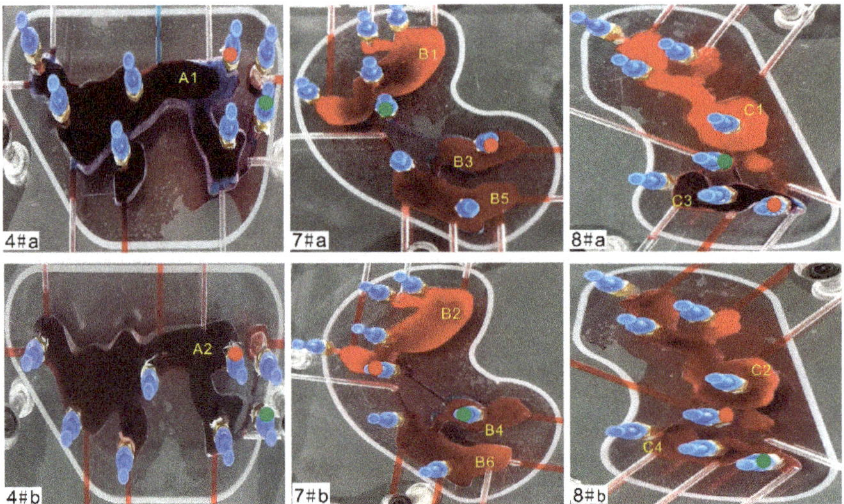

Figure 9. The comparisons between (**a**) FIIVO and (**b**) FOIVI in the residual oil distributions of Models 4#, 7#, and 8#.

However, both the water displacement efficiency (Figure 8a) and the residual oil distributions (Figure 9) show that FOIVI is better in Model 8#. This phenomenon was due to different amounts of main flow channels. During water flooding, only one main flow channel (yellow circle) was observed in FIIVO (Figure 10a), while three main flow channels (yellow circles) appeared in FOIVI (Figure 10b). In FIIVO, only cave C3 was affected by water drive (Figure 10a). However, in FOIVI, both cave C2 and C4 were swept by water (Figure 10b). So compared with FIIVO, FOIVI had the bigger sweep area of water flooding (Figure 9). Nevertheless, in Models 4# and 7# (Figure 9), the main flow channel was unique and constant in water flooding. So, the comparison of two relationships in Model 8# may be insignificant for seeking laws of well placements design.

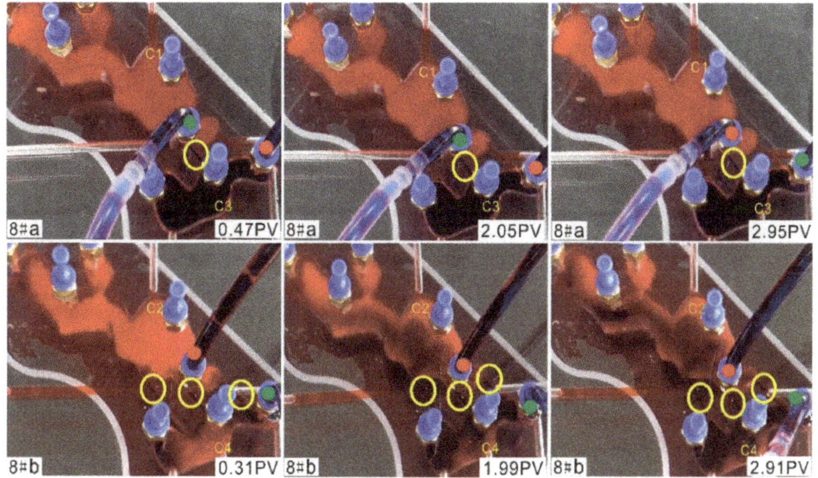

Figure 10. The flow channels in Model 8# under (**a**) FIIVO and (**b**) FOIVI.

Wells Designed on Fractures or Small-Sized Vugs

In order to study the well placements on fractures and caves further, the small scale caves and the fractures were also combined to form FISVO and FOSVT.

In three models, the comparisons in the displacement efficiency indicate that FISVO was superior to FOSVT (Figure 11a). Moreover, the averaged injection volume of the former also exceeded the latter at the difference of 0.09 PV during the period of water-free oil production (Figure 11b). The small sizes of the caves were attributed to this small difference. In addition, after water flooding, the remaining oil in cave A1, B1, and C1 was less than cave A2, B2, and C2 (Figure 12). Thus, FISVO would benefit the effect of water flooding.

Because of the low flow resistance, some caves apart from the main flow channel also were affected by water drive, such as cave A3, B3, B5, B7, C3, and C5 in Models 10#a, 11#a, and 12#a (Figure 12). However, the same caves marked with A4, B4, B6, C4, and C6 were almost not driven by water. Meanwhile, the oil in some caves also was not displaced completely in FISVO and FOSVT, though these caves were connected with the main flow channel.

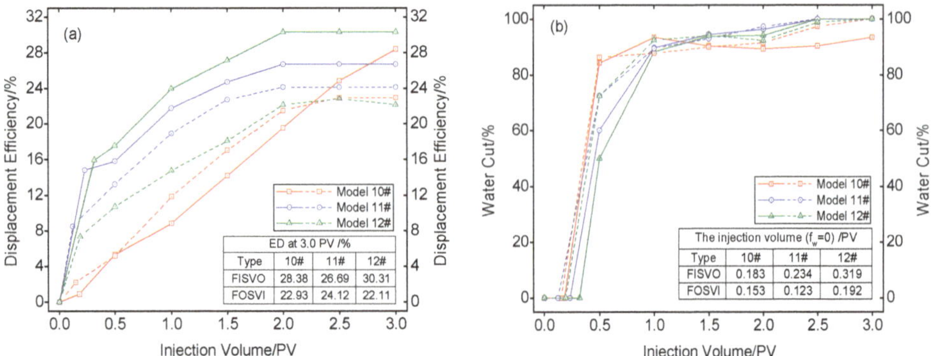

Figure 11. The curves of the displacement efficiency (**a**) and the water cut (**b**) for Models 10#–12#. The solid and dash lines represent FISVO and FOSVT, respectively.

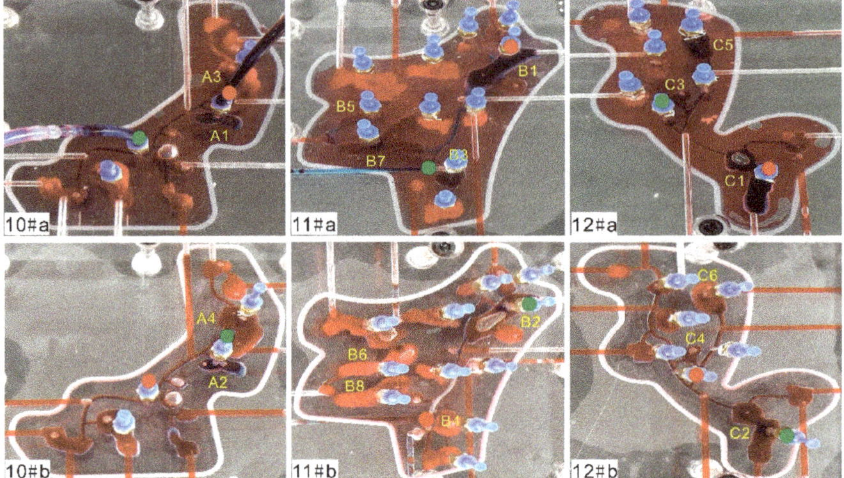

Figure 12. The comparisons between (**a**) FISVO and (**b**) FOSVI in the residual oil distributions of Models 10#–12#.

Wells Designed on Various-Scaled Vugs

The wells can also be placed on various scale caves. The averaged water displacement efficiency of SVIIIV at 3.0 PV was 24.85% higher than this of SVOLIVI in Models 2#, 4#, and 9# (Figure 13a). In SVIIIV and SVOLIVI, the averaged injection volume is 0.47 PV and 0.16 PV, respectively (Figure 13b), both of which indicate the exit was easier to undergo water logging in the SVOLIVI. In addition, the residual oil in cave A3, B1, and C1 was less than in cave A4, B2, and C2 (Figure 14). Hence, the water flooding performance would be better, as the entrance was placed in the smaller caves, and the exit was arranged in the larger caves.

Meanwhile, since the structures of the caves were complex in Models 2# and 9#, cave A1, A2, C5, and C6 were approximately not flooded under SVIIIV and SVOLIVI (Figure 14).

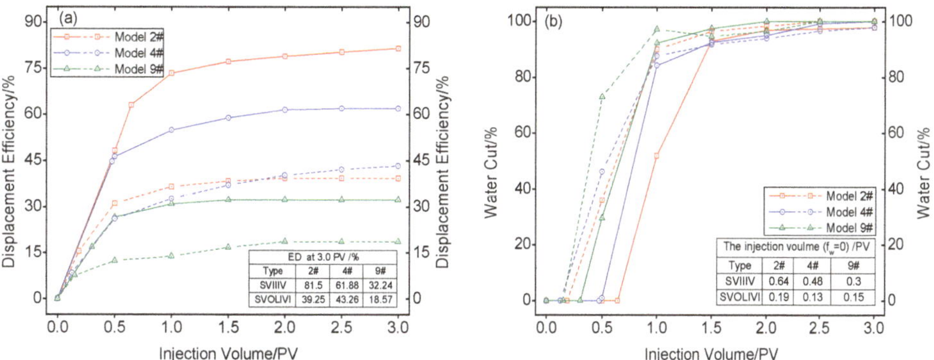

Figure 13. The curves of the displacement efficiency (**a**) and the water cut (**b**) for Models 2#, 4#, and 9#. The solid and dash lines represent SVIIIV and SVOLIVI, respectively.

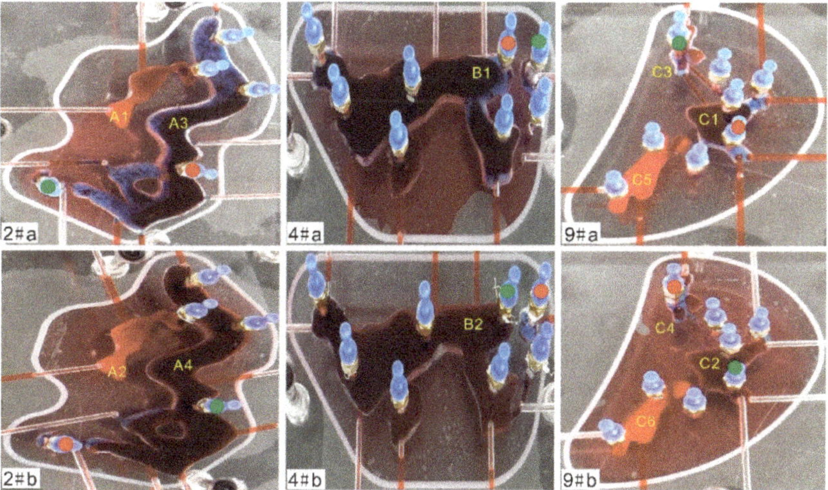

Figure 14. The comparisons between (**a**) SVIIIV and (**b**) SVOLIVI in the residual oil distributions of Models 2#, 4#, and 9#.

Analysis of Injection/Production Relationships

In fact, six injection/production relationships represent six injection and production patterns in fractured-vuggy carbonate reservoirs. Various fractured-cave structures have exhibited different water flooding behaviors under these well placements. However, the gravity differentiation between oil and water was considered as the mechanism governing these behaviors. As shown in Figures 15 and 16, the dynamic processes in Model 4# (Figure 9) and in Model 2# (Figure 14) were analyzed to reveal this mechanism. The injection/production relationship in Model 4# is both placing the injection well in the fracture while placing the production well in the intermediate scale cave (FIIVO) and placing the injection well in the intermediate scale cave while placing the production well in the fracture (FOIVI). The injection/production relationship in Model 2# is both placing the injection well in the small scale cave while placing the production well in the intermediate scale cave (SVIIIV) and placing the injection well in the intermediate scale cave while the production well in the small scale cave (SVOLIVI).

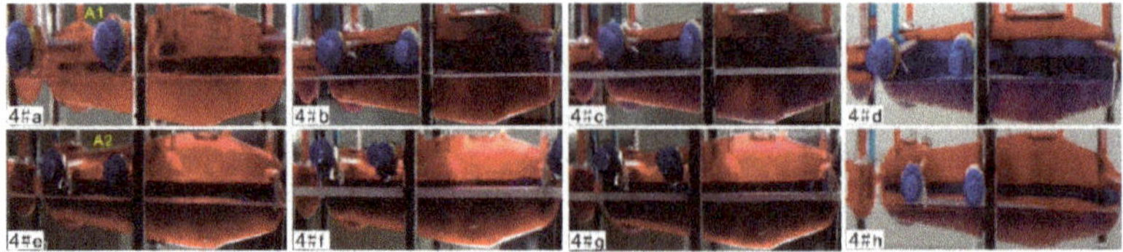

Figure 15. The oil/water interface changes in Model 4# under (**a–d**) FIIVO and (**e–h**) FOIVI.

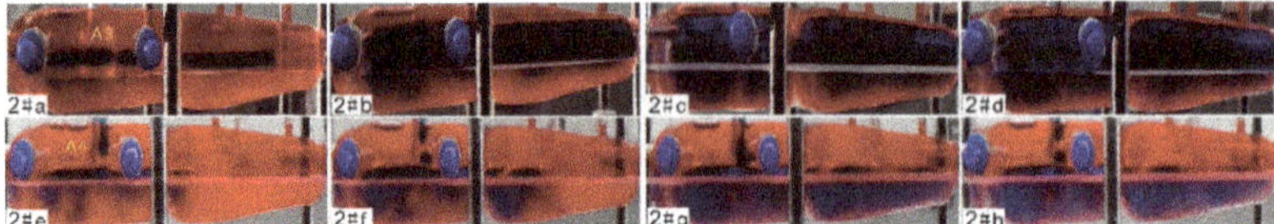

Figure 16. The oil/water interface changes in Model 2# under (**a–d**) SVIIIV and (**e–h**) SVOLIVI.

Before the water entered the fracture, the water occupied the small cave quickly in SVIIIV. In FIIVO and SVIIIV, the drive force further pushed water to balance the additional force caused by wettability in the fracture. After water entering cave A1 (Figure 15a) and cave A3 (Figure 16a), the oil/water interface was advanced horizontally at first. Then, the water was accumulated on the upper of fillings because of the gravity differentiation. The oil/water interface was elevated vertically in cave A1 (Figure 15b) and in cave A3 (Figure 16b). The water gradually displaced the oil in the area without fillings. Meanwhile, water also partly immersed the fillings, but the oil was displaced a little in the fillings (Figures 15b and 16b).

After the oil/water interface arrived at the outlets, the water was observed at once in the exits. The oil/water interface almost remained constant in the following process (Figure 15c,d and Figure 16c,d). The oil could not be displaced above the oil/water interface again and remained in the cave. Furthermore, because of the gravity differentiation, the water also got over the capillary force in the pores of the fillings featured with the water wettability and further expelled the oil in the fillings (Figure 15c,d and Figure 16c,d). Finally, as the oil was driven out enough from the fillings, the water cut reached 100% in the exits (Figures 15d and 16d). Nevertheless, some oil still remained in the fillings.

FOIVI and SVOLIVI behaved with different characteristics in two models. Since the fractures were carved on the horizontal plane, the raising oil/water interface in the vertical direction was almost not observed in cave A2 (Figure 15h) and in cave A4 (Figure 16h), and its corresponding height was lower than that in cave A1 (Figure 15d) and cave A2 separately (Figure 16d). The oil was not displaced above the oil/water interface. Meanwhile, as a drive, the gravitational force immersed the water much earlier in the fillings (Figures 15f and 16f). So, the oil was gradually driven out from the fillings (Figures 15g and 16g). However, there was still abundant residual oil in cave A2 and A4 (Figures 15h and 16h).

In these two injection/production relationships, the heights of entrances and exits are 0.5 and 36 mm in Model 4# while 7 and 42 mm in Model 2#, respectively. In FIIVO and SVIIIV, the heights of the entrances (0.5 and 7 mm) were lower than those of the exits (36 and 42 mm). Under the gravity differentiation, two relationships elevated the oil/water interfaces in intermediate or large caves, so a better water flooding effect was achieved. In FOIVI and SVOLIVI, the heights of the entrances (36 and 42 mm) were higher than those of the exits (0.5 and 7 mm). These relationships hindered the gravity differentiation from upraising oil/water interfaces. Hence, FOIVI and SVOLIVI are not proposed in well placement design.

In essence, the gravity differentiation must be used adequately for a reasonable injection/production relationship in fractured-vuggy carbonate reservoirs.

3.1.3. High/Low Positions of the Reservoir Body

A reservoir body features with high and low parts in the spatial distributions. For a reservoir body, there were four injection and production patterns, including placing the injection well in the high part while placing the production well in the high part (HIHO), placing the injection well in the low part while placing the production well in the high part (LIHO), placing the injection well in the high part while placing the production well in the low part (HILO), and placing the injection well in low part while placing the production well in the low part (LILO). Their water flooding performances behaved differently in Model 2#, 6#, and 11# (Figures 17 and 18).

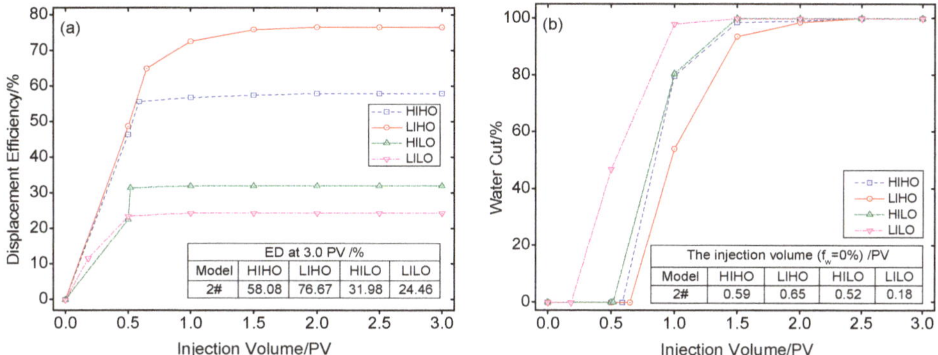

Figure 17. The curves of the displacement efficiency (**a**) and the water cut (**b**) for Model 2# in HIHO, LIHO, HILO, and LILO.

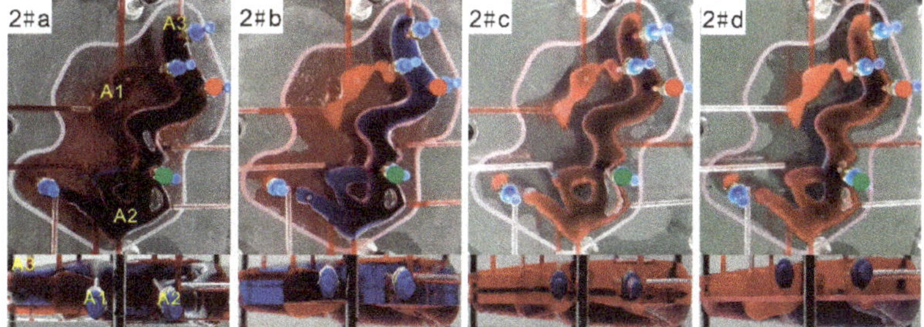

Figure 18. The comparisons in the residual oil distributions among HIHO (**a**), LIHO (**b**), HILO (**c**), and LILO (**d**) of Model 2#, respectively.

Large Scale Reservoir Body

In Model 2#, LIHO achieved the highest water displacement efficiency (Figure 17a) and the longest water-free oil production time (Figure 17b).

Under four injection and production patterns, a unique oil/water interface with different height was observed to exist in the whole reservoir body (Figure 18a–d). Since production wells were placed at the high part, the oil/water interfaces were also high in cave A2 and A3 (Figure 18a,b). At the same time, it was also observed that more oil in the fillings was displaced by LIHO instead of HIHO. Thus, LIHO was more advantageous to use gravitational differentiation. Due to using the low part to arrange the production well in HILO and LILO, their oil/water interfaces are apparently lower than those in LIHO and HIHO (Figure 18c,d). Because locating the injection well in the high position (Figure 18c) could sweep the oil a little in the unfilled zone of cave A2 and A3, the oil/water interface in the HILO

(Figure 18c) was slightly higher than LILO (Figure 18d). However, under HILO and LILO, the gravity differentiation did not function well in the area without fillings, with much residual oil remaining.

In addition, the oil in cave A1 could hardly be driven out under four situations (Figure 18a–d), due to the complex vuggy structure and the deviation of the main flow channel.

Intermediate Scale Reservoir Body

In Model 6#, cave B1 and B2 connect to each other by a fracture. In cave B1 and B2, the vertical height of two locations as the entrance or the exit is 42 and 44 mm, respectively. Figure 19a,b indicates that LIHO still had the best water flooding performance like in Model 2#.

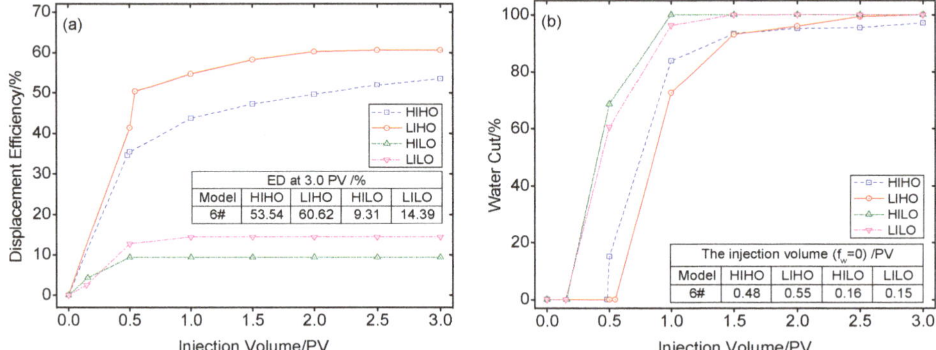

Figure 19. The curves of the displacement efficiency (a) and the water cut (b) for Model 6# in HIHO, LIHO, HILO, and LILO.

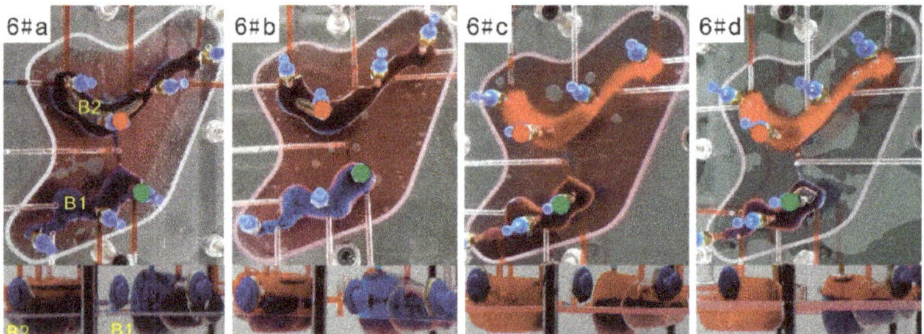

Figure 20. The comparisons in the residual oil distributions among HIHO (a), LIHO (b), HILO (c), and LILO (d) of Model 6#, respectively.

Nevertheless, several different characteristics were still found in this case.

Firstly, the oil/water interface in cave B1 is higher than in cave B2 (Figure 20a–d). In other words, there did not exist a unique oil/water interface in the whole model. These oil/water interfaces were not only governed by the locations of the entrance and the exit but the orientations of fractures.

Secondly, arranging the production well in the high position lifted the oil/water interface in cave B1 and B2 (Figure 20a,b). Thus, using LIHO almost did not retain the oil in cave B1 (Figure 20b).

Thirdly, since the production well was placed in the low position, the oil in the fillings of cave B2 was displaced (Figure 20c,d). In cave B1, despite almost the same oil/water interfaces in HILO and LILO, the water flowed mainly along with the upper of fillings and entered the fracture easily in HILO, so oil was partly trapped in the fillings (Figure 20c), causing the smallest displacement efficiency (Figure 19a). In LILO, the oil in the fillings was mainly displaced (Figure 20d).

Therefore, the use of the gravity differentiation was still vital to achieve better performances of water flooding in intermediate scale reservoir bodies.

Small Scale Reservoir Body

Compared with Models 2# and 6#, the sizes of the caves scaled down apparently in Model 11#. Using cave C1 and C6 was to arrange the injection well and production well. The height of their locations is merely 12 and 10 mm separately. The fracture does connect cave C1 to C6. LIHO still possessed the highest water flooding efficiency (Figure 21a), but HILO had the longest water breakthrough time (Figure 21b).

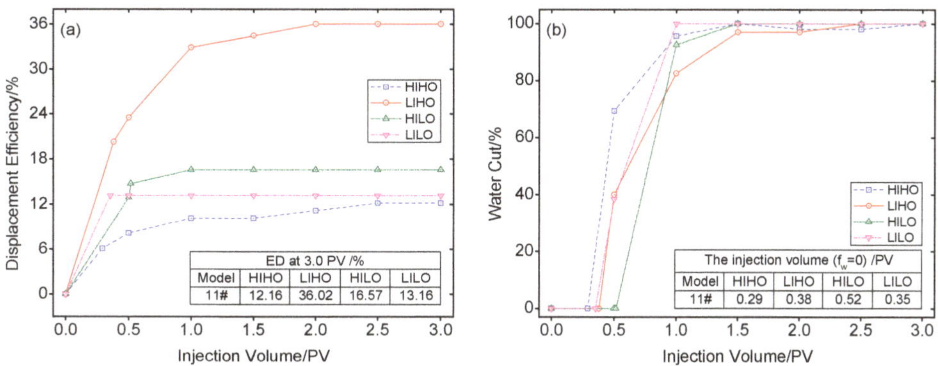

Figure 21. The curves of the displacement efficiency (**a**) and the water cut (**b**) for Model 11# in HIHO, LIHO, HILO, and LILO.

In cave C1, placing the injection well in the low position used the gravity differentiation fully, so the remaining oil was not approximately observed (Figure 22b,d). Because placing the entrance in a high position made the water enter the fractures quickly, much of the oil was locked in the fillings (Figure 22a,c).

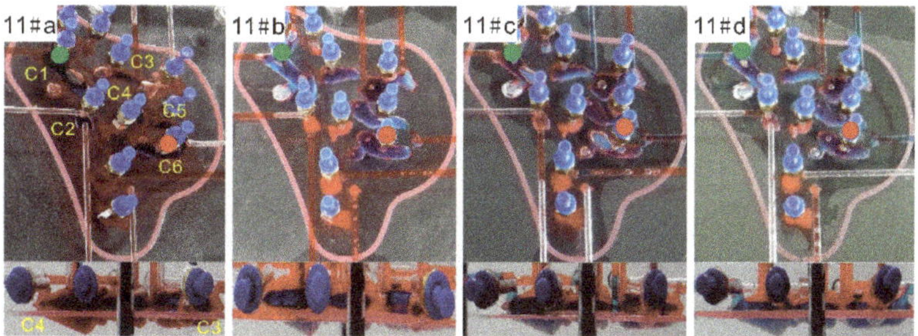

Figure 22. The comparisons in the residual oil distributions among HIHO (**a**), LIHO (**b**), HILO (**c**), and LILO (**d**) of Model 11#, respectively.

In cave C6, the locations of the exits appeared unrelated to the remaining oil distributions. Although the exits of both HIHO and LIHO were placed in high positions, the remaining oil in HIHO was less than in LIHO (Figure 22a,b). This similar circumstance also occurred in HILO and LILO (Figure 22c,d). Various residual oil distributions in four situations indicate that the gravity differentiation was not the main factor to displace oil in cave C6.

Meanwhile, from cave C2 to C5 (Figure 22a–d), the residual oil was distributed irregularly in every cave. The residual oil of HIHO was most in cave C3 and C4 (Figure 22a).

In addition, in cave C1 to C6, the small sizes of these caves caused the tough observation of their changing oil/water interfaces, though there was still gravity differentiation in them.

For a small scale reservoir body, the effect of water flooding was not only controlled by the gravity differentiation but fillings, fluid properties, flow channels, and fractured-cave structures. Given the water displacement efficiency (Figure 21a) and water breakthrough time (Figure 21b), LIHO may benefit the water drive to sweep these caves connected with the main flow channel. However, unlike the use of gravity differentiation to improve the effect of water flooding in Models 2# and 6#, sweeping more small caves was assumed as a possible mechanism controlling the water flooding effect in Model 11#. Hence, a high/low position was not considered as a key factor of well placement design in this kind of reservoir body.

3.2. Adjusting Injection and Production Pattern

As shown in Table 4, the effect of the measures on not changing locations or amounts of wells (Schemes 1 and 2) was far inferior to that of the measures of changing locations or amounts of wells (Schemes 3–6). In Schemes 1 and 2, the locations and amounts of wells had not been adjusted. In Figure 23a, an example is combined of Models ⑦③ in Scheme 1. The main flow channel had not been altered during water flooding. Nonetheless, a certain effect was still achieved by this measure. The reason is that massive oil was observed to remain in the model, including the injection well (Figure 24), so oil close to the dominant flow path was partly flushed out as the old injection well was used as the production well. In Figure 23b, there was not any effect in the combined Models ⑦⑪ in Scheme 2, and water injection had not changed flow direction. Thus, only adding amounts of connected channels for flow was not supposed to enlarge the sweeping area of water flooding. In Schemes 3–6, some changes may occur in the main flow channels due to changing locations and amounts of wells. For instance, closing the old production well ensured water to move towards the new one in Scheme 3 (Figure 25), so the directions of the main flow channels were altered to improve the water flooding effect. The above comparisons show that the alteration of the main flow channels benefited the improvement of the water flooding effect.

Table 4. The effect of water flooding under various adjusting schemes.

Adjusting Measures	Combined Models	Displacement Efficiency/%		Effect Evaluation		Averaged Data	
		Before Adjustment	After Adjustment	Displacement Efficiency Increase/%	Effective Injection Volume/PV	Displacement Efficiency Increase/%	Effective Injection Volume/PV
Scheme 1	⑦③	68.55	69.82	1.27	0.22	1.39	0.175
	⑦⑧	40.86	42.37	1.51	0.13		
Scheme 2	⑥④	47.09	47.09	0	0	0.26	0.013
	⑦⑪	33.72	33.72	0	0		
	⑪④	46.89	47.67	0.78	0.04		
Scheme 3	③④	41.9	57.91	16.01	0.27	10.91	0.5
	⑥⑦	22.25	37.1	14.85	0.73		
	⑥⑪	29.66	31.54	1.88	0.5		
Scheme 4	③⑦	48.28	49.23	0.95	0.5	4.46	0.6
	⑧⑦	30.83	33.09	2.26	0.88		
	⑧⑩	38.52	48.68	10.16	0.42		
Scheme 5	⑦⑪	29.22	44.25	15.03	1.82	9.29	1.37
	⑪⑦	51.51	53.64	2.13	0.3		
	⑧⑪	39.45	50.17	10.72	2		
Scheme 6	⑥③	57.77	66.27	8.5	0.43	12.34	0.57
	⑪⑦	31.81	54.68	22.87	0.98		
	⑪③	62.28	67.95	5.67	0.3		

However, it does not mean that all measures on changing locations and amounts of wells must cause the change of the main flow channels, such as in Scheme 4 (Figure 26). After adjustment, the new production well named Z had produced nothing during water flooding in combined Models ③⑦, and the main flow channel was the same as that before 3.0 PV. In combined Models ⑧⑦, the new production well called S74 stopped production after a simulated oil bulk of 6 mL having been driven out. Meanwhile, the main flow pathway recovered like that at 3.0 PV. By contrast, in combined

Models ⑧⑩, the old production well TK659 had no production, and the new production well marked as A started to produce continuously, so the new main flow pathway was formed. The effect of water drive also did behave much better, as shown in Table 4. This further implies that the adjusting effect of water flooding was determined by whether new flow channels could be formed.

Moreover, the adjusting effect of water flooding also depended on the sizes and residual oil distributions of caves selected to arrange new wells. Take Scheme 6 as an example (Figure 27). At 3.0 PV, a lot of non-displaced oil remained in combined Models ⑥③ and ⑩⑦ separately, while great attic oil was distributed in combined Models ⑪③ (Figure 27). At the same time, the sizes of the caves in the new production well TK617 and Z are larger than those in the new production well marked as C (Figure 27). Although the altered flow channels were observed in three combined models, both displacement efficiency increases and the effective injection volume of combined Models ⑪③ were lower than those in the other two combinations. Hence, new formed flow channels should pass through relative large-sized caves containing enough residual oil. This identical mechanism also appeared in combined Models ⑥⑪ of Scheme 3 and combined Models ⑩⑦ of Scheme 5.

In addition, the injection and production pattern was "one injection and two productions" in Scheme 4 and was "two injections and one production" in Schemes 5 and 6, separately. As shown in Table 4, from the view of averaged data, the effect of using Schemes 5 and 6 is superior to that of using Scheme 4. Hence, "two injections and one production" was better to favor the enhanced effect of water flooding in these experiments.

On the whole, forming new flow channels was supposed to control the effect of the injection and production pattern adjustment under water flooding. Both scales of cave chosen to arrange new wells and residual oil distribution also influenced the final adjusting effect. The effective measures included shutting down old production wells while adding new production wells (Scheme 3), adding new injection wells (Scheme 5), or switching production wells into injection wells while adding new production wells (Scheme 6).

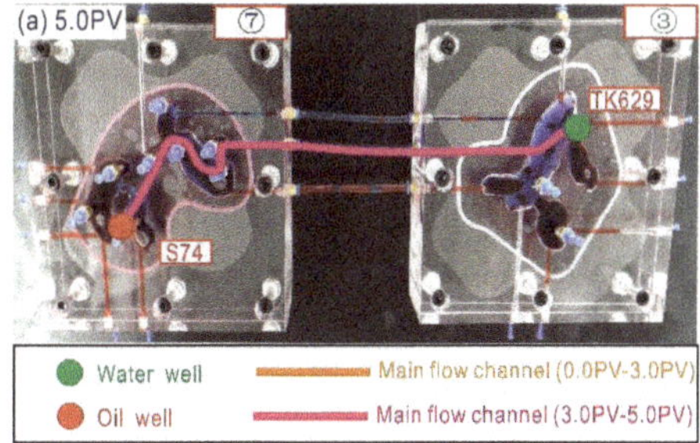

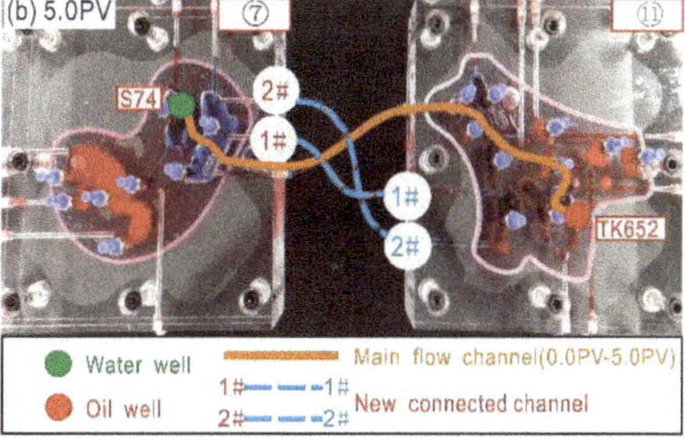

Figure 23. Changes of main flow channels in the combined Models ⑦③ (**a**) in Scheme 1 and ⑦⑪ in Scheme 2 (**b**).

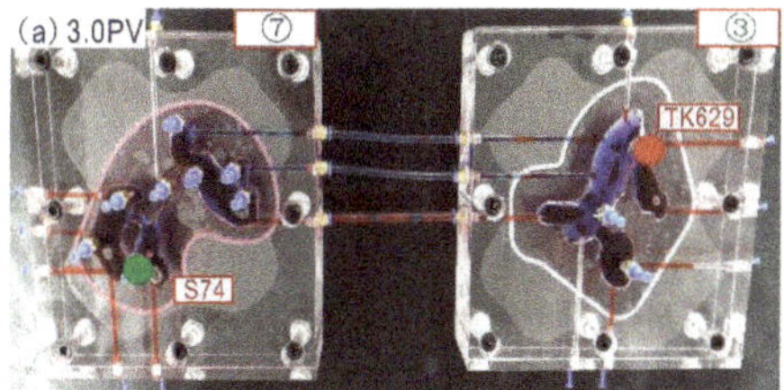

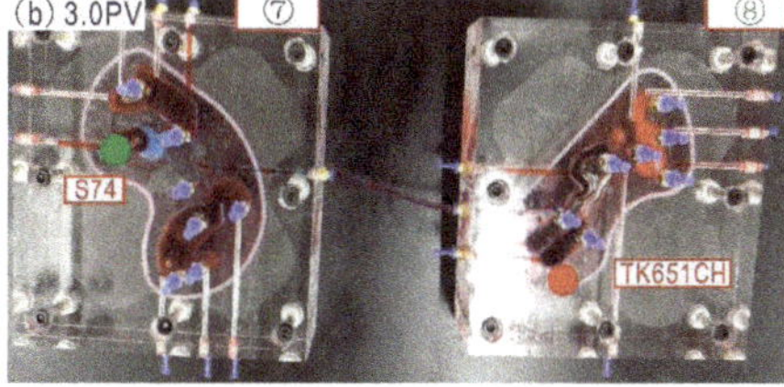

Figure 24. Residual oil distributions of the combined model ⑦③ (**a**) and ⑦⑧ (**b**) in Scheme 1 at 3.0 PV.

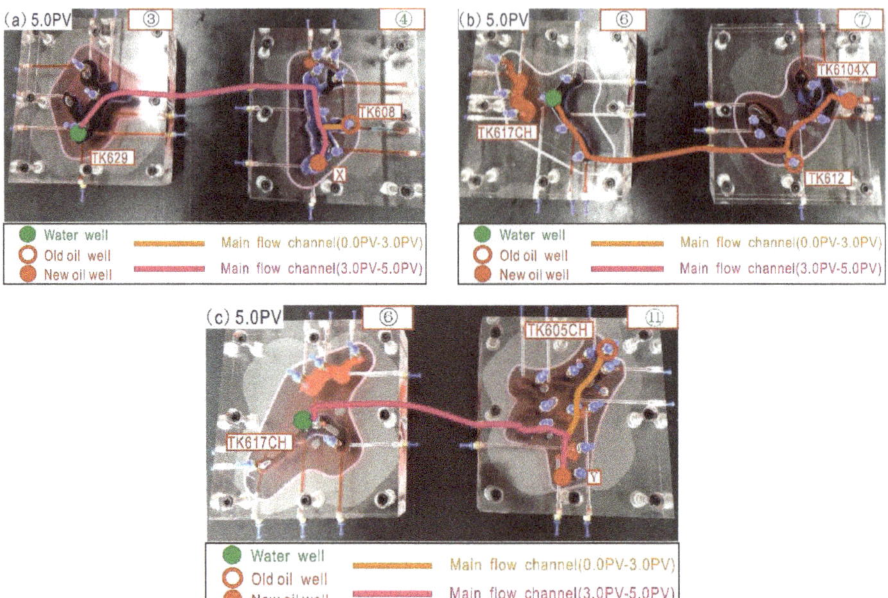

Figure 25. Main flow channel changes of the combined model ③④ (**a**), ⑥⑦ (**b**), and ⑥⑪ (**c**) in scheme 3, respectively.

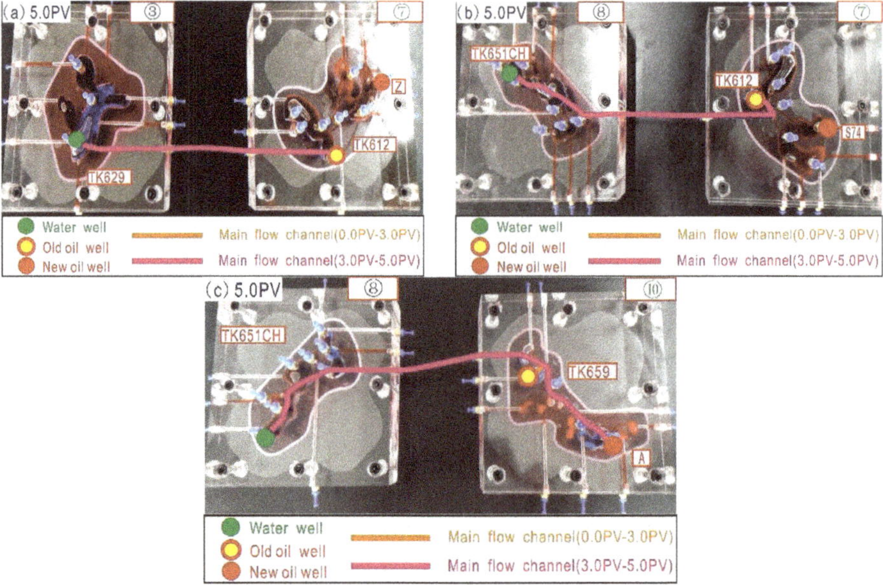

Figure 26. Main flow channel changes of the combined model ③⑦ (**a**), ⑧⑦ (**b**), and ⑧⑩ (**c**) in scheme 4, respectively.

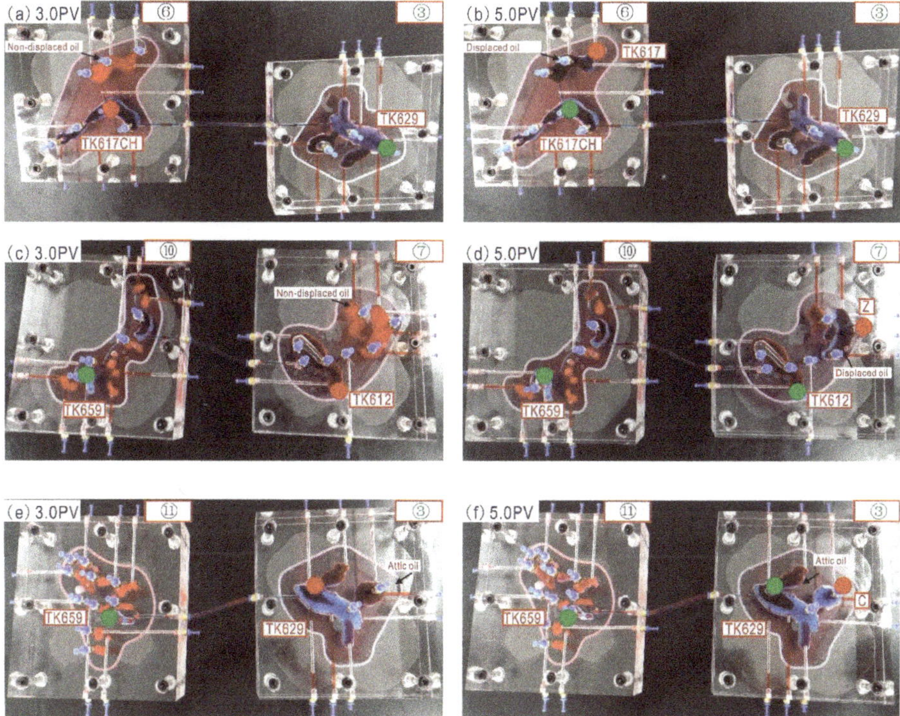

Figure 27. Residual oil distributions of the combined model ⑥③ (**a** vs. **b**), ⑩⑦(**c** vs. **d**), and ⑪③ (**e** vs. **f**) in Scheme 6 at 3.0 PV and 5.0 PV, respectively.

For fractured-vuggy reservoirs, different emphases existed in constituting and adjusting injection and production patterns separately. The former was to make full use of gravitational differentiation between oil and water, while the latter was to ameliorate the water driving effect for the caves containing enough remaining oil through forming new flow channels. In fact, both of them surpassed their original contents. An example is the adjusting measure on opening the old oil well while opening the new oil well (Scheme 4). In spite of adjusting the injection and production pattern, a relatively inferior water flooding effect was achieved due to no formation of newly effective flow channels. This result implies that the validity of flow channels was crucial to improving the effect of water flooding. In our understanding, this validity of flow channels was embodied in two aspects that are to form new ones and to make these new ones work on intermediate or large scaled caves with enough residual oil. Nevertheless, in actual reservoirs, it is tough to meet these two aspects simultaneously by directly adjusting flow channels, particularly in letting new flow channels pass through caves with a large quantity of oil still remaining. According to experimental results, if caves with sufficient residual oil were arranged by adding a new injection well or production well and at the same time, the present flow channels were adjusted directly, the adjusting effect of water flooding may be controllable. In other words, the adjustment of both well location and flow channel were proposed to adjust together. Consequently, the contents of injection and production patterns were supposed to include flow channel adjustment.

On the other hand, although there are different main missions in constituting and adjusting injection and production patterns, their common objectives are to utilize gravitational differentiation well. Therefore, the measures on the former also play a significant role in the latter. For instance, in adjusting the injection and production pattern, apart from residual oil distribution, using gravitational differentiation reasonably also should be taken into consideration as appropriate caves are chosen to adjust. Physical simulation was employed to study our topic in the laboratory. Although we have combined this method partly with actual production characteristics in the Tahe oilfield [45], some specific methods had not yet been developed to put the experimental measures into practice. This awaits our further investigation.

4. Conclusions

Due to extremely strong heterogeneity, constituting and adjusting injection and production patterns in fractured-vuggy reservoirs differ from those in traditional reservoirs. This difference lies in how to use gravity segregation and how to form effective flow channels. Through the results of physical simulation in this work, the former can be achieved by selecting reservoir bodies with intermediate and large scales preferentially to exploit, or arranging injection well in fractures or small caves while placing production well in large caves, or putting injection/production wells in low/high parts of a reservoir body with intermediate and large scales separately. The latter can be solved by shutting down an old production well while adding new a production well, adding a new injection well, or switching a production well into an injection well while adding a new production well. Considering the complex connectivity in actual fractured-vuggy reservoirs, a suggestion is to adjust the injection and production pattern with the flow channel together.

Author Contributions: All the authors contributed to this paper. J.H., D.L., and R.T. conceived and designed the experiments; J.H., A.L., and S.W. performed the experiments and wrote the paper; J.H., Y.L. and X.L. prepared the experiments and analyzed the data. All authors have read and agreed to the published version of the manuscript.

Funding: This research was funded by the National Major Science and Technology Projects of China, grant number "2016ZX05053-002-003".

Acknowledgments: This research is supported by the National Major Science and Technology Projects of China (2016ZX05053-002-003).

References

1. Li, Y.; Hou, J.; Li, Y. Features and hierarchical modeling of carbonate fracture-cavity reservoirs. *Pet. Explor. Dev.* **2016**, *43*, 600–606. [CrossRef]
2. Hu, X.; Li, Y.; Quan, L.; Kong, Q.; Wang, Y.; Lye, X. Three-dimensional geological modeling of factured-vuggy carbonate reservoirs: A case from the Ordovician reservoirs in Tahe-IV block, Tahe oilfield. *Oil Gas Geol.* **2013**, *34*, 387–393.
3. Tangen, G.I.; Smaaskijaer, G.; Bergseth, E.; Clark, A.; Fossli, B.; Claudey, E.; Qiang, Z. Experience from Drilling a Horizontal Well in a Naturally Fractured and Karstified Carbonate Reservoir in the Barents Sea Using a CML MPD System. In Proceedings of the IADC/SPE Managed Pressure Drilling and Underbalanced Operations Conference and Exhibition, Amsterdam, The Netherlands, 9–10 April 2019.
4. Liao, M.; Pei, Y.; Chen, P.; Liu, X.; He, J. Formation and controlling factors of karst fracture-cave reservoir in the 4th block of Tahe Oilfield. *J. Southwest Pet. Univ.* **2013**, *35*, 1–8.
5. Li, Y.; Fan, Z. Developmental pattern and distribution rule of the fracture-cavity system of Ordovician carbonate reservoirs in the Tahe Oilfield. *Acta Petrolei Sin.* **2011**, *32*, 101–106.
6. Li, Y. Ordovician carbonate fracture-cavity reservoirs identification and quantitative characterization in Tahe Oilfield. *J. China Univ. Pet.* **2012**, *36*, 1–7.
7. Li, Y.; Kang, Z.; Xue, Z.; Zheng, S. Theories and practices of carbonate reservoirs development in China. *Pet. Explor. Dev.* **2018**, *45*, 669–677. [CrossRef]
8. Wang, L.; Dou, Z.; Lin, T.; Zhao, H.; Luo, J. Study on the visual modeling of water flooding in carbonate fracture-cavity reservoir. *J. Southwest Pet. Univ.* **2011**, *33*, 121–124.
9. Li, J.; Chen, Z.; Gao, S. Microcosmic experiment modeling on water-driven-oil mechanism in fractured-vuggy reservoirs. *Pet. Geol. Exp.* **2009**, *31*, 637–642.

10. Cruz-Hernández, J.; Islas-Juárez, R.; Pérez-Rosales, C.; Rivas-Gómez, S.; Pineda-Muñoz, A.; González-Guevara, J.A. Oil Displacement by Water in Vuggy Fractured Porous Media. In Proceedings of the SPE Latin American and Caribbean Petroleum Engineering Conference, Buenos Aires, Argentina, 25–28 March 2001.

11. Li, S.; Li, Y. An experimental research on water injection to replace the oil in isolated caves in fracture-cavity carbonate rock oilfield. *J. Southwest Pet. Univ.* **2010**, *32*, 117–120.

12. Li, A.; Zhang, D.; Yao, J.; Gao, C.; Lyu, A. Physical simulation of water flooding in fractured-vuggy unit. *J. China Univ. Pet.* **2012**, *36*, 130–135.

13. Zheng, Z.; Zhu, T.; Hou, J.; Luo, M.; Gao, Y.; Wu, X.; Qu, M. Visible research on remaining oil after nitrogen flooding in fractured-cavity carbonate reservoir. *Pet. Geol. Recover. Effic.* **2016**, *23*, 93–97.

14. Zheng, X.; Sun, L.; Wang, L.; Yi, M.; Yang, D.; Dong, G. Large-scale visible water/oil displacing physical modeling experiment and mechanism research of fracture-vuggy reservoir. *Geol. Sci. Technol. Inf.* **2010**, *29*, 77–81.

15. Zhao, Q.; Zhang, J.; Ding, B.; Wang, Y. Physical simulation of gas enhanced oil recovery for fractured-vuggy carbonate reservoirs in Tahe Oilfield. *Sci. Technol. Eng.* **2017**, *17*, 55–62.

16. Yuan, D.; Hou, J.; Song, Z.; Wang, Y.; Luo, M.; Zheng, Z. Residual oil distribution characteristic of fractured-cavity carbonate reservoir after water flooding and enhanced oil recovery by N2 flooding of fractured-cavity carbonate reservoir. *J. Pet. Sci. Eng.* **2015**, *129*, 15–22. [CrossRef]

17. Yuan, D.; Hou, J.; Song, Z.; Luo, M.; Zheng, Z.; Qu, M. Optimization of water injection methods and N2 flooding for EOR in Tahe fractured-vuggy carbonate reservoir. *J. Northeast Pet. Univ.* **2015**, *39*, 102–110.

18. Wang, J.; Liu, H.; Ning, Z.; Zhang, H.; Hong, C. Experiments on water flooding in fractured-vuggy cells in fractured-vuggy reservoirs. *Pet. Explor. Dev.* **2014**, *41*, 67–73. [CrossRef]

19. Qu, M.; Hou, J.; Qi, P.; Zhao, F.; Ma, S.; Churchwell, L.; Wang, Q.; Li, H.; Yang, T. Experimental study of fluid behaviors from water and nitrogen floods on a 3-D visual fractured-vuggy model. *J. Pet. Sci. Eng.* **2018**, *166*, 871–879. [CrossRef]

20. Lyu, X.; Liu, Z.; Hou, J.; Lyu, T. Mechanism and influencing factors of EOR by N 2 injection in fractured-vuggy carbonate reservoirs. *J. Nat. Gas Sci. Eng.* **2017**, *40*, 226–235. [CrossRef]

21. Ji, X.; Pu, W.; Jin, F.; Zhang, J.; Zhao, S.; Zhao, Q. Physical simulation experiment research on enhancing "attic oil" recovery in fracture-cavity reservoir. *Fault Block Oil Gas Field* **2016**, *23*, 375–379.

22. Guo, P.; Yuan, H.; Li, X.; Ma, H. Experiments on gas injection mechanisms in carbonate fracture-cavity reservoir using microvisual model. *J. China Univ. Pet.* **2012**, *36*, 89–93.

23. Zhang, D. Study of Production Mechanism for Fractured-Vuggy Carbonate Reservoirs. Ph.D. Thesis, China University of Petroleum, Beijing, China, 2012.

24. Wang, J.; Liu, H.; Zhang, J.; Zhao, W.; Huang, Y.; Kang, Z.; Zheng, S. Experiments on the influences of well pattern on water flooding characteristics of dissolution vug-cave reservoir. *Pet. Explor. Dev.* **2018**, *45*, 1035–1042. [CrossRef]

25. Li, X.; Rong, Y. Optimization research on reasonable injection-production pattern in Tahe Oilfield fractured carbonate reservoir. *Drill. Prod. Technol.* **2013**, *36*, 47–52.

26. Lu, X.; Rong, Y.; Li, X.; Wu, F. Construction of injection-production well pattern in fractured-vuggy carbonate reservoir and its development significance: A case study from Tahe oilfield in Tarim Basin. *Oil Gas Geol.* **2017**, *38*, 658–664.

27. Li, Y. The theory and method for development of carbonate fractured-cavity reservoirs in Tahe oilfield. *Acta Pet. Sin.* **2013**, *34*, 115–121.

28. Peter, P.; Bi, L.; Yalchin, E.; Richard, E.E.; Qin, G.; Li, J.; Ren, Y. Multiphysics and Multiscale Methods for Modeling Fluid Flow Through Naturally Fractured Vuggy Carbonate Reservoirs. In Proceedings of the SPE Annual Technical Conference and Exhibition, Anaheim, CA, USA, 11–14 November 2007.

29. Nwachukwu, A.; Jeong, H.; Pyrcz, M.; Lake, L.W. Fast evaluation of well placements in heterogeneous reservoir models using machine learning. *J. Pet. Sci. Eng.* **2018**, *163*, 463–475. [CrossRef]

30. Nasrabadi, H.; Morales, A.; Zhu, D. Well placement optimization: A survey with special focus on application for gas/gas-condensate reservoirs. *J. Nat. Gas Sci. Eng.* **2012**, *5*, 6–16. [CrossRef]

31. Liu, C.; Wei, X.; Xu, S.; Fu, Z. The overview of geophysical techniques in prediction of carbonate rock reservoir. *Progress Geophys.* **2007**, *22*, 1815–1822.

32. Kang, S.; Datta-Gupta, A.; John Lee, W. Impact of natural fractures in drainage volume calculations and optimal well placement in tight gas reservoirs. *J. Pet. Sci. Eng.* **2013**, *109*, 206–216. [CrossRef]

33. Hamida, Z.; Azizi, F.; Saad, G. An efficient geometry-based optimization approach for well placement in oil fields. *J. Pet. Sci. Eng.* **2017**, *149*, 383–392. [CrossRef]

34. Forouzanfar, F.; Reynolds, A.C.; Li, G. Optimization of the well locations and completions for vertical and horizontal wells using a derivative-free optimization algorithm. *J. Pet. Sci. Eng.* **2012**, *86*, 272–288. [CrossRef]

35. Cheng, Y.; McVay, D.A.; John Lee, W. A practical approach for optimization of infill well placement in tight gas reservoirs. *J. Nat. Gas Sci. Eng.* **2009**, *1*, 165–176. [CrossRef]

36. Bagherinezhad, A.; Boozarjomehry Bozorgmehry, R.; Pishvaie, M.R. Multi-criterion based well placement and control in the water-flooding of naturally fractured reservoir. *J. Pet. Sci. Eng.* **2017**, *149*, 675–685. [CrossRef]

37. Arinkoola, A.O.; Onuh, H.M.; Ogbe, D.O. Quantifying uncertainty in infill well placement using numerical simulation and experimental design: Case study. *J. Pet. Explor. Prod. Technol.* **2015**, *6*, 201–215. [CrossRef]

38. Ariadji, T.; Haryadi, F.; Rau, I.T.; Aziz, P.A.; Dasilfa, R. A novel tool for designing well placements by combination of modified genetic algorithm and artificial neural network. *J. Pet. Sci. Eng.* **2014**, *122*, 69–82. [CrossRef]

39. Wang, Z.; Wang, S.; Tang, W. Seismic response and frequency difference analysis of cavern-type reservoirs in carbonates in Tahe oilfield. *Oil Gas Geol.* **2004**, *25*, 93–98.

40. Wang, S.; Lu, X. Prediction techniques for deep carbonate reservoirs in Tahe Oilfield. *Geophys. Prospect. Pet.* **2004**, *36*, 153–158.

41. Wang, D.; Li, Y.; Hu, Y.; Li, B.; Deng, X.; Liu, Z. Integrated dynamic evaluation of depletion-drive performance in naturally fractured-vuggy carbonate reservoirs using DPSO–FCM clustering. *Fuel* **2016**, *181*, 996–1010. [CrossRef]

42. Rong, Y.; Pu, W.; Zhao, J.; Li, K.; Li, X.; Li, X. Experimental research of the tracer characteristic curves for fracture-cave structures in a carbonate oil and gas reservoir. *J. Nat. Gas Sci. Eng.* **2016**, *31*, 417–427. [CrossRef]

43. Ni, X.; Shen, A.; Pan, W.; Zhang, R.; Yu, H.; Dong, Y.; Zhu, Y.; Wang, C. Geological modeling of excellent fracture-vug carbonate reservoirs: A case study of the Ordovician in the northern slope of Tazhong palaeouplift and the southern area of Tabei slope, Tarim Basin, NW China. *Pet. Explor. Dev.* **2013**, *40*, 444–453. [CrossRef]

44. Liu, Y.; Rong, Y.; Yang, M. Detailed classification and evaluation of reserves in fracture-cavity units for carbonate fracture-cavity reservoirs. *Pet. Geol. Exp.* **2018**, *40*, 431–439.

45. Yang, B.; He, J.; Lyu, D.; Tang, H.; Zhang, J.; Li, X.; Zhao, J. Production optimization for water flooding in fractured-vuggy carbonate reservoir—From laboratory physical model to reservoir operation. *J. Pet. Sci. Eng.* **2020**, *184*, 106520. [CrossRef]

Article

Visual Experimental Study on Gradation Optimization of Two-Stage Gravel Packing Operation in Unconventional Reservoirs

Xingbang Meng [1], Minhui Qi [1], Zhan Meng [2,*], Tong Li [3] and Zhongxiao Niu [3]

[1]   Petroleum Engineering, China University of Petroleum (East China), Qingdao 266000, China;
      mengxingbang@upc.edu.cn (X.M.); b17020077@s.upc.edu.cn (M.Q.)
[2]   Petroleum Systems Engineering, Faculty of Engineering and Applied Science, University of Regina, Regina,
      SK S4S 0A2, Canada
[3]   PetroChina Huabei Oilfield Company, Cangzhong 062552, China; cy3_lt@petrochina.com.cn (T.L.);
      cy3_nzx@petrochina.com.cn (Z.N.)
*   Correspondence: Zhan.Meng@uregina.ca; Tel.: +001-306-501-7159

Received: 2 April 2019; Accepted: 18 April 2019; Published: 22 April 2019

Abstract: During the development of unconventional reservoirs with high sand production rate and fine silt content such as heavy oil and hydrate reservoirs, silt sand blockage problem is a serious issue. A two-stage gravel-packing sand control technique is applied to solve the silt sand blockage now. However, traditional experiments on this technique could not obtain the dynamic distribution law of intrusive sand in the gravel pack. In this study, a new visualization experiment based on hydrodynamic similarity criterion for studying particle blockage in gravel packs was conducted. Real-time monitoring of sand particle migration in the gravel pack could be achieved. Also, the stable penetration depth and the distributing disciplinarian of invaded particles could be determined. The results show that when the gravel-to-sand median size ratio of gravel bed I is less than five, the sand bridge can be formed at the front end of the gravel pack. This could prevent sand from further intruding. As the grain size of gravel bed II is increased, the flow velocity is reduced. Thus, the sand invading into gravel bed II tends to settle at the interface. A large amount of sand intrusion can happen to gravel pack II when the pore filling front breaks through the gravel bed I.

Keywords: unconventional reservoirs; gravel pack; sand control; gradation optimization; visual experiment

---

1. Introduction

Sand production has become one of the main factors restricting the efficient development of unconventional reservoirs such as hydrate and heavy oil reservoirs. It is mainly reflected in the damage to the near region of the wellbore, artificial lift, and surface equipment. A high-rate gravel-packing sand control technique has been widely applied to solve this problem. This technique has high sand control efficiency, a long validity period, and alleviation of the productivity impairment caused by the sand blocking material [1–3]. However, after a period of production, the gravel pack could be blocked by silty sand. Sparlin studied the permeability of common size gravel with different percentages of silty sand [4]. The study showed that the permeability of the gravel would be seriously decreased after mixing a small amount of silty sand. When the gravel pack blockage occurred in tubing-casing annulus and perforation holes, the additional pressure in the gravel layer could be raised significantly. Thus, well productivity would be decreased, and unnecessary reservoir energy loss would occur.

For unconventional reservoirs with high sand production rate and fine silt content, it is necessary to adopt appropriate sand control technology to achieve effective exploitation [5]. Two-stage gravel packing sand control technology is one of the most mature appropriate sand control technologies.

This technology can effectively solve the sand blockage problem in high argillaceous and silty sand reservoirs. Gravel bed I with smaller grain size is packed to prevent the invasion of particles of larger grain size. The bigger grain size of gravel bed II allows the silt particles transported to gravel bed II to be discharged with the produced fluid. This can effectively avoid the blockage, and the permeability could be impaired, as shown in Figure 1. Good application efficiency was achieved [6,7]. The advantage of this sand-control technique is that it can effectively alleviate the blockage of the gravel layer and increase the effective period of the gravel pack.

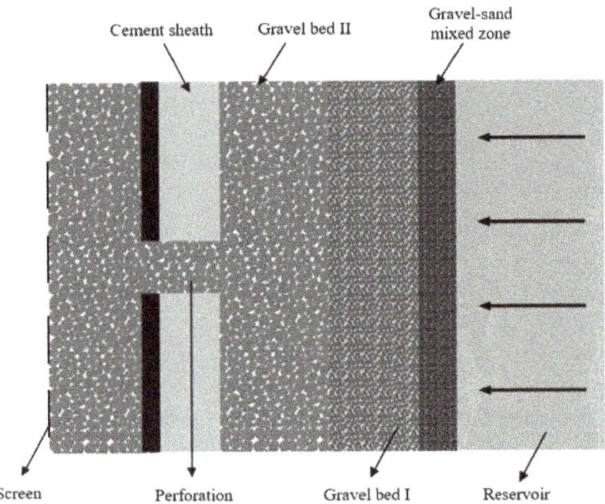

Figure 1. Diagram of two-stage gravel packing well.

The main causes of permeability damage of a gravel pack are particle migration, deposition, and blockage in gravel pores. Clarifying the plugging mechanism of a gravel pack is of great significance when optimizing sand control design. Also, it can help to predict the productivity of sand control wells more accurately. Due to the heterogeneity of pore throat and particle size distribution in the gravel pack, it is difficult to describe the clogging process accurately by analytical and theoretical methods only. It is necessary to conduct physical simulation research and revise the theoretical model based on the experimental results of studying particle transport in porous media. Many scholars have conducted a series of studies on particle migration in porous media and obtained many useful results. Bouhroum and Civan studied the pore structural changes of the gravel pack due to the clogging process of intruding particles by visualizing laboratory investigation. They concluded that the particulates migration process is characterized by a discontinuity of particulates concentration [8]. Bigno et al. studied the trends of gravel pack permeability reduction as a function of pore blocking and identified five prevailing pore blocking mechanisms [9]. To estimate the amount of deposited material quantitatively, Ail et al. used an on-line linear X-ray apparatus to study the deposition profile in deep bed filtration during produced water re-injection [10]. Deep penetration can be distinguished from external filter cake buildup by this method. Shirinabadi et al. studied the effect of gravel pack size on sand production experimentally and numerically [11]. Villarroel et al. built a gravel pack prototype of the two most common mesh sizes (16/20 and 20/40) and tested the influence of the gravel size under high in-situ stress contrast [12]. Lawal et al. carried out a series of studies on porous media permeability impairment by incorporating the kinetics of dynamical particle deposition into the classic deep-bed-filtration theory [13]. Gravelle et al. reproduced particle generation and particle transport in porous media and detachment after ionic strength reduction by laboratory experiments [14]. Li et al. studied the blockage rule of particles in a gravel pack experimentally and established the damage evaluation model of physical parameters of the gravel filling layer using the deposition model of particles in the liquid phase [15,16]. In addition to the above, many other scholars have characterized the deposition and plugging mechanism of particles in porous media through laboratory experiments and theoretical study [17–24].

Traditional experimental studies of gravel packs are usually conducted in invisible metal sand-filling pipes. The results typically obtained from the experiments convey little pressure distribution along the pipes [25–27]. The sand front interface and the penetration depth in a gravel pack cannot be obtained from the experiments. Also, simulation studies with a one-dimensional phenomenological model based on the average particle concentration obtained by experiments have been conducted [28–32]. However, these simulation studies cannot characterize the particle distribution. These studies cannot provide an effective guidance for gravel packing design. To meet the need of the two-stage gravel packing design, visualization experiments need to be conducted to determine the penetration depth of intrusive particles in a gravel pack. In this paper, newly designed visualization experiments were conducted to study the dynamic distribution of intrusive particles in the two-stage gravel pack. The influence of the two-stage gravel size and the packing ratio was analyzed based on our experimental results. The design optimizations of the two-stage gravel packing process are described herein.

2. Experimental Design

2.1. Methodology

There is no chromatic aberration between the packed gravel and the sand. Thus, the visualization cannot be achieved. Moreover, it is difficult to separate sand from gravel after the experiment, and the measurement of retained particles is hard to carry out. In this experiment, the prototype was enlarged. Glass beads were used instead of gravel to meet the experimental requirements of visibility and measurability. In the process of hydrodynamic and solid-liquid two-phase flow experiments, it was necessary to ensure a certain flow similarity between the model and the prototype. This required a fixed proportional relationship of physical quantities (geometrical, kinematical, and dynamical). To fit the motion modes of the sand particle in an actual stratum, similarity criteria needed to be established to determine the similarity scale of time, velocity, and flow in the visualization model. For the solid-liquid two-phase flow, flow and particle Reynolds criterion, Stokes criterion, and Freud criterion were the four dominant similarity criteria in designing the experiments.

$$\frac{l_p \rho_p u_p}{\mu_{fp}} = \frac{l_m \rho_m u_m}{\mu_{fm}} \quad (Re)_p = (Re)_m \tag{1}$$

$$\frac{d_{sp} \rho_{sp} u_{rp}}{\mu_{fp}} = \frac{d_{sm} \rho_{sm} u_{rm}}{\mu_{fm}} \quad (Re_s)_p = (Re_s)_m \tag{2}$$

$$\frac{d_{sp}^2 \rho_{sp} u_{rp}}{\mu_{fp} l_p} = \frac{d_{sm}^2 \rho_{sm} u_{rm}}{\mu_{fm} l_m} \quad (Stk)_p = (Stk)_m \tag{3}$$

$$\frac{\rho_p l_p^2 u_p^2}{\rho_p g l_p^3} = \frac{\rho_m l_m^2 u_m^2}{\rho_m g l_m^3} \quad (Fr)_p = (Fr)_m \tag{4}$$

where subscript p represents prototype and subscript m represents model; Re and Re_s are the flow and particle Reynolds number, dimensionless; Stk is the Stokes number, dimensionless; Fr is the particle Freud number, dimensionless; l is the geometric characteristic length, m; d_s is the particle diameter, m; ρ and ρ_s are the fluid and particle density, kg/m^3; u and u_r are the fluid velocity and relative velocity between solid and liquid phases, m/s; μ is the fluid viscosity, N·s/m^2.

In the experiment, the prototype and the model could not satisfy all the above similarity criteria at the same time [33,34]. The approximation model was adapted to ignore the secondary factors. By ignoring the difference of surface roughness between gravel and glass beads, the fluids and the particles used in the experiment could be regarded as the same as those in natural conditions. When the

Reynolds number of the prototype and the model was in the same self-modeling region, the similarity with criterion number Re could be ignored. For the flow in the sand-packing tube, the vertical distance that the gravity acted on was relatively short, thus the Froude number (Fr) that reflected the gravity action could be ignored. After the above simplification, it was concluded that the prototype and the model should have obeyed the particle Reynolds and Stokes similarity criteria for designing the experiment.

The following scale relations were obtained according to the established modeling law:

$$\lambda_v = \lambda_l^{-1}$$
$$\lambda_t = \frac{\lambda_l}{\lambda_v} = \lambda_l^2 \tag{5}$$

where λ_l is the length similarity coefficient equal to l_p/l_m; λ_v is the velocity similarity coefficient equal to u_p/u_m; λ_t is the time similarity coefficient equal to t_p/t_m.

According to the selected similarity criterion, the velocity magnification of the experiment was determined to be equal to the λ_l. For example, if the quartz sand with ten times enlarged median grain size was selected to simulate the intrusive sand in the real situation, the injection rate of the visual experiment should be set as ten times the traditional displacement experiments.

2.2. Experiment Materials

The experimental set-up is shown in Figure 2. The Kamoer Lab UIP-S25-6 peristaltic pump was selected as the displacement pump with a range of 1–1300 mL/min. The liquid injection caliber was 8 mm, and the sand production was measured by the container placed at the outlet of the pipe. To achieve the visualization of sand migration and blockage in the gravel bed, a clear cylindrical PMMA (poly methyl methacrylate) pipe with a diameter of 40 mm and a length of 300 mm was selected as the sand-packing tube. The gravel used in traditional experiments were replaced by scale-up glass beads to monitor particle transportation. The injection rate was determined according to the hydrodynamic similarity criterion discussed before, thus, the motion of sand particles in the simulation bed was similar to the real situation. During the experiment, the movement and the blockage of particles in the gravel layer were monitored by real-time photography. When the gravel-sand interface was stable, the glass beads and the retained sand were taken out and weighed separately. Thus, the plugging degree of each section could be obtained.

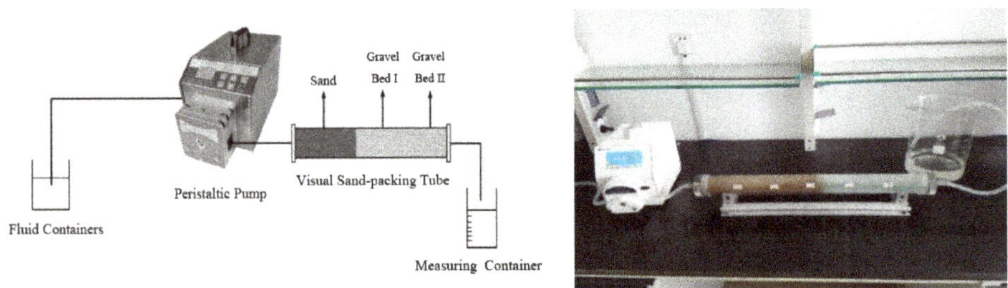

(**a**) Schematic of experiment   (**b**) Equipment of the experiment

Figure 2. The flowchart and set-up of the experimental device.

According to the selected similarity criterion, the geometric magnification of the experiment was determined to be ten times. Taking sand with 0.1 mm median grain size as an example, the quartz sand with ten times enlarged median grain size was selected to simulate the intrusive sand in a real reservoir. The glass beads with specific particle size were selected to simulate gravel packs to ensure the visibility and measurability of intrusive particles. During the experiment, the temperature was kept at 20 degrees and the pressure was 1 atm. The injection fluid was the glycerol mixture with the viscosity of 5 cP.

2.3. Experiment Procedure

A schematic of the experiment is illustrated with Figure 2. The general procedure of the experiment is described as follows:

(1)   Clean up the tube and place the metal gauze at the outlet joint to prevent the glass beads from discharging at the outlet. This can help to ensure the flow stability during the experiment.

(2)   The end of the tube is sealed, and the composite simulated quartz sand sample is filled to one side. After packing to the design depth, the sand sample is compacted. Then, pack glass beads into the pipe from the other end and compact slightly. During packing, the quartz sand should be kept at the bottom to prevent a large amount of pre-intruding sand before the experiment.

(3)   After calibrating the flow rate of the peristaltic pump, the displacement container, the peristaltic pump, and the sand-packing tube are connected. The depth of sand invasion, the movement of the gravel-sand interface, and the sand distribution characteristics are recorded every 10 min during the experiment.

(4)   After displacement, unfold the sand-packing tube from the exit end. The glass beads and the retained sand are extracted from segments. The mixture is filtered, and the glass beads and the retained sand are recovered separately. The amount of retained sand of this packing section can be obtained by weighing the dried recycled sand.

2.4. Experiment Scheme

In order to investigate the influence of the two-stage gravel size and packing length on the two-stage gravel bed, the experimental scheme was designed as shown in Table 1.

Table 1. Experimental scheme.

Group Number	Sand Median Grain size (mm)	Gravel Bed I Grain Size (mm)	Gravel Bed II Grain Size (mm)	Displacement Rate (mL/min)	Packing Length (I)/cm	Packing Length (II)/cm
1#	0.85	3.5	7	200	10	20
2#	0.85	4	7	200	10	20
3#	0.85	4.5	7	200	10	20
4#	0.85	5	7	200	10	20
5#	0.85	4.5	6	200	10	20
6#	0.85	4.5	8	200	10	20
7#	0.85	4.5	7	200	5	25
8#	0.85	4.5	7	200	7	23
9#	0.85	4.5	7	200	15	15

The experimental design adopted in this work was based on the principle of moderate sand control. Moderate sand control should raise the sand retention rate of gravel bed I to avoid the blockage of gravel bed II. The design of gravel bed I grain size was based on the criterion proposed by Saucier. After taking both conductivity and sand retention rate into account, Saucier proposed that the gravel-sand size ratio should be 5–6 [35]. The selection of gravel bed II grain size was mainly based on field experience and the commonly used size of industrial gravel derived from the operation of two-stage gravel packing in Shengli Oilfield, China [7]. The design of the injection rate was based on the experimental data of Li et al. in the conventional gravel packing displacement experiments [15,16].

3. Results and Discussions

3.1. Particle Grain Size of Gravel Bed I

The grain size of packing gravel in gravel bed I is one of the most critical factors impacting the permeability of the two-stage gravel packs. Currently, the selection criteria for gravel grain size of field operation mainly refer to the research of Saucier, which suggested a five to six times gravel-size median grain size ratio. By our visual experimental study, the sand penetration depth of four groups of different gravel-sand grain size ratio was obtained. This could provide an accurate and quantitative basis for the design of two-stage packing proportion. The experimental scheme is illustrated with groups 1#–4#, Table 1.

In the four groups of experiments, the sand-to-gravel grain size ratio of gravel bed I was conformed to the design criterion of field operation. By conducting the visual experiments, the penetration depth and the distribution characteristics of the intruding particles of different gravel size was obtained. Thus, it could help to design the packing gravel grain size and the effective packing depth of gravel bed I. It could effectively prevent deep intrusion of sand into gravel bed II. Also, in this way, overall gravel pack permeability could be maintained or increased. During the experiment, the dynamic settlement and the blockage of intrusive sand in the pore space of the gravel were observed and regularly recorded, as shown in Figure 3.

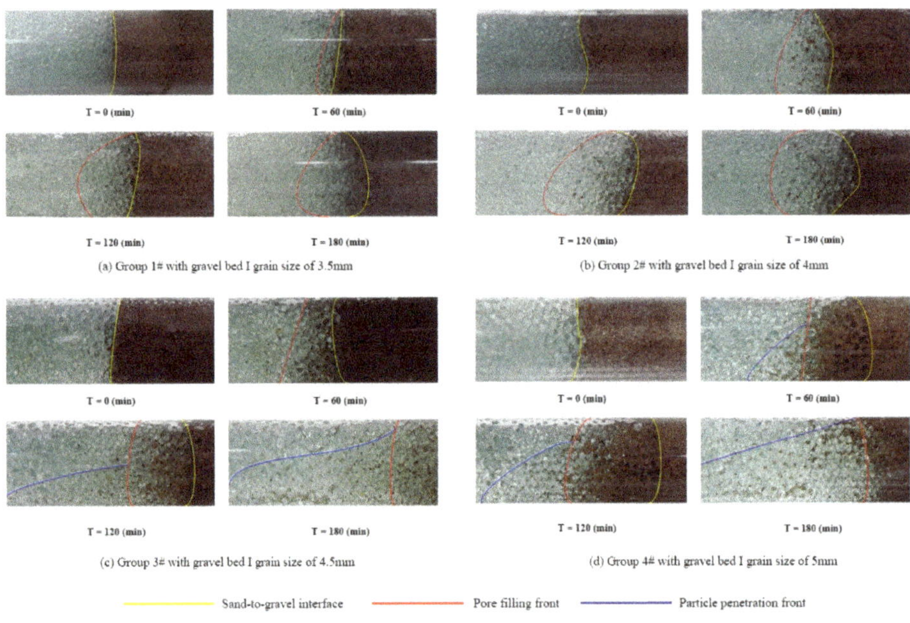

Figure 3. Sand distribution in gravel packs affected by grain size of gravel bed I.

The sand-gravel interface, the pore filling front, and the sand penetration front are shown in Figure 3. The pore filling zone refers to the gravel-sand mixing area where the pore throat was blocked severely at the gravel-sand interface due to the massive invasion of the sand particle. The sand penetration front indicates the longest distance of intrusive sand migrated into the gravel pack. It can be seen that the grain size of grade I gravel had an obvious influence on sand migration and blockage in formation.

The influence of gravel bed I grain size on sand migration and blockage in the gravel pack was obviously observed during the experiment. When gravel grain size was 3.5–4 mm, gravel bed I had a better sand controlled efficiency compared with other groups. This resulted in the shorter sand filling length, the slower expansion speed, and the lower level of the gravel pack plugging the degree near the sand-gravel interface. Also, a stable sand bridge could be formed in the pore space of the packed gravel. A stable sand bridge could restrain the expansion of the particle penetration front. The amount of retained particles of the gravel pack could be increased substantially with the enlargement of packed grain size. With the increase of pore space, it took a longer time to form a sand bridge that could exclude further particle intrusive. Thus, the penetration depth of particles was increased obviously.

It can be seen from Figure 4a that when the gravel grain size was less than five times that of the median grain size of the formation sand (group 1# and group 2#), the gravel bed I entirely excluded the further intrusion of the production sand. The variation characteristics of pore filling depth were similar in groups 1# and 2#. Both of them had a stable pore filling depth of 3–4 cm, which was nearly two times smaller than that of groups 3# and 4#. When the gravel-to-sand grain size ratio was more than five according to the Saucier criterion (group 3#), the depth of the gravel-sand mixing zone was increased significantly. However, when the grain size continued to increase to 5 mm, there was no obvious change in the trend of the pore filling front. The stable filling depth was slightly increased from 7 to 7.5 cm.

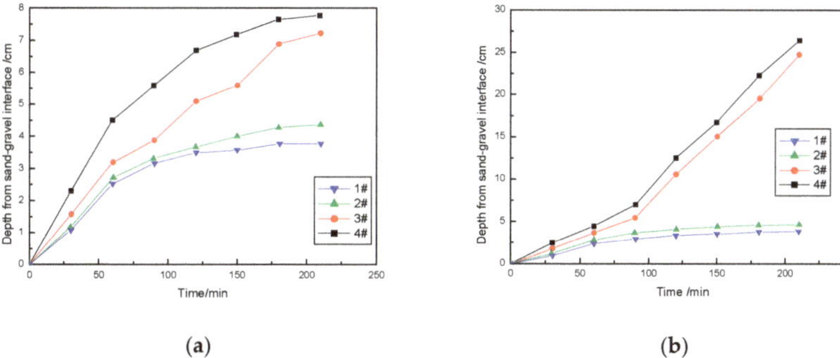

(a)   (b)

Figure 4. The dynamic diversification chart of the pore filling and the particle penetration depth influenced by grain size of gravel bed I: (**a**) pore filling depth; (**b**) particle penetration depth.

The particle penetration front of the gravel pack could be divided into two groups based on whether it was five times larger than the median grain size of the intruding sand, as shown in Figure 4b. In groups 1# and 2# with gravel bed I grain sizes of 3.5 mm and 4 mm (gravel-sand size ratio less than five), a stable sand bridge existed that could prevent the particles from further intrusion into the gravel-sand mixing zone. With the increase of packing gravel size, a large number of intrusive sands migrated to the deep place. Both of the experiments for group 1# and group 2# had a particle penetration depth of up to 20 cm. Additionally, because the overall blockage in the gravel pack was not critical, this part of the sand could usually be settled in the lower part of the pipe wall due to the action of gravity, as shown in Figure 3.

Figure 5 shows the distribution of sand retention in the two-stage gravel beds. The sand retention of the two groups of experiments with smaller grain size could only be detected in gravel bed I, which was mainly concentrated at 3–4 cm, where the pore fill took place. The mass of retained sand all over the gravel beds of groups 1# and 2# was about 7.8 g, which was much less than the other two groups. The amount of sand retained in the first 3 cm of the experimental group with gravel grain of 4.5 mm was similar to that in the experimental group with 3.5 mm and 4 mm diameters. However, it covered deeper gravel bed in longitudinal depth since there was retained sand detected in the gravel pore with a distance of 27 cm from the interface. Although the pore filling and the penetration depth of the two experimental groups with particle sizes of 4.5 mm and 5 mm were similar, the overall blockage amount of the gravel pack was much larger because of the larger pore inlet area and the accumulation space. The mass of retained sand all over the two gravel beds in group 4# was 26 g, which was much larger than that in group 3#.

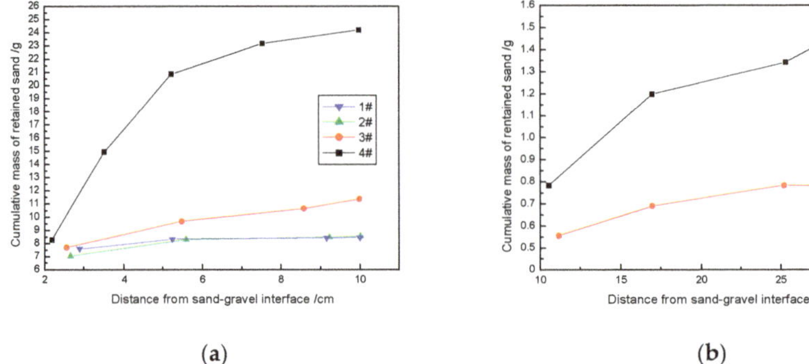

Figure 5. The cumulative mass of retained sand in two-stage gravel beds influenced by grain size of gravel bed I: (**a**) gravel bed I (**b**) gravel bed II.

By analyzing the obtained experimental results, it can be concluded that when the gravel-sand grain size ratio was less than five, sand bridges that formed in the pore space of the gravel pack could restrain further penetration of particles intrusion. The depth of the gravel-sand mixing zone was decreased, and there were few particles entering gravel bed II. Therefore, the gravel-sand grain ratio of gravel bed I could be controlled within five. At the same time, the packing length of gravel bed I should have been reduced, and the packing grain size of gravel bed II should have been increased to improve the gravel pack's overall permeability.

3.2. Particle Grain Size of Gravel Bed II

Until now, there had been no reliable criterion for packing gravel selection of gravel bed II. To increase the validity period and the permeability of the gravel pack, the following principles should be followed in designing packing gravel in gravel pack II. Packed gravel should effectively prevent the gravel in bed I from further migration.

In the three experimental groups, the median grain size ratio of gravel bed II to gravel bed I ranged from 1.33 to 1.78, which covered the possible combination range of industrial gravel selection of the field operation. The dynamic blockage in the gravel pack of the experiment is shown in Figure 6. The pore filling front of the three groups had similar regularity, while the penetration depth was quite different. With the increase in packing grain size of gravel bed II, the gravel pack's pore space was increased, and the fluid velocity in the pore of the gravel layer was decreased. This resulted in a significant increase in the sedimentation sand at the interface between gravel beds I and II.

It can be seen from Figure 7 that these three groups had similar change regulation. Also, the steady-state values of the pore filling depth were all about 7 cm. The grain size of gravel bed II mainly affected the dynamic development law of the particle penetration depth. During the first 90 min of the experiment, the intrusive particles mainly migrated in gravel bed I, thus the three groups' penetration depths had similar development characteristics. As the frontier of penetrated sand migrated to the junction of gravel beds I and II, the increase in gravel size led to more sand intrusion. Meanwhile, because the fluid flow velocity in the pore throat became slower, the experimental group of the larger gravel bed II gravel had more sand settlement at the two-stage gravel pack interface. Compared with group 5# with gravel grain size of 6 mm, the penetration depth of group 3# with grain size of 7 mm was increased significantly by 8 cm. Meanwhile, when the grain size was changed from 7 mm to 8 mm in group 6#, the penetration depth was only changed by 2 cm.

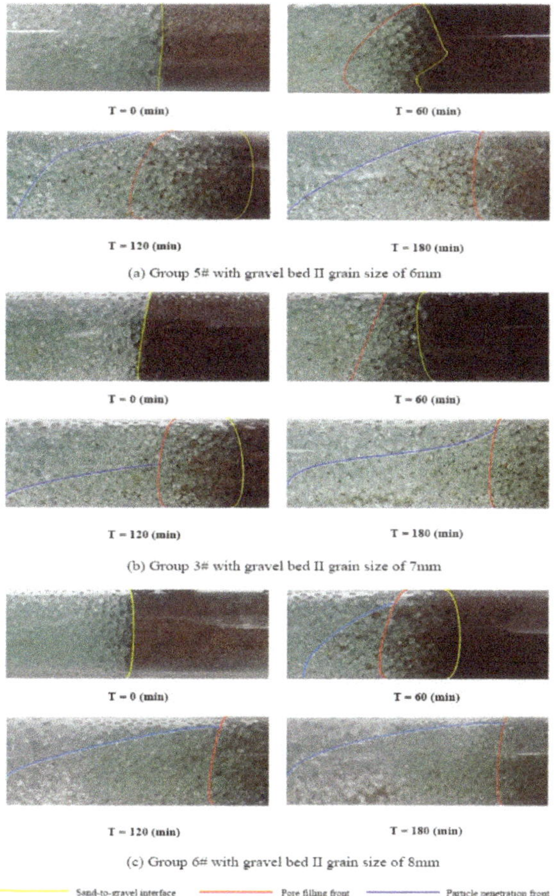

Figure 6. Sand distribution in gravel packs affected by grain size of gravel bed II.

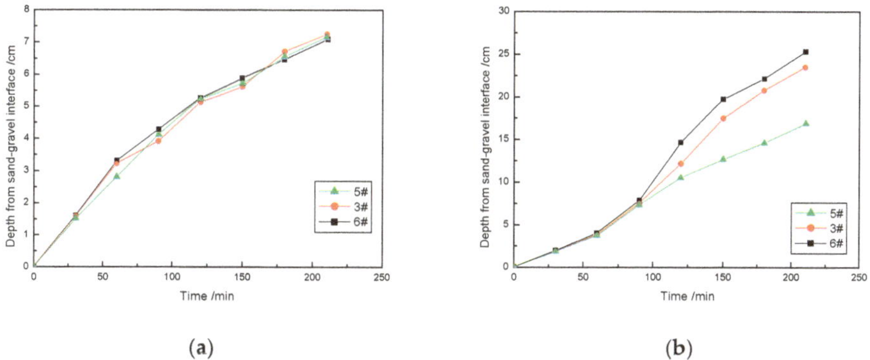

Figure 7. The dynamic diversification chart of the pore filling and particle penetration depth influenced by grain size of gravel bed II: (**a**) pore filling depth (**b**) particle penetration depth.

In group 5# with a gravel bed II grain size of 6 mm, the channel for blocked sand migration at the interface was relatively small, thus it was more difficult for particles to further enter gravel bed II, and this reduced the amount of retained particles, as shown in Figure 8.

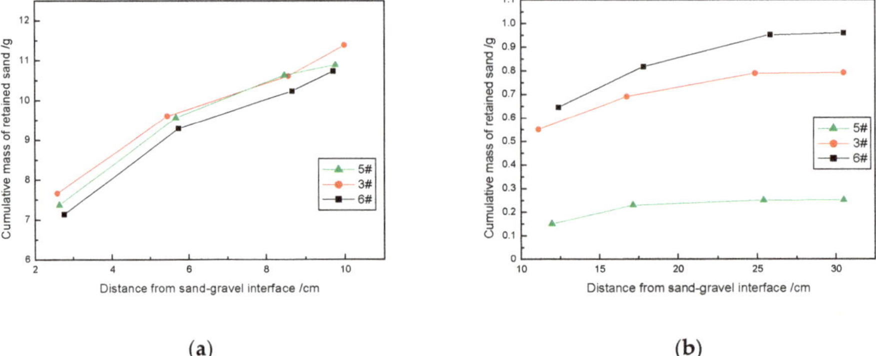

(a)   (b)

Figure 8. The cumulative mass of retained sand in two-stage gravel beds influenced by grain size of gravel bed II: (**a**) gravel bed I (**b**) gravel bed II.

When the grain size of gravel bed II was increased to 7 mm, the amount of sand retained at the interface was increased by 0.4 g compared with the group of 6 mm. The reason is that the amount of sand intruded was increased by the increase of the pore throat and the particle settlement caused by the decrease of flow velocity at the interface. However, the sand deposited at the interface did not further migrate, thus no blockage occurred to gravel bed II. The experimental group of packing grain size with 8 mm had a similar regularity in retained sand distribution law as the group with 7 mm, with only a minor increase of 0.2 g.

The following conclusions can be drawn from the above study: the grain size of gravel bed II had little effect on the length of the pore filling zone and the amount of retained particles of gravel beds I and II. It mainly affected the amount of retained sand at the two-stage interface and the particle penetration depth. During the experiment, the amount of retained sand at the interface of the experimental group with a grain size of 7 mm was larger than that of the experimental group with 6 mm, but it was a bit smaller than that of the experimental group with a grain size of 8 mm. From the perspective of maximizing overall permeability, 8 mm should be selected as the grain size of packed gravel in gravel bed II.

3.3. Two-Stage Gravel Packing Length

The two-stage packing length ratio is a critical factor that influences the sand control effect of the gravel packs. However, because of the limitations of objective factors such as the length of the sand packing tube and the distribution of pressure measuring points, very few conventional gravel packing experiments had the pack length ratio taken into account. In this experiment, because the packing quantity of the two-stage gravel could be controlled, the influence of packing length was investigated in this study.

It can be observed from Figure 9 that the sand pore filling phenomenon only existed in gravel bed I, while sedimentation occurred when particles intruded into the gravel bed II layer instead of pore filling. For all four groups, in the first 60 min, the intrusive sand was filled in the pore space of gravel bed I with the same expansion regularity. From 60 to 180 min, the pore filling depth of group 7# exceeded 5 cm, which was the packing length of gravel bed I. The sand fell into gravel bed II, and this resulted in sand sedimentation on the pipe wall and a large amount of retained sand at the two-stage interface. The pore filling front of groups 8# and 3# broke through gravel bed I around 120 min. Sand settlement took place at the two-stage gravel interface as well. However, the amount of retained sand was quite a bit smaller than group 7#. Intrusive sand in group 9# migrated within gravel bed I all through the displacement process, and no visible penetration front formed.

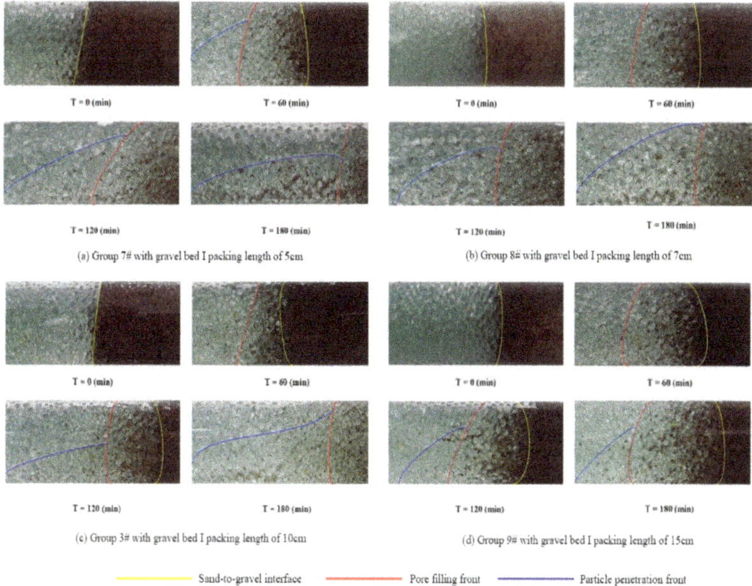

Figure 9. Sand distribution in gravel packs affected by two-stage packing length.

From Figure 10, it can be seen that the pore filling of the intrusion sand was only formed in gravel bed I. After the pore filling of gravel bed I was completed in group 7# with a packing length of 5 cm, the pore filling front had not been extended to gravel bed II. Sand particles only migrated in gravel bed I of group 9#, thus the pore filling depth of group 9# with a packing length of 15 cm was only slightly less than 7 cm and 10 cm. However, since gravel bed I had a lower porosity, particle penetration depth was apparently less than that of the other two groups, with a value of 10 cm. Because the stability value of pore filling depth under the experimental setting was more than 7 cm but less than 10 cm, the sand intrusion depth of group 3# was significantly reduced compared with that of group 8#.

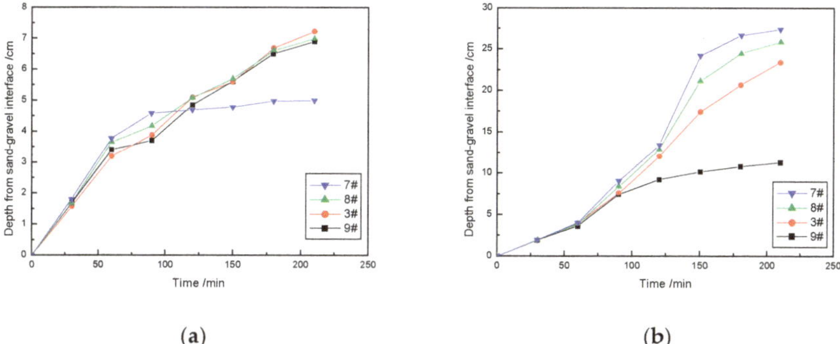

Figure 10. The dynamic diversification chart of the pore filling and particle penetration depth influenced by two-stage packing length: (**a**) pore filling depth; (**b**) particle penetration depth.

As shown in Figure 11, in group 9#, intrusive sand was only distributed in gravel bed I, while in the first 10 cm or so of gravel bed I, the degree of blockage was similar to that of group (c), but no sand intrusion occurred after 10 cm. In the other three groups of experiments, the retained sand was all distributed along the whole of the two-stage gravel packs. The masses of retained particles in gravel bed I of groups 7# and 8# were 16 g and 15 g, respectively, which were significantly higher than 11.5 g and 10.4 g in groups 3# and 9#. A large amount of sand migrated into gravel bed II in group 7# and intruded to the end front of the gravel pack, consequently forming sand-production. Retained sand in gravel bed II of group 3# was about 0.5 g and was mainly concentrated at the two-stage interface, while the sand blockage in the middle of gravel bed II was less serious compared to that of group 8#, which had 1.6 g of sand retention.

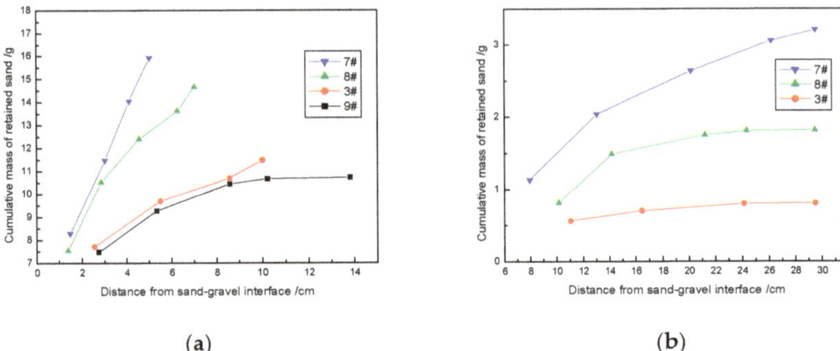

(a)   (b)

Figure 11. The cumulative mass of retained sand in two-stage gravel beds influenced by two-stage packing length: (**a**) gravel bed I (**b**) gravel bed II.

Through the above experiments, the following conclusions can be drawn: after the completion of pore filling in gravel bed I of groups (a) and (b) with packing lengths of 5 cm and 7 cm, no pore filling phenomenon occurred in gravel bed II, however, a large amount of sand settlement formed at the two-stage interface. Under the experimental conditions, the pore filling front of group (c) would not break through gravel bed I, and the majority of intrusive sand settled at the two-stage interface instead of further penetration. It is suggested that the packing depth of gravel bed I should be 10 cm in this case.

4. Conclusions

In this study, particle migration and blockage in a two-stage gravel pack were obtained by conducting our newly designed visualization experiment. The effect of the two-stage gravel size and the packing ratio on pore filling length and particle penetration depth was analyzed. Also, the particle retained in each gravel pack section was analyzed. By obtaining the stable value of pore filling depth and pore plugging degree, the basis for gradation and grain size selection in sand control operations of two-stage gravel packing was proposed. Our work can provide guidance for designing a two-stage gravel pack during the development of unconventional reservoirs with high sand production rates and fine silt content. The following conclusions can be drawn from this work:

(1) The pore filling depth could be shorter in the cases where grain ratio of gravel bed I to sand was less than five. The reason is that the sand bridge formed in the pore space of the gravel pack could restrain further particle penetration.

(2) The grain size of gravel bed II had little influence on the length of the pore filling zone and the amount of retained particles in gravel beds I and II. It mainly affected the amount of retained sand at the two-stage interface and the particle penetration depth.

(3) After completion of pore filling in gravel bed I in cases with packing lengths of 5 cm and 7 cm, no pore filling phenomenon occurred in gravel bed II. However, a large amount of sand settlement formed at the two-stage interface.

Author Contributions: Conceptualization, X.M.; Methodology, X.M., M.Q. and Z.M.; Formal analysis, Z.M. and T.L.; Investigation, X.M., Z.M. and M.Q.; Writing—original draft preparation, X.M.; Writing—review and editing, T.L., Z.N., and M.Q.; Funding Acquisition, X.M.

Funding: This research was funded by Shandong Natural Science Foundation Funded Project (Grant ZR2018BEE011), Fundamental Research Funds for the Central Universities National Science (Grant No. 18CX02027A), and Technology Major Project of the Ministry of Science and Technology of China (Grant No. 2016ZX05011004-003).

Acknowledgments: The Sand Control Lab in China University of Petroleum (East China) is highly appreciated.

Conflicts of Interest: The authors declare no conflict of interest.

References

1. Chen, Y.; Deng, J.; Zhao, W. Research on solutions of the influences of high pressure repeating packing sand control on productivity. *Pet. Sci. Technol.* **2013**, *31*, 88–93. [CrossRef]
2. Deng, F.; Feng, Y.; Yan, C.; Lin, H.; Gong, N.; Wang, J. Experimental investigation of factors affecting gravel pack efficiency for thermal recovery wells in Bohai Bay, China. *J. Pet. Sci. Eng.* **2017**, *156*, 835–844. [CrossRef]
3. Mirshekari, B.; Dadvar, M.; Modares, H.; Dabir, B. Modelling and simulation of gravel-pack damage due to particle transport by single phase flow in cased hole completion. *Int. J. Oil Gas Coal Technol.* **2014**, *7*, 152–168. [CrossRef]
4. Sparlin, D.D. Sand and Gravel—A Study of Their Permeabilities. In Proceedings of the SPE Symposium on Formation Damage Control, New Orleans, LA, USA, 30 January–2 February 1974. [CrossRef]
5. Li, Y.; Liu, L.; Sun, J.; Ye, Y.; Chen, Q. Sanding prediction and sand-control technology in hydrate exploitation: A review and discussion. *Mar. Geol. Front.* **2016**, *32*, 36–43.
6. Deng, J.; Tan, Q.; Sun, Y.; Chen, Y.; Duan, Z.; Wei, L.; Wang, E. Influence of high pressure pack repeat sand control on productivity and solution. *Fault Block Oil Gas Field* **2010**, *17*, 494–496.
7. Chen, Y.; Deng, J.; Zhao, W. Research of gravel size optimization in sand control with variable particle diameter packing. *J. Oil Gas Technol.* **2010**, *6*, 319–321.
8. Bouhroum, A.; Civan, F. A Study of particulates migration in gravel pack. In Proceedings of the SPE Formation Damage Control Symposium, Lafayette, LA, USA, 7–10 February 1994.
9. Bigno, Y.; Oyeneyin, M.B.; Peden, J.M. Investigation of pore-blocking mechanism in gravel packs in the management and control of fines migration. In Proceedings of the SPE Formation Damage Control Symposium, Lafayette, LA, USA, 7–10 February 1994.
10. Ali, M.A.J.; Currie, P.K.; Salman, M.J. Measurement of the particle deposition profile in deep-bed filtration during produced water re-injection. In Proceedings of the SPE Middle East Oil and Gas Show and Conference, Manama, Kingdom of Bahrain, 12–15 March 2015.
11. Shirinabadi, R.; Moarefvaand, P.; Goshtasbi, K.; Ahangari, K. The physical and numerical modeling of sand production and gravel pack in oil wells by designing and manufacturing the machine and presenting a relation estimating sand production rate. *J. Min. Sci.* **2016**, *52*, 300–312. [CrossRef]
12. Villarroel, F.M.G.; Vargas, E.A.J.; Bloch, M. Experimental analysis of gravel pack mesh size effect on the screen deformation under high in situ stress contrast. *J. Am. Chem. Soc.* **2010**, *95*, 4634–4639.
13. Lawal, K.A.; Vesovic, V.; Boek, E.S. Modeling permeability impairment in porous media due to asphaltene deposition under dynamic conditions. *Energy Fuels* **2011**, *25*, 5647–5659. [CrossRef]
14. Gravelle, A.; Peysson, Y.; Tabary, R.; Egermann, P. Controlled release of colloidal particles and remediation: Experimental investigation and modelling. In Proceedings of the SPE International Symposium on Oilfield Chemistry, the Woodlands, TX, USA, 11–13 April 2011.
15. Li, Y.; Li, M.; Wang, L.; Meng, W. A Productivity Prediction Model for the Gravel-packed Horizontal Well. *Liq. Fuels Technol.* **2013**, *31*, 633–642. [CrossRef]
16. Li, Y.; Qin, G.; Li, M.; Wang, W. Numerical simulation study on gravel-packing layer damage by integration of innovative experimental observations. In Proceedings of the SPE Heavy Oil Conference Canada, Calgary, AB, Canada, 12–14 June 2012.
17. Civan, F. Reservoir formation damage: Fundamentals, modeling, assessment, and mitigation. *Herd Health Environ. Res. Des. J.* **2000**, *7*, 60–77.
18. Keller, A.A.; Auset, M. A review of visualization techniques of biocolloid transport processes at the pore scale under saturated and unsaturated conditions. *Adv. Water Resour.* **2007**, *30*, 1392–1407. [CrossRef]
19. Jacobsen, O.H.; Moldrup, P.; Larsen, C.; Petersen, L.W. Particle transport in macropores of undisturbed soil columns. *J. Hydrol.* **1997**, *196*, 185–203. [CrossRef]
20. Benamar, A.; Ahfir, N.D.; Wang, H.Q.; Alem, A. Particle transport in a saturated porous medium: Pore structure effects. *Comptes Rendus Géosci.* **2007**, *339*, 674–681. [CrossRef]

21. Moghadasi, J.; Müller-Steinhagen, H.; Jamialahmadi, M.; Sharif, A. Theoretical and experimental study of particle movement and deposition in porous media during water injection. *J. Pet. Sci. Eng.* **2004**, *43*, 163–181. [CrossRef]

22. Rolle, M.; Hochstetler, D.; Chiogna, G.; Kitanidis, P.K.; Grathwohl, P. Experimental Investigation and Pore-Scale Modeling Interpretation of Compound-Specific Transverse Dispersion in Porous Media. *Transp. Porous Media* **2012**, *93*, 347–362. [CrossRef]

23. Parker, S.E.; Lee, W.W. A fully nonlinear characteristic method for gyrokinetic simulation. *Phys. Fluids B Plasma Phys.* **1993**, *5*, 77–86. [CrossRef]

24. Shin, C. Numerical simulation for particle penetration depth distribution in deep bed filtration. *Chem. Eng. Technol.* **2006**, *29*, 905–909. [CrossRef]

25. Ma, S.; Dong, M.; Li, Z.; Shirif, E. Evaluation of the effectiveness of chemical flooding using heterogeneous sandpack flood test. *J. Pet. Sci. Eng.* **2007**, *55*, 294–300. [CrossRef]

26. Liu, Q.; Cui, X.; Zhang, C.; Huang, S. Experimental investigation of suspended particles transport through porous media: Particle and grain size effect. *Environ. Technol. Lett.* **2016**, *37*, 11. [CrossRef]

27. Bedrikovetsky, P.; Marchesin, D.; Shecaira, F.; Souza, A.L.; Milanez, P.V.; Rezende, E. Characterisation of deep bed filtration system from laboratory pressure drop measurements. *J. Pet. Sci. Eng.* **2001**, *32*, 167–177. [CrossRef]

28. Bedrikovetsky, P. Upscaling of stochastic micro model for suspension transport in porous media. *Transp. Porous Media* **2008**, *75*, 335–369. [CrossRef]

29. Kim, Y.S.; Whittle, A.J. Filtration in a porous granular medium: 1. simulation of pore-scale particle deposition and clogging. *Transp. Porous Media* **2006**, *65*, 53–87. [CrossRef]

30. Guedes, R.G.; Al-Abduwani, F.A.H.; Bedrikovetsky, P.; Currie, P.K. Deep-bed filtration under multiple particle-capture mechanisms. *SPE J.* **2009**, *14*, 477–487. [CrossRef]

31. Araújo, J.A.; Santos, A. Analytic model for DBF under multiple particle retention mechanisms. *Transp. Porous Media* **2013**, *97*, 135–145. [CrossRef]

32. Kuhnen, F.; Barmettler, K.; Bhattacharjee, S.; Elimelech, M.; Kretzschmar, R. Transport of iron oxide colloids in packed quartz sand media: Monolayer and multilayer deposition. *J. Colloid Interface Sci.* **2000**, *231*, 32–41. [CrossRef]

33. Carlson, D.R.; Widnall, S.E.; Peeters, M.F. A flow-visualization study of transition in plane Poiseuille flow. *J. Fluid Mech.* **1982**, *121*, 487–505. [CrossRef]

34. Zamani, A.; Maini, B. Flow of dispersed particles through porous media—Deep bed filtration. *J. Pet. Sci. Eng.* **2009**, *69*, 71–88. [CrossRef]

35. Saucier, R.J. Considerations in gravel pack design. *J. Pet. Technol.* **1974**, *26*, 205–212. [CrossRef]

Article

Study of Downhole Shock Loads for Ultra-Deep Well Perforation and Optimization Measures

Qiao Deng [1,2,*], Hui Zhang [1,*], Jun Li [1], Xuejun Hou [3] and Hao Wang [1]

[1] College of Petroleum Engineering, China University of Petroleum-Beijing, 18 Fuxue Road, Changping, Beijing 102249, China

[2] Department of Civil and Environmental Engineering, University of Pittsburgh, 3700 O'Hara St, Pittsburgh, PA 15261, USA

[3] School of Petroleum Engineering, Chongqing University of Science & Technology, 20 East Road, University City, Shapingba District, Chongqing 401331, China

* Correspondence: qid15@pitt.edu (Q.D.); zhanghuicup2018@163.com (H.Z.); Tel.: +(1)412-478-8135 (Q.D.)

Received: 13 June 2019; Accepted: 15 July 2019; Published: 17 July 2019

Abstract: Ultra-deep well perforation is an important direction for the development of unconventional oil and gas resources, the security with shock loads is a difficult technical problem. Firstly, the theoretical analysis of perforated string is carried out, the dynamics models of which are established in the directions of axial, radial and circumferential. Secondly, the process of perforating with hundreds of bullets is simulated by using the software of LS-DYNA (ANSYS, Inc, Pennsylvania, USA). The propagation attenuation model of shock loads is established, and a calculation model to predict shock loads at different positions of the tubing interval has been fitted by considering multiple factors. The dynamic response of perforated string is studied, and the vulnerable parts of which are found out. Thirdly, the optimization measures are put forward for ultra-deep well perforation by the design of shock adsorption and safety distance of the packer. Finally, the field case of an ultra-deep well shows that the research method in this paper is practical, and the optimization measures are reasonable and effective. This study can provide important guidance to reduce shock damage and improve security for ultra-deep well perforation.

Keywords: ultra-deep well; shock loads; perforated string; safety analysis; optimization measures

---

1. Introduction

With the rapid development of petroleum exploration and development, unconventional oil and gas resources have become the focus of global oil fields, the field operation of which is getting harder and harder with the increase in deep and ultra-deep well exploitation [1]. Ultra-deep wells are characterized by deep reservoir burial, and a number of ultra-deep wells with depths of 8000 m have been drilled in China, the maximum formation pressure of which can be close to 140 MPa. With the ultra-high pressure conditions, the potential safety risk of perforation is getting higher, particularly in order to maximize well productivity and recoup the higher cost for ultra-deep wells, in recent years higher-shot densities, propellants, and larger perforating guns have become widely adopted and developed rapidly for field application. With the use of such systems comes the additional explosive load that could cause perforated string to burst, collapse, bend, buckle, and shear, as well as the packer seals to fail as perforating guns are detonated [2]. Figure 1 shows tubing damage (breaking and buckling) after field perforation of an ultra-deep well, resulting in the failure of perforation, and affecting the progress of oil testing operations with huge economic losses [3]. Predicting the magnitude and transient behavior of perforating shock loads is a critical step for ultra-deep well perforation that can avoid damage or destruction to tool strings and production equipment.

(a)   (b)

Figure 1. Tubing damage caused by perforating shock loads: (**a**) tubing breaking; (**b**) tube buckling.

Due to the recent advances in well design and production techniques, tubing-conveyed perforating (TCP)-combined well testing is being increasingly used in challenging ultra-deep well completions [4]. A series connection of the perforating gun, tubing string, shock absorbers, packers, and other instruments is suspended into the downhole casing, as shown in Figure 2. When the shape charges detonate, the hollow carriers deform due to internal gas pressure and debris impacting the inner side of the carrier, the perforating jets puncture the hollow carrier wall, casing, cement, and formation. At the same time, a huge generated detonation wave will be released into the long and narrow space downhole with the packer setting. On the one hand, part of the shock loads will directly act on the perforating gun, which transfers to the tubing pipe, shock absorbers, packers, screen liner, the other connected components, resulting in a strong shock vibration of the perforated string system. On the other hand, the detonation gas inside the gun interacts with the wellbore fluid. The pressure difference between the gun and the wellbore produces shock waves in the wellbore fluid, propagating radially and axially up and down in a short time, leading to large fluid deformation and high-speed violent movement. The large-amplitude pressure waves that produce very large loads on the equipment, affecting the structural stability of the string system [5].

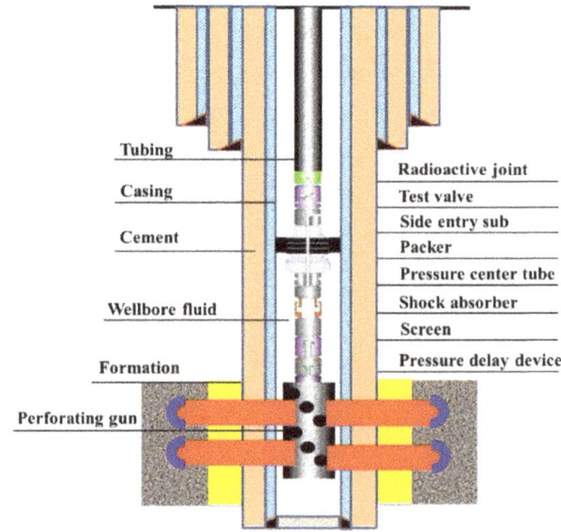

Figure 2. Structure of downhole tubing-conveyed perforation.

As there are potentially more perforation engineering problems in field operation, and the influence of perforating shock loads on the perforation safety attracted many researchers' attention. At present, theoretical analysis, laboratory test, and numerical simulation are mainly used to study the shock damage during perforating.

Some scholars have carried out early relevant research. Lu et al. established the pressure pulsation equation under actual perforation conditions by using superposition fundamentals. However, only theoretical deduction and qualitative analysis are carried out due to the fact the equation is very complex [6]. The method of static theory analysis was used to study the perforated string by Yin et al., and it was concluded that that the high pressure formed in the sealing section during perforating is one of the reasons for the vibration of the string [7]. Yu et al. applied the finite clearance element theory method to study the contact problem of perforated strings in well testing, which is to use the finite element to discretize the string into several spatial straight beam elements along the axis, and the clearance elements are set on the nodes of the beam elements [8]. Through a field case of a complex well perforation project, the strength safety of perforated string has been theoretically studied by Dou et al., and the influence of perforation section length and downhole "pocket" length was obtained [9]. Xu et al. theoretically analyzed the influence of the shooting density and charge quantity on the strength security of perforated string by a field case of gas well completion [10]. According to the detonation theory, the peak pressure in the wellbore was obtained by calculating the energy of the detonation gas by Yang et al., and the strength of the string was analyzed based on the buckling theory [11]. Zhang et al. established the calculation model of perforating shock loads based on the theory of underwater explosion. The influence of the number of perforation bullets, detonation interval time, charge quantity, physical condition of artificial wellbore bottom, the depth of "pocket" on the perforating shock loads was analyzed [12].

In laboratory research, it's hard to stimulate all of the real underground working conditions, and the cost for real tests is too high. Therefore, there are few studies in this field at present. Zhou et al. developed a comprehensive dynamic load test system, the acquired data of strain and outer ring pressure of perforated string were obtained in the perforation section [13]. A ground simulation test system was developed by Lu et al., in which the pressure and acceleration responses at the end of the perforated string were tested by experiments. A preliminary understanding of the dynamic response characteristics of perforated string with shock loads were obtained [14].

In the aspect of numerical simulation, many studies have emerged in this field in recent years. Chen et al carried out a mechanical analysis of the perforated string in ultra-deep wells by using the ANSYS/LS-DYNA finite element software. The results showed that the longer the operating string length is, the greater the perforation depth is, and the longer the distance from the restraint end is, the smaller the effective stress of the perforated string is, the more stable the whole downhole string system is, and some optimization methods for effectively alleviating the impact of perforated string are proposed [15]. The influencing factors of perforating shock loads in ultra-deep wells were studied by Chen et al., it was found that the axial force of perforated string was affected by factors such as "pocket" length, well fluid density, string length, perforation depth, well depth and perforating gun loading strategy [16]. Baumann et al. comprehensively summarized all influencing factors of perforation design, including the length of perforated interval, types, and sizes of perforated string, shaped charges, cable, and conveyance, rathole length, installation position of the packer, type and number of shock absorbers, wellbore and formation pressure with fluid properties [17]. The numerical simulation of the perforating process with a single perforation bullet was carried out by Teng, and the effects of shock loads on the perforated string were explored by static analysis [18]. Kang et al. studied the dynamic response of the perforating gun under different perforation conditions using the ANSYS/LS-DYNA software, whereby the influence of the axial and radial shock loads on the structural strength of the perforating was analyzed [19]. Cai et al. established a dynamic model of perforated string by using the method of space beam element and spring element with considering the annulus between tubing and casing, and the dynamic response process of which was simulated [20]. The perforated string was divided into space beam elements to simulate the dynamic response process by Yang, and the effects of the perforating shock loads, size and collapsing strength of perforated string, impact time on the maximum stress of the string for a horizontal well were studied [21]. Zhang et al. obtained the vibration displacement, velocity, acceleration and equivalent stress of perforated string by establishing a finite element model. The influence of the length and thickness of the string on the stress intensity was studied [22]. Li et al. reviewed that the shock vibration of the perforated string is caused by perforating bullets and bumper jar shock. The variation and distribution of shock pressure along the string with time were not clear [23]. Based on the AUTODYN software (ANSYS, Inc, Pennsylvania, USA), the variation law of density, velocity and pressure fluctuation of perforation fluid were analyzed by Li, and the effects of charge quantity and density on perforated string were studied [24]. Li et al. analyzed perforating pressure fluctuation of annular by software simulation based on field measured data, the vibration velocity and acceleration of perforated string were obtained. The results show that the stress concentration occurs near the packer, where the stress value is the largest [25]. The professional commercial software has been developed by some major oil companies (Schlumberger, Halliburton, etc.), which can simulate dynamic downhole conditions of ultra-deep wells and model all relevant aspects of well perforation, including gun loading, wellbore pressure waves and related fluid movement [26–30].

These studies set a foundation for further research on the shock damage of perforating shock loads. However, there is a lack of a model to accurately calculate the shock loads during perforating under different conditions, especially for the high formation pressure condition for ultra-deep wells. These numerical simulation studies are often aimed at a single perforating bullet, which is not consistent with the actual working conditions in the field with hundreds of bullets. Meanwhile, the vulnerable position of the perforated string needs to be found, and the propagation law of perforation shock load in the wellbore is not clear. In this paper, through theoretical and numerical simulation analysis, the effects of the downhole perforating shock loads on the perforated string system are analyzed for ultra-deep well, and the propagation law of perforating shock loads in the wellbore is explored. A model for predicting the magnitude of shock loads under different perforation conditions is established, the dynamic response process of the perforated string is studied, and the relevant optimization measures are put forward, including the design of shock absorption and safety distance of the packer. Finally, the research results are applied to the perforation of an ultra-deep well case.

2. Mechanical Model of a Perforated String

As the initial output unit of perforating shock loads, analyzing the dynamic response behavior of perforated string system is the basis of studying the shock damage for ultra-deep wells. The security of the perforated string and its correlative problems gradually turned into the investigative object for researchers in this field in recent years. In this study, the dynamics models of perforated string have been established in the directions of axial, radial and circumferential, the displacement of perforated string can be calculated by numerical integration method.

2.1. Axial, Radial and Circumferential Model of Perforated String

In the process of perforation operation, due to the asymmetry of the perforating charge structure, the spiral distribution pattern and the coupling effect of the sequential explosion, perforating shock loads on the string can be divided into in axial, radial and circumferential directions simultaneously. Taking the perforated string below the packer as the research object, which the string can be assumed to be a cantilever with one fixed end and the other end applied with shock loads. According to the structure of the perforation string system and the operation condition, the following assumptions are made by the authors: the material of the perforated string is an isotropic homogeneous continuous linear elastic body, satisfying Hooke's law; the string is a continuous uniform-section straight thin rod and the effects of wellbore fluid and casing are ignored; the shock loads are all applied on the bottom of the string. A rectangular coordinate system is established with the center of packer as the origin, the down direction as the x positive direction and right direction as the y positive direction. The mechanical model of the perforated string can be described in Figure 3.

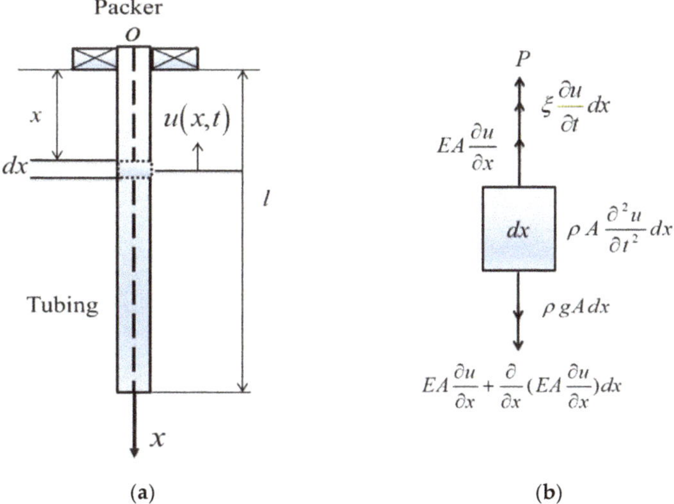

Figure 3. Simplified physical model of perforated string: (**a**) sketch of mechanical analysis; (**b**) mechanical model of perforated string under axial shock loads.

According to D'Alambert's principle, Equation (1) can be written:

$$\rho A dx \frac{\partial^2 u}{\partial t^2} = \left(EA\frac{\partial u}{\partial x} + \frac{\partial}{\partial x}(EA\frac{\partial u}{\partial x})dx \right) + \rho g A dx - P dx - EA\frac{\partial u}{\partial x} - \xi\frac{\partial u}{\partial t}dx \tag{1}$$

where ρ is the density of the string, kg/m^3; A is the cross sectional area of the string, in m^2; E is the elasticity modulus, Pa; $u(x,t)$ is the axial displacement of the string at time t and position x, in m; g is the gravitational acceleration, m/s^2; ξ is the damping coefficient of the string in perforating fluid; P is the perforating impact load, MPa.

Considering the effect of the gravitation, and assuming the external loads uniformly distribute along the string, the force acting on perforated string can be established in Equation (2):

$$F = -P\delta(x-l) + \rho Ag \tag{2}$$

where δ is the unit impulse function. By simplification, the differential equation of perforated string vibration under axial shock loads can be obtained in Equation (3):

$$\frac{\partial^2 u}{\partial t^2} = \frac{E}{\rho}\frac{\partial^2 u}{\partial x^2} + \frac{1}{\rho A}F - \frac{\xi}{\rho A}\frac{\partial u}{\partial t} \tag{3}$$

To build the dynamic model of perforated string under radial shock loads, the string is simplified as a cantilever with a fixed end and a free end, like the condition of axial shock loads. Following assumptions are made by authors: main inertia axes of cross sections of the string are in the same plane where the string moves radially; shock loads are applied on the lowest part of the string; bend is the main deformation of the string. Like the axial analysis of the perforated string, a unit of dx in the radial direction of the string is shown in Figure 4a. Similarly, the cross sections of the string are assumed to be planes with axial and circumferential vibrations. Like the coordinate system of the string under axial shock loads, the down direction along the string was assumed positive. A unit of the perforated string in the circumferential direction is shown in Figure 4b.

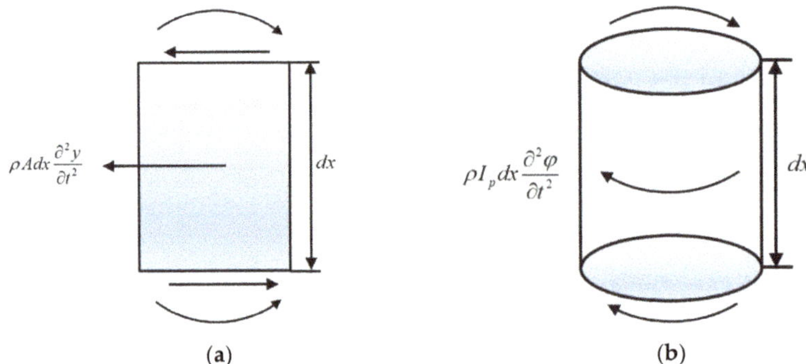

(a)   (b)

Figure 4. Mechanical analysis of a perforated string element: (**a**) sketch of mechanical analysis under radial shock loads; (**b**) sketch of mechanical analysis under circumferential shock loads.

The radial vibration differential equation of the string can be established in Equation (4):

$$EI\frac{\partial^4 y}{\partial x^4} + \rho A\frac{\partial^2 y}{\partial t^2} = F \tag{4}$$

where I the inertia moment of the perforating string, m⁴; $y(x,t)$ is the radial displacement of the perforating string at position x, in m, and time t.

The circumferential vibration equation can be expressed by Equation (5):

$$\frac{\partial u}{\partial t}\frac{\partial^4 \varphi}{\partial t^2} = \frac{G}{\rho}\cdot\frac{\partial^2 \varphi}{\partial x^2} + \frac{1}{\rho I_p}\cdot F \tag{5}$$

2.2. Displacement of Perforated String

Due to the fixed restriction, the displacement of the perforated string is zero at the position of the packer. Meanwhile, the internal force on the end face of the string is also zero. Therefore, boundary conditions at the fixed end and the free end of the string can be demonstrated in Equation (6):

$$\begin{cases} EA\dfrac{\partial u(x,t)}{\partial x}\bigg|_{x=l} = 0 \\ u(x,t)\big|_{x=0} = 0 \end{cases} \tag{6}$$

where $\varphi(x,t)$ is the angular displacement of the tubing at position x and time t, rad; G is the shear modulus, Pa; I_p is the polar moment of inertia of a string cross section, m⁴.

Before the explosion of the perforation charge, the perforated string remains still and its initial velocity is zero. As the effect of gravity of the string on its displacement is ignored, the initial displacements of every string unit can be regarded as zero. The initial conditions can be obtained using Equation (7):

$$\begin{cases} u(x,t)\big|_{t=0} = 0 \\ \dfrac{\partial u(x,t)}{\partial t}\bigg|_{t=0} = 0 \end{cases} \tag{7}$$

Based on the above boundary and initial conditions, the longitudinal free vibration equation of the perforated string can be established in Equation (8):

$$\frac{\partial^2 u(x,t)}{\partial t^2} = \frac{E}{\rho}\frac{\partial^2 u(x,t)}{\partial x^2} \tag{8}$$

The solution process of the longitudinal free vibration equation of perforated string is shown in Appendix A. According to the solving process of the axial mechanical model, the infinite series of sinusoidal vibration modes can be used to express the displacement response of perforated string under axial shock loads, as shown in Equation (9):

$$u(x,t) = \sum_{i=1}^{\infty} \frac{1}{2\pi f_i} \cdot U_i(x) \cdot \int_0^t g_i(\tau)\sin 2\pi f_i(t-\tau)d\tau \tag{9}$$

Through the above theoretical analysis, the dynamic response of perforated string has been studied. However, it is difficult to make a comprehensive analysis for the complicated shock waves loading rules and the dynamic response laws of perforated strings through theoretical models, especially for the dynamic fluid-structure interaction. In the laboratory research field, it is difficult to fully stimulate the real downhole working conditions, especially for the complex underground environments found in ultra-deep wells. What's more, the experimental data obtained is limited and the cost of obtaining it is too high. As for numerical simulations, if the modeling and meshing are reasonable, they can fully present the perforating explosion process and simulate various working conditions for ultra-deep wells, with a comprehensive analysis of the dynamic response of perforated string by considering the dynamic fluid-structure interaction. In this way, the perforating shock loads under different perforation conditions can be obtained, and the propagation laws of which in the wellbore can be studied. At last, the optimization design and proposal can be proposed, which can provide an important theoretical basis for field perforating operations.

3. Numerical Simulations

The process of bullet explosion, jet formation, and penetration during perforating have been simulated by researchers by combining the arbitrary Lagrange-Euler (ALE) method with the self-adaptive mesh technique (AMR) [31–33]. However, due to the complexity of modeling and meshing, a single perforating bullet or a few were simulated in these studies. In addition, the effects of high formation pressure for ultra-deep wells were not considered, the actual downhole perforation conditions cannot be presented accurately, affecting the results of numerical simulations. In this study, hundreds of perforating bullets have been simulated with considering the actual ultra-deep well perforation environment by using the large computers. The physical models are created by ANSYS/WORKBENCH, meshed by HYPERMESH, and ANSYS/LS-DYNA is adopted for numerical simulations. A number of numerical simulations are carried out with extracting calculation results of different perforating parameters and simulating various working conditions.

3.1. Modeling and Meshing

Due to the plastic deformation of the pipe string and the damage behaviors such as bending and fracture often occur below the packer, the tubing in a well section below which can be taken as a specific research object. In the actual perforation operation for ultra-deep wells, the length of the tubing varies from tens of meters to several hundred meters under different perforating conditions [34]. A 3D physical model can be established by simplification, which mainly consists of perforating gun, tubing and casing by ignoring the thread of each connector. The upper end of the tubing is radially restrained by the packer, the lower portion is restricted by the well bottom, and the surrounding is confined by the casing, as shown in Figure 5a. The perforating gun length is 9 m, tubing length is 20 m, rathole length is 4 m; the steel grade of the perforated string is N80, the size of the gun, tubing and casing are 177.80/152.53 mm, 73.02/62.00 mm, 244.40/220.50 mm, respectively. The yield limit of the casing is 460 MPa, and that of tubing is 536 MPa. The number of perforating bullets is 180, which are distributed in the perforating gun with a phase angle is 90°. The charge per hole is 45 g, the charge type is Royal Demolition Explosive (RDX). The remaining space inside the gun is filled with air, the tubing and annulus are filled with wellbore fluid, the density of which is 1.78 g/cm^3. The formation pressure is 130 MPa, the wellbore initial pressure is 125 MPa. The time range is from 0 to 5000 μs.

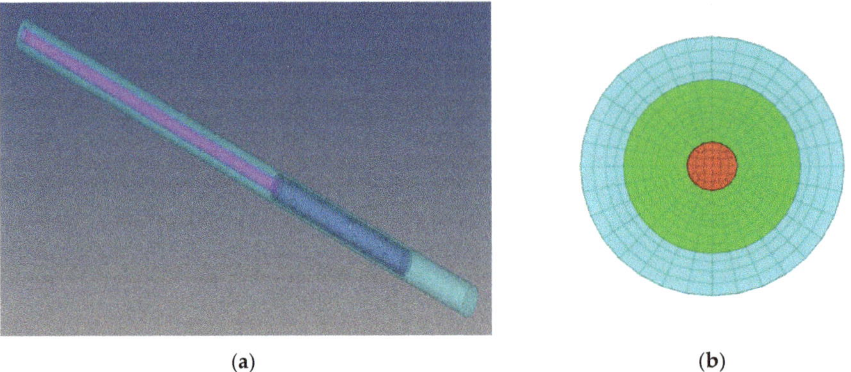

(a)　　　　　　　　　　　　　　　　　　(b)

Figure 5. Modeling and meshing: (**a**) model for numerical simulation; (**b**) cross-section view of mesh model.

Due to the highly nonlinear nature of the explosion, the grid of full hexahedron must be used in the fluid area and partially encrypted at the charge, as shown in Figure 5b. Each part of the material on the joined interface must have the common mesh node, which can effectively capture the movement and deformation of the construction of the material, and ensure energy between the partial grid effective transmit. The Lagrange algorithm is used in the perforating gun, tubing and casing. The Arbitrary Lagrange-Euler (ALE) algorithm is used in charge, air, and fluid, in which the space position of the ALE grid remains unchanged, and the material flows among the grids. The mean grid spacing is 4–5 mm and the total number of grids is about 1 million. The non-linear plastic kinematic hardening model (MAT_PLASTIC_KENEMATIC) is adopted for the material of the pipe string, as shown in Equation (10):

$$\sigma = \left[1 + \left(\frac{\varepsilon}{Q_1}\right)^{\frac{1}{Q_2}}\right]\left(\sigma_0 + \alpha E_p \varepsilon_P^{eff}\right) \tag{10}$$

where σ is the yield stress; ε is the strain rate; Q_1 and Q_2 are the parameters of strain rate; σ_0 is the initial yield stress; E_p is the plastic hardening modulus; ε_P^{eff} is the effective plastic strain.

The fluid is coupled to the solid interface and the material model of the charge is high-energy dynamite (High_Explosive_Burn), the state equation is EOS_JWL, as shown in Equation (11).

$$p = C_1\left(1 - \frac{w}{R_1 V_1}\right)e^{-R_1 V_1} + C_2\left(1 - \frac{w}{R_2 V_1}\right)e^{-R_2 V_1} + \frac{w E_1}{V_1} \tag{11}$$

where V_1 is the relative volume; C_1, C_2, w, R_1, R_2 are the explosive physical parameters; E_1 is the initial internal energy of unit explosive volume.

The unit algorithm adopts the constant stress unit algorithm of SOLID164 unit. The parameters of the RDX explosive are shown in Table 1.

Table 1. Explosive parameters.

ρ_0 (g/cm^3)	D (cm/μs)	B (GPa)	C (GPa)	R_1	R_2	w	E_1	V_1
1.69	0.8310	850	18	4.60	1.3	0.38	0.1	1.0

The final model can be imported to the LS-DYNA in the form of K file, which the model parameters are defined by keywords. *INCLUDE and * INCLUDE TRANSFORM can be used to import the meshed models, using *INITIAL_DETONATION to define the detonation point and initiation time, using *ALE_MULTI - MATERIAL_GROUP to define the material of fluid field grids which can flow in each other, using *CONTACT_ERODING_SURFACE_TO_SURFACE to define the erosive face to face contact algorithm, using the keyword *CONSTRAINED_LAGRANGE_IN_SOLID and *SECTION SOLID ALE to define the fluid-solid coupling, using *CONTROL_TERMINATION and *CONTROL_TIMESTEP to define the simulation time and the output step of the model calculation [35].

3.2. Computing Results

Based on the above modeling and parameters setting, numerical simulations are carried out on a large computer, with the data analysis by post-processing software of LS-PrePost (LSTC, California, USA). The unit during simulation is cm-g-μs, the pressure nephogram unit is 10^{11} Pa.

3.2.1. Perforating Dynamic Pressure

Figure 6 shows the pressure variation of perforation gun section after the perforating bullets detonate from $t = 49.9$ μs to $t = 249.9$ μs, and the pressure waves propagate from top to bottom along wellbore with the explosives explode gradually, which will reflect when reaching the bottom of the wellbore. The generated pressure waves in the gun continue to propagate to the upper tubing interval in the wellbore, acting on the perforated string and interacting with the wellbore fluid. This is the formation of dynamic shock loads in the wellbore, which has an impact on wellbore safety.

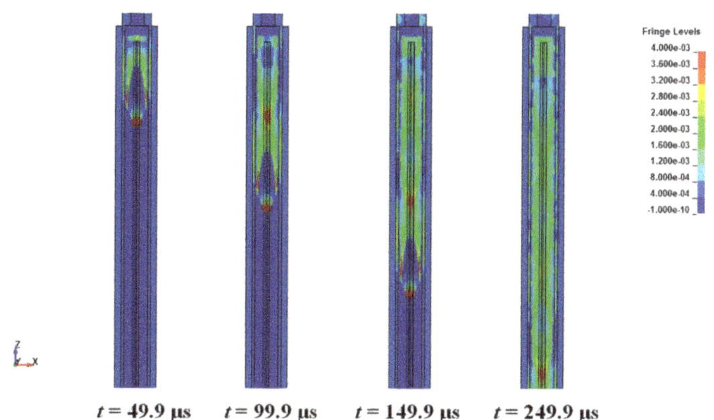

Figure 6. Pressure variation of perforation gun section (units for contour plots: 10^{11} Pa).

At present, there are several methods for accurately obtaining the perforation pressure: tracked and recorded by downhole high-speed P-T testing instruments in the whole process of downhole perforation, calculated by perforation software by entering specific parameters, predicted based on the empirical formula of underwater explosion, and obtained by laboratory perforation experiments. The field measured data are very limited with specific parameters and the data obtained by special software for perforation are relatively single, both of them are very accurate, which can be used as an important reference basis for verifying numerical simulation calculation. The results obtained from empirical formula calculation of underwater explosion are often inaccurate and in the laboratory research field, it's hard to stimulate real underground working condition and it can't really reflect the characteristics of asymmetric dynamic load. Therefore, the perforating pressure can be obtained by extracting numerical simulation data, which can be verified by special perforation software.

The pressure-time curve can be drawn by extracting data from the annulus of the wellbore according in the unit. Figure 7 shows the wellbore bottom pressure-time curves, in which the blue solid line is the wellbore bottom pressure-time curve by extracting data from the simulated result, and the red dashed line is calculated by the perforation software by inputting relevant parameters. The blue solid line shows that as the pressure wave arrives at the bottom of the wellbore, the wellbore pressure increases sharply, rapidly reaching its peak value (168.72 MPa) within hundreds of microseconds. As the pressure wave continues to propagate upward, its pressure value drops instantaneously and shows a trend of oscillation attenuation, as a result of the reflection in the wellbore. When the pressure value drops to the same level as the formation pressure, it tends to be stable gradually. This pressure-time curve is basically consistent with the variation law of pressure with time in the actual downhole perforation process. The red dashed line tends to rise first and then decrease, which the rising stage is very steep, and it drops rapidly after reaching the peak pressure (172.5 MPa), and when it falls near formation pressure, it tends to be gentle. The fluctuation range of the two curves is similar, and the curve simulated by perforation software are more regular, both of them can basically reflect the law of perforation pressure changing with time. The difference between the peak pressure values obtained by the two methods is within a reasonable range, which indicates that the modeling, meshing, and numerical simulation calculation are accurate, effective and reasonable. Therefore, a large number of numerical simulation based on the above simulation process can be carried out by changing different model parameters.

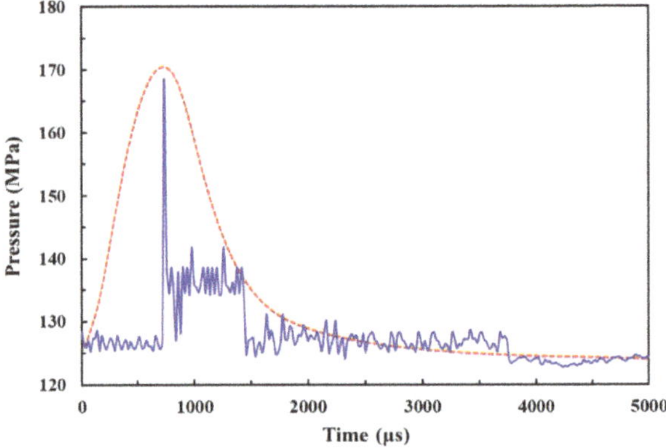

Figure 7. Wellbore bottom pressure-time curves.

3.2.2. Propagation of Perforating Shock Loads

At present, there is a lack of research on the propagation law of perforation shock loads in the wellbore. However, there are many studies on underwater explosion shock wave propagation, and the theory of underwater explosion is mature. The explosive overpressure was studied in the book "Explosives in Water", and the physical effects, basic laws and experimental methods of underwater explosion are reported. The theory of explosive-induced shock wave proposed is still widely used at present [36]. When the explosive explodes underwater, the detonation products of high temperature and high pressure are formed in the volume of the charge, and the pressure is far greater than the static pressure of the surrounding medium, resulting in the shock wave and bubble pulsation in water. With the propagation of the shock wave in water, the pressure and velocity of wave front decrease rapidly, and the waveform widens continuously [37]. As the low compressibility and high density of water, which can be regarded as an incompressible medium under overpressure. The propagation of shock wave and reflection wave in water can be approximately regarded as conforming to the law of acoustic theory, the attenuation of its propagation obeys exponential attenuation, which is verified by the data of underwater explosion experiment, as shown in Figure 8 [38]. Similarly, the detonation products formed by the explosion of the perforating charge rapidly expand in the perforation fluid in the form of gas. Since its initial pressure is far greater than the static pressure of the surrounding medium, the shock wave is formed in the wellbore, and the attenuation model of which in perforation fluid basically accords with the exponential attenuation mode.

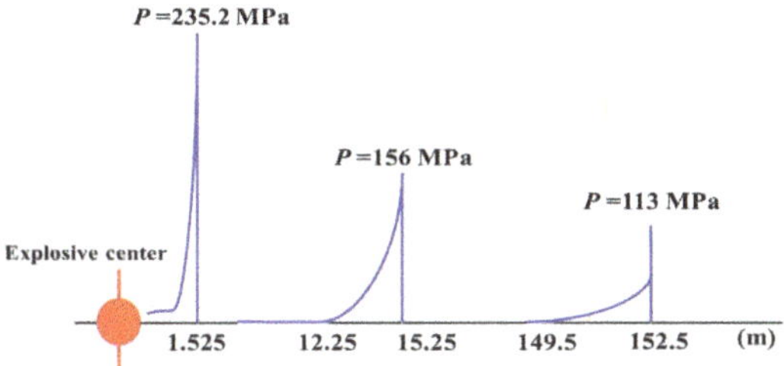

Figure 8. Data regularity curves of shock wave disseminating experiment tested in water.

According to the method of undetermined coefficients, the attenuation formula of the perforating shock wave in the wellbore can be set as Equation (12):

$$P_s = a \times P_t \times e^{\beta R} \tag{12}$$

where P_s is the perforating peak pressure after attenuation; P_t is the perforating peak pressure at the bottom of tubing interval; β is the attenuation index; a is the unknown coefficient; R is the distance from the position to the bottom of the tubing.

In order to solve the undetermined coefficients of the formula, the perforating pressure data at different positions in the wellbore can be extracted, as shown in Figure 9. These pressure curves vary similarly with different positions of the wellbore, which show that the perforation pressure at where increases first to the peak when the shock wave arrives, and as the shock wave continues to propagate upward, its pressure value drops instantaneously and shows a trend of oscillation attenuation until reaches the approximate stable state. The previous study shows that the effect of perforating shock wave on the perforated string is mainly considered the influence of peak overpressure. Figure 9 shows that the peak pressure attenuates with the increase of the upward distance along the wellbore from the initial position, which is the bottom of the tubing.

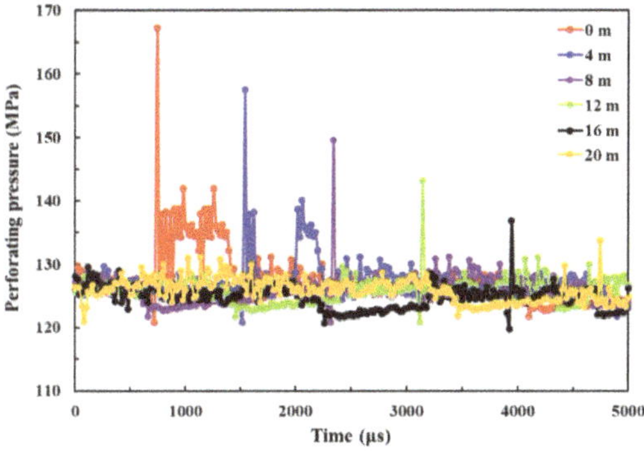

Figure 9. Pressure-time curves at different positions of the wellbore.

Based on the above numerical simulation data, the undetermined coefficients of the formula can be solved by using least squares fitting, as shown in Equation (13):

$$P_s = \frac{249}{252} \times P_t \times e^{-\frac{11}{1000}R} \tag{13}$$

In order to obtain the perforating shock loads at the bottom of the tubing, a large number of numerical simulations can be carried out by changing the number of perforating bullets, charge per hole, tubing length, rathole length, formation pressure, and wellbore initial pressure. According to the simulation calculation results, a database can be established. The basic expression form of the function containing several unknowns can be established in Equation (14):

$$P_t = f(x_1, x_2, x_3, x_4, x_5, x_6) \tag{14}$$

Based on the principle of the least square method, the modified multivariate nonlinear regression model has been established through MATLAB calculation code, the fitting formulas of the perforating shock loads at the bottom of the tubing interval can be obtained as Equation (15):

$$P_t = \frac{91}{50} \frac{p_i^{\frac{27}{100}} \times p_f^{\frac{17}{100}} \times Ln(N) \times Ln(m)}{L_t^{\frac{7}{50}} L_r^{\frac{3}{25}}} \tag{15}$$

where p_i is the wellbore initial pressure; p_f is the formation pressure; L_t is the parameters of strain rate; L_r is the rathole length; N is the number of perforating bullets; m is the charge per hole.

The final calculation model to predict the perforating shock loads at different positions of the tubing interval can be obtained, as shown in Equation (16):

$$P_s = \frac{1079}{600} \frac{p_i^{\frac{27}{100}} \times p_f^{\frac{17}{100}} \times Ln(N) \times Ln(m)}{L_t^{\frac{7}{50}} L_r^{\frac{3}{25}}} \times e^{-\frac{11}{1000}R} \tag{16}$$

The formula analysis shows that the more the number of perforating bullets are, the larger the charge per hole is, the stronger the perforating shock loads is. The longer the tubing length and rathole length are, the weaker perforating shock loads are, this is because the larger wellbore space provides more room for explosive energy release, the formed shock loads are smaller. The higher the wellbore initial pressure and formation pressure are, the larger the peak perforating pressure is, and with the increase of wellbore initial pressure, the peak pressure increases obviously, the formation pressure increases, and the peak pressure increases relatively smaller, as shown in Figure 10a. This is because the hydrostatic pressure of wellbore fluid provides the initial load for the perforating dynamic load, which the wellbore initial pressure is an important factor affecting the perforating shock loads.

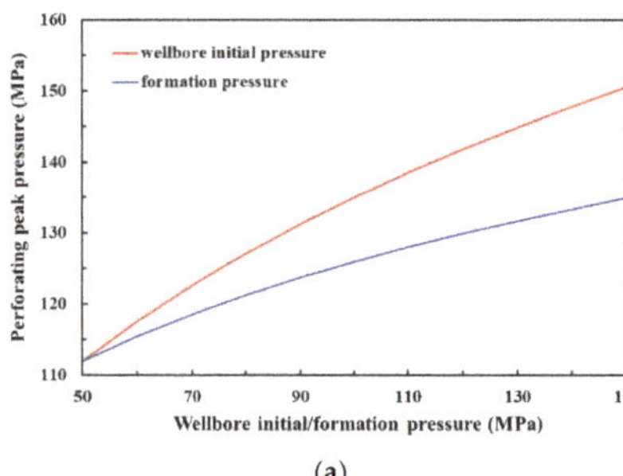

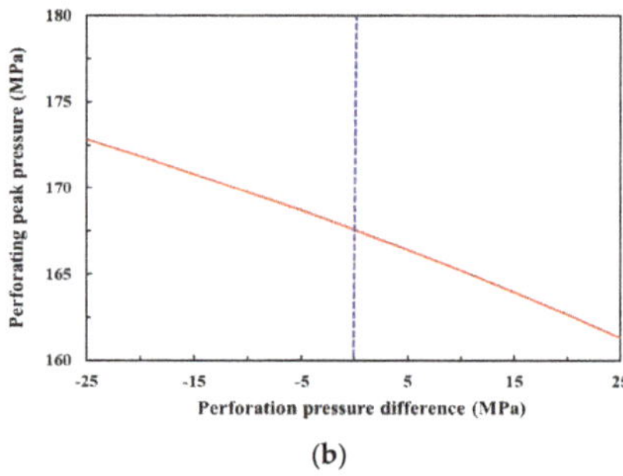

Figure 10. Formula analysis: (**a**) Influence of wellbore initial and formation pressure on perforating pressure; (**b**) Influence of perforation pressure difference on perforating pressure.

Figure 10b shows that as the pressure difference from negative to positive, the peak pressure of perforating gradually decreases, which means that the peak pressure during underbalanced perforation is slightly larger than that of underbalanced or balanced conditions. This is because that the explosive energy will be more hardly spread to the formation during underbalanced, which can increase the shock loads in the wellbore. It proves that most wells are susceptible to perforating shock damage under the condition of underbalance perforation for ultra-deep wells.

3.2.3. Dynamic Response of Perforated String

With the strong shock loads and fluid-structure interaction, perforated pipe string will be in a very complex state of stress and strain for ultra-deep wells, some research was carried out to study the failure mechanism of perforated string with shock loads [39,40]. In order to present the dynamic response process of the string more clearly, the nephograms of displacement change during perforating at different times are grouped together in one, as shown in Figure 11. It can be seen that when the shock waves come to the tubing interval in the wellbore, the shock loads act on the bottom of the string and transfers upward, with the displacement occurring at the bottom part of the string and accumulating gradually. The shock loads are transmitted upward the string and the energy is gradually absorbed by the string, which finally becomes the strain energy of the string. The maximum displacement appears at the bottom of the string, the shock loads cannot propagate further as the top restricted, with no displacement occurring at the top of the string.

Figure 11. Displacement of string during perforating.

The maximum equivalent stress changing over time curve can be drawn by extracting data from a structural block (part) at the perforated string, as shown in Figure 12.

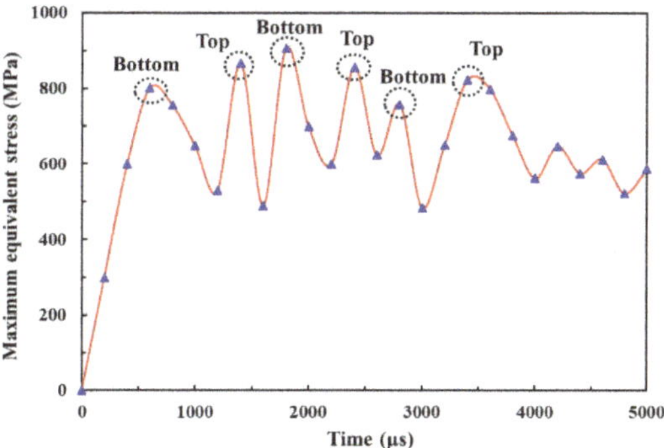

Figure 12. Maximum equivalent stress time-varying curve.

It can be seen the equivalent stress of the string presents a periodic change, the bottom of the string responds firstly after the detonation of the perforating shaped charge, with the first peak point (802 MPa). Then the equivalent stress reaches the extreme value at the top of the string where is the fixed end, with the second peak point (868 MPa). Subsequently, the shock waves reflect downwards and the maximum value (908 MPa) appears at the bottom, which is one cycle. The shock waves reflect at the bottom of the string and will enter the next cycle. It can be concluded that the maximum equivalent stress appears at the bottom and top of the string, which the bottom is the initial output unit of perforating shock loads, and the top is restricted with the loads reflected and enhanced. From the aspect of the changing trend of equivalent stress, it can be known that the maximum equivalent stress still shows periodic variation, which fluctuates back and forth between the bottom and the top of the string, with the maximum value repeatedly appears at the bottom and top of the string, which can be assessed as vulnerable parts. In a well sealing section, the bottom of the tubing can be used as the initial position of the perforating shock loads acting on the string, the top of the tubing is the position of the packer.

A structural block (part) at the bottom of the string is built to obtain the data of dynamic response of the perforated string. Table 2 shows the acceleration peaks and maximum displacements of the perforated string in the axial, radial X and radial Y directions at the bottom of the perforated string respectively. The greatest dynamic response appears in the axial direction, which the acceleration peak and maximum displacement are both much greater than the other two directions. These in the radial X direction are slightly greater than that in the radial Y direction, while there are little differences between radial X and radial Y directions. Assuming the shock loads (Figure 7) act on the string in the axial direction. The axial displacement of perforated string can be calculated by Equation (14), the result shows that the maximum displacements are ±6.8 cm, much greater than simulation results. The reason is that the string material has the ability to absorb the shock loads of perforation, and the theoretical calculations often do not take this into account.

Table 2. Acceleration peaks of perforated string in axial, radial X and radial Y.

Direction	Peak (+)/10,000g	Peak (−)/10,000g	Displacement (+)/cm	Displacement (−)/cm
Axial	6.32	−6.17	4.54	−3.92
Radial X	2.56	−2.28	0.24	−0.18
Radial Y	2.43	−2.19	0.16	−0.11

Through the above analysis, with such the strong cyclical axial shock loads, the perforated string will show buckling instability or even fracture in the macroscopic, which usually oscillates due to the changing radial load, resulting in a significant shear load on the string. Therefore, it is necessary to install the axial shock absorbers below the packer to reduce the impact vibration on it and the upper instruments.

4. Optimization Measures

Common shock absorbers are mostly connected by spring or rubber elements in series, or the combination of the two elements for shock absorption. The shock absorbers based on the principle of rubber cylinder is equipped with two rubber cylinders, and the elastic function of rubber is used to achieve the purpose of shock absorption. As the outer diameter of the rubber drum is much smaller than the inner diameter of the oil casing, which the shock absorption effect is not good. For the ultra-deep well perforation, the temperature and pressure at the bottom of the wellbore are relatively high, which make the rubber components in the shock absorber more easily for serious damage, and the maintenance is troublesome with no reused. In order to ensure the shock absorption effect in such an environment, the shock absorber based on spring shock absorption principle is adopted, which can close to the inner wall of the casing and greatly reduce the axial vibration caused by perforation shock loads.

4.1. Design of Shock Absorption

In order to achieve the best shock absorption effect, the installation position of shock absorbers can be optimized by numerical simulation, which can be simplified as a mechanical spring element adding the axial shock to the numerical model. The distance between the shock absorber and the perforating gun is R_1 and the ratio of the distance between the shock absorber and the perforating gun to the distance of the packer is R_1/R, as shown in Figure 13. The position proportional parameters are shown in Table 3.

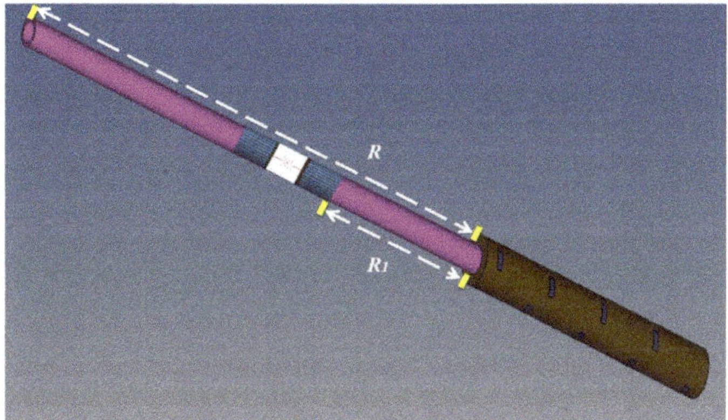

Figure 13. Model for numerical simulation with shock absorber.

Table 3. Installation position of shock absorber.

R/m	R_1/m	R_1/R
20	0	0
20	2	0.1
20	4	0.2
20	6	0.3
20	8	0.4
20	10	0.5
20	12	0.6
20	14	0.7
20	16	0.8
20	18	0.9
20	20	1

According to the method of modeling, meshing and numerical simulation in Section 3.1, several groups of numerical simulation calculations are carried out. Based on the simulation results, the curves of peak pressure on packer during perforating with different numbers of shock absorbers under different installation positions can be obtained, as shown in Figure 14.

The red dashed line represents the peak pressure (133.79 MPa) on the packer without the installation of shock absorbers. Three solid lines of different colors represent the peak pressure of the packer with different numbers of shock absorbers. The colorful area represents the reduction of the peak pressure on the packer, which decreases significantly with the installation of one shock absorber, the effect of shock absorption is obvious. When the numbers of shock absorbers are two or three, the peak pressure on the packer continues to decrease, but the reduction is smaller. When the installation positions of shock absorbers are in the colorful area ($R_1/R = 0.4$–0.6), the peak pressure on the packer decreases the most and the shock absorption effect is the best. It shows that the best shock absorption effect can be achieved by optimizing the installation positions of shock absorbers, which is the middle of the perforated string connecting the packer and perforating gun. In order to meet the needs of the practical application, the formula for reducing the peak pressure on the packer can be fitted with condensing different installation positions and different number of shock absorbers, as shown in Equation (17):

$$\begin{cases} \Delta P_1 = 14.76x^3 - 87.88x^2 + 79.27x + 39.86 \\ \Delta P_2 = 12.82x^3 - 88.58x^2 + 81.32x + 33.97 \\ \Delta P_3 = 3.89x^3 - 80.30x^2 + 85.33x + 23.98 \end{cases} \tag{17}$$

where ΔP_1, ΔP_2, ΔP_3 are the reduction values of perforating peak pressure on the packer with one, two, three shock absorbers, respectively.

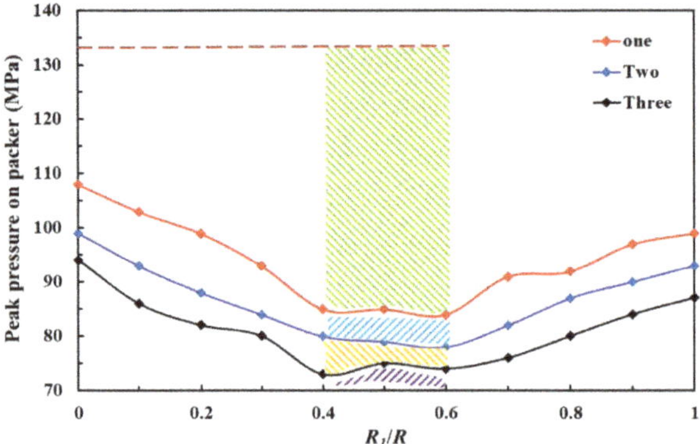

Figure 14. Perforating pressure on packer with different shock absorber installation positions.

With the installation of shock absorbers, the peak pressure on the packer is still large, which may exceed the pressure range of the general packer and pose a serious threat to the safety of the packer. Therefore, it is necessary to analyze the safety of the packer during perforating for ultra-deep wells.

4.2. Safety Analysis of Packer

The dynamic response process of the packer under perforating shock loads is a very complex process. In this study, only the downhole sealing capacity of packer is studied, which mainly bears the function of maintaining pressure. By regarding the wellbore fluid as the medium of pressure transmission, without considering the fluid-solid interaction between the fluid and the solid structure of packer, according to the technical index of packer products, the safety of the packer can be studied. In addition to its own bearing capacity of the packer load level, the upper-end surface of the packer also needs to withstand the pressure of the fluid column in the well, and the lower end surface needs to withstand the perforating shock loads, as shown in the left of Figure 15. As the packer is placed in the fluid medium, there are two interfaces, and when the shock loads propagate in the wellbore fluid, it will reflect and transmit when encountering the packer, as shown in the right of Figure 15.

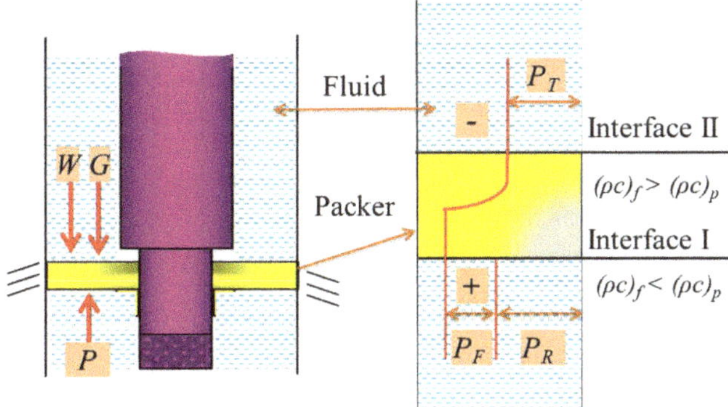

Figure 15. Perforating pressure on packer with different installation positions of shock absorbers.

The carrying capacity of the packer can be expressed as Equation (18):

$$\Delta P = P - \frac{W}{S} - G \tag{18}$$

where W is the bearing capacity of the packer; S is the cross section area of the packer; G is the liquid column gravity on the packer.

As the reflection of the packer will increase the pressure, the perforating pressure on the packer is the difference between the overpressure of the pressure wave and the pressure of transmitted wave [41], which can be expressed in Equation (19):

$$P = P_s + P_F - P_T = 2P_s \times \frac{(\rho c)_p \left[(\rho c)_p - (\rho c)_f \right]}{\left[(\rho c)_f + (\rho c)_p \right]^2} \tag{19}$$

where P_F is the reflected pressure by the packer; P_T is the transmission pressure by the packer; $(\rho c)_f$ is the impact resistance parameters of water medium at normal temperature and pressure; $(\rho c)_p$ is the impact resistance parameters of rubber medium at normal temperature and pressure.

The incident pressure can be calculated by Equation (16), and the final peak pressure on the packer after during perforating can be obtained by Equation (20):

$$P_p = \frac{1079}{300} \frac{p_i^{\frac{27}{100}} \times p_f^{\frac{17}{100}} \times Ln(N) \times Ln(m)}{L_t^{\frac{7}{50}} L_r^{\frac{3}{25}}} \times e^{-\frac{11}{1000}R} \times \frac{(\rho c)_p \left[(\rho c)_p - (\rho c)_f \right]}{\left[(\rho c)_f + (\rho c)_p \right]^2} \tag{20}$$

From the above analysis, combining Equation (17) with Equation (20), the pressure difference between the upper and lower end of the packer can be calculated by Equation (21):

$$\Delta P = \frac{1079}{300} \frac{p_i^{\frac{27}{100}} \times p_f^{\frac{17}{100}} \times Ln(N) \times Ln(m)}{L_t^{\frac{7}{50}} L_r^{\frac{3}{25}}} \times e^{-\frac{11}{1000}R} \times \frac{(\rho c)_p \left[(\rho c)_p - (\rho c)_f \right]}{\left[(\rho c)_f + (\rho c)_p \right]^2} - \frac{W}{S} - G \tag{21}$$

5. Case Study

The field case is an ultra-deep well located in the western part of China. The well depth is 8000 m; perforation interval is 7965–7980 m, the operation parameters during perforating are shown in Table 4. The type of the packer is static type packer, which has strong pressure resistance and the rated working pressure is 70 MPa. Based on these actual parameters of the field example, the method presented in this paper can be used to analyze and optimize the perforation safety.

Table 4. Perforation operation parameters.

Casing Inner Diameter/m	Tubing Outer Diameter/m	Tubing Length/m	Wellbore Fluid Density/(Kg/m^3)	Formation Pressure/MPa	Wellbore Initial Pressure/MPa	Perforating Bullets	Single Charge/g	Rathole Length/m	Wellbore Fluid Height/m
0.22	0.073	45	1790	131	120	276	53	10	7060

According to the strength check theory of the pipe string, the maximum allowable peak pressure can be calculated by Equation (22). The internal pressure strength of the tubing is 79 MPa, the safety coefficient of internal pressure strength is 1.25, the minimum external pressure of perforated string is provided by wellbore fluid, which is calculated as water:

$$P_{\max} = \frac{P_{pi}}{K_{rpi}} + P_o \tag{22}$$

where $P_{\max}$ is the permissible maximum peak pressure on the tubing; P_{pi} is the internal pressure strength of the tubing; K_{rpi} is the safety coefficient of internal pressure strength; E_p is the plastic hardening modulus; P_0 is the local pressure outside the tubing.

The permissible maximum peak value of perforation calculated by Equation (22) is 132.83 MPa. Through the analysis of Section 3.2, considering the bottom of the tubing and the packer as the object of the safety analysis of the perforated string system. The perforating peak pressure at the bottom of the tubing interval calculated by Equation (15) is 150.84 MPa, which exceeds the maximum peak pressure calculated above. The result shows that the tubing will be damaged due to excessive negative pressure difference. The peak pressure propagating to the top of the tubing (packer) after attenuation calculated by Equation (16) is 90.86 MPa, the pressure difference between the upper and lower end of the packer calculated by Equation (21) is 117.17 MPa, which exceeds the pressure range of the packer (70 MPa) and poses a serious threat to the safety of the packer.

Based on the optimization method of shock absorption proposed in this paper, the design of shock absorption is carried out for the case, as shown in Table 5. It shows that when the shock absorbers are installed, the peak pressure at the bottom of the tubing is reduced to the allowable peak range, the safety of perforated string is ensured. When three shock absorbers are installed at 22.5 m from the top of the perforating gun, the shock absorption effect is the best, the value is reduced to the lowest (117.51 MPa). However, when the pressure difference between the upper and lower of the packer is reduced to the lowest (84.19 MPa), the value still exceeds the pressure-bearing capacity of the packer. Therefore, the further optimization is needed.

Table 5. Shock absorption with different number of shock absorptions.

Shock Absorption	Tubing Bottom Pressure/MPa			Packer Pressure Difference/MPa		
R_0/m	One	Two	Three	One	Two	Three
18	125.21	120.97	118.02	91.88	87.64	84.7
22.5	124.35	120.44	117.51	91.02	87.12	84.19
27	124.32	120.69	117.73	90.99	87.36	84.41

The method of optimizing the set distance of the packer is used to ensure the safety of the packer, which the safe distance of packer can be calculated by Equation (26), the final optimization design is shown in Figure 16. Three solid lines of different colors represent pressure difference on the packer with different numbers of shock absorbers. The black horizontal dashed line represents the maximum pressure-bearing capacity of the packer (70 MPa). The black vertical dashed line is an auxiliary line. The colorful area represents the packer is safe when the parameters are within a reasonable range, which the set distance of the packer is 69.46–85 m with two or three shock absorbers installing from the distance 35–42.5 m to the perforating gun.

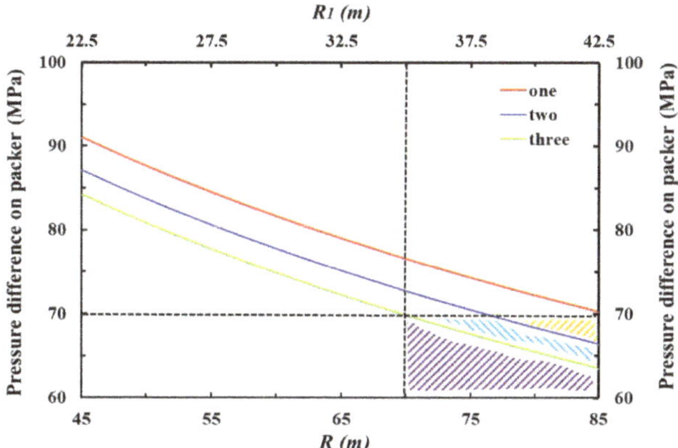

Figure 16. Field case optimization design.

Through the above analysis, the following optimization scheme was adopted in the field case: the packer is set 75 m away from the perforating gun; two shock absorbers are mounted in series in the middle of the tubing, which is 37.5 m away from the perforating gun. On this basis, the perforation test operation proceeded smoothly, and there was no safety problem of perforation string system and packer, which have a good effect on field application and improve the safety of perforation. In addition to the above optimization measures, it is suggested that better shock resistant materials should be used for pipe string, packer, and other instruments, with thicker wall thickness.

6. Conclusions

By combining theory with numerical simulation, this study proposes a new method for the study of perforating shock loads and the effects on the perforated string and packer for ultra-deep wells, the related optimization measures are put forward, which can provide important guidance for the design of field perforation operations and to improve security. Based on the analysis, some conclusions are reached:

(1) Through mechanical analysis, dynamic models in the axial, radial and circumferential directions have been established preliminarily, by which the displacement of perforated strings under axial shock loads can be calculated.

(2) The propagation attenuation law of shock loads in the wellbore is obtained, a multi-factors prediction model of which is presented, which shows that the wellbore initial pressure provides the basis for the perforating dynamic pressure, and the shock damage is more obvious with negative perforating pressure.

(3) It is found that the vulnerable parts of the perforated string system are the bottom of the tubing and the position of the packer, and the axial dynamic response of which is the largest with shock loads.

(4) A shock absorption design based on optimizing the installation position and number of shock absorbers is proposed, and the pressure difference on the packer can be calculated.

(5) The case study shows that the optimization methods proposed in this paper are practical, as the shock damage can be greatly reduced by combining shock absorption with a safe distance of the downhole packer.

Author Contributions: All the authors conceived and designed the study. Formal analysis, Q.D. and H.Z.; Software, Q.D. and H.Z.; Writing—original draft, Q.D. and H.Z.; Writing—review & editing, J.L., X.H. and H.W.

Acknowledgments: The authors gratefully acknowledge the Natural Science Foundation of China (Grant No. U1762211, Grant No. 51734010, Grant No. 51574262, Grant No.51774304, Grant No. 51774063), National Oil and

Gas Major Project (Grant No. 2017ZX05009), the Foundation for Innovative Research Groups of the National Natural Science Foundation of China (Grant No. 51521063), the State Key Laboratory of Petroleum Resources and Engineering.

Conflicts of Interest: The authors declare that there is no conflict of interests regarding the publication of this paper.

Nomenclature

ρ	Density of perforated string
A	Cross sectional area of perforated string
E	Elasticity modulus of perforated string
ξ	Damping coefficient of perforated string in the fluid
F	Force acting on perforated string
P	Perforating pressure
δ	Unit impulse function
I	Inertia moment of perforated string
G	Shear modulus of perforated string
I_p	Polar moment of inertia of a cross section of perforated string
k	Constant
ω	Positive number
$i = 1, 2, 3, 4 \ldots \ldots$	Positive integer
f_i	Natural frequencies
τ	Time integral variable
ω_i	Intrinsic angular frequency
$y(x, t)$	Radial displacement of perforated string
$u(x, t)$	Axial displacement of perforated string
$\varphi(x, t)$	Angular displacement of perforated string
$T(t)$	Function of time
$U(x)$	Longitudinal vibration amplitude of the section from the origin of pipe string
σ	Yield stress
σ_0	Initial yield stress
ε	Strain rate
ε_P^{eff}	Effective plastic strain
Q_1, Q_2	Parameters of strain rate
V_1	Relative volume
E_1	Initial internal energy of unit explosive volume
C_1, C_2, w, R_1, R_2	Physical parameters of explosive
P_s	Perforating peak pressure after attenuation
P_t	Perforating peak pressure at the bottom of tubing interval
β	Attenuation index
R	Distance from the position to the bottom of the tubing
a	Unknown coefficient
p_i	Wellbore initial pressure
p_f	Formation pressure
L_t	Tubing length
L_r	Rathole length
N	Number of perforating bullets
m	Charge per hole
$\Delta P_1, \Delta P_2, \Delta P_3$	Perforating peak pressure reduction on the packer with one, two, three shock absorbers
W	Bearing capacity of the packer
S	Cross section area of the packer
G	Liquid column gravity on the packer
P_F	Reflected pressure by the packer
P_T	Transmission pressure by the packer
$(\rho c)_f$	Impact resistance parameters of water medium at normal temperature and pressure

$(\rho c)_p$	Impact resistance parameters of rubber medium at normal temperature and pressure
P_{max}	Permissible maximum peak pressure on the tubing
P_{pi}	Internal pressure strength of the tubing
K_{rpi}	Safety coefficient of internal pressure strength
P_o	Local pressure outside tubing

Appendix A

Based on the model of perforated string with the boundary and initial conditions in Section 2, the longitudinal free vibration equation of the perforated string can be established in Equation (A1):

$$\frac{\partial^2 u(x,t)}{\partial t^2} = \frac{E}{\rho}\frac{\partial^2 u(x,t)}{\partial x^2} \tag{A1}$$

Assuming that all the points on the perforated string move synchronously, the method of separating variables can be used to assume in Equation (A2);

$$u(x,t) = U(x)T(t) \tag{A2}$$

where $T(t)$ is the function of time; $U(x)$ is the longitudinal vibration amplitude of the section at the point of x from the origin of the pipe string. Equtaion (A2) can be changed into Equation (A3).

$$U(x)'' + \frac{k\rho}{E}U(x) = 0 \tag{A3}$$

where k is a constant. If $k\rho/E > 0$, Equation (A3) has a non-zero solution. The natural frequencies and principal modes of perforated pipe string can be obtained by solving a linear differential equation with constant coefficients. Making $k = \omega^2$, the natural frequency and the main modes of the perforated pipe string can be obtained in Equation (A4) and Equation (A5):

$$f_i = \frac{\omega_i}{2\pi} = \frac{2i-1}{4l}\sqrt{\frac{E}{\rho}} \tag{A4}$$

$$U_i(x) = \sqrt{\frac{2}{\rho Al}} \cdot \sin\left(\frac{2i-1}{2l} \cdot \pi x\right) \tag{A5}$$

where ω is the positive number; $i = 1,2,3,4\ldots\ldots$ are positive integers; f_i is the natural frequencies, Hz; ω_i is the intrinsic angular frequency, rad/s.

For solving the axial mechanical model, assuming the shock loads of perforation act on the string in the axial direction. The generalized force can be expressed in Equation (A6):

$$g_i(t) = -\sqrt{\frac{2}{\rho Al}} \cdot F(t) \cdot \sin\left(\frac{2i-1}{2} \cdot \pi\right) + \frac{2g}{(2i-1)\pi}\sqrt{2\rho Al} \cdot \left(1 - \cos\frac{2i-1}{2} \cdot \pi\right) \tag{A6}$$

As the coordinate transformation does not affect the initial condition, which is still zero. The infinite series of sinusoidal vibration modes can be used to express the displacement response of perforated string under axial shock loads, as shown in Equation (A7);

$$u(x,t) = \sum_{i=1}^{\infty} \frac{1}{2\pi f_i} \cdot U_i(x) \cdot \int_0^t g_i(\tau)\sin 2\pi f_i(t-\tau)d\tau \tag{A7}$$

where τ is the time integral variable.

References

1. Zhen, X.; Moan, T.; Gao, Z.; Huang, Y. Risk Assessment and Reduction for an Innovative Subsurface Well Completion System. *Energies* **2018**, *11*, 1306. [CrossRef]
2. Schatz, J.F.; Folse, K.C.; Fripp, M.; Dupont, R. High-Speed Pressure and Accelerometer Measurements Characterize Dynamic Behavior during Perforating Events in Deepwater Gulf of Mexico. In Proceedings of the SPE Annual Technical Conference and Exhibition, Houston, TX, USA, 26–29 September 2004. SPE-90042-MS. [CrossRef]

3. Xu, C. Numerical Analysis and Model Establishment of Dynamic Response of Perforating String. Master's Thesis, Xi'an Shiyou University, Xi'an, China, 2017.

4. Gilliat, J.; Bale, D.; Satti, R.P.; Li, C.; Howard, J.J. The Importance of Pre-Job Shock Modeling as a Risk Mitigation Tool in TCP Operations. In Proceedings of the SPE Deepwater Drilling and Completions Conference, Galveston, TX, USA, 10–11 September 2014. SPE-170260-MS. [CrossRef]

5. Baumann, C.E.; Dutertre, A.; Martin, A.; Williams, H. Risk Evaluation Technique for Tubing-Conveyed Perforating. In Proceedings of the SPE Europec/EAGE Annual Conference, Copenhagen, Denmark, 4–7 June 2012. [CrossRef]

6. Lu, D.T.; Xu, G.M.; Kong, X.Y. Calculating Transient Pressure of Wellbore Taking under Perforation Condition with Parallel Computation Method. *Univ. Sci. Technol. China* **1999**, *29*, 638–643.

7. Yin, H.D.; Zhang, J.J.; Li, S.Y. Mechanics Analysis of Perforating Combined Well Testing String and Protection Technology of Downhole Instrument. *Oil. Drilling. Prod. Techno.* **2003**, *25*, 61–63.

8. Yu, Z.D.; Li, Y. Finite Clearance Element Analysis of Perforation Tube String for Testing for Oil. *Chinese J. Appl. Mech.* **2003**, *20*, 73–77.

9. Dou, Y.H.; Li, M.F.; Zhang, F.X.; Yang, X.T. An Analysis of the Effect of the Well Bore Structure on the Strength Safety of the Tubing String at the Perforation Section. *China Pet. Mach.* **2012**, *40*, 27–29.

10. Xu, F.; Li, M.F.; Dou, Y.H.; Zhang, F.X.; Yang, X.T. The Analysis of the Influence of Perforating Parameters on the Strength Security of Perforation String. 12. *Appl. Mech. Mater.* **2013**, *268*, 514–517. [CrossRef]

11. Yang, X.T.; Zhang, F.X.; Li, M.F.; Dou, Y.H. Analysis of strength safety of perforated string considering detonation parameters. *Adv. Mater. Res.* **2013**, *634*, 3573–3576. [CrossRef]

12. Zhang, W.B.; Xie, S.; Lu, Q. Influencing Analysis of Perforation Impact Load on Combined String of Perforation-acidification-test. *Well Test.* **2016**, *25*, 8–11.

13. Zhou, H.F.; Ma, F.; Chen, H.B.; Xu, Y.X.; Xi, LX.; W, S.S. Comprehensive dynamic load test for the column in perforation section. *Well Logging Technol.* **2014**, *38*, 247–250.

14. Lu, X.; Wang, S.S.; Ma, F.; Zhou, H.F. Dynamic responses test of perforating string section under explosion impact. *Sci. Technol. Eng.* **2014**, *14*, 53–56.

15. Chen, F.; Chen, H.B.; Tang, K. Influence of Perforating Impact Load on the Operating String and the Countermeasures. *Nat. Gas Ind.* **2010**, *30*, 61–65.

16. Chen, H.B.; Tang, K.; Ren, G.H. String Dynamic Mechanics Analysis on Ultra-Deep Perforation. *Well. Logging Technol.* **2010**, *34*, 487–491.

17. Baumann, C.E.; Pesantes, E.; Guerra, J.; William, A.; Williams, H. Reduction of Perforating Gunshock Loads. *SPE Drill. Complet.* **2012**, *27*, 65–74. [CrossRef]

18. Teng, Y.S. Dynamic Simulation for Perforating and String Safety Evaluation. Master's Thesis, China University of Petroleum (East China), Beijing, China, 2014.

19. Kang, K.; Ma, F.; Zhou, H.F. Study on Dynamic Numerical Simulation of String Damage Rules in Oil-gas Well Perforating Job. *Procedia Eng.* **2014**, *84*, 898–905.

20. Cai, L.Z.; Zhao, H.; Xue, S.F.; Yang, Z.Z. Dynamic Response Analysis of String Structure during Perforating Process. *Oil Field Equip.* **2015**, *44*, 26–30.

21. Yang, Z.Z. Dynamic Response of String under Perforating Pulse Load. Master's Thesis, China University of Petroleum (East China), Beijing, China, 2015.

22. Zhang, W.; Xu, C.; Li, M.F.; Zhang, L.; Wang, G.Q. Transient Response and Strength Analysis of Perforating String. *China Pet. Mach.* **2017**, *45*, 90–94.

23. Li, Z.F.; Zhang, C.Y.; Song, G.M. Research Advances and Debates on Tubular Mechanics in Oil and Gas Wells. *J. Pet. Sci. Eng.* **2017**, *151*, 194–212. [CrossRef]

24. Li, B.Y. Analysis of Strength and Safety of Packer Central Tube under Perforating Impact Loading. Master's Thesis, Xi'an Shiyou University, Xi'an, China, 2018.

25. Li, M.F.; Xu, F.; Dou, Y.H. Measurement of Perforating Column Vibration Parameters and ALE-based Numerical Simulation. *Chin. J. Appl. Mech.* **2019**, *36*, 458–465.

26. Baumann, C.E. An HP-Adaptive Discontinuous Finite Element Method for Computational Fluid Dynamics. Ph.D. Dissertation, the University of Texas at Austin, Austin, TX, USA, 1997.

27. Baumann, C.E.; Oden, J.T. An Adaptive-order Discontinuous Galerkin Method for the Solution of the Euler Equations of Gas Dynamics. *Int. J. Num. Methods Eng.* **2000**, *47*, 61–73. [CrossRef]

28. Burman, J.; Schoener-Scott, M.F.; Le, C.V. Predicting Wellbore Dynamic-Shock Loads Prior to Perforating. In Proceedings of the Brasil Offshore Conference and Exhibition, Macaé, Brazil, 14–17 September 2011. SPE-143787-MS. [CrossRef]

29. Bale, D.; Ji, M.; Satti, R.; Gilliat, J. Advances in Numerical Modeling of Downhole Dynamics for Perforated Well Completions. In Proceedings of the SPE Annual Caspian Technical Conference and Exhibition, Astana, Kazakhstan, 12–14 November 2013. SPE-172308-MS. [CrossRef]

30. Bale, D.; Satti, R.; Ji, M. A Next-generation Shock-capturing, Multi-phase Flow Simulator for Perforating Applications in HPHT Environment. In Proceedings of the SPE Deepwater Drilling & Completions Conference, Galveston, TX, USA, 14–15 September 2016. SPE-180283-MS. [CrossRef]

31. Katayama, M.; Kibe, S. Numerical Study of the Conical Shaped Charge for Space Debris Impact. *Int. J. Impact Eng.* **2001**, *26*, 357–368. [CrossRef]

32. Molinari, J.F. Finite Element Simulation of Shaped Charges. *Finite Elem. Anal. Des.* **2002**, *38*, 921–936. [CrossRef]

33. Liu, J.; Guo, X.Q.; Liu, Z.J.; Liu, X.; Liu, Q.Y. Pressure Field Investigation into Oil & Gas Wellbore during Perforating Shaped Charge Explosion. *J. Pet. Sci. Eng.* **2019**, *172*, 1235–1247.

34. Deng, Q.; Zhang, H.; Li, J.; Wang, H.; Cai, Z.Y.; Tan, T.Y.; Hou, X.J. A Model for Estimating Penetration Length under Different Conditions. In Proceedings of the 52nd US Rock Mechanics/Geomechanics Symposium, Seattle, WA, USA, 17–20 June 2018; ARMA-2018-344. American Rock Mechanics Association. Available online: https://www.onepetro.org/conference-paper/ARMA-2018-344 (accessed on 1 June 2019).

35. Deng, Q.; Zhang, H.; Li, J.; Hou, X.; Wang, H. Analysis of Impact Load on Tubing and Shock Absorption during Perforating. *Open Phys.* **2019**, *17*, 214–221. [CrossRef]

36. Cole, R.H. *Underwater Explosion*; LISA Princeton University Press: Princeton, NJ, USA, 1948.

37. Gao, Y.D.; Wang, S.H.; Wang, X.C.; Liu, Z.Y.; Guo, S.S.; Tang, J.J. Assessment Packer in Perforating String in Safety Distance. *Well Logging Technol.* **2016**, *40*, 382–384.

38. Yu, Y.; R, F.Y. Application of Packer in Fracture Perforating. *J. Xian Univ. Sci. Technol.* **2007**, *27*, 423–425.

39. Qiu, Q.; Cui, L. Reliability Evaluation Based on a Dependent Two-stage Failure Process with Competing Failures. *Appl. Math. Model.* **2018**, *64*, 699–712. [CrossRef]

40. Qiu, Q.; Cui, L. Optimal Mission Abort Policy for Systems Subject to Random Shocks Based on Virtual Age Process. *Reliab. Eng. Syst. Safe* **2019**, *189*, 11–20. [CrossRef]

41. Deng, Q.; Zhang, H.; Li, J.; Hou, X.; Wang, H. Safety Distances of Packers for Deep-water Tubing-conveyed Perforating. In Proceedings of the Offshore Technology Conference, Houston, TX, USA, 5–7 November 2018. OTC-28770-MS. [CrossRef]

Article

An Automatic Classification Method of Well Testing Plot Based on Convolutional Neural Network (CNN)

Hongyang Chu [1,2,†], **Xinwei Liao** [1,2], **Peng Dong** [1,2,†], **Zhiming Chen** [1,2,*], **Xiaoliang Zhao** [1,2] **and Jiandong Zou** [1,2]

1   College of Petroleum Engineering, China University of Petroleum, Beijing 102249, China
2   State Key Laboratory of Petroleum Resources and Engineering, Beijing 102249, China
*   Correspondence: zhimingchn@cup.edu.cn
†   These authors contributed equally to this work.

Received: 16 June 2019; Accepted: 21 July 2019; Published: 24 July 2019

Abstract: The precondition of well testing interpretation is to determine the appropriate well testing model. In numerous attempts in the past, automatic classification and identification of well testing plots have been limited to fully connected neural networks (FCNN). Compared with FCNN, the convolutional neural network (CNN) has a better performance in the domain of image recognition. Utilizing the newly proposed CNN, we develop a new automatic identification approach to evaluate the type of well testing curves. The field data in tight reservoirs such as the Ordos Basin exhibit various well test models. With those models, the corresponding well test curves are chosen as training samples. One-hot encoding, Xavier normal initialization, regularization technique, and Adam algorithm are combined to optimize the established model. The evaluation results show that the CNN has a better result when the ReLU function is used. For the learning rate and dropout rate, the optimized values respectively are 0.005 and 0.4. Meanwhile, when the number of training samples was greater than 2000, the performance of the established CNN tended to be stable. Compared with the FCNN of similar structure, the CNN is more suitable for classification of well testing plots. What is more, the practical application shows that the CNN can successfully classify 21 of the 25 cases.

Keywords: convolutional neural network; well testing; tight reservoirs; pressure derivative; automatic classification

---

1. Introduction

Well testing generally has two major categories: Transient rate analysis and transient pressure analysis. For the transient pressure analysis, its main purpose is to identify the type of target reservoir and further quantitatively determine the reservoir properties. Muskat [1] first proposed a method of estimating the initial reservoir pressure and parameters using a buildup test plot. Due to the fact that compressibility of the formation fluid is difficult to study, this method only can qualitatively analyze the results. Van Everdingen and Hurst [2] used the Laplace integral method to obtain the analytical solution of the transient diffusion equation, which gives the mathematical theoretical basis of well testing. Based on this truth, Horner et al. [3] developed a classic "semi-log" analysis method, which can determine the permeability, skin factor, productivity index, and other parameters. These methods make full use of the mid and late period data in well testing, but a common disadvantage is that the early data of the well testing is ignored.

In order to make reasonable use of the early data in well testing, Ramey et al. [4] first proposed a "plate analysis method" of the log–log type plot. Further, Gringarten et al. [5] extended this method to various well test models such as the dual-porosity model and fractured well model, and a combination of different parameters were used to greatly reduce the difficulty in curve classification and interpretation, which indicated that the well testing interpretation was widely used around the world. Bourdet et al. [6] found that different types of reservoirs had distinct responses in the pressure derivative curve, so the pressure derivative curve was introduced into the "plate analysis method". Compared to the pressure dynamic, the application of the pressure derivative curve makes the classification of reservoir types, and the overall curve fitting, easier. Therefore, the pressure derivative plot is the most critical part of the large-scale application of well testing interpretation methods.

Recently, with the advancement in machine learning technology and the vast datasets in the petroleum industry, the broad prospects of machine learning technology in the petroleum industry have gradually been proven, and it has been applied to different aspects of the petroleum industry [7–13].

Awoleke et al. [12] combined self-organizing maps, the k-means algorithm, the competitive-learning-based network (CLN), and the feed-forward neural network (FFNN) to predict the well water production in Barnett shale. The expected misclassification error was about 10% for CLN and the average prediction error was between 10% and 26% for FFNN, which depended on the quality of the training data set.

Akbilgic et al. [14] used a neural network-based model to predict the steam-to-oil ratio in oil sands reservoirs. Porosity, permeability, oil saturation, reservoir depth, and thickness characterized by well logging and core data were used as data sets for the models.

With deep neural networks (DNNs), Wang et al. [11] used production data from 2919 wells in Bakken shale reservoirs to forecast well productivity. Results show that the predicted oil production of DNNs for both six months and 18 months was acceptable and the average proppant placed per stage was the most important factor in affecting productivity.

In numerous research studies about machine learning in the petroleum industry, Al-Kaabi and Lee [15] firstly used a three-layer FCNN to determine the well test interpretation model. In their work, the pressure derivative and corresponding time were entered into the FCNN with 60 input nodes. Additionally, different well test models were exported, and the accuracy of the prediction was verified by two field examples. This meaningful work demonstrates meaningful guidelines for later work on well test plot identification by neural networks. Following that, a series of researchers [16–18] utilized a more complex network structure and data set to optimize FCNN's recognition result of well testing curves.

Although a large number of scholars have done some meaningful research, due to the previous limitations of CPU performance and mathematical theory basis, there are also several shortcomings as follows: (a) The number of training samples and input nodes in present neural network models are relatively insufficient, which greatly restricts the generalization ability of neural network models. (b) There is no corresponding method to overcome the over-fitting and local minimum problem, which is the phenomenon that exists widely in the fitting process of neural network models. (c) Almost all the current research is limited to the FCNN, and the newly proposed CNN has not been considered in research.

Nowadays, CNN is one of the most popular methods in the field of machine learning. Compared with FCNN, the CNN has a better performance in the domain of image recognition [19–22]. Since the different forms of pressure derivative curves represent various reservoir types, flow regimes, and outer boundary properties, in this paper, an automatic classification method of well testing curves is proposed based on CNN. By summarizing the buildup test data in low permeability reservoirs, the vertically fractured well model, dual-porosity model, and radial composite model were selected as the base model, which were used to generate 2500 theoretical curves of five different types. To overcome the problem of overfitting and local minimum, the regularization technique, Adam optimization algorithm, ReLU activation function, and mini batch technique were used to optimize the established CNN. The model input nodes were 488, which ensured that the information of the curve is completely input. Further, we compared the training performance of CNN and FCNN. The analysis of confusion matrix showed that the *Score* for CNN and FCNN on the validation set were 0.91 and 0.81 respectively, which means that the CNN had a better prediction result than FCNN. Finally, 25 buildup test curves from Ordos Basin were used to verify the generalization ability of the CNN noted above.

2. Background

The Ordos Basin is the second largest sedimentary basin in China and it contains abundant oil and gas reserves. In terms of geology, the Ordos Basin is a large-scale multicycle craton basin with simple tectonics, which is made up of the Yimeng uplift, Weibei uplift, the Western margin thrust belt, the Tianhuan depression, and the Jinxi flexure belt [23,24]. This Basin is a typical low-permeability formation with an average permeability of less than 1 mD. Except for the Chang 6 reservoir with developed horizontal bedding [25,26], the horizontal stress in most areas of the basin is greater than the vertical stress, which means that the fractures generated by hydraulic fracturing are mainly vertical fractures [27–30].

3. Theory

3.1. Concept of CNN

Traditional neural networks (like the FCNNs) use the matrix multiplication to describe the connection between input nodes and output nodes. Wherein, each individual weight of the weights matrix describes the interaction between an input unit and an output unit. For traditional neural networks, when the number of input nodes is quite large, the number of weights will also become very huge, and the training efficiency will drop drastically. To address this issue, the convolution method of reducing the number of weights is needed to reduce training costs. The two main advantages of the convolution method are weight sharing and sparse connections, which effectively improve the situation. The calculation process of convolution method is shown in Figure 1. The filter contains the weights to be optimized, and the forward propagation process of the filter is a process of calculating the output data by using the inner product matrix multiplication between the weights in the filter and the input data. In a CNN, the filter weights used by each convolutional layer (CONV layer) are the same. The sharing of filter weights can make image content unaffected by local feature and reduce the number of weights. Then, the data are convolved and output through the activation function.

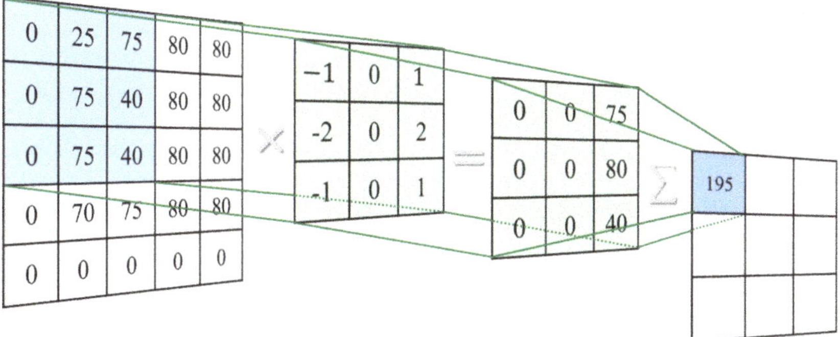

Figure 1. Schematic diagram of convolution in convolutional neural network (CNN) (the elements in the matrix represent the pixel values of the input data and weights).

In addition to the CONV layer, the network often uses a pooling layer, which can adjust the output structure of the layer and reduce the size of the model. With the pooling layer, the calculation speed and the robustness of the model are improved. The pooling layer usually includes a max-pooling layer and an average-pooling layer, as shown in Figure 2, which is used to output the maximum value and the average value in the filter area. So, no weights exist in the filter of pooling layer.

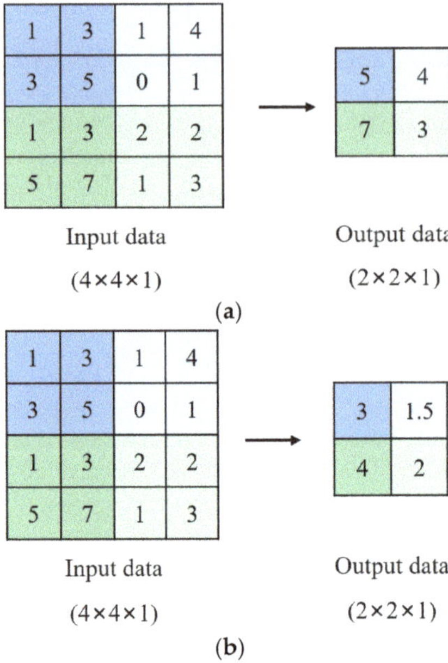

Figure 2. Schematic diagram of pooling layer calculation process (**a**) max-pooling layer (**b**) average-pooling layer (the elements in the matrix are various pixels).

To achieve different test tasks, different layers need to be connected to form a CNN. The AlexNet is a typical CNN proposed by Krizhevsky et al. [31], which has a simple model structure but accurate image recognition rate. The AlexNet fully demonstrates the superior performance of CNN in dealing with complex data. As shown in Figure 3, the structure of AlexNet is 8 layers with weights, including 5 layers of CONV layers and 3 layers of fully connected layers (FC layers). Three max-pooling layers are utilized to adjust the output shape. Additionally, to reduce the dimension of curve probabilistic prediction data, a flatten method is used before the FC layer. Finally, the FC layers are used to achieve the data dimensional reduction and output the final results. In the calculation process of FC layers, the softmax function is usually chosen to calculate the probability of the data after dimension reduction. The picture with the highest probability is the final result of CNN. Equation (1) gives the mathematical expression of the softmax function.

$$softmax(l) = \frac{e^{a_l}}{\sum_{k=1}^{c} e^{a_k}} \tag{1}$$

where a_l is the output value of the lth node of the output layer, c is the total number of sample classes.

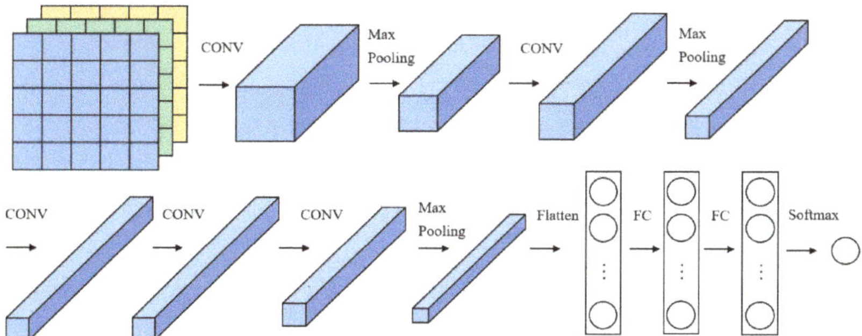

Figure 3. Schematic diagram of the AlexNet structure.

3.2. Model of CNN

3.2.1. Sample Obtaining

The type curve of well testing is the log–log plot, which is based on the analysis of the time, pressure, and its derivative in the log–log coordinates. The reservoir types are determined by different shapes of the curve, and they are very critical to well testing interpretation results. Due to the non-uniqueness in interpretation results, it is difficult to quickly and accurately determine the reservoir type corresponding to a large amount of interpretation data. Automatic identification of well test curve types based on CNN can significantly reduce the workload of identification, and it provides a reliable basis for accurate parameter inversion.

Production wells in unconventional reservoirs represented by the Ordos Basin are generally hydraulically fractured, so the vertically fractured model is one of the commonly used well test interpretation models in unconventional reservoirs. At the same time, due to hydraulic fracturing, the natural fractures in the formation are activated, and the resulting considerable amounts of buildup test data of Ordos basins are characterized by a dual-porosity model. On the other hand, large-scale hydraulic fracturing significantly improves the permeability of the near-well region, which means that the radial composite model is also used as the reservoir model for well test interpretation in unconventional reservoirs. The mathematic expressions of the above well testing models are given in the Appendices B–D. With these mathematic expressions, Figure 4 shows that the typical well test curves for the above models can be roughly divided into the following categories. In the same reservoir conditions, there is no doubt that the radial composite model with mobility ratio >1 and dispersion ratio >1 in the five well test models has the greatest productivity. The reason is that this model assumes that the area around the production well has been adequately stimulated by the hydraulic fracturing, so an inner zone of high permeability is formed around the production well, which contributes to the largest productivity.

In this paper, the training set included 2725 well test curves for five well test models, and 25 field buildup test cases were used to evaluate the generalization ability of CNN. The pressure derivative-time curve data for each training sample was used for classification. There were 545 curves of each well test model type and Table 1 shows the range of corresponding parameters for five well test models.

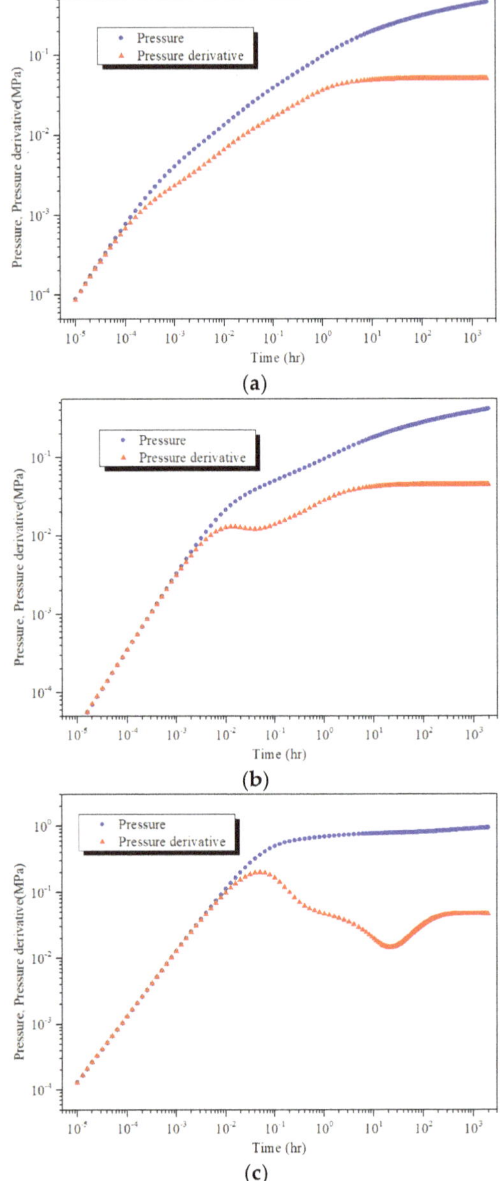

Figure 4. *Cont.*

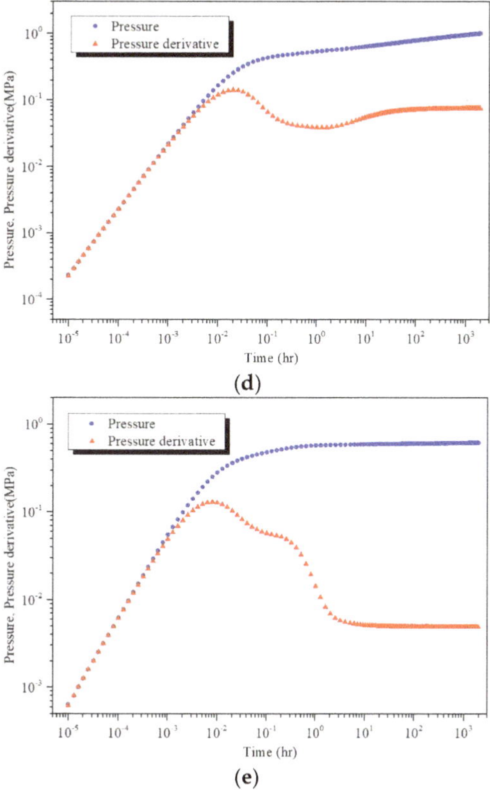

Figure 4. The typical well test curves for models used in this work. (**a**) Infinite-conductivity vertically fractured model without skin factor (Model 1); (**b**) infinite-conductivity vertically fractured model with skin factor (Model 2); (**c**) dual-porosity model with pseudo-steady state (Model 3); (**d**) radial composite model with mobility ratio >1 and dispersion ratio >1 (Model 4); (**e**) radial composite model with mobility ratio <1 and dispersion ratio >1 (Model 5).

Table 1. The range of model parameters of various well test models in this paper.

Model	1	2	3	4	5
Wellbore storage coefficient (m^3/MPa)	0–0.25	0–0.25	0–0.25	0–0.25	0–0.25
Skin factor	0–0.05	0.05–2	0–1	0–1	0–1
Fracture half length (m)	20–80	20–80	/	/	/
Initial pressure (MPa)	15–35	15–35	15–35	15–35	15–35
Permeability (mD)	0.10–50	0.10–50	0.10–50	0.10–50	0.10–50
Thickness (m)	9.14	9.14	9.14	9.14	9.14
Porosity	0.10	0.10	0.10	0.10	0.10
Omega	/	/	0.01–0.60	/	/
lambda	/	/	10^{-6}–10^{-9}	/	/
Mobility ratio	/	/	/	1–20	0–1
Dispersion ratio	/	/	/	1–20	0–1
Composite radius (m)	/	/	/	10–200	10–200

Before training, improving, and evaluating the CNN model for well test plots, it was necessary to divide the training data sets into training set, validation set, and test set. Their quantities respectively accounted for 90.909%, 8.182%, and 0.909% of the total number of samples. The primary role of the validation set was to compare the performance of different neural network models. The test set was used to verify the generalization ability of the model based on the mine data. The validation set and test set were not involved in the training process of the network. The first time they were entered into the network was in the process of network verification. In total, 2500 of the theoretical curves were determined as training sets, and the remaining 225 curves were chosen as the validation set. Additionally, 25 field buildup test cases from the Ordos Basin were used as a test set. Figure 5 is a schematic diagram of training data partition.

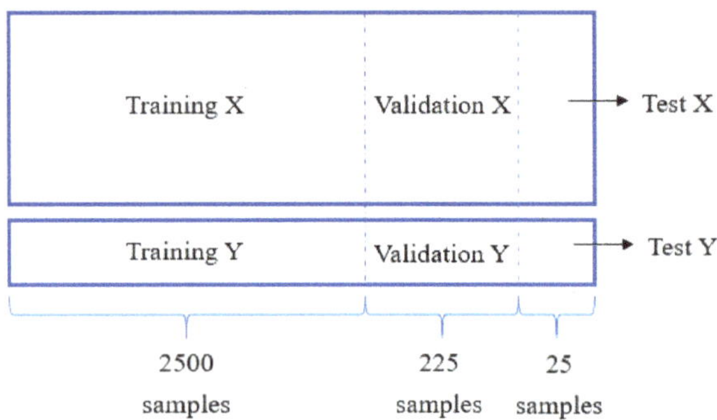

Figure 5. Data partition including training set, validation set, and test set.

3.2.2. Structure of Neural Network Model

The neural network model has strong ability of nonlinear representation, as its basic unit is a neuron. Through the design of different numbers of neurons and different layers, various mapping relations can be characterized.

Model Building of CNN

The CNN constructed in this paper was a five-layer deep network with weights, in which three layers were CONV layers and two layers were FC layers, as shown in the Figure 6. There was also the max-pooling layer and average-pooling layer between various CONV layers, which was used to compress the input data and reduce overfitting problem. Table 2 shows the number of network weights in the different layers and the total number of weights was 76,583. In order to minimize the weights number of CNN, the input layer of the CNN was the data point of pressure derivative time plot, rather than the curve picture. Since the input data point of pressure derivative time curve was one-dimensional data with respect to time, we used the layers containing one-dimensional (1D) filter, including CONV1D, max-pooling1D. Layers containing two-dimensional (2D) filters (such as CONV2D, max-pooling2D, and average-pooling2D) were used to transform 1D data into 2D data needed for convolutional calculations. In the final layer, flatten method and softmax activation function were used to output the result of CNN.

Table 2. The layer shape and weights number of CNN.

Layer	Layer Shape (Output Shape)	Weights Number
Input	(2,244)	0
Conv1D	(38,80)	418
Max-Pooling1D	(38,38)	0
Conv2D	(17,17,64)	1664
Max-Pooling2D	(5,5,64)	0
Conv2D	(2,2,128)	73856
Average-Pooling2D	(1,1,128)	0
Flatten	128	0
FC (Output)	5	645

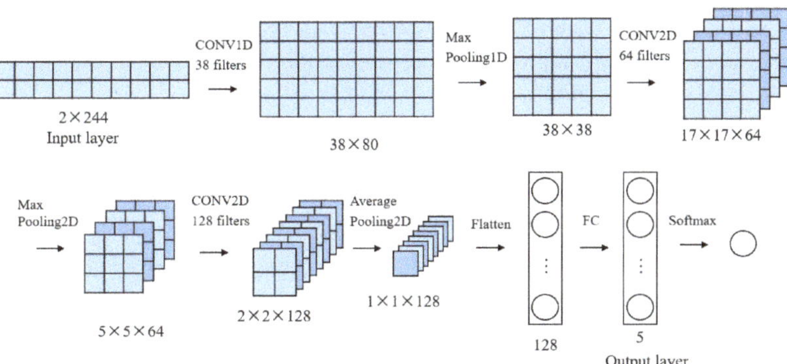

Figure 6. Schematic diagram of CNN structure.

Model Building of FCNN

In contrast, a FCNN was established, which had a similar number of weights as CNN. The input layer, hidden layer, and output layer for FCNN had 488, 106, and 5 neurons respectively. Figure 7 shows that the input layer consisted of 488 nodes that accepted the 244 data points (t, dP). Table 3 demonstrates that the FCNN had a total of 76,575 weights.

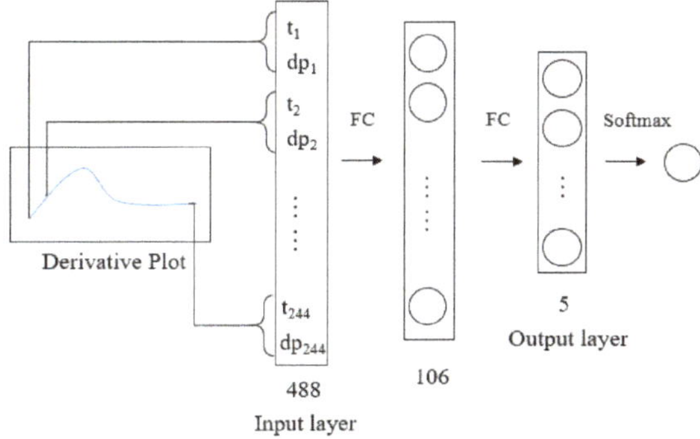

Figure 7. Schematic diagram of fully connected neural networks (FCNN) structure.

Table 3. The layer shape and weights number of FCNN.

Layer	Layer Shape (Output Shape)	Weights Number
Input	488	0
FC	106	75,795
FC (Output)	5	780

Evaluation Results for the CNN and FCNN

During the training process of the FCNN and CNN, the maximum value of the output corresponded to the type of curves being predicted, which was recorded as $\hat{y}$. In order to optimize the weights in models, the cross-entropy function values of the predicted and the theoretical curve type were calculated. As shown in Equation (2), the cross-entropy function is usually recorded as the loss function L. When the loss function value is the smallest, the ratio of the number of accurately predicted training samples to the total number of samples (called accuracy) is the largest, which means that the network model has the highest performance.

$$L(\hat{y}, y) = -\sum_{i=1}^{m} y_i \log \hat{y}_i \tag{2}$$

where y_i the type of the ith training samples, $\hat{y}_i$ is the probability of ith training samples, and m is the number of training samples. In order to obtain a robust and fast CNN, the newly proposed one-hot encoding, Xavier normal initialized model, ReLU activation function, L2 regularization method, Adam optimization algorithm, and mini batch technique were combined to further construct the CNN.

3.2.3. One-Hot Encoding

In the training tasks of machine learning, the variety of sources of training data led to more complex data types. The training data can be roughly divided into type data and numerical data. The training process of the neural network model was performed on the basis of numerical data. Therefore, in the classification task, the type data needed to be converted into numerical data and were further used to train the neural network model. The one-hot encoding method is a commonly used method of encoding type data into numerical data, which encodes the type data into a binary vector with at most one valid value. As shown in Table 4, each column represents each category in the training sample data and the unit containing "1" represents the category to which the sample data belongs.

Table 4. The schematic diagram of one-hot encoding.

	Class1	Class2	Class3	Class4	Class5
Sample1	0	1	0	0	0
Sample2	1	0	0	0	0
Sample3	0	0	1	0	0
			...		
Sample2724	0	0	1	0	0
Sample2725	0	0	0	0	1

3.2.4. Determination of Model Initialization

During the training process of the neural network model, proper initialization of the weights was essential to establish a robust neural network model. The proper initialization of the weights will cause the weights to be distributed over a reasonable range. If the initial weight value is small, the effective information in the backpropagation process will be ignored and the training process of neural network model maybe invalidated. When the initial weight values are large, the weight fluctuations in the backpropagation process will increase, which may lead to the instability or even collapse of the model. The commonly used initialization methods of neural network model include four categories and they are as follows: (1) Randomnormal method; (2) Randomuniform; (3) Xavier normal; (4) Xavier uniform [32]. In this work, we compared the effects of four initialization methods on the training results. After 100 iterations of the model, the Xavier normal initialized model had the highest accuracy in the training set and the validation set (Figure 8).

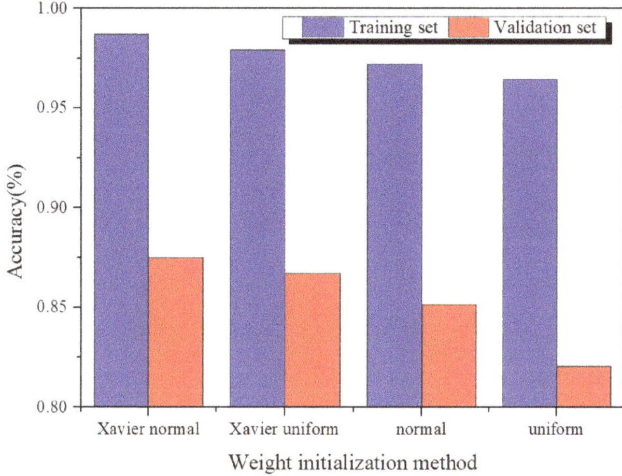

Figure 8. Comparison results of CNN on training set and validation set under different initialization methods.

3.2.5. Selection of Activation Functions

The activation function of the neural network has a significant impact on the prediction effect of the model. When no activation function is used (i.e., $f(x) = x$), the input of each node is a linear function of the output result of the node in upper layer. In this case, regardless of the number of layers in the neural network, the output result only is a linear combination of input results, and the hidden layer does not work. Only when the neural network model uses a nonlinear activation function are the output results no longer a linear combination of input results and it can approximate an arbitrary function. Table 5 shows the five commonly used activation functions (i.e., linear, tanh, sigmoid, ELU, and ReLU). As shown in Figure 9, the comparative results showed that the neural network model had a better effect when the ReLU function was used in the middle layer.

Table 5. The mathematical expression of five commonly activation functions.

Type	Equation
linear	$f(x) = x$
tanh	$f(x) = \dfrac{e^x - e^{-x}}{e^x + e^{-x}}$
sigmoid	$f(x) = \dfrac{1}{1 + e^x}$
ELU	$f(x) = \begin{cases} e^x - 1 & , \quad x < 0 \\ x & , \quad x \geq 0 \end{cases}$
ReLU	$f(x) = max(0, x)$

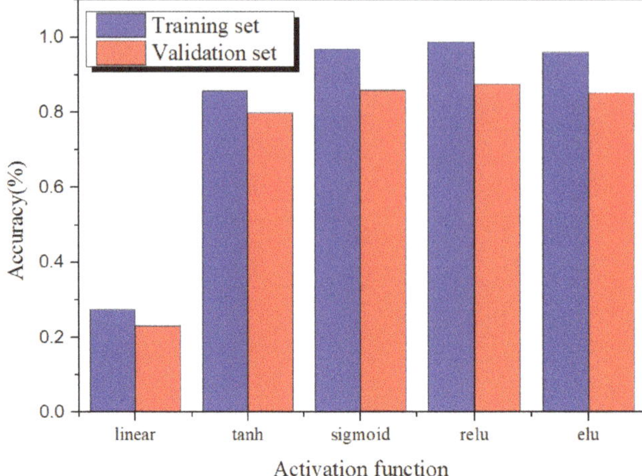

Figure 9. Comparison results of CNN on training set and validation set under different activation functions.

3.2.6. Regularization Technique

The overfitting is a common problem in the training process of neural network models and it greatly reduces the generalization ability of neural network models [33]. The main reasons for the overfitting problem are insufficient training samples and a complex structure of networks. To overcome this problem, the dropout method and L2 regularization method are used to dynamically adjust the network structure, which can effectively avoid the overfitting problem. (1) In the process of forward propagation, the dropout method can make various nodes stop working in a certain probability p (called dropout rate) and the relative importance of each node is balanced. After the introduction of the dropout method, each node of the neural network model contributes more equally to the output results and avoids the situation where a few high-weight nodes fully control the output results. (2) For the L2 regularization method, the sum of the squared value for weight squares is added to the loss function, which can constrain the size of the weight values to reduce complexity of the model. Therefore, Equation (2) is rewritten as Equation (3).

$$L(\hat{y}, y) = -\sum_{i=1}^{m} y_i \log \hat{y}_i + \lambda \sum_{j=1}^{n} w_j^2 \tag{3}$$

where λ is the super-parameter, which is used to control the level of weight decay. n is the amount of weights. In contrast, the results of the classification accuracy of the model with dropout method, L2 regularization method, and without regularization method are compared. It can be seen from Figure 10 that the model had the highest accuracy in the validation set when using the dropout method and the accuracy in validation set was close to the that for the training set.

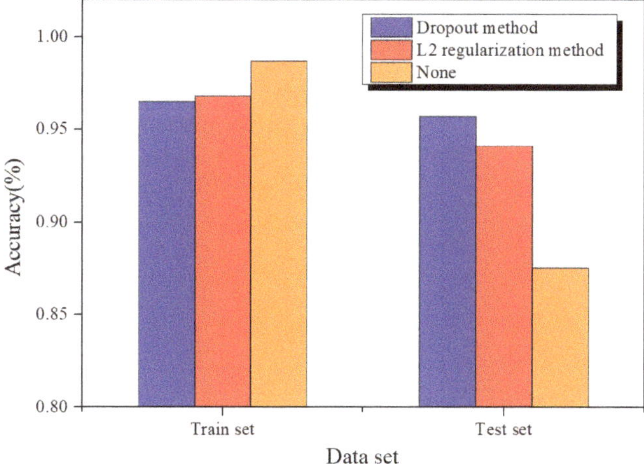

Figure 10. Comparison results of CNN on training set and validation set under different regularization techniques.

3.2.7. Adam Optimization Algorithm

To obtain the minimum value of loss function of the model, the weights in the network model need to be updated at each iteration step. Among various optimization algorithms for weight updating, the Adam optimization algorithm proposed by Kingma and Ba, [34] has the highest performance [35,36]. Compared to the classical gradient descent method, this method can avoid the oscillation of the loss function and the model with Adam optimization algorithm has a higher convergence speed. The Adam optimization algorithm updates the network model weights in the form of Equation (4).

$$w_j^t = w_j^{t-1} - \frac{\eta^t}{\sqrt{v_j^t} + \varepsilon} \omega_j^t \tag{4}$$

where η^t is the learning rate at the tth time step, w_j^t is the network weight of the jth feature of the training sample data under the tth time step, ε is a small constant to avoid a zero denominator.

In Equation (5),

$$\omega_j^t = \frac{\beta_1 \omega_j^{t-1} + (1-\beta_1) g_j^t}{1 - (\beta_1)^t} \tag{5}$$

$$v_j^t = \frac{\beta_2 v_j^{t-1} + (1-\beta_2)(g_j^t)^2}{1 - (\beta_2)^2} \tag{6}$$

where β_1 and β_2 are the exponential decay rates for the moment estimates, g_j^t is the gradient in the jth parameter under the tth time step. In this work, we used the parameters recommended by Kingma and Ba, [34]: $\beta_1 = 0.9$, $\beta_2 = 0.999$, $\omega_j^0 = 0$, $v_j^0 = 0$ and $\varepsilon = 10^{-8}$.

3.2.8. Mini Batch Technique

The premise of machine learning is that it requires a huge sample size. During each iteration of the model, the optimization algorithm needs to fit the model to all training samples at once, so the requirement for CPU will be enormous. In order to reduce the requirements for CPU and improve computational efficiency, the mini batch technique was utilized as it can randomly select a small portion of the training samples in the training set for each iteration process of the model. Meanwhile, the random selection of training samples also made the mini batch technology effectively avoid the neural network model falling into local minimum problems during the training process. For the mini batch technique, the gradient g^t in Equations (5) and (6) is as follows:

$$g^t = \frac{1}{b}\sum_{k=1}^{b} g_k^t \tag{7}$$

$$g_k^t = \frac{1}{s}\sum_{r=1}^{s} \nabla L(w^{t-1}; x_{i_r}; y_{i_r}) \tag{8}$$

$$b = \left[\frac{m}{s}\right] \tag{9}$$

where b represents the number of iterative step of the mini batch method from $t - 1$th time step to tth time step. g_k^t is the gradient of the kth iterative step from t-1th time step to tth time step. w^{t-1} is the weights at the t-1th time step. s is the number of training samples in one mini batch. $i_1, \ldots, i_s$ are a random number between 1 and m. x_{i_r} is the pressure derivative-time curve data of the i_r training sample. y_{i_r} the type of the i_r training sample. m is the total number of training samples.

4. Results and Discussions

4.1. Comparison of Classification Performance for FCNN and CNN

We used the same techniques (including regularization technique, Adam optimization algorithm, activation functions, and initialization methods) to optimize the FCNN model. Table 6 and Figure 11 compare the errors of different models. The error of the test set verified the performance of the well test plot classification in the field buildup test cases and demonstrated the generalization ability of the CNN. For the FCNN, after 100 iterations, the loss function value was 0.44, and the classification accuracy was 91.2%. In the validation set, its accuracy was 89.8%. For the CNN, the loss function and accuracy for training set and validation set respectively were 0.19, 96.6%, and 95.6%. The comparing results of FCNN and CNN showed that the CNN had a higher accuracy when the number of weights of two models were close (76,583 weights and 76,575 weights).

As shown in Figures 12–14, the confusion matrix analysis is a method for judging the classification performance of neural network model, which shows the accuracy result of classification. Mukhanov et al. [37] used the confusion matrix to evaluate the classification result of the waterlogging curve by the support vector machine technology. The confusion matrix separately calculates the number of misclassification classes and the number of correct classification classes in the model. It can be seen from Figures 12 and 13, and Tables 7 and 8 that the FCNN had different classification capabilities for various types of well test curves in the training set and validation set. For the CNN, its classification results of the 5 well test models were 0.98, 0.94, 0.97, 0.95, 0.98 for training set and 0.97, 0.93, 0.96, 0.93, 0.98 for validation set, which were generally better than the FCNN results. Figure 13 and Table 8 also showed that the FCNN forecasting results of class1, class3, and class5 in the validation set were basically correct, but there was a large error for the curves of class2 and class4. For CNN, the stability of forecasting result was high, and the prediction errors of various types of curves were almost the same, indicating the reliability of CNN. Through the confusion matrix, the recall rate (Equation (11)) and precision rate (Equation (10)) of the model could be calculated. The precision rate represents the number of correctly predicted samples in a class (TP) to all actually retrieved items (the sum of TP and FP). The recall rate refers to the TP as a percentage of all items (the sum of TP and FN) that should be retrieved. F_1 value is the harmonic mean of the precision rate and recall rate. The average of the F_1 values for all training samples is determined as *Score* (Equation (13)). Table 8 summarizes the performance of different network models in validation sets. It can be seen that the *Score* of the FCNN model was 0.81 and the value of CNN was 0.91, indicating that the overall performance of CNN was better than FCNN.

$$Precision = \frac{TP}{TP + FP} \tag{10}$$

$$Recall = \frac{TP}{TP + FN} \tag{11}$$

$$F_1 = 2\frac{Precison \times Recall}{Precison + Recall} \tag{12}$$

$$Score = \left[\frac{1}{c}\sum_{k=1}^{c}(F_1)_k\right]^2 \tag{13}$$

where TP is the number of correctly predicted samples in a class. For a certain type of training sample, FN is the difference between the number of successfully predicted training samples and the total number of training samples, FP is the difference between the number of successfully predicted training samples and the predicted number of samples for a certain type, c is the total number of sample classes.

Finally, the classification ability of CNN was verified on 25 field buildup test data, among which 21 samples are successfully classified. Table 9 and Figure 14 show the confusion matrix of the model in the test set, and its *Score* was 0.69. Appendix A shows the data of the 25 field buildup test data.

Table 6. Model prediction accuracy.

	Loss Function	**Accuracy (%)**
CNN train set	0.19	96.6
CNN validation set	/	95.6
FCNN train set	0.44	91.2
FCNN validation set	/	89.8

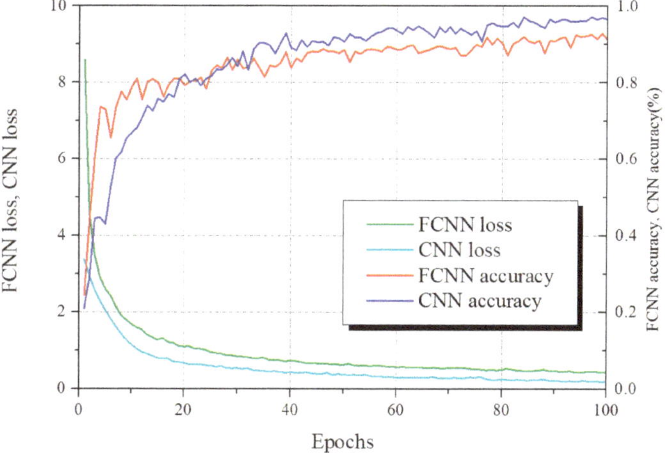

Figure 11. The changes in the accuracy and loss function curve for FCNN and CNN on training set as the number of iterations increases.

CNN	Predict 1	2	3	4	5
Ture 1	495	0	5	0	0
2	4	453	12	25	6
3	10	0	489	0	1
4	0	13	0	483	4
5	0	1	0	4	495

(a)

FCNN	Predict 1	2	3	4	5
Ture 1	461	13	26	0	0
2	9	411	24	41	15
3	26	17	456	0	1
4	0	25	8	463	4
5	0	8	0	3	489

(b)

Figure 12. The confusion matrix of CNN and FCNN on training set. (**a**) CNN confusion matrix; (**b**) FCNN confusion matrix.

CNN	Predict 1	2	3	4	5
Ture 1	43	2	0	0	0
2	0	41	0	4	0
3	0	0	43	0	2
4	0	0	2	43	0
5	0	0	0	0	45

(a)

FCNN	Predict 1	2	3	4	5
Ture 1	39	5	0	1	0
2	1	33	0	11	0
3	0	0	45	0	0
4	0	4	1	40	0
5	0	0	0	0	45

(b)

Figure 13. The confusion matrix of CNN and FCNN on validation set. (**a**) CNN confusion matrix; (**b**) FCNN confusion matrix.

CNN	Predict 1	2	3	4	5
Ture 1	3	1	0	0	0
2	0	5	0	1	0
3	0	0	2	0	0
4	0	1	0	4	0
5	0	0	1	0	7

Figure 14. The confusion matrix of CNN on test set.

Table 7. The evaluation result of FCNN and CNN on training set.

Model	Index	Class1	Class2	Class3	Class4	Class5	*Score*
FCNN	*Precision* (%)	92.94	86.71	88.72	91.32	96.07	0.83
	Recall (%)	92.20	82.20	91.20	92.60	97.80	
	F_1 *Score*	0.93	0.84	0.90	0.92	0.97	
CNN	*Precision* (%)	97.25	97.00	96.64	94.34	97.83	0.93
	Recall (%)	99.00	90.60	97.80	96.60	99.00	
	F_1 *Score*	0.98	0.94	0.97	0.95	0.98	

Table 8. The evaluation result of FCNN and CNN on validation set.

Model	Index	Class1	Class2	Class3	Class4	Class5	*Score*
FCNN	*Precision (%)*	97.50	78.57	97.83	76.92	100	
	Recall (%)	86.67	73.33	100	88.89	100	0.81
	F_1 *Score*	0.92	0.76	0.99	0.82	1.00	
CNN	*Precision (%)*	100	95.35	95.56	91.49	95.74	
	Recall (%)	95.56	91.11	95.56	95.56	100	0.91
	F_1 *Score*	0.97	0.93	0.96	0.93	0.98	

Table 9. The evaluation result of test set for CNN.

Index	Model 1	Model 2	Model 3	Model 4	Model 5	*Score*
Recall	75	83.3	100	80	87.5	
Precision	100	71.4	66.7	80	100	0.69
F_1 *Score*	0.86	0.77	0.80	0.80	0.93	

4.2. Effects of Parameters on Classification Results

Sensitivity analysis is a key step in testing CNN performance and determining the impact of input parameters on the predictive results [38–43]. In order to study the influence of a series of CNN parameters on the prediction results and further optimize the established CNN, a sensitivity analysis was conducted.

4.2.1. Effect of the Learning Rate

The loss function is a function of weights, and the learning rate determines the update speed of the weights in the CNN and determines the value of the loss function. If the learning rate is too large, it will cause the loss function to oscillate and the CNN is hard to converge. When the learning rate is a small value, the updated value of the weights also is small and the model converges slowly. Therefore, there is an optimized learning rate for each CNN. In order to determine the optimal learning rate, we changed the value of the learning rate from 0.0001 to 0.03 and remaining values were constant. Figure 15 shows that the CNN had the highest accuracy on both the validation set and the training set when the learning rate was 0.005.

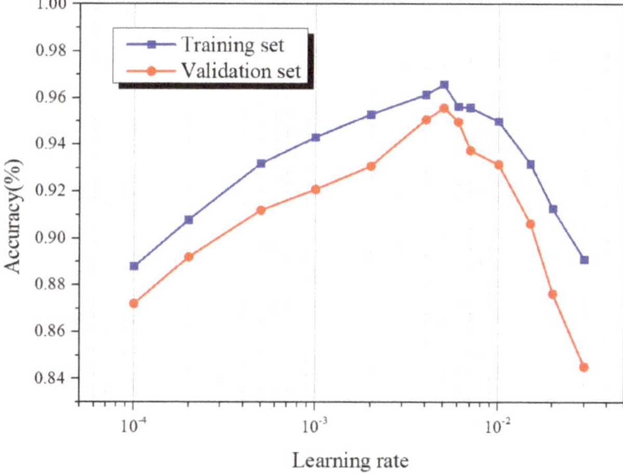

Figure 15. Comparison of the training results of CNN on training set and validation set under different learning rates.

4.2.2. Effect of the Dropout Rate

The key point of the dropout method to prevent the overfitting problem is to make some nodes of the CNN to stop working in a probability of dropout rate. Therefore, the value of the dropout rate has a significant impact on the training effect of the CNN. As shown in Figure 16, as the dropout rate increased, the accuracy of CNN on the training set continued to decrease. For the value for the validation set, it increased firstly and then decreased when the dropout rate increased. Meanwhile, the accuracy difference of the training set and the validation set was large in the case of small dropout value, indicating that the overfitting phenomenon had occurred. As the dropout rate became larger, the accuracy difference was very small, meaning that the CNN had not been well fitted to the training data. For the CNN in this paper, the optimal value of the dropout rate was 0.4.

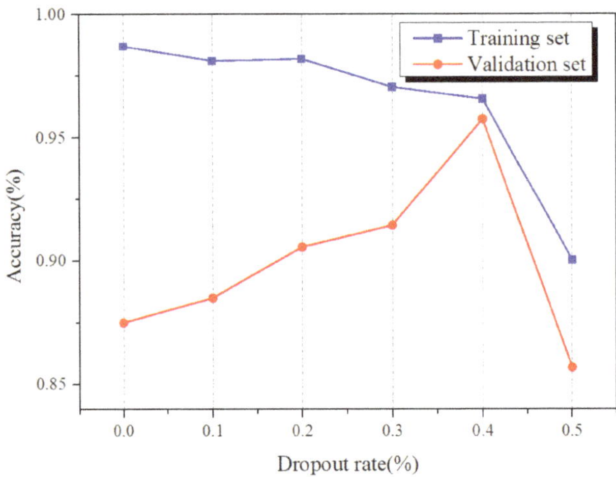

Figure 16. Comparison of the training results of CNN on training set and validation set under different dropout rates.

4.2.3. Effect of the Number of Training Samples

The performance of CNN is strongly controlled by the number of training samples. A small sample size can make the training process of the CNN difficult to converge. In general, CNN require a pretty large training sample size and the negative effect is that this large sample size usually increases the requirement for CPU. To select as few samples as possible while ensuring the best performance of CNN, the impact of sample size on CNN learning curves was investigated. Figure 17 shows that as the number of training samples increased, the accuracy of CNN model on the validation set also increased. When the number of training samples was greater than 2000, the increase of the accuracy on the validation set tended to be flat. Meanwhile, the CNN had a similar accuracy on both the training set and the validation set. Therefore, the number of training samples was finally determined to be 2500.

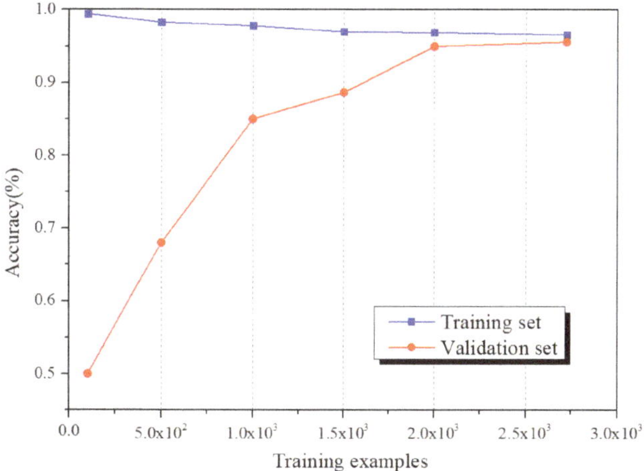

Figure 17. Comparison of the training results of CNN on training set and validation set under different numbers of training samples.

5. Conclusions

In this paper, a CNN model was developed to classify the well testing curves. In order to obtain the best test curve classification effect, before the training, we optimized the CNN model from several aspects such as regularization technology, activation function, and optimization algorithm. The results show that the Xavier normal initialization worked best in the four optimization methods. Among the five activation functions, the CNN model had the best performance when the activation function of the convolution layer and the output layer was chosen as the ReLU function. Compared to the L2 regularization method, the dropout method had a better performance in avoiding overfitting problem. In addition, the utilization of mini batch technique and Adam optimization algorithm made the model not fall into local minimum and fast convergence. Further, the impacts of key parameters in the CNN model on work performance were studied. It was found that when the learning rate was 0.005, the CNN had the highest precision in the validation set and the training set. For the dropout rate, CNN could better fit the training data without over-fitting phenomenon in the case of 0.4. The analysis of training sample numbers showed that the accuracy difference between the training set and the validation set could be ignored when the number of training samples was 2500. Finally, the classification results of CNN and FCNN with similar structures on well testing curves were compared. For the validation set, the *Score* of FCNN and CNN were 0.81 and 0.91, respectively, indicating that the CNN had a more robust performance in the classification results of the well test curve. The 25 field cases from the Ordos Basin showed that the trained CNN could successfully classify 21 cases and the robustness of the model was further proved.

Author Contributions: Every author has contributed to this work. Conceptualization, C.H. and D.P.; Methodology, C.H.; Software, D.P.; Validation, C.H., D.P. and L.X.; Formal Analysis, L.X.; Investigation, L.X.; Resources, L.X. and C.Z; Data Curation, L.X.; Writing-Original Draft Preparation, C.H.; Writing-Review & Editing, C.H., D.P. and C.Z; Visualization, Z.J.; Supervision, Z.X.; Project Administration, L.X.; Funding Acquisition, L.X., Z.X. and C.Z.

Funding: This research was funded by the Joint Funds of the National Natural Science Foundation of China (Grant No. U1762210), National Science and Technology Major Project of China (Grant No.2017ZX05009004-005), Science Foundation of China University of Petroleum, Beijing (Grant No. 2462018YJRC032), and Post-doctoral Program for Innovation Talents (BX20180380).

Conflicts of Interest: The authors declare no conflict of interest. The funders had no role in the design of the study; in the collection, analyses, or interpretation of data; in the writing of the manuscript, and in the decision to publish the results.

Nomenclature

CNN	Convolutional Neural Network
FCNN	Fully Connected Neural Network
CONVlayer	Convolutional Layer
FC layer	Fully Connected Layer
1D	One Dimensional
2D	Two Dimensional
TP	True Positive
FP	False Positive
FN	False Negative
w	Network Weight
g	Gradient
n	Number of Network Weight
b	Number of Iterative Step in Mini Batch Technique
s	Number of Training Samples in Mini Batch Technique
c	Number of Sample Classes
x	Sample Matrix
y	Real Sample Label Matrix
$\hat{y}$	Predictive Sample Label Matrix
a	Output Value of Neural Network
m	Number of Training Samples

Greek

η	Learning Rate
β_1, β_2	Exponential Decay Rates in Adam Algorithm
ω, υ	Momentum in Adam Algorithm
ε	Constant
λ	L2 Regularization Parameter

Subscript

i	i-th Sample
j	j-th Feature
k	Iteration

Superscript

t	t-th Time Step

Appendix A. Field Cases Used in This Work

Table A1. Parameter distribution of field cases used in this work.

	Thickness (m)	Porosity (%)	Permeability (mD)	Initial Pressure (MPa)	Wellbore Storage Coefficient	Skin Factor	Composite Radius (m)	Mobility Ratio	Dispersion Ratio	Fracture Half Length (m)	Omega	Lambda	Curve Type
Case1	9.4	10.94	0.82	15.06	0.19	0.05	/	/	/	23.1	/	/	1
Case2	9.56	13.12	0.02	13.56	0.12	−5.88	/	/	/	59	/	/	1
Case3	13.12	9.08	7.53	14.02	2.12	0.02	/	/	/	112	/	/	1
Case4	5.68	11.04	0.5	17.2	0.01	0.01	/	/	/	76	/	/	1
Case5	16	11.61	0.2	14.28	0.03	0.63	/	/	/	11.1	/	/	2
Case6	4.3	13.68	0.41	10.29	0.6	0.11	/	/	/	46	/	/	2
Case7	13.7	12.56	0.09	26.82	0.07	0.18	/	/	/	16.4	/	/	2
Case8	9.7	10.8	1.14	20.05	1.96	0.26	/	/	/	128	/	/	2
Case9	13.3	11.28	0.17	12.29	0.33	0.21	/	/	/	56.5	/	/	2
Case10	8.5	11.77	1.08	22.61	0.12	0.3	/	/	/	126	/	/	2
Case11	9.67	10.13	1	19.13	0.01	0.02	/	/	/	/	0.29	1.4×10^{-8}	3
Case12	10.78	13.03	0.27	40.55	0.07	−0.81	/	/	/	/	0.08	9.4×10^{-4}	3
Case13	13.2	11.1	0.84	17.08	0.24	−3.77	40.6	7.36	3.61	/	/	/	4
Case14	5.4	12.3	0.34	13.47	0.16	−4.58	12.3	7.71	12.8	/	/	/	4
Case15	16.3	10.95	0.13	12.41	0.14	−3.54	31	2.32	1.4	/	/	/	4
Case16	11.6	13.12	0.25	16.02	0.15	−3.64	31	2	3.06	/	/	/	4
Case17	11.2	9.63	0.78	13.24	0.14	−2.23	13.3	9.56	8.03	/	/	/	4
Case18	8.2	11.25	0.25	20.87	0.15	−2.82	33.26	0.82	0	/	/	/	5
Case19	9.2	14.06	0.91	18.52	0.66	−1.37	52.1	0.44	0	/	/	/	5
Case20	27.2	13.12	0.4	24.64	0.32	−1.72	13.9	0.36	0	/	/	/	5
Case21	9.7	10.8	1.14	20.05	1.96	0.26	94.3	0.75	0.1	/	/	/	5
Case22	8.2	13.36	0.42	25.82	0.89	−3.78	59	0.51	0	/	/	/	5
Case23	11.2	10.31	0.1	23.21	0.26	−3.11	41	0.03	0	/	/	/	5
Case24	7.3	14.6	1.35	24.68	0.61	−3.56	92.2	0.72	0.01	/	/	/	5
Case25	7.6	12.74	0.66	23.66	1.13	−3.45	18.9	0.98	0	/	/	/	5

B. Infinite-Conductivity Vertically Fractured Model

At first, a series of dimensionless variables need to be defined:

$$p_{wD} = \frac{kh\Delta p}{1.842 \times 10^{-3} q\mu B} \tag{A1}$$

$$t_D = \frac{3.6kt}{\varphi\mu C_t L^2} \tag{A2}$$

$$C_D = \frac{0.1592C}{\varphi h C_t L^2} \tag{A3}$$

$$y_D = \frac{y}{L} \tag{A4}$$

$$x_D = \frac{x}{L} \tag{A5}$$

$$r_D = \frac{r}{L} \tag{A6}$$

where k is the permeability, φ is the porosity, C is the wellbore storage coefficient, L is the reference length, C_t is the compressibility, t refers to time, μ refers to the viscosity, B is the volume factor, q refers to the flux rate, p is the pressure, h is the formation thickness. After dimensionless treatment, the diffusion equation in Laplace domain can be expressed as:

$$\frac{d^2\bar{p}_D}{dr_D^2} + \frac{1}{r_D}\frac{d\bar{p}_D}{dr_D} = u\bar{p}_D \tag{A7}$$

where $\bar{p}_D$ is the dimensionless pressure in Laplace domain, u refers to the Laplace variable, and r_D is the dimensionless distance. The initial condition is:

$$\bar{p}_D(r_D, 0) = 0. \tag{A8}$$

The internal boundary condition and exterior boundary respectively are:

$$\lim_{r_D \to 0}\left[r_D\frac{d\bar{p}_D}{dr_D}\right] = -1 \tag{A9}$$

$$\bar{p}_D(\infty, t_D) = 0. \tag{A10}$$

The general solution of Equation (A7) is shown as following:

$$\bar{p}_D = \frac{1}{u}K_0\left(r_D\sqrt{u}\right) \tag{A11}$$

where K_0 is the first class zero order Bessel function. With pressure superposition method, the pressure solution of the infinite conductivity vertically fractured model is obtained:

$$\bar{p}_D = \frac{1}{u}\int_{-1}^{1} K_0\left(\sqrt{(x_D - x_i)^2 + y_D^2}\sqrt{u}\right)da \tag{A12}$$

C. Dual-Porosity Model with Pseudo-Steady State

In the dual-porosity model, the corresponding dimensionless variables are:

$$p_{wD} = \frac{k_f h\Delta p}{1.842 \times 10^{-3} q\mu B} \tag{A13}$$

$$t_D = \frac{3.6kt}{(\varphi C_t)_{f+m}\mu L^2} \tag{A14}$$

$$C_D = \frac{0.1592C}{(\varphi C_t)_{f+m}\mu L^2} \tag{A15}$$

$$\lambda = aL^2 \frac{k_m}{k_f} \tag{A16}$$

$$\omega = \frac{(\varphi C_t)_f}{(\varphi C_t)_f (\varphi C_t)_m} \tag{A17}$$

where subscript f is the natural fracture system, subscript m refers to the matrix system, w is the wellbore system, λ is the interporosity flow coefficient, ω refers to the storage ratio. The diffusion equation of the pseudo-steady state in Laplace domain can be expressed as:

$$\frac{d^2 \bar{p}_{fD}}{dr_D^2} + \frac{1}{r_D}\frac{d\bar{p}_{fD}}{dr_D} = \lambda\left(\bar{p}_{fD} - \bar{p}_{mD}\right) + \omega u \bar{p}_D \tag{A18}$$

$$(1-\omega)u\bar{p}_{mD} = \lambda\left(\bar{p}_{fD} - \bar{p}_{mD}\right). \tag{A19}$$

The initial condition is:

$$\bar{p}_{fD}(r_D,0) = \bar{p}_{mD}(r_D,0) = 0. \tag{A20}$$

The boundary conditions are:

$$C_D u d\bar{p}_{wD} - \left.\frac{d\bar{p}_{fD}}{dr_D}\right|_{r_D=1} = \frac{1}{u} \tag{A21}$$

$$\bar{p}_{wD} = \left.\left(\bar{p}_{fD} - S\frac{d\bar{p}_{fD}}{dr_D}\right)\right|_{r_D=1} \tag{A22}$$

$$\bar{p}_{fD} = \bar{p}_{mD} = 0\big|_{r_D\to\infty} \tag{A23}$$

where S is the skin factor and C_D is the dimensionless wellbore storage coefficient. Combining Equations (A18) and (A19), the general solution is determined as [44–46]:

$$\bar{p}_{wD} = \frac{K_0\left(\sqrt{f(u)u}\right) + S\sqrt{f(u)u}K_1\left(\sqrt{f(u)u}\right)}{u\sqrt{f(u)u}K_1\left(\sqrt{f(u)u}\right) + C_D u^2 K_0\left(\sqrt{f(u)u}\right) + SC_D u^2 \sqrt{f(u)u}K_1\left(\sqrt{f(u)u}\right)}. \tag{A24}$$

In Equation (A24),

$$f(u) = \frac{\omega(1-\omega)\times u + \lambda}{(1-\omega)\times u + \lambda}. \tag{A25}$$

D. Radial Composite Model

For the radial composite model, the dimensionless variables are defined as follows:

$$p_{irD} = \frac{k_{ir}h\Delta p}{1.842\times10^{-3}q\mu_{ir}B} \tag{A26}$$

$$p_{erD} = \frac{k_{er}h\Delta p}{1.842\times10^{-3}q\mu_{er}B} \tag{A27}$$

$$p_{wD} = \frac{k_{ir}h\Delta p_{wf}}{1.842\times10^{-3}q\mu_{ir}B} \tag{A28}$$

$$t_D = \frac{3.6k_{ir}t}{\varphi\mu_{ir}C_t L^2} \tag{A29}$$

$$C_D = \frac{0.159C}{\varphi h C_t L^2} \tag{A30}$$

$$r_D = \frac{r}{L} \tag{A31}$$

$$r_{fD} = \frac{r_f}{L} \tag{A32}$$

$$M = \frac{(k/\mu)_{ir}}{(k/\mu)_{er}} \tag{A33}$$

$$W = \frac{(\varphi C_t)_{ir}}{(\varphi C_t)_{er}} \tag{A34}$$

where p_{irD} and p_{erD} are the dimensionless pressure in inner region and outer regions, r_{fD} is the dimensionless radius of the interface, M is the mobility ratio, W is the dispersion ratio. The diffusion equations in Laplace domain for the inner and outer regions of the composite model can be written as:

$$\frac{1}{r_D}\frac{d}{dr_D}\left(r_D\frac{d\overline{p}_{irD}}{dr_D}\right) = u\overline{p}_{irD} \tag{A35}$$

$$\frac{1}{r_D}\frac{d}{dr_D}\left(r_D\frac{d\overline{p}_{erD}}{dr_D}\right) = \frac{W}{M}u\overline{p}_{erD}. \tag{A36}$$

Correspondingly, the inner and outer boundary conditions are:

$$C_D u d\overline{p}_{wD} - \left.\frac{d\overline{p}_{irD}}{dr_D}\right|_{r_D=1} = \frac{1}{u} \tag{A37}$$

$$\overline{p}_{wD} = \left.\left(\overline{p}_{irD} - S\frac{d\overline{p}_{irD}}{dr_D}\right)\right|_{r_D=1} \tag{A38}$$

$$\left.\overline{p}_{erD}\right|_{r_D\to\infty} = 0. \tag{A39}$$

There is an interface between the inner and outer regions. For this interface, the pressure and pressure derivative meet the following requirements:

$$\overline{p}_{irD} = \left.\overline{p}_{erD}\right|_{r_D=r_{fD}} \tag{A40}$$

$$\frac{d\overline{p}_{irD}}{dr_D} = \left.\frac{1}{M}\frac{d\overline{p}_{erD}}{dr_D}\right|_{r_D=r_{fD}}. \tag{A41}$$

Therefore, the solution can be determined as:

$$\overline{p}_{irD} = AK_0\left(r_D\sqrt{u}\right) + BI_0\left(r_D\sqrt{u}\right). \tag{A42}$$

To satisfy the conditions of the interface, the value of A and B can be obtained:

$$A = \overline{q}_D \tag{A43}$$

$$B = \frac{\overline{q}_D MK_0\left(r_{irD}\sqrt{Mu/W}\right)K_1\left(r_{irD}\sqrt{u}\right)\sqrt{U} - \overline{q}_D MK_1\left(r_{irD}\sqrt{Mu/W}\right)K_0\left(r_{irD}\sqrt{u}\right)\sqrt{Mu/W}}{I_0\left(r_{irD}\sqrt{u}\right)K_1\left(r_{irD}\sqrt{Mu/W}\right)\sqrt{Mu/W} - MI_1\left(r_{irD}\sqrt{u}\right)K_0\left(r_{irD}\sqrt{Mu/W}\right)\sqrt{u}}. \tag{A44}$$

References

1. Muskat, M. The flow of homogeneous fluids through porous media. *Soil Sci.* **1938**, *46*, 169. [CrossRef]
2. Van Everdingen, A.F.; Hurst, W. The application of the Laplace transformation to flow problems in reservoirs. *J. Pet. Technol.* **1949**, *1*, 305–324. [CrossRef]
3. Horner, D.R. Pressure build-up in wells. In Proceedings of the 3rd World Petroleum Congress, The Hague, The Netherlands, 28 May–6 June 1951.
4. Ramey, H.J., Jr. Short-time well test data interpretation in the presence of skin effect and wellbore storage. *J. Pet. Technol.* **1970**, *22*, 97–104. [CrossRef]
5. Gringarten, A.C.; Ramey, H.J., Jr.; Raghavan, R. Unsteady-state pressure distributions created by a well with a single infinite-conductivity vertical fracture. *Soc. Pet. Eng. J.* **1974**, *14*, 347–360. [CrossRef]
6. Bourdet, D.; Ayoub, J.A.; Pirard, Y.M. Use of pressure derivative in well test interpretation. *SPE Form. Eval.* **1989**, *4*, 293–302. [CrossRef]

7. Zhou, Q.; Dilmore, R.; Kleit, A.; Wang, J.Y. Evaluating gas production performances in Marcellus using data mining technologies. *J. Nat. Gas. Sci. Eng.* **2014**, *20*, 109–120. [CrossRef]

8. Ma, Z.; Leung, J.Y.; Zanon, S.; Dzurman, P. Practical implementation of knowledge-based approaches for steam-assisted gravity drainage production analysis. *Expert Syst. Appl.* **2015**, *42*, 7326–7343. [CrossRef]

9. Lolon, E.; Hamidieh, K.; Weijers, L.; Mayerhofer, M.; Melcher, H.; Oduba, O. Evaluating the relationship between well parameters and production using multivariate statistical models: A middle Bakken and three forks case history. In Proceedings of the SPE Hydraulic Fracturing Technology Conference, The Woodlands, TX, USA, 9–11 February 2016.

10. Wang, S.; Chen, S. Insights to fracture stimulation design in unconventional reservoirs based on machine learning modeling. *J. Pet. Sci. Eng.* **2019a**, *174*, 682–695. [CrossRef]

11. Wang, S.; Chen, Z.; Chen, S. Applicability of deep neural networks on production forecasting in Bakken shale reservoirs. *J. Pet. Sci. Eng.* **2019b**, *179*, 112–125. [CrossRef]

12. Awoleke, O.; Lane, R. Analysis of data from the Barnett shale using conventional statistical and virtual intelligence techniques. *SPE Reserv. Eval. Eng.* **2011**, *14*, 544–556. [CrossRef]

13. Chu, H.; Liao, X.; Zhang, W.; Li, J.; Zou, J.; Dong, P.; Zhao, C. Applications of Artificial Neural Networks in Gas Injection. In Proceedings of the SPE Russian Petroleum Technology Conference, Moscow, Russia, 15–17 October 2018.

14. Akbilgic, O.; Zhu, D.; Gates, I.D.; Bergerson, J.A. Prediction of steam-assisted gravity drainage steam to oil ratio from reservoir characteristics. *Energy* **2015**, *93*, 1663–1670. [CrossRef]

15. Al-Kaabi, A.U.; Lee, W.J. Using artificial neural networks to identify the well test interpretation model (includes associated papers 28151 and 28165). *SPE Form. Eval.* **1993**, *8*, 233–240. [CrossRef]

16. Sultan, M.A.; Al-Kaabi, A.U. Application of neural network to the determination of well-test interpretation model for horizontal wells. In Proceedings of the SPE Asia Pacific Oil and Gas Conference and Exhibition, Melbourne, VIC, Australia, 8–10 October 2002.

17. Kharrat, R.; Razavi, S.M. Determination of reservoir model from well test data, using an artificial neural network. *Sci. Iran.* **2008**, *15*, 487–493.

18. AlMaraghi, A.M.; El-Banbi, A.H. Automatic Reservoir Model Identification using Artificial Neural Networks in Pressure Transient Analysis. In Proceedings of the SPE North Africa Technical Conference and Exhibition, Cairo, Egypt, 14–16 September 2015.

19. Schroff, F.; Kalenichenko, D.; Philbin, J. Facenet: A unified embedding for face recognition and clustering. In Proceedings of the IEEE Conference on Computer Vision and Pattern Recognition, Boston, MA, USA, 7–12 June 2015; pp. 815–823.

20. Redmon, J.; Divvala, S.; Girshick, R.; Farhadi, A. You only look once: Unified, real-time object detection. In Proceedings of the IEEE Conference on Computer Vision and Pattern Recognition, Las Vegas, NV, USA, 27–30 June 2016; pp. 779–788.

21. Gatys, L.A.; Ecker, A.S.; Bethge, M. A neural algorithm of artistic style. *arXiv* **2015**, arXiv:1508.06576. [CrossRef]

22. He, K.; Zhang, X.; Ren, S.; Sun, J. Deep residual learning for image recognition. In Proceedings of the IEEE Conference on Computer Vision and Pattern Recognition, Las Vegas, NV, USA, 27–30 June 2016; pp. 770–778.

23. Liu, D.; Li, Y.; Agarwal, R.K. Numerical simulation of long-term storage of CO_2 in Yanchang shale reservoir of the Ordos basin in China. *Chem. Geol.* **2016**, *440*, 288–305. [CrossRef]

24. Chu, H.; Liao, X.; Chen, Z.; Zhao, X.; Liu, W. Estimating carbon geosequestration capacity in shales based on multiple fractured horizontal well: A case study. *J. Pet. Sci. Eng.* **2019**, *181*, 106179. [CrossRef]

25. Meng, X. Horizontal Fracture Seepage Model and Effective Way for Development of Chang 6 Reservoir. Ph.D. Thesis, Southwest Petroleum University, Chengdu, China, 2018.

26. Chu, H.; Liao, X.; Chen, Z.; Zhao, X.; Liu, W.; Dong, P. Transient pressure analysis of a horizontal well with multiple, arbitrarily shaped horizontal fractures. *J. Pet. Sci. Eng.* **2019**, *180*, 631–642. [CrossRef]

27. Jingli, Y.; Xiuqin, D.; Yande, Z.; Tianyou, H.; Meijuan, C.; Jinlian, P. Characteristics of tight oil in Triassic Yanchang formation, Ordos Basin. *Pet. Explor. Dev.* **2013**, *40*, 161–169.

28. Hua, Y.A.N.G.; Jinhua, F.; Haiqing, H.; Xianyang, L.I.U.; Zhang, Z.; Xiuqin, D.E.N.G. Formation and distribution of large low-permeability lithologic oil regions in Huaqing, Ordos Basin. *Pet. Explor. Dev.* **2012**, *39*, 683–691.

29. Li, Y.; Song, Y.; Jiang, Z.; Yin, L.; Luo, Q.; Ge, Y.; Liu, D. Two episodes of structural fractures: Numerical simulation of Yanchang Oilfield in the Ordos basin, northern China. *Mar. Pet. Geol.* **2018**, *97*, 223–240. [CrossRef]
30. Guo, P.; Ren, D.; Xue, Y. Simulation of multi-period tectonic stress fields and distribution prediction of tectonic fractures in tight gas reservoirs: A case study of the Tianhuan Depression in western Ordos Basin, China. *Mar. Pet. Geol.* **2019**, *109*, 530–546. [CrossRef]
31. Krizhevsky, A.; Sutskever, I.; Hinton, G.E. Imagenet classification with deep convolutional neural networks. In Proceedings of the 25th International Conference on Neural Information Processing Systems, Lake Tahoe, NV, USA, 3–6 December 2012; pp. 1097–1105.
32. Glorot, X.; Bengio, Y. Understanding the difficulty of training deep feedforward neural networks. In Proceedings of the Thirteenth International Conference on Artificial Intelligence and Statistics, Sardinia, Italy, 13–15 March 2010; pp. 249–256.
33. Ghaderi, A.; Shahri, A.A.; Larsson, S. An artificial neural network based model to predict spatial soil type distribution using piezocone penetration test data (CPTu). *Bull. Eng. Geol. Environ.* **2018**, 1–10. [CrossRef]
34. Kingma, D.P.; Ba, J. Adam: A method for stochastic optimization. *arXiv* **2014**, arXiv:1412.6980.
35. Chen, X.; Liu, S.; Sun, R.; Hong, M. On the convergence of a class of adam-type algorithms for non-convex optimization. *arXiv* **2018**, arXiv:1808.02941.
36. Reddi, S.J.; Kale, S.; Kumar, S. On the convergence of adam and beyond. *arXiv* **2019**, arXiv:1904.09237.
37. Mukhanov, A.; Arturo Garcia, C.; Torres, H. Water Control Diagnostic Plot Pattern Recognition Using Support Vector Machine. In Proceedings of the SPE Russian Petroleum Technology Conference, Moscow, Russia, 15–17 October 2018.
38. Hamdia, K.M.; Silani, M.; Zhuang, X.; He, P.; Rabczuk, T. Stochastic analysis of the fracture toughness of polymeric nanoparticle composites using polynomial chaos expansions. *Int. J. Fract.* **2017**, *206*, 215–227. [CrossRef]
39. Shahri, A.A.; Asheghi, R. Optimized developed artificial neural network-based models to predict the blast-induced ground vibration. *Innov. Infrastruct. Solut.* **2018**, *3*, 34. [CrossRef]
40. Saltelli, A.; Ratto, M.; Andres, T.; Campolongo, F.; Cariboni, J.; Gatelli, D.; Saisana, G.; Tarantola, S. *Global Sensitivity Analysis: The Primer*; John Wiley & Sons: Chichester, UK, 2008; ISBN 9780470725184.
41. Saltelli, A. Sensitivity analysis for importance assessment. *Risk Anal.* **2002**, *22*, 579–590. [CrossRef]
42. Vu-Bac, N.; Lahmer, T.; Zhuang, X.; Nguyen-Thoi, T.; Rabczuk, T. A software framework for probabilistic sensitivity analysis for computationally expensive models. *Adv. Eng. Soft.* **2016**, *100*, 19–31. [CrossRef]
43. Shahri, A.A. An optimized artificial neural network structure to predict clay sensitivity in a high landslide prone area using piezocone penetration test (CPTu) data: A case study in southwest of Sweden. *Geotech. Geol. Eng.* **2016**, *34*, 745–758. [CrossRef]
44. Chen, Z.; Liao, X.; Sepehrnoori, K.; Yu, W. A Semianalytical Model for Pressure-Transient Analysis of Fractured Wells in Unconventional Plays With Arbitrarily Distributed Discrete Fractures. *SPE J.* **2018**, *23*, 2041–2059. [CrossRef]
45. Chen, Z.; Liao, X.; Zhao, X.; Lyu, S.; Zhu, L. A comprehensive productivity equation for multiple fractured vertical wells with non-linear effects under steady-state flow. *J. Pet. Sci. and Eng.* **2017**, *149*, 9–24. [CrossRef]
46. Zongxiao, R.; Xiaodong, W.; Dandan, L.; Rui, R.; Wei, G.; Zhiming, C.; Zhaoguang, T. Semi-analytical model of the transient pressure behavior of complex fracture networks in tight oil reservoirs. *J. Nat. Gas Sci. Eng.* **2016**, *35*, 497–508. [CrossRef]

Article

Study on the Impacts of Capillary Number and Initial Water Saturation on the Residual Gas Distribution by NMR

Tao Li [1], Ying Wang [1,2,*], Min Li [1,*], Jiahao Ji [1], Lin Chang [1,2] and Zheming Wang [2]

[1] State Key Laboratory of Oil and Gas Reservoir Geology and Exploitation, Southwest Petroleum University, Chengdu 610500, China

[2] Pacific Northwest National Laboratory, 902 Battelle Boulevard, P.O. Box 999, MSIN K8-96, Richland, WA 99352, USA

* Correspondence: ywang@swpu.edu.cn (Y.W.); hytlxf@swpu.edu.cn (M.L.); Tel.: +86-028-83034269 (Y.W. & M.L.)

Received: 14 May 2019; Accepted: 12 July 2019; Published: 16 July 2019

Abstract: The determination of microscopic residual gas distribution is beneficial for exploiting reservoirs to their maximum potential. In this work, both forced and spontaneous imbibition (waterflooding) experiments were performed on a high-pressure displacement experimental setup, which was integrated with nuclear magnetic resonance (NMR) to reveal the impacts of capillary number (Ca) and initial water saturation (S_{wi}) on the residual gas distribution over four magnitudes of injection rates ($Q = 0.001, 0.01, 0.1$ and 1 mL/min), expressed as Ca ($\log Ca = -8.68, -7.68, -6.68$ and -5.68), and three different S_{wi} ($S_{wi} = 0\%, 39.34\%$ and 62.98%). The NMR amplitude is dependent on pore volumes while the NMR transverse relaxation time (T_2) spectrum reflects the characteristics of pore size distribution, which is determined based on a mercury injection (MI) experiment. Using this method, the residual gas distribution was quantified by comparing the T_2 spectrum of the sample measured after imbibition with the sample fully saturated by brine before imbibition. The results showed that capillary trapping efficiency increased with increasing S_{wi}, and above 90% of residual gas existed in pores larger than 1 μm in the spontaneous imbibition experiments. The residual gas was trapped in pores by different capillary trapping mechanisms under different Ca, leading to the difference of residual gas distribution. The flow channels were mainly composed of micropores (pore radius, $r < 1$ μm) and mesopores ($r = 1$–10 μm) at $\log Ca = -8.68$ and -7.68, while of mesopores and macropores ($r > 10$ μm) at $\log Ca = -5.68$. At both $S_{wi} = 0\%$ and 39.34%, residual gas distribution in macropores significantly decreased while that in micropores slightly increased with $\log Ca$ increasing to -6.68 and -5.68, respectively.

Keywords: capillary number; initial water saturation; capillary trapping; residual gas distribution; nuclear magnetic resonance

---

1. Introduction

Gas reservoir with aquifers is one of the most common type of reservoirs, in which edge-bottom water entering and fingering often occur, leading to residual gas becoming trapped [1]. A major difficulty in studying gas-water flow in reservoirs is the emergence of trapped phases [2]. Faced with increasing global energy demand, further understanding of gas-water migration characteristics in gas reservoirs is not only beneficial for ensuring the trapping mechanism, but also for optimizing production parameters to exploit reservoirs to their maximum potential [3,4].

Because of the strong compressibility and good fluidity (related to the smaller viscosity of gas) of gas, the experiment for gas-water flow usually encounters more difficulties than liquid flow [5]. Many experimental studies have been carried out on imbibition (waterflooding) in reservoir samples, including both spontaneous imbibition [6,7] and forced imbibition [8,9]. The relationship between the amount of water into samples and the time it takes are first concerned in the imbibition experiments [10], while the water content is usually evaluated by the weighing method. The gas-phase saturation remaining in porous media is referred to as residual gas saturation (S_{gr}) as the gas phase relative permeability reduces to zero [3]. Numerous relationships between S_{gr} and other petrophysical properties, such as porosity [6,11], permeability [7], clay mineral components [6], pore types [12], heterogeneity [13], S_{wi} (namely water-phase saturation in porous media at the beginning of experiment) [12], etc., have been established. Among them, the S_{wi} of reservoirs refers to the percentage of water volume in reservoir rock to their total pore volume before exploitation, which is an important parameter to evaluate the geological reserves and flow capacity of reservoirs. However, some of the above correlations were found to behave consistently under various rock conditions, few are universally applicable. Also, the residual gas distribution in sample cannot be obtained via the weighing method.

Subsequently, the gas-water imbibition was conducted in micromodels [1]. The process of water-gas migration and the residual gas distribution in micromodels were observed visually via μ-CT [5,14], X-ray [10,15,16] or microscope technique experiments [8]. During the process of imbibition, the variation of water-gas interface geometry in porous media are extremely complicated [1], and some of the gas is immobile due to capillary forces [5,17]. The capillary trapping strongly depends on the core-scale capillary number (expressed as Ca, characterizing the ratio of viscous forces to capillary forces, $Ca = \frac{\mu_w v_w}{\phi \sigma_{gw}}$) [18–21], where μ is water viscosity (Pa·s); v_w is the flow velocity of water, (m/s); ϕ is rock porosity; σ_{gw} represents the interface tension of gas and water (N/m). At different development stages of gas reservoirs, Ca and water saturation are changing. Meanwhile, the capillary forces and gas production in water-bearing samples are related to the water content [22]. Based on micro-CT technology, Mohammadian et al. [23] studied the magnitude and structure of the residual gas phase at the pore scale. With X-ray microtomography, Buchgraber et al. [24] studied the saturation of trapped CO_2 in 2D micromodels for Ca from 10^{-8} to 10^{-5} and found it decreasing continuously as Ca increased, Geistlinger et al. [2] quantified the gas-water interface and the different shapes of isolated gas clusters. Also, with 2D micromodels, Geistlinger et al. [18,25] illustrated that the linear surface-volume relationship of trapped gas clusters, and found that the morphology and number of trapped gas clusters change with Ca. In addition, Herring et al. [26] found that there was a linear relationship between the initial gas phase connectivity and capillary trapping efficiency. However, the residual gas distribution in the pore structure of porous media at different Ca and S_{wi} remain ambiguous. Also, these micromodels cannot completely characterize the complex pore structure of reservoir rocks.

Recently, NMR relaxometry measurement was introduced to analyze the influences of micropores and initial imbibition rate on S_{gr} [6], and the accuracy of NMR measurement have been fully verified. T_2 relaxation time is related to pore properties, and a corresponding relationship between T_2 relaxation time and pore size can be theoretically derived [27]. The T_2 distribution of a sample saturated by water can be used to characterize its porous structure and evaluate its pore size distribution [28]. Also, the initial water saturation in rock sample can be precisely quantified at the beginning of experiment by NMR measurement [29,30]. Moreover, combining with mercury intrusion (MI) data, the conversion coefficient between T_2 relaxation time and pore size can be further determined [31,32]. Fluid migration in porous media can be monitored at the pore scale in real time by applying NMR. So far, NMR has been successfully applied to analyze the oil recovery ratio under different displacement volumes [33], the residual oil distribution in rock samples [29] and the effect of micro structures on imbibition [32], etc. However, few studies have looked into the microscopic residual gas distribution within rock samples at different S_{wi} and Ca. The interaction between S_{wi} and Ca during imbibition processes is still unknown.

Therefore, spontaneous and forced imbibition experiments are carried out on reservoir sandstone samples with a NMR relaxation system in this work. The NMR transverse relaxation time (T_2)

spectrum was obtained and recorded to monitor the gas saturation changes and their saturation distribution changes during the imbibition processes in samples. The effects of S_{wi} and *Ca* on residual gas distribution are studied. Combining with mercury intrusion measurements, the NMR T_2 spectrum as a function of pore sizes is determined based on the modified linear relationship of pore size and T_2. Then, the residual gas distribution in pores of rock sample, including micropores ($r < 1$ μm), mesopores (1–10 μm) and macropores ($r > 10$ μm), is quantitatively analyzed. Finally, the trapping mechanisms of gas phase in sandstone samples and the interaction between S_{wi} and *Ca* during imbibition processes are discussed. Results from this study provide an improved understanding on the processes that affect the development of water-drive gas reservoirs.

2. Imbibition Experiments Theories and Methods

2.1. Material

In this study, two similar samples with similar properties were selected from the same position of a gas reservoir core (defined as samples #1-1 and #1-2). Samples #1-1 and #1-2 were each further cut into two pieces (defined as samples #1-1F and #1-1M, #1-2F and #1-2M). Samples #1-1F and #1-2F are longer, and are used for spontaneous and forced imbibition experiments, while samples #1-1M and #1-2M are shorter and used for mercury intrusion experiments. The size, components, porosity and permeability of each sample are listed in Table 1. The microscopic properties of samples are shown in Figure 1. There appears to be very little difference of the pore structures between samples #1-1F and #1-1M (or samples #1-2F and #1-2M). The range of primary grain diameter is from 80 to 120 μm. Then, the rock samples were salt-washed and dried in an oven at 105 °C until their weights became constant [9,11]. Also, the wettability of rock samples was assumed to be constant before and after waterflooding experiments and drying treatments. Nitrogen gas (99.99%) (termed gas throughout the text) and synthetic brine (termed brine) were selected as the fluid media. The physical properties and composition of the brine are presented in Table 2.

Table 1. Sizes, components and properties of the sandstone samples.

Sample	Mineral Component	Diameter (cm)	Length (cm)	Porosity (%)	Permeability (mD)
#1-1F	55% quartz + 35% feldspar	2.440	4.534	24.01	82.26
#1-1M	+ 10% cement		2.500		
#1-2F	58% quartz + 34% feldspar	2.451	4.580	22.72	92.22
#1-2M	+ 8% cement		2.500		

Figure 1. The scanning electron microscopy photos of samples: (**a**) #1-1; (**b**) #1-2.

Table 2. Physical properties and composition of the brine.

Composition	Concentration (mg/L)	Room Temperature and Brine Properties	
Sodium	7863.25	Room temperature (°C)	20.0 ± 1.0
Chlorine	17,717.50	Density (g/mL)	1.03
Potassium	2617.45	Viscosity (mPa·s)	1.0
Calcium	1801.80	Interfacial tension (mN/m)	72.9

2.2. Setup

A high-pressure displacement experimental system (Figure 2) was built to inject fluid into the core samples while real-time NMR images and corresponding data were obtained. The system is primarily composed of the following three parts: (i) a high-pressure fluid delivery system consisting of a ISCO pump (100DX, Teledyne, Thousand Oaks, CA, USA), two storage cylinders (one filling with brine and the other filling with nitrogen) that can reach a certain S_{wi} in samples or inject brine into samples, and a core holder (Niumag, China); (ii) a confining pressuring system including a syringe pump (Series III, Scientific Systems, USA) and a container filled with fluorocarbon oil (supplied by Niumag, China). The syringe pump transports the fluorocarbon oil from the container into the core holder to produce the confining pressure; and (iii) an NMR image and data acquisition system including a computer and an NMR spectrometer (MesoMR23-60H-I, Niumag, China) that can scan and image the entire sample and record the T_2 spectrum at real-time during the processes of waterflooding experiments. The outlet of the core holder is connected to a valve that opens to the ambient atmosphere and the effluent is collected by a graduated cylinder. The basic framework of the NMR device has been described elsewhere by Blümich et al. [34] To avoid interference of the solvent protons in the process of NMR measurement, fluorocarbon oil, composed of only three elements including carbon, fluorine, and oxygen, is used as the confining pressure liquid. In the core holder, polymer heat-shrinkable tube was utilized to seal the samples [29]. The inlet and outlet are covered with copper mesh to avoid magnetic interferences. Moreover, a mercury intrusion apparatus (PoreMaster60, Quantachrome, USA) is used to conduct the mercury intrusion measurements by injecting mercury in a range of pressure from 0 to 413.79 MPa (i.e., 0–60,000 psi).

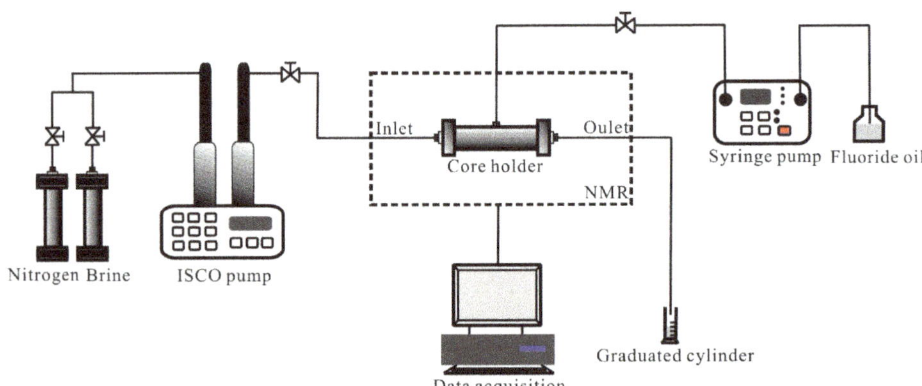

Figure 2. Experimental setup for the forced and spontaneous imbibition experiments.

2.3. Procedures

The experimental procedure of the forced imbibition experiment at $S_{wi} = 0$ is as following: (i) the dried sample, sealed with a heat-shrinkable tube, is placed in the core holder; (ii) the core holder with the sample is placed in the NMR setup. The fluorocarbon oil is injected into the core holder to produce a confining pressure of 2 MPa on the sample; (iii) the brine is injected into the sample at a specified flowrate (Q = 0.001, 0.01, 0.1, and 1 mL/min, corresponding to 0.28, 2.83, 28.29 and 282.89 pore volumes per day, respectively). The T_2 spectrum is acquired in real-time. The inlet pore pressure and the outlet flow volume are measured simultaneously; (iv) the experiment was stopped when T_2 spectrum showed no further changes. For the forced imbibition experiments at S_{wi} = 39.34 ± 1.57 and 62.98 ± 2.22% (in this work, S_{wi} = 0, 39.34 ± 1.57 and 62.98 ± 2.22% are expressed as S_{wi1}, S_{wi2} and S_{wi3} respectively), the experimental procedures are as following: (i) the dried rock sample was initially immersed into the brine and then vacuumed at 10 MPa over 10 h (i.e., the sample is fully saturated); (ii) then the fully saturated sample was sealed with heat-shrinkable tube and placed in the core holder into the NMR setup with confining pressure of 2 MPa; (iii) the T_2 spectrum of the fully saturated sample was acquired; (iv) then, gas is injected into the sample to displace the brine with continuously increasing flow rates from 0.01 to 1 mL/min and stopped until the desired water saturation (S_{wi2} or S_{wi3}) was reached (the experimental setup was shown in Figure 2); (v) the brine is injected into the sample at a specified flowrate (Q = 0.001, 0.01, 0.1, and 1 mL/min) and the T_2 spectrum is acquired in real-time. The inlet pore pressure and the outlet flow volume are measured simultaneously; (vi) the experiment was stopped as the total amplitudes of the T_2 spectrum are constant. More than two pore volumes of water was injected during imbibition experiments. In each experiment, NMR waiting time is set to 5000 ms, echo time is set to 301 µs, echo number is set to 12,000, scanning number is set to 32 and gain number (the magnification times of signals) is set to 20. After each experiment, the sample was taken out and re-dried in the oven at 105 °C until their weights became constant (over 12 h) [9,11,29]. For the spontaneous imbibition experiment, the experimental procedures are almost the same as the forced imbibition experiment for each corresponding S_{wi}, except for the step of brine injection. The brine was introduced from the upstream end of the sample surface to saturate while T_2 spectrum was continuously recorded until the T_2 spectrum became constant.

A total of 15 imbibition experiments, 12 forced imbibition and three spontaneous imbibition, were carried out at three different S_{wi} (S_{wi1}, S_{wi2} and S_{wi3}) and four magnitudes of Q (0.001, 0.01, 0.1 and 1 mL/min) (Table 3). Samples #1-1F and #1-2F were used for the forced imbibition and the spontaneous imbibition experiments, respectively. All the imbibition experiments were conducted at room temperature (20.0 ± 1.0 °C). The corresponding Darcy's velocity (v), Ca and logCa of the injection rate of brine in the forced imbibition experiments were also shown in Table 3. In addition, samples #1-1M and #1-2M were measured the pore size distribution by mercury intrusion method, respectively.

Table 3. Experimental conditions in spontaneous imbibition and forced imbibition experiments.

S_{wi}(%)	Spontaneous Imbibition	Forced Imbibition				
	Sample	Sample	Q (mL/min)	v (m/d)	Ca	logCa
0			0.001	0.0129	2.09×10^{-9}	−8.68
39.34 ± 1.57	#1-1F	#1-1F	0.01	0.129	2.09×10^{-8}	−7.68
62.98 ± 2.22		#1-2F	0.1	1.29	2.09×10^{-7}	−6.68
			1	12.9	2.09×10^{-6}	−5.68

2.4. Water/Gas Saturation Measurement Theories and Methods

With the NMR technique, the T_2 spectrum is acquired to illustrate the pore size distribution. The hydrogen protons, one of the two elements of water, can produce nuclear magnetic resonance signals under the adscititious magnetic field, and the signal decay speed is described as relaxation time in NMR physics. The relaxation time denotes the loss of the transverse magnetization with T_2 and the increase of longitudinal magnetization with T_1 [9]. Thus, each sample pore space, occupied by water, can be described by T_2 displayed by the following Equation (1) [32].

$$\frac{1}{T_2} = \frac{1}{T_{2,bulk}} + \frac{1}{T_{2,surface}} + \frac{1}{T_{2,diffusion}} \tag{1}$$

where $T_{2,\text{bulk}}$ is the bulk relaxation time (ms), $T_{2,\text{surface}}$ is the surface relaxation time (ms) and $T_{2,\text{diffusion}}$ is the relaxation time induced by diffusion (ms). In this study, the brine flows through the sandstone sample, $T_{2,\text{bulk}}$ and $T_{2,\text{diffusion}}$ can be neglected [29]. Then, T_2 is mainly dependent on $T_{2,\text{surface}}$, which is associated with the specific surface area of a pore. $T_{2,\text{surface}}$ can be expressed as follows [29].

$$\frac{1}{T_{2,\text{surface}}} = \rho_2 \left(\frac{S}{V}\right)_{\text{pore}} \tag{2}$$

where ρ_2 is the surface relaxivity (μm/ms), S is the pore surface area (μm^2) and V is the pore volume (μm^3). The ratio of S/V can be expressed as the ratio of the dimensionless shape factor for the pore (F_s) and pore radius (μm), as follows [31].

$$\frac{S}{V} = \frac{F_s}{r} \tag{3}$$

Combining Equations (2) and (3), we have

$$T_{2,\text{surface}} = \frac{1}{\rho_2 F_s} r \tag{4}$$

For a given sample, its surface relaxivity and shape factor can be assumed to be constant. Thus,

$$T_2 = Cr \tag{5}$$

where $C = 1/(\rho_2 F_s)$. C is a constant conversion coefficient (ms/μm) and can be determined by combining with mercury intrusion method.

In the NMR experiment, the T_2 decay curve, which consists of many exponential decay components can be obtained directly. As for the rock sample composed of various sizes pores that correspond to various decay components, the different pore size is exponential to different T_2. And the sum of magnetization vector is superimposed by magnetization vectors of pore radius [34]:

$$M(t) = \sum_i M_i e^{-t/T_{2i}} \tag{6}$$

where $M(t)$ is the sum of the magnetization vector measured at time t, M_i is the ith magnetization vector corresponding to transverse relaxation time T_{2i}.

Using the inversion technology, the share of different T_2, i.e., the T_2 spectrum of a sample can be calculated based on Equation (6). The abscissa of the T_2 spectrum is the transverse relaxation time (T_2), and the ordinate is the amplitude. and the total amplitude of the T_2 spectrum represents the total pore volume in rock samples. Combining Equations (5) and (6), the amplitude of the T_2 spectrum can be converted into the share of different pores expressed as A_i. The larger the pore size is, the longer the relaxation time T_2 is. Thus, the volumetric percentage of the pores occupied by water in a rock sample after water flooding can be expressed as $A_i/A_{s\,i} \times 100\%$, while the rest volumetric percentage, $(1 - A_i/A_{si}) \times 100\%$, is gas occupying. Where A_{si} is the amplitude of the T_2 spectrum for the sample that is fully saturated by water. Therefore, the S_w in sample during water flooding can be expressed as the percentage of the sum of the T_2 spectrum amplitude from the sum of the T_2 spectrum amplitude for a fully saturated sample. S_{gr} of sample after water flooding is calculated as

$$S_{gr} = 100\% - S_w = 100\% - \frac{\Sigma A_i}{\Sigma A_{si}} \times 100\% \tag{7}$$

Combining Equations (5) and (7), the residual gas distribution in the pore space of a sample can be analyzed. In this study, the pore size distribution curves of samples #1-1M and #1-2M were measured by mercury intrusion method. The maximum pressure of the mercury intrusion was 186 MPa, corresponding to a pore size of 0.004 μm. The conversion coefficient C was obtained by comparing the weighted average value of the T_2 spectrum of each sample with the corresponding mercury intrusion curve, calculated by the following

$$C = \frac{\sum\limits_{i} T_{2i}A_i / \sum\limits_{i} A_i}{\sum\limits_{i} r_i S_i / \sum\limits_{i} S_i} \tag{8}$$

where T_{2i} is the NMR transverse relaxation time, A_i is the amplitude of the T_2 spectrum at T_{2i}, r_i is the sample pore radius obtained by mercury intrusion method, and S_i is the mercury injection saturation corresponding to r_i on the mercury intrusion curve, and $\sum\limits_{i} S_i$ is the maximum mercury injection saturation of sample in mercury intrusion experiment.

3. Results and Discussion

3.1. The T_2 Spectrum

The T_2 spectra of sample #1-1F acquired in spontaneous imbibition experiments at S_{wi1}, S_{wi2} and S_{wi3} are shown in Figure 3a–c, respectively. The measured T_2 spectra represents the properties of the entire rock sample, which reflects water distribution in all of pores in the whole rock sample. In Figure 3a–c, curve I shows the T_2 spectrum of the sample that is fully saturated with brine; curve II shows the T_2 spectrum of the sample at three S_{wi} conditions; and curve III shows the T_2 spectrum of the sample that is completely taken with spontaneous imbibition. To compare curves I and II, most of the initial gas is found existing in the range of 1–100 ms. Comparing curves II and III, it is observed that most of the residual gas is trapped in the range of relaxation time from 10 ms to 100 ms. According to the characteristics of water imbibing into the rock samples during the processes of spontaneous imbibition, the rock samples can be assumed to be strongly water-wet in this work. Subsequently, the S_{gr} in spontaneous imbibition experiments at S_{wi1}, S_{wi2} and S_{wi3} are obtained as 34.02%, 38.84% and 30.89%, respectively. In addition, the relative errors of samples porosity obtained by NMR testing and a routine gas porosity measurement method, which is the base of Boyle's law, are less than 5.33%, which fully validates the accuracy of NMR testing.

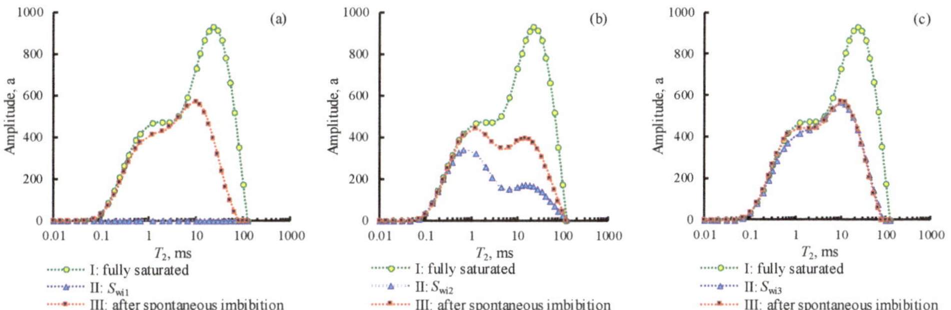

Figure 3. The T_2 spectrum of sample #1-1F that measured in the spontaneous experiments with three S_{wi}: (**a**) $S_{wi1} = 0$; (**b**) $S_{wi2} = 39.34 \pm 1.57\%$; (**c**) $S_{wi3} = 62.98 \pm 2.22\%$.

The T_2 spectrums of samples #1-1F and #1-2F acquired in forced imbibition experiments at the same three S_{wi} conditions (0, 39.34 ± 1.57 and 62.98 ± 2.22%) and four waterflooding flow rates ($\log Ca$ = −8.68, −7.68, −6.68, −5.68) are shown in Figure 4a–c, respectively. In Figure 4, curve I and curve II are the same as those in Figure 3, while curves III, IV, V and VI show that the T_2 spectra of samples #1-1F and #1-2F displaced by the brine with flowrate of $\log Ca$ = −8.68, −7.68, −6.68 and −5.68, respectively. From Figure 4, the amplitude of the T_2 spectrum increases significantly at $\log Ca$ = −6.68 and $\log Ca$ = −5.68, indicating different capillary trapping mechanisms for the sample under different Ca conditions. Then, the S_{gr} in forced imbibition experiments at the same three S_{wi} conditions was calculated.

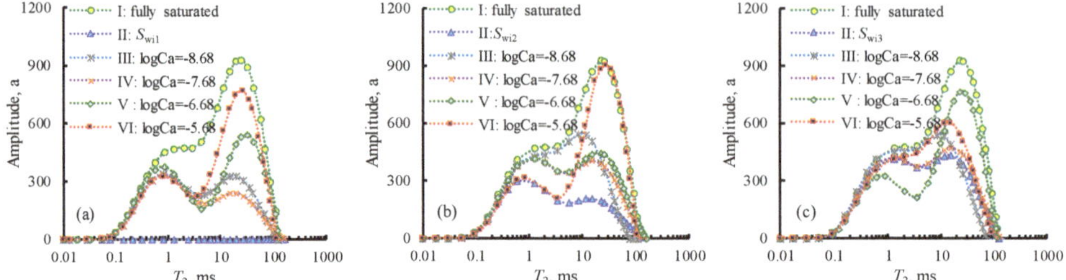

Figure 4. T_2 spectrum of samples #1-1F and #1-2F that measured in the forced imbibition experiments with three S_{wi} and four different flowrates: (**a**) S_{wi1} = 0; (**b**) S_{wi2} = 39.34 ± 1.57%; (**c**) S_{wi3} = 62.98 ± 2.22%. The curve I and curve II in the above pictures only presents the results of sample #1-1F for concise. ((**a**): curves III, IV and VI corresponding to sample #1-2F, curve V corresponding to sample #1-1F; (**b**): curves III, IV, V corresponding to sample #1-1F, curve VI corresponding to sample #1-2F; (**c**) curves III, IV, VI corresponding to sample #1-1F, curve V corresponding to sample #1-2F).

3.2. Pore Radius and Pore Size Distribution

Since each rock sample has its intrinsic conversion coefficient, C, it is impossible to describe the distribution of residual gas in the pore structure of a sample quantitatively by using the T_2 spectrum alone [29]. With the mercury intrusion technique, the pore size distribution of the samples #1-1F and #1-2F, or the samples #1-1M and #1-2M, can be obtained. The pore radius was in the range of 0.004 μm–43 μm with an average of 6.20 μm for sample #1-1F, while the pore radius was between 0.004 μm and 37.17 μm with an average of 6.26 μm for sample #1-2F. Using Equation (8), the conversion coefficients can be calculated. The conversion coefficients of samples #1-1F and #1-2F were 2.87 ms/μm and 3.32 ms/μm, respectively. In this work, the sizes of the sample pores (pore radius) are divided into three scales; < 1 μm, 1–10 μm and > 10 μm, to represent micro-, meso- and macro-pores, respectively [29].

3.3. Initial Water, Produced Gas and Residual Gas Distribution

Based on the induced conversion coefficients and Equations (5) and (8), the pore distribution of initial water, residual gas and produced gas in micro-, meso- and macro-pores in spontaneous imbibition are calculated and plotted in Figure 5, along with their corresponding percentages of pore distribution in each range of pore size. As shown in Figure 5, the sum of initial water, residual gas and produced gas in micro-, meso- and macro-pores represents the total pore space (expressed as 100%). The grey volume represents the percentage of produced gas from the pores of the corresponding size, while the orange is residual gas and the blue is initial water. In spontaneous imbibition experiments: (i) at S_{wi1}, most of gas in micropores (89.62%) and mesopores (70.46%) were produced, while most of the gas in macropores (82.25%) were retained; (ii) at S_{wi2}, most of the micropores (70.49%) were occupied by the initial water that was kept in the micropores all the time, and most of the residual gas remained in meso- and macro-pores; (iii) at S_{wi3}, most of the residual gas were retained in meso- and macro-pores, and most of the initial water were kept in micro- and meso-pores. Namely, gas can be produced from all sizes of pores at S_{wi1} and S_{wi2}. However, no gas was produced from meso- or macro-pores at S_{wi3} (Figure 5). More than 90% of residual gas existed in meso- and macro-pores at all three S_{wi} conditions.

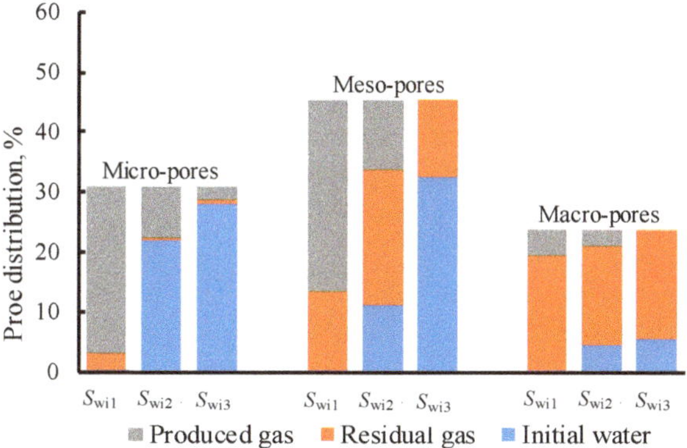

Figure 5. The pore distribution of initial water, residual gas and produced gas in micro-, meso- and macro- pores in spontaneous imbibition experiments of sample #1-1F at different S_{wi}.

In the same way, the pore distribution of the initial gas, residual gas and produced gas in micro-, meso- and macro-pores in forced imbibition at the four $\log Ca$ and three S_{wi} conditions are also plotted in Figure 6a–c. From Figure 6a, in forced imbibition at S_{wi1} condition, most of the gas in micropores (81.46–87.90%) were produced under all $\log Ca$ conditions, while most of residual gas was found in meso- (64.91–74.07%) and macro-pores (70.94–74.55%) at low $\log Ca$ ($\log Ca = -8.68, -7.68$). Meanwhile, most of the gas was produced from meso- (68.73%) and macro-pores (76.90%) at high $\log Ca$ ($\log Ca = -5.68$). From Figure 6b, in forced imbibition at S_{wi2} condition, most of the initial water was trapped in micropores (66.44–77.36%), some in mesopores, and a small amount in macropores at any $\log Ca$. About half of the gas was trapped in mesopores (44.59–48.26%), and a small amount in micropores. More than half of the macropores (58.00–75.22%) stored the residual gas at low $\log Ca$ ($\log Ca = -8.68, -7.68$) and about half of the gas in macropores could be produced at high $\log Ca$ ($\log Ca = -5.68$). In the case of forced imbibition in the S_{wi3} condition (Figure 6c), most of the initial water was kept in micro- (64.72–92.05%) and meso-pores (55.60–72.75%) at all $\log Ca$ values, while much of the residual gas were trapped in meso- (25.32–40.73%) and macro-pores (34.66–75.04%) at any $\log Ca$. A small amount of gas can be produced from pores in the S_{wi3} condition. Therefore, S_{wi} and Ca significantly affect the distributions of residual gas in micro-, meso- and macro-pores, and the amount of gas that can be produced from each size of pores. In addition, we also noticed nonuniform initial conditions for S_{wi3} in forced imbibition experiments. It is very difficult to build identical S_{wi} in the four groups of forced imbibition experiments at the same Ca. Therefore, S_{wi} built in rock samples is usually a range for imbibition experiments at the same Ca, which can lead to the nonuniform initial conditions. In addition, the hysteresis and capillary end effects can also cause the nonuniform initial conditions [23,35]. However, their influence on the residual gas distribution in rock samples can be negligible compared with Ca. The hysteresis and capillary end effects are neglected in this work.

3.4. Impacts of S_{wi} on Residual Gas Distribution

In the spontaneous imbibition experiments, capillary forces dominate the imbibition process [32]. The initial water imbibition rate, defined as the ratio of water imbibition volume and imbibition time from the beginning to the quasi-stable state of spontaneous imbibition experiment, is determined by capillary forces, which are related to the pore structure of the sample and S_{wi} [36]. In this work, the initial water imbibition rate at S_{wi1}, S_{wi2}, and S_{wi3} were obtained as 0.0141 mL/min, 0.0024 mL/min and 0.0008 mL/min, respectively. That is to say, the larger the S_{wi}, the smaller the initial water imbibition rate.

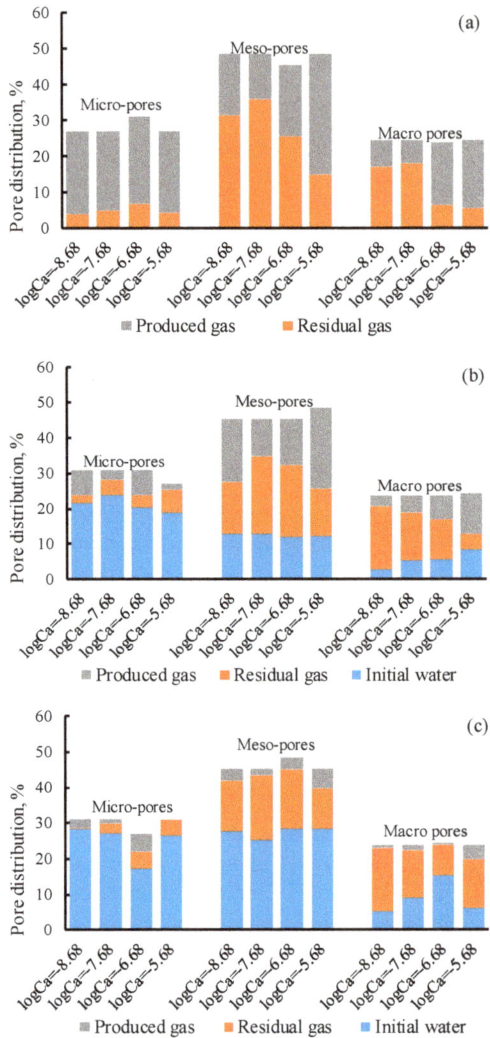

Figure 6. The pore distribution of initial water, residual gas and produced gas in micro-, meso- and macro- pores in forced imbibition experiments of samples #1-1F and #1-2F at different *Ca* and different S_{wi} (the details of experiment samples are consistent with Figure 4): (**a**) $S_{wi1} = 0$; (**b**) $S_{wi2} = 39.34 \pm 1.57\%$; (**c**) $S_{wi3} = 62.98 \pm 2.22\%$.

During the spontaneous imbibition processes, the gas porosity distribution (A_{gi}) in sample at S_{wi1} and S_{wi2}, which is calculated by Equation (9), are shown in Figure 7. In Figure 7, the curves labeled as pore distribution show the porosity distribution of the sample with pore radius, and the sum of porosity distribution represents the total porosity (ϕ) of the sample. The curves labeled with imbibition time (min) represent the gas porosity distribution (A_{gi}) in the sample during spontaneous imbibition experiments, while the curves labeled as residual gas represents the porosity distribution of residual gas. From Figure 7a, a significant decrease of spectral amplitude at the imbibition time of 7 min and 23 min indicated that much of the gas initially found in micro-pores was forced out. Then, the spectral amplitude corresponding to gas in micro- and meso-pores decreased simultaneously at 87 min, 135 min and 167 min. This explains that there is an obvious thin-film flow process ahead of the bulk flow for the experiment at S_{wi1} (such as 7 min and 23 min). Namely, water first fills the small pores and forms water film on the surfaces of large pores, while the thickness of water film depends on the surface roughness, wettability and interfacial tension [2,36]. Subsequently, the bulk water flow occurs, and bulk flow channels are composed of micro- and meso- pores in samples (87 min, 135 min and 167 min in Figure 7a). Some large gas clusters, initially gathered from thin-film flow, finally fill into the macro-pores surrounded by the bulk water phase. In the end, 90.61% of residual gas remains in the meso- and macro-pores for S_{wi1}, which agrees well with the claim that 85% of residual gas remains as larger gas clusters [18]. Whereas for S_{wi2}, the connectivity of initial gas phase (especially in micro- and meso-pores) is worse. The thin-film flow rarely occurs, as initial water has existed in samples with S_{wi2} (Figure 5), while the bulk water flow slowly occurs during the entire process of the spontaneous imbibition experiment (Figure 7b). The snap-off trapping is more significant with the lower initial water imbibition rate of 0.0024 mL/min [2,24], which leads to more gas (59.19%) trapped in mesopores at S_{wi2}. Although the initial water imbibition rate (0.0008 mL/min) is minimum for S_{wi3}, 90.84% of micropores and 71.47% of mesopores have been occupied by initial water, which indicates that water flow channels have existed in samples from inlet to outlet before the spontaneous imbibition processes (Figure 5). Finally, only 5.42% of the original gas is produced at S_{wi3}, while 97.34% of residual gas exists in meso- and macro-pores (Figure 5). Therefore, residual gas primarily remains in macropores for S_{wi1} (51.52%) and S_{wi3} (54.81%) under spontaneous imbibition, while in mesopores for S_{wi2} (59.19%). However, the residual gas distribution in the pore structure of sample is basically consistent for three S_{wi} in the spontaneous imbibition experiments (Figure 5).

$$A_{gi} = \left(\frac{A_{si} - A_i}{\Sigma A_{si}}\right) \times \phi \tag{9}$$

$$CTE = \frac{S_{gr}}{1 - S_{wi}} \times 100\% \tag{10}$$

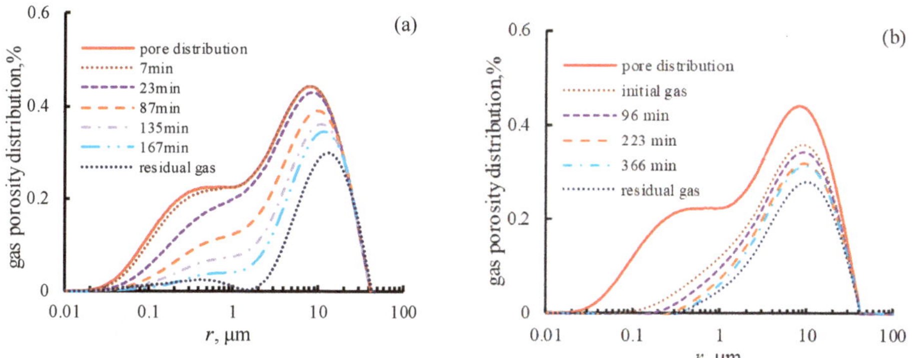

Figure 7. The gas porosity distribution in sample #1-1F during the spontaneous imbibition experiments: (a) $S_{wi1} = 0$; (b) $S_{wi2} = 39.34 \pm 1.57\%$.

3.5. Impacts of Ca on Residual Gas Distribution

In the forced imbibition experiments, viscous forces increase with the increasing Ca [8]. Figure 8 depicts the capillary trapping efficiency (*CTE*) of the gas phase (calculated by Equation (10)) and the experimental time for experiments to reach a stable state in the imbibition experiments. In Figure 8, the green, blue and red solid points represent the spontaneous imbibition experiments (*Ca* is calculated based on the initial water imbibition rate) at S_{wi1}, S_{wi2} and S_{wi3} respectively, while green, blue and red hollow points represent the forced imbibition experiments at S_{wi1}, S_{wi2} and S_{wi3} respectively. *Ca* in spontaneous imbibition experiments is larger when the S_{wi} is smaller. If *Ca* in forced imbibition experiment is less than that in the spontaneous imbibition experiment at the same condition of S_{wi}, capillary forces will dominate the waterflooding. Otherwise, viscous forces and capillary forces will together control the waterflooding. Figure 8a indicates that the *CTE* increases as S_{wi} increases, and *CTE* at S_{wi3} is 60% higher than that at S_{wi1} in the spontaneous imbibition experiments. Since a small amount of gas (only about 6% of the initial gas) was produced at S_{wi3}, showing a bewildering welter with log*Ca* (Figure 8a), we will focus on S_{wi1} and S_{wi2} conditions in the forced imbibition experiments. Based on the NMR T_2 spectrum (Figure 4), it can also be found that the bulk flow channels are mainly composed of micro- and meso- pores for S_{wi1} and S_{wi2} at log*Ca* = −8.68 and −7.68. At log*Ca* = −6.68, the bulk flow in meso- and macro- pores begins to become obvious. For log*Ca*= −5.68, the T_2 spectral amplitude of meso- and macro-pores most significantly increases, especially for macro-pores. This fact indicates that under the action of viscous forces, the bulk flow channels are mainly composed of meso- and macro- pores for the smaller resistance, and the residual gas in macropores is minimum with log*Ca* = −5.68 for S_{wi1} and S_{wi2} (Figure 6). Meanwhile, comparing Figures 5 and 6, it can be observed that S_{wi} has little influence on the residual gas distribution in spontaneous imbibition experiments, but *Ca* has greater effects (especially in macro- pores) at conditions of S_{wi1} and S_{wi2}. It implies that water flow preferential channels were formed in samples going through the inlet and outlet before imbibition starts to work, *Ca* will not affect the residual gas distribution. The critical *Ca* was not observed from experimental results. In this work, the confining pressure is set as 2 MPa and the pore pressure is less than 0.1 MPa during imbibition experiments. Meanwhile, rock samples used in this work are of high permeability and porosity. Therefore, the influences of rock stress sensitivity and gas slippage effect on the experimental results are negligible. In addition, Figure 8b indicates that the experimental time to reach a stable state increases with increasing S_{wi} in the spontaneous imbibition experiments, which is contrary to the forced imbibition experiments. Meanwhile, the water injection pore volumes (PVs) to reach a stable state are 0.66–1.98, 0.43–1.73 and 0.11–1.08 for S_{wi1}, S_{wi2}, and S_{wi3} in the forced imbibition experiments, respectively, and the PVs increases with increasing *Ca*. The water injection pore volumes to reach a stable state are 0.97, 0.37 and 0.12 for S_{wi1}, S_{wi2}, and S_{wi3} in the spontaneous imbibition experiments, respectively.

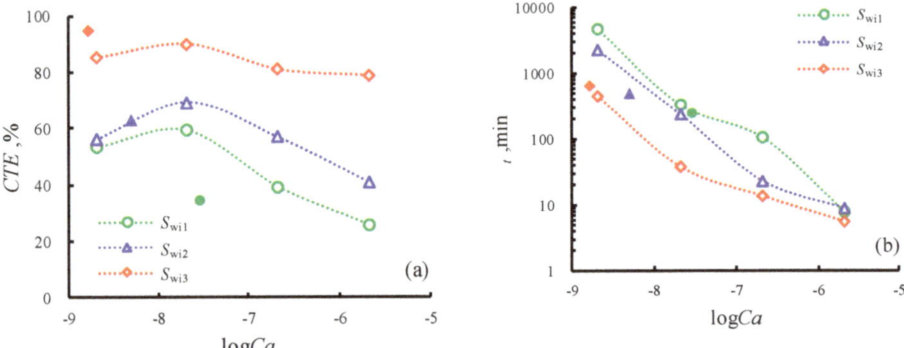

Figure 8. The capillary trapping efficiency and time for experiments to reach a stable state in the imbibition experiments (solid points represent the spontaneous imbibition experiments and hollow points represent the forced imbibition experiments, while the details of experiment samples refer to Figures 3 and 4): (**a**) capillary trapping efficiency; (**b**) the time for experiments to reach stable state.

In conclusion, during spontaneous and forced imbibition processes, most of the residual gas is trapped in larger pores as a continuous or discontinuous phase. *Ca* and S_{wi} jointly determine the residual gas distribution characteristics in rock samples. For a gas reservoir with a lower S_{wi} (such as S_{wi1} and S_{wi2}, which are related to the pore structure and wettability of reservoir rocks), the initial water phase is separately distributed in pores. With reasonable adjustment of gas production rate, the residual gas in larger pores of reservoir rocks and the *CTE* of reservoir rocks can effectively be reduced so as to raise recovery efficiency of gas reservoirs. However, if S_{wi} in reservoir rocks is at a higher level (such as S_{wi3}), continuous and stable water flow channels will be formed in reservoir rocks, which leads to a high *CTE*. Accordingly, it rarely works to decrease the *CTE* and increase the recovery of gas reservoirs by adjusting gas production rate. The measurement of drainage gas recovery in gas wells should be carried out to improve the recovery of gas reservoirs.

4. Conclusions

In this work, the impacts of *Ca* and S_{wi} on the microscopic residual gas distribution in core samples were investigated by both forced and spontaneous imbibition coupled with NMR. From this study, the following conclusions were drawn.

The conversion coefficients of samples were 2.87 ms/μm and 3.32 ms/μm, respectively. The larger the S_{wi}, the lower the initial imbibition rate. In the spontaneous imbibition experiments, the residual gas distribution is basically consistent for all three S_{wi} (0, 39.34 ± 1.57%, and 62.98 ± 2.22%), and above 90% of residual gas exists in pores larger than 1 μm. In addition, the capillary trapping efficiency of the gas phase increases as S_{wi} increases. The experimental time to reach a stable state increases with increasing S_{wi} in the spontaneous imbibition experiments, which is contrary to the forced imbibition experiments.

Ca significantly impacts the residual gas distribution in the pores of sample. The bulk flow channels are mainly composed of pores smaller than 10 μm at log*Ca* = −8.68 and −7.68. As log*Ca* increases to −5.68, the bulk flow channels are mainly composed of the pores larger than 1 μm. With log*Ca* increasing to −6.68 and −5.68 at S_{wi} = 0 and 39.34 ± 1.57%, the residual gas distribution in pores larger than 10 μm significantly decreases, and that in pores smaller than 1 μm slightly increases. The impact of *Ca* on the residual gas distribution is greater than the S_{wi} for S_{wi} = 0 and 39.34 ± 1.57%, while *Ca* and S_{wi} have little effect on the residual gas distribution for S_{wi} = 62.98 ± 2.22%. This work improves the understanding on the trapping mechanism and water-gas migration in reservoirs.

Author Contributions: Each author has made contributions to the present paper. Conceptualization, methodology and formal analysis, M.L., Y.W. and T.L.; investigation, T.L., J.J. and L.C.; data curation and writing, T.L., Y.W. and Z.W.; supervision and funding acquisition, M.L.

Funding: The project is supported by the open fund (PLN201601) of the State Key Laboratory of Oil and Gas Reservoir Geology and Exploitation (Southwest Petroleum University), the joint fund of the National Natural Science Foundation of China (No. U1562217) and the National Natural Science Foundation of China (No. 41304081).

Conflicts of Interest: The authors declare no conflict of interest.

References

1. Lu, W.Y.; Huang, B.X. Numerical simulation of migration characteristics of the two-phase interface in water-gas displacement. *Energy Explor. Exploit.* **2018**, *36*, 246–264. [CrossRef]
2. Geistlinger, H.; Mohammadian, S.; Schlueter, S. Quantification of capillary trapping of gas clusters using X-ray microtomogrephy. *Water Resour. Res.* **2014**, *50*, 4514–4529. [CrossRef]
3. Meng, Q.B.; Liu, H.Q.; Wang, J. Entrapment of non-wetting phase during co-current spontaneous imbibition. *Energy Fuels* **2015**, *29*, 686–694. [CrossRef]
4. Qian, K.; Yang, S.L.; Dou, H.G.; Wang, Q.; Wang, L.; Huang, Y. Experimental investigation on microscopic residual oil distribution during CO_2 Huff-and-Puff process in tight oil reservoirs. *Energies* **2018**, *11*, 2843. [CrossRef]
5. Wang, L.; Yang, S.L.; Peng, X.; Deng, H.; Meng, Z.; Qian, K.; Wang, Z.L.; Lei, H. An improved visual investigation on gas–water flow characteristics and trapped gas formation mechanism of fracture–cavity carbonate gas reservoir. *J. Nat. Gas Sci. Eng.* **2018**, *49*, 213–226. [CrossRef]
6. Hamon, G.; Suzanne, K.; Billiotte, J.; Trocme, V. Field-wide variations of trapped gas saturation in heterogeneous sandstone reservoirs. In Proceedings of the SPE Annual Technical Conference and Exhibition, New Orleans, LA, USA, 30 September–3 October 2001.

7. Wang, S.; Li, Y.; Sun, L.; Wei, Y.; Wei, X.; Zhang, F. Measurements of residual gas saturation by unidirectional spontaneous imbibitions. *Pet. Sci. Technol.* **2014**, *32*, 2815–2821. [CrossRef]

8. Hekmatzadeh, M.; Dadvar, M.; Emadi, M.A. Visual investigation of residual gas saturation in porous media. *Int. J. Oil Gas Coal Technol.* **2015**, *10*, 161–178. [CrossRef]

9. Ding, M.; Kantzas, A. Capillary number correlations for gas-liquid systems. *J. Can. Pet. Technol.* **2007**, *46*, 27–32. [CrossRef]

10. Lin, H.; Zhang, S.C.; Wang, F.; Pan, Z.Q.; Mou, J.Y.; Zhou, T.; Ren, Z.X. Experimental investigation on imbibition-front progression in shale based on nuclear magnetic resonance. *Energy Fuels* **2016**, *30*, 9097–9105.

11. Suzanne, K.; Hamon, G.; Billiotte, J.; Trocmé, V. Experimental relationships between residual gas saturation and initial gas saturation in heterogeneous sandstone reservoirs. In Proceedings of the SPE Annual Technical Conference and Exhibition, Denver, CO, USA, 5–8 October 2003.

12. Batycky, J.; Irwin, D.; Fish, R. Trapped gas saturations in leduc-age reservoirs. *J. Can. Pet. Technol.* **1998**, *37*, 32–39. [CrossRef]

13. Rezaee, M.; Rostami, B.; Pourafshary, P. Heterogeneity effect on non-wetting phase trapping in strong water drive gas reservoirs. *J. Nat. Gas Sci. Eng.* **2013**, *14*, 185–191. [CrossRef]

14. Kumar, M.; Knackstedt, M.A.; Senden, T.J. Visualizing and quantifying the residual phase distribution in core material. *Petrophysics* **2010**, *51*, 323–332.

15. Suekane, T.; Zhou, N.; Hosokawa, T. Direct observation of trapped gas bubbles by capillarity in sandy porous media. *Transp. Porous Media* **2010**, *82*, 111–122. [CrossRef]

16. Iglauer, S.; Paluszny, A.; Blunt, M.J. Simultaneous oil recovery and residual gas storage: A pore-level analysis using in situ X-ray micro-tomography. *Fuel* **2013**, *103*, 905–914. [CrossRef]

17. Pentland, C.H.; Itsekiri, E.; Al Mansoori, S.K.; Iglauer, S.; Branko, B.; Blunt, M.J. Measurement of nonwetting-phase trapping in sandpacks. *SPE J.* **2010**, *15*, 274–281. [CrossRef]

18. Geistlinger, H.; Mohammadian, S. Capillary trapping mechanism in strongly water wet systems: Comparison between experiment and percolation theory. *Adv. Water Resour.* **2015**, *79*, 35–50. [CrossRef]

19. Tiab, D.; Donaldson, E.C. *Petrophysics*, 2nd ed.; Gulf Professional Publishing: Houston, TX, USA, 2004; p. 352.

20. Guo, H.; Dou, M.; Wang, H.Q.; Wang, F.Y.; Gu, Y.Y.; Yu, Z.Y.; Wang, Y.S.; Li, Y.Q. Proper Use of Capillary Number in Chemical Flooding. *J. Chem.* **2017**, *4307368*, 1–11. [CrossRef]

21. Zendehboudi, S.; Chatzis, I. Experimental Study of Controlled Gravity Drainage in Fractured Porous Media. *J. Can. Pet. Technol.* **2011**, *50*. [CrossRef]

22. Shen, Y.H.; Ge, H.K.; Meng, M.M.; Jiang, Z.X.; Yang, X.Y. Effect of water imbition on shale permeability and its influence on gas production. *Energy Fuels* **2017**, *31*, 4873–4980. [CrossRef]

23. Mohammadian, S.; Geistlinger, H.; Vogel, H.J. Quantification of Gas-Phase Trapping within the Capillary Fringe Using Computed Microtomography. *Vadose Zone J.* **2015**, *14*. [CrossRef]

24. Buchgraber, M.; Kovscek, A.R.; Castanier, L.M. A study of microscale gas trapping using etched silicon micromodels. *Transp. Porous Med.* **2012**, *95*, 647–668. [CrossRef]

25. Geistlinger, H.; Ataei-Dadavi, I.; Vogel, H.J. Impact of surface roughness on capillary trapping using 2D-micromodel visualization experiments. *Transp. Porous Med.* **2016**, *112*, 207–227. [CrossRef]

26. Herring, A.L.; Andersson, L.; Schlüter, S.; Sheppard, A.; Wildenschild, D. Efficiently engineering pore-scale processes: The role of force dominance and topology during nonwetting phase trapping in porous media. *Adv. Water Resour.* **2015**, *79*, 91–102. [CrossRef]

27. Yuan, Y.; Rezaee, R.; Verrall, M.; Hu, S.Y.; Zou, J.; Testmanti, N. Pore characterization and clay bound water assessment in shale with a combination of NMR and low-pressure nitrogen gas adsorption. *Int. J. Coal Geol.* **2018**, *194*, 11–21. [CrossRef]

28. Slijkerman, W.F.J.; Hofman, J.P. Determination of surface relaxivity from NMR diffusion measurements. *Magn. Reson. Imaging* **1998**, *16*, 541–544. [CrossRef]

29. Yang, P.; Guo, H.K.; Yang, D.Y. Determination of residual oil distribution during waterflooding in tight oil formations with NMR relaxometry measurements. *Energy Fuels* **2013**, *27*, 5750–5756. [CrossRef]

30. Testamanti, M.N.; Rezaee, R. Determination of NMR T2 cut-off for clay bound water in shales: A case study of Carynginia Formation, Perth Basin, Western Australia. *J. Pet. Sci. Eng.* **2017**, *149*, 497–503. [CrossRef]

31. Xu, H.J.; Fan, Y.R.; Hu, F.L.; Li, C.X.; Yu, J.; Liu, Z.C.; Wang, F.Y. Characterization of pore throat size distribution in tight sandstones with nuclear magnetic resonance and high-pressure mercury intrusion. *Energies* **2019**, *12*, 1528. [CrossRef]

32. Lyu, C.H.; Wang, Q.; Ning, Z.F.; Chen, M.Q.; Li, M.Q.; Chen, Z.L.; Xia, Y.X. Investigation on the application of NMR to spontaneous imbibition recovery of tight sandstones: An experimental study. *Energies* **2018**, *11*, 2359. [CrossRef]
33. Chen, T.; Yang, Z.M.; Luo, Y.T.; Lin, W.; Xu, J.X.; Ding, Y.H.; Niu, J.L. Evaluation of displacement effects of different injection media in tight oil sandstone by online nuclear magnetic resonance. *Energies* **2018**, *11*, 2836. [CrossRef]
34. Blümich, B.; Casanova, F.; Appelt, S. NMR at low magnetic fields. *Chem. Phys. Lett.* **2009**, *477*, 231–240. [CrossRef]
35. Andersen, P.Ø.; Standnes, D.C.; Skjæveland, S.M. Waterflooding oil-saturated core samples-Analytical solutions for steady-state capillary end effects and correction of residual saturation. *J. Pet. Sci. Eng.* **2017**, *157*, 364–379. [CrossRef]
36. Kibbey, T.C.G. The configuration of water on rough natural surfaces: Implications for understanding air-water interfacial area, film thickness, and imaging resolution. *Water Resour. Res.* **2013**, *49*, 4765–4774. [CrossRef]

Article

Numerical Investigation of Downhole Perforation Pressure for a Deepwater Well

Qiao Deng [1], Hui Zhang [1,*], Jun Li [1], Xuejun Hou [2] and Binxing Zhao [3]

[1] College of Petroleum Engineering, China University of Petroleum-Beijing, 18 Fuxue Road, Changping 102249, China; bridges0307@hotmail.com (Q.D.); lijuncup2018@163.com (J.L.)

[2] School of Petroleum Engineering, Chongqing University of Science & Technology, 20 East Road, University City, Shapingba District, Chongqing 401331, China; xuejunhou1973@163.com

[3] School of Optoelectronic Science and Engineering, University of Electronic Science and Technology of China, Chengdu 610054, China; zhaobxing@uestc.edu.cn

* Correspondence: zhanghuicup2018@163.com; Tel.: +86-199-1988-1457

Received: 29 August 2019; Accepted: 4 October 2019; Published: 8 October 2019

Abstract: During the production of deepwater wells, downhole perforation safety is one of the key technical problems. The perforation fluctuating pressure is an important factor in assessing the wellbore safety threat. Due to difficulty in describing the downhole perforation pressure by using the existing empirical correlations, a prediction model based on data fitting of a large number of numerical simulations has been proposed. Firstly, a numerical model is set up to obtain the dynamic data of downhole perforation, and the exponential attenuation model of perforation pressure in the wellbore is established. Secondly, a large number of numerical simulations have been carried out through orthogonal test design. The results reveal that the downhole perforation pressure is logarithmic to the total charge quantity, increases linearly to the wellbore initial pressure, shows an exponential relationship with downhole effective volume for perforation, and has a power relationship with the thickness of casing and cement as well as formation elastic modulus. Thirdly, the prediction of perforation peak pressure at different positions along the wellbore agrees well with the field measurement within a 10% error. Finally, the results of this study have been applied in the field case, and an optimization scheme for deepwater downhole perforation safety has been put forward.

Keywords: deepwater well; perforation safety; peak pressure; numerical model; orthogonal test

---

1. Introduction

With the improvement of oil and gas resources exploitation technology, more and more offshore oil and gas resources have been exploited in recent years [1,2]. The exploitation of deepwater oil and gas resources has increased gradually, and the safety of downhole perforation is an important technical problem [3]. The shaped charge jet perforation has been widely used in deepwater completion operations, the purpose of which is to form a channel between reservoir and wellbore that can effectively transport oil and gas [4,5]. On the one hand, the energy released by the explosion of perforating charge is converted into the effective energy to penetrate casing. On the other hand, part of the energy will be released into the wellbore, resulting in the pulsation of perforation pressure in the wellbore. This is the beginning of the wellbore dynamic effect in the process of perforation, in which the pulsating pressure of perforation will produce strong impact loads on the downhole tools and affect the stability of the whole perforation system [6–8]. Predicting the magnitude of downhole perforation pressure is a critical step for the safety analysis of deepwater perforation.

Perforation testing refers to the process of connecting perforating gun, upper operating string, and packer into a cluster for downward entry. The downhole perforation system for the deepwater wells is shown in Figure 1, which shows that the wellbore is filled with fluid after the packer is set. With the increase of water depth, the perforation process of deepwater wells is more complex and difficult. This is bound to greatly increase the downhole perforation pressure and wellbore safety risk. The normal exploitation of oil and gas will be affected, and even some irreparable damage may be caused.

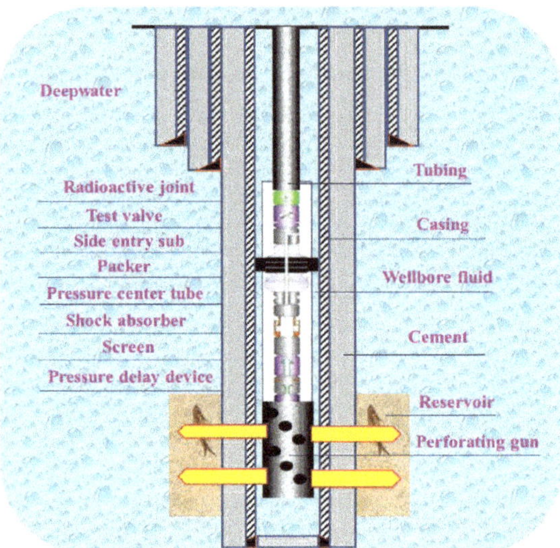

Figure 1. Downhole perforation system.

The fluctuating pressure of perforation mainly comes from the asymmetric dynamic loads, which are produced by the explosion of many perforating bullets arranged in a certain phase in the perforating gun. Progress has been made in this field in recent years. Lu et al. [9] proposed a perforation pressure equation based on reservoir conditions by using the superposition principle, with preliminary theoretical derivation and qualitative analysis. Zhao et al. [10,11] qualitatively studied the characteristics of perforation pressure fluctuation in the process of composite perforation by combining theoretical research with experimental tests. Chen et al. [12] analyzed the dynamic load of downhole perforation by using theoretical empirical formula and simulation software. Sanders et al. [13] proposed that the pulsating pressure of downhole perforation would increase the risk of downhole equipment damage in deepwater well. Yang et al. [14] indirectly calculated the peak pressure of downhole perforation by calculating the energy of detonation gas based on the detonation theory. Zhou et al. [15] used the load test system to measure perforation pressure data downhole. Bale et al. [16] developed a commercial perforation software to simulate the dynamic downhole perforation pressure of deepwater wells. Li et al. [17] pointed out that the change and distribution of perforation impact pressure in the wellbore with time is not clear and needs further investigation. Liu et al. [18] set up a pressure field model of shaped charge to study the downhole perforation impact pressure. These studies have promoted the research progress of downhole perforation pressure. However, the functional relationships between downhole perforation pressure and different influencing factors are not clear. The distribution of perforation pressure along the wellbore is not clear. There is a lack of an accurate method to predict the magnitude of perforation pressure by considering multiple downhole influencing factors.

There are three main methods to study the dynamic pressure of downhole perforation, which are the empirical formula calculation, laboratory test, and numerical simulation [19]. However, due to the asymmetric load of perforating bullets and the complexity of the downhole dynamic environment, it is difficult to comprehensively describe the whole perforation process by relying solely on the existing empirical formulas. The data obtained by the laboratory test for perforation is limited. Therefore, the numerical simulation method of non-linear dynamics can be applied to describe the three-dimensional dynamic process of downhole perforation.

In this study, the parameters setting of the model materials and the state equations have been explored. By setting the fluid-structure coupling, the modeling method of multiple perforating bullets is formed. The complex perforating conditions are successfully simulated by LS-DYNA software on a large computer. The physical process of perforation with hundreds of perforating bullets has been innovatively simulated. Based on the simulated data, the functional relationships between perforation peak pressure and different influencing factors have been obtained. A prediction model to calculate the downhole perforating peak pressure has been proposed. The perforation peak pressure at different positions in the wellbore can be obtained.

2. Empirical Formulas of Perforation Pressure

At present, the calculation of downhole perforation pressure is mainly based on the empirical formulas of the explosive explosion. One of them is referring to the method of measuring explosive performance in an isolated container. It is assumed that the perforating charge is put in the sealed casing and the remaining space of which is filled with air. The empirical formula of explosive explosion pressure is used to calculate perforation pressure, as shown in [20]:

$$P = \frac{f\omega}{V - \alpha\omega} \tag{1}$$

where P is the explosive pressure, MPa; f is the explosive specific energy, J/kg; ω is the perforating charge quantity, kg; V is the effective volume of the casing, L; $\alpha = 0.001v$, v is the Hexogen specific volume, L/kg.

By equating the perforating charge explosion with the underwater explosion test, the perforating pressure at different locations is calculated by [21]:

$$P = K\left(\frac{\omega^{1/3}}{R}\right)^{\beta} \tag{2}$$

where K is the tested coefficient, 44.5 for spherical TNT explosive, dimensionless; ω is the explosive quality, kg; R is the distance from the detonation center, m; β is the attenuation coefficient, 1.5 for spherical TNT explosive, dimensionless.

Based on the detonation theory, the energy of detonation products is determined according to the detonation parameters of the perforating charge. The explosion pressure of perforation is obtained by [22]:

$$P = 1.558\rho^2 N \sqrt{MQ} \tag{3}$$

where ρ is the explosive density, g/cm^3; N is the molar number of the detonation gas products per unit mass of explosive, mol/g; M is the average molar mass of detonation gas components, g/mol; Q is the detonation heat, KJ/mol.

Assuming that a certain proportion of explosive energy will be converted into wellbore fluid internal energy, a certain increment of pressure can be produced in the wellbore. After perforation, the wellbore pressure is obtained as [23]:

$$P = P_0 + \varphi(\delta - 1)\frac{n}{V}\frac{m}{M}Q \tag{4}$$

where P_0 is the wellbore initial pressure, MPa; φ is the explosive energy transfer rate, dimensionless; δ is the imaginary gas-liquid ratio in the wellbore, dimensionless; n the number of perforating bullets; m is the charge per hole, g; M is the average molar mass of detonation gas components, g/mol; V is the wellbore effective volume, m^3; Q is the detonation heat, KJ/mol.

The idea of the above empirical formulas is to calculate the explosion pressure of perforation by the equivalent explosion of explosives. It is assumed that other calculation parameters remain constant: charge density is 1.80 g/cm^3; inner diameter of casing is 0.157 m; casing length is 2.5 m; inner diameter of tubing is 0.059 m; tubing length is 1 m; inner diameter of perforating gun is 0.105 m; gun length is 1 m; residual volume of charge is 0.83 L/kg; explosive specific energy is 1181861 J/kg; molecular mass of explosive is 222.12 g/mol; molar number of explosive per unit mass is 0.03377; detonation heat is 6205.23 KJ/mol. The initial wellbore pressure is 10 MPa, with 30% of explosive energy converted into wellbore pressure during perforation. The total charge quantity of perforation is calculated as a variable by Equations (1)–(4), which is changed as 2 kg, 4 kg, 6 kg, 8 kg, 10 kg. The calculation results are shown in Figure 2.

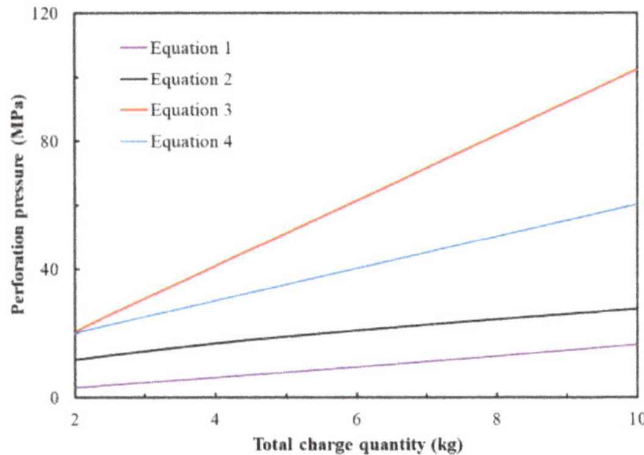

Figure 2. Calculation results from different empirical formulas.

Based on the above calculation results, it can be seen that the perforation pressure increases with the increase of the total charge quantity. The pressure values of perforation calculated Equations (1)–(4) are quite different. The actual downhole perforation conditions are complex, with many factors affecting the fluctuating pressure of perforation. The above empirical formulas cannot consider the factors comprehensively, especially for the influence of formation conditions. The calculation results of the above empirical formulas are inaccurate. It is important to account for the reservoir condition since it can act as an absorber or reflector (based on porosity, permeability, etc.) of shock waves that eventually affects the propagation and strength of shock waves within the wellbore [24,25]. Therefore, these theoretical empirical formulas can only be used as a preliminary reference for qualitative analysis of downhole perforation pressure. The above analysis shows that the charge of perforation, the effective space of the wellbore, the wellbore initial pressure, and so on are the important factors affecting the explosion pressure of perforation. Therefore, further research is needed to explore a more accurate calculation model for perforating pressure.

Due to high flexibility and ability, the technologies of numerical simulation have been widely applied for the study of complex non-linear problems in recent years. The simulation of perforation completion has undergone a process from simple to complex, and the perforation model has gradually developed from two-dimensional to three-dimensional. At present, the perforation simulation analysis has entered the stage of finite element simulation calculation with the further development of computer technologies. Some complex non-linear problems with high calculation accuracy have been solved with the premise of adapting to various complex mathematical models. The numerical simulation has many advantages and has become an effective means of piercing engineering analysis, in which the whole dynamic process can be presented by numerical simulation [26,27].

3. Modeling and Meshing

The whole process of perforation includes charge explosion, jet formation, casing penetration, and fluid-solid coupling. The sequential explosion of perforating bullets has an extremely complex coupling superposition effect with increasing shock loads in the wellbore. At present, the finite element software of LS-DYNA or AUTODYNA has been widely used to simulate the whole dynamic process. Lee [28] simulated the process of jet formation and penetration into the casing wall based on the 2D Euler coding, the foundation of perforation numerical simulation was laid. Han et al. [29] studied the process of explosive explosion forming jet and penetrating casing wall during perforation by numerical simulation. Kang et al. [30] used the software of LS-DYNA to study the dynamic response of the perforating gun, and few numbers of perforating bullets were simulated in three-dimensional. Yang [31] simulated a single perforating bullet of the horizontal well perforation by the software of LS-DYNA, and the state equations and model parameters during perforation were explored. Li [32] simulated the fluctuation of perforation pressure by the software of AUTODYN using spherical explosive instead of perforating bullets. Yan et al. [33] studied the process of perforation to cement damage by the software of LS-DYNA, and the ALE (Arbitrary Lagrange–Euler) algorithm was used to describe the fluid-structure coupling phenomenon in the perforation process. The above researches have laid the foundation for modern perforation completion simulation and promoted its rapid development. The selection of state equations, the setting of model parameters, and the fluid-structure coupling phenomenon in the perforation process have been preliminarily studied. However, previous work often focused on a single or a few numbers of perforation bullets due to the complexity of perforation modeling and meshing with huge workloads. Without considering the influence of multiple factors on downhole perforation pressure, the simulation cannot reflect the actual working conditions of perforation.

In this section, the perforation model, with hundreds of perforating bullets, has been innovatively established. Several trial calculations have been carried out by exploring the parameters of the model materials and state equation and setting the fluid-solid coupling interface. The numerical simulations have been carried out smoothly.

3.1. Physical Model

A number of perforating bullets are distributed in the gun at a certain phase. The downhole perforation system of deepwater wells has many components, as shown in Figure 1. The physical model needs to be simplified for the convenience of simulation calculation. Make the following assumptions: Firstly, the cement or reservoir is assumed as isotropic materials by ignoring their heterogeneity; Secondly, all the downhole components in the perforation string system are regarded as isotropic tubing string materials. Through the above assumptions, a 3D physical model can be established by simplification, as shown in Figure 3. The perforating gun, tubing string, and casing are involved in the model with 300 perforating bullets with phase 45°, 63 g charge per hole. The air fills inside the remaining space of the gun. The wellbore fluid fills the annulus space of tubing and casing with the initial pressure of 15 MPa. The packer restrains the upper end of the string radially, and the circumference of the string is restrained by the casing with reservoir surrounding.

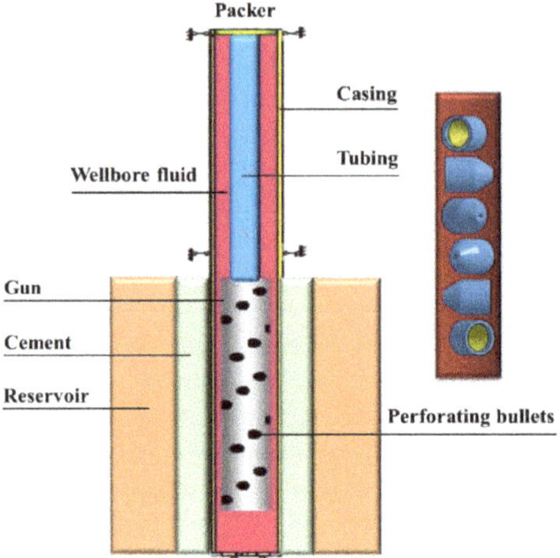

Figure 3. 3D physical model.

3.2. Model Meshing

The physical process of perforation includes complex fluid-structure interaction with a high strain rate, which is commonly solved by the BEM (boundary element method) and FEM (finite element method). The BEM is suitable for solving linear homogeneous problems, while FEM is suitable for solving nonlinear problems with complex boundaries [34,35]. Therefore, the finite element method can be used to simulate perforation. In LS-DYNA, there are mainly ALE (Arbitrary Lagrange–Euler) algorithm and SPH (Smoothed Particle Hydrodynamics) algorithm for fluid-structure interaction calculation. The mesh deformation caused by meshless can be avoided for the SPH algorithm [36]. However, the ALE algorithm is more stable under perforation conditions, accurately describing the deformation response of perforated string, cement, and reservoir without mesh non-convergence. The ALE algorithm can accurately simulate the formation and penetration of a shaped charge jet with a high strain rate and large deformation, which can be used in RDX (Royal Demolition Explosive) explosives, air, and fluid. The spatial position of the ALE grid remains unchanged with the material flows among the grids. The Lagrange algorithm is used in the downhole string. The Lagrange algorithm can be firstly executed by the ALE algorithm at each time step [37].

Since the model meshing is complex, the grid is built on the cylindrical boundary to make the whole area transition into a cuboid structure. The remaining 3/4 models can be obtained by rotating the replication command. The fluid area model of the perforating bullet and the torsion model are connected by twisting the air area model between the cylindrical segments, in which different perforating bullets are located. The twisting model is connected by replicating rotation. The fluid area model of several perforating bullets can be obtained. Through the above methods, the geometric parameters of the liner, charge, the height of the upper end face of the cuboid, and the number of corresponding meshes can be adjusted. Different sizes of the perforating bullets can be quickly established by changing the parameters. Multiple perforating bullets can be obtained by simple replication with the distance that can be quickly adjusted. Figure 4a shows the vertical section of the mesh model, and the vertical section of the mesh model of the perforating bullet is shown in Figure 4b.

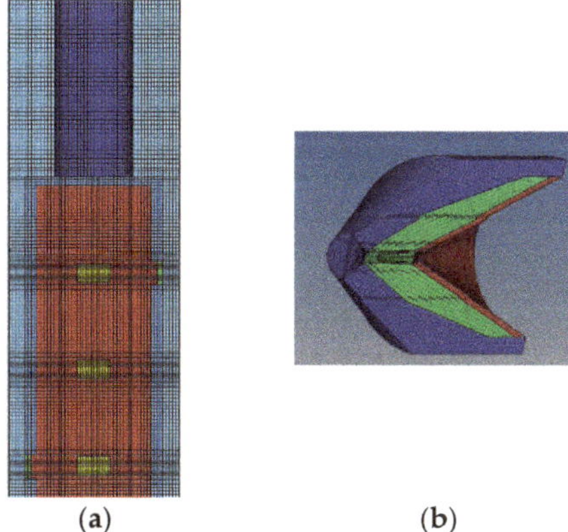

(a)   (b)

Figure 4. Meshing: (**a**) vertical section of mesh model; (**b**) mesh model of perforating bullet.

3.3. Material Model and State Equation

The selection of material parameters and the related equation of state have been briefly introduced in the section. The state equation of EOS_JWL (Jones-Wilkins-Lee) is adopted for RDX explosive, and the expression is shown in [38]:

$$P = A(1 - \frac{\omega}{R1V})e^{-R1V} + B(1 - \frac{\omega}{R2V})e^{-R2V} + \frac{\omega E}{V} \tag{5}$$

where p is the pressure of detonation product, MPa; $V = \frac{1}{\rho}$ is the specific volume, cm^3/g; $A, B, \omega, R1, R2$ are material constants; E is the explosive internal energy of unit volume, J/m^3. The parameters' values are listed in Table 1.

Table 1. Parameters' values of RDX (Royal Demolition Explosive).

ρ (g/cm^3)	D (cm/μs)	A (GPa)	B (GPa)	R_1	R_2	ω	E (J/m^3)
1.78	0.65	3.71	0.074	4.15	0.95	0.3	9×10^9

The MAT_NULL material model and state equation of EOS_GRUNEISEN are used for describing the wellbore fluid (water) dynamic response. The equation is shown in [39]:

$$P = \frac{\rho_0 C^2 \mu \left[1 + \left(1 - \frac{\gamma_0}{2}\right)\mu - \frac{a_1}{2}\mu^2 \right]}{\left[1 - (S_1 - 1)\mu - S_2^2\frac{\mu^2}{\mu^2+1} - S_3\frac{\mu^3}{(\mu+1)^2} \right]} + (\gamma_0 + a_1\mu)E \tag{6}$$

where C is the sound velocity in water, cm/μs; $S1, S2, S3$ are the constant coefficients, dimensionless; γ_0 is the Gruneisen constant, dimensionless; $\mu = \frac{\rho}{\rho_0} - 1$ is the water compressibility, dimensionless; a_1 is the first-order volume correction of γ_0, μ, dimensionless. The main parameter values are shown in Table 2.

Table 2. Water parameters of the Gruneisen state equation.

C(cm/μs)	S_1	S_2	S_3	γ_0	a_1	A	E_0	V_0
0.165	1.92	−0.0096	0	0.35	0.47	0	0	0.9925

The EOS of material air is shown as [40]:

$$P = (C_0 + C_1\mu + C_2\mu^2 + C_3\mu^3) + (C_4 + C_5\mu + C_6\mu^2)E \tag{7}$$

where $C_0, C_1, C_2, C_3, C_4, C_5, C_6$ are the EOS parameters, dimensionless. The values of the parameters are shown in Table 3.

Table 3. Air parameters.

ρ (g/cm^3)	C_0	C_1	C_2	C_3	C_4	C_5	C_6
0.001225	0	0	0	0	0.4	0.4	0

The Cowper–Symonds model is adopted for the shell and perforated string (gun, tubing, casing), as shown in [41]:

$$\sigma_y = \left[1 + \left(\frac{\varepsilon}{Q}\right)^{\frac{1}{P}}\right]\left(\sigma_0 + \beta E_p \varepsilon_P^{eff}\right) \tag{8}$$

where ε is the strain rate of materials, dimensionless; Q and P are the strain rate parameters, dimensionless; σ_0 is the initial value of yield stress, MPa; $E_P = \frac{E_{tan}E}{E-E_{tan}}$ is the plastic hardening modulus of materials, GPa; E_{tan} is the tangent modulus of materials, GPa; E is the young modulus of materials, GPa; ε_P^{eff} is the effective plastic strain of materials, dimensionless. According to the API (American Petroleum Institute) standard, the material parameters of the perforated string are shown in Table 4 [42].

Table 4. Parameters of perforated string.

Parameters	Steel Grade -	Length (m)	Density (kg/m^3)	Size (mm)	Yield Stress (MPa)	Elastic Modulus (GPa)	Poisson's Ratio -
Tubing	N80	20	7800	93/62	536	206	0.3
Casing	N80	35	7800	245/221	460	206	0.25
Gun	N80	9	-	178/153	550	206	0.3

The cement and reservoir can be described by the material model of Holmquist–Johnson–Cook model [38]:

$$\sigma = f_c\left[Q_1(1 - Q_2) + Q_3 P^{*N}\right]\left[1 + Q_4 \ln(\varepsilon^*)\right] \tag{9}$$

where f_c is the uniaxial compression strength, MPa; P^{*N} is the normalized pressure value, dimensionless; ε^* is the normalized strain rate, dimensionless; Q_1, Q_2, Q_3 are the constants of the material, dimensionless; Q_4 is the damage value, dimensionless. In order to reduce the amount of calculation, the cement and reservoir model is only established near the perforation section. The modeling parameters are shown in Table 5.

Table 5. Cement and reservoir parameters.

Parameters	Density (kg/m^3)	Length (m)	Thickness (m)	Inner Diameter (mm)	Elastic Modulus (GPa)	Poisson's Ratio -
Cement	2560	20	0.03	245	206	0.3
Reservoir	2210	20	3	545	1.27	0.35

The numerical model is input into the LS-DYNA program in the form of a K file for calculation. The parameters of numerical simulation are defined by keywords. * INCLUDE and * INCLUDE TRANSFORM: importing the grid model information; * INITIAL_DETONATION: defining the initiation point and time of explosives; * ALE_MULTI-MATERIAL_GROUP: making the material flows with each other in the fluid domain grid; * CONSTRAINED_LAGRANGE_IN_SOLID: realizing the fluid-solid coupling; * SECTION_SOLID and * SECTION_SOLID_ALE: setting the grid cell algorithms; * CONTROL_TERMINATION and *CONTROL_TIMESTEP: controlling the simulation time and the output step of the model calculation [37].

4. Numerical Analysis

In this section, numerical simulation has been carried out on a large computer based on the above modeling and calculation methods. The simulation data can be extracted through post-processing. The total explosion time during perforation of 0–5000 μs has been simulated.

4.1. Simulation Results

The perforation pressure in the wellbore distribution from 800 μs to 1400 μs with an interval of 100 microseconds can be obtained, as shown in Figure 5. When the perforator has not been detonated, there is no fluctuation of the pressure in the wellbore. When the shock wave is formed by the explosion of perforating bullets, the perforation pressure in the wellbore increases sharply upward along the wellbore. When the shock wave reaches the position of the packer at the top of the wellbore, it reaches the peak value, which is caused by the reflection of the packer. The reflected shock wave gradually propagates downward along the wellbore, showing a trend of attenuation. The wellbore pressure-time curve can be drawn by extracting the simulated data, as shown in Figure 6.

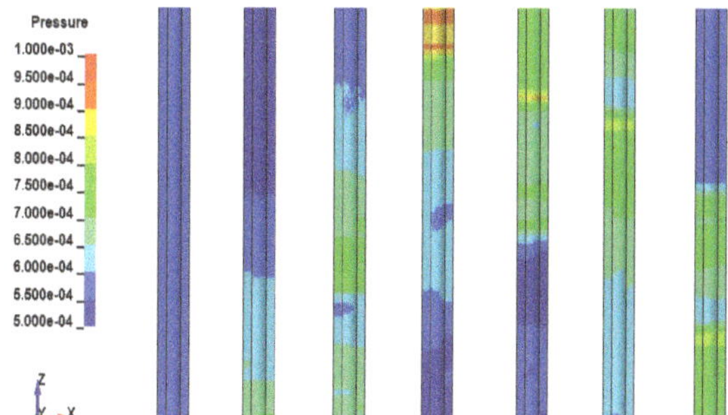

Figure 5. Perforation pressure distribution in the wellbore (unit: 10^{11} Pa).

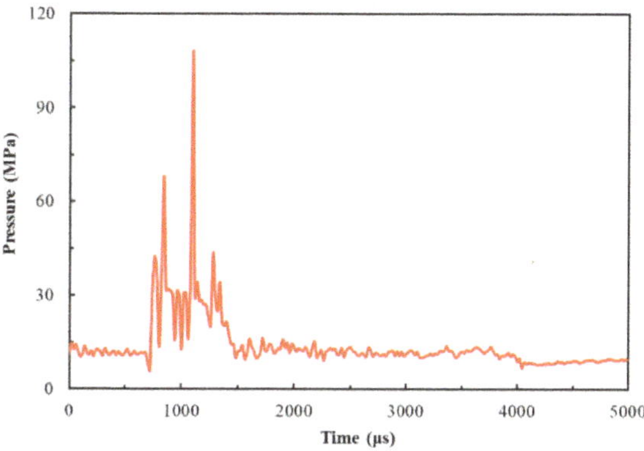

Figure 6. Simulated perforation pressure-time curve.

It can be seen from the simulated perforation pressure-time curve that the wellbore pressure reaches peak instantaneously with the detonation of the perforating bullets. Then, the oscillation occurs when shock waves are reflected back and forth in the wellbore. The attenuation occurs with propagation in the wellbore fluid. Finally, the pressure value tends to be stable after perforation.

The changing trend is similar to the field measured data curve, as shown in Section 5. It can be seen from Figure 6 that the peak pressure of deepwater perforation is as high as 110 MPa, which is a very important index for downhole wellbore safety. The high peak perforation pressure will cause the direct damage of downhole tools, such as the fracture of perforated string, the failure of the packer, etc. The peak value of perforation pressure can be used as the test index of the downhole safety for deepwater wells [43,44]. Before the shock wave reaches the packer, the contour of annulus pressure is captured along the different positions of the wellbore from bottom to top, as shown in Figure 7.

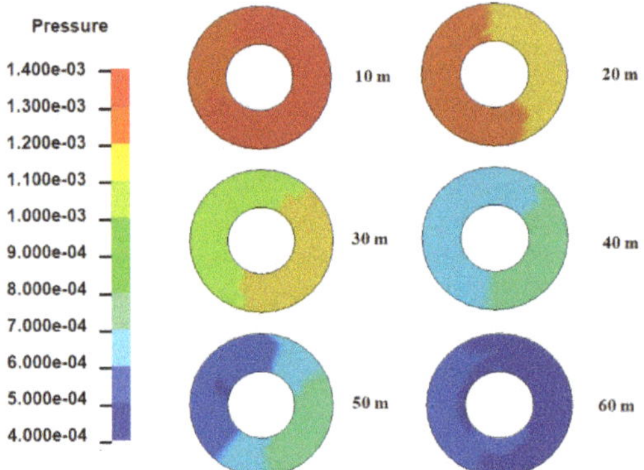

Figure 7. Perforation pressure distribution at different positions of the wellbore (unit: 10^{11} Pa).

The position of perforating gun is regarded as the starting position of perforating explosive load. From Figure 7, it can be seen that the farther away from the source of the explosion, the lower the perforating pressure is. It means that the propagation of perforating pressure in wellbore fluid shows an attenuation trend. The peak perforation pressure at different positions in the wellbore can be extracted [45–49]. By using the extremum method, the simulated data can be dimensionless. The change trend curve of peak perforation pressure at different wellbore positions has been obtained, as shown in Figure 8.

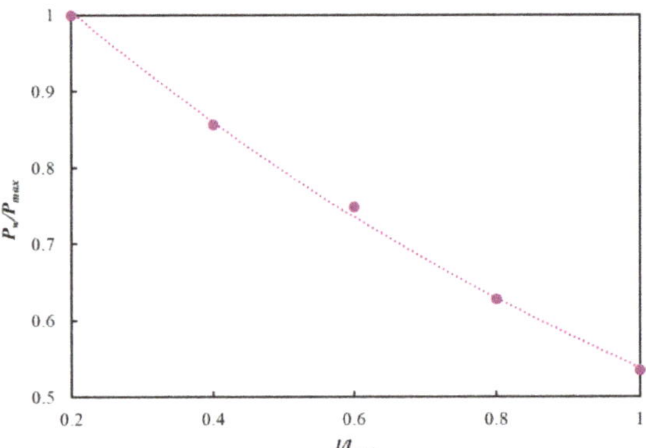

Figure 8. Perforation peak pressure at different positions of the wellbore. The abscissa l/l_{max} is the ratio of the distance along the wellbore from the source of the explosion and the distance from the packer to the source of the explosion. The ordinate $Pw/Pmax$ is the ratio of the peak pressure of perforation and the maximum peak pressure of perforation.

The ultra-high pressure detonation products formed by perforation bullets rapidly expand in wellbore fluid in the form of gas. Since the pressure of the detonation products is far greater than the static pressure of surrounding wellbore fluid, the shock waves and perforation pressure pulsation have been formed in the wellbore. There will be a decaying trend with the propagation of shock waves in the wellbore fluid. The physical process is very similar to the underwater explosion. The mathematical model of the underwater pressure attenuation is usually obtained by fitting the experimental data. Cole [50] first proposed the theory to describe the pressure attenuation of underwater explosion; it is considered that the propagation attenuation law of underwater explosion accords with exponential attenuation. Through the above analysis, the attenuation coefficient model of the downhole perforation pressure in the wellbore is proposed by the undetermined coefficient method:

$$Pw = K \times Ps \times e^{\beta k} \tag{10}$$

where Pw is the peak pressure of perforation at different positions of the wellbore, MPa; Ps is the initial pressure of the explosive source of perforation bullets, MPa; K is the undetermined coefficient, dimensionless; β is the attenuation index, dimensionless; $k = l/l_{max}$ is the ratio of the distance along the wellbore from the source of the explosion and the distance from the packer to the source of the explosion, dimensionless.

Based on the numerical simulation data in Figure 8, the model of the undetermined coefficient in the equation can be solved, $K = 1.18$ and $\beta = -0.37$. In order to obtain the pressure of the source of the explosion of perforating bullets, a large number of numerical simulation analysis is carried out for different factors based on the orthogonal test.

4.2. Orthogonal Test

The orthogonal test is a scientific, effective, and simple method to study the influence of multiple factors, which is applicable to a certain index affected by these factors, with each factor taking more than one value. As we know, there are many factors affecting the pulse pressure of perforation, such as perforation-related parameters, pressure in the initial wellbore fluid, wellbore size, the thickness of casing and cement sheath penetrated, reservoir conditions, and so on. In this study, in order to obtain the variation law of downhole perforation pressure under different perforation conditions, the orthogonal test method has been used to carry out a series of numerical simulation calculations on a large computer for the influence of different factors on perforation pressure. Five correlative factors have been considered: total charge quantity, wellbore initial pressure, downhole effective volume for perforation, the thickness of casing and cement, formation elastic modulus. The designed orthogonal test table is shown in Table 6.

Table 6. Orthogonal test level table.

Orthogonal Test Level	Total Charge Quantity (kg)	Wellbore Initial Pressure (MPa)	Downhole Perforation Effective Volume (m^3)	Thickness of Casing and Cement (m)	Formation Elastic Modulus (MPa)
Level 1	4	15	1.5	0.03	5
Level 2	8	50	2	0.05	10
Level 3	12	85	2.5	0.07	15
Level 4	16	120	3	0.09	20

The perforating peak pressure is the index affected by the factors; the results of orthogonal numerical simulation tests can be obtained in Table 7. Through range analysis, the order of importance of each factor to test index can be judged. The range can be calculated by the formula, as shown in [28]:

$$R_i = \{\text{Max}(K_i) - \text{Max}(K_i)\}/n \tag{11}$$

where R_i is the range formula of the ith column, dimensionless; K_{ij} is the sum of the test indices corresponding to the ith column (factor) and jth level (value), dimensionless; $\overline{K}_{ij}$ is the average value of the test indices corresponding to the ith column (factor) and jth level (value), which can be used to obtain the optimum combination of factors to make the test index, dimensionless.

Table 7. Results of orthogonal tests.

Sequence Number (-)	Total Charge (kg)	Wellbore Initial Pressure (MPa)	Downhole Effective Volume (m^3)	Thickness of Casing and Cement (m)	Formation Elastic Modulus (MPa)	Peak Value of Perforating Pressure (MPa)
1	4	15	3	0.09	15	52.6
2	4	50	2.5	0.07	20	89.88
3	4	85	2	0.05	10	149.22
4	4	120	1.5	0.03	5	206.2
5	8	15	2.5	0.05	5	85.33
6	8	50	3	0.03	10	148.55
7	8	85	1.5	0.09	20	160.75
8	8	120	2	0.07	15	208
9	12	15	2	0.03	20	96.22
10	12	50	1.5	0.05	15	140.74
11	12	85	3	0.07	5	189.87
12	12	120	2.5	0.09	10	215.17
13	16	15	1.5	0.07	10	145.52
14	16	50	2	0.09	5	195.68
15	16	85	2.5	0.03	15	248.9
16	16	120	3	0.05	20	276.65

Through range analysis, the order of importance of each factor to the perforating peak pressure can be obtained. The larger the range value, the greater the influence of the factor on the peak perforation pressure is. The range value of the total charge quantity, wellbore initial pressure, downhole effective volume for perforation, the thickness of casing and cement, formation elastic modulus is 90.3, 106.52, 24.27, 29.1, 19.79, respectively. Therefore, the order of influence of each factor on the peak value of perforation pressure is as follows: wellbore initial pressure > total charge quantity > thickness of casing and cement > downhole effective volume for perforation > formation elastic modulus.

4.3. Analysis of Influence Factors

In order to study the correlation between influence factors and peak perforation pressure, the dimensionless analysis is adopted for the simulated data. Five dimensionless normalized forms of the factors of the total charge quantity, wellbore initial pressure, downhole effective volume for perforation, the thickness of casing and cement, formation elastic modulus are represented by X_1, X_2, X_3, X_4, X_5, respectively. The dimensionless normalized form of peak perforation pressure can be represented by Y. The downhole wellbore perforation peak pressure function can be expressed in the dimensionless form:

$$Y = f(X_1, X_2, X_3, X_4, X_5) \tag{12}$$

where $X_1 = M_0/M_{\max}$, $X_2 = P_0/P_{0\max}$, $X_3 = V_0/V_{\max}$, $X_4 = L_0/L_{\max}$, $X_5 = G_0/G_{\max}$ are the ratios of the total charge quantity, wellbore initial pressure, downhole effective volume for perforation, the thickness of casing and cement, formation elastic modulus to the maximum value of that, respectively, dimensionless.

Based on the dimensionless simulation data, taking X_1 as abscissa and Y as a longitudinal coordinate, the trend curve and the influence law can be obtained, as shown in Figure 9.

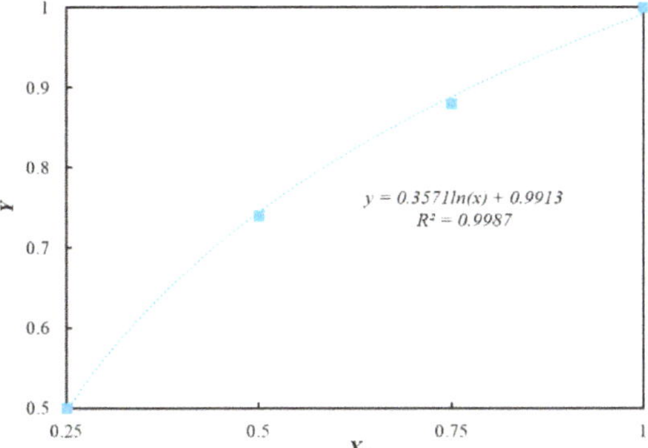

Figure 9. The influence of total charge quantity on downhole perforation peak pressure.

Figure 9 indicates that the peak value of perforation pressure and the total charge quantity show a logarithm function. The perforating peak pressure increases with the total charge quantity increasing. The reason is that the fluctuating pressure of downhole perforation mainly comes from the explosion source of perforated bullets with the shaped charge. The larger the single charge and the more the number of perforation bullets are, the more the total charge quantity is, the higher the peak pressure of perforation pulsation is.

Similarly, in order to investigate the effects of wellbore initial pressure on the pulsating pressure of perforation, the relationship between the wellbore initial pressure and the peak value of downhole perforation can be obtained in Figure 10.

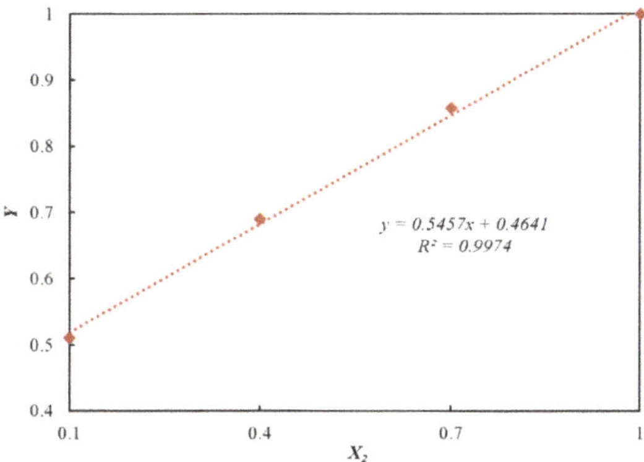

Figure 10. The influence of wellbore initial pressure on downhole perforation peak pressure.

As shown in Figure 10, the peak perforation pressure increases linearly with the increase of wellbore initial pressure. The reason is that the initial wellbore pressure provides the basis for the pulsating pressure of perforation. When the explosive energy produced by the explosion of the perforation bullets is released into the wellbore, the larger the wellbore initial pressure is, the higher the perforating explosion pressure is.

Figure 11 illustrates the relationship between the downhole perforation effective volume and the peak value of perforation. It can be seen that the peak value of perforation and the downhole perforation effective volume shows an exponential function. The peak value of perforation gradually becomes smaller with the increase of the downhole perforation effective volume. This is because as the packer is seated, the downhole wellbore is in a closed space. With the increase of downhole perforation effective volume, the energy generated by the perforation explosion has more space to release. Less energy can be converted into the perforating pulsating pressure.

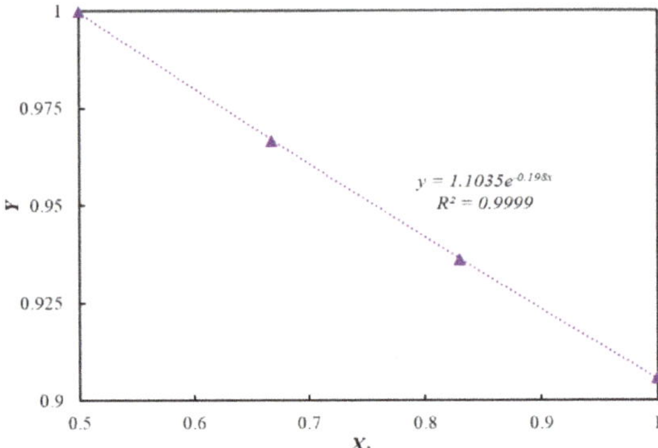

Figure 11. The influence of downhole perforation effective volume on perforation peak pressure.

The relationship between the thickness of casing and cement and the peak value of downhole perforating pressure can be drawn, as shown in Figure 12. It can be seen that the peak value of downhole perforating pressure and the thickness of casing and cement show a power relation function. The peak value of downhole perforating pressure gradually becomes smaller with the increase of the thickness of casing and cement. The reason is that with the increase of energy used to perforate casing and cement, the explosive energy released in wellbore decreases correspondingly. The pulsating pressure of perforation decreases accordingly.

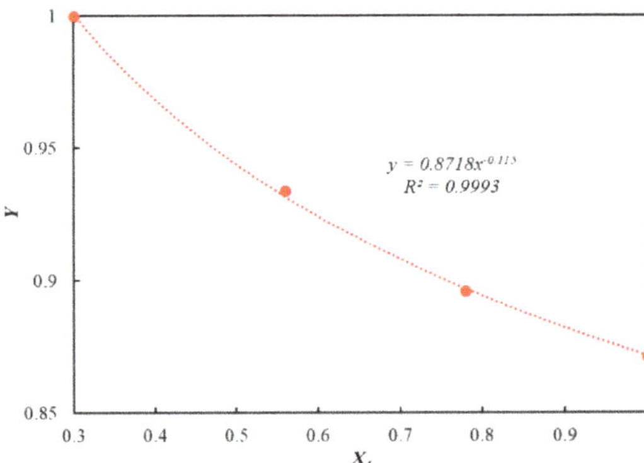

Figure 12. The influence of the thickness of casing and cement on downhole perforation peak pressure.

Figure 13 illustrates the relationship between the formation elastic modulus and the peak value of perforation pressure. It can be seen that the peak value of perforation pressure and the formation elastic modulus show a power function. The peak value of perforation pressure gradually becomes smaller with the increase of the formation elastic modulus. The reason is that the greater the formation elastic modulus is, the stronger the ability to resist elastic deformation is. More energy is required for perforation to penetrate the formation. Less energy will be released in the wellbore with the smaller perforating pressure.

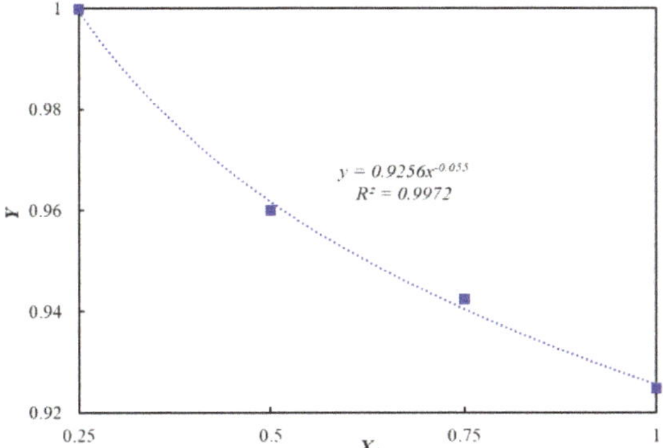

Figure 13. The influence of formation elastic modulus on downhole perforation peak pressure.

Based on the above analysis, the influence laws of the factors on the downhole perforation peak pressure have been obtained. In order to obtain a prediction model that can be used to calculate the peak pressure of perforation pulsation accurately in engineering, it can be fitted based on a large number of numerical simulation data by considering multi-factor changes. Equation (12) can be written further as:

$$Y = \frac{\sqrt{K_1 Ln(X_1) + K_2}}{e^{K_3 X_3} X_4{}^{K_4} X_5{}^{K_5}} + K_6 X_2 + K_7 \tag{13}$$

where $K_1, K_2, K_3, K_4, K_5, K_6, K_7$ are the undetermined coefficients, which can be obtained by fitting the numerical simulation data, dimensionless.

Combining with the attenuation coefficient model of the downhole perforation pressure in the wellbore of Equation (10), the final prediction model of perforation peak pressure at different positions of the wellbore can be obtained in:

$$Pw = P_{\max} \left(\frac{\sqrt{ALn(X_1) + B}}{e^{0.33X_3} X_4{}^{0.02} X_5{}^{0.01}} + CX_2 + D \right) e^{0.037k} \tag{14}$$

where A, B, C, D are the undetermined coefficients, which can be solved, dimensionless $A = 0.13$, $B = 0.35, C = 0.33, D = 0.27$.

Through the above analysis, a calculation model for the prediction of the fluctuating peak pressure of downhole deepwater perforation has been obtained, which can be applied to the field perforation case study.

5. Field Application

The field case is a deepwater well located in the South China Sea [51]. The well's depth is 1295 m, the length of the perforating gun, rathole length, and tubing is, respectively, 9 m, 19.14 m, 35 m. The packer adopts a static type with rated working pressure of 30 MPa. The specific operation parameters of the field well are shown in Table 8. The measured perforation pressure-time curve can be obtained by the test instrument installed on the perforation gun, as shown in Figure 14.

Table 8. Field case parameters.

Parameters	Data
Perforation bullets	336
Charge per hole	40 g
Casing size	244.40/220.50 mm
Tubing size	73.02/62.00 mm
Gun size	177.80/152.53 mm
Wellbore initial pressure	10 MPa
Fluid density	1030 kg/m3
Confining pressure	12 MPa
Cement thickness	0.06 m
Formation elastic modulus	1.27 MPa
Charge type	RDX
Perforation type	TCP (Tubing Conveyed Perforation)

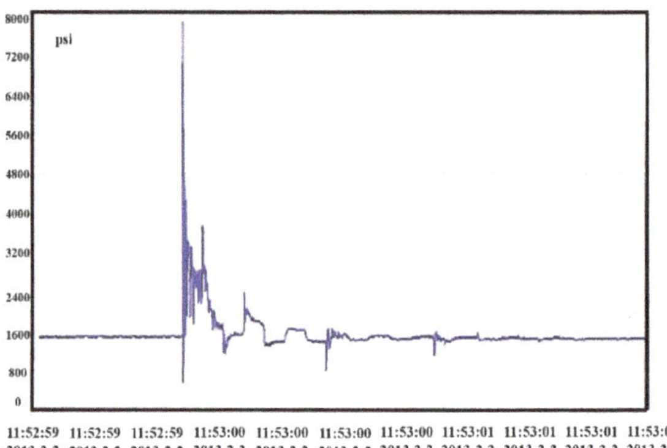

Figure 14. Measured perforation pressure-time curve.

It can be seen that the measured and simulated pressure-time curves of Figures 6 and 14 show similar trends, in which the perforation pressure reaches peak instantaneously with the detonation of the perforating bullets. Then, the oscillation occurs when shock waves are reflected back and forth in the wellbore, and the attenuation occurs due to propagation in the wellbore fluid. Finally, the pressure value tends to be stable after perforation. The measured peak pressure during perforating is 7900 psi (54.57 MPa). According to the prediction model of perforation peak pressure of Equation (13), the result can be obtained. The downhole effective volume for perforation can be calculated in:

$$V_0 = V_t + V_g + V_r \tag{15}$$

where V_t, V_g, V_r are the volume of tubing section, perforating gun section, and pocket section, respectively, m^3. The predicted peak perforation pressure by this study (Equation (13)) has been found to be 56.85 MPa; the error between the predicted value and the measured value can be calculated by:

$$\delta = \frac{56.85 - 54.57}{54.57} \times 100\% = 4.18\% \tag{16}$$

Similarly, the results of different empirical formulas (Equations (1)–(4)) in previous reports for calculating the peak pressure of perforation and the errors in comparison with the measured data can be obtained, as shown in Table 9.

Table 9. Calculation results of different empirical formulas.

Formula Number	Calculation Result (MPa)	Measured Result (MPa)	Relative Error
Equation (1)	19.89	54.57	63.55%
Equation (2)	39.54	54.57	27.54%
Equation (3)	85.68	54.57	57.01%
Equation (4)	71.2	54.57	30.47%
Equation (13)	56.85	54.57	4.18%

By comparing the results of this study with those of previous studies, it has been found that the error in comparison with the measured data by Equation (13) is within 10%. The result calculated by the prediction model accords with the accuracy needs of actual perforation operation. The safety of downhole perforation can be analyzed, in which the safety of the packer is an important index to measure the safety of downhole wellbore. When the packer is seated, it is in a fixed state. When the perforation pressure wave is transmitted to the lower end of the packer, it will produce reflection and transmission. According to the principle of reflection and transmission of the wave, the pressure value acting on the packer can be calculated by:

$$P_1 = P_2 + P_3 - P_4 = 2P_2 \times \frac{(\rho c)_2[(\rho c)_2 - (\rho c)_1]}{[(\rho c)_1 + (\rho c)_2]^2} \tag{17}$$

where P_1 is the final pressure value acting on the packer, MPa; P_2 is the perforation pressure of the lower end of the packer, MPa; P_3 is the reflection pressure, MPa; P_4 is the transmission pressure, MPa; $\frac{(\rho c)_1}{(\rho c)_2} = \frac{1}{5}$ is the ratio of shock impedance between wellbore fluid and the packer, dimensionless.

Combining with the prediction model of perforation peak pressure at different positions of the wellbore of Equation (14), the final pressure value acting on packer can be obtained as 51.37 MPa. The value is beyond the range of a packer (30 MPa), and the packer will be damaged with such perforating peak pressure. To ensure the safety of perforation while keeping the perforation strength is an important challenge. It means that the number of perforating bullets and charge quantity cannot be reduced. The material properties, type, and size of the downhole string cannot be changed, and the reservoir conditions are determined. The perforating pressure can be reduced by adjusting the installation distance of the packer and the installation of shock absorbers. The shock absorbers are usually installed below the packer to ensure the safety of the whole perforation process. The shock absorber is usually composed of spring or rubber parts. For deepwater wells, the rubber parts in the shock absorber are more easily damaged due to the complex bottom-hole environment, which causes great trouble to perforation operation. So, the spring shock absorber is often used in the perforation of a deepwater well. According to the previous analysis, the axial dynamic response of the downhole string under perforation impact is the most obvious. The perforating peak pressure can be alleviated by installing the longitudinal spring shock absorber. It is proposed that the shock absorbers installed in the middle of the string can achieve the best shock absorption effect [5].

The shock absorber is simulated by adding spring elements connected in series to the string in the numerical model of Figure 3. The numerical models with shock absorbers installed in the middle of the string have been used to carry out the simulation calculation. Through the numerical simulations, the date of perforating peak pressure on the packer with a different number of shock absorbers under different packer setting distances can be obtained, as shown in Figure 15.

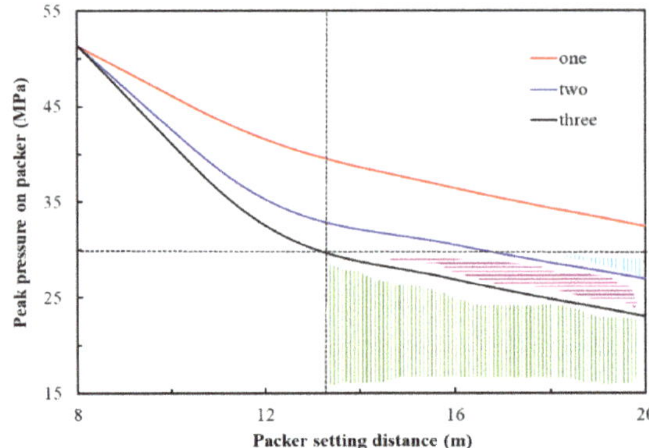

Figure 15. Safety optimization scheme.

Figure 15 shows the reduction of the peak value of perforating pressure on the downhole packer with different numbers of shock absorbers. The more the number of shock absorbers is, the larger the reduction is, and the better shock absorption effect can be achieved. With the increase of packer installation distance, the peak value of the perforating pressure on the downhole packer decreases. When one shock absorber is installed, the minimum value is reduced to 32.5 MPa, which still exceeds the range of a packer (30 MPa). When the optimal installation position cannot meet the requirements of shock absorption, the number of shock absorbers can be increased. With the increase of the number of shock absorbers, the downhole packer can be ensured in a safe state combined with the optimization of packer installation distance. The optimization range refers to the color region in Figure 15. Through the above analysis, the specific optimization scheme is put forward. In the field operation, the distance of the packer is 20 m, and two shock absorbers are installed in series in the middle of the string. After safety optimization, there is no damage to the downhole packer after perforation.

6. Conclusions

In this paper, the numerical model with hundreds of perforation bullets is set up to study the downhole perforation pressure for deepwater wells. The downhole dynamic perforation process is clearly described. The propagation attenuation model of perforation impact pressure in wellbore fluid is established based on the theory of underwater explosion. According to the results of orthogonal numerical simulation tests, the order of influence of each factor on the peak value of perforation pressure has been obtained: wellbore initial pressure > total charge quantity > thickness of casing and cement > downhole effective volume for perforation > formation elastic modulus. The results show that the downhole perforation peak pressure, logarithmically related to the total charge quantity, is linearly related to the wellbore initial pressure, has an exponential relationship with the downhole effective volume for perforation, and is a power relationship with the casing and cement thickness and the formation elastic modulus. The model to predict the perforation peak pressure considering multiple factors at different positions of the wellbore has been obtained. The proposed correlation shows accurate performance based on a field case under consideration, which demonstrates promising capability as a predictive tool. With the field application, an optimized scheme is proposed to improve the perforation safety for the deepwater well.

Author Contributions: All the authors conceived and designed the study. Investigation, Q.D. and H.Z.; Writing-original draft, Q.D. and H.Z.; Writing-review and editing, J.L., X.H., and B.Z.

Funding: This research received no external funding.

Acknowledgments: The authors gratefully acknowledge the Natural Science Foundation of China (Grant No. U1762211, Grant No. 51734010, Grant No. 51574262, Grant No.51774304, Grant No. 51774063), National Oil and Gas Major Project (Grant No. 2017ZX05009), the Foundation for Innovative Research Groups of the National

Natural Science Foundation of China (Grant No. 51521063), the State Key Laboratory of Petroleum Resources and Engineering.

Conflicts of Interest: The authors declare that there is no conflict of interest regarding the publication of this paper.

References

1. Brockway, P.E.; Owen, A.; Brand-Correa, L.I.; Hardt, L. Estimation of Global Final Stage Energy Return-On-Investment for Fossil Fuels with Comparison to Renewable Energy Sources. *Nat. Energy* **2019**, *4*, 612–621. [CrossRef]
2. Xi, Y.; Li, J.; Liu, G.; Li, J.; Jiang, J. Mechanisms and Influence of Casing Shear Deformation near the Casing Shoe, Based on MFC Surveys during Multistage Fracturing in Shale Gas Wells in Canada. *Energies* **2019**, *12*, 372. [CrossRef]
3. Gilliat, J.; Bale, D.; Satti, R.P.; Li, C.; Howard, J.J. The Importance of Pre-Job Shock Modeling as a Risk Mitigation Tool in TCP Operations. In Proceedings of the SPE Deepwater Drilling and Completions Conference, Galveston, TX, USA, 10–11 September 2014.
4. McLeod, H.O., Jr. The Effect of Perforating Conditions on Well Performance. *J. Pet. Technol.* **1983**, *35*, 31–39. [CrossRef]
5. Giunta, R.; Le, C.; Schoener-Scott, M.; Anderson, R.; Glass, J. Modeling Dynamic Shock Loads for High-Pressure/Deep-Water Tubing-Conveyed Perforating Operations in the Gulf of Mexico. In Proceedings of the SPE Deepwater Drilling and Completions Conference, Galveston, TX, USA, 20–21 June 2012.
6. Deng, Q.; Zhang, H.; Li, J.; Hou, X.; Wang, H. Study of Downhole Shock Loads for Ultra-Deep Well Perforation and Optimization Measures. *Energies* **2019**, *12*, 2743. [CrossRef]
7. Qiu, Q.; Cui, L. Reliability Evaluation Based on a Dependent Two-stage Failure Process with Competing Failures. *Appl. Math. Model.* **2018**, *64*, 699–712. [CrossRef]
8. Qiu, Q.; Cui, L. Optimal Mission Abort Policy for Systems Subject to Random Shocks Based on Virtual Age Process. *Reliabil. Eng. Syst. Saf.* **2019**, *189*, 11–20. [CrossRef]
9. Lu, D.T.; Xu, G.M.; Kong, X.Y. Calculating Transient Pressure of Wellbore Taking under Perforation Condition with Parallel Computation Method. *Univ. Sci. Technol. China* **1999**, *29*, 638–643.
10. Zhao, X.; Liu, G.H. Theoretical Modeling of the Movement of the Top Killing Fluids in Complex Perforation. *Drill. Fluid Complet. Fluid* **2008**, *25*, 7–9.
11. Zhao, X.; Liu, G.H.; Li, J. Experimental Study of the Movement of Control Fluid in Compound Perforation. *Pet. Sci.* **2009**, *6*, 389–394. [CrossRef]
12. Chen, H.B.; Tang, K.; Ren, G.H. String Dynamic Mechanics Analysis on Ultra-Deep Perforation. *Well. Logging Technol.* **2010**, *34*, 487–491.
13. Sanders, W.; Baumann, C.E.; Williams, H.A.R.; de Moraes, F.D.; Shipley, J.; Bethke, M.E.; Ogier, S. Efficient Perforation of High-Pressure Deepwater Wells. In Proceedings of the Offshore Technology Conference, Houston, TX, USA, 2–5 May 2011.
14. Yang, X.T.; Zhang, F.X.; Li, M.F.; Dou, Y.H. Analysis of strength safety of perforated string considering detonation parameters. *Adv. Mater. Res.* **2013**, *634*, 3573–3576. [CrossRef]
15. Zhou, H.F.; Ma, F.; Chen, H.B.; Xu, Y.X.; Xi, L.X.; W, S.S. Comprehensive dynamic load test for the column in perforation section. *Well Logging Technol.* **2014**, *38*, 247–250.
16. Bale, D.; Satti, R.; Ji, M. A Next-generation Shock-capturing, Multi-phase Flow Simulator for Perforating Applications in HPHT Environment. In Proceedings of the SPE Deepwater Drilling & Completions Conference, Galveston, TX, USA, 14–15 September 2016.
17. Li, Z.F.; Zhang, C.Y.; Song, G.M. Research Advances and Debates on Tubular Mechanics in Oil and Gas Wells. *J. Pet. Sci. Eng.* **2017**, *151*, 194–212. [CrossRef]
18. Liu, J.; Guo, X.Q.; Liu, Z.J.; Liu, X.; Liu, Q.Y. Pressure Field Investigation into Oil & Gas Wellbore during Perforating Shaped Charge Explosion. *J. Pet. Sci. Eng.* **2019**, *172*, 1235–1247.
19. Lu, D.W. *Perforating Technology for Oil and Gas Wells*; Petroleum Industry Press: Beijing, China, 2012.
20. Chen, F.; Chen, H.B.; Tang, K. Influence of Perforating Impact Load on the Operating String and the Countermeasures. *Nat. Gas Ind.* **2010**, *30*, 61–65.
21. Ning, J.G.; Wang, C.; Ma, T.B. *Explosion and Shock Dynamics*; National Defense Industry Press: Beijing, China, 2010.

22. Kamlet, M.J.; Jacobs, S.J. Chemistry of Detonations. I. A Simple Method for Calculating Detonation Properties of C–H–N–O Explosives. *J. Chem. Phys.* **1968**, *48*, 23–35. [CrossRef]
23. Deng, Q.; Zhang, H.; Li, J.; Hou, X.; Wang, H. Analysis of Impact Load on Tubing and Shock Absorption during Perforating. *Open Phys.* **2019**, *17*, 214–221. [CrossRef]
24. Gao, Y.D.; Wang, S.H.; Wang, X.C.; Liu, Z.Y.; Guo, S.S.; Tang, J.J. Assessment Packer in Perforating String in Safety Distance. *Well Logging Technol.* **2016**, *40*, 382–384.
25. Baumann, C.E.; Pesantes, E.; Guerra, J.; William, A.; Williams, H. Reduction of Perforating Gunshock Loads. *SPE Drill. Complet.* **2012**, *27*, 65–74. [CrossRef]
26. Baumann, C.E.; Oden, J.T. An Adaptive-Order Discontinuous Galerkin Method for the Solution of the Euler Equations of Gas Dynamics. *Int. J. Numer. Methods Eng.* **2002**, *47*, 61–73. [CrossRef]
27. Bale, D.; Ji, M.; Satti, R.; Gilliat, J. Advances in Numerical Modeling of Downhole Dynamics for Perforated Well Completions. In Proceedings of the SPE Annual Caspian Technical Conference and Exhibition, Astana, Kazakhstan, 12–14 November 2013. SPE-172308-MS. [CrossRef]
28. Lee, W.H. Oil Well Perforator Design Using 2D Eulerian Code. *Int. J. Impact Eng.* **2002**, *27*, 535–559. [CrossRef]
29. Han, X.Q.; Cao, L.N.; Cao, Y.X.; Zhang, Y.G. Numerical Simulation of the Shaped Charge Jet Formation and Penetration through the Wall of Perforating Gun. *Eng. Sci. Technol.* **2009**, *9*, 6960–6964.
30. Kang, K.; Ma, F.; Zhou, H.F. Study on Dynamic Numerical Simulation of String Damage Rules in Oil-gas Well Perforating Job. *Procedia Eng.* **2014**, *84*, 898–905.
31. Yang, Z.Z. Dynamic Response of String under Perforating Pulse Load. Master's Thesis, China University of Petroleum (East China), Qingdao, China, 2015.
32. Li, B.Y. Analysis of Strength and Safety of Packer Central Tube under Perforating Impact Loading. Master's Thesis, Xi'an Shiyou University, Xi'an, China, 2018.
33. Yan, Y.; Guan, Z.; Xu, Y.; Yan, W.; Wang, H. Numerical Investigation of Perforation to Cementing Interface Damage Area. *J. Pet. Sci. Eng.* **2019**, *179*, 257–265. [CrossRef]
34. Mirzaei, M.; Harandi, A.; Karimi, R. Finite Element Analysis of Deformation and Fracture of an Exploded Gas Cylinder. *Eng. Fail. Anal.* **2009**, *16*, 1607–1615. [CrossRef]
35. Villavicencio, R.; Soares, C.G. Numerical Modelling of the Boundary Conditions on Beams Stuck Transversely by a Mass. *Int. J. Impact Eng.* **2011**, *38*, 384–396. [CrossRef]
36. Ma, S.; Zhang, X.; Qiu, X.M. Comparison Study of MPM and SPH in Modeling Hyper-Velocity Impact Problems. *Int. J. Impact Eng.* **2009**, *36*, 272–282. [CrossRef]
37. Elshenawy, T.; Elbeih, A.; Li, Q. Influence of Target Strength on the Penetration Depth of Shaped Charge Jets into RHA Targets. *Int. J. Mech. Sci.* **2018**, *136*, 234–242. [CrossRef]
38. Chen, C.Y.; Shiuan, J.H.; Lan, I.F. The Equation of State of Detonation Products Obtained from Cylinder Expansion test. *Propellants Explos. Pyrotech.* **1994**, *19*, 9–14. [CrossRef]
39. Elshenawy, T.; Li, Q. Influences of Target Strength and Confinement on the Penetration Depth of an Oil Well Perforator. *Int. J. Impact. Eng.* **2013**, *54*, 130–137. [CrossRef]
40. LS-DYNA. *Keyword User's Manual*; Version 971; Livermore Software Technology: Livermore, CA, USA, 2013.
41. Wei, Q.; Jones, N. A Failure Criterion for Beams under Impulsive Loading. *Int. J. Impact Eng.* **1992**, *12*, 101–121.
42. API. *Technical Report on Equations and Calculations for Casing, Tubing and Line Pipe Used as Casing or Tubing; and Performance Properties Tables for Casing and Tubing, TR 5C3*; API: Washington, DC, USA, 2008.
43. Yu, Y.; R, F.Y. Application of Packer in Fracture Perforating. *J. Xi'an Univ. Sci. Technol.* **2007**, *27*, 423–425.
44. Deng, Q.; Zhang, H.; Li, J.; Hou, X.; Wang, H. Safety Distances of Packers for Deep-water Tubing-conveyed Perforating. In Proceedings of the Offshore Technology Conference, Houston, TX, USA, 5–7 November 2018.
45. Cheraghian, G.; Hemmati, M.; Bazgir, S. Application of TiO_2 and Fumed Silica Nanoparticles and Improve the Performance of Drilling Fluids. *AIP Conf. Proc.* **2014**, *1590*, 266–270.
46. Cheraghian, G. Application of Nano-Particles of Clay to Improve Drilling Fluid. *Int. J. Nanosci. Nanotechnol.* **2017**, *13*, 177–186.
47. Deng, Q.; Zhang, H.; Li, J.; Wang, H.; Cai, Z.Y.; Tan, T.Y.; Hou, X.J. A Model for Estimating Penetration Length under Different Conditions. In Proceedings of the 52nd US Rock Mechanics/Geomechanics Symposium, Seattle, WA, USA, 17–20 June 2018; Available online: https://www.onepetro.org/conference-paper/ARMA-2018-344 (accessed on 1 June 2019).

48. Sun, T.; Zhang, H.; Gao, D.; Liu, S.; Cao, Y. Application of the Artificial Fish Swarm Algorithm to Well Trajectory Optimization. *Chem. Technol. Fuels Oils* **2019**, *55*, 213–218. [CrossRef]

49. Cheraghian, G.; Wu, Q.; Mostofi, M.; Li, M.C.; Afrand, M.; Sangwai, J.S. Effect of a Novel Clay/Silica Nanocomposite on Water-Based Drilling Fluids: Improvements in Rheological and Filtration Properties. *Colloid Surf. A* **2018**, *555*, 339–350. [CrossRef]

50. A Cole, R.H. *Underwater Explosion*; LISA Princeton University Press: Princeton, NJ, USA, 1948.

51. Teng, Y.S. Dynamic Simulation for Perforating and String Safety Evaluation. Master's Thesis, China University of Petroleum (East China), Beijing, China, 2014.

Article

The Effect of Supercritical CO$_2$ on Shaly Caprocks

Pooya Hadian and Reza Rezaee *

Department of Petroleum Engineering, Western Australian School of Mines: Minerals, Energy and Chemical Engineering, Curtin University, Kensington, WA 6151, Australia; pooya.hadian@postgrad.curtin.edu.au
* Correspondence: R.Rezaee@curtin.edu.au

Received: 14 November 2019; Accepted: 23 December 2019; Published: 27 December 2019

Abstract: The effect of supercritical CO$_2$ on the shaly caprocks is one of the critical issues to be considered in CO$_2$ sequestration programs. Shale-scCO$_2$ interactions can alter the seal integrity, leading to environmental problems and bringing into question the effectiveness of the program altogether. Several analytical studies were conducted on samples from Jurassic Eneabba Basal Shale and claystone rich facies of the Triassic Yalgorup Member (725–1417 m) in the Harvey CO$_2$ sequestration site, Western Australia, to address the shale-scCO$_2$ interactions and their effect on the petrophysical properties of the caprock. Shale samples saturated with NaCl brine were exposed to scCO$_2$ under the reservoir condition (T = 60 °C, P = 2000 psi) for nine months and then tested to determine their altered mineralogical, petrophysical and geochemical properties. The experimental study examined changes to the mineralogical composition, capillary threshold pressure, and pore size distribution (PSD) of samples. The X-ray diffraction (XRD) results showed several changes in mineralogy because of rock-brine-CO$_2$ reactions. Quartz, feldspars, kaolinite, and goethite were dissolved in most samples and muscovite, and halite were precipitated in general. Nuclear magnetic resonance (NMR), low-pressure nitrogen adsorption (LPNA), and mercury injection capillary pressure (MICP) tests indicate an increase in pore volume, except for relatively tighter, clay-rich samples. A reduction in capillary threshold pressures of samples after exposure to scCO$_2$ is observed.

Keywords: carbon dioxide sequestration; caprock integrity; shale alteration; rock-water-CO$_2$ interaction; lab tests under reservoir condition

---

1. Introduction

The geosequestration of anthropogenic CO$_2$ has been suggested as a solution for resolving the problem of increasing greenhouse gas emissions which are responsible for global warming [1]. CO$_2$ is the major greenhouse gas, resulting from fossil fuel combustion for domestic and industrial purposes [2]. The injection of anthropogenic CO$_2$ deep underground instead of releasing it to the atmosphere is the basic concept in this method [3]. CO$_2$, under the storage condition of target reservoirs (below 800 m), is in the critical state (CO$_2$ critical point: 31.8 °C and 7.38 MPa). Saline aquifers are the most common target for the injection of CO$_2$, due to their abundancy and proximity to the source [4,5]. The excessive amount of CO$_2$ in the geologic formation, which also contain water, modifies the chemical equilibrium of the existing system and induces a series of reactions. The interactions between rock-forming minerals, brine, and injected CO$_2$ in deep brine aquifers alter the natural petrophysical properties of CO$_2$ geosequestration sites [6].

The mitigation of ScCO$_2$ into the subsurface initiates the CO$_2$-water-rock interactions in geologic storage strata that, in time, render the pH of the environment into an acidic condition, which, in turn, is expected to be buffered by reactions with the silicate/oxide/carbonate phases. Increasing the pH with time, due to this buffering effect, may result in carbonates or other phases re-precipitating in the pore system that had been dissolved initially [7,8]. Even after pH buffering, due to carbonate dissolution, the acidity of the brine will still be sufficient to attack alumino-silicate minerals (e.g., clays and feldspars) [9]. Subsequently, such reactions are often followed by alterations to the petrophysical properties of the rock. Several laboratory investigations on mineral transformations in CO_2 storage reservoirs indicated that some changes of caprock's fluid transport properties and mineralogical composition may occur [10–17]). Credoz et al. (2009) suggested that potential pathways for CO_2 leakage through the caprock could be induced as the result of geochemical alteration occurring along with the reservoir-caprock interface with the CO_2-brine mixture [18]. However, Busch et al. (2008) suggested that the depth of influence of such reactions is only limited to the lower part of the caprock in the vicinity of the CO_2 plume [7].

CO_2 containment and long-term safety for both humans and the environment are crucial factors to consider before implementing large scale CO_2 storage. Containment of the injected CO_2 for hundreds to thousands of years by trapping mechanism determines the success of geologic CO_2 sequestration as a large-scale carbon management strategy. Interaction of scCO$_2$-shale and the subsequent shale alteration can cause seal failure and leakage of CO_2 to the upper formations and the atmosphere [9]. Assessing the CO_2-rock interaction is an important part of such studies, as these potentially affect physical properties. There have been few studies to address the caprock sealing properties' variation during the geosequestration of CO_2 [8,11,12,14,15,19–23]. This study aims to assess the directions of the geochemical reactions and the petrophysical alteration as a result of dissolution/precipitation mechanisms in potential shaly caprock in the South West Hub, Western Australia under in situ conditions. This research is aimed at reducing uncertainties in the efficacy of shaly caprocks in a CO_2 storage system.

Laboratory experiments of rock samples with CO_2 and brine have been frequently applied as a strategy for studying the potential geochemical reactions. Most of the existing experiments have been conducted for less than three months and at very high temperatures to increase the chemical reaction rates. Some drawback of such batch experiments cannot be fully resolved, including short laboratory time scales and the increase of reactive surface area and reaction temperature, variation in the brine/rock ratio and the potential formation of experimental artefacts during reactor depressurisation and cooling [24]. However, the mineral-brine-CO_2 reaction is a long-term process, and short time-scale experiments are less accurate in predicting field conditions. Therefore, there is a necessity for experiments conducted over long-term timescales at low temperatures to represent actual reservoir conditions [1].

Short laboratory time scales of batch experiments are one drawback, which is why we tried to expand the reaction time to the maximum possible of nine months as opposed to several weeks to several months, as in previous studies. Elevated temperature is one other drawback which we avoided by selecting the in situ temperature of 60 °C. Samples were in different forms according to the requirement of each method. Core plugs, disks, rock fragments, and powdered samples were used in the reactor to avoid unnecessary changes to the surface area.

Ten samples out of the fifteen original samples were chosen to be used in this research. In this study, several methodologies for assessing CO_2-rock interactions are discussed and applied. These ten samples were analysed with X-ray diffraction (XRD) and low-pressure nitrogen adsorption (LPNA) methods in powder form. Only five plugs were recovered for nuclear magnetic resonance (NMR) measurement. The same samples were also analysed with the mercury intrusion porosimetry (MICP) method. The overall experimental procedure started with testing the samples in their original state, then exposing them to scCO$_2$, and finally testing them again after the exposure phase. The NMR samples were also tested midway through the exposure phase. The results show changes in mineralogical composition and alterations in pore size distribution, total pore volume, average pore width, total pore surface area, and capillary pressure of the samples, because of the reactivity between scCO$_2$-brine-minerals. The porosity increase was more significant in larger pores and samples which lacked clay minerals (kaolinite). The direction of the changes was to enhance the transport properties, as the pore volumes increased generally, and capillary threshold pressure decreased. Future works include a geochemical reaction path modelling to simulate the results of this study, and to further supplement the results of this study for longer periods.

2. Geological Setting

There has been a focus on the Carbon Capture and Storage (CCS) program in Australia in recent years, as the country is one of the top twenty countries in CO_2 emission per person. With its stationary energy-intensive industries, Australia is predicted to contribute to 20% of CO_2 emissions by the end of 2020, Currently, several projects have been proposed or already established in Australia for CCS purposes, such as South West Hub, CarbonNet, Otway (Victoria), Surat, Gippsland (SA), and Gorgon [25,26].

In Western Australia, two suitable mainland locations for the geosequestration of carbon dioxide have been identified. Carbon Capture and Storage in Gorgon and Borrow Island in the North West are active projects. The South West Hub project has been introduced as a potential mainland location for the geosequestration of carbon dioxide. This potential CCS system has suitable storage capacity in its saline aquifer and efficient seal properties in the shaly caprock. In addition, the area is located relatively close to the biggest CO_2 emitter of the state, the Kwinana Power Plant. It is planned to collect CO_2 from several industrial emitters, where the injection masses are expected to be in the order of 6.5 million tons per year for the 40 years of project activity [27].

The potential geosequestration site is in a deep saline aquifer within the Lesueur Sandstone. A total of four wells have been drilled in the Harvey region to investigate the suitability of the Lesueur Formation of Southern Perth basin for storage of industrial-scale CO_2 emissions. The results have identified significant differences in sedimentology and petrophysical properties of the Upper and Lower Members of the Lesueur Sandstone. The Lower Member of the Triassic Lesueur (Wonnerup) is a laterally extensive and thick layer of sandstone which represents the targeted reservoir, whereas the Upper Lesueur (Yalgorup) is far more heterogeneous, due to the mixed nature of the mudstone intervals and the thick continuous clean sandstone succession [28].

The targeted seals are identified in a series of rock baffles in the Triassic Yalgorup Member (Triass) and a thick interval of paleosol section (Jurassic Basal Eneabba Shale). The Yalgorup Member consists of mixed-thickness, interbedded high- to low-energy channel-fill facies, and swampy/overbank deposits, and paleosols. The Wonnerup Member consists of thick, continuous, high-energy channel-fill facies, with minor intercalations of moderate- to low-energy channel-fill/stacked rippleforms and rare swampy deposits. This is an ideal lithology for a CO_2 reservoir [27].

3. Samples and Methods

3.1. Samples

Relatively fresh samples of shale intervals were selected from the Harvey3 well cores in Perth's Core Library, Western Australia, to achieve the objectives of this study. A total of fifteen core samples from Eneabba Basal Shale and claystone rich facies of the Yalgorup Member (Lesueur Sandstone), depth ranging from 741 to 1218 m, were collected in September 2016.

The shale samples were characterised with a full suite of non-destructive petrophysical methods. The samples were analysed before and after they were dynamically exposed to supercritical CO_2 under in situ reservoir conditions. X-ray diffraction (XRD) analyses were used to examine the compositional changes of the seal rock mineralogy. XRD analysis indicated whether any dissolution or precipitation has taken place [29]. Pore size distributions were measured using low-pressure nitrogen adsorption [7], and nuclear magnetic resonance (NMR) [30–32] methods. Low-pressure nitrogen adsorption measured the surface area and pore volume to check the occurrence of any changes to this critical rock parameter [33,34]. The pore size distribution was also obtained from the mercury injection capillary pressure method. MICP also provided more data, such as the threshold pressure and the height of the column of CO_2.

3.2. Exposure to scCO2

The selected shale samples were required to be saturated with synthetic brine before being placed into the pressure–volume–temperature (PVT) cell at storage conditions and exposed dynamically to scCO$_2$ for nine months. Simplified artificial pore water (brine) of 30,000 ppm (512 mmol/L NaCl) was prepared from ROWE ultrapure water and ROWE NaCl (CS10307) and used for all the batch reactions. This concentration was chosen to be equivalent to the available groundwater data of the Harvey3 Lesueur formation fluid sampling from the Rockwater Hydrological and Environmental Consultants Report [35]. This synthetic water chemistry was selected so that there would be minimal risk of damage to the core materials before exposing them to scCO$_2$. Also, the dissolution process was encouraged by simple NaCl brine as it lacked the divalent cations, which have a pH-buffering effect and decrease the acid-induced reactions [36]. On the other hand, the cations needed for mineral precipitation can only be derived from the reactions with rock-forming minerals, instead of brine [12]. The selected shale samples were pressure-saturated at 2000 psi with 30,000 ppm NaCl solution before being placed into the PVT cell.

Shale samples in various forms of core-plugs, small rock fragments and crushed rock were tested in this study for pore accessibility and conformance with measurement methods [37]. We placed the selected shale samples into a PVT cell (Figure 1) under the reservoir condition of 2000 psi and 60 °C, related to the deepest sample recovered from the depth of 1400 m, for nine months. During the exposure, high-purity (99.9 mole%) scCO$_2$ mixed with deionised water was injected continuously using a high-accuracy continuous pump with a constant rate into the PVT cell. The high-performance liquid chromatography (HPLC) injection pump (SHIMADZU LC-20AT) allowed the pressure and flow-rate values to be set at 2000 psi and 0.7 cm^3/h (i.e., 0.01 mL/min), respectively, which could be controlled and recorded with the accuracies of 1 psi, and 0.001 mL/min, respectively. The injection rate must be realistic so that we can expect the same experimental results as the real-life injection process. We adopted the same method of rate calculation as [38]. The accumulator connected to the input of the samples cell was filled with CO$_2$ and pressurised with an air-driven compressor about every 40 days.

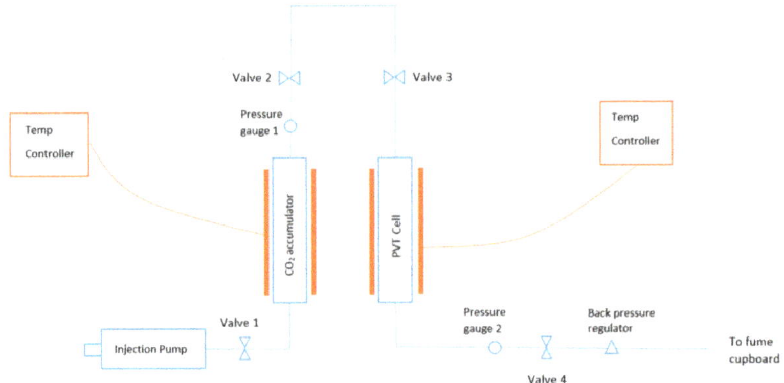

Figure 1. Schematic of exposure set-up.

To make the CO$_2$ saturated with water, 50 mL of water was injected inside the accumulator initially. Under the high pressure and temperature of the cell, water dissolved in the scCO$_2$ and made it wet. The outlet end of the samples cell was connected to a Swagelok KPB back-pressure regulator, which maintained the pressure inside the cell equal to 2000 psi (in situ pore pressure). The temperature inside the cell was set to 60 °C and maintained using a Proportional-Integral-Derivative (PID) Temperature Controller (omega CSi8DH) and heating jackets (omega SRFG series) with an accuracy of 0.04 °C. The accumulator and exposure cell with the attached heating jacket were wrapped in insulation pads to minimise the temperature fluctuation. All wetted parts of the apparatus are made from Hastelloy, super duplex stainless steel or titanium to minimise any fluid contamination or damages caused by corrosion when exposed to highly corrosive materials (e.g., carbonated water or high salinity brine) under high pressures and temperatures for a prolonged period. At the end of the exposure, the pressure and temperature were reduced gradually to room conditions and samples were examined again to assess any rock chemical and petrophysical alterations.

3.3. Experimental Approach

The composition of the shale samples in the unreacted sample was measured by XRD before conducting the batch experiments. For XRD analysis, the samples were crushed, homogenised and divided into two aliquots to avoid the heterogeneity of samples effecting the results. Each patch was then mixed with methanol and further ground in a micronizing mill to around 1–5 μm. The solution was dried overnight at 40 °C in a fume cupboard. Once the samples had dried, they were packed in sample holders and an XRD pattern was obtained. X-ray diffraction (XRD) analyses of the samples were performed using a Bruker D8 Advance automated powder diffractometer with Bragg-Brentano configuration at the John de Laeter Centre of Curtin University. It used a LynxEye detector and a copper X-ray tube at 20 kV and 5 mA. The whole-rock samples were analysed over an angular range of 7 to 120 degrees 2θ. The instrument was complemented with Bruker DIFFRAC EVA software for search/match analysis, and the XRD quantitative results were calculated using the Rietveld method in the comprehensive full pattern data analysis programme, TOPAS.

The 2 MHz Magritek bench-top laboratory NMR spectrometer was applied to record T_2 relaxation time spectrum of shale core plug samples before and after exposure to CO_2 to estimate the porosity and pore size distribution. The magnet operated at a stable temperature of 30 °C and atmospheric pressure, with a frequency of approximately 2 MHz. Shale core plugs of 1.5-inch diameter and about 1.5-inch length were measured using a P54 probe, which allowed a minimum CPMG (Carr-Purcel-Meiboom-Gill) echo-spacing of 100 μs [39,40]. In this work, the T_2 relaxation time measurements were conducted on six types of samples. Dry samples, as received from the core library, and then the samples saturated with 30,000 ppm NaCl brine were measured. The T_2 relaxation time measurements were conducted again after samples were exposed to scCO$_2$ for four and nine months, both in a dry and saturated state.

The BET surface area and pore size distribution of the core samples were determined by the low-pressure nitrogen adsorption method based on the quantity of gas that adsorbs as a single layer of molecules on the rock surface. Approximately 0.6 g of each sample was loaded into a glass sample tube. To ensure a surface free of traces of gas and water molecules, it was necessary to remove moisture content and degas the samples before pore structure analysis [41]. The samples were degassed for 8 hours at 110 °C before starting the main analysis. Finally, the nitrogen adsorption and desorption isotherms were collected at a constant temperature of 77 K (−196 °C) using a Micromeritics® TriStar II 3020 apparatus [42]. The adsorption and desorption isotherm curves were measured using between 40 and 60 points. The amount of gas adsorbed depended on the adsorbent and the temperature and pressure, and was measured by the instrument at different relative pressures (P/P_0) where P is the gas vapour pressure in the system and P_0 is the saturation pressure of the adsorbent. The adsorption isotherm is the point by point measurement of the quantity of nitrogen adsorbed against the equilibrium pressure, while desorption isotherms are the measured quantity of nitrogen from the sample as the relative pressure is lowered [42]. Pore size distributions (PSD) were obtained based on the density functional theory (DFT) model in this study. The DFT model provides a much more accurate approach for pore size analysis of micropores and even smaller mesopores [38]. It overcomes the oversimplified approach of previous methods which underestimated the pore sizes and yields more reliable pore size results over the full nanopore range by describing the adsorbate configuration at the molecular level [43,44].

The MICP experimental data were measured using a Micromeritics Autopore IV 9500, Capillary pressure curves were collected for pressure points in the range 0.5–60,000 psi, which allowed for the assessment of equivalent pore sizes between 3 and 3.5×10^5 nm approximately. Ten small rock fragments weighing 1.8–4.1 g for five samples were used for tests. Like for low-pressure adsorption measurement, the samples were evacuated before the test to remove moisture and possible gas content. Before mercury was injected into the chamber, the shale samples were degassed at a 25 µmHg vacuum pressure for 10 min. In all cases, the experimental data were collected using a reference equilibration rate of at least 0.001 µL/g/s with a mercury filling pressure of 0.53 psi. The pore throat sizes were calculated based on the Washburn equation [45] for a mercury surface tension of 485 dynes/cm, with an advancing contact angle of 130°.

$$d = -\frac{4\gamma \cos \theta}{P_C} \times C.$$

P_C = capillary pressure (psi), γ = interfacial tension (dynes/cm), θ = wetting angle, d-pore diameter (microns), C = 0.145 (constant to convert to psi). Considering $\gamma_{Hg/air}$ = 485 dynes/cm and $\theta_{Hg/air}$ = 130°, we have the following equation:

$$d = \frac{180.8}{P_C}.$$

4. Results

4.1. Mineralogical Changes

Bulk X-ray diffraction analysis was performed on selected samples to characterise their mineral composition before and after nine months. The collected samples were quartz-rich mudstone with significant Illite/mica and kaolinite and sub-dominant feldspars (Table 1). The average detrital quartz and feldspar grains were 30 and 9 weight percent, respectively. The major clay types were kaolinite and Illite-mica, with an average of 27% and 31%, respectively. Samples H1, H6, H9, H10, H11, and H16 were the clay-rich samples with a clay content of more than 60%. The most quartz-rich sample was H8, with a maximum of 56.7%, and the highest amount of kaolinite was in samples H1 (43%) and H11 (45%). Minor amounts of goethite, pyrite, hematite, and calcite also occurred in some of the samples analysed. Sample H1 had the highest amount of goethite (9%) among the samples; a lesser amount of goethite (<2%) was found in most of the samples. The samples were free from carbonates. Only a minor amount of hematite and pyrite was found in some samples.

Table 1. XRD semi-quantitative mineralogy for Harvey3 seal rocks in weight%, before and after exposure.

Samples	Quartz		Microcline		Albite		Anorthite		Illite/Mica		Kaolinite		Chlorite		Goethite		Hematite		Halite	
	Before	After	Before	After	Before	After	Before	After	Before	After	Before	After	Before	After	Before	After	Before	After	Before	After
H1	18.40	17.50	3.91	4.00	0.35	0.48	0.00	0.00	24.94	28.48	43.06	37.86	0.00	0.00	9.33	9.34	0.00	0.00	0.00	2.34
H2	39.49	30.04	1.31	3.64	0.31	4.22	0.00	0.00	17.69	27.53	38.39	30.73	0.00	0.00	2.48	1.13	0.00	0.00	0.33	2.70
H4	21.01	18.78	8.59	7.59	12.24	11.71	2.24	1.65	33.53	36.44	21.66	19.82	0.00	0.00	0.67	0.53	0.00	0.00	0.20	3.48
H5	36.93	32.90	7.02	7.07	0.00	0.00	0.00	0.00	25.09	28.46	20.25	19.94	10.47	7.50	0.24	0.88	0.00	0.00	0.06	3.26
H8	58.78	49.23	8.14	8.29	0.10	0.28	0.00	0.00	18.19	19.98	13.47	15.72	1.28	2.68	0.01	0.00	0.00	0.00	0.00	3.82
H9	30.39	27.29	8.67	8.55	0.00	0.00	0.00	0.00	34.42	36.56	26.43	23.27	0.00	0.12	0.08	0.35	0.00	0.00	0.05	3.87
H10	22.37	21.87	5.08	5.06	0.01	0.38	0.00	0.00	41.93	40.55	27.79	26.83	0.00	0.00	0.07	0.18	2.74	2.78	0.00	2.34
H13	43.84	39.52	9.64	9.48	0.00	0.00	0.00	0.00	31.35	30.97	14.91	16.40	0.00	0.00	0.16	1.42	0.00	0.05	0.02	2.17
H14	34.83	32.04	8.53	8.38	0.47	0.54	0.00	0.00	27.77	32.01	26.25	22.72	0.00	0.03	0.00	0.04	2.10	1.79	0.00	2.45
H15	25.00	23.45	7.26	6.99	0.30	0.13	0.00	0.00	33.55	33.87	28.85	26.42	1.41	2.27	2.46	1.85	1.18	1.28	0.01	3.75

Table 1 summarises the quantitative amount of minerals in each sample before and after four and nine months of exposure to $scCO_2$. The quartz content is constantly reduced in all the samples, which is a sign of quartz dissolution. However, an increase in the amount of quartz is attributed to the dissolution of feldspar and clay minerals. Microcline is decreased in some samples (H4, H9, H13, H14, and H15) and increased in samples H1 and H2 after exposure. For samples H5, H8, and H10, there is a mixed trend in the amount of microcline. Albite and anorthite rich H4 show a decrease in both minerals post-exposure, but a sharp increase in albite amount in H2. A significant increase in albite and microcline content is observed in sample H2 after nine months.

For clay minerals, kaolinite decreases, and muscovite increases in most samples (H1, H2, H4, H9, H10, and H14). The clay content in the samples follows an interesting trend. In the samples where the kaolinite content is continuously decreased, an increase in muscovite content is constantly evident. Halite is increased in all samples as a result of brine desiccation. Carbonate minerals, which are of great significance in the mineral trapping of CO_2 through the reaction with magnesium and calcium ions provided by silicate minerals, are mostly absent in the samples analysed.

4.2. Nuclear Magnetic Resonance

Nuclear magnetic resonance (NMR) experiments were carried out on a low field Magritek bench-top Rock Core Analyzer, located in the Department of Petroleum Engineering at Curtin University. A total of seven core plugs were recovered from the samples received from the core library. The other samples were either too soft or were broken, and plugs could not be taken from them. The study involved six types of T_2 relaxation time measurements: (1) dry state (as received samples from core store) before exposure to $scCO_2$ (Before_Dry); (2) brine-saturated state before exposure to $scCO_2$ (Before_Sat); (3) after exposure to $scCO_2$ for four months (4M_Dry); (4) brine re-saturated state after exposure to $scCO_2$ for four months (4M_Sat); (5) after exposure to $scCO_2$ for nine months (9M_Dry); and (6) brine re-saturated state after exposure to $scCO_2$ for nine months (9M_Sat). Sample H8 and H10 collapsed later during saturation stages.

Figure 2 lists the NMR porosity for the samples analysed. There is a notable increase in porosity for the samples that were exposed to $scCO_2$. The average NMR porosities are 7.62%, 19.55%, 14.44%, 20.68%, 14.76%, and 20.46% for Before_Dry, Before_Sat, 4M_Dry, 4M _Sat, 9M_Dry, and 9M_Sat respectively. There is an increase in the average porosity of samples exposed to $scCO_2$.

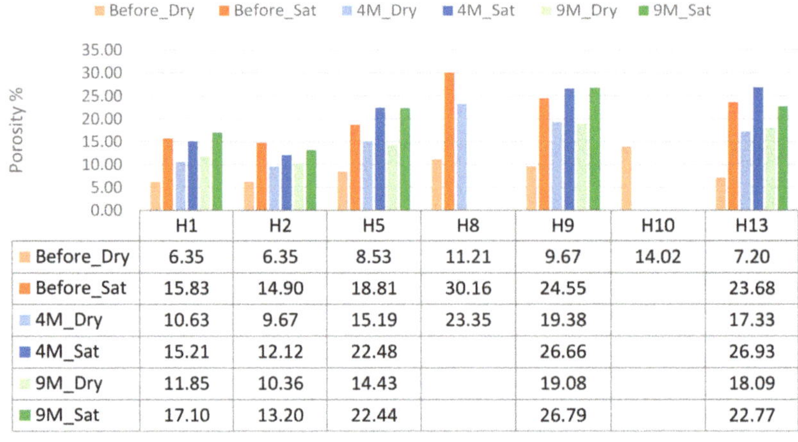

	H1	H2	H5	H8	H9	H10	H13
Before_Dry	6.35	6.35	8.53	11.21	9.67	14.02	7.20
Before_Sat	15.83	14.90	18.81	30.16	24.55		23.68
4M_Dry	10.63	9.67	15.19	23.35	19.38		17.33
4M_Sat	15.21	12.12	22.48		26.66		26.93
9M_Dry	11.85	10.36	14.43		19.08		18.09
9M_Sat	17.10	13.20	22.44		26.79		22.77

Figure 2. A comparison of porosities for samples before and after exposure to $scCO_2$.

Figure 3 illustrates the combined T_2 distribution for all Before_Dry, Before_Sat, 4M_Dry, 4M_Sat, and 9M_Dry core plugs respectively. The T_2 distributions of the Before_Dry samples, which is peaked at the lower T_2 region, illustrate a shape that corresponds to clay surface water. After the saturation, the dominant peak of the T_2 distribution of the Before_Sat state is shifted toward higher T_2 times, which likely corresponds to a combination of both clay surface water and capillary pores. The T_2 distribution of the samples after exposure to $scCO_2$ show a slight reduction in the height of dominant peak.

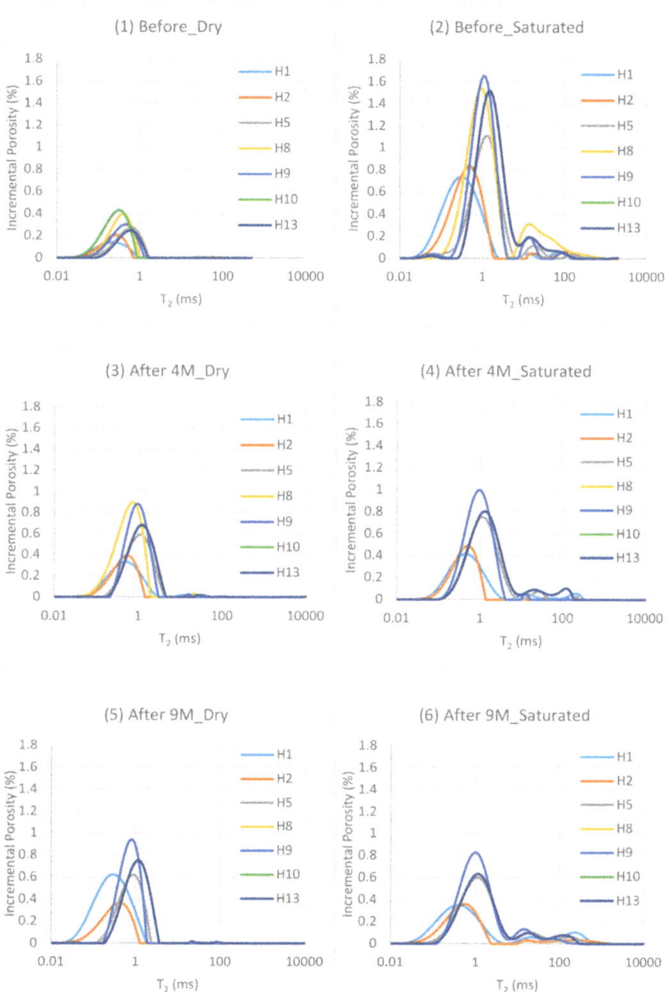

Figure 3. T_2 distribution of (**1**) dry state (as received samples from core store) before exposure to $scCO_2$, (**2**) brine-saturated state before exposure to $scCO_2$, (**3**) directly after 4 months exposure to $scCO_2$, (**4**) brine re-saturated state after exposure to $scCO_2$, (**5**) directly after 9 months exposure to $scCO_2$, (**6**) brine re-saturated state after 9 months exposure to $scCO_2$. Sample H10 failed after first saturation and sample H8 failed after second saturation.

Some of the water in the samples was removed after exposure to scCO$_2$ (Figure 3) which resulted in reductions in the NMR signals. Exposure to scCO$_2$ could induce rock dehydration in the samples.

The bimodal T_2 distribution observed in the saturated samples is indicative of clay surface water and water in the interstitial pores. Samples H10 and H8 failed during saturation process and could not be analysed.

4.3. Low-Pressure Nitrogen Absorption

A low-pressure nitrogen Adsorption (<18.4 psi, 77 K) test was applied on ten samples from Harvey3 before and after they were exposed to scCO$_2$. The values for the surface area and pore volume and their distribution along the pore width are reported and compared here. The pore size (diameter) is calculated from these values and included in this study. The results indicate some significant changes in the nanopore structure system of the samples after scCO$_2$ and shale interactions. However, the variation trend of the pore structure parameter was quite different after exposure to scCO$_2$, which is related to the discrepancies in the mineralogical and geochemical properties of the samples.

Tables 2 and 3 summarise the collected results from low-pressure adsorption measurement for the samples before and after exposure to scCO$_2$. The average pore width for the Before and After data was found to be 3.90 and 4.5 nm, respectively. The average pore volume obtained from LPNA tests for Before and After samples is about 4 and 4.4 cm^3/100 g, which is characterised by a very high contribution of the mesopore size, 76.63% and 83.48% of total porosity, respectively. The average BET surface area is found to be 43.42 m^2/g, and 39.42 m^2/g (Before and After). Figure 4 compares the total pore volume in samples studied. Figures 5 and 6 compare the BET surface area and pore size of the analysed samples before and after exposure to scCO$_2$.

Table 2. Summary of low-pressure nitrogen adsorption results before exposure to scCO$_2$.

Samples	BET Surface Area (m^2/g)	Average Pore Width (nm)	DFT Model			
			Micropore Vol. (cm^3/100 g)	Mesopore Vol. (cm^3/100 g)	Macropore Vol. (cm^3/100 g)	Total Pore Vol. (cm^3/100 g)
H1	32.96	6.868	0.347	4.859	0.453	5.659
H2	29.71	5.467	0.347	3.500	0.214	4.060
H4	50.29	3.718	1.226	3.186	0.263	4.675
H5	29.29	3.521	0.332	2.186	0.061	2.578
H8	23.88	2.889	0.405	1.299	0.021	1.725
H9	41.34	3.406	0.731	2.720	0.070	3.520
H10	51.96	3.157	0.987	3.067	0.047	4.101
H13	36.54	4.210	0.620	3.033	0.193	3.846
H14	67.61	2.904	1.654	3.205	0.049	4.908
H15	70.63	2.844	1.495	3.504	0.023	5.022

Table 3. Summary of low-pressure nitrogen adsorption results after exposure to scCO$_2$.

Samples	BET Surface Area (m^2/g)	Average Pore Width (nm)	DFT Model			
			Micropore Vol. (cm^3/100 g)	Mesopore Vol. (cm^3/100 g)	Macropore Vol. (cm^3/100 g)	Total Pore Vol. (cm^3/100 g)
H1	33.12	4.889	0.315	3.614	0.119	4.048
H2	30.69	4.354	0.344	2.949	0.051	3.344
H4	26.19	4.264	0.409	2.319	0.064	2.792
H5	29.59	5.754	0.324	3.724	0.208	4.256
H8	25.45	3.325	0.332	1.760	0.023	2.115
H9	41.77	4.775	0.573	4.203	0.210	4.986
H10	46.1	4.063	0.775	3.812	0.094	4.682
H13	42.4	5.207	0.548	4.702	0.270	5.520
H14	63.21	3.700	1.363	4.309	0.175	5.847
H15	55.69	4.638	1.088	5.096	0.272	6.457

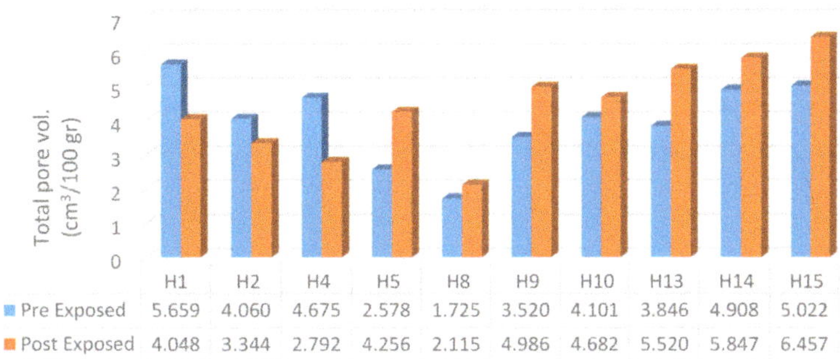

Figure 4. Histogram showing total pore volume derived from maximum pressure for the analysed samples.

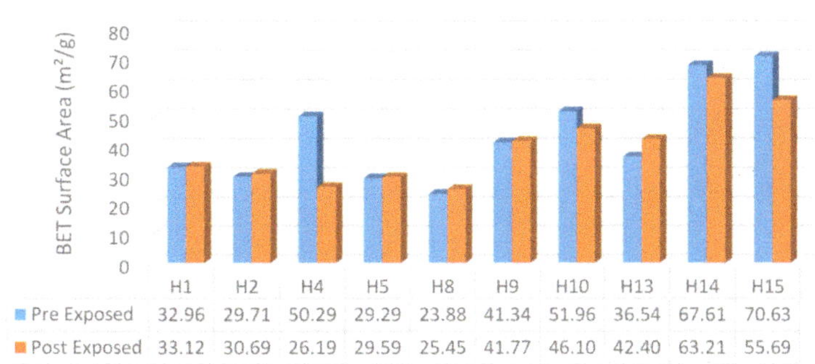

Figure 5. Histogram showing the measured BET surface area for the analysed samples.

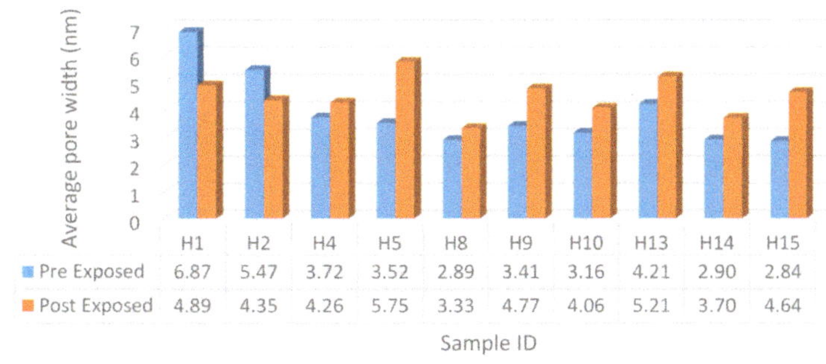

Figure 6. Histogram showing the average pore size of the analysed samples.

The low-pressure nitrogen adsorption analysis shows that the pore structure changed after the shale samples were exposed to scCO$_2$. For most of the samples studied there was an increase in the pore volume (H5–H15). However, there was a decrease in the pore volume for samples H1, H2, and H4 (Figure 4). The porosity increased significantly in samples H5, H9, and H13, where the post-exposed pore size distribution (PSD) curve sits on top of the preexposed curve along the whole pore width range (Figure 7). An average increase of 21% is observed in samples H8, H10, H14, and H15, where the post-exposed curve is increased only in the lower pore width range (<20 nm).

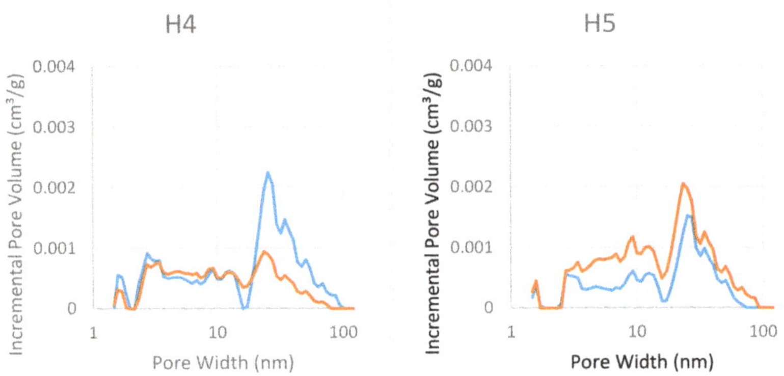

Figure 7. The pore size distribution of sample H4, representing the group of samples where the total pore area was reduced, and sample H5, representing the group of samples where the total pore area was enhanced.

Figures 8 and 9 show the pore volume distribution of the micropores, mesopores, and part of macropores before and after the exposure. As can be seen for H8 and H10, there are no significant changes in the PSD after exposure to $scCO_2$ for pores larger than 20 nm. The pore volume has slightly increased for pores smaller than 20 nm of width in these two samples. Whereas, for samples H5, H9, and H13, the PSD shows an increase in the pore volume in the whole pore size spectrum, but the general trend follows the PSD of the sample before exposure to CO_2. Samples H14 and H15 show a significant increase in pore volume in pore widths smaller than 20 nm, which is followed by a pore volume decrease in pores larger than 20 nm. Samples H1, H2, and H4 have a more complex increase/decrease trend. They show a large decrease in the pore volume of the larger pore size range (> 18 nm) after exposure to CO_2, which results in total pore volume decrease in these three samples (Figure 4) In pore sizes smaller than 18 nm, a mixed pattern is observed. Samples H1 and H2 show an increase then decrease, but sample H4 shows a decrease then increase in pore volume in the lower mesopore range.

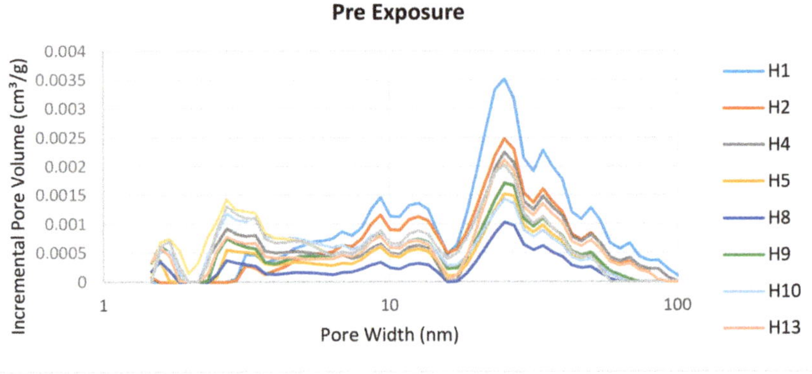

Figure 8. Pore size distributions for the samples analysed before exposure to $scCO_2$.

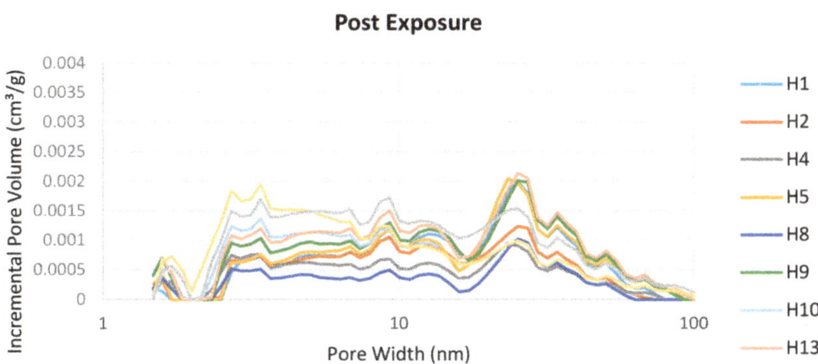

Figure 9. Pore size distributions for the samples analysed after exposure to scCO$_2$.

In summary, pores with a diameter of 18–20 nm are the turning point for most of the samples. In the first three samples, the substantial decrease of pore volume above this turning point results in an overall pore volume reduction of samples. For the remaining samples (H5 to H15), the total pore volume is increased, but the contribution of pore sizes above or below the turning point is different. Pore size below the turning point plays the deciding role in samples H8, H10, H14, and H15, as the pore volume increase in this range is so large that it either covers the reduction in larger pore sizes, or there is no pore volume reduction in the larger pore size range. On the other hand, in samples H5, H9, and H13, the pore volume is increased in the whole 2–100 nm range and no turning point is observed.

In general, the surface area was increased in most of the samples after exposure, especially in mesopore size ranges. As can be seen in Figure 5, the BET surface area in samples H1, H2, H5, H8, H9, and H13 is increased, and in samples H4, H10, H14, and H15 is decreased after exposure. Among the samples, H8 has the lowest surface area, which is the most quartz-rich sample (57%). Samples H4, H10, H14, and H15 have a surface area in the range of 50–70 m^2/g and samples H1, H2, H5, H9, and H13 have a surface area in the range of 30–40 m^2/g. The pore diameter was calculated using the volume to surface area ratio and shows the same trend as pore volume changes. The calculated average pore diameter follows the same decreasing/increasing trend of pore volume, except for sample H4 after exposure to CO$_2$ (Figure 6).

4.4. Mercury Injection Capillary Pressure

Table 4 presents the results obtained from MICP measurements on five samples before and after exposure to scCO$_2$. Variation can be observed, comparing the petrophysical parameters in the tested samples. The porosity is increased in post-exposed samples H5, H9, and H13, and decreased in H1 and significantly decreased in H2. The pore areas in samples H1, H2 and H5 are decreased after exposure and increased in samples H9 and H13. The pore diameter in all the samples is enhanced after the exposure.

Table 4. Summary of results from the MICP test for samples before and after exposure to $scCO_2$.

Sample		Porosity (%)	Total Pore Area (m^2/g)	Peak Diameter (nm)	Median Pore Diameter (Volume) (nm)	Median Pore Diameter (Area) (nm)	Average Pore Diameter (4V/A) (nm)	Threshold Pressure (psi)
H1	Before	9.59	14.61	8.74	11.3	9.6	11.3	8349
	After	9.21	12.28	12.5	14.8	11.4	14.2	6708
H2	Before	10.65	18.53	12.5	13.4	9.6	12.7	6310
	After	4.59	4.17	15.1	21.4	11.9	20.4	1736
H5	Before	10.12	10.26	21.1	23.6	9.1	16.2	1897
	After	10.80	6.35	21.1	64.1	12.1	28.1	316
H9	Before	13.93	10.89	21.1	43.6	8.9	22	412
	After	16.32	17.2	18.1	63.5	9.4	24.1	392
H13	Before	14.09	13.6	21.1	30	9	17.9	1179
	After	19.14	18.32	18.1	31.4	9.2	19.6	1090

Porosity ranges of 9.59–14.09% for the Before samples and 4.17–18.53% for the After samples were recorded. The porosity was reduced for samples H1 and H2, from an average of 10.12% to 6.90%, but was enhanced from an average of 12.72% to 15.42% in the case of samples H5, H9, and H13. The general peak pore throat diameter of samples H1 and H2 show Before at 10.62 nm and After at 13.80 nm, smaller than samples H5, H9, and H13, which have values of 21.10 nm and 19.10 nm for the before- and after-exposure samples respectively (Figure 10). The average peak diameter of 10.62 nm was increased to 13.80 nm after exposure for samples H1 and H2. However, the higher average peak diameter of 21.10 nm for the before-exposure samples H5, H9, and H13 was decreased to 19.10 nm after the exposure.

Capillary pressure curves (Figure 11) summarize the relationship between the pressure applied to the samples at different stages and the volume of intruded mercury for each tested sample. Samples H1 and H2 have sigmoidal capillary pressure curves. Samples H5, H9, and H13 show distinctly bimodal pore-size distributions. All the samples exhibited 100% mercury saturation close to 60,000 psi. The highest entry pressures are observed for samples H1 and H2, whereas H9 display the lowest ones. Samples H1 and H2 are likely to have the most uniform pore size distributions, due to the more abrupt increment at low normalised mercury saturation, without further visible changes to the rate of intrusion (Figure 11). Since most of the intrusion occurs at higher pressures, H1 and H2 are also presumed to have the largest proportion of narrow pore sizes. This is verified by pore size distribution curves in Figure 10. In contrast, the more stepwise intrusion observed in the H5, H9, and H13 curves (Figure 11) is indicative of significant compression beyond the initial entry point, which is related to the presence of a wider and continuous range of pore sizes. The capillary pressure curves of after exposure samples in Figure 11 show a smoother slope. The intrusion of CO_2 into the samples alters the pore sizes and pore throats in the samples, resulting in mercury intrusion starting at lower pressures compared with the before exposure samples.

Figure 10. Incremental Intrusion vs Pore Size.

The pore size distributions of the samples (Figure 10) demonstrate a dual trend. The first two samples exhibit narrow-distributed modals with higher peak values. A major throat size in the range of 6–18 nm is observed for samples H1 and H2; larger mesopores are very limited in terms of volume. The other three samples (H5, H9, and H13) demonstrate lower peaks and a wider range of pore throats spreading in the mesopore and macropore range.

In this work, the MICP threshold pressures for shale samples have been determined graphically at the point of inflection in the cumulative curves upon which the first rapid increase is observed. The same trend as capillary pressure curves and pore size distribution is also observed in the threshold pressure, where the two first samples have the highest and the other three samples have the lowest threshold pressure. However, after exposure to scCO$_2$, samples H2 and H5 show the greatest decrease in their $P_{Threshold}$ of 72% and 83%, respectively, where all the samples' threshold pressures are decreased. The height of the CO$_2$ column is calculated assuming a contact angle of 130° and a Hg surface tension of 485 dynes/cm (Table 5 and Figure 12).

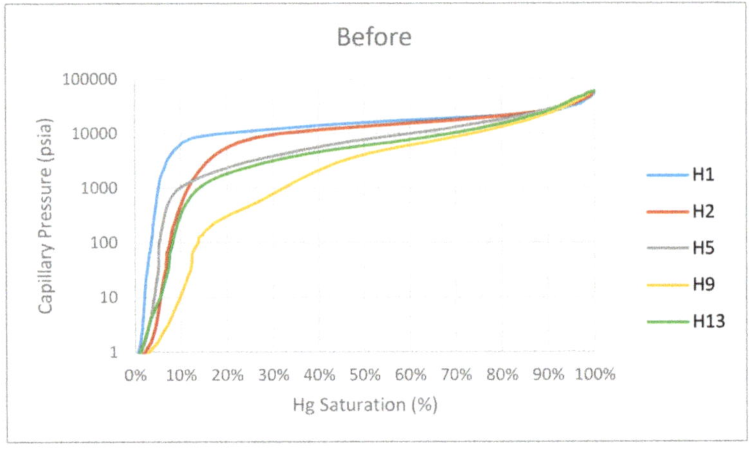

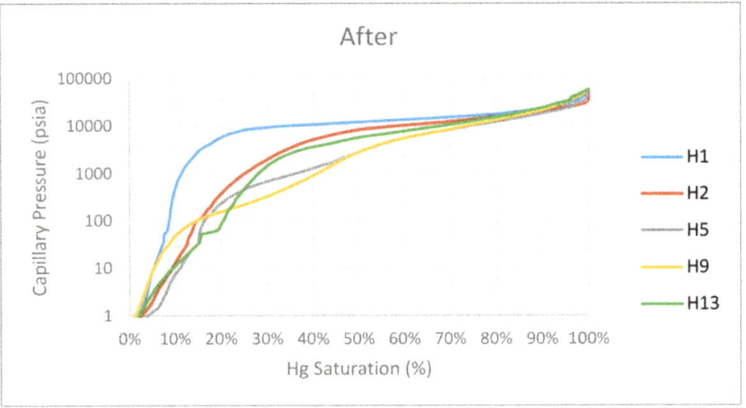

Figure 11. Capillary pressure vs Hg saturation before and after exposure.

Table 5. Parameters used for the calculation of maximum CO_2 column heights retained by different shale intervals before and after exposure to $scCO_2$.

Sample	Depth (m)	P Pore (psi)	T (°C)	ρ_{CO2} (g/cc)	ρ_{brine} (g/cc)	$\gamma\,b$, CO_2 (mN/m)	$P_{Threshold}$ (psi) before	$P_{Threshold}$ (psi) after	CO_2 Column Height (m), before	CO_2 Column Height (m), after
H1	1417.5	2055	64.6	0.494	0.990	28.06	8349	6708	1067.3	857.5
H2	1412	2047	64.5	0.493	0.990	28.07	6310	1736	805.4	221.6
H5	1187.6	1722	56.9	0.443	0.993	28.68	1897	316	223.3	37.2
H9	916.3	1329	47.7	0.326	0.996	30.94	412	392	43.0	40.9
H13	766.1	1111	42.6	0.229	0.998	34.01	1179	1090	117.7	108.8

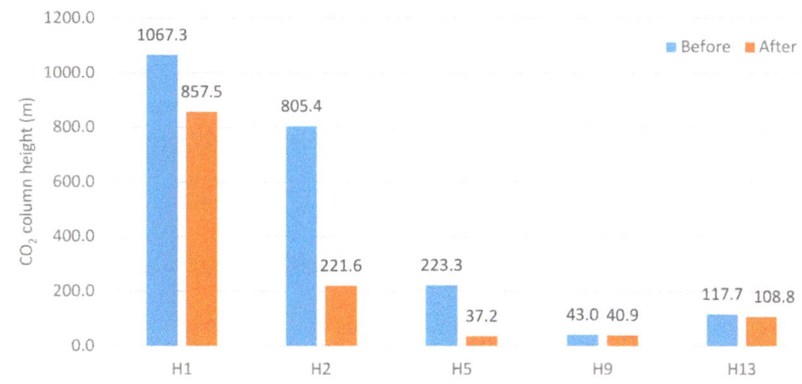

Figure 12. Bar chart showing CO_2 column height (m) for shale samples before and after exposure to $scCO_2$.

5. Discussion

Assessing CO_2-rock interaction is an important part of seal integrity studies, as these potentially affect physical properties through highly coupled processes. The driving process of CO_2-rock interactions includes dissolution of CO_2 in brines, acid-induced reactions, reactions due to brine concentration, clay desiccation, pure CO_2-rock interactions and reactions induced by other gasses than CO_2 [9]. The observed changes in the petrophysical properties of the after-exposure samples are generally attributed to the three mechanisms of mineral dissolution, mineral precipitation and physical compaction [46]. The fine migration of susceptible minerals such as kaolinite is another important mechanism which can alter the properties of rock samples. The migration of clay fines generated and released by mineral dissolution is believed to have occurred in the exposure experiment conducted in this work. The occurrence and the extent of the effects of these mechanisms on a sample depend on the exact mineral composition, presence of core-scale heterogeneities and the original textural features of the sample and therefore, may vary from one sample to the next. Much of the discussions presented in the upcoming sections of this section are to determine how CO_2-induced dissolution and precipitation reactions affect the pore space evolution and thus the physical properties of after-exposure samples, as revealed by the complementary measurements conduced.

5.1. The Effect on Mineralogy

The XRD analysis of the original caprock samples showed that samples H8 and H13 are quartz-rich, samples H1, H9, H10, and H15 are clay-rich (total clay > 60%), and samples H2, H4, H5, H13, and H14 have more clay content than quartz. The siliceous samples investigated in the current study are marked by a strong component of clay minerals, with up to 75% kaolinite and Illite. Samples are rich in both kaolinite (average: 26% and maximum 43%) and Illite (average: 29% and maximum 42%). Quartz content is 33% on average, with a maximum of 59%. The dominant feldspar in the samples is microcline (average 7% and maximum 10%). Samples H14 and H15 originated from the Eneabba shale formation, and samples H1–H13 originated from the same formation (Yalgroup member of Lesueur formation), but different localities. The different mineralogical composition of the samples leads to different reaction responses to the CO_2-brine and brine systems.

XRD analysis of samples was carried out to measure the exact weight percentage of each mineral phase. The results were compared to the after-exposure results to monitor any changes in the amount of mineral phases. The X-ray diffraction (XRD) results show several changes in mineralogy because of rock-brine-CO_2 reactions (Figure 13). Quartz, kaolinite, and goethite were dissolved in most samples and muscovite and halite were precipitated in general. Feldspars showed mixed behaviour across the samples.

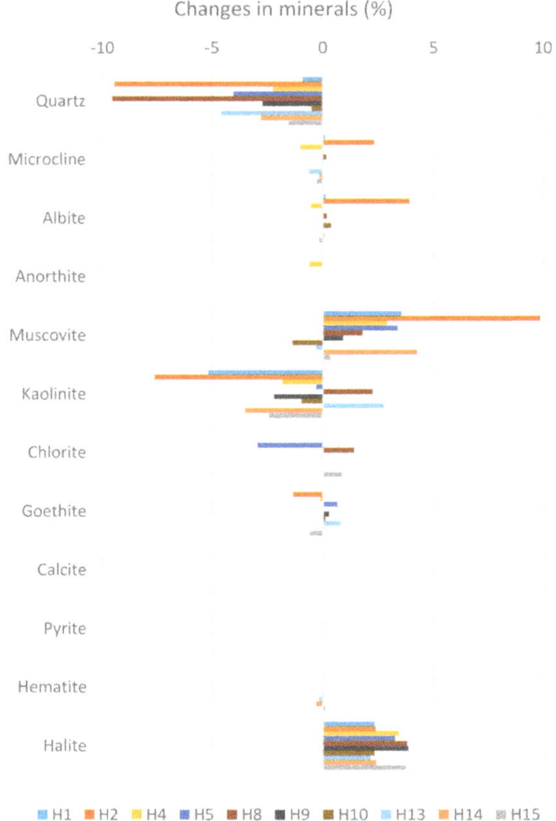

Figure 13. Changes in mineral content after exposure in percentage.

Previous experimental work [11] showed some reactions occurring on a measurable time scale only at high temperatures of 200 °C. Other works [16] noted only a limited reaction of CO_2 with pure mineral phases (anorthite and glauconite) at 50 °C and 150 °C and a low pressure. Changes in caprock fluid transport properties may happen regardless of the extent and scale of the mineralogical composition alteration. Wollenweber et al. (2010) reported carbonate dissolution and re-precipitation, leading to the alteration of shaly caprock transport parameters. In this case, no detectable mineral alterations were observed. The loss of quartz [47], feldspars [14], kaolinite [18], and chlorite [21] has been reported in other studies, resulting in an increase in porosity and opening up pore throats, leading to an increase in permeability.

The observed mineralogical alterations can be explained by the generation of carbonic acid (H_2CO_3). Changes in pH control the dissolution of minerals, release of chemical species, and formation of new compounds and minerals [7,8,10,17,48,49]. In the absence of calcite, reactive minerals such as feldspars, kaolinite and chlorite start to dissolve. The dissolution of these minerals releases Si, Al, Na, K, Fe, and Ca, which can form new compounds. The dissolution of feldspars and kaolinite results in the formation of muscovite/Illite.

The decrease in the quartz content of the samples, observed in previous experiments, is attributed to the dissolution of quartz itself. Rathnaweera et al. (2016) reported significant quartz dissolution under the 40 °C after a 1.5-year interaction with scCO$_2$ and suggested that over a short period and under low temperature, quartz dissolution is too minor to be observed [50]. Szabó et al. (2016) also reported moderate dissolution of quartz, however in durations ranging 28–57 days [24]. Kaszuba et al. (2003) also observed dissolution of quartz along with oligoclase and biotite dissolution during their batch experiment at an elevated temperature and pressure (2000 °C and 200 bars) [11]. Other cases when quartz content was increased have been reported in other studies [22]. In such cases, the additional quartz is the product of the dissolution of other minerals such as feldspars and clay minerals. In an experiment with two types of shale caprocks, Alemu, et al. (2011) reported no major mineralogical alteration in the case of clay-rich shale, whereas in the case of carbonate-rich shale, plagioclase and clay minerals were dissolved, carbonates dissolved and re-precipitated, and smectite was formed [51]. Credoz et al. (2009) used purified clay minerals to evaluate their reactivity and found kaolinite to be the most reactive clay mineral in the long-term experiments [18].

Carbonate precipitation, such as the precipitation of dawsonite, siderite, calcite, dolomite, and magnesite, has been reported in several studies [52,53]. However, carbonate formation was not observed during this study. Since the shale sample used in this experimental study did not contain any carbonates and a very limited amount (2%) of Ca-feldspar in just one of the samples, a potential source for Ca was not available to form calcite and kaolinite assemblage. Also, low cation concentrations with an acidic pH were observed, which showed that dissolution prevailed over carbonate precipitation [17].

The formation of halite was interpreted as an experimental artefact. NaCl precipitated from Na and Cl coming from the solution, since the solid was not rinsed with deionised water before analysis [18].

5.2. The Effect on Petrophysical Properties

The low-pressure nitrogen adsorption method showed a slight increase in porosity observed in seven out of ten samples (Figure 4), except for samples H1, H2, and H4. Samples recorded a pore volume range of 1.7–5.7 cm^3/g before exposure, which was changed to 2.1–6.4 cm^3/g after exposure to CO$_2$. There was an average 1.17 cm^3/g increase in the pore volume of ten samples and an average 1.4 cm^3/g decrease in the pore volume of three other samples tested.

The porosities were also estimated with the MICP and NMR methods for five samples. Samples H1 and H2 recorded a reduction in their porosity in both of these methods and samples H5, H9, and H13 show an enhancement in their porosities. The MICP porosities range is 9.6–16.3% for pre-exposure samples and 4.6–19.1% for after-exposure samples (Table 4). There was an average 2.7% increase in MICP porosity for three samples and an average 3.2% decrease in MICP porosity for two other samples tested.

The NMR porosities were measured in both dry and saturated states. The dry samples before exposure (as received) had porosities in the range of 6.35–9.67% and the range of 9.67–19.38% after four months of treatment and 10.36–19.08% after nine months of CO$_2$ treatment. NMR porosities in saturated states are generally higher than dry states and their ranges for before, after four months, and after nine months of CO$_2$ exposure are 14.90–24.55%, 12.12–26.93%, and 13.20–26.79%, respectively (Figure 2). While the NMR porosity of samples in the dry state increased across the sample by about 7.7%, the NMR porosity of saturated samples increased by an average 3% in three samples and decreased by an average 1.7% in two other samples tested. The increased porosity after the initial saturation stage is simply because the NMR method measures the response of hydrogen nuclei and then converts it to porosity. The received samples from storage were relatively dry and showed smaller porosities, which was significantly increased after the saturation. Each following exposure to CO$_2$ reduced the porosities because of the evaporation of some of the water content. The following re-saturation stages did not increase the porosity as much as the initial saturation because the samples contained much of the water from the first saturation stage.

The received samples have a systematic monomodal distribution, with a relaxation time (T_2) centred around 0.3–0.7 ms. After saturation, the previous population showed a shift toward a longer T_2, centred between 9–1.6 ms, except for samples H1 and H2 which stayed centred between 0.33–0.46 ms. This first population is defined as the short relaxation time. A second population defined by long relaxation time was also recorded for the samples at around 13.5 ms. This second population is more likely the effect of the macropores filled by brine during the saturation process. The saturated samples show porosity values significantly higher than partially saturated samples.

The average pore width results of the LPNA method show an increase in most samples (H5, H8, H9, H10, H13, H14, and H15 and a decrease in three samples, H1, H2, and H4 (Figure 6). The average pore width range according to this method is 2.9–6.9 nm for pre-exposure samples, which is changed to 3.3–5.8 nm after exposure. However, the MICP average pore width is increased across all samples. The average pore diameter range of before-exposure samples is 11.3–22 nm, and for after-exposure samples the range is 12.7–28.1 nm (Table 4).

The leaching of the mineral constituents of the shale is the main reason for porosity increase. Mineral dissolution and precipitation can induce changes in porosity and permeability. In this study, an exposure period of nine months was selected, based on the reaction time required to complete the kinetically slow reactions of existing minerals with CO_2 and brine to identify the fate of these major rock minerals. However, the timescales required to complete the reactions are certainly more than one year, up to hundreds or even thousands of years [54,55]. Therefore, considering the time frame available for this study and in order to record the maximum impact possible, nine months was selected as being a reasonable period for the CO_2-brine-rock interactions. However, this period is not nearly enough to capture all the possible mineral reactions that occur in CO_2 and brine environment, such as precipitation of feldspars and secondary precipitation of calcite and quartz. Therefore, it would be possible to see the dominant reactions such as the initial dissolution of quartz and kaolinite and the salt drying-out effect [50]. The significant increase in porosity observed in most experimental samples would potentially deteriorate the sealing capacity of the caprock.

The bulk porosities from LPNA, MICP, and NMR present some differences (Figure 14). The porosities from NMR, averaging about 20% for saturated samples, are much higher than the LPNA and MICP porosities, which are about 12% and 4 cm^3/g, respectively. This scale of differences has also been investigated by [56] for mudstone samples. Possible explanations include: 1—the ability of NMR to measure both connected and isolated pores (pores located within the grains and clay-bound water spaces) in contrast with MICP that only measures the connected pores; 2—NMR samples are saturated with brine which make them prone to clay swelling and cracking, while MICP samples were dried which induced potential clay shrinkage [57].

Two groups of samples were identified from the capillary pressure profiles (Figure 11). The capillary pressure curves in samples H1 and H2 start to plateau earlier than the other three samples and remain mostly flat for a considerably larger portion of the curves. This suggests a unimodal type of pores in the first two samples compared with more complex pore sizes in the last three samples covering a large range of pore sizes. The general peak pore throat radius shows, in the H1 and H2 samples, a value of 12.5 nm, smaller than H5, H9, and H13 samples which have values around 21.1 nm (Table 5 and Figure 11). More specifically, pore throat distribution reveals a second minor population in the second group of samples that have a pore throat size >250 nm that is easily invaded by mercury injection at low pressure. Samples H1 and H2 (group one) have the smallest average pore diameter at 11.3–12.7nm with low permeability at around 160–430 nD, while other samples are 16.2–22 nm.

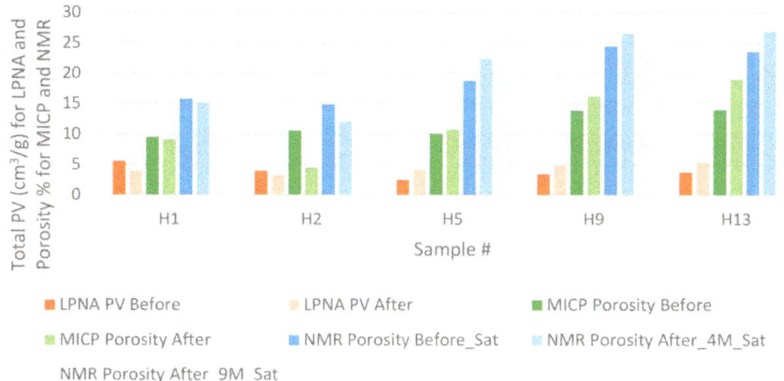

Figure 14. Porosity results from low-pressure nitrogen adsorption (LPNA), mercury injection capillary pressure (MICP), and NMR-Sat methods. LPNA pore volume values are in cm^3/g, and MICP and NMR values are porosity in percentage.

Similarly, threshold pressure shows distinctive values in these two groups. The group one threshold pressures are in the range of 6700–8350 psi and the group two threshold pressure range is 412–1900 psi, which is significantly lower than group one (Table 4). The threshold pressure is reduced in all five samples. However, the decrease is significant in the case of samples H2, where it is decreased from 6310 psi to 1736 psi.

When the buoyancy pressure, due to an accumulated CO_2 plume, dominates the capillary pressure of caprock, the plume intrudes into the pore throats, and the occurrence of capillary leakage is inevitable in the caprock. The capillary pressure of the CO_2-brine can be equated to the buoyancy pressure of the injected CO_2 column [38]. Here, the experimental investigation of capillary pressure was conducted, and the result was used to calculate the height of the CO_2 column. Mercury injection capillary pressure (MICP) indicated a reduction in the capillary pressure and the calculated maximum column heights of CO_2.

There is a slight positive relationship between MICP porosity and the average pore diameter. However, a strong relationship exists between MICP porosity and peak pore throat diameter and threshold pressure (Figure 15). The general trend follows the expected relationship between these three parameters. Smaller porosities and peak pore diameter (group one) correspond to higher threshold pressure and vice versa (group two).

These two groups are distinctive in their content of kaolinite clays. Group one, compromising samples H1 and H2, with high kaolinite content, records a high threshold pressure due to low pore throat diameter. The second group of samples, H5, H9, and H13, with lower kaolinite content, has a lower entry pressure and more diverse range of pore throat diameters.

The total area of pores is estimated from both the LPNA and MICP methods. Ten samples were tested with LPNA before and after exposure. However, only five of them were measured with the MICP method (H1, H2, H5, H9, and H13). The LPNA surface area shows slight changes in most samples (Figure 16). A slight positive change is observed in samples H1, H2, H5, H8, H9 and a slight negative change in sample H10. However, in samples H5 and H15, the total pore area is decreased significantly, and, in the case of sample H13, it is increased significantly. The MICP total pore area shows more dramatic changes. The total pore area is decreased in samples H1, H2, and H5 and is increased in samples H9 and H13. In the case of sample H2, the reduction is very significant (−77%). MICP confirms the LPNA total pore area results in terms of the direction of change in samples H9 and H13 where both methods show an enhancement in total surface area. However, the slight LPNA pore area increase in samples H1, H2, and H5 contradicts the MICP significant decrease after exposure to scCO$_2$.

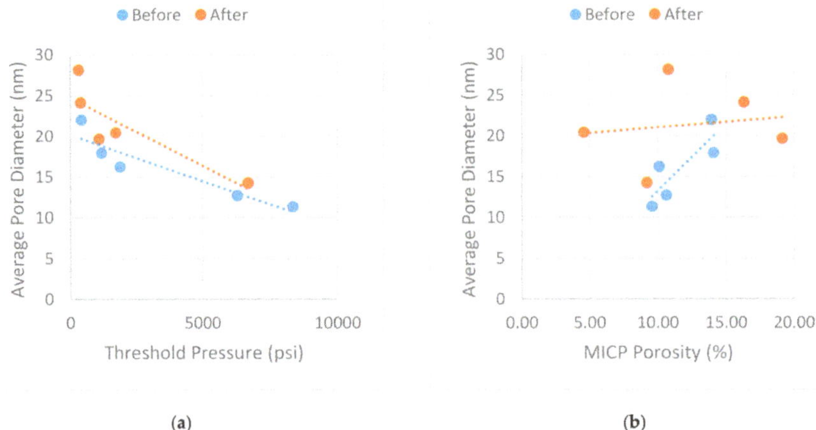

Figure 15. Pore throat diameter versus (**a**) threshold pressure and (**b**) MICP porosity.

The BET surface areas were found to be 23.9–70.6 m^2/g (Table 2). There was a direct relationship found between the BET surface area and the occurrence of Illite/mica (Figure 17). Group one of samples has a lower percentage of Illite/mica clays but higher kaolinite percentages compared with group two of the samples. The second group of samples with a higher kaolinite content has higher threshold pressure due to its low pore throat diameter and lower porosities (Figures 15 and 17). Furthermore, the presence of kaolinite influences the NMR response (Figure 17). As the kaolinite content increases, the T_2 relaxation time peak tends to decrease, with corresponding smaller pore sizes or a restricted environment.

The high presence of total clay (mostly kaolinite) could cause the blocking of the pore throats and could give access to the neighbouring larger pores during the saturation process, leading to lower T_2 amplitude values. The long T_2 in group two of the samples is an indication of macropores, and potentially new cracks induced by artificial saturation under pressure and brine reactivity with shales, during the sample recovery and preparation steps.

There is no strong relationship between the LPNA average pore width and either total pore area or total pore volume (Figure 18). Average pore width shows a strong relationship with fractions of micro-, meso-, and macropores. Before and after exposure samples show a decrease in micropore percentage and an increase in mesopore and macropore percentage with increasing pore diameter (Figures 19 and 20). Based on IUPAC pore classification, LPNA pore volume showed a pore range from 65.3% to 86.2% mesopores, 6.1% to 33.7% micropores, and a small portion of 0.5% to 8% macropores.

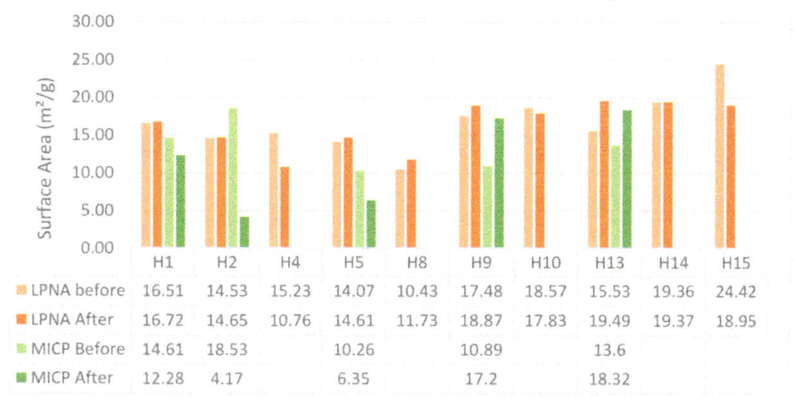

Figure 16. Surface area estimates before and after exposure to scCO$_2$ from LPNA (orange) and MICP (green).

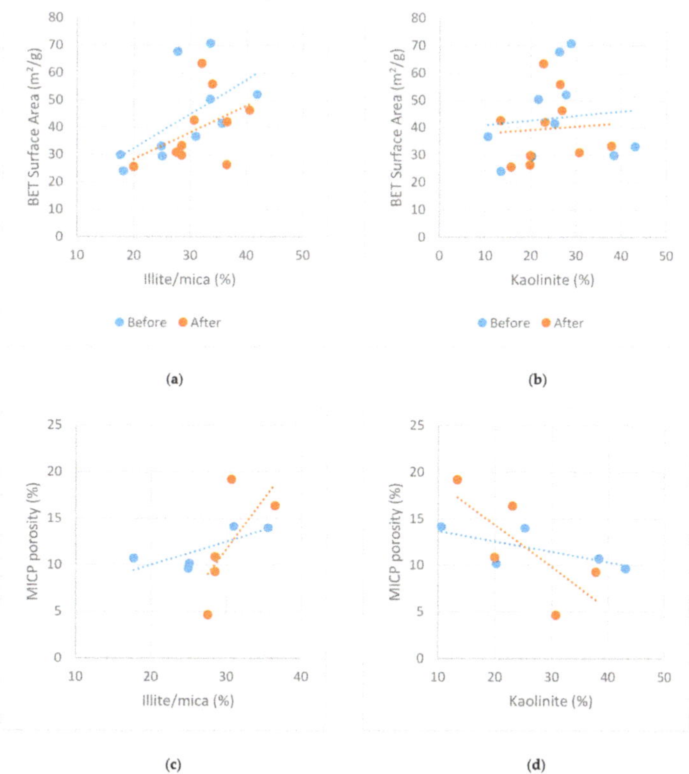

Figure 17. *Cont.*

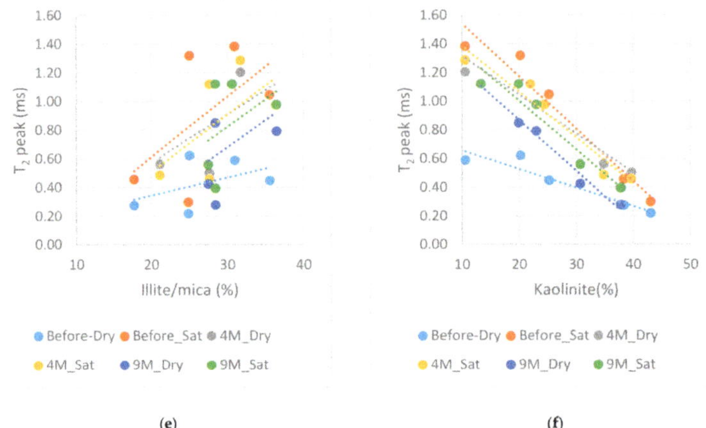

Figure 17. Influence of Illite/mica and kaolinite on various parameters. (**a**) BET Surface Area vs. Illite/mica (**b**) BET Surface Area vs. kaolinite (**c**) MICP porosity vs. Illite/mica (**d**) MICP porosity vs. kaolinite (**e**) T_2 peak vs. Illite/mica (**f**) T_2 peak vs. kaolinite.

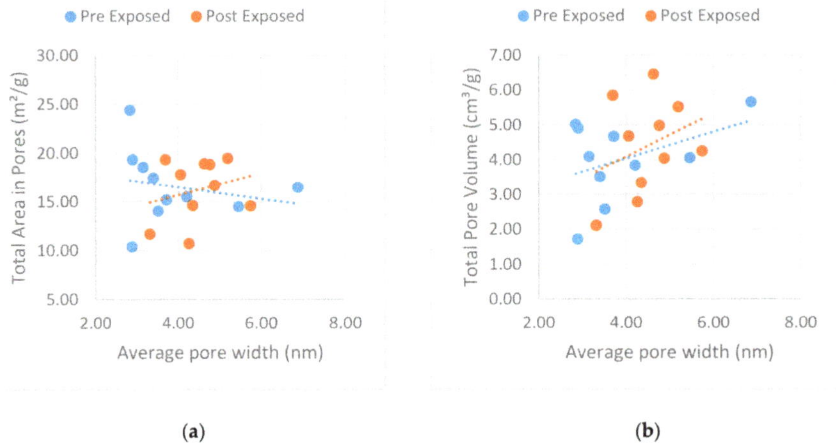

Figure 18. Relationship between average pore width versus total pore area (**a**), and total pore volume (**b**) from LPNA measurements.

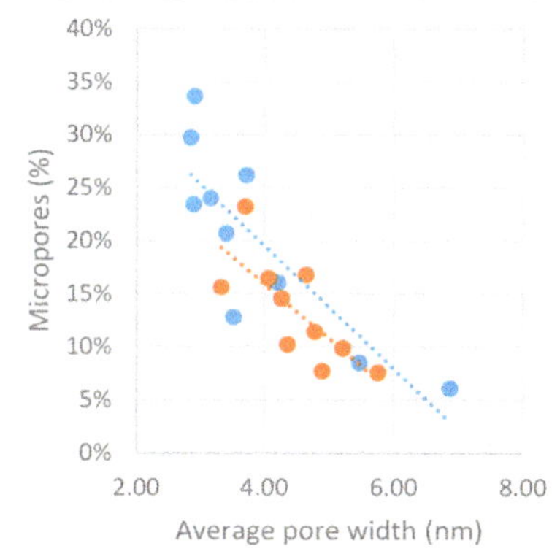

Figure 19. Relationship between average pore width vs the percentage of micropores.

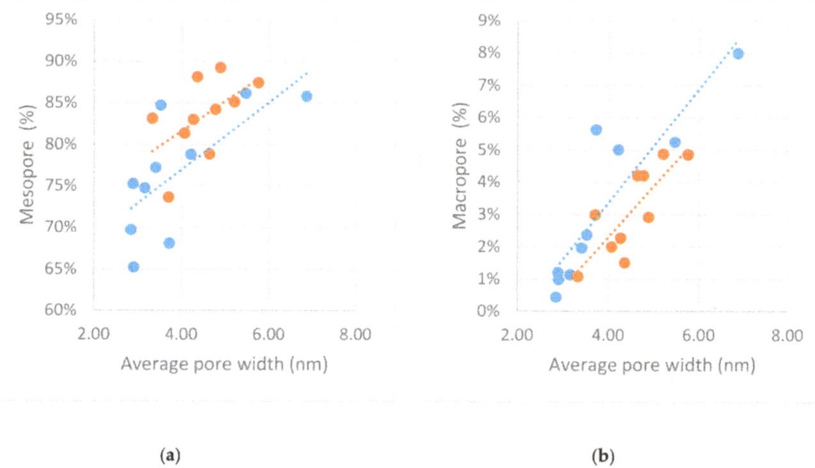

(a)   (b)

Figure 20. Relationship between average pore width vs mesopore percentage (**a**), and macropore percentage (**b**).

PSD Comparisons

The LPNA results show different changes in the sample PSD. After nine months of exposure, an increase was observed in the pore volume of micropores and mesopores smaller than about 8 nm in samples H1, H2, and H4. Sample H4 is different in this matter to the other two samples in that the pore volume of micropore range was decreased (Figure 7). However, the pore volume of pores larger than about 8 nm is decreased in samples H1, H2, and H4. In samples H5, H8, H9, and H13, the pore volume is increased across all pore size range. Samples H10, H14, and H15 follow a deferent pattern. Their pore volume is increased in pore sizes smaller than about 20 nm and is decreased bellow this turning point. Their after-exposure isotherms are lower than the before-exposure isotherms.

The MICP results are two-fold too. MICP shows a decrease in the porosity of samples H1 and H2 and an increase in samples H5, H9, and H13. Although H5 porosity increased, which puts it in the second group, its total pore volume is decreased from 10.26 m^2/g to 6.35 m^2/g (Table 4). The MICP PSD did not include the micropore range as the LPNA did. MICP advocates substantial pore volume in the meso- and macropores range. Samples H1 and H2 obtained from MICP analysis have the largest mesopore volume and the lowest macropore volume. The slight inconsistency between MICP and LPNA is because MICP only quantifies pore throat sizes and not the pore bodies, whereas LPNA quantifies both of them [57].

The grouping mentioned above is also obvious in the MICP pore size distribution curves (Figure 10). Samples H1 and H2 with the most uniform pore size distributions have the largest proportion of their pores in narrow pore sizes (peak at about 10 nm) which is reduced after exposure. A dual behaviour is observed in the pore volume distribution of samples H5, H9, and H13. The pore sizes are reduced below around 200 nm and enhanced above this point. PSD analysis using the NMR and MICP methods gives similar results, with pore distribution made of meso- and macropores. The pressure injection of mercury in the MICP method is not enough to override the strong capillary pressure of pores <2 nm and some of the nanopore signal associated with micro-porosity is not detectable by low-field NMR [57]. The same trend is also observed in the NMR results (Figure 3). The NMR pore size distribution of samples H1 and H2 are in the smaller pore size range and show a lower porosity than other samples (H5, H8, H9, H13).

6. Conclusions

Four laboratory techniques (MICP, LPNA, low-field NMR, and XRD) have been utilised to assess the alteration of petrophysical and chemical properties of shale caprocks under the influence of supercritical CO_2. The following conclusions can be reached about the effect of supercritical CO_2 on the shale samples in this study.

The experimental study of shaly caprock samples under in situ reservoir conditions ($T = 60\,^\circ$C, $P = 2000$ psi) has shown that the injection of scCO_2 into saline formation water of deep saline aquifers influences their chemical character and their sealing efficiency. Reactions of the mixture of scCO_2 and brine are documented by changes in mineral composition of exposed samples relative to the initial samples confirmed by XRD examinations. The injection of scCO_2 into the brine resulted in the dissolution of quartz, aluminosilicates such as K-feldspar and clay minerals such as kaolinite and the precipitation of muscovite (Illite).

The dissolution and reprecipitation of minerals are responsible for the changes in the pore structure properties of the samples. The results of this study indicate that chemical/mineralogical alterations of the shale samples, after exposure to scCO_2, have measurable effects on the porosity, and sealing properties of shales, with a tendency to enhance the transport properties. An increase in the porosity of most samples is observed in the NMR, MICP, LPNA results because of the mineralogical alterations. A noticeable increase in the T_2 relaxation time is observed in the samples exposed to scCO_2. MICP capillary pressure also shows a distinct shift toward smaller values after exposure to scCO_2. This agrees with a general shift toward larger pore throat sizes in the MICP PSD curves of the samples analysed. Also, the calculated maximum column height of CO_2 retention is reduced for samples exposed to scCO_2 as a result of the reduction of the threshold pressure.

Samples are grouped based on the clay content and pore sizes. The higher kaolinite contents which are present in samples H1 (43.06%) and H2 (38.39%) along with lower pore size ranges would contribute to their anomalous behaviour in terms of petrophysical alterations. Shale samples with higher kaolinite and lower quartz contents demonstrated uniform but poorly connected pores, with most pore sizes in the smaller pore size range. Their porosity was reduced after exposure to CO_2 when the smaller pores were decreased. By contrast, for shale samples of lower kaolinite and higher quartz contents (H5, H9, and H13) with unevenly distributed but well-connected pores, the porosity was enhanced. The overall larger pore size allowed CO_2 to penetrate more easily and to interact with pores' minerals. The dissolution in connected larger pores could play a major role in porosity enhancement in these samples.

A shale composition such as group one studied here is considered geochemically suitable caprock for CO_2 geological storage, as it contains clay minerals and matrix permeability is close to zero. Matrix solution transport would not be significant, and dissolution could not lead to increased porosity and permeability. In other words, seal integrity is maintained in such a case. Even when the petrophysical properties are enhanced, substantial leakage problems are unlikely through the undisturbed matrix of massive shale sequences [7]. Future works include a geochemical reaction path modelling of the experiment to further supplement the results of this study for longer periods.

Author Contributions: Formal analysis, P.H.; Investigation, P.H.; Methodology, P.H.; Supervision, R.R.; Visualization, P.H.; Writing—original draft, P.H.; Writing—review & editing, R.R. All authors have read and agreed to the published version of the manuscript.

Funding: This research received no external funding.

Acknowledgments: The authors wish to acknowledge the Department of Mines, Industry Regulation and Safety for providing the rock samples (Approval No. S32833). The authors would like to acknowledge the contribution of an Australian Government Research Training Program Scholarship in supporting this research. This study was also supported by a Curtin Research Scholarship (CRS). Part of this research was undertaken using the EM/XRD instrumentation (ARC LE0775553/LE0775551) at the John de Laeter Centre, Curtin University.

Conflicts of Interest: The authors declare no conflict of interest.

List of Acronyms

CPMG	Carr-Purcell-Meiboom-Gill
DFT	Density Functional Theory
HPLC	High-performance Liquid Chromatography
IPV	Incremental Pore Volume
LPNA	Low-pressure Nitrogen Adsorption
MICP	Mercury Intrusion Porosimetry
NMR	Nuclear Magnetic Resonance
PID	Proportional-Integral-Derivative
PSD	Pore Size distribution
PV	Pore Volume
PVT	Pressure-Volume-Temperature
SA	Surface Area
XRD	X-ray Diffraction

References

1. De Silva, G.P.D.; Ranjith, P.G.; Perera, M.S.A. Geochemical aspects of CO_2 sequestration in deep saline aquifers: A review. *Fuel* **2015**, *155*, 128–143. [CrossRef]
2. Kweon, H.; Deo, M. The impact of reactive surface area on brine-rock-carbon dioxide reactions in CO_2 sequestration. *Fuel* **2017**, *188*, 39–49. [CrossRef]
3. IPCC. *Climate Change 2013, The Physical Science Basis*; Contribution of Working Group I to the Fifth Assessment Report of the Intergovernmental Panel on Climate Change; Stocker, T.F., Qin, D., Plattner, G.K., Tignor, M., Allen, S.K., Boschung, J., Nauels, A., Xia, Y., Bex, V., Midgley, P.M., Eds.; Cambridge University Press: Cambridge, UK; New York, NY, USA, 2013; p. 1535.
4. Bachu, S.; Gunter, W.; Perkins, E. Aquifer disposal of CO_2, Hydrodynamic and mineral trapping. *Energy Convers. Manag.* **1994**, *35*, 269–279. [CrossRef]
5. Gunter, W.D.; Wiwchar, B.; Perkins, E.H. Aquifer disposal of CO_2-rich greenhouse gases: Extension of the time scale of experiment for CO_2-sequestering reactions by geochemical modelling. *Mineral. Petrol.* **1997**, *59*, 121–140. [CrossRef]
6. Chalbaud, C.; Robin, M.; Lombard, J.M.; Martin, F.; Egermann, P.; Bertin, H. Interfacial tension measurements and wettability evaluation for geological CO_2 storage. *Adv. Water Resour.* **2009**, *32*, 98–109. [CrossRef]
7. Busch, A.; Alles, S.; Gensterblum, Y.; Prinz, D.; Dewhurst, D.N.; Raven, M.D.; Stanjek, H.; Krooss, B.M. Carbon dioxide storage potential of shales. *Int. J. Greenh. Gas Control* **2008**, *2*, 297–308. [CrossRef]
8. Wollenweber, J.; Alles, S.; Busch, A.; Krooss, B.M.; Stanjek, H.; Littke, R. Experimental investigation of the CO_2 sealing efficiency of caprocks. *Int. J. Greenh. Gas Control* **2010**, *4*, 231–241. [CrossRef]
9. Gaus, I. Role and impact of CO_2-rock interactions during CO_2 storage in sedimentary rocks. *Int. J. Greenh. Gas Control* **2010**, *4*, 73–89. [CrossRef]

10. Bertier, P.; Swennen, R.; Laenen, B.; Lagrou, D.; Dreesen, R. Experimental identification of CO_2–water–rock interactions caused by sequestration of CO_2 in Westphalian and Buntsandstein sandstones of the Campine Basin (NE-Belgium). *J. Geochem. Explor.* **2006**, *89*, 10–14. [CrossRef]

11. Kaszuba, J.P.; Janecky, D.R.; Snow, M.G. Carbon dioxide reaction processes in a model brine aquifer at 200 °C and 200 bars: Implications for geologic sequestration of carbon. *Appl. Geochem.* **2003**, *18*, 1065–1080. [CrossRef]

12. Kaszuba, J.P.; Janecky, D.R.; Snow, M.G. Experimental evaluation of mixed fluid reactions between supercritical carbon dioxide and NaCl brine: Relevance to the integrity of a geologic carbon repository. *Chem. Geol.* **2005**, *217*, 277–293. [CrossRef]

13. Kharaka, Y.K.; Cole, D.R.; Thordsen, J.J.; Kakouros, E.; Nance, H.S. Gas–water–rock interactions in sedimentary basins: CO_2 sequestration in the Frio Formation, Texas, USA. *J. Geochem. Explor.* **2006**, *89*, 183–186. [CrossRef]

14. Liu, F.; Lu, P.; Griffith, C.; Hedges, S.W.; Soong, Y.; Hellevang, H.; Zhu, C. CO_2–brine–caprock interaction: Reactivity experiments on Eau Claire shale and a review of relevant literature. *Int. J. Greenh. Gas Control* **2012**, *7*, 153–167. [CrossRef]

15. Lu, J.; Kharaka, Y.K.; Thordsen, J.J.; Horita, J.; Karamalidis, A.; Griffith, C.; Hakala, J.A.; Ambats, G.; Cole, D.R.; Phelps, T.J.; et al. CO_2–rock–brine interactions in Lower Tuscaloosa Formation at Cranfield CO_2 sequestration site, Mississippi, U.S.A. *Chem. Geol.* **2012**, *291*, 269–277. [CrossRef]

16. Rosenbauer, R.J.; Koksalan, T.; Palandri, J.L. Experimental investigation of CO_2–brine–rock interactions at elevated temperature and pressure: Implications for CO_2 sequestration in deep-saline aquifers. *Fuel Process. Technol.* **2005**, *86*, 1581–1597. [CrossRef]

17. Wigand, M.; Carey, J.W.; Schütt, H.; Spangenberg, E.; Erzinger, J. Geochemical effects of CO_2 sequestration in sandstones under simulated in situ conditions of deep saline aquifers. *Appl. Geochem.* **2008**, *23*, 2735–2745. [CrossRef]

18. Credoz, A.; Bildstein, O.; Jullien, M.; Raynal, J.; Pétronin, J.C.; Lillo, M.; Pozo, C.; Geniaut, G. Experimental and modeling study of geochemical reactivity between clayey caprocks and CO_2 in geological storage conditions. *Energy Procedia* **2009**, *1*, 3445–3452. [CrossRef]

19. Liu, M. Behaviour of Shale Cap Rock under Exposure to Supercritical CO_2. Master's Thesis, Department of Civil and Environment Engineering, University of Alberta, Edmonton, AB, Canada, 2013; p. 108.

20. Schlomer, S.; Krooss, B.M. Experimental characterisation of the hydrocarbon sealing efficiency of cap rocks. *Mar. Pet. Geol.* **1997**, *14*, 563–578. [CrossRef]

21. Armitage, P.J.; Faulkner, D.R.; Worden, R.H. Caprock corrosion. *Nat. Geosci.* **2013**, *6*, 79–80. [CrossRef]

22. Yin, H.; Zhou, J.; Jiang, Y.; Xian, X.; Liu, Q. Physical and structural changes in shale associated with supercritical CO_2 exposure. *Fuel* **2016**, *184*, 289–303. [CrossRef]

23. Rezaee, R. Shale alteration after exposure to supercritical CO_2. *Appl. Polym. Sci.* **2014**, *62*, 91–99. [CrossRef]

24. Szabó, Z.; Hellevang, H.; Király, C.; Sendula, E.; Kónya, P.; Falus, G.; Török, S.; Szabó, C. Experimental-modelling geochemical study of potential CCS caprocks in brine and CO_2-saturated brine. *Int. J. Greenh. Gas Control* **2016**, *44*, 262–275. [CrossRef]

25. Alajmi, M.S. *Feasibility of Seismic Monitoring Methods for Australian CO_2 Storage Projects*; Department of Exploration Geophysics, Curtin University: Perth, Australia, 2015; p. 235.

26. Gibson-Poole, C.M.; Svendsen, L.; Underschultz, J.; Watson, M.N.; Ennis-King, J.; Van Ruth, P.J.; Nelson, E.J.; Daniel, R.F.; Cinar, Y. Site characterisation of a basin-scale CO_2 geological storage system: Gippsland Basin, southeast Australia. *Environ. Geol.* **2007**, *54*, 1583–1606. [CrossRef]

27. Delle Piane, C.; Olierook, H.K.H.; Timms, N.E.; Saeedi, A.; Esteban, L.; Rezaee, R.; Mikhaltsevitch, V.; Lebedev, M. *Facies-Based Rock Properties Distribution along the Harvey 1 Stratigraphic Well*; CSIRO Report Number EP133710; CSIRO: Perth, Australia, 2013.

28. Olierook, H.K.; Delle Piane, C.; Timms, N.E.; Esteban, L.; Rezaee, R.; Mory, A.J.; Hancock, L. Facies-based rock properties characterization for CO_2 sequestration: GSWA Harvey 1 well, Western Australia. *Mar. Pet. Geol.* **2014**, *50*, 83–102. [CrossRef]

29. Soldal, M. *Caprock Interaction with CO_2 Geomechanical and Geochemical Effects*; Department of Geosciences, University of Oslo: Oslo, Norway, 2008.

30. Manalo, F.; Ding, M.; Bryan, J.; Kantzas, A. *Separating the Signals from Clay Bound Water and Heavy Oil in NMR Spectra of Unconsolidated Samples*; Society of Petroleum Engineers: Calgary, AB, Canada, 2003.

31. Matteson, A.; Tomanic, J.P.; Herron, M.M.; Allen, D.F.; Kenyon, W.E. NMR Relaxation of Clay-Brine Mixtures. In Proceedings of the SPE Annual Technical Conference and Exhibition, New Orleans, LA, USA, 27–30 September 1998.

32. Bouton, J.C.; Drack, E.D.; Gardner, J.S.; Prammer, M.G. Measurements of Clay-Bound Water and Total Porosity by Magnetic Resonance Logging. In *SPE Annual Technical Conference and Exhibition*; Society of Petroleum Engineers: Denver, CO, USA, 1996.

33. Amann, A.; Waschbüsch, M.; Bertier, P.; Busch, A.; Krooss, B.M.; Littke, R. Sealing rock characteristics under the influence of CO_2. *Energy Procedia* **2011**, *4*, 5170–5177. [CrossRef]

34. Labani, M.M.; Rezaee, R.; Saeedi, A.; Al Hinai, A. Evaluation of pore size spectrum of gas shale reservoirs using low pressure nitrogen adsorption, gas expansion and mercury porosimetry: A case study from the Perth and Canning Basins, Western Australia. *J. Pet. Sci. Eng.* **2013**, *112*, 7–16. [CrossRef]

35. Rockwater Hydrogelogical and Environmetal Consultants. *Harvey 3 Lesueur Formation Fluid Sampling*; Southwesthub Geosequestration Project; Department of Mines and Petroleum: Perth, Australia, 2015; p. 24.

36. Wang, K.; Xu, T.; Wang, F.; Tian, H. Experimental study of CO_2–brine–rock interaction during CO_2 sequestration in deep coal seams. *Int. J. Coal Geol.* **2016**, *154*, 265–274. [CrossRef]

37. Comisky, J.T.; Santiago, M.; McCollom, B.; Buddhala, A.; Newsham, K.E. *Sample Size Effects on the Application of Mercury Injection Capillary Pressure for Determining the Storage Capacity of Tight Gas and Oil Shales*; Society of Petroleum Ezngineers: Calgary, AB, Canada, 2011.

38. Rezaee, R.; Saeedi, A.; Iglauer, S.; Evans, B. Shale alteration after exposure to supercritical CO_2. *Int. J. Greenh. Gas Control* **2017**, *62*, 91–99. [CrossRef]

39. Carr, H.Y.; Purcell, E.M. Effects of Diffusion on Free Precession in Nuclear Magnetic Resonance Experiments. *Phys. Rev.* **1954**, *94*, 630–638. [CrossRef]

40. Meiboom, S.; Gill, D. Modified Spin-Echo Method for Measuring Nuclear Relaxation Times. *Rev. Sci. Instrum.* **1958**, *29*, 688–691. [CrossRef]

41. Bustin, R.M.; Clarkson, C.R. Geological controls on coalbed methane reservoir capacity and gas content. *Int. J. Coal Geol.* **1998**, *38*, 3–26. [CrossRef]

42. Rezaee, R.; Saeedi, A.; Iglauer, S.; Evans, B. *CarbonNet Dynamic Seal Capacity*; Cooperative Research Centre for Greenhouse Gas Technologies Canberra, Curtin University: Perth, Australia, 2013.

43. Thommes, M.; Kaneko, K.; Neimark, A.V.; Olivier, J.P.; Rodriguez-Reinoso, F.; Rouquerol, J.; Sing, K.S. Physisorption of gases, with special reference to the evaluation of surface area and pore size distribution (IUPAC Technical Report). *Pure Appl. Chem.* **2015**, *87*, 1051–1069. [CrossRef]

44. Testamanti, M.N. Assessment of Fluid Transport Mechanisms in Shale Gas Reservoirs. Ph.D. Thesis, Curtin University, Perth, Australia, 2018.

45. Washburn, E.W. Note on a Method of Determining the Distribution of Pore Sizes in a Porous Material. *Proc. Natl. Acad. Sci. USA* **1921**, *7*, 115. [CrossRef] [PubMed]

46. Izgec, O.; Demiral, B.; Bertin, H.; Akin, S. CO_2 injection into saline carbonate aquifer formations I: Laboratory investigation. *Transp. Porous Media* **2008**, *72*, 1–24. [CrossRef]

47. Rathnaweera, T.D.; Ranjith, P.G.; Perera, M.S.; Haque, A. Influence of CO_2-Brine Co-injection on CO_2 Storage Capacity Enhancement in Deep Saline Aquifers: An Experimental Study on Hawkesbury Sandstone Formation. *Energy Fuels* **2016**, *30*, 4229–4243. [CrossRef]

48. Lahann, R.; Mastalerz, M.; Rupp, J.A.; Drobniak, A. Influence of CO_2 on New Albany Shale composition and pore structure. *Int. J. Coal Geol.* **2013**, *108*, 2–9. [CrossRef]

49. Ketzer, J.M.; Iglesias, R.; Einloft, S.; Dullius, J.; Ligabue, R.; De Lima, V. Water-rock-CO_2 interactions in saline aquifers aimed for carbon dioxide storage: Experimental and numerical modeling studies of the Rio Bonito Formation (Permian), southern Brazil. *Appl. Geochem.* **2009**, *24*, 760–767. [CrossRef]

50. Rathnaweera, T.D.; Ranjith, P.G.; Perera, M.S.A. Experimental investigation of geochemical and mineralogical effects of CO_2 sequestration on flow characteristics of reservoir rock in deep saline aquifers. *Sci. Rep.* **2016**, *6*, 19326. [CrossRef]

51. Alemu, B.L.; Aagaard, P.; Munz, I.A.; Skurtveit, E. Caprock interaction with CO_2, A laboratory study of reactivity of shale with supercritical CO_2 and brine. *Appl. Geochem.* **2011**, *26*, 1975–1989. [CrossRef]

52. Gunter, W.D.; Perkins, E.H.; Hutcheon, I. Aquifer disposal of acid gases: Modelling of water–rock reactions for trapping of acid wastes. *Appl. Geochem.* **2000**, *15*, 1085–1095. [CrossRef]

53. Moore, J.; Adams, M.; Allis, R.; Lutz, S.; Rauzi, S. Mineralogical and geochemical consequences of the long-term presence of CO_2 in natural reservoirs: An example from the Springerville-St. Johns Field, Arizona, and New Mexico, U.S.A. *Chem. Geol.* **2005**, *217*, 365–385. [CrossRef]

54. Gunter, W.D.; Bachu, S.; Benson, S.M. The role of hydrogeological and geochemical trapping in sedimentary basins for secure geological storage of carbon dioxide. *Geol. Soc. Spec. Publ.* **2004**, *233*, 129–145. [CrossRef]

55. Davis, M.C.; Wesolowski, D.J.; Rosenqvist, J.; Brantley, S.L.; Mueller, K.T. Solubility and near-equilibrium dissolution rates of quartz in dilute NaCl solutions at 398-473 K under alkaline conditions. *Geochim. Cosmochim. Acta* **2011**, *75*, 401–415. [CrossRef]

56. Hildenbrand, A.; Urai, J.L. Investigation of the morphology of pore space in mudstones—First results. *Mar. Pet. Geol.* **2003**, *20*, 1185–1200. [CrossRef]

57. Hinai, A.A.; Rezaee, R. Pore Geometry in Gas Shale Reservoirs. In *Fundamentals of Gas Shale Reservoirs*; John Wiley & Sons, Inc.: Hoboken, NJ, USA, 2015; pp. 89–116.

MDPI

St. Alban-Anlage 66

4052 Basel

Switzerland

Tel. +41 61 683 77 34

Fax +41 61 302 89 18

www.mdpi.com

Energies Editorial Office

E-mail: energies@mdpi.com

www.mdpi.com/journal/energies